# Radiopharmaceuticals

# Radiopharmaceuticals

Editor

**Svend Borup Jensen**

Basel • Beijing • Wuhan • Barcelona • Belgrade • Novi Sad • Cluj • Manchester

*Editor*
Svend Borup Jensen
Department of Nuclear
Medicine
University Hospital Aalborg
Aalborg C
Denmark

*Editorial Office*
MDPI
St. Alban-Anlage 66
4052 Basel, Switzerland

This is a reprint of articles from the Special Issue published online in the open access journal *Molecules* (ISSN 1420-3049) (available at: www.mdpi.com/journal/molecules/special_issues/radiopharmacy).

For citation purposes, cite each article independently as indicated on the article page online and as indicated below:

Lastname, A.A.; Lastname, B.B. Article Title. *Journal Name* **Year**, *Volume Number*, Page Range.

ISBN 978-3-03928-620-1 (Hbk)
ISBN 978-3-03928-619-5 (PDF)
doi.org/10.3390/books978-3-03928-619-5

© 2024 by the authors. Articles in this book are Open Access and distributed under the Creative Commons Attribution (CC BY) license. The book as a whole is distributed by MDPI under the terms and conditions of the Creative Commons Attribution-NonCommercial-NoDerivs (CC BY-NC-ND) license.

# Contents

About the Editor .................................................................. ix

Preface ............................................................................ xi

**Svend Borup Jensen**
Radioactive Molecules 2021–2022
Reprinted from: *Molecules* **2024**, *29*, 265, doi:10.3390/molecules29010265 ............... 1

**Amarnath Challapalli, Tara D. Barwick, Suraiya R. Dubash, Marianna Inglese, Matthew Grech-Sollars, Kasia Kozlowski, et al.**
Bench to Bedside Development of [$^{18}$F]Fluoromethyl- (1,2-$^2$H$_4$)choline ([$^{18}$F]D4-FCH)
Reprinted from: *Molecules* **2023**, *28*, 8018, doi:10.3390/molecules28248018 ............... 12

**Stefan Schmitl, Julia Raitanen, Stephan Witoszynskyj, Eva-Maria Patronas, Lukas Nics, Marius Ozenil, et al.**
Quality Assurance Investigations and Impurity Characterization during Upscaling of [$^{177}$Lu]Lu-PSMA$^{I\&T}$
Reprinted from: *Molecules* **2023**, *28*, 7696, doi:10.3390/molecules28237696 ............... 28

**Monika Skulska and Lise Falborg**
A Simple Kit for the Good-Manufacturing-Practice Production of [$^{68}$Ga]Ga-EDTA
Reprinted from: *Molecules* **2023**, *28*, 6131, doi:10.3390/molecules28166131 ............... 42

**Lin Shao**
Optimization of Deuteron Irradiation of $^{176}$Yb for Producing $^{177}$Lu of High Specific Activity Exceeding 3000 GBq/mg
Reprinted from: *Molecules* **2023**, *28*, 6053, doi:10.3390/molecules28166053 ............... 57

**Luis Michel Alonso Martinez, Nabil Naim, Alejandro Hernandez Saiz, José-Mathieu Simard, Mehdi Boudjemeline, Daniel Juneau and Jean N. DaSilva**
A Reliable Production System of Large Quantities of [$^{13}$N]Ammonia for Multiple Human Injections
Reprinted from: *Molecules* **2023**, *28*, 4517, doi:10.3390/molecules28114517 ............... 79

**Viktoriya V. Orlovskaya, Olga S. Fedorova, Nikolai B. Viktorov, Daria D. Vaulina and Raisa N. Krasikova**
One-Pot Radiosynthesis of [$^{18}$F]Anle138b—5-(3-Bromophenyl)-3-(6-[$^{18}$F]fluorobenzo[*d*][1,3] dioxol-5-yl)-1*H*-pyrazole—A Potential PET Radiotracer Targeting α-Synuclein Aggregates
Reprinted from: *Molecules* **2023**, *28*, 2732, doi:10.3390/molecules28062732 ............... 90

**Flavio Cicconi, Alberto Ubaldini, Angela Fiore, Antonietta Rizzo, Sebastiano Cataldo, Pietro Agostini, et al.**
Dissolution of Molybdenum in Hydrogen Peroxide: A Thermodynamic, Kinetic and Microscopic Study of a Green Process for $^{99m}$Tc Production
Reprinted from: *Molecules* **2023**, *28*, 2090, doi:10.3390/molecules28052090 ............... 109

**Anna Pees, Melissa Chassé, Anton Lindberg and Neil Vasdev**
Recent Developments in Carbon-11 Chemistry and Applications for First-In-Human PET Studies
Reprinted from: *Molecules* **2023**, *28*, 931, doi:10.3390/molecules28030931 ............... 128

**Elisabeth Plhak, Christopher Pichler, Edith Gößnitzer, Reingard M. Aigner and Herbert Kvaternik**
Development of in-House Synthesis and Quality Control of [$^{99m}$Tc]Tc-PSMA-I&S
Reprinted from: *Molecules* **2023**, *28*, 577, doi:10.3390/molecules28020577 ............... 149

**Amanda J. Boyle, Emily Murrell, Junchao Tong, Christin Schifani, Andrea Narvaez, Melinda Wuest, et al.**
PET Imaging of Fructose Metabolism in a Rodent Model of Neuroinflammation with 6-[$^{18}$F]fluoro-6-deoxy-D-fructose
Reprinted from: *Molecules* **2022**, *27*, 8529, doi:10.3390/molecules27238529 . . . . . . . . . . . . . . 158

**Fedor Zhuravlev, Arif Gulzar and Lise Falborg**
Recovery of Gallium-68 and Zinc from HNO$_3$-Based Solution by Liquid–Liquid Extraction with Arylamino Phosphonates
Reprinted from: *Molecules* **2022**, *27*, 8377, doi:10.3390/molecules27238377 . . . . . . . . . . . . . . 172

**Travis S. Laferriere-Holloway, Alejandra Rios, Giuseppe Carlucci and R. Michael van Dam**
Rapid Purification and Formulation of Radiopharmaceuticals via Thin-Layer Chromatography
Reprinted from: *Molecules* **2022**, *27*, 8178, doi:10.3390/molecules27238178 . . . . . . . . . . . . . . 186

**Hishar Hassan, Muhamad Faiz Othman, Hairil Rashmizal Abdul Razak, Zainul Amiruddin Zakaria, Fathinul Fikri Ahmad Saad, Mohd Azuraidi Osman, et al.**
Preparation, Optimisation, and In Vitro Evaluation of [$^{18}$F]AlF-NOTA-Pamidronic Acid for Bone Imaging PET
Reprinted from: *Molecules* **2022**, *27*, 7969, doi:10.3390/molecules27227969 . . . . . . . . . . . . . . 202

**Johan Hygum Dam, Niels Langkjær, Christina Baun, Birgitte Brinkmann Olsen, Aaraby Yoheswaran Nielsen and Helge Thisgaard**
Preparation and Evaluation of [$^{18}$F]AlF-NOTA-NOC for PET Imaging of Neuroendocrine Tumors: Comparison to [$^{68}$Ga]Ga-DOTA/NOTA-NOC
Reprinted from: *Molecules* **2022**, *27*, 6818, doi:10.3390/molecules27206818 . . . . . . . . . . . . . . 227

**Mohammad B. Haskali, Peter D. Roselt, Terence J. O'Brien, Craig A. Hutton, Idrish Ali, Lucy Vivash and Bianca Jupp**
Effective Preparation of [$^{18}$F]Flumazenil Using Copper-Mediated Late-Stage Radiofluorination of a Stannyl Precursor
Reprinted from: *Molecules* **2022**, *27*, 5931, doi:10.3390/molecules27185931 . . . . . . . . . . . . . . 236

**Kelly D. Orcutt, Kelly E. Henry, Christine Habjan, Keryn Palmer, Jack Heimann, Julie M. Cupido, et al.**
Dosimetry of [$^{212}$Pb]VMT01, a MC1R-Targeted Alpha Therapeutic Compound, and Effect of Free $^{208}$Tl on Tissue Absorbed Doses
Reprinted from: *Molecules* **2022**, *27*, 5831, doi:10.3390/molecules27185831 . . . . . . . . . . . . . . 248

**Caroline Stokke, Monika Kvassheim and Johan Blakkisrud**
Radionuclides for Targeted Therapy: Physical Properties
Reprinted from: *Molecules* **2022**, *27*, 5429, doi:10.3390/molecules27175429 . . . . . . . . . . . . . . 258

**Yiwei Wang, Daiyuan Chen, Ricardo dos Santos Augusto, Jixin Liang, Zhi Qin, Juntao Liu and Zhiyi Liu**
Production Review of Accelerator-Based Medical Isotopes
Reprinted from: *Molecules* **2022**, *27*, 5294, doi:10.3390/molecules27165294 . . . . . . . . . . . . . . 278

**Suliman Salih, Ajnas Alkatheeri, Wijdan Alomaim and Aisyah Elliyanti**
Radiopharmaceutical Treatments for Cancer Therapy, Radionuclides Characteristics, Applications, and Challenges
Reprinted from: *Molecules* **2022**, *27*, 5231, doi:10.3390/molecules27165231 . . . . . . . . . . . . . . 298

**Valentina Di Iorio, Stefano Boschi, Cristina Cuni, Manuela Monti, Stefano Severi, Giovanni Paganelli and Carla Masini**
Production and Quality Control of [$^{177}$Lu]Lu-PSMA-I&T: Development of an Investigational Medicinal Product Dossier for Clinical Trials
Reprinted from: *Molecules* **2022**, *27*, 4143, doi:10.3390/molecules27134143 . . . . . . . . . . . . . . 314

**Svend Borup Jensen, Lotte Studsgaard Meyer, Nikolaj Schandorph Nielsen and Søren Steen Nielsen**
Issues with the European Pharmacopoeia Quality Control Method for $^{99m}$Tc-Labelled Macroaggregated Albumin
Reprinted from: *Molecules* **2022**, *27*, 3997, doi:10.3390/molecules27133997 . . . . . . . . . . . . . . 329

**Costantina Maisto, Michela Aurilio, Anna Morisco, Roberta de Marino, Monica Josefa Buonanno Recchimuzzo, Luciano Carideo, et al.**
Analysis of Pros and Cons in Using [$^{68}$Ga]Ga-PSMA-11 and [$^{18}$F]PSMA-1007: Production, Costs, and PET/CT Applications in Patients with Prostate Cancer
Reprinted from: *Molecules* **2022**, *27*, 3862, doi:10.3390/molecules27123862 . . . . . . . . . . . . . . 338

**James M. Kelly, Thomas M. Jeitner, Nicole N. Waterhouse, Wenchao Qu, Ethan J. Linstad, Banafshe Samani, et al.**
Synthesis and Evaluation of $^{11}$C-Labeled Triazolones as Probes for Imaging Fatty Acid Synthase Expression by Positron Emission Tomography
Reprinted from: *Molecules* **2022**, *27*, 1552, doi:10.3390/molecules27051552 . . . . . . . . . . . . . . 347

# About the Editor

**Svend Borup Jensen**

Svend Borup Jensen received his BSc in Chemistry and Biochemistry from Aarhus University, Denmark (1992), a MPhil in Chemistry from the University of Strathclyde, Scotland (1994), and a PhD in Chemistry from the University of Paisley, Scotland (1998).

Following graduation, Svend worked as an organic chemist and as an analysis chemist for private companies. However, his main career has been in radiochemistry, first at Aarhus University Hospital as head of production, and since 2009, he has been head of radiochemistry/QP (Qualified Person) at Aalborg University Hospital. Svend's responsibility as a QP revolves around the production and release of radiopharmaceutical drugs used in humans for diagnosis. He has also been strongly involved in designing and qualifying new and heavily enlarged cleanroom facilities for the Dept. of Nuclear Medicine at "Nyt Aalborg Universitetshospital" (NAU) to ensure that the facilities are in compliance with the legislation for the production of sterile pharmaceuticals and contain suitable apparatus for radioactive drug production.

Research in the field of radioactive drugs is very versatile. Most of Svend's research has revolved around the synthesis of new radiopharmaceutical drugs or the optimization of known syntheses as the focal point, but the analysis and identification of radioactive drugs and by-products have also received considerable attention.

As a researcher, Svend has authored or co-authored 49 research papers in peer-reviewed journals, 3 book chapters, and 11 articles in Danish professional journals with a more educational angle.

# Preface

The purpose of this Special Issue was to host research and review papers on Radiopharmaceuticals in line with the three previous Special Issues that I have Guest Edited for *Molecules*. This subject may vary from radioactive isotope production, the synthesis of precures, radioactive labeling reactions, to the purification of radiopharmaceuticals to quality control and regulatory efforts, which are required prior to using radiopharmaceuticals in humans.

This Special Issue contains 17 original research papers, 1 communication, 5 reviews, and 1 editorial review all within the area of radiochemistry.

I sincerely hope that you enjoy reading this Special Issue.

**Svend Borup Jensen**
*Editor*

*Editorial*

# Radioactive Molecules 2021–2022

Svend Borup Jensen [1,2]

1. Department of Nuclear Medicine, Aalborg University Hospital, 9000 Aalborg, Denmark; svbj@rn.dk
2. Department of Chemistry and Biochemistry, Aalborg University, 9220 Aalborg, Denmark

In 2020 I was invited to write an editorial review on radioactive molecules published in *Molecules* in 2019 and 2020 [1]. The aim of the review was to identify all the papers published in the journal that applied radioactive isotopes. The present review has the same aim, but for the years 2021 and 2022. To make the data comparable, the articles are examined and categorized in the same manner as in the first review.

## 1. The Search Criteria

As in the previous editorial review, a search in *Molecules* was performed using the keyword "radio*". In the 2021 to 2022 period, the search resulted in 276 articles, compared to 181 for 2019–2020, an increase of 52%.

A closer look at the identified articles revealed that 109 (60%) of those published in 2019–2020 were in fact dealing with radioactive isotopes, with this figure being 172 (62%) for 2021–2022. Again, approximately 40% of the articles from the search were not considered relevant to the review. I conducted an investigation into which search terms caused the articles not relevant to this review to appear in the search results and found that many of these articles deal with radioactive therapy and protection against radioactivity. The words found to occur more than 10 times in the 104 "irrelevant" articles were: radiotherapy, radioresistance/radio resistance, radiosensitivity, radio sensitization/radiosensitization, radiosensitizer(s), radio frequency/radiofrequency, and radiolysis.

Taking the above identified words into account, I attempted to apply othersearch terms to optimize the search, see Table 1. However, as can be seen from Table 1, the application of "radioa*", "radioactive", "isotope", and "isotop*", resulted in the search failing to find all of the relevant articles, so this search strategy was discarded.

**Table 1.** Search terms applied.

|  | 2019 | 2020 | 2021 | 2022 |
|---|---|---|---|---|
| Radioa* | 12 | 16 | 12 | 23 |
| Radioactive | 11 | 15 | 12 | 26 |
| Isotope | 46 | 42 | 51 | 56 |
| Isotop* | 46 | 42 | 50 | 54 |
| Radio* | 68 | 117 | 126 | 150 |
| Relevant papers |  | 109 | 78 | 94 |
| Not relevant |  | 76 | 48 | 56 |

## 2. Main Categories

### 2.1. Non-Medical Applications

Only 23 of the 172 selected articles pertained to non-medical subjects, typically dealing with pollution and nuclear waste.

As an example of the subjects covered, two interesting papers dealing with the synthesis of compounds able to capture iodine [L1, L2]. In both papers, the aim was to develop

compounds capable of removing radioactive iodine from nuclear power plant waste. The capture of iodine is, in fact, a hot research topic [2–5]. Compounds capable of removing radioactive iodine may also be medically relevant, as hospitals use iodine I-131 (with a half-life of 8 days) for treatment of some thyroid disorders, resulting in the patients urine becoming radioactive, resulting in it being categorized as radioactive waste. Hospitals handle this problem by using storage tanks for toilet waste to postpone its release into the environment and/or by mixing the toilet waste with other wastewater to dilute the radioactivity below the threshold level. Methods designed to extract I-131 from urine may indeed prove be more economically advantageous compared to storing toilet wastewater, especially if the iodine capture is reversible such that the extraction material can be re-used.

The other articles focusing on non-medical use of radioactive isotopes constitute a very diverse group. They cover a range of topics from the examination of naturally occurring isotopes by applying in vivo isotope degradation [L3], to thermo-degradation experiments [L4], and $^{14}$C-dating of a Roman Egyptian mummy portrait [L5].

*2.2. Medical Applications*

The number of articles covering medical uses of radioactive molecules was 149. Excluding my own editorial review [1] from 2021 leaves a total of 148 articles dealing, in one way or the other, with the medical uses of radioactive isotopes, to be further examined.

As in the previous review, the articles were divided into subgroups: diagnostics, therapy, synthesis, and nuclide preparation, and it was noted whether the article was a review (see Table 2). An article may appear in multiple subgroups.

**Table 2.** Division of the articles dealing with the medical use of isotopes into subgroups.

|  | Diagnostic | Therapy | Synthesis | Nuclide Preparation | Review | Articles Concerning Medical Use of Radioactive Isotopes |
|---|---|---|---|---|---|---|
| 2019–2020 | 73 articles (71%) | 24 articles (23%) | 41 articles (40%) | 9 articles (9%) | 21 articles (20%) | 103 articles (100%) |
| 2021–2022 | 129 articles (87%) | 57 articles (39%) | 96 articles (65%) | 21 articles (14%) | 42 articles (28%) | 148 articles (100%) |

Table 2 shows that the number of articles dealing with medical uses of radioactive isotopes has increased from 103 to 148 (an increase of 36%). In general, the use of radioactive molecules in clinical settings at hospitals has been increasing for many years, which can to some extent explain the rising research articles on this topic.

The percentage of reviews has increased from 20% to 28%. It is worth noting that all of the percentages have increased from 2019–2020 to 2021–2022. Part of the explanation for this is that the number of reviews has increased, and review articles often touch upon several topics such as diagnostics, therapy, synthesis, and nuclide preparation. However, the tendency of a greater number of articles covering topics from more than one of the subgroups cannot solely be put down to the increase in reviews. This increase supports a general sense I have when reading through the articles that, although an article focuses on, for example, synthesis, it also often contains information about how the radioactive substances are used in diagnostics and/or therapy. In other words, articles are increasingly covering more subjects (diagnostic, therapy, synthesis, and/or nuclide preparation) than was the case merely two years ago.

### 3. Use of Isotopes

The 148 articles on medical uses for radioactive isotopes, and the 23 on non-medical uses, deal with or mention many different isotopes, as can be seen in Tables 3–7.

The placement of the isotopes in Tables 3–7 depends on how the isotope mainly decays—i.e., whether the main form of its decay is via positron [6–9], gamma [10], beta [11], Auger [12], or alpha emissions [13]. The decay of an isotope will often take place in several ways; however, this review will focus on the main form of decay. The categorization of the isotopes and their location in Tables 3–7 follows the description of the isotopes primary decay in the reviewed articles, which moreover in broad terms are identical to what is found in the literature published elsewhere.

In Tables 3 and 4, the articles are classified into three groups: diagnostic/use, synthesis/chelation, and isotope preparation. In Tables 5–7, the diagnostic column has been replaced by therapy/detection. Each article is placed in at least one of these groups but may appear in more than one group. The column entitled "Total Number of Articles" is the sum of the other columns but with duplicates subtracted.

The column "Selected Articles" lists articles that support and illustrate what I aim to express, furthermore it provide the reader with the opportunity to seek additional knowledge from the original literature.

Table 3. Positron emitters.

|  | Total Number of Articles | Diagnostic | Synthesis/ Chelation | Isotope Preparation | Selected Articles |
|---|---|---|---|---|---|
| $^{11}$C | 26 | 26 | 9 | - | [L6–L12] |
| $^{13}$N | 7 | 7 | - | - | [L8, L9, L11, L13] |
| $^{15}$O | 7 | 7 | - | - | [L7–L9, L13] |
| $^{18}$F | 59 | 57 | 25 | - | [L6–L20] |
| $^{22}$Na | 2 | 2 | - | - | - |
| $^{43}$Sc | 1 | 1 | - | - | - |
| $^{44}$Sc | 10 | 8 | 5 | 3 | [L20, L21] |
| $^{52}$Mn | 5 | 3 | 3 | 3 | [L8, L20] |
| $^{55}$Co | 2 | 2 | - | 1 | - |
| $^{61}$Cu | 2 | 2 | - | 1 | [L13, L22] |
| $^{64}$Cu | 33 | 32 | 7 | 4 | [L8, L9, L11, L13, L15, L18, L20, L22] |
| $^{66}$Ga □ |  | 3 | 1 | 1 | [L7] |
| $^{68}$Ga | 51 | 48 | 22 | 4 | [L7–L9, L11, L13, L18–L20, L23–L25] |
| $^{82}$Rb | 3 | 3 | - | - | [L9, L13] |
| $^{86}$Y/$^{86g}$Y | 4 | 3 | 2 | 1 | [L20] |
| $^{89}$Zr | 19 | 17 | 8 | 3 | [L8, L13, L15, L20] |
| $^{124}$I | 13 | 11 | 6 | 1 | [L8, L10, L13, L19, L26] |

Positron emitter isotopes mentioned only in one article: $^{43}$Sc, $^{152}$Tb, $^{44}$Ti, $^{76}$Br, and $^{120}$I

(□ both a positron and a gamma emitter).

As was the case with the 2021 review [1], $^{18}$F is the isotope mentioned in most articles. There has subsequently been an increase in papers mentioning $^{18}$F, but the increase has been much bigger for several of other isotopes. The second most used PET isotope in 2021/2022 was $^{68}$Ga, yet there were more than 2.5 as many articles on $^{18}$F than $^{68}$Ga in 2019/2020. At present, the numbers of articles covering these isotope are more closely aligned, there being only 1.16 more articles on $^{18}$F than $^{68}$Ga. The same trend is seen for a number of the other isotopes such as, for example, $^{11}$C, $^{64}$Cu, and $^{89}$Zr. The numbers of articles on $^{11}$C has risen from 6 to 26, on $^{64}$Cu the number is up from 2 to 33, and the number of articles on $^{89}$Zr has risen from 5 to 20. Furthermore, a significant number of new PET isotopes were discussed in the 2021–2022 issues of *Molecules*, which was not the case in 2019–2020; examples include $^{13}$N, $^{22}$Na, $^{52}$Mn, $^{61}$Cu, $^{66}$Ga, $^{82}$Rb, and $^{86}$Y.

Table 4. Gamma emitters.

| | Total Number of Articles | Diagnostic | Synthesis/ Chelation | Isotope Preparation | Selected Articles |
|---|---|---|---|---|---|
| $^{57}$Co | 2 | 2 | - | 1 | [L20] |
| $^{62}$Zn | 2 | 2 | - | - | [L18, L20] |
| $^{66}$Ga | 3 | 3 | 1 | 1 | [L7] |
| $^{67}$Ga [a] | 10 | 10 | 2 | 1 | [L7, L11, L13, L18, L22] |
| $^{99m}$Tc | 47 | 43 | 20 | 4 | [L6, L8, L9, L13, L18, L20, L25, L27-L31] |
| $^{105}$Rh | 2 | 2 | - | - | [L18] |
| $^{111}$In | 24 | 23 | 3 | 2 | [L7, L9, L11, L13, L15, L18, L20, L24] |
| $^{123}$I | 16 | 14 | 4 | - | [L8, L10, L11, L13, L18, L19, L26] |
| $^{125}$I | 18 | 17 | 6 | - | [L8, L10, L13, L18, L26, L32, L33] |
| $^{133}$Xe | 2 | 2 | - | - | [L13] |
| $^{201}$Tl | 2 | 2 | - | - | [L13] |

Isotopes mentioned only in one article: $^{51}$Cr [L13], $^{51}$Mn, and $^{155}$Tb.
Reference [L19] is the only article to mention: $^{76}$As, $^{77}$Br, $^{82}$Br, $^{86}$Rb, $^{90}$Nb

([a] both a positron and gamma emitter).

There has been a considerable rise in the number of articles that mention gamma emitters. $^{99m}$Tc and $^{125}$I are mentioned approximately three times as often in 2021–2022 compared to 2019–2020. For the other isotopes, the increase is, in many cases, much higher. This supports my point that articles generally have an increased focus on explaining how their findings contribute to the general state of knowledge. In order to illustrate what their contribution means in a broader context, the authors must, to a higher degree, describe what other researchers have achieved. For that reason, a greater number of tracers and more isotopes are mentioned.

Additionally, there is an awareness of that developing and identifying the best possible tracer often means that a nuclide replacement is performed to unroll the full potential of a radioactive drug, this of course also contributes to more isotopes being mentioned in a given article.

As already mentioned, an isotope can often decay by more than one route; for example, $^{177}$Lu emits both beta and gamma radiation during its decay and can, therefore, be used for therapy as well as diagnostics. I have placed the isotopes in Tables 3–7 depending on how they are normally referred to. As $^{177}$Lu is normally referred to as a beta emitter, it is placed in Table 5.

A very interesting article by Stokke et al. [L18], mentions almost 100 different isotopes. For some of these, it was not clear in which of Tables 3–7 they should be placed. I have taken the liberty of not including those isotopes, as they are only mentioned in reference [L18] and they decay in several different ways—for example, as gamma and beta emitters, as well as by electron capture.

Theranostic is becoming more and more widespread in nuclear medicine. Theranostic is a two-stage process to diagnosing and treating cancers using radiotracers. The notion underpinning theranostics is that two almost identical radiopharmaceutical can first be used to diagnose/identify a cancer and then subsequently to treat the cancer. The difference between the diagnostic use and the therapeutic use is the replacement of the radioactive isotope: a positron or gamma emitter is used for diagnostics and, for treatment, a beta emitter, Auger electron emitter, or alpha emitter. It makes sense to know exactly where a beta, alpha, or Auger electron emitter goes before it is injected into a human subject.

Table 5. Beta emitters.

| | Total Number of Articles | Therapy/ Detection | Synthesis/ Chelation | Isotope Preparation | Selected Articles |
|---|---|---|---|---|---|
| $^{3}$H | 15 | 14 | 5 | - | [L9] |
| $^{14}$C | 1 | 1 | - | - | - |
| $^{32}$P | 5 | 4 | 1 | 2 | [L18, L34] |
| $^{35}$S | 3 | 3 | 1 | 1 | - |
| $^{47}$Sc | 8 | 7 | 1 | - | [L18, L20, L22] |
| $^{67}$Cu * | 9 | 8 | 2 | 1 | [L18, L20, L22] |
| $^{89}$Sr | 3 | 3 | - | - | [L18] |
| $^{90}$Y | 18 | 17 | 4 | 1 | [L8, L13, L18, L25, L29, L34] |
| $^{117m}$Sn | 2 | 2 | - | - | [L13, L18] |
| $^{131}$I * | 25 | 18 | 11 | - | [L8, L13, L18, L19, L26, L29, L32, L34] |
| $^{153}$Sm | 9 | 9 | - | 1 | [L13, L25, L34] |
| $^{161}$Tb | 3 | 3 | - | - | [L18] |
| $^{166}$Ho | 4 | 4 | 1 | - | [L13, L18, L29, L34] |
| $^{177}$Lu | 42 | 39 | 7 | 2 | [L8, L13, L15, L18–L20, L22–L25, L34] |
| $^{186}$Re | 7 | 7 | 1 | 2 | [L13, L18, L22, L34] |
| $^{188}$Re | 12 | 12 | 1 | 2 | [L13, L18, L25, L29, L34] |
| $^{198}$Au | 3 | 3 | - | 1 | [L18, L34] |

Isotopes mentioned only in one article: $^{14}$C, $^{33}$P, $^{90}$Sr, $^{129}$I, $^{210}$Pb, $^{169}$Er [L34]
Reference [L19] is the only article to mention: $^{24}$Na, $^{33}$P, $^{54}$Mn, $^{76}$As, $^{77}$As, $^{77}$Br, $^{80}$Br, $^{91}$Y, $^{109}$Pd, $^{111}$Ag, $^{121}$Sn, $^{135}$La, $^{142}$Pr, $^{143}$Pr, $^{165}$Dy, $^{185}$W, $^{212}$Pb

* $^{67}$Cu and $^{131}$I are both beta and gamma emitters. One article mentions that $^{67}$Cu is also a gamma emitter and 10 articles mention that $^{131}$I is also a gamma emitter.

Table 6. Auger electron emitters.

| | Total Number of Articles | Therapy | Synthesis/ Chelation | Isotope Preparation | Selected Articles |
|---|---|---|---|---|---|
| $^{58}$Co/$^{58m}$Co | 2 | 2 | - | 1 | [L18] |
| $^{165}$Er | 2 | 2 | 1 | 1 | [L18] |
| $^{193m}$Pt | 2 | 2 | 1 | - | [L18] |
| $^{195m}$Pt | 2 | 2 | 1 | - | [L18] |

Isotopes mentioned in only one article: $^{114m}$In, $^{134}$Ce [L18]

Table 7. Alpha emitters.

| | Number of Articles | Therapy | Synthesis/ Chelation | Isotopes Preparation | Selected Articles |
|---|---|---|---|---|---|
| $^{149}$Tb | 2 | 2 | - | - | [L18] |
| $^{211}$At | 6 | 6 | - | - | [L18, L34] |
| $^{212}$Bi | 2 | 2 | - | - | [L18] |
| $^{213}$Bi | 5 | 4 | 1 | - | [L18, L34] |
| $^{223}$Ra | 7 | 7 | - | - | [L13, L18, L34] |
| $^{225}$Ac | 14 | 13 | 2 | 1 | [L13, L15, L18, L25, L31, L34] |
| $^{227}$Th | 2 | 2 | 1 | 1 | |

Reference [L18] is the only article to mention: $^{224}$Ra, $^{255}$Fm

## 4. Development of New Radiopharmaceuticals for Clinical Use

The development of new radiopharmaceuticals is performed in a similar manner to non-radioactive medicine, with synthesis followed by ex vivo examination, then in vivo animal examination, and finally by human clinical trials. Synthesis was discussed in 96 articles (65%, see Table 2). Following synthesis there is often some initial ex vivo testing of the radiopharmaceuticals. This was referred to in 58 of the 148 articles [L6, L8, L9, L10, L12, L16, L24, L25, L30, L31] on isotopes for medical use. Ex vivo tests are often simple stability tests of the compound or analysis to find log P (the partition coefficient: measuring how an analyte will partition between an aqueous and organic phase).

The next step is to perform in vivo tests with the radiopharmaceuticals in animals. Such tests are performed to examine the distribution of the radiopharmaceuticals, there in vivo stability, and in many cases, to examine their uptake in, for example, tumour cells. The preferred animals are mice: 51 articles covered the use of mice [L7–L13, L15, L16, L20, L24, L25, L28, L31, L32], 16 the use rats [L7, L13, L17, L20], 6 pigs [L7, L29], 3 baboons/monkeys, 2 rabbits, 1 dogs, and 1 sheep.

There were nine preliminary human studies and five clinical studies focusing on tracers. This is higher than in 2019–2020. The increase is probably due to more and more radiopharmaceuticals being approved for clinical trials. The amount of work a radiochemist must invest in a radiopharmaceutical being approved for clinical trials is significant, and that is probably also part of the reason why we see an increased number of publications on preliminary human studies focusing on the radiopharmaceuticals.

## 5. Diagnosing or Treating a Specific Disease

Of the 148 articles on isotopes for medical uses, 78 concerned cancers (Table 8), 24 brain disorders (Table 9), 15 inflammation and infection (Table 10), 15 other diseases, or which do not discuss a specific disease. The "other diseases" category includes, for example, liver diseases, diabetes mellitus, sclerosis, hyperthyroidism, myocardial injury, invasive fungal, renal protection, traumatic brain injury, spinal cord injury, and pulmonary disease. Also placed in this category are the articles which focus on imaging for, for example, lung perfusions, bone formation and bone imaging [L13], and sentinel lymph node detection. There was also one article on analgesia, one on liposomes, and one on the influence of caffeine on glucose uptake.

The column "Selected Articles" in Tables 8–10 lists articles that support and illustrate what I aim to express, furthermore they provide the reader with an opportunity to seek additional knowledge from the original literature.

Table 8. Cancers.

| | Number of Articles | Selected Articles |
|---|---|---|
| All articles concerning cancer | 78 | [L8, L9, L12, L13, L15, L16, L18, L19, L24, L31, L32, L34] |
| Prostate cancer | 14 | [L8, L9, L12, L13, L31, L34] |
| Lymphoma (including B-cell) | 2 | [L8] |
| Neuroendocrine | 13 | [L8, L9, L13, L24] |
| Breast cancer | 20 | [L8, L9, L13, L31] |
| Gastric cancer | 3 | [L8] |
| Leukaemia | 7 | [L8, L12, L13, L34] |
| Renal cell carcinoma | 4 | [L8, L9] |
| Solid tumour/multiple myeloma | 7 | [L9, L12, L13] |
| Non-Hodgkin's lymphoma | 7 | [L8, L9, L18, L34] |
| Pancreatic cancer | 9 | [L8, L9, L15, L24] |
| Ovarian cancer | 12 | [L9, L18, L24, L31, L34] |
| Bone-metastases, -pain, and -cancer | 15 | [L8, L9, L13, L18, L34] |
| Thyroid cancer | 11 | [L8, L9, L13, L19, L24, L34] |
| Non-small-cell lung cancer/lung cancer | 12 | [L12, L24] |
| Skin cancer | 8 | [L8, L12, L18, L34] |
| Bladder cancer | 3 | [L8] |
| Liver cancer | 11 | [L8, L9, L12, L13, L31, L34] |
| Colon/colorectal cancer | 8 | [L8, L9, L13, L16, L32] |
| Hypoxia | 3 | [L9, L13] |
| Head and neck cancer | 2 | [L9] |
| Oesophagus cancer/oesophageal squamous cell carcinoma | 3 | [L9] |
| Sarcoma cancer/Ewing sarcoma | 2 | [L9] |

Types of cancer mentioned only once: stomach cancer [L8], $\alpha_v\beta_3$ integrin (examination of tumour cells to survive during therapy), sentinel lymph node, neuroblastoma, cancer immunotherapy [L16], thymus cancer [L9], circulating cancer cells (CCC) [L9], cancer of unknown primary (CUP) [L9], oral cancer, small intestine cancer [L9], and pleural mesothelioma

Cancers is by far the largest category. In 2021–2022, there were 78 articles mentioning cancers, compared to 39 in 2019–2020; i.e., an increase of 61%. The most common cancer type in 2019–2020 was prostate with eight articles, which increased to 14 articles in 2021–2022. Although the number of articles dealing with prostate cancer increased by 75%, it was overtaken by other types of cancer when it comes to the number of articles published on a specific cancer type: for example, the number of articles on breast cancer increased from two articles in 2019–2020 to 20 articles in 2021–2022. Other significant increases were for bone cancer, which increased from 2 to 15, lung cancer which was up from 1 to 12, and liver cancer rising from 0 to 11. The reason why articles on prostate cancer did not increase as much as, for example, breast, liver, and lung cancer, is that many articles dealing with the PSMA analogue had already been published [14,15]. It is only natural that the research focus turned to other types of cancers where good radiopharmaceuticals are not yet as available as is the case for prostate cancer.

Table 9. Brain disorders.

|  | Number of Articles | Selected Articles |
|---|---|---|
| All articles concerning brain disorders/brain function | 24 | [L10, L11, L13, L14, L17] |
| Alzheimer's disease | 8 | [L10] |
| Parkinson's disease | 9 | [L13, L14] |
| Anxiety | 2 | [L11] |
| Regulates/chronic pain treatment | 2 |  |
| Opioid receptor | 3 |  |
| Substance use disorder | 2 | [L17] |
| Compulsive behaviour disorder like feeding disorders/obesity, gambling, addition | 2 | [L11] |
| Neurodegenerative | 2 | [L11] |
| Epilepsy | 2 | [L17] |

Type of brain disorders/brain functions mentioned only once: neurological disorders, serotonin neurotransmitter, mood disorder, spinal cord injury, CNS-modulating agents, hallucinogens, Huntington's disease, bipolar disorder, dementia, benzodiazepine receptors, monoamine oxidase-B (MAO-B), schizophrenia [L17], autism [L17], metabolic diseases [L11], and depression

Table 10. Inflammation and infection.

|  | Number of Articles | Selected Articles |
|---|---|---|
| All articles concerning inflammation/infection | 15 | [L7, L13] |
| Osteomyelitis | 3 | [L7] |
| Neuroinflammation | 4 |  |
| Bone infection | 5 | [L7, L13] |
| Rheumatoid arthritis | 2 | [L13] |

Types of inflammation/infection mentioned only once: vascular inflammation, inflammatory joint disease, and arthritis.

The increase in articles concerning brain disorders/brain function was by 33%, from 18 articles in 2019–2020 to 24 in 2021–2022. In 2019–2020, Alzheimer's disease was the most mentioned subject with six articles, this increased to eight articles in 2021–2022. However, in 2021–2022, Alzheimer's disease was overtaken by studies on Parkinson's disease, which went up from 3 to 9 articles.

The increase of articles concerning inflammation/infection was by 114%, from 7 articles in 2019–2020 to 15 in 2021–2022. Bone infection/osteomyelitis increased from two articles in 2019–2020 to seven in 2021–2022, and neuroinflammation increased from one article in 2019–2020 to four in 2021–2022.

## 6. Overall Summary

In summary, the number of articles which apply radioactive isotopes in *Molecules* has increased by 58% over the last 2 years, from 109 articles in 2019–2020 to 176 articles in 2021–2022. Although a lot of interesting articles are being published in the area of non-medical research, articles with radioactive molecules are still mainly published in medicine for diagnosing or treating diseases. The number of different isotopes applied (see Tables 3–7) and the different diseases covered (see Tables 8–10) have increased considerably over the last two years.

**Conflicts of Interest:** The author declares no conflict of interest.

**List of Contributions**

L1. Yan, Z.; Qiao, Y.;Wang, J.; Xie, J.; Cui, B.; Fu, Y.; Lu, J.; Yang, Y.; Bu, N.; Yuan, Y.; et al. An Azo-Group-Functionalized Porous Aromatic Framework for Achieving Highly Efficient Capture of Iodine. *Molecules* **2022**, *27*, 6297. https://doi.org/10.3390/molecules2719629.

L2. Tian, P.; Ai, Z.; Hu, H.; Wang, M.; Li, Y.; Gao, X.; Qian, J.; Su, X.; Xiao, S.; Xu, H.; et al. Synthesis of Electron-Rich Porous Organic Polymers via Schiff-Base Chemistry for Efficient Iodine Capture. *Molecules* **2022**, *27*, 5161. https://doi.org/10.3390/molecules27165161.

L3. Zuo, B.; Cao, M.; Tao, X.; Xu, X.; Leng, H.; Cui, Y.; Bi, K. Metabolic Study of Tetra-PEG-Based Hydrogel after Pelvic Implantation in Rats. *Molecules* **2022**, *27*, 5993. https://doi.org/10.3390/molecules27185993.

L4. Planche, C.; Chevolleau, S.; Noguer-Meireles, M.-H.; Jouanin, I.; Mompelat, S.; Ratel, J.; Verdon, E.; Engel, E.; Debrauwer, L. Fate of Sulfonamides and Tetracyclines in Meat during Pan Cooking: Focus on the Thermodegradation of Sulfamethoxazole. *Molecules* **2022**, *27*, 6233. https://doi.org/10.3390/molecules27196233.

L5. Dal Fovo, A.; Fedi, M.; Federico, G.; Liccioli, L.; Barone, S.; Fontana, R. Correction: Dal Fovo et al. Multi-Analytical Characterization and Radiocarbon Dating of a Roman Egyptian Mummy Portrait. *Molecules* **2021**, *26*, 5268. *Molecules* **2022**, *27*, 3822. https://doi.org/10.3390/molecules27123822.

L6. Wodtke, R.; Pietzsch, J.; Löser, R. Solid-Phase Synthesis of Selectively Mono-Fluorobenz(o)ylated Polyamines as a Basis for the Development of 18F-Labeled Radiotracers. *Molecules* **2021**, *26*, 7012. https://doi.org/10.3390/molecules26227012.

L7. Jødal, L.; Afzelius, P.; Alstrup, A.K.O.; Jensen, S.B. Radiotracers for Bone Marrow Infection Imaging. *Molecules* **2021**, *26*, 3159. https://doi.org/10.3390/molecules26113159.

L8. Kumar, K.; Ghosh, A. Radiochemistry, Production Processes, Labeling Methods, and ImmunoPET Imaging Pharmaceuticals of Iodine-124. *Molecules* **2021**, *26*, 414. https://doi.org/10.3390/molecules26020414.

L9. Lin, M.; Coll, R.P.; Cohen, A.S.; Georgiou, D.K.; Manning, H.C. PET Oncological Radiopharmaceuticals: Current Status and Perspectives. *Molecules* **2022**, *27*, 6790. https://doi.org/10.3390/molecules27206790.

L10. Nguyen, G.A.; Liang, C.; Mukherjee, J. [124I]IBETA: A New A Plaque Positron Emission Tomography Imaging Agent for Alzheimer's Disease. *Molecules* **2022**, *27*, 4552. https://doi.org/10.3390/molecules27144552.

L11. Fonseca, I.C.F.; Castelo-Branco, M.; Cavadas, C.; Abrunhosa, A.J. PET Imaging of the Neuropeptide Y System: A Systematic Review. *Molecules* **2022**, *27*, 3726. https://doi.org/10.3390/molecules27123726.

L12. Wang, L.; Zhou, Y.;Wu, X.; Ma, X.; Li, B.; Ding, R.; Stashko, M.A.; Wu, Z.;Wang, X.; Li, Z. The Synthesis and Initial Evaluation of MerTK Targeted PET Agents. *Molecules* **2022**, *27*, 1460. https://doi.org/10.3390/molecules27051460.

L13. Holik, H.A.; Ibrahim, F.M.; Elaine, A.A.; Putra, B.D.; Achmad, A.; Kartamihardja, A.H.S. The Chemical Scaffold of Theranostic Radiopharmaceuticals: Radionuclide, Bifunctional Chelator, and Pharmacokinetics Modifying Linker. *Molecules* **2022**, *27*, 3062. https://doi.org/10.3390/molecules27103062.

L14. Campoy, A.-D.T.; Liang, C.; Ladwa, R.M.; Patel, K.K.; Patel, I.H.; Mukherjee, J. [18F]Nifene PET/CT Imaging in Mice: Improved Methods and Preliminary Studies of α4β2*Nicotinic Acetylcholinergic Receptors in Transgenic A53T Mouse Model of -Synucleinopathy and Post-Mortem Human Parkinson's Disease. *Molecules* **2021**, *26*, 7360. https://doi.org/10.3390/molecules26237360.

L15. Handula, M.; Chen, K.-T.; Seimbille, Y. IEDDA: An Attractive Bioorthogonal Reaction for Biomedical Applications. *Molecules* **2021**, *26*, 4640. https://doi.org/10.3390/molecules26154640.

L16. Khanapur, S.; Yong, F.F.; Hartimath, S.V.; Jiang, L.; Ramasamy, B.; Cheng, P.; Narayanaswamy, P.; Goggi, J.L.; Robins, E.G. An Improved Synthesis of N-(4-[18F]Fluorobenzoyl)-Interleukin-2 for the Preclinical PET Imaging of Tumour-Infiltrating T-cells in CT26 and MC38 Colon Cancer Models. *Molecules* **2021**, *26*, 1728. https://doi.org/10.3390/molecules26061728.

L17. Haskali, M.B.; Roselt, P.D.; O'Brien, T.J.; Hutton, C.A.; Ali, I.; Vivash, L.; Jupp, B. Effective Preparation of [18F]Flumazenil Using Copper-Mediated Late-Stage Radiofluorination of a Stannyl Precursor. *Molecules* **2022**, *27*, 5931. https://doi.org/10.3390/molecules27185931.

L18. Stokke, C.; Kvassheim, M.; Blakkisrud, J. Radionuclides for Targeted Therapy: Physical Properties. *Molecules* **2022**, *27*, 5429. https://doi.org/10.3390/molecules27175429.

L19. Sakulpisuti, C.; Charoenphun, P.; Chamroonrat, W. Positron Emission Tomography Radiopharmaceuticals in Differentiated Thyroid Cancer. *Molecules* **2022**, *27*, 4936. https://doi.org/10.3390/molecules27154936.

L20. Pyrzynska, K.; Kilian, K.; Pęgier, M. Porphyrins as Chelating Agents for Molecular Imaging in Nuclear Medicine. *Molecules* **2022**, *27*, 3311. https://doi.org/10.3390/molecules27103311.

L21. Larenkov, A.A.; Makichyan, A.G.; Iatsenko, V.N. Separation of 44Sc from 44Ti in the Context of A Generator System for Radiopharmaceutical Purposes with the Example of [44Sc]Sc-PSMA-617 and [44Sc]Sc-PSMA-I&T Synthesis. *Molecules* **2021**, *26*, 6371. https://doi.org/10.3390/molecules26216371.

L22. Mou, L.; Martini, P.; Pupillo, G.; Cieszykowska, I.; Cutler, C.S.; Mikołajczak, R. 67Cu Production Capabilities: A Mini Review. *Molecules* **2022**, *27*, 1501. https://doi.org/10.3390/molecules27051501.

L23. Moon, E.S.; Van Rymenant, Y.; Battan, S.; De Loose, J.; Bracke, A.; Van der Veken, P.; De Meester, I.; Rösch, F. In Vitro Evaluation of the Squaramide-Conjugated Fibroblast Activation Protein Inhibitor-Based Agents AAZTA5.SA.FAPi and DOTA.SA.FAPi. *Molecules* **2021**, *26*, 3482. https://doi.org/10.3390/molecules26123482.

L24. Hörmann, A.A.; Plhak, E.; Klingler, M.; Rangger, C.; Pfister, J.; Schwach, G.; Kvaternik, H.; von Guggenberg, E. Automated Synthesis of 68Ga-Labeled DOTA-MGS8 and Preclinical Characterization of Cholecystokinin-2 Receptor Targeting. *Molecules* **2022**, *27*, 2034. https://doi.org/10.3390/molecules27062034.

L25. Trujillo-Benítez, D.; Luna-Gutiérrez, M.; Ferro-Flores, G.; Ocampo-García, B.; Santos-Cuevas, C.; Bravo-Villegas, G.; Morales-Ávila, E.; Cruz-Nova, P.; Díaz-Nieto, L.; García-Quiroz, J.; et al. Design, Synthesis and Preclinical Assessment of 99mTc-iFAP for In Vivo Fibroblast Activation Protein (FAP) Imaging. *Molecules* **2022**, *27*, 264. https://doi.org/10.3390/molecules27010264.

L26. Kumar, K.; Woolum, K. A Novel Reagent for Radioiodine Labeling of New Chemical Entities (NCEs) and Biomolecules. *Molecules* **2021**, *26*, 4344. https://doi.org/10.3390/molecules26144344.

L27. Martini, P.; Uccelli, L.; Duatti, A.; Marvelli, L.; Esposito, J.; Boschi, A. Highly Efficient Micro-Scale Liquid-Liquid In-Flow Extraction of 99mTc from Molybdenum. *Molecules* **2021**, *26*, 5699. https://doi.org/10.3390/molecules26185699.

L28. Papasavva, A.; Shegani, A.; Kiritsis, C.; Roupa, I.; Ischyropoulou, M.; Makrypidi, K.; Pilatis, I.; Loudos, G.; Pelecanou, M.; Papadopoulos, M.; et al. Comparative Study of a Series of 99mTc(CO)3 Mannosylated Dextran Derivatives for Sentinel Lymph Node Detection. *Molecules* **2021**, *26*, 4797. https://doi.org/10.3390/.

L29. d'Abadie, P.; Hesse, M.; Louppe, A.; Lhommel, R.; Walrand, S.; Jamar, F. Microspheres Used in Liver Radioembolization: From Conception to Clinical Effects. *Molecules* **2021**, *26*, 3966. https://doi.org/10.3390/molecules26133966.

L30. Jensen, S.B.; Meyer, L.S.; Nielsen, N.S.; Nielsen, S.S. Issues with the European Pharmacopoeia Quality Control Method for 99mTc-Labelled Macroaggregated Albumin. *Molecules* **2022**, *27*, 3997. https://doi.org/10.3390/molecules27133997.

L31. Hernández-Jiménez, T.; Ferro-Flores, G.; Morales-Ávila, E.; Isaac-Olivé, K.; Ocampo-García, B.; Aranda-Lara, L.; Santos-Cuevas, C.; Luna-Gutiérrez, M.; De Nardo, L.; Rosato, A.; et al. 225Ac-rHDL Nanoparticles: A Potential Agent for Targeted Alpha-Particle Therapy of Tumors Overexpressing SR-BI Proteins. *Molecules* **2022**, *27*, 2156. https://doi.org/10.3390/molecules27072156.

L32. Cruz-Nova, P.; Ocampo-García, B.; Carrión-Estrada, D.A.; Briseño-Diaz, P.; Ferro-Flores, G.; Jiménez-Mancilla, N.; Correa-Basurto, J.; Bello, M.; Vega-Loyo, L.; Thompson-Bonilla, M.d.R.; et al. 131I-C19 Iodide Radioisotope and Synthetic I-C19 Compounds as K-Ras4B–PDE6 Inhibitors: A Novel Approach against Colorectal Cancer—Biological Characterization, Biokinetics and Dosimetry. *Molecules* **2022**, *27*, 5446. https://doi.org/10.3390/molecules27175446.

L33. Lau, J.; Lee, H.; Rousseau, J.; Bénard, F.; Lin, K.-S. Application of Cleavable Linkers to Improve Therapeutic Index of Radioligand Therapies. *Molecules* **2022**, *27*, 4959. https://doi.org/10.3390/molecules27154959.

L34. Salih, S.; Alkatheeri, A.; Alomaim, W.; Elliyanti, A. Radiopharmaceutical Treatments for Cancer Therapy, Radionuclides Characteristics, Applications, and Challenges. *Molecules* **2022**, *27*, 5231. https://doi.org/10.3390/molecules27165231.

## References

1. Jensen, S.B. Radioactive Molecules 2019–2020. *Molecules* **2021**, *26*, 529. [CrossRef] [PubMed]
2. Xie, Y.; Pan, T.; Lei, Q.; Chen, C.; Dong, X.; Yuan, Y.; Shen, J.; Cai, Y.; Zhou, C.; Pinnau, I.; et al. Ionic Functionalization of Multivariate Covalent Organic Frameworks to Achieve Exceptionally High Iodine Capture Capacity. *Angew. Chem. Int. Ed.* **2021**, *133*, 22606–22614. [CrossRef]
3. Muhire, C.; Reda, A.T.; Zhang, D.; Xu, X.; Cui, C. An overview on metal Oxide-based materials for iodine capture and storage. *Chem. Eng. J.* **2022**, *431*, 133816. [CrossRef]
4. Dai, D.; Yang, J.; Zou, Y.-C.; Wu, J.-R.; Tan, L.-L.; Wang, Y.; Li, B.; Lu, T.; Wang, L.B.; Yang, Y.-W. MacrocyclicArenes-Based Conjugated Macrocycle Polymers for Highly Selective $CO_2$ Capture and Iodine Adsorption. *Angew. Chem. Int. Ed.* **2021**, *60*, 8967–8975. [CrossRef] [PubMed]
5. He, L.; Chen, L.; Dong, X.; Zhang, S.; Zhang, M.; Dai, X.; Liu, X.; Lin, P.; Li, K.F.; Chen, C.L.; et al. A nitrogen-rich covalent organic framework for simultaneous dynamic capture of iodine and methyl iodide. *Chem* **2021**, *7*, 699–714. [CrossRef]
6. Boros, E.; Packard, A.B. Radioactive Transition Metals for Imaging and Therapy. *Chem. Rev.* **2019**, *119*, 870–901. [CrossRef] [PubMed]
7. Kyriakou, I.; Sakata, D.; Tran, H.N.; Perrot, Y.; Shin, W.G.; Lampe, N.; Zein, S.; Bordage, M.C.; Guatelli, S.; Villagrasa, C.; et al. Review of the Geant4-DNA Simulation Toolkit for Radiobiological Applications at the Cellular and DNA Level. *Cancers* **2022**, *14*, 35. [CrossRef] [PubMed]
8. Damont, A.; Roeda, D.; Dollé, F. The potential of carbon-11 and fluorine-18 chemistry: Illustration through the development of positron emission tomography radioligands targeting the translocator protein 18 kDa. *J. Label. Compd. Radiopharm.* **2013**, *56*, 96–104. [CrossRef] [PubMed]
9. Campbell, E.; Jordan, C.; Gilmour, R. Fluorinated carbohydrates for 18F-positron emission tomography (PET). *Chem. Soc. Rev.* **2023**, *52*, 3599–3626. [CrossRef] [PubMed]
10. Schillaci, O.; Spanu, A.; Danieli, R.; Madeddu, G. Molecular breast imaging with gamma emitters. *Q. J. Nucl. Med. Mol. Imaging* **2013**, *57*, 340–351. [PubMed]
11. Waksman, R.; Raizner, A.; Popma, J.J. Beta emitter systems and results from clinical trials. state of the art. *Cardiovasc. Radiat. Med.* **2003**, *4*, 54–63. [CrossRef] [PubMed]
12. Pirovano, G.; Wilson, T.C.; Reiner, T. Auger: The future of precision medicine. *Nucl. Med. Biol.* **2021**, *96–97*, 50–53. [CrossRef]
13. Kunikowska, J.; Królicki, L. Targeted α-Emitter Therapy of Neuroendocrine Tumors. *Semin. Nucl. Med.* **2020**, *50*, 171–176. [CrossRef] [PubMed]
14. Rachel Levine, R.; Krenning, E.P. Clinical History of the Theranostic Radionuclide Approach to Neuroendocrine Tumors and Other Types of Cancer: Historical Review Based on an Interview of Eric P. Krenning by Rachel Levine. *J. Nuc. Med.* **2017**, *58*, 3S–9S. [CrossRef]
15. Alati, S.; Singh, R.; Pomper, M.G.; Rowe, S.P.; Banerjee, S.R. Preclinical Development in Radiopharmaceutical Therapy for Prostate Cancer. *Semin. Nucl. Med.* **2023**, *53*, 663–686. [CrossRef] [PubMed]

**Disclaimer/Publisher's Note:** The statements, opinions and data contained in all publications are solely those of the individual author(s) and contributor(s) and not of MDPI and/or the editor(s). MDPI and/or the editor(s) disclaim responsibility for any injury to people or property resulting from any ideas, methods, instructions or products referred to in the content.

*Review*

# Bench to Bedside Development of [¹⁸F]Fluoromethyl-(1,2-²H₄)choline ([¹⁸F]D4-FCH)

Amarnath Challapalli [1,2], Tara D. Barwick [1,3], Suraiya R. Dubash [1], Marianna Inglese [1], Matthew Grech-Sollars [1,†,‡], Kasia Kozlowski [1], Henry Tam [3], Neva H. Patel [3], Mathias Winkler [4], Penny Flohr [5], Azeem Saleem [6,7], Amit Bahl [2], Alison Falconer [4], Johann S. De Bono [5], Eric O. Aboagye [1,*] and Stephen Mangar [4,*]

[1] Department of Surgery and Cancer, Imperial College London, Hammersmith Hospital Campus, Du Cane Road, London W12 0NN, UK; amarnath.challapalli@uhbw.nhs.uk (A.C.); tara.barwick@nhs.net (T.D.B.); suraiya.dubash@nhs.net (S.R.D.); marianna.inglese17@imperial.ac.uk (M.I.); m.grech-sollars@ucl.ac.uk (M.G.-S.); kaskoz@yahoo.com (K.K.)

[2] Department of Clinical Oncology, Bristol Haematology and Oncology Center, Horfield Road, Bristol BS2 8ED, UK; amit.bahl@uhbw.nhs.uk

[3] Department of Radiology & Nuclear Medicine, Imperial College Healthcare NHS Trust, Hammersmith Hospital, Du Cane Road, London W12 0HS, UK; henry.tam@nhs.net (H.T.); neva.patel1@nhs.net (N.H.P.)

[4] Department of Urology, Imperial College Healthcare NHS Trust, Charing Cross Hospital, London W6 8RF, UK; mathias.winkler@nhs.net (M.W.); alison.falconer1@nhs.net (A.F.)

[5] Division of Clinical Studies, The Institute of Cancer Research and Royal Marsden Hospital, Cotswold Road, Sutton SM2 5NG, UK; penny.flohr@icr.ac.uk (P.F.); johann.debono@icr.ac.uk (J.S.D.B.)

[6] Invicro, A Konica Minolta Company, Burlington Danes Building, Hammersmith Hospital, Du Cane Road, London W12 0NN, UK; azeem.saleem@hyms.ac.uk

[7] Hull York Medical School, University of Hull, Cottingham Road, Hull HU6 7RX, UK

* Correspondence: eric.aboagye@imperial.ac.uk (E.O.A.); s.mangar@imperial.ac.uk (S.M.)

† Current address: Centre for Medical Image Computing, Department of Computer Science, University College London, London WC1N 3BG, UK.

‡ Current address: Lysholm Department of Neuroradiology, National Hospital for Neurology and Neurosurgery, University College London Hospitals NHS Foundation Trust, London WC1N 3BG, UK.

Citation: Challapalli, A.; Barwick, T.D.; Dubash, S.R.; Inglese, M.; Grech-Sollars, M.; Kozlowski, K.; Tam, H.; Patel, N.H.; Winkler, M.; Flohr, P.; et al. Bench to Bedside Development of [¹⁸F]Fluoromethyl-(1,2-²H₄)choline ([¹⁸F]D4-FCH). *Molecules* **2023**, *28*, 8018. https://doi.org/10.3390/molecules28248018

Academic Editor: Svend Borup Jensen

Received: 15 November 2023
Revised: 5 December 2023
Accepted: 6 December 2023
Published: 8 December 2023

Copyright: © 2023 by the authors. Licensee MDPI, Basel, Switzerland. This article is an open access article distributed under the terms and conditions of the Creative Commons Attribution (CC BY) license (https://creativecommons.org/licenses/by/4.0/).

**Abstract:** Malignant transformation is characterised by aberrant phospholipid metabolism of cancers, associated with the upregulation of choline kinase alpha (CHKα). Due to the metabolic instability of choline radiotracers and the increasing use of late-imaging protocols, we developed a more stable choline radiotracer, [¹⁸F]fluoromethyl-[1,2-²H₄]choline ([¹⁸F]D4-FCH). [¹⁸F]D4-FCH has improved protection against choline oxidase, the key choline catabolic enzyme, via a $^1H/^2D$ isotope effect, together with fluorine substitution. Due to the promising mechanistic and safety profiles of [¹⁸F]D4-FCH in vitro and preclinically, the radiotracer has transitioned to clinical development. [¹⁸F]D4-FCH is a safe positron emission tomography (PET) tracer, with a favourable radiation dosimetry profile for clinical imaging. [¹⁸F]D4-FCH PET/CT in lung and prostate cancers has shown highly heterogeneous intratumoral distribution and large lesion variability. Treatment with abiraterone or enzalutamide in metastatic castrate-resistant prostate cancer patients elicited mixed responses on PET at 12–16 weeks despite predominantly stable radiological appearances. The sum of the weighted tumour-to-background ratios (TBRs-wsum) was associated with the duration of survival.

**Keywords:** [¹⁸F]Fluoromethyl-(1,2-²H₄)choline; positron emission tomography; choline; cell membrane metabolism; in vivo; in vitro; dosimetry; metastatic castrate-resistant prostate cancer

## 1. Introduction

Choline is one of the components of phosphatidylcholine (PC), an essential element of phospholipids in the cell membrane [1] and is required for structural stability and cell proliferation. Tumours show a high proliferation and increased cell membrane components that will lead to an increased uptake of choline. The progression of normal cells to malignant phenotypes is associated with altered membrane choline phospholipid metabolism [2].

Choline kinase α (CHKα), the first enzyme in the Kennedy pathway [3], is responsible for the de novo synthesis of PC and the generation of phosphorylcholine (PCho) from its precursor, choline. CHKα has been extensively linked to cell proliferation and human carcinogenesis [4–6]. A strong correlation between CHKα activity and cancer onset has been established by providing evidence that CHK dysregulation is a frequent event occurring in a variety of human tumours, such as breast, lung, colorectal, and prostate tumours [6,7]. Therefore, choline kinase activity represents a potential biomarker for diagnostic use in oncology [8,9]. The use of positron emission tomography (PET) to measure choline transport and CHKα activity is the subject of this review article.

Nature-identical radiolabelled choline represented the first probe to be established for measuring choline tumour uptake by using PET; the main biological variable of interest being the maximum voxel Standardised Uptake Value ($SUV_{max}$). Hara et al. in the first report of the use of [$^{11}$C]choline in prostate cancer, noted that the $SUV_{max}$ of prostate cancer was three-fold greater than that of normal prostate tissue or of benign prostatic hyperplasia (BPH) [10]. Although [$^{11}$C]choline injection is approved for use in men with recurrent prostate cancer by the US Food and Drug Administration, the short half-life of carbon-11 ($t_{1/2}$ = 20.1 min) limits the application of [$^{11}$C]choline to centres with an on-site cyclotron. As a result, fluorine-18 ($t_{1/2}$ = 109.8 min)-labelled choline analogues were independently developed by Hara et al. [11] ([$^{18}$F]fluoroethylcholine (FEC)) and by DeGrado et al. [12] ([$^{18}$F]fluoromethylcholine (FMC)). In theory, the longer half-life of [$^{18}$F] (109.8 min) is potentially advantageous in permitting the late imaging of tumours when sufficient clearance of the parent tracer in systemic circulation has occurred.

## 2. Deuterated Choline

[$^{11}$C]Choline and [$^{18}$F]FMC are readily oxidized by choline oxidase mainly in kidney and liver tissues, with metabolites detectable in the plasma soon after injection of the radiotracer [13–15]. This causes difficulty in highlighting the relative contributions of parent radiotracers and catabolites to the PET signal when a late-imaging protocol is used. A more metabolically stable fluorocholine analogue, [$^{18}$F]fluoromethyl-[1,2-$^{2}$H$_4$]choline ([$^{18}$F]D4-FCH), has been developed based on the deuterium isotope effect [16]. The simple substitution of deuterium ($^2$D) for hydrogen ($^1$H) and the presence of $^{18}$F reduces the degradation of the parent tracer and improves stability [15,17,18]. This modification is hypothesised to increase the net availability of the parent tracer for phosphorylation and intracellular trapping leading to a better signal-to-background contrast, thus improving sensitivity. The translational studies of [$^{18}$F]D4-FCH from bench to bedside are summarised in Table 1. In this article, we elaborate the development of the more stable choline radiotracer, [$^{18}$F]D4-FCH (Figure 1a; [$^{18}$F]D4-FCH has improved protection against choline oxidase, the key choline catabolic enzyme, via a $^1$H/$^2$D isotope effect, together with fluorine substitution). We discuss the steps in the translation of the new radiotracer from in vitro discovery to preclinical imaging, in first in-human studies, in an evaluation in lung cancer patients, and then finally in an evaluation of its role in response assessment in metastatic prostate cancer patients. The initial part of the paper reviews the various preclinical and clinical studies of [$^{18}$F]D4-FCH. In the section on response assessment, we discuss our findings regarding the role of [$^{18}$F]D4-FCH in the response assessment to novel antiandrogens in metastatic prostate cancer patients.

**Table 1.** Summary of [$^{18}$F]D4-FCH studies: Translation from bench to bedside.

| Setting | Findings/Comments |
|---|---|
| In vitro stability [15], 2011 | • Chromatographic analyses showed that [$^{18}$F]fluoro-[1,2-$^2$H$_4$]choline ([$^{18}$F]D4-FCH) was ~80% intact after treatment for 1 h with potassium permanganate, which oxidizes choline to betaine. Conversely, only 40% of [$^{18}$F]Fluorocholine ([$^{18}$F]FCH) was intact.<br>• Therefore, [$^{18}$F]D4-FCH is more stable than [$^{18}$F]FCH against oxidation. |

**Table 1.** Cont.

| Setting | Findings/Comments |
|---|---|
| In vivo biodistribution [15], 2011 | • [$^{18}$F]D4-FCH exhibited higher accumulation in the tumours than [$^{18}$F]FCH, particularly at later time points.<br>• The distribution of the three radiotracers, [$^{18}$F]FCH, [$^{18}$F]fluoro-[1-$^2$H$_2$]choline, and [$^{18}$F]D4-FCH, showed a similar uptake profile in most organs, with prominent radioactivity in the kidneys and liver.<br>• The plasma concentrations of both tracers were <0.1% ID/g at 2 min after injection, with <10% of [$^{18}$F]FCH intact and ~50% of [$^{18}$F]D4-FCH intact as predicted from the deuterium isotope effect.<br>• A reduced uptake of [$^{18}$F]D4-FCH, relative to [$^{18}$F]FCH and [$^{18}$F]fluoro-[1-$^2$H$_2$]choline, in lung tissue may make imaging of thoracic tumours using this radiotracer superior to [$^{18}$F]FCH. |
| In vivo response assessment [17], 2009 | At day 10 after drug treatment, compared with the pretreatment group, the following observations were seen:<br>• Reduction in tumour size by 12%.<br>• Marked reduction in radiotracer retention in the treated tumours.<br>• Decrease in all imaging variables.<br>[$^{18}$F]D4-FCH can be used for response assessment even under conditions where large changes in tumour size reduction are not seen. |
| In vitro and in vivo comparison [18], 2012 | • Deuteration and fluorination combine to provide protection against choline oxidation in vivo.<br>• Uptake of [$^{18}$F]D4-choline was the same in three different tumour types, suggesting that [$^{18}$F]D4-choline has utility for cancer detection irrespective of histologic type. |
| Human biodistribution [19], 2014 | • [$^{18}$F]D4-FCH was well tolerated, with no radiotracer-related serious adverse events reported.<br>• The mean effective dose averaged over both males and females (±SD) was estimated to be 0.025 ± 0.004 (male 0.022 ± 0.002; female 0.027 ± 0.002) mSv/MBq.<br>• Highest-absorbed dose (mGy/MBq) was in the kidneys (0.106 ± 0.03), liver (0.094 ± 0.03), pancreas (0.066 ± 0.01), urinary bladder wall (0.047 ± 0.02), and adrenals (0.046 ± 0.01). Elimination was through the renal and hepatic systems.<br>• [$^{18}$F]D4-FCH is a safe radiotracer with a dosimetry profile comparable to other common [$^{18}$F]PET tracers. |
| First in-patient evaluation (lung cancer) [20], 2020 | • Oxidation of [$^{18}$F]D4-FCH to [$^{18}$F]D4-fluorobetaine was suppressed, confirming the slow catabolism of [$^{18}$F]D4-FCH.<br>• Early (5 min) and late (60 min) images showed specific uptake of tracer in all 51 lesions (tumours, lymph nodes, and metastases) from 17 patients analysed.<br>• [$^{18}$F]D4-FCH-derived uptake ($SUV_{max}$) in index primary lesions ($n = 17$) ranged between 2.87 and 10.13; lower than that of [$^{18}$F]FDG-PET [6.89, 22.64].<br>• Mathematical modelling demonstrated net irreversible uptake of [$^{18}$F]D4-FCH at steady-state, and parametric mapping of the entire tumour showed large intratumoural heterogeneity in radiotracer retention, which highlights the potential for radiotherapy dose delivery and treatment response monitoring. |
| Impact of hypoxia on D4-FCH kinetics [21], 2021 | • The export of phosphorylated [$^{18}$F]D4-FCH and [$^{18}$F]D4-FCHP via HIF-1α-responsive efflux transporters, including ABCB4, when the HIF-1α level is augmented.<br>• This is supported by a graphical analysis of human data with a compartmental model (M2T6k + $k_5$) that accounts for the efflux.<br>• Hypoxia/HIF-1α increases the efflux of phosphorylated radiolabelled choline species, thus supporting the consideration of efflux in the modelling of radiotracer dynamics. |
| Prostate cancer response assessment (Current Study) | • Metastatic castrate-resistant prostate cancer (mCRPC) patients ($n = 9$) prospectively recruited for abiraterone/enzalutamide therapy. [$^{18}$F]D4-FCH was performed at baseline, 4–6 and 12–16 weeks and compared to prostate-specific antigen (PSA), Prostate cancer working group 3 (PCWG3) response criteria and survival duration.<br>• Heterogeneity of response: A wide-ranging response profile is seen in both bone and soft-tissue lesions. Changes in [$^{18}$F]D4-FCH PET/CT variables at 12–16 weeks are likely to reflect an escape of individual lesions from selective pressures of therapy. We show bioinformatically, that the choline transporters may also contribute to the cholinic phenotype in addition to the phenotype being a proliferation-independent phenotype in advanced prostate cancer.<br>• Strong association between [$^{18}$F]D4-FCH-detectable cholinic phenotype (sum of weighted tumour-to-background ratios (TBRs-wsum)) and PSA response.<br>• TBR-wsum could reflect disease burden: Patients with larger TBR-wsum changes survived longer, with a TBR-wsum >30% giving a PFS advantage of >25 months. Superiority of TBR-wsum to PSA change or PCWG3 classification is demonstrated by inability of these latter routine measures to predict survival. TBR-wsum should be explored in future studies (together with individual TBR values signifying heterogeneity) to account for aggregate cholinic phenotype from multiple lesions. |

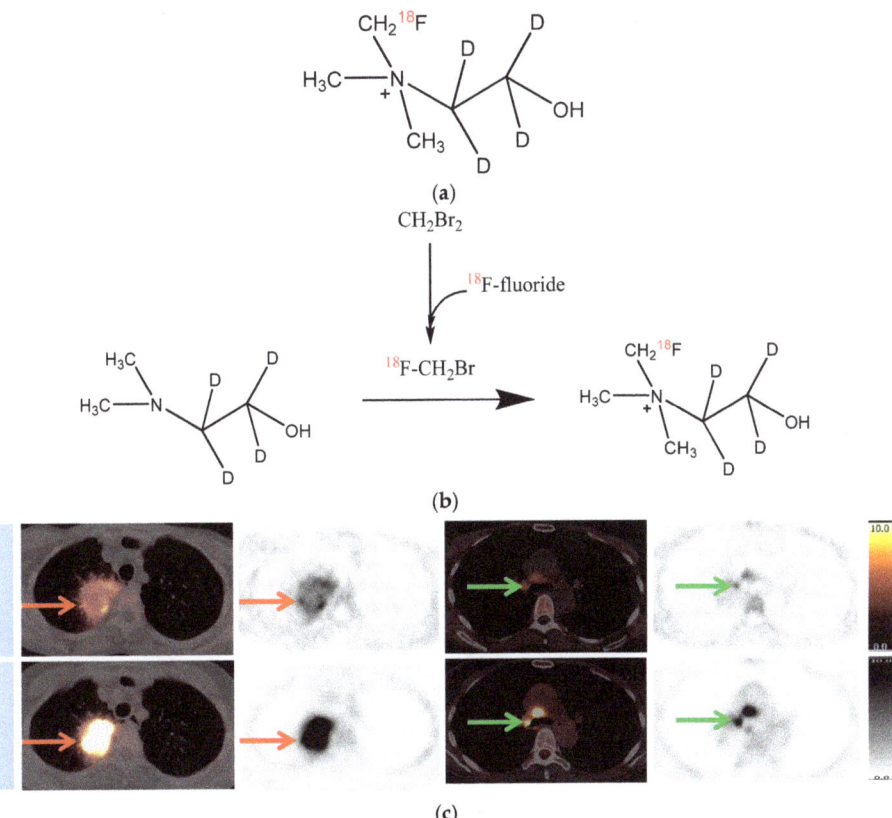

**Figure 1.** Chemical structure and synthesis of [$^{18}$F]D4-FCH, and images of [$^{18}$F]D4-FCH-PET in patients with non-small cell lung cancer (NSCLC). (**a**) Chemical structure of [$^{18}$F]D4-FCH. (**b**) Synthesis of [$^{18}$F]D4-FCH. No-carrier-added [$^{18}$F]D4-FCH is synthesised by reacting [$^{18}$F]fluorobromomethane, synthesised from dibromomethane by [$^{18}$F]fluoride-bromide substitution, with D4-N,N-dimethylaminoethanol precursor. (**c**) [$^{18}$F]D4-FCH (top row) and [$^{18}$F]FDG PET/CT images (bottom row) in a patient with NSCLC right upper lobe primary (red arrows) demonstrating tumour heterogeneity, and right paratracheal lymph node (green arrows).

## 3. Synthesis and Stability

[$^{18}$F]D4-FCH was prepared in a two-step reaction with a 12% overall yield, a final radiochemical purity of >99%, and a total synthesis time of 150 min. Alkylation with [$^{18}$F]fluoromethyl tosylate proved to be the most reliable radiosynthetic route [15].

The effect of deuterium substitution on bond strength was tested by evaluation of the chemical oxidation pattern using potassium permanganate. Chromatographic analyses showed that [$^{18}$F]D4-FCH was ~80% intact after treatment for 1 h with potassium permanganate, which oxidizes choline to choline betaine. Conversely, only 40% of nondeuterated FCH was intact. In the presence of choline oxidase, [$^{18}$F]D4-FCH was 29 ± 4% intact after treatment for 1 h, whereas only 11 ± 8% of FCH was intact. Therefore, [$^{18}$F]D4-FCH is more stable than FCH against oxidation [15].

## 4. In Vivo Biodistribution

In a murine HCT116 human colon xenograft model, in vivo biodistribution patterns of [$^{18}$F]D4-FCH, and [$^{18}$F]FCH were compared at 2, 30, and 60 min after injection [15].

[$^{18}$F]D4-FCH exhibited a higher accumulation in the tumours than [$^{18}$F]FCH. The distribution of all FCH radiotracers showed a similar uptake profile in most organs, with prominent radioactivity in the kidneys and liver. A pronounced increase in tumour uptake of [$^{18}$F]D4-FCH at the later time points was evident. A reduced uptake of [$^{18}$F]D4-FCH, relative to [$^{18}$F]FCH, in lung tissue may make imaging of thoracic tumours using this radiotracer superior to [$^{18}$F]FCH.

## 5. In Vivo Metabolic Stability

[$^{18}$F]D4-FCH was validated for imaging tumours preclinically and was found to be a very promising, metabolically stable radiotracer for imaging choline metabolism in tumours [17,18]. The improved stability (protection against oxidation by choline oxidase) is conferred by a $^{1}$H/$^{2}$D isotope effect, together with fluorine substitution [17,18].

## 6. In Vivo Response Assessment

The use of [$^{18}$F]D4-FCH for imaging response to CHK inhibition represents the most direct biomarker investigation. Trousil et al. showed that [$^{18}$F]D4-FCH reports reduction in phosphocholine formation when mice are treated with the CHK inhibitor ICL-CCIC-0019 [22]. Furthermore, Leyton et al. have demonstrated a marked reduction in radiotracer retention (all imaging variables) in the tumours treated with a mitogenic extracellular kinase inhibitor, despite a reduction in tumour size by only 12% after 10 days of treatment [17]. This suggests that [$^{18}$F]D4-FCH can be used to detect treatment response early during the course of therapy prior to any potential tumour size reduction. More recent studies highlight the role of hypoxia in [$^{18}$F]D4-FCH tumour biomarker response, suggesting that HIF-1α-responsive efflux transporters, including ABCB4, can modulate [$^{18}$F]D4-FCH uptake; consideration of efflux is supported in a mathematical analysis of [$^{18}$F]D4-FCH PET data [20]. Incidentally, an RNA-Seq analysis has identified an upregulation of ABCB1 and ABCB4 transporters as the main mechanisms involved in therapy resistance to CHK inhibitors in pancreatic cancer [23].

## 7. In Vitro and In Vivo Comparison of [$^{11}$C] and [$^{18}$F]choline Analogues

Witney et al. carried out biodistribution, metabolism, small-animal PET studies, and kinetic analysis of [$^{11}$C]choline, [$^{11}$C]methyl-[1,2-$_2$H$_4$]-choline ([$^{11}$C]D4-choline), and [$^{18}$F]D4-FCH uptake in human colon (HCT116), melanoma (A375), and prostate cancer (PC3-M) xenograft-bearing mice [18]. All tracers were converted intracellularly to their respective phosphocholine analogues. Their analyses have confirmed that deuteration and fluorination combine to provide protection against choline oxidation in vivo. Uptake of [$^{18}$F]D4-FCH was quantitatively similar in three tumours—HCT116, A375, and PC3-M—with similar radiotracer delivery ($K_1$) and CHKα expression, suggesting that [$^{18}$F]D4-FCH may have utility for cancer detection irrespective of histologic type.

## 8. Healthy Volunteer Biodistribution

In order to extend the pharmacokinetic aspects of the preclinical findings into human application, a first in-human study of [$^{18}$F]D4-FCH was performed on eight healthy volunteers to evaluate dosimetry, biodistribution, and safety [19]. [$^{18}$F]D4-FCH was well tolerated in all subjects, with no radiotracer-related serious adverse events reported. No significant changes in vital signs, clinical laboratory blood tests, or electrocardiograms were observed. The mean effective dose averaged over both males and females ($\pm$SD) was estimated to be 0.025 $\pm$ 0.004 (male 0.022 $\pm$ 0.002; female 0.027 $\pm$ 0.002) mSv/MBq, which is comparable with the ED of [$^{18}$F]FDG (0.019 mSv/MBq) [24]. The five organs receiving the highest-absorbed dose (mGy/MBq) were the kidneys (0.106 $\pm$ 0.03), liver (0.094 $\pm$ 0.03), pancreas (0.066 $\pm$ 0.01), urinary bladder wall (0.047 $\pm$ 0.02), and adrenals (0.046 $\pm$ 0.01). Elimination was through the renal and hepatic systems. The human biodistribution study has concluded that [$^{18}$F]D4-FCH is a safe PET radiotracer with a dosimetry profile comparable to other common [$^{18}$F] PET tracers.

## 9. First In-Patient (Lung Cancer) Evaluation

Given the low background uptake of [$^{18}$F]D4-FCH in the thorax [19], [$^{18}$F]D4-FCH was studied in 17 newly diagnosed non-small cell lung cancer (NSCLC) patients to evaluate tumour heterogeneity [20]. PET/CT scans were acquired concurrently with radioactive blood sampling to permit mathematical modelling of the blood-tissue transcellular rate constants. Comparisons were made with diagnostic [$^{18}$F]fluorodeoxyglucose (FDG) scans. The [$^{18}$F]D4-FCH-derived uptake (SUV$_{max}$) in index primary lesions ($n$ = 17) ranged between 2.87 and 10.13, which was lower than that of [$^{18}$F]FDG-PET (6.89 and 22.64, respectively). Mathematical modelling demonstrated the net irreversible uptake of [$^{18}$F]D4-FCH at steady-state, and parametric mapping of the entire tumour showed large intratumoural heterogeneity in radiotracer retention (Figure 1b). This highlights the potential for radiotherapy dose escalation.

As aforementioned, hypoxia is known to regulate CHKα activity and choline transport through hypoxia-inducible factor-1α (HIF-1α). This may confound the uptake of [$^{18}$F]D4-FCH. Moreover, in dynamic PET scans, most tumours exhibit a rapid radiotracer uptake phase followed by a plateaued time versus radioactivity (TAC) curve, regardless of their oxidative stability [15,18]. This suggests that other mechanisms are at play in tumour-radiolabelled choline biodistribution dynamics [20]. Li et al.'s work relating to the effect of hypoxia on [$^{18}$F]D4-FCH uptake [21] highlighted the export of phosphorylated [$^{18}$F]D4-FCH and [$^{18}$F]D4-FCHP via HIF-1α-responsive efflux transporters, including ABCB4. These findings are supported by the graphical analysis of NSCLC patient data with a compartmental modelling that accounts for the efflux [20].

## 10. Longitudinal Case Study in a Brain Tumour Patient

[$^{18}$F]FMC has been previously used to evaluate choline metabolism in patients with primary brain tumours, with one study showing that a combination of [$^{18}$F]FMC uptake on PET/CT and MR spectroscopy correlated with the tumour grade [25]. In order to assess the utility of choline uptake in the prediction of high-grade transformation of low-grade gliomas, a longitudinal case study was carried out using [$^{18}$F]D4-FCH. A patient with a low-grade glioma was recruited to the study following ethical approval and consent. The patient had two PET and MRI scans, 6 months apart. At baseline, a faint tracer uptake was visible in the lesion (Figure 2), which was stable on imaging 6 months later. There has been no history of transformation on routine follow-up.

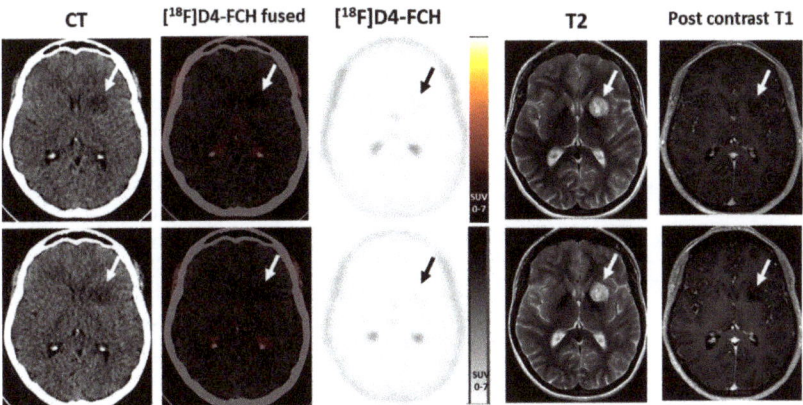

**Figure 2.** [$^{18}$F]D4–FCH PET/CT in a left thalamic low-grade glioma 6 months apart (**top row**, baseline; **bottom row**, 6 months later) showing minimal tracer uptake above background in the lesion, stable between scans, high signal on T2–weighted MRI imaging with no significant enhancement post-contrast. (Note physiological activity in the choroid plexus). Stable on continued follow-up 6 years later (not shown).

## 11. Response Evaluation in Metastatic Castrate-Resistant Prostate Cancer (mCRPC): Proof of Concept Study

Limitations of RECIST 1.1 [26], particularly in mCRPC patients with bone metastases [27], has led to the investigation of metabolic PET imaging methods. Other studies using [$^{18}$F]FCH have demonstrated the potential of choline tracers to monitor treatment to abiraterone or enzalutamide at early (3–6 weeks) times post-treatment [28,29]. In this original research, we examined temporal variations of [$^{18}$F]D4-FCH in mCRPC patients on abiraterone or enzalutamide in a prospective nonrandomized feasibility study.

mCRPC patients due to receive abiraterone (1000 mg with twice-daily 5 mg prednisolone in 28-day cycles)/enzalutamide (daily at 160 mg) were enrolled. PET/CT analyses were conducted at baseline, 4–6 weeks, and 12–16 weeks (detailed methodology in Supplementary Table S1). In brief, index lesions ≥10 mm with increased tracer uptake above background structures were outlined. For the follow-up visits, the same index lesions were outlined. For lesions no longer visible above the background, the background activity at the site of the lesion was outlined. Changes in the semiquantitative Standardized Uptake Value (SUV), both $SUV_{max}$ and $SUV_{mean}$, were calculated. In addition, for index lesions, tumour-to-background ratios (TBRs) were documented ($SUV_{max}$ lesion/$SUV_{mean}$ background) using background muscle for nodes and soft-tissue lesions and background bone for bone metastases, for both the early 4–6 week post-therapy scan and the 12–16 week post-therapy scan. For follow-up data, per-patient and per-lesion analyses were performed. [$^{18}$F]D4-FCH PET changes were compared to prostate-specific antigen (PSA) levels, prostate cancer working group 3 (PCWG3) response criteria [30], and survival duration. All statistical tests were run in Matlab (Mathworks, R2018b).

*Patient Characteristics and Optimal Imaging Time for [$^{18}$F]D4-FCH PET in mCRPC*

In this 'no-benefit trial', nine patients underwent a baseline PSA measurement, conventional staging, and [$^{18}$F]D4-FCH scan (Table 2). Of these, five patients had an early post-treatment scan at 4–6 weeks post-therapy (median 5.5 weeks; range 3.9–6.4) and seven patients completed midtreatment time points (12–16 weeks) post-therapy (median 15.3 weeks; range 13.1–17.3) (Figure 3). Six patients had stable disease, and one was a partial responder. A total of 55 lesions were documented. Of these, only 8 out of the 55 were measurable by RECIST 1.1 (two peritoneal, five nodes, and one bone, which had a soft-tissue component). Imaging features of mCRPC demonstrate significant response heterogeneity within and between patients (per-lesion analyses).

We found substantial temporal response heterogeneity in patients having either abiraterone or enzalutamide treatment. Significant reduction in [$^{18}$F]D4-FCH uptake on PET was seen in lesions, which were stable on conventional imaging (Figure 4). An initial decrease in TBR was seen in most bone lesions and soft-tissue lesions at visit 2, with subsequent increases in TBR in most lesions by visit 3. This could represent a 'flare effect' in the bone lesions. Another pattern was a mixed response (Figure 5). The heatmaps in Figure 6 show that the absolute uptake variable, as opposed to % change, was moderate (at TBR of 6.7) at visit 3 in patient (#2) who recorded a partial response; the bone lesion in this patient recorded a low uptake (TBR of 2.4) at visit 3. Notably, patients # 2, 4, and 5 all had one or more individual lesions with TBR values > 5 at visit 3 (Figure 6) signifying an escape from therapeutic pressures. The aggregate lesion uptake correlates with baseline PSA and associates with survival (per-patient analyses).

**Table 2.** Patient characteristics of the mCRPC response study.

| Pt No. | Age (year) | Metastatic Site (Number) | Drug Used | Baseline Parameters PSA (ng/mL) | TBR 5 min p.i. Sum | TBR 5 min p.i. Sumw | TBR 30 min p.i. Sum | TBR 30 min p.i. Sumw | 3-Month Parameters PSA (ng/mL) | TBR 5 min p.i. Sum | TBR 5 min p.i. Sumw | TBR 30 min p.i. Sum | TBR 30 min p.i. Sumw | PCWG3 (Subsequent Treatment) | Alive/Dead | PFS (m) | OS (m) |
|---|---|---|---|---|---|---|---|---|---|---|---|---|---|---|---|---|---|
| 1 | 75 | left SV, right iliac bone. | Enzalutamide | 95.25 | 28.45 | 14.31 | 38.83 | 19.42 | 1.08 | 3.80 | 1.98 | 4.89 | 1.34 | SD (Docetaxel) | Alive | 34.63 | 48.43 |
| 2 | 73 | node (1), T8 bone. | Enzalutamide | 10.54 | 19.80 | 13.98 | 17.48 | 10.81 | 2.22 | 11.37 | 5.55 | 8.76 | 1.18 | PR ¶ (Docetaxel) | Dead | 11.93 | 29.17 |
| 3 | 58 | bone (sacrum and scapula) | Abiraterone | 9.67 | 13.42 | 7.87 | 13.56 | 7.73 | 3.02 | 6.17 | 3.74 | 8.91 | 1.73 | SD (Docetaxel) | Alive | 47.73 | 53.80 |
| 4 | 63 | multiple bone mets (19), nodes (3) | Abiraterone | 33.44 | 170.51 | 8.65 | 191.21 | 9.52 | 14.5 | 181.36 | 9.60 | 211.00 | 0.81 | SD (Carboplatin/ Etoposide) | Dead | 9.13 | 29.53 |
| 5 | 80 | left SV, peritoneal lesions (2), C3 bone (1) | Abiraterone | 198 | 27.91 | 7.75 | 23.91 | 6.45 | 63.67 | 30.91 | 8.29 | 27.90 | 1.06 | SD (Cabazitaxel) | Dead | 14.47 | 38.67 |
| 6 | 74 | bone mets (7), node (1) | Enzalutamide | 228.82 | 18.51 | 3.34 | 21.25 | 3.26 | 92.95 | 14.88 | 2.33 | 14.88 | 2.17 | SD (Docetaxel) | Dead | 12.13 | 28.10 |
| 7 | 84 | right iliac bone (1). | Abiraterone | 27.19 | 2.62 | 2.62 | 4.10 | 4.10 | 8.17 | 1.38 | 1.38 | 1.69 | 1.69 | SD (‡ Nil) | Dead | 25.60 | 25.80 |
| 8 | 74 | nodes (2) | Enzalutamide | 22.54 | 7.70 | 3.50 | 6.62 | 3.50 | 14.90 | N/A | N/A | N/A | N/A | SD (Enzalutamide) | Alive | 36.00 | 36.00 |
| 9 | 69 | bone (2) | Enzalutamide | 94.47 | 97.32 | 8.56 | 98.73 | 8.77 | 32.78 | N/A | N/A | N/A | N/A | N/A († Treatment break) | Alive | 34.00 | 36.00 |

SV, seminal vesicle; TBR, tumour-to-background ratio; p.i., post-injection; PSA, prostate-specific antigen; N/A, not available as overseas. ¶ Had nodal disease that responded to treatment on the 3-month follow-up. Subsequently, patient showed multiple (routine PSMA-avid) bone metastases and died from their disease. ‡ Nil further due to Dementia. † Treatment break abroad (in COVID-19 lockdown). Then, progressive disease on return. Radium-223 and radiotherapy.

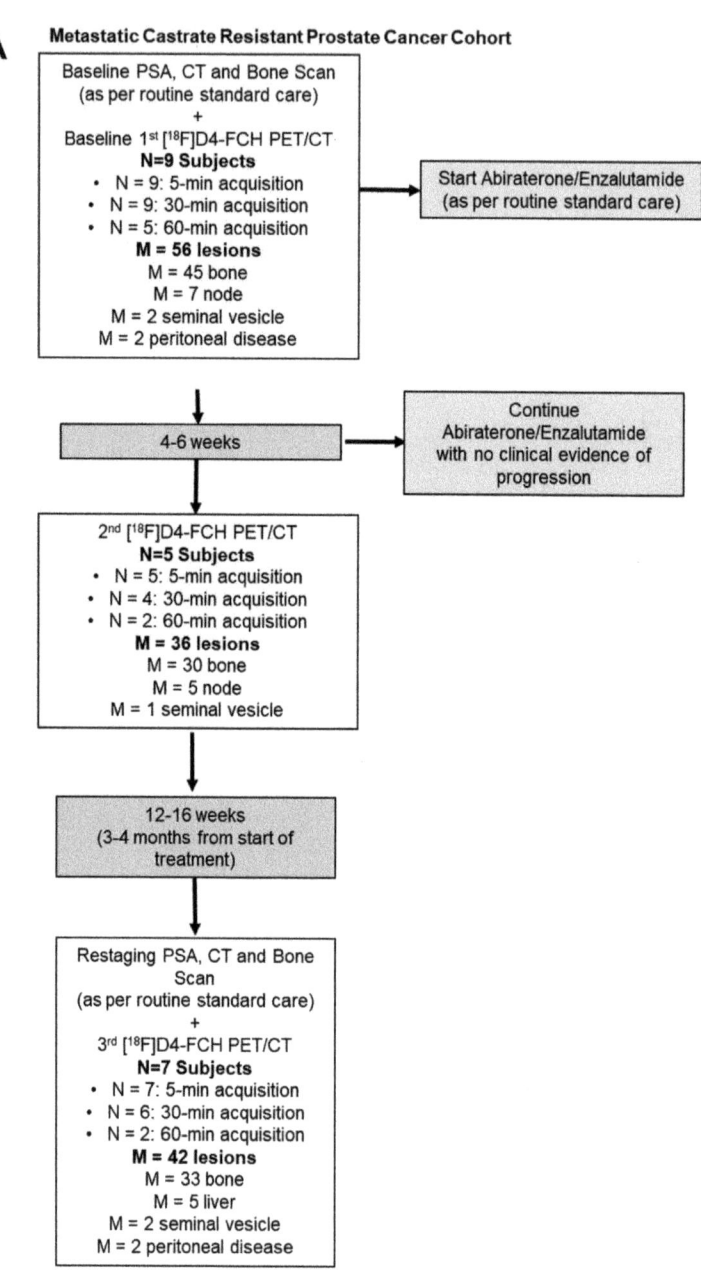

**Figure 3.** Cont.

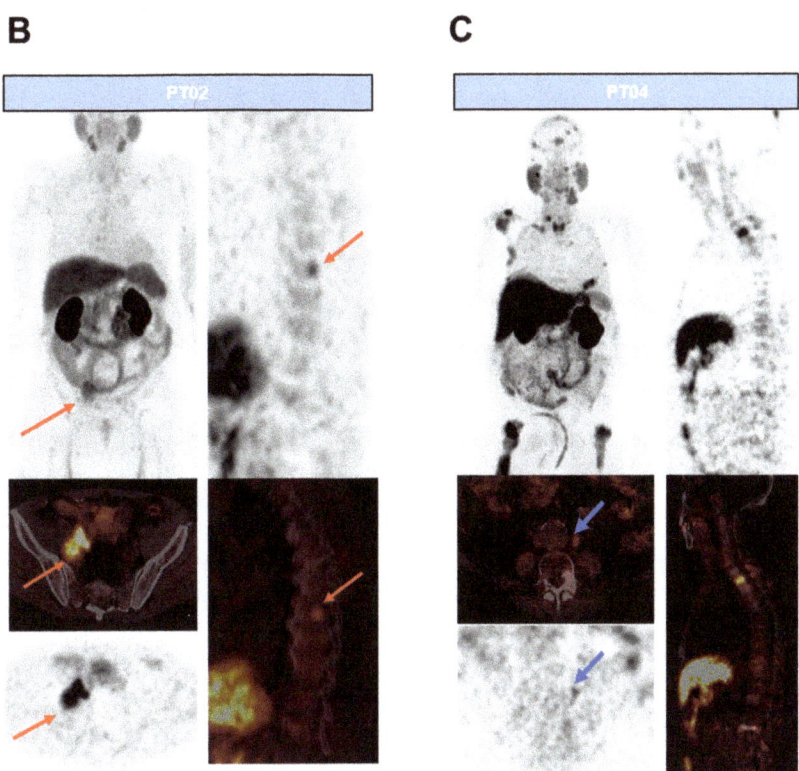

**Figure 3.** Study design and PET images. (**A**) Details of the PET study and summary of patients and lesions analysed. (**B**) Typical [$^{18}$F]D4-FCH uptake. MIP, axial, and sagittal views of PET and PET/CT of PT02, showing right external iliac node and T8 bone metastasis (red arrows). (**C**) Typical [$^{18}$F]D4-FCH uptake. MIP, axial, and sagittal views of PET and PET/CT of PT04, showing multiple bone metastases and a left paraaortic nodal metastasis (blue arrows).

TBR-wsum ensured that changes in lesions with low and high cholinic phenotypes alike were adequately captured. Baseline PSA or PSA at different time points was not associated with TBR-wsum variables (Supplementary Figure S1A,B). In contrast, change in PSA at 3 months compared to baseline correlated with similar changes in TBR-wsum (Supplementary Figure S1C). Generally, patients with larger TBR-wsum changes survived longer, with the exception being patient# 2 who had one lesion that showed TBR > 5 at visit 3 (Figure 6); this was not the case for PSA or PCWG3 (Supplementary Figure S2A–D). With the exception of patient# 2, TBR-wsum >30% gave a PFS advantage of >25 months, while <30% gives a PFS of <14 months (Supplementary Figure S2A–D). We did not have patients with PD or CR at 3 m post-therapy initiation; hence, the interpretation of these results should be restricted to patients with SD and PR. Neither baseline PSA nor TBR-wsum predicted survival ($p > 0.05$).

In this first use of [$^{18}$F]D4-FCH PET/CT in patients with prostate cancer, we show large interlesion heterogeneity, together with temporal heterogeneity in mCRPC patients undergoing treatment with abiraterone or enzalutamide. The treatment elicited mixed responses on PET at 12–16 weeks despite predominantly stable radiological appearances. We identify lesions that have escaped the selective pressures of cancer therapy and show that aggregate measures of [$^{18}$F]D4-FCH PET/CT from multiple lesions correlate with PSA and are associated with survival, thus providing a quantitative visual reflection of response heterogeneity.

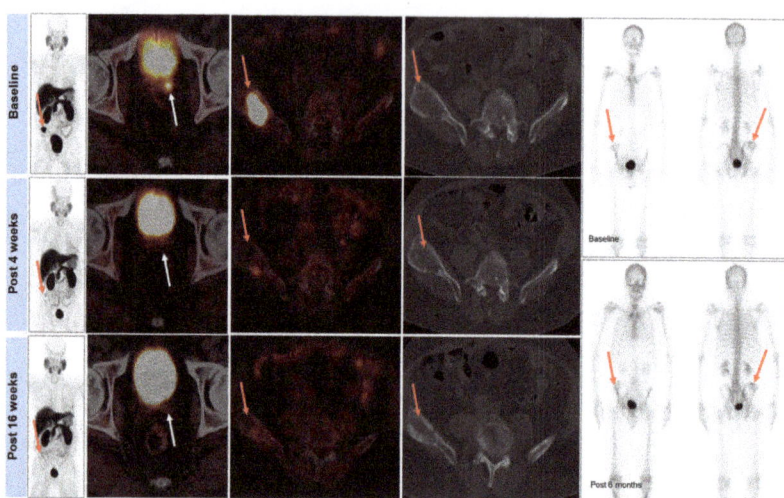

**Figure 4.** Response heterogeneity—changes in activity of lesions within individual patients at different times—detected by using [$^{18}$F]D4-FCH PET/CT. Representative PET images acquired at baseline (t1), early post-treatment (4–6 weeks; t2), and midtherapy (14–16 weeks; t3) demonstrating decreasing radiotracer uptake in a left seminal vesicle lesion (white arrows) and right iliac bone metastasis (red arrows).

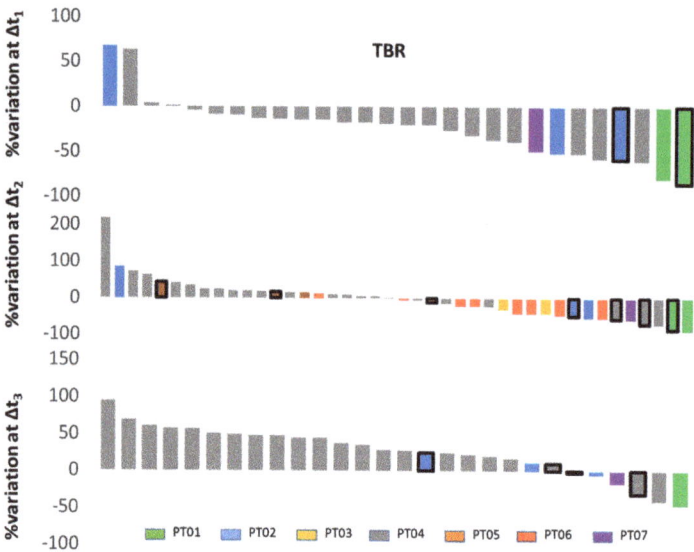

**Figure 5.** Relative lesional radiotracer uptake at the different time points represented as 'waterfall' plots of the percentage variation (%Δ1/2/3) of the tumour-to-background ratio (TBR). We only included patients who completed at least visit 1 and 3 scans within this analysis to avoid bias. The variations were evaluated as follows: %Δt1 = ((t2 − t1)/t1) × 100, n = 28 lesions; %Δt2 = ((t3 − t1)/t1) × 100, n = 48 lesions; %Δt3 = ((t3 − t2)/t2) × 100, n = 28 lesions. TBRBoneMet = SUV$_{max}$BoneMet/SUV$_{mean}$BackgBone. TBRSoftTissueMet = SUV$_{max}$SoftTissueMet/SUV$_{mean}$BackgMuscle (Backg: background). Soft-tissue metastases are identified with black-outlined bars (rest are bone metastases). Subject P02 is classified as partial responder based on PCWG3.

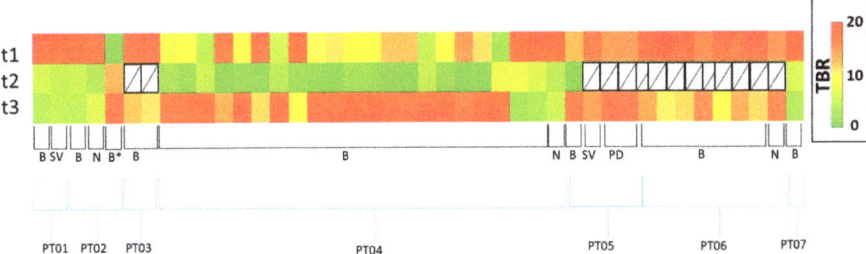

**Figure 6.** Response plots of the absolute TBR (per-lesion analysis) evaluated at the three time points, represented as 'heatmaps'. Of the seven patients who had both t1 and t3 scans, those who did not have t2 scans were excluded from the analysis. N: Node; SV: Seminal vesicle; PD: Peritoneal Disease; B: Bone. *: progressive bone metastasis in patient #2.

## 12. Discussion

Malignant transformation is characterised by aberrant phospholipid metabolism of cancers [6], associated with an upregulation of CHKα [2]. Choline metabolism has been studied utilizing magnetic resonance spectroscopy (MRS) of tumour tissue biopsies, as well as noninvasive tissue measurement [31]. PET-labelled choline tracers provide improved sensitivity when compared to MRS and enable dynamic measurements of the early steps of choline metabolism. Due to the metabolic instability of choline radiotracers and the increasing use of late-imaging protocols (~60 min, to permit elimination of nonspecific metabolites), we developed a more stable choline radiotracer, [$^{18}$F]D4-FCH [15]. Preclinical studies (Table 1) showed that [$^{18}$F]D4-FCH has improved protection against choline oxidase, the key choline catabolic enzyme, via a $^{1}$H/$^{2}$D isotope effect, together with fluorine substitution [15,17,18].

To date, [$^{11}$C]choline and [$^{18}$F]FCH have been extensively used for the clinical imaging of prostate, brain, breast, hepatocellular (HCC), and esophageal carcinomas [32,33]. Choline metabolism is altered in gliomas, and pilot clinical studies have shown a differential uptake of choline radiotracers between glioma and normal brain tissue and between gliomas and other disease processes. Sollars et al. have shown that [$^{18}$F]FMC PET/CT differentiated WHO (World Health Organization) grade IV from grade II and III tumours. Tumoural [$^{18}$F]FMC PET-CT uptake was higher than in normal-appearing white matter across all grades and markedly elevated within regions of contrast enhancement. This uptake was independent of choline kinase expression [25].

Tumour differentiation is a major predictive factor of post-operative recurrence in HCC [34]. However, the histological analysis of tumour differentiation, which remains the gold standard, is currently carried out only in atypical cases. Conventional imaging is essential for the management of HCC patients [35], but its limited value motivated the use of PET imaging although still not consensually recommended. As [$^{18}$F]FDG shows limited sensitivity to detect HCC, choline PET has been proposed as a complementary diagnostic tool [36]. The [$^{18}$F]FDG/[$^{18}$F]choline dual-tracer PET behaviour of uptake shows a high overlap between well- and less-differentiated HCC, making the characterization of tumours challenging based on such a PET combination [9]. Using [$^{18}$F]D4-FCH may potentially improve the sensitivity and is worth further evaluation.

We have shown previously that the choline-based radiotracer, [$^{11}$C]choline, which correlates with CHKα expression and represents a proliferation-independent phenotype in prostate cancer [37], decreased predictably following androgen deprivation [38], interpreted as a reduction in choline transport/metabolism or loss of cell viability (similar directionality of change) [39].

Recently, prostate-specific membrane antigen (PSMA)-PET has taken centre stage in functional imaging of prostate cancer and is superseding choline PET. A recent meta-analysis of a head-to-head comparison of detection rates (DRs) between radiolabelled choline and PSMA PET/CT has shown that the overall DR of radiolabelled PSMA PET/CT on a per-patient-based analysis is higher compared to that of radiolabelled choline PET/CT (78% versus 56%, respectively) in the setting of biochemical relapse [40]. This is due to the higher contrast and tumour uptake of PSMA than radiolabelled choline.

In rationalising the use of choline and PSMA PET, it is appreciated that PSMA expression is inversely related to androgen response; androgen deprivation, or abiraterone treatment of human castration-resistant PCa cell line VCaP-stimulated PSMA expression and increased PSMA-based radiotracer uptake [41]. On the other hand, a loss of cell viability will be expected to decrease radiotracer uptake; thus, while human studies with PSMA-PET to monitor therapy response following arbiraterone or enzalutamide have been reported [37,42], the interpretation of the data is challenging [38].

[$^{18}$F]D4-FCH PET/CT shows large lesion variability in response to abiraterone or enzalutamide, suggesting an escape from selective pressures of therapy. PET variables including TBR-wsum predicted the duration of survival, depending on individual lesion response. Future studies should elaborate how this variable, together with progression (escape of the resistant lesion(s)) on PET influences progression-free and overall survival in a larger patient cohort and encourages exploration of choline-transport-targeted theranostics as recently suggested [43].

Choline PET/CT imaging is gathering momentum in the localisation of parathyroid adenomas to guide parathyroid surgery with high detection rates in patients with primary hyperparathyroidism [44]. However, in the more challenging persistent/recurrent primary hyperparathyroidism cases, detection rates are lower and the potential improved sensitivity of [$^{18}$F]D4-FCH PET could be explored in this setting.

Paired [$^{18}$F]FCH and [$^{18}$F]FDG PET have been shown to predict a 6-month response to $^{90}$Y-transarterial radioembolisation in hepatocellular carcinoma (HCC) [45]. It also has the potential to identify metabolically active tumour remnants after $^{90}$Y-transarterial radioembolisation in HCC [46].

As CHKα is overexpressed in a number of cancers and has a role in the onset and progression of some cancers, choline kinase inhibitors have been proposed as novel therapeutic targets [47]. CHKα inhibitors are being evaluated in a first in-human phase 1 trial in patients with advanced solid tumours [48]. [$^{18}$F]D4-FCH PET imaging may provide a noninvasive biomarker for developing and assessing the mechanism of action of future choline kinase inhibitors.

## 13. Conclusions

[$^{18}$F]D4-FCH has improved protection against choline oxidase, the key choline catabolic enzyme, via a $^{1}$H/$^{2}$D isotope effect, together with fluorine substitution. Due to the promising mechanistic and safety profiles of [$^{18}$F]D4-FCH in vitro and preclinically in vivo, the radiotracer has transitioned to clinical development. [$^{18}$F]D4-FCH is a safe PET tracer, with a favourable radiation dosimetry profile for clinical imaging. [$^{18}$F]D4-FCH PET-CT in lung and prostate cancers has shown highly heterogeneous intratumoural distribution and large lesion variability suggesting a use for potential radiotherapy dose escalation and treatment response monitoring.

**Supplementary Materials:** The following supporting information can be downloaded at: https://www.mdpi.com/article/10.3390/molecules28248018/s1, Figure S1: Patient level comparison of PSA with [18F]D4-FCH PET/CT; Figure S2: Patient level prediction of progression-free survival using current approaches and by [18F]D4-FCH PET/CT. Table S1: Methodology of [18F]D4-FCH studies.

**Author Contributions:** Conceptualization: A.C., T.D.B., S.R.D., E.O.A. and S.M. Methodology: A.C., T.D.B., S.R.D., M.I., M.G.-S., K.K., H.T., N.H.P., M.W., P.F., A.S., A.B., A.F., J.S.D.B., E.O.A. and S.M. Formal Analysis: A.C., T.D.B., S.R.D., M.I., K.K., H.T., N.H.P. and A.B. Resources: M.G.-S., M.W., P.F., A.S., A.F., J.S.D.B., E.O.A. and S.M. Data Curation: A.C., T.D.B., S.R.D., E.O.A. and S.M. Writing—Original Draft Preparation: A.C., T.D.B., S.R.D., M.I., M.G.-S., K.K., H.T., N.H.P., M.W., P.F., A.S., A.B., A.F., J.S.D.B., E.O.A. and S.M. Writing—Review and Editing: A.C., T.D.B., S.R.D. and

E.O.A. Supervision: E.O.A. Funding Acquisition: E.O.A. All authors have read and agreed to the published version of the manuscript.

**Funding:** These studies were made possible by funds from the UK Medical Research Council award MR/N020782/1 and the National Institute for Health Research (NIHR) Biomedical Research Centre award to the Imperial College Healthcare NHS Trust. The authors would like to acknowledge support from the Imperial College Experimental Cancer Medicines Centre, and Cancer Research UK funded National Cancer Imaging Translational Accelerator award (NCITA; C2536/A28680).

**Institutional Review Board Statement:** All procedures performed in the studies involving human participants were in accordance with the ethical standards of the institutional research committee and with the 1964 Helsinki declaration and its later amendments or comparable ethical standards. The institutional review board (including Health Research Authority, Research Ethics, Research and Development, and Administration of Radioactive Substances Advisory Committee) approved the study under Integrated Research Application System IRAS ID: 202727.

**Informed Consent Statement:** Informed consent was obtained from all subjects involved in the study.

**Data Availability Statement:** Data supporting the conclusions of this manuscript are included within the article and the Supplementary Figures.

**Acknowledgments:** The authors would like to thank all patients who selflessly contributed to this research and staff at Invicro Ltd. who supported radiotracer production and PET studies.

**Conflicts of Interest:** Abiraterone was first designed and synthesized at The Institute of Cancer Research (PF and JSDB are employees), which has a commercial interest in the drug. The authors declare no conflict of interest.

## References

1. Zeisel, S.H. Dietary choline: Biochemistry, physiology, and pharmacology. *Annu. Rev. Nutr.* **1981**, *1*, 95–121. [CrossRef] [PubMed]
2. Aboagye, E.O.; Bhujwalla, Z.M. Malignant transformation alters membrane choline phospholipid metabolism of human mammary epithelial cells. *Cancer Res.* **1999**, *59*, 80–84. [PubMed]
3. Gibellini, F.; Smith, T.K. The Kennedy pathway--De novo synthesis of phosphatidylethanolamine and phosphatidylcholine. *IUBMB Life* **2010**, *62*, 414–428. [CrossRef]
4. Yoshimoto, M.; Waki, A.; Obata, A.; Furukawa, T.; Yonekura, Y.; Fujibayashi, Y. Radiolabeled choline as a proliferation marker: Comparison with radiolabeled acetate. *Nucl. Med. Biol.* **2004**, *31*, 859–865. [CrossRef] [PubMed]
5. Ramirez de Molina, A.; Penalva, V.; Lucas, L.; Lacal, J.C. Regulation of choline kinase activity by Ras proteins involves Ral-GDS and PI3K. *Oncogene* **2002**, *21*, 937–946. [CrossRef]
6. Ramirez de Molina, A.; Rodriguez-Gonzalez, A.; Gutierrez, R.; Martinez-Pineiro, L.; Sanchez, J.; Bonilla, F.; Rosell, R.; Lacal, J. Overexpression of choline kinase is a frequent feature in human tumor-derived cell lines and in lung, prostate, and colorectal human cancers. *Biochem. Biophys. Res. Commun.* **2002**, *296*, 580–583. [CrossRef] [PubMed]
7. Ramirez de Molina, A.; Banez-Coronel, M.; Gutierrez, R.; Rodriguez-Gonzalez, A.; Olmeda, D.; Megias, D.; Lacal, J.C. Choline kinase activation is a critical requirement for the proliferation of primary human mammary epithelial cells and breast tumor progression. *Cancer Res.* **2004**, *64*, 6732–6739. [CrossRef]
8. Alongi, P.; Laudicella, R.; Lanzafame, H.; Farolfi, A.; Mapelli, P.; Picchio, M.; Burger, I.A.; Iagaru, A.; Minutoli, F.; Evangelista, L. PSMA and Choline PET for the Assessment of Response to Therapy and Survival Outcomes in Prostate Cancer Patients: A Systematic Review from the Literature. *Cancers* **2022**, *14*, 1770. [CrossRef]
9. Ghidaglia, J.; Golse, N.; Pascale, A.; Sebagh, M.; Besson, F.L. $^{18}$F-FDG/$^{18}$F-Choline Dual-Tracer PET Behavior and Tumor Differentiation in HepatoCellular Carcinoma. A Systematic Review. *Front. Med.* **2022**, *9*, 924824. [CrossRef] [PubMed]
10. Hara, T.; Kosaka, N.; Kishi, H. PET imaging of prostate cancer using carbon-11-choline. *J. Nucl. Med.* **1998**, *39*, 990–995. [PubMed]
11. Hara, T.; Kosaka, N.; Kishi, H. Development of $^{18}$F-fluoroethylcholine for cancer imaging with PET: Synthesis, biochemistry, and prostate cancer imaging. *J. Nucl. Med.* **2002**, *43*, 187–199.
12. DeGrado, T.R.; Coleman, R.E.; Wang, S.; Baldwin, S.W.; Orr, M.D.; Robertson, C.N.; Polascik, T.J.; Price, D.T. Synthesis and evaluation of $^{18}$F-labeled choline as an oncologic tracer for positron emission tomography: Initial findings in prostate cancer. *Cancer Res.* **2001**, *61*, 110–117.
13. Bansal, A.; Shuyan, W.; Hara, T.; Harris, R.A.; Degrado, T.R. Biodisposition and metabolism of [$^{18}$F]fluorocholine in 9L glioma cells and 9L glioma-bearing fisher rats. *Eur. J. Nucl. Med. Mol. Imaging* **2008**, *35*, 1192–1203. [CrossRef] [PubMed]
14. Roivainen, A.; Forsback, S.; Gronroos, T.; Lehikoinen, P.; Kahkonen, M.; Sutinen, E.; Minn, H. Blood metabolism of [methyl-11C]choline; implications for in vivo imaging with positron emission tomography. *Eur. J. Nucl. Med.* **2000**, *27*, 25–32. [CrossRef]
15. Smith, G.; Zhao, Y.; Leyton, J.; Shan, B.; Nguyen, Q.D.; Perumal, M.; Turton, D.; Arstad, E.; Luthra, S.K.; Robins, E.G.; et al. Radiosynthesis and pre-clinical evaluation of [$^{18}$F]fluoro-[1,2-(2)H(4)]choline. *Nucl. Med. Biol.* **2011**, *38*, 39–51. [CrossRef] [PubMed]

16. Gadda, G. pH and deuterium kinetic isotope effects studies on the oxidation of choline to betaine-aldehyde catalyzed by choline oxidase. *Biochim. Biophys. Acta* **2003**, *1650*, 4–9. [CrossRef] [PubMed]
17. Leyton, J.; Smith, G.; Zhao, Y.; Perumal, M.; Nguyen, Q.D.; Robins, E.; Arstad, E.; Aboagye, E.O. [$^{18}$F]fluoromethyl-[1,2-2H4]-choline: A novel radiotracer for imaging choline metabolism in tumors by positron emission tomography. *Cancer Res.* **2009**, *69*, 7721–7728. [CrossRef] [PubMed]
18. Witney, T.H.; Alam, I.S.; Turton, D.R.; Smith, G.; Carroll, L.; Brickute, D.; Twyman, F.J.; Nguyen, Q.D.; Tomasi, G.; Awais, R.O.; et al. Evaluation of deuterated $^{18}$F- and $^{11}$C-labeled choline analogs for cancer detection by positron emission tomography. *Clin. Cancer Res.* **2012**, *18*, 1063–1072. [CrossRef] [PubMed]
19. Challapalli, A.; Sharma, R.; Hallett, W.A.; Kozlowski, K.; Carroll, L.; Brickute, D.; Twyman, F.; Al-Nahhas, A.; Aboagye, E.O. Biodistribution and radiation dosimetry of deuterium-substituted $^{18}$F-fluoromethyl-[1,2-2H4]choline in healthy volunteers. *J. Nucl. Med.* **2014**, *55*, 256–263. [CrossRef] [PubMed]
20. Dubash, S.; Inglese, M.; Mauri, F.; Kozlowski, K.; Trivedi, P.; Arshad, M.; Challapalli, A.; Barwick, T.; Al-Nahhas, A.; Stanbridge, R.; et al. Spatial heterogeneity of radiolabeled choline positron emission tomography in tumors of patients with non-small cell lung cancer: First-in-patient evaluation of [$^{18}$F]fluoromethyl-(1,2-(2)H(4))-choline. *Theranostics* **2020**, *10*, 8677–8690. [CrossRef]
21. Li, Y.; Inglese, M.; Dubash, S.; Barnes, C.; Brickute, D.; Braga, M.C.; Wang, N.; Beckley, A.; Heinzmann, K.; Allott, L.; et al. Consideration of Metabolite Efflux in Radiolabelled Choline Kinetics. *Pharmaceutics* **2021**, *13*, 1246. [CrossRef] [PubMed]
22. Trousil, S.; Kaliszczak, M.; Schug, Z.; Nguyen, Q.D.; Tomasi, G.; Favicchio, R.; Brickute, D.; Fortt, R.; Twyman, F.J.; Carroll, L.; et al. The novel choline kinase inhibitor ICL-CCIC-0019 reprograms cellular metabolism and inhibits cancer cell growth. *Oncotarget* **2016**, *7*, 37103–37120. [CrossRef] [PubMed]
23. Mazarico, J.M.; Sanchez-Arevalo Lobo, V.J.; Favicchio, R.; Greenhalf, W.; Costello, E.; Carrillo-de Santa Pau, E.; Marques, M.; Lacal, J.C.; Aboagye, E.; Real, F.X. Choline Kinase Alpha (CHKalpha) as a Therapeutic Target in Pancreatic Ductal Adenocarcinoma: Expression, Predictive Value, and Sensitivity to Inhibitors. *Mol. Cancer Ther.* **2016**, *15*, 323–333. [CrossRef] [PubMed]
24. Radiation dose to patients from radiopharmaceuticals (addendum 2 to ICRP publication 53). *Ann. ICRP* **1998**, *28*, 1–126.
25. Grech-Sollars, M.; Ordidge, K.L.; Vaqas, B.; Davies, C.; Vaja, V.; Honeyfield, L.; Camp, S.; Towey, D.; Mayers, H.; Peterson, D.; et al. Imaging and Tissue Biomarkers of Choline Metabolism in Diffuse Adult Glioma: 18F-Fluoromethylcholine PET/CT, Magnetic Resonance Spectroscopy, and Choline Kinase alpha. *Cancers* **2019**, *11*, 1969. [CrossRef] [PubMed]
26. Eisenhauer, E.A.; Therasse, P.; Bogaerts, J.; Schwartz, L.H.; Sargent, D.; Ford, R.; Dancey, J.; Arbuck, S.; Gwyther, S.; Mooney, M.; et al. New response evaluation criteria in solid tumours: Revised RECIST guideline (version 1.1). *Eur. J. Cancer* **2009**, *45*, 228–247. [CrossRef]
27. Bubendorf, L.; Schopfer, A.; Wagner, U.; Sauter, G.; Moch, H.; Willi, N.; Gasser, T.C.; Mihatsch, M.J. Metastatic patterns of prostate cancer: An autopsy study of 1589 patients. *Hum. Pathol.* **2000**, *31*, 578–583. [CrossRef]
28. Challapalli, A.; Barwick, T.; Tomasi, G.; Doherty, M.O.; Contractor, K.; Stewart, S.; Al-Nahhas, A.; Behan, K.; Coombes, C.; Aboagye, E.O.; et al. Exploring the potential of [$^{11}$C]choline-PET/CT as a novel imaging biomarker for predicting early treatment response in prostate cancer. *Nucl. Med. Commun.* **2014**, *35*, 20–29. [CrossRef]
29. Inazu, M. Choline transporter-like proteins CTLs/SLC44 family as a novel molecular target for cancer therapy. *Biopharm. Drug Dispos.* **2014**, *35*, 431–449. [CrossRef] [PubMed]
30. Scher, H.I.; Morris, M.J.; Stadler, W.M.; Higano, C.; Basch, E.; Fizazi, K.; Antonarakis, E.S.; Beer, T.M.; Carducci, M.A.; Chi, K.N.; et al. Trial Design and Objectives for Castration-Resistant Prostate Cancer: Updated Recommendations From the Prostate Cancer Clinical Trials Working Group 3. *J. Clin. Oncol.* **2016**, *34*, 1402–1418. [CrossRef]
31. Glunde, K.; Bhujwalla, Z.M. Metabolic tumor imaging using magnetic resonance spectroscopy. *Semin. Oncol.* **2011**, *38*, 26–41. [CrossRef]
32. Treglia, G.; Giovannini, E.; Di Franco, D.; Calcagni, M.L.; Rufini, V.; Picchio, M.; Giordano, A. The role of positron emission tomography using carbon-11 and fluorine-18 choline in tumors other than prostate cancer: A systematic review. *Ann. Nucl. Med.* **2012**, *26*, 451–461. [CrossRef] [PubMed]
33. Umbehr, M.H.; Muntener, M.; Hany, T.; Sulser, T.; Bachmann, L.M. The Role of 11C-Choline and 18F-Fluorocholine Positron Emission Tomography (PET) and PET/CT in Prostate Cancer: A Systematic Review and Meta-analysis. *Eur. Urol.* **2013**, *64*, 106–117. [CrossRef] [PubMed]
34. Cucchetti, A.; Piscaglia, F.; Caturelli, E.; Benvegnu, L.; Vivarelli, M.; Ercolani, G.; Cescon, M.; Ravaioli, M.; Grazi, G.L.; Bolondi, L.; et al. Comparison of recurrence of hepatocellular carcinoma after resection in patients with cirrhosis to its occurrence in a surveilled cirrhotic population. *Ann. Surg. Oncol.* **2009**, *16*, 413–422. [CrossRef] [PubMed]
35. Llovet, J.M.; Lencioni, R. mRECIST for HCC: Performance and novel refinements. *J. Hepatol.* **2020**, *72*, 288–306. [CrossRef]
36. Park, J.W.; Kim, J.H.; Kim, S.K.; Kang, K.W.; Park, K.W.; Choi, J.I.; Lee, W.J.; Kim, C.M.; Nam, B.H. A prospective evaluation of $^{18}$F-FDG and $^{11}$C-acetate PET/CT for detection of primary and metastatic hepatocellular carcinoma. *J. Nucl. Med.* **2008**, *49*, 1912–1921. [CrossRef]
37. Zukotynski, K.A.; Emmenegger, U.; Hotte, S.; Kapoor, A.; Fu, W.; Blackford, A.L.; Valliant, J.; Benard, F.; Kim, C.K.; Markowski, M.C.; et al. Prospective, Single-Arm Trial Evaluating Changes in Uptake Patterns on Prostate-Specific Membrane Antigen-Targeted $^{18}$F-DCFPyL PET/CT in Patients with Castration-Resistant Prostate Cancer Starting Abiraterone or Enzalutamide. *J. Nucl. Med.* **2021**, *62*, 1430–1437. [CrossRef]

38. Hofman, M.S.; Hicks, R.J.; Maurer, T.; Eiber, M. Prostate-specific Membrane Antigen PET: Clinical Utility in Prostate Cancer, Normal Patterns, Pearls, and Pitfalls. *Radiographics* **2018**, *38*, 200–217. [CrossRef]
39. Contractor, K.; Challapalli, A.; Barwick, T.; Winkler, M.; Hellawell, G.; Hazell, S.; Tomasi, G.; Al-Nahhas, A.; Mapelli, P.; Kenny, L.M.; et al. Use of [$^{11}$C]choline PET-CT as a noninvasive method for detecting pelvic lymph node status from prostate cancer and relationship with choline kinase expression. *Clin. Cancer Res.* **2011**, *17*, 7673–7683. [CrossRef]
40. Treglia, G.; Pereira Mestre, R.; Ferrari, M.; Bosetti, D.G.; Pascale, M.; Oikonomou, E.; De Dosso, S.; Jermini, F.; Prior, J.O.; Roggero, E.; et al. Radiolabelled choline versus PSMA PET/CT in prostate cancer restaging: A meta-analysis. *Am. J. Nucl. Med. Mol. Imaging* **2019**, *9*, 127–139.
41. Meller, B.; Bremmer, F.; Sahlmann, C.O.; Hijazi, S.; Bouter, C.; Trojan, L.; Meller, J.; Thelen, P. Alterations in androgen deprivation enhanced prostate-specific membrane antigen (PSMA) expression in prostate cancer cells as a target for diagnostics and therapy. *EJNMMI Res.* **2015**, *5*, 66. [CrossRef] [PubMed]
42. Oruc, Z.; Guzel, Y.; Ebinc, S.; Komek, H.; Kucukoner, M.; Kaplan, M.A.; Oruc, I.; Urakci, Z.; Isikdogan, A. Efficacy of 68Ga-PSMA PET/CT-derived whole-body volumetric parameters in predicting response to second-generation androgen receptor axis-targeted therapy, and the prognosis in metastatic hormone-refractory prostate cancer patients. *Nucl. Med. Commun.* **2021**, *42*, 1336–1346. [CrossRef] [PubMed]
43. Svec, P.; Novy, Z.; Kucka, J.; Petrik, M.; Sedlacek, O.; Kuchar, M.; Liskova, B.; Medvedikova, M.; Kolouchova, K.; Groborz, O.; et al. Iodinated Choline Transport-Targeted Tracers. *J. Med. Chem.* **2020**, *63*, 15960–15978. [CrossRef]
44. Quak, E.; Cavarec, M.; Ciappuccini, R.; Girault, G.; Rouzier, R.; Lequesne, J. Detection, resection and cure: A systematic review and meta-analysis of $^{18}$F-choline PET in primary hyperparathyroidism. *Q. J. Nucl. Med. Mol. Imaging* **2023**, *67*, 122–129. [CrossRef] [PubMed]
45. Reizine, E.; Chalaye, J.; Mule, S.; Regnault, H.; Perrin, C.; Calderaro, J.; Laurent, A.; Amaddeo, G.; Kobeiter, H.; Tacher, V.; et al. Utility of Early Posttreatment PET/CT Evaluation Using FDG or $^{18}$F-FCH to Predict Response to $^{90}$Y Radioembolization in Patients with Hepatocellular Carcinoma. *AJR Am. J. Roentgenol.* **2022**, *218*, 359–369. [CrossRef]
46. Filippi, L.; Bagni, O.; Notarianni, E.; Saltarelli, A.; Ambrogi, C.; Schillaci, O. PET/CT with $^{18}$F-choline or $^{18}$F-FDG in Hepatocellular Carcinoma Submitted to $^{90}$Y-TARE: A Real-World Study. *Biomedicines* **2022**, *10*, 2996. [CrossRef]
47. Lacal, J.C.; Zimmerman, T.; Campos, J.M. Choline Kinase: An Unexpected Journey for a Precision Medicine Strategy in Human Diseases. *Pharmaceutics* **2021**, *13*, 788. [CrossRef] [PubMed]
48. Study of Intravenous TCD-717 in Patients with Advanced Solid Tumors. Available online: https://clinicaltrials.gov/ct2/show/NCT01215864 (accessed on 10 April 2023).

**Disclaimer/Publisher's Note:** The statements, opinions and data contained in all publications are solely those of the individual author(s) and contributor(s) and not of MDPI and/or the editor(s). MDPI and/or the editor(s) disclaim responsibility for any injury to people or property resulting from any ideas, methods, instructions or products referred to in the content.

Article

# Quality Assurance Investigations and Impurity Characterization during Upscaling of [$^{177}$Lu]Lu-PSMA$^{I\&T}$

Stefan Schmitl [1], Julia Raitanen [1,2,3,4], Stephan Witoszynskyj [1], Eva-Maria Patronas [1], Lukas Nics [1], Marius Ozenil [1], Victoria Weissenböck [1], Thomas L. Mindt [2,3,5], Marcus Hacker [1], Wolfgang Wadsak [1], Marie R. Brandt [2,3,5,*] and Markus Mitterhauser [1,2,3,5]

[1] Department of Biomedical Imaging and Image-Guided Therapy, Division of Nuclear Medicine, Medical University of Vienna, 1090 Vienna, Austria
[2] Ludwig Boltzmann Institute Applied Diagnostics, AKH Wien c/o Sekretariat Nuklearmedizin, 1090 Vienna, Austria
[3] Department of Inorganic Chemistry, Faculty of Chemistry, University of Vienna, 1090 Vienna, Austria
[4] Vienna Doctoral School of Chemistry (DoSChem), University of Vienna, 1090 Vienna, Austria
[5] Joint Applied Medicinal Radiochemistry Facility, University of Vienna & Medical University of Vienna, 1090 Vienna, Austria
\* Correspondence: marie.brandt@univie.ac.at

**Abstract:** [$^{177}$Lu]Lu-PSMA$^{I\&T}$ is widely used for the radioligand therapy of metastatic castration-resistant prostate cancer (mCRPC). Since this kind of therapy has gained a large momentum in recent years, an upscaled production process yielding multiple patient doses in one batch has been developed. During upscaling, the established production method as well as the HPLC quality control were challenged. A major finding was a correlation between the specific activity and the formation of a pre-peak, presumably caused by radiolysis. Hence, nonradioactive reference standards were irradiated with an X-ray source and the formed pre-peak was subsequently identified as a deiodination product by UPLC-MS. To confirm the occurrence of the same deiodinated side product in the routine batch, a customized deiodinated precursor was radiolabeled and analyzed with the same HPLC setup, revealing an identical retention time to the pre-peak in the formerly synthesized routine batches. Additionally, further cyclization products of [$^{177}$Lu]Lu-PSMA$^{I\&T}$ were identified as major contributors to radiochemical impurities. The comparison of two HPLC methods showed the likelihood of the overestimation of the radiochemical purity during the synthesis of [$^{177}$Lu]Lu-PSMA$^{I\&T}$. Finally, a prospective cost reduction through an optimization of the production process was shown.

**Keywords:** [$^{177}$Lu]Lu-PSMA$^{-I\&T}$; radiolysis; radioligand therapy; upscaling; quality assurance; HPLC; UPLC-MS

## 1. Introduction

[$^{177}$Lu]Lu-PSMA$^{I\&T}$ (INN: Lutetium ($^{177}$Lu) zadavotide guraxetan) [1] is one of the main PSMA-directed radiopharmaceuticals that are successfully used for the treatment of metastatic castration-resistant prostate cancer in nuclear medicine departments around the world [2]. Lutetium-177 decays via $\beta^-/\gamma$ emission to its daughter nuclide hafnium-176 ($^{176}$Hf). With a physical half-life of 6.647 days and a maximum tissue penetration range of <2 mm of its emitted β particles, it displays beneficial properties for the deposition of high radiation doses to tumor lesions and metastases while sparing the surrounding tissue [3].

The success of PSMA-directed radioligand therapy is not only reflected by the very recent approval of [$^{177}$Lu]Lu-PSMA-617 (Pluvicto®) for mCRPC in the US and EU, but also by the increasing clinical demand for comparable agents [2,4,5]. In the literature, administered activities ranging from 3.7 to 9.3 GBq were described for [$^{177}$Lu]Lu-PSMA therapy in metastatic castration-resistant prostate cancer (mCRPC) [6]. In our clinic, 7.4 GBq of

[$^{177}$Lu]Lu-PSMA$^{I\&T}$ every six weeks represents the current standard in mCRPC treatment, based on the marketing authorization of Pluvicto®. However, lower levels have been administered in earlier stages of prostate carcinoma in recent pilot studies [7,8].

In the Division of Nuclear Medicine at the General Hospital of Vienna, the number of mCRPC patients referred to [$^{177}$Lu]Lu-PSMA$^{I\&T}$ therapy under named patient use has constantly increased throughout recent years. This development is in accordance with a recent study that showed an increasing prevalence of mCRPC in the United States, with stable incidence rates between 2010 and 2017, mostly attributable to the introduction of novel therapies (e.g., taxane-based chemotherapy, androgen deprivation therapy) [9]. To ensure the supplying of all our patients with this treatment using the available resources, it became necessary to upscale our production process from two patients per batch to at least four patients per batch (18 to 36 GBq starting activity/batch). Due to the increasing amount of radioactivity, radiolysis was suspected to constitute a major problem with regard to radiochemical purity results during upscaling.

The degradation of a radiopharmaceutical under the influence of its incorporated radionuclide (radiolysis) is a phenomenon known to occur during the production of [$^{177}$Lu]Lu-PSMA$^{I\&T}$ and other radiopharmaceuticals labeled with α or β$^-$ emitters. The process is mediated predominantly by solvated electrons, radicals and highly reactive molecules that result from the irradiation of water and thereby formed reactive oxygen species (ROS). To a lower extent, radiolysis is mediated by direct effects of radiation on the radiopharmaceutical. Different radiolysis quenchers have been investigated to maintain the radiochemical purity of therapeutic radiopharmaceuticals during storage [10]. The major radiolysis product of [$^{177}$Lu]Lu-PSMA$^{I\&T}$ was recently suggested to be the result of deiodination of the iodotyrosin moiety (see Figure 1) [11].

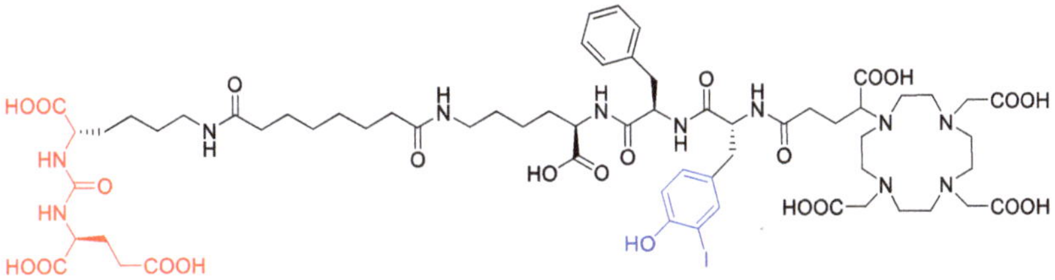

**Figure 1.** Molecular structure of PSMA$^{I\&T}$ with Glu-urea-Lys-binding motif (red, left) and iodotyrosine moiety (center, blue).

For [$^{177}$Lu]Lu-PSMA$^{I\&T}$, the radiolabeling with up to 30 GBq in the presence of gentisic acid, ascorbic acid and sodium acetate followed by the dilution of the reaction mixture with an ascorbic acid solution showed excellent radiochemical purity results of >97% after storage for 30 h [12]. The effectiveness of ascorbic acid/gentisic acid combination buffers can be explained by the oxidation of gentisic acid via ROS and the subsequent reduction of the primary radicals via ascorbic acid, hence sparing the radiopetide [13].

Apart from radiolysis, the chemical rearrangement of the product was also considered to have a possible influence on the formation of radioactive side products, although not necessarily related to the increased radioactivity amounts. Recently, the heat-dependent formation of hydantoins via cyclization of the PSMA binding motif was shown to occur during the radiosynthesis of [$^{177}$Lu]Lu-PSMA-617. These hydantoins showed no binding affinity to PSMA and were rapidly excreted by the kidneys. It was shown that these byproducts constitute >2% of the sum of radiochemical impurities, even under optimized conditions [14]. To the best of our knowledge, the formation of hydantoins in the PSMA binding motif has not been described for [$^{177}$Lu]Lu-PSMA$^{I\&T}$ yet. Given the likelihood of the formation of these by-products in [$^{177}$Lu]Lu-PSMA-I&T, we suspect that an overestimation of radiochemical purity could be prevalent throughout the existing literature, possibly due to the use of unsuitable HPLC methods for the detection of the cyclization products.

The aim of this study was therefore to investigate the formation and identity of radiolysis and cyclization products during the upscaling of [$^{177}$Lu]Lu-PSMA$^{I\&T}$ production.

## 2. Results and Discussion

### 2.1. Investigation of the Production Process of [$^{177}$Lu]Lu-PSMA$^{I\&T}$

Due to the recent publications on radiolysis-generated side products [10,12,15] in Lutetium-177-labeled radiopharmaceuticals, the radiochemical purity (RCP) depending on the radioactivity concentration was defined as a key parameter to be investigated and prioritized in the upscaling process. Hence, several batches with different starting activity concentrations (AC, 0.5–2.5 GBq/mL) were analyzed with HPLC method 1 (see chapter "Methods") for possible radiolysis products. The batches were produced according to the method referred to as "original method" (see Section 3.4). A pre-peak ($R_t$ = 3.88 min) was detected in all cases and all activity concentrations, with a slightly shorter retention time than the main product ($R_t$ = 4.20 min; see Figure 2). Its integral was below the in-house limit of 5% by that time for ACs up to 1 GBq/mL. No other peaks were detected in the radioactivity channel.

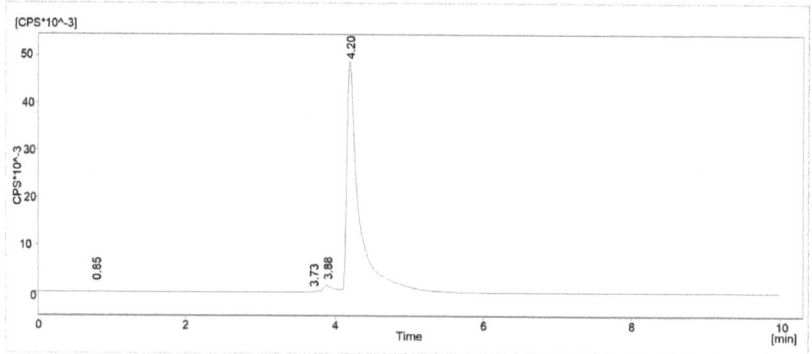

**Figure 2.** Typical γ-HPLC chromatogram (method 1) of [$^{177}$Lu]Lu-PSMA$^{I\&T}$ with main product ($R_t$ = 4.20 min) and radiolysis-induced side product ($R_t$ = 3.88 min).

Depending on the AC, the integral of the impurity was rising, revealing a correlation between the pre-peak and the AC (see Figure 3).

### 2.2. HPLC Optimization Studies and HPLC Validation

In the next step, our HPLC quality control methods were revised, as the relatively poor separation of the product from the radiolysis pre-peak was identified as problematic. Hence, the HPLC method described as "method C" by Martin et al. [14] was slightly modified and the resulting method will be called "method 2" throughout this text [14] (for details see the chapter "Methods"). The method was originally used to detect cyclization products of [$^{177}$Lu]Lu-PSMA-617 in patient urine samples. For comparison, simultaneous measurements of test batches in two identical HPLC setups were conducted to compare the performance of method 1 and method 2. The direct comparison showed significant underestimations of the presence of impurities by method 1 (see Figure 4).

The validation results of HPLC method 2 are displayed in Table 1.

To improve the radiochemical purity of the product, the need for a modified production process and storage formulation to suppress the suspected radiolysis became obvious. As described in the section "Methods: Radiosyntheses", a manual process described by Di Orio et al. [12] was modified and automatized (called "adapted method" in this text, whereas "original method" describes our previously used procedure, for details see Section 3.4).

The implementation of the new production method combined with HPLC method 2 showed a maximum of ≤10% total impurity for syntheses with a maximum AC of 1.8 GBq/mL in the product solution, as displayed in Figure 5.

It is also shown that, in addition to the baseline separated radiolysis product, hydrophobic substances eluting after the main product additionally appeared in HPLC method 2 (contribution of 2.87 ± 0.85% to total impurities) and furthermore formed independently from the activity concentration. A typical HPLC chromatogram acquired with method 2 is displayed in Figure 6.

Stability measurements were performed on three batches containing 8.6, 26.3 and 26.5 GBq. All batches showed >95% radiochemical purity (HPLC) after 4 h. Long term stability studies were not conducted since no shipment to other facilities or commercial use was intended.

In the next step, the radiochemical by-products were identified.

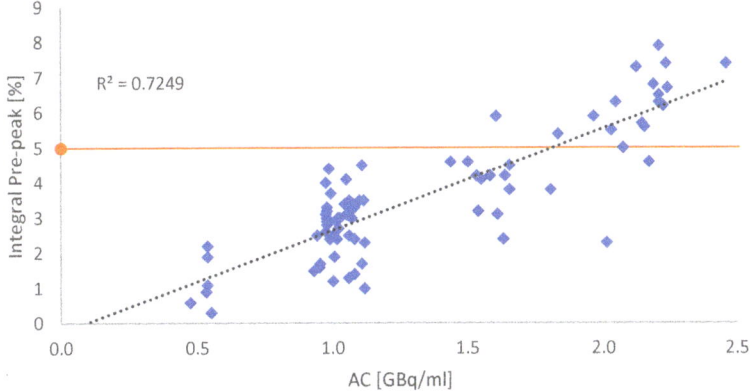

**Figure 3.** Graphical correlation between the integral of the suspected radiolysis-induced side product and the AC. Each blue dot refers to a single production batch. The orange line indicates the in-house specification benchmark of 5% impurities. All dots above the red line indicate batches with RCP < 95%.

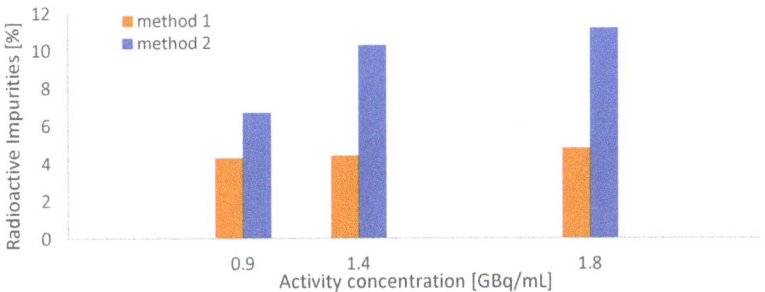

**Figure 4.** Comparison between HPLC methods 1 and 2 through simultaneous measurements of [$^{177}$Lu]Lu-PSMA$^{I\&T}$ batches; three different batches with increasing activities (18–36 GBq, AC = 1–2 GBq/mL, N = 1) were synthesized according to the original synthetic procedure.

**Table 1.** Validation of HPLC method 2.

| Parameter | Radiodetector | | Results |
|---|---|---|---|
| | Radiochemical Identity | Radiochemical Purity | |
| | | Acceptance Criteria | |
| Precision (repeatability) | CV% ≤ 5% | | complies |
| Specificity (radiolysis product) | Rs ≥ 1.5 | | complies |
| Linearity | R2 ≥ 0.99 | | complies |
| | UV detector | | |
| Parameter | Acceptance Criteria | | Results |
| Precision (repeatability) | CV% ≤ 5% | | complies |
| LOD ([$^{nat}$Lu]Lu-PSMA$^{I\&T}$) | Based on calibration curve | | 0.0361 µg/µL |
| LOD (PSMA$^{I\&T}$) | Based on calibration curve | | 0.0149 µg/µL |
| LOQ ([$^{nat}$Lu]Lu-PSMA$^{I\&T}$) | Based on calibration curve | | 0.1269 µg/µL |
| LOQ (PSMA$^{I\&T}$) | Based on calibration curve | | 0.0531 µg/µL |
| Linearity [$^{nat}$Lu]Lu-PSMA$^{I\&T}$ | R2 ≥ 0.99 | | complies |
| Linearity PSMA$^{I\&T}$ | R2 ≥ 0.99 | | complies |

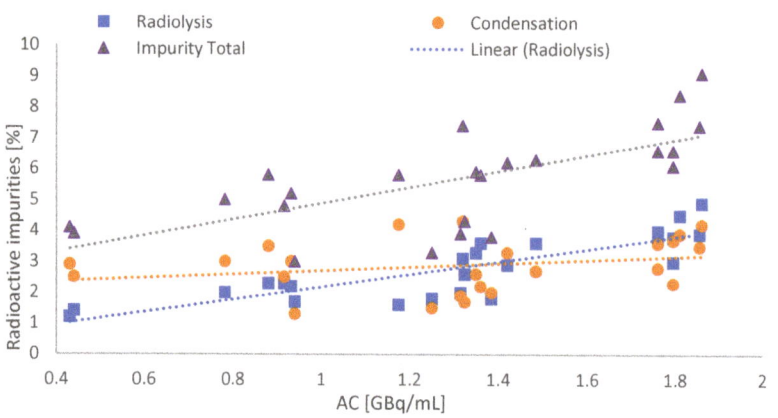

**Figure 5.** Total amount of radioactive impurities by using the new production and HPLC method in dependence of AC (blue dots: radiolysis pre-peak; orange circle: condensation side products; purple triangles: total impurities).

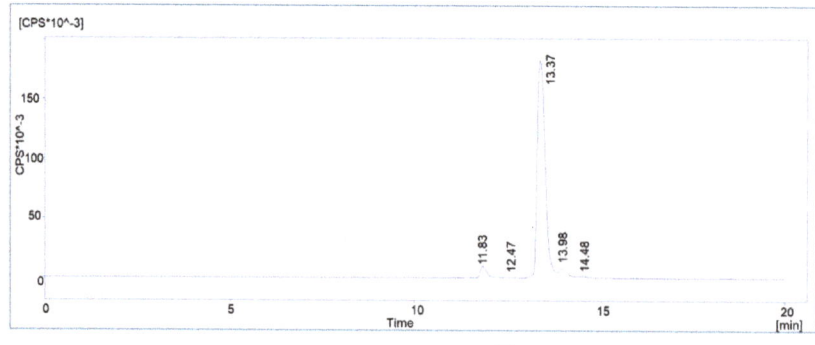

**Figure 6.** Representative chromatogram of a routine batch of [$^{177}$Lu]Lu-PSMA$^{I\&T}$ with HPLC method 2.

## 2.3. Identification of the Radiolysis By-Products: Irradiation Experiments and Coelution

At this stage of the project, the identity of the pre-peak in [$^{177}$Lu]Lu-PSMA$^{I\&T}$ was not yet clear (a study by Kraihammer et al. with a similar experimental setup has been published since then [11,13]). Due to the almost linear correlation between the intensity of the by-product and the radioactivity concentrations as described above, radiolysis was highly likely to be the cause of the formation. The dissociation of the iodine atom in the iodotyrosine moiety of the molecule (see Figure 7), which was shown to be caused by reactive oxygen species (ROS) induced by external radiation in aqueous media [16], was suspected to be the most probable cause.

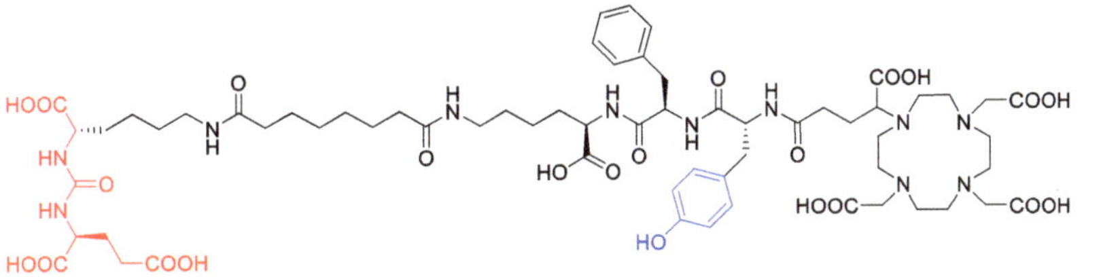

**Figure 7.** Molecular structure of the deiodinated PSMA$^{I\&T}$ by-product with the intact binding motif and missing iodine atom on the tyrosine side chain.

Hence, radiolysis was investigated by simulation. The intact Gd-PSMA$^{I\&T}$ was at first subjected to LC-MS and showed a retention time of 7.721 min and [M+2H]$^{2+}$ was found to be 828.30 (calculated: 828.31).

Then, the non-radioactive reference compound Gd-PSMA$^{I\&T}$ was irradiated in both the reaction buffer (ascorbate buffer from POLATOM) as well as in the reaction buffer plus ethanol with our in-house X-ray irradiation device (see "methods: Radiosyntheses"). Due to the similar chemical behavior and lower cost, a gadolinium salt was used as a surrogate for lutetium.

First, a simplified dose calculation was performed under the following four assumptions: 1. The β$^-$ emitter Lutetium-177 was causing substance damage comparable to X-rays. 2. The radionuclide was evenly distributed throughout the reaction vessel in 19.6 mL total volume. 3. Only ß$^-$ particles contributed to the total dose. 4. The storage time was 45 min. The physical dose for the ACs of 0.4, 0.9, 1.8 and 2 GBq/mL, corresponding to batch activities of 7, 18, 36 and 40 GBq per batch (for the calculation, see the chapter "Dose Calculations" in the Section 3), was calculated. Then, the samples were irradiated with the calculated doses and analyzed via LC-MS to identify potentially formed side products. In the samples without ethanol, two side products could be detected, whose intensities increased with rising dose equivalents (Figure 8) and were barely present in the ethanol containing samples kept for the same time in the same buffer.

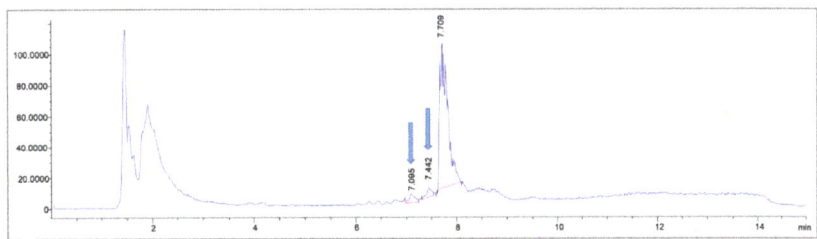

**Figure 8.** MS trace of an exemplary LC-MS measurement of a sample without buffer adjustments, irradiated with X-rays equivalent to 130 Gy. The blue arrows indicate the pre-peaks newly formed due to the irradiation.

In the corresponding MS spectra, one of the pre-peaks ($R_t$ = 7.095 min) could in fact be identified as the deiodinated Gd-PSMA$^{I\&T}$ (see Figure 9) with a molecular mass of 1527 ($m/z$ = 763.8 [M−I+H]$^{+2}$). The other side product, with a retention time of $R_t$ = 7.442 min, showed multiple mass peaks ($m/z$ = 104.4; 740.8; 521.2) and could not be identified unambiguously (Figure 9). The main peak, however, showed intact Gd-PSMA$^{I\&T}$ ($R_t$ = 7.709 min; $m/z$ = 828.8 [M+2H]$^{2+}$; 839.1 [M+H+Na]$^{2+}$; see Figures 8 and 9).

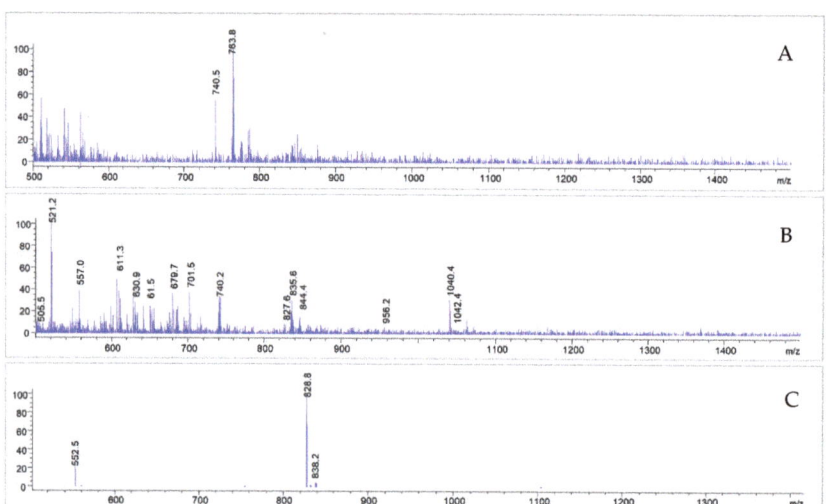

**Figure 9.** MS spectra corresponding to the impurities detected ($R_t$ = 7.095 (**A**); 7.442 (**B**) and 7.709 (**C**)); after irradiation of Gd-PSMA$^{I\&T}$ without adjusted buffer system, as described further above.

To quantify the effect of the irradiation on Gd-PSMA$^{I\&T}$ in the two different reaction buffers, the peak of the identified deiodinated compound was integrated and correlated with the radiation dose (Figure 10). The effect of the radiolysis quencher EtOH was shown to be very effective in the dose ranges up to 117 Gy (equivalent to 36 GBq or 1.8 GBq/mL). In the batch with the highest irradiation dose, the deiodinated compound could be detected in both buffer formulations. Since the primary goal was to identify the major radiolysis product, the irradiation experiment was performed only once (in duplicates).

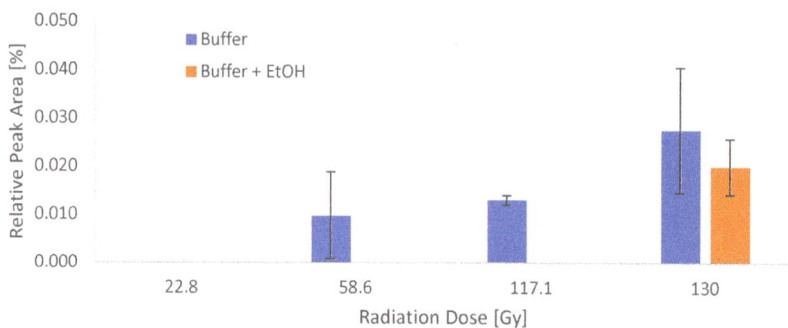

**Figure 10.** Integrals of the deiodinated compound determined via UPLC-MS in correlation with the applied dose in the two different reaction buffers. N = 1, performed in duplicate.

To further prove the suspicion of deiodination as a radiolysis mechanism, customized "deiodinated" PSMA$^{I\&T}$ was labeled with Lutetium-177. Using the optimized synthesis method, the overlay chromatogram showed highly comparable retention times, $R_t$ = 11.967

vs. 11.950 min, for the radiolabeled custom peptide and the radiolysis peak of the [$^{177}$Lu]Lu-PSMA$^{I\&T}$ routine batch from the same day (Figure 11).

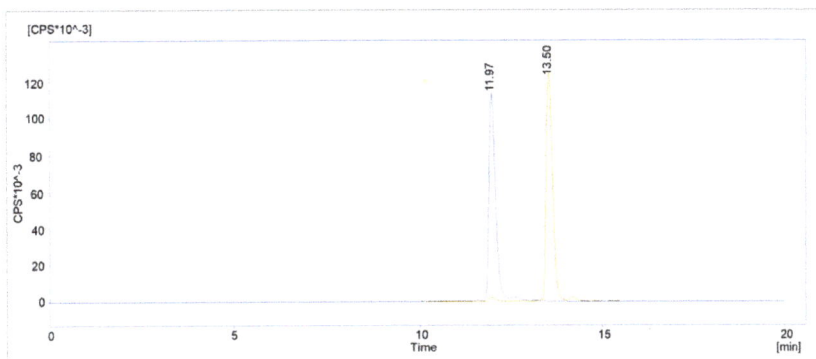

**Figure 11.** Overlay chromatogram of deiodinated [$^{177}$Lu]Lu-PSMA$^{I\&T}$ (blue) and a routine [$^{177}$Lu]Lu-PSMA$^{I\&T}$ batch (orange) from the same day.

Next to the radiolysis side products, a side product in the peak tail of the main product was also found by UPLC-MS. The area showed a mass difference of 18 in comparison to the parent compound, indicating a cyclization process under the dissociation of water (Figure 12).

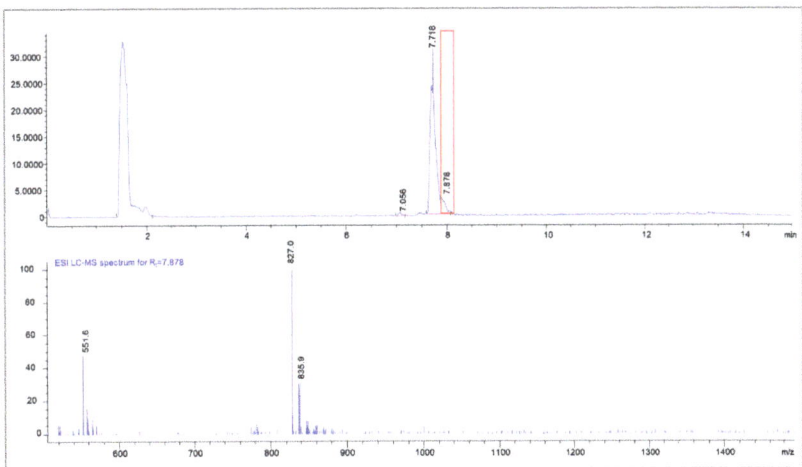

**Figure 12.** UPLC-MS chromatogram of [$^{nat}$Lu]Lu-PSMA$^{I\&T}$ with the MS spectrum of the peak tail (Rt = 7.878 min); clearly visible are condensation products of $^{nat}$Lu-PSMA ($m/z$ = 835.0 + [M+2H]$^{2+}$) with the mass of $m/z$ = 827 [M-H$_2$O+2H]$^{2+}$.

This product was attributed to the heat-dependent condensations of the PSMA-binding motif, as published earlier by Martin et al. [14] for [$^{177}$Lu]Lu-PSMA-617, resulting in three possible side products (Figure 13). As already mentioned in Section 2.2, the condensation products did not correlate with the activity concentration.

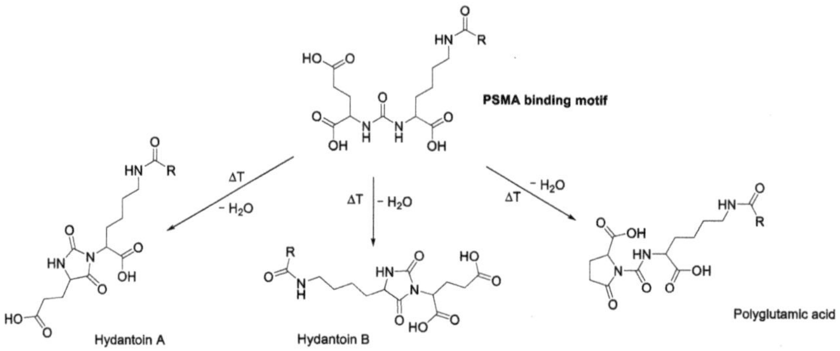

**Figure 13.** The PSMA binding motif and its possible heat-dependent condensation products, as proposed by Martin et al. [14].

### 2.4. Economical Optimization

Lastly, a radiochemical yield analysis of the new production method yielded a possible reduction in starting activity per patient from 9 to 8 GBq. Given the daily limit of 42 GBq Lutetium-177 in our department, serving up to 5 patients per batch became possible due to the successful upscaling process, reducing relative radiation exposure for the involved staff, overall costs and the relative formation of the deiodination product at the same time (Table 2).

**Table 2.** Retrospective analysis of 26 routine batches of [$^{177}$Lu]Lu-PSMA$^{I\&T}$ with average yields shows excellent yields with the possibility of reduction in starting activity; SA = starting activity; * 1 patient receives a therapeutic dose of 7.4 GBq.

| Patients */Batch | SA Original Method [GBq] | SA Adapted Method [GBq] | av. RCY [%] | av. Absolute Yield [GBq] | Minimum Final Activity [GBq] |
|---|---|---|---|---|---|
| 5 | 40 | - | 97.7 ± 0.5 | 41.5 ± 0.2 | 37.0 |
| 4 | 36 | 32 | 96.3 ± 1.2 | 36.8 ± 0.8 | 29.6 |
| 3 | 27 | 24 | 96.5 ± 2.6 | 27.4 ± 2.0 | 22.2 |
| 2 | 18 | 16 | 93.6 ± 5.3 | 17.6 ± 1.3 | 14.8 |

## 3. Methods

### 3.1. Chemicals

All chemicals and substances were used as received without further purification. GMP-grade PSMA$^{I\&T}$ was purchased from SCINTOMICS (Fürstenfeldbruck, Germany). Custom-synthesized Glu-CO-Lys[(Sub)DLys-DPhe-DTyr-DOTAGA] trifluoroacetate was obtained from piCHEM Forschungs- und Entwicklungs GmbH (Raaba-Grambach, Austria). Sodium ascorbate buffer (Polatom; kit ASC-01 containing 50 mg ascorbic acid and 7.9 mg NaOH) was received from Polatom (Warsaw, Poland) and diluted as indicated with Trace Select™ water (Honeywell Austria, Vienna, Austria). N.c.a. [$^{177}$Lu]LuCl$_3$ was purchased either from Isotope Technologies Munich (Munich, Germany) or Isotopia (Petah Tikva, Israel) in GMP quality. For dilutions and HPLC, deionized water generated from a MilliQ device (Merck Millipore) or Aqua ad injectabilia (B. Braun, Maria Enzersdorf, Austria) was used.

### 3.2. Chromatography

#### 3.2.1. UPLC-MS Measurements

Analytical UPLC-MS runs were performed on an Agilent 1260 Infinity II system equipped with a flexible pump, an Agilent 1260 UV detector with variable wavelength (λ = 220 nm) and an LC/MSD mass detector, in combination with either a Chromolith performance RP-18 (100 × 4.6 mm, flow = 1 mL/min) or an Acquity BEH C18 column, 1.7 µm, 3.0 × 50 mm (flow = 0.6 mL/min). A binary mobile phase of $H_2O$ + 0.1% TFA (A) and acetonitrile + 0.1% TFA (B) was used.

#### 3.2.2. HPLC Measurements

All HPLC measurements of the radioactive products were performed on a VWR Hitachi Chromaster System which included a column oven and UV detection unit, equipped with a Ramona Star Beta radiation detector (Elysia Raytest, Straubenhardt, Germany). HPLC measurements were performed at RT with a UV detection wavelength of 250 nm with a binary mobile phase of $H_2O$ + 0.1% TFA (A) and acetonitrile + 0.1% TFA (B).

For method 1, a chromolith performance RP-18 (100 × 4.6 mm, flow = 2 mL/min) column (Merck, Darmstadt, Germany) was applied. The respective gradient was 5% to 50% B for 5.5 min.

For method 2, a Jupiter Proteo 4 µm RP (250 × 4.6 mm, flow = 1 mL/min, Phenomenex, Phenomenex Inc., Torrance, CA, USA) was used with the gradient displayed in Table 3. The validation of HPLC method 2 was analyzed via Validat® software (v1) (GUS LAB GmbH, Gera, Germany) and based on the current ICH Q2(R1) and EANM guidelines [17,18].

**Table 3.** Solvent gradient for HPLC method 2 (flow = 1 mL/min).

| Time [min] | A (%) | B (%) |
|---|---|---|
| 1 | 90 | 10 |
| 2 | 88 | 12 |
| 3 | 84 | 16 |
| 5 | 80 | 20 |
| 7 | 76 | 24 |
| 8 | 74 | 26 |
| 9 | 72 | 28 |
| 10 | 71 | 29 |
| 11 | 70.5 | 29.5 |
| 12 | 70 | 30 |
| 14 | 69.5 | 30.5 |
| 17 | 5 | 95 |
| 18 | 95 | 5 |
| 20 | 95 | 5 |

### 3.3. Validation of HPLC Method 2

Calculations were performed automatically via Validat® software. Intermediate blank injections were performed between measurements and checked for residuals of the respective test compounds. Injection volume was 20 µL for all measurements. Acceptance criteria are displayed in Table 1. Specificity regarding [$^{177}$Lu]Lu-PSMA$^{I\&T}$ and the deiodinated product was calculated by Validat® according to the retention times of the substances and the resulting resolution.

For the determination of linearity and LOD/LOQ of the precursor, seven different concentrations between 1 µg/µL and 0.005 µg/µL were prepared by dilution in water for injection. Each concentration was measured as triplicate.

Precision measurement of the reference standard was performed as follows: 10 µL (1 µg/µL) reference standard were mixed with 10 µL water for injection and the measurement was repeated 6 times.

For linearity and LOD/LOQ of the reference standard, seven different concentrations between 1 µg/µL and 0.005 µg/µL were prepared by dilution in water for injection. Each concentration was measured as triplicate. For the determination of the linearity of the radiodetetor, 5 different activities of Lutetium-177 between 3.780 and 0.034 MBq were injected. To determine the repeatability of the radiodetector, a sample of 3.8 MBq was injected 5 times and total peak areas were analyzed.

### 3.4. Radiosyntheses

Radiosyntheses were performed on a Modular-Lab PharmTracer (Eckert & Ziegler GmbH, Berlin, Germany) using the corresponding single-use disposable cassettes.

Before each synthesis, an automated cassette pressure test was performed to ensure the leak-proofness of the synthesis cassette. The production was performed automatically and step-wise by the Modular Lab (version 6.2). The steps are displayed in Table 4.

**Table 4.** Automated radiosynthesis procedure; the reaction buffer, radiosynthesis conditions in step 3 and the final formulation buffer and volume in step 6 vary between the original and adapted method and are described in Table 5.

| # | Step |
|---|---|
| 1 | Conditioning of the Sep-Pak® C18 Plus cartridge with water/ethanol (50/50 mixture) followed by the formulation buffer |
| 2 | Transfer of radioactivity to reactor and rinsing of the activity vial with 1.4 mL of reaction buffer |
| 3 | Radiosynthesis |
| 4 | Transfer of the reaction mixture to the Sep-Pak® C18 Plus Cartridge |
| 5 | Elution of the product with EtOH/$H_2$O (2.5 mL, 50/50 mixture) |
| 6 | Formulation of the product to a final volume of 20 mL with formulation buffer |

**Table 5.** Differences between the original and the adapted synthetic procedure.

| Parameter | Original Method | Adapted Method |
|---|---|---|
| Product vial preparation | Sterile filtration of 0.15 mL (=30 mg DTPA) Ditripentat-Heyl® (200 mg/mL) solution into product vial | Sterile filtration of 0.15 mL (=30 mg DTPA) Ditripentat-Heyl® (200 mg/mL) solution and 1.4 mL reaction buffer into product vial |
| Reaction buffer | 35.7 mg/mL L (+)-ascorbic acid, 11.1 mg/mL NaOH (commercial buffer kit, Polatom®) in 1.4 mL Trace Select® water | 92.1 mg/mL L (+)-ascorbic acid, 111.4 mg/mL sodium acetate trihydrate, 34 mg/mL gentisic acid in 1.4 mL water for injection, adjusted to pH 5.2 with 2 N NaOH |
| Precursor amount | 125 µg/~9 GBq n.c.a. Lutetium-177 | |
| EtOH present during synthesis | 200 µL+ 0.5 µL per µg precursor | 0.5 µL per µg precursor |
| T [°C] | 90 | 95 |
| t [min] labelling reaction | 10 | 30 |
| Formulation buffer | 16 mL phys NaCl 0.9% | 24 mg/mL sodium ascorbate + 2.4 mg/mL L (+)-ascorbic acid in 16 mL phys NaCl 0.9% |
| Total volume EOS [mL] | 18.6 | 20.0 |

After each synthesis, a fully automated filter test of the product sterile filter, as programmed in the software, was performed. Differences between the old and new synthesis method are displayed in Table 5.

## 3.5. Syntheses of $^{nat}$Lu-PSMA$^{I\&T}$ and $^{nat}$Gd-PSMA$^{I\&T}$

Briefly, 250 µg (167 nmol) PSMA$^{I\&T}$ was dissolved in 200 µL of sodium ascorbate buffer (1 M, pH 4.5). Then, 10 equivalents (1.67 µmol) of either Gadolinium(III) nitrate or Lutetium(III) chloride in 1 mM aqueous solution were added. The vial containing the mixture was heated at 95 °C in an aluminum heating block for 10 min. After cooling to RT, qualitative LC-MS was performed without further workup. LC-MS conditions were RP-LCMS, Chromolith performance; 5–50% MeCN + 0.1% TFA over 10 min.

## 3.6. Dose Calculations

To estimate the radiation dose absorbed in a sample, the contributions of all particles resulting from any decay have to be considered. $^{177}$Lu decays to $^{177}$Hf by four possible β$^-$ transitions. The relative intensities of these transitions are 11.6%, 0.016%, 8.89% and 79.44% [19]. The average energies <$E_\beta$> of the corresponding β particles are 47.23 keV, 78.12 keV, 111.20 keV and 148.84 keV, respectively. During the subsequent transition of the resulting excited state to the ground state additional conversion electrons, Auger electrons, γ photons and X-rays are emitted. The total energy of β particles and electrons is 146 keV (2.34 × 10$^{-14}$ J) per decay and the total energy of photons is 33 keV (5.29 × 10$^{-15}$ J) per decay. If a sample is sufficiently large, the energy of all β$^-$ particles and electrons is deposited within the sample. For the dose estimation, the dose of the photons will be neglected, since only a small fraction of the energy will be absorbed within the sample. Thus, the absorbed dose $D$ is given by

$$D = \frac{NE_{e^-}}{m} \quad (1)$$

where $N$ is the number of decays, $E_{e^-}$ is the total energy per decay of β particles and electrons and $m$ is the mass of the sample.

The number of disintegrations within a time interval $t$ is given by

$$N = \int_0^T A e^{-\frac{t \ln 2}{t_{1/2}}} dt = A \frac{\ln 2}{T_{1/2}} \left(1 - e^{-\frac{t \ln 2}{t_{1/2}}}\right) \quad (2)$$

with $A$ being the activity and $t_{1/2}$ the half-life.

If $t$ is large with respect to the half-life $t_{1/2}$, $N$ approaches $N = A \frac{\ln 2}{t_{1/2}}$. For $t$ much shorter than $t_{1/2}$, $N$ can be approximated as $N \approx A T$ using the Taylor series of the exponential function. Assuming an approximate density of 1 g/cm$^3$ and t much shorter than $t_{1/2}$, the dose can be calculated as

$$D = t \times A_c \times 2.34 \times 10^{-2} \text{Gy} * \text{mL} \quad (3)$$

where $t$ is the time in seconds, $D$ is the dose in Gy and $A_c$ is the activity concentration in GBq/mL.

## 3.7. Irradiations

After the preparation of one batch of Gd-PSMA$^{I\&T}$ and another batch of Gd-PSMA$^{I\&T}$ mixed with 280 µL of ethanol, each batch was divided into 8 portions. The samples were irradiated in duplicates with 4 different doses of X-rays, according to Table 6.

For that, an YXLON reference irradiator (Maxishot, YXLON International GmbH, Hamburg, Germany) was used, as previously described [20]. In brief, irradiation was performed at 200 kV, 20 mA with a focus size of 5.5 mm, using a 0.5 mm copper filter and a 3 mm aluminum filter. The average dose rate was about 1.1 Gy/min. After irradiation, the samples were subjected to LC-MS measurements.

Table 6. Irradiation parameters (actual dose and calculated activity equivalents) of the irradiated samples.

| Sample Number Gd-PSMA$^{I\&T}$ | Sample Number Gd-PSMA$^{I\&T}$ + EtOH | Dose [Gy] | Activity Equivalent for 45 min Storage Time [GBq] |
|---|---|---|---|
| 1.1 | 2.1 | 22.8 | 7 |
| 1.2 | 2.2 | 58.6 | 18 |
| 1.3 | 2.3 | 117.1 | 36 |
| 1.4 | 2.4 | 130 | 40 |

*3.8. Conclusions*

In this study, a thorough analysis of the quality control parameters of [$^{177}$Lu]Lu-PSMA$^{I\&T}$ was performed. Major optimizations in HPLC analyses and irradiation experiments followed by UPLC-MS studies revealed a significant contribution of cyclization and radiolysis products to the amount of radioactive impurities in the production of [$^{177}$Lu]Lu-PSMA$^{I\&T}$.

The pre-peak, now known to consist of deiodinated [$^{177}$Lu]Lu-PSMA$^{I\&T}$ and induced by radiolysis, is expected to show altered pharmacokinetics and might reduce the overall tumor dose, but still demonstrated cell uptake in a previous study [15]. Our findings regarding the identity of the pre-peak are in accordance with a recently published study [11] and corroborate them by the approach of radiolabeling of the suspected radiolysis product and comparing the HPLC retention time with that of a [$^{177}$Lu]Lu-PSMA$^{I\&T}$ routine batch.

The cyclization products were already shown to lack tumor cell binding, due to the altered PSMA binding motif. Further, NMR studies were performed to elucidate the definite structure of the cyclization products [14]. In our study, the conclusion was drawn based on the combination of UPLC-MS, HPLC elution profile and independency from activity concentration.

A risk-based approach resulted in the extension of our radiochemical purity thresholds for the sum of hydantoins and deiodinated product for 90%, given the fact that the cyclization products constitute the most critical impurity with respect to therapeutic efficacy and only represented 2.87 ± 0.85% of the total radioactivity in the samples. As expected, the formation of the main radiolysis product was shown to depend on the activity concentration and represented 2.8 ± 0.1% of the total radioactivity in the samples.

To our knowledge, this is the first time that the discussed cyclization peaks were shown to contribute to the overall amount of radiochemical impurities in [$^{177}$Lu]Lu-PSMA$^{I\&T}$ and the question of how adequate HPLC methods should be used to detect them was addressed.

Further, an overall improvement of processes for the routine supply of [$^{177}$Lu]Lu-PSMA$^{I\&T}$ was achieved and it was possible to keep up with the increasing demand for this radioligand therapy. Due to the reduction in starting activity, a possible cost reduction through quality assurance procedures was also shown in this work.

**Author Contributions:** Conceptualization, S.S., J.R., E.-M.P. and M.R.B.; Formal analysis, S.S. and S.W.; Investigation, S.S., J.R., S.W. and M.R.B.; Project administration, T.L.M., M.H. and M.M.; Resources, L.N., M.O. and V.W.; Supervision, E.-M.P., M.O., T.L.M., M.H., W.W., M.R.B. and M.H.; Validation, S.S., E.-M.P., L.N., M.O. and M.R.B.; Visualization, S.S. and J.R.; Writing—original draft, S.S., J.R. and M.R.B.; Writing—review and editing, J.R., S.W., E.-M.P., V.W., T.L.M., W.W., M.R.B. and M.M. All authors have read and agreed to the published version of the manuscript.

**Funding:** Open Access Funding by the University of Vienna.

**Institutional Review Board Statement:** Not applicable.

**Informed Consent Statement:** Not applicable.

**Data Availability Statement:** Data can be provided on an individual basis by the authors.

**Acknowledgments:** The authors thank the radiopharmaceutical routine team, especially R. Bartosch for the production and quality control of [$^{177}$Lu]Lu-PSMA$^{I\&T}$. We also thank C. Vraka, S. Mairinger and J. Cardinale for scientific discussions.

**Conflicts of Interest:** The authors declare no conflict of interest.

## References

1. Pubchem Entry for [177Lu]Lu-PSMA-I&T. Available online: https://pubchem.ncbi.nlm.nih.gov/compound/Unii-G5B860B0G1#section=NCI-Thesaurus-Tree (accessed on 15 September 2023).
2. Sadaghiani, M.S.; Sheikhbahaei, S.; Werner, R.A.; Pienta, K.J.; Pomper, M.G.; Solnes, L.B.; Gorin, M.A.; Wang, N.Y.; Rowe, S.P. A Systematic Review and Meta-analysis of the Effectiveness and Toxicities of Lutetium-177-labeled Prostate-specific Membrane Antigen-targeted Radioligand Therapy in Metastatic Castration-Resistant Prostate Cancer. *Eur. Urol.* **2021**, *80*, 82–94.
3. Emmett, L.; Willowson, K.; Violet, J.; Shin, J.; Blansby, A.; Lee, J. Lutetium (177) PSMA radionuclide therapy for men with prostate cancer: A review of the current literature and discussion of practical aspects of therapy. *J. Med. Radiat. Sci.* **2017**, *64*, 52–60. [CrossRef]
4. European Medicines Agency. Pluvicto Entry European Medicines Agency. Available online: https://www.ema.europa.eu/en/medicines/human/EPAR/pluvicto (accessed on 30 May 2023).
5. U.S. Food and Drug Administration. FDA Approves Pluvicto for Metastatic Castration-Resistant Prostate Cancer 2022. Available online: https://www.fda.gov/drugs/resources-information-approved-drugs/fda-approves-pluvicto-metastatic-castration-resistant-prostate-cancer (accessed on 16 September 2023).
6. Kratochwil, C.; Fendler, W.P.; Eiber, M.; Baum, R.; Bozkurt, M.F.; Czernin, J.; Delgado Bolton, R.C.; Ezziddin, S.; Forrer, F.; Hicks, R.J.; et al. EANM procedure guidelines for radionuclide therapy with (177)Lu-labelled PSMA-ligands ((177)Lu-PSMA-RLT). *Eur. J. Nucl. Med. Mol. Imaging* **2019**, *46*, 2536–2544.
7. Privé, B.M.; Peters, S.M.; Muselaers, C.H.; van Oort, I.M.; Janssen, M.J.; Sedelaar, J.M.; Konijnenberg, M.W.; Zámecnik, P.; Uijen, M.J.; Schilham, M.G.; et al. Lutetium-177-PSMA-617 in Low-Volume Hormone-Sensitive Metastatic Prostate Cancer: A Prospective Pilot Study. *Clin. Cancer Res.* **2021**, *27*, 3595–3601. [CrossRef]
8. Golan, S.; Frumer, M.; Zohar, Y.; Rosenbaum, E.; Yakimov, M.; Kedar, D.; Margel, D.; Baniel, J.; Steinmetz, A.P.; Groshar, D.; et al. Neoadjuvant (177)Lu-PSMA-I&T Radionuclide Treatment in Patients with High-risk Prostate Cancer Before Radical Prostatectomy: A Single-arm Phase 1 Trial. *Eur. Urol. Oncol.* **2022**, *6*, 151–159.
9. Wallace, K.L.; Landsteiner, A.; Bunner, S.H.; Engel-Nitz, N.M.; Luckenbaugh, A.N. Increasing prevalence of metastatic castration-resistant prostate cancer in a managed care population in the United States. *Cancer Causes Control* **2021**, *32*, 1365–1374. [CrossRef]
10. Larenkov, A.; Mitrofanov, I.; Pavlenko, E.; Rakhimov, M. Radiolysis-Associated Decrease in Radiochemical Purity of 177Lu-Radiopharmaceuticals and Comparison of the Effectiveness of Selected Quenchers against This Process. *Molecules* **2023**, *28*, 1884. [CrossRef]
11. Kraihammer, M.; Garnuszek, P.; Bauman, A.; Maurin, M.; Alejandre Lafont, M.; Haubner, R.; von Guggenberg, E.; Gabriel, M.; Decristoforo, C. Improved Quality Control of [177Lu]Lu-PSMA I&T. *EJNMMI Radiopharm. Chem.* **2023**, *8*, 1–3.
12. Di Iorio, V.; Boschi, S.; Cuni, C.; Monti, M.; Severi, S.; Paganelli, G.; Masini, C. Production and Quality Control of [(177)Lu]Lu-PSMA-I&T: Development of an Investigational Medicinal Product Dossier for Clinical Trials. *Molecules* **2022**, *27*, 4143.
13. Joshi, R.; Gangabhagirathi, R.; Venu, S.; Adhikari, S.; Mukherjee, T. Antioxidant activity and free radical scavenging reactions of gentisic acid: In-vitro and pulse radiolysis studies. *Free Radic. Res.* **2012**, *46*, 11–20. [CrossRef]
14. Martin, S.; Tonnesmann, R.; Hierlmeier, I.; Maus, S.; Rosar, F.; Ruf, J.; Holland, J.P.; Ezziddin, S.; Bartholoma, M.D. Identification, Characterization, and Suppression of Side Products Formed during the Synthesis of [(177)Lu]Lu-PSMA-617. *J. Med. Chem.* **2021**, *64*, 4960–4971. [CrossRef]
15. Hooijman, E.L.; Ntihabose, C.M.; Reuvers, T.G.; Nonnekens, J.; Aalbersberg, E.A.; van de Merbel, J.R.; Huijmans, J.E.; Koolen, S.L.; Hendrikx, J.J.; de Blois, E. Radiolabeling and quality control of therapeutic radiopharmaceuticals: Optimization, clinical implementation and comparison of radio-TLC/HPLC analysis, demonstrated by [(177)Lu]Lu-PSMA. *EJNMMI Radiopharm. Chem.* **2022**, *7*, 29. [CrossRef]
16. Das, T.N.; Priyadarsini, K.I. Characterization of Transients Produced in Aqueous Medium by Pulse Radiolytic Oxidation of 3,5-Diiodotyrosine. *J. Phys. Chem.* **1994**, *98*, 5272–5278. [CrossRef]
17. Gillings, N.; Todde, S.; Behe, M.; Decristoforo, C.; Elsinga, P.; Ferrari, V.; Hjelstuen, O.; Peitl, P.K.; Koziorowski, J.; Laverman, P.; et al. EANM guideline on the validation of analytical methods for radiopharmaceuticals. *EJNMMI Radiopharm. Chem.* **2020**, *5*, 7. [CrossRef]
18. ICH Guideline Q2(R2) on Validation of Analytical Procedures Step 2b European Medicines Agency. 2022. Available online: https://www.ema.europa.eu/en/documents/scientific-guideline/ich-guideline-q2r2-validation-analytical-procedures-step-2b_en.pdf (accessed on 15 January 2023).
19. Kondev, F.G. Nuclear Data Sheets for A=177☆. *Nuclear Data Sheet* **2019**, *159*, 1–412.
20. Raitanen, J.; Barta, B.; Hacker, M.; Georg, D.; Balber, T.; Mitterhauser, M. Comparison of Radiation Response between 2D and 3D Cell Culture Models of Different Human Cancer Cell Lines. *Cells* **2023**, *12*, 360.

**Disclaimer/Publisher's Note:** The statements, opinions and data contained in all publications are solely those of the individual author(s) and contributor(s) and not of MDPI and/or the editor(s). MDPI and/or the editor(s) disclaim responsibility for any injury to people or property resulting from any ideas, methods, instructions or products referred to in the content.

Article

# A Simple Kit for the Good-Manufacturing-Practice Production of [$^{68}$Ga]Ga-EDTA

Monika Skulska and Lise Falborg *

Department of Nuclear Medicine, Gødstrup Hospital, 7400 Herning, Denmark
* Correspondence: lisefalb@rm.dk

**Abstract:** Glomerular filtration rates for individual kidneys can be measured semi-quantitatively by a gamma camera using [$^{99m}$Tc]Tc-DTPA, with limited diagnostic accuracy. A more precise measurement can be performed on a PET/CT scanner using the radiotracer [$^{68}$Ga]Ga-EDTA, which has been validated in animal studies. The purpose of this study was to develop an easy kit-based synthesis of [$^{68}$Ga]Ga-EDTA that is compliant with good manufacturing practice (GMP) and applicable for human use. The production of the cold kit and its labeling were validated, as were the radiochemical purity measurement and analytical procedures for determining the Na$_2$EDTA dihydrate content in the kits. In this study, we validated a GMP kit for the simple production of [$^{68}$Ga]Ga-EDTA, with the intention of applicability for human use.

**Keywords:** [$^{68}$Ga]Ga-EDTA; PET; renography; GMP-production; Kit; $^{68}$Ga-colloids

## 1. Introduction

With the current widespread distribution of positron emission tomography (PET) scanners, the interest in gallium-68 ($^{68}$Ga)-labeled radiopharmaceuticals has increased. $^{68}$Ga's high positron-emission fraction (89% maximum energy; 1899 keV) and its 67.71 min half-life provide sufficient levels of radioactivity for high-quality images, while minimizing the radiation dose that is given to patients [1]. The parent radionuclide, germanium-68 ($^{68}$Ge), with a half-life of 271 days, provides an easily available method of producing $^{68}$Ga, with a shelf-life of approximately one year, from an efficient and medically approved $^{68}$Ge/$^{68}$Ga generator [2].

Early $^{68}$Ge/$^{68}$Ga-generators were eluted using ethylenediaminetetraacetic acid (EDTA) and provided the direct production of [$^{68}$Ga]Ga-EDTA for use in brain imaging as well as the quantitative assessment of blood–brain barrier abnormalities that are associated with multiple sclerosis [3,4]. However, since the equilibrium constant for the formation of the complex is high ($K_{ML} = 7.9 \times 10^{18}$), the [$^{68}$Ga]Ga-EDTA complex has high thermodynamic stability and, therefore, its decomposition is difficult [5]. This drastically limited the development of other $^{68}$Ga-labelled radiopharmaceuticals for these early radionuclide generators. Consequently, modern $^{68}$Ge/$^{68}$Ga-generators use acidic eluent (hydrochloric acid) and provide $^{68}$Ga in cationic form to enable further labeling chemistry [5].

Therefore, $^{68}$Ga is also used to label ligands, such as peptides, antibodies, or hormones, which can be targeted to specific biologically accessible proteins, such as receptors, that are over-expressed by tumor cells. Examples include [$^{68}$Ga]Ga-DOTATATE, [$^{68}$Ga]Ga-DOTANOC, and [$^{68}$Ga]Ga-DOTATOC, all of which play important roles in the diagnosis of neuroendocrine tumors due to their affinity with somatostatin receptors [6], or $^{68}$Ga-PSMA, which is used for the clinical imaging of prostate cancer [7].

[$^{68}$Ga]Ga-EDTA is known to be cleared from the blood via the kidneys with a rate that depends on the renal glomerular filtration function [8]. In 2016, Hofman et al. proved that [$^{68}$Ga]Ga-EDTA can be used as a substitute for [$^{99m}$Tc]Tc-DTPA, which is used in conventional gamma camera single-photon nuclear medical imaging for a wide variety of clinical indications [9]. The quantitative capabilities of PET, combined with its inherent ability to perform 3D tomographic imaging, provide major advantages over conventional planar imaging, as has been shown in recent animal studies [10,11]. However, earlier methods for the production of [$^{68}$Ga]Ga-EDTA, as, described in the literature, cannot be directly transferred to a GMP-compliant production method for human use. Therefore, we aimed to implement a local kit-based synthesis, analogous to the standard $^{99}$Mo/$^{99m}$Tc-generator/kit preparation of radiopharmaceuticals used for gamma-camera and single-photon emission computed tomography (SPECT) examinations.

The goal of the present work was to develop, establish, and validate a kit-based production of [$^{68}$Ga]Ga-EDTA, in which labeling takes place in a simple one-pot synthesis, where the generator is eluted into a vial containing disodium EDTA dihydrate. We developed a kit (hereinafter referred to as the EDTA kit) containing EDTA and the necessary buffer system. The production of an EDTA kit and the $^{68}$Ga-labeling reaction to obtain [$^{68}$Ga]Ga-EDTA must be GMP-compliant and in accordance with the national regulations of the Danish Medicine Agency (DMA).

## 2. Results
### 2.1. EDTA Kits
#### 2.1.1. EDTA-Kit Composition

The composition of EDTA kits was designed with respect to the following requirements: appropriate amounts of reagents to provide the correctly labeled product without toxic effects; an appropriate ion strength and pH for intravenous injection and the correct pH to avoid possible side reactions during the labeling reaction. At pH values higher than 3, $^{68}$Ga$^{3+}$-ions form oxide or hydroxide species of the $^{68}$Ga$^{3+}$-ion with low solubility, some of which form insoluble colloids [12]. The introduction of buffers that act as stabilizing ligands in the reaction mixture prevents the formation of colloids and supports complexation with the intended ligand, which, in our case, is EDTA. Bauwens et al. showed that the optimal buffer choices for the radiosynthesis of $^{68}$Ga-Dotatoc are HEPES, acetate, or succinate with a pH of 3.5–5.0 [13]. The colloids are impurities, which are hereinafter referred to as $^{68}$Ga-colloids. The composition of a single EDTA kit is presented in Table 1. Such kits are stored in a freezer. The constraint requirements for kit design are further described in the Discussion section of this article.

**Table 1.** Composition of an EDTA kit.

| Reagent | Amount per EDTA Kit |
| --- | --- |
| Disodium EDTA dihydrate | 1.86 mg |
| Sodium acetate trihydrate | 136 mg |
| NaOH | 7.99 mg |
| Sterile water | Up to 3.00 mL |

#### 2.1.2. EDTA Kit Validation

A GMP-compliant production of three batches of the EDTA kit was performed. A comparison of the measured parameters with the pharmaceutical/chemical specifications showed that the production of the EDTA kit using the described method was robust and highly reproducible (Table 2). Stability studies of the EDTA kits were performed over a period of up to 14 months by executing repeated measurements of pH and full quality control (QC) programs for the labeling of the three validation batches.

#### 2.1.3. Na$_2$EDTA Dihydrate Content Determination in EDTA Kits

The amount of Na$_2$EDTA dihydrate (Table 2) in the EDTA kits was determined using a complexation reaction with Fe$^{3+}$ followed by HPLC analysis, Figure 1. The peak at 1.5 min corresponds to Fe$^{3+}$ ions, the peak at 1.9 min to Fe(OAc)$_3$ and the peak at 3.0 min to Fe-EDTA [14–16]. A calibration curve was produced by analyzing solutions with various known concentrations of disodium EDTA dihydrate. The curve was constructed by plotting the area under the Fe-EDTA peak as a function of $Na_2EDTA$ dihydrate concentration (Figure 2). The linearity of the investigated EDTA concentrations was validated for the range 0.01–0.1 mg/mL. Samples of EDTA kits were diluted by a factor 10 prior to complexation with $Fe^{3+}$, followed by HPLC analysis. Thus, using the slope of the standard curve, the concentrations of $Na_2EDTA$ dihydrate in the three batches of EDTA kit were 0.64, 0.69, and 0.69 mg/mL, respectively.

Table 2. Specifications and validation of three productions of EDTA kits.

| Test (Method) * | Specifications | EDTA Kit Batch 1 | EDTA Kit Batch 2 | EDTA Kit Batch 3 |
| --- | --- | --- | --- | --- |
| pH (pH meter) | 12.0–13.0 | 12.4 | 12.3 | 12.4 |
| Filter integrity (Manual) | Intact | Intact | Intact | Intact |
| Sterility | No growth | No growth | No growth | No growth |
| Volume (Visual) | 3.0 ± 0.5 mL | 3.0 | 3.0 | 3.0 |
| Appearance (Visual) | Clear without particles | Clear without particles | Clear without particles | Clear without particles |
| Identity (Fe-EDTA) (HPLC) | 2.5–3.5 min | 3.0 | 3.0 | 3.0 |
| $Na_2EDTA \cdot 2H_2O$ (HPLC) | 0.62 g/mL ± 20% (0.50–0.74 mg/mL) | 0.64 | 0.69 | 0.69 |
| Labelling (full QC program) | Comply | Comply | Comply | Comply |

* Results are from analyses performed immediately after production, except the HPLC results, which are obtained after 14 months.

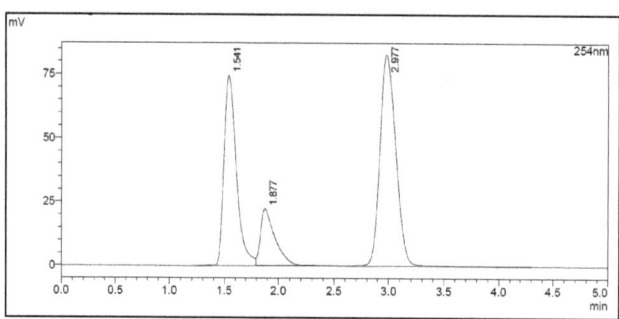

Figure 1. HPLC chromatogram.

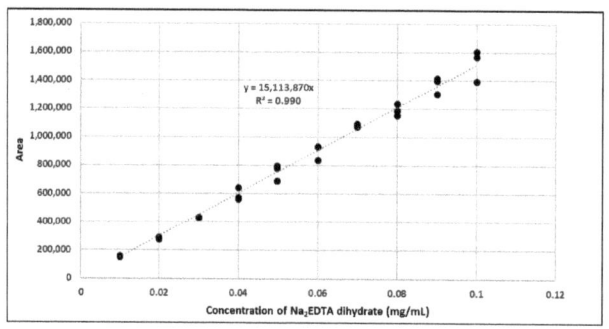

Figure 2. Calibration curve with area of Fe-EDTA peak as a function of concentration of $Na_2EDTA$ dihydrate.

### 2.1.4. Buffer Capacity: Labelling Process' Robustness

The buffer capacity of the kit, both during and after elution of the generator, is important for the robustness of the overall labelling process. Thus, the challenge was to design an EDTA kit and labelling process that did not lead to stable insoluble $^{68}$Ga-colloid production as a radiochemical impurity whilst eluting the generator into the EDTA kit. Since the EDTA kit itself had a high pH value and the formation of $^{68}$Ga-colloid is known to take place at moderately acidic-to-basic conditions, the elution needs to be fast enough to obtain a low enough pH in time to avoid the formation of these impurities. The quality control results of the labelled product show that this was achieved.

Additionally, using the syringe module of the PharmTracer, the generator is eluted with 7 mL 0.1 N HCl, 2 mL/min. However, small leaks in the cassette can cause a reduced volume of 0.1 N HCl. Therefore, the pH of the solution in the final product vial after complete elution of the generator should be stable and robust for varying volumes of 0.1 N HCl. Figure 3 illustrates the pH profile w.r.t. the addition of different volumes of 0.1 N HCl into an EDTA kit, which mimics the pH of [$^{68}$Ga]Ga-EDTA with different volumes of 0.1 N HCl added from the generator during the elution of the generator into the vial. A pH of 4.65, which is optimal for a good labelling reaction as well as avoiding the formation of $^{68}$Ga-colloid and having an appropriate pH for i.v. injection, was obtained by the addition of 7 mL. The labelling process robustness showed a resultant pH range of 4.5–5.0 in the situation where the volume of eluent differed due to possible variations of ±2 mL in the automatic dispensing of the eluent while eluting the generator.

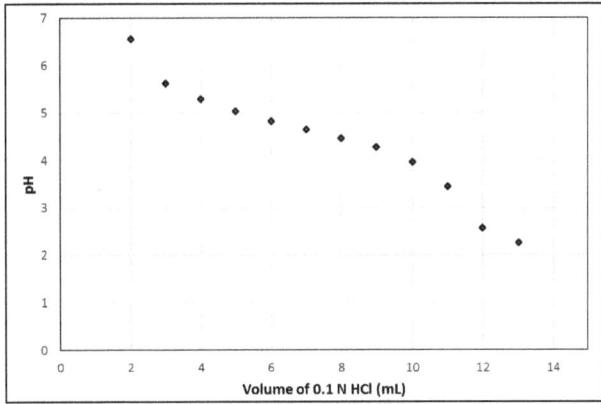

**Figure 3.** pH of EDTA kit during the addition of different volumes of 0.1 N HCl.

## 2.2. [$^{68}$Ga]Ga-EDTA

Nine labelling procedures were performed using Modular-Lab PharmTracer: three for validation of the radiolabelled product, three for bioburden testing and three for stability studies. Three different batches of EDTA kits were used for each of the above procedures.

### 2.2.1. EDTA Kit Labelling with $^{68}$Ga

$^{68}$Ge/$^{68}$Ga-generator qualification and validation were conducted prior to use. The identity of $^{68}$Ga was confirmed by radionuclide purity testing (half-life = 68.7 min; $^{68}$Ge-breakthrough = 0.00003% and gamma spectrum analysis (only 511 keV and 1077 keV photons characteristic to $^{68}$Ga were detected). The microbiological testing of eluate (sterility and endotoxin levels) detected no microbial contamination. The generator was eluted a minimum of 24 hours prior to labeling.

During synthesis, 7 mL of $^{68}$Ga-eluate was automatically transferred by the PharmTracer directly to the EDTA kit, where the conjugation reaction proceeded immediately.

Bioburden testing showed no growth in the controlled batches. Specifications and results of the three validation runs are presented in Table 3.

Table 3. Specifications and validation for three productions of [$^{68}$Ga]Ga-EDTA.

| QC (Method) * | Specifications | [$^{68}$Ga]Ga-EDTA Batch 1 | [$^{68}$Ga]Ga-EDTA Batch 2 | [$^{68}$Ga]Ga-EDTA Batch 3 |
|---|---|---|---|---|
| Radioactivity (dose calibrator) | ≤1373 MBq at EOS | 1363 | 1373 | 1329 |
| Volume (visual) | 9.0–11.0 mL | 9.0 | 9.5 | 9.5 |
| Appearance (visual) | Clear without particles | Clear without particles | Clear without particles | Clear without particles |
| Filter integrity (manual) | Intact | Intact | Intact | Intact |
| pH (indicator paper) | 4.0-6.0 | 4.7 | 4.7 | 4.7 |
| $^{68}$Ga-colloid (paper chromatography) | <3% | 0.3 | 0.3 | 0.2 |
| RCP (paper chromatography) | >95% | 99.8 | 99.7 | 99.8 |
| Identity (paper chromatography) | 0.7 < Rf < 1.3 | 1.1 | 1.0 | 1.1 |
| Radionuclidic purity (Gamma counter) | <0.001% activity from $^{68}$Ge | <0.00001 | <0.00001 | <0.00001 |
| Endotoxin (EndoSafe) | <17.5 EU/mL | <5.00 | <5.00 | <5.00 |
| Sterility | No growth | No growth | No growth | No growth |

* Displayed results are from analyses performed immediately after production.

The stability studies proved stability up to 2 hours after end of synthesis (EOS). The specification for bioburden was less than one colony-forming unit (CFU) per 10 mL (<1 CFU/10 mL) of the product. The test for bioburden in the three batches resulted in 0 CFU.

2.2.2. Paper Chromatography of [$^{68}$Ga]Ga-EDTA

Quality control of [$^{68}$Ga]Ga-EDTA included the analytical procedures shown in Table 3. The paper chromatography method for the determination of the radiochemical purity of [$^{68}$Ga]Ga-EDTA was performed using Whatman Grade 1 Chr paper as the stationary phase and 0.9% sodium chloride as the mobile phase. The plate was 2 cm × 12 cm, and the sample was added 2 cm from the bottom edge and developed to 8 cm from the bottom. Typical chromatograms of $^{68}$Ga-colloid and [$^{68}$Ga]Ga-EDTA are shown in Figure 4. The method is specific, precise and robust.

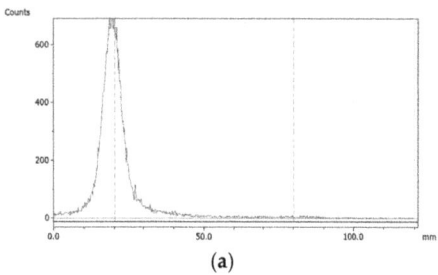
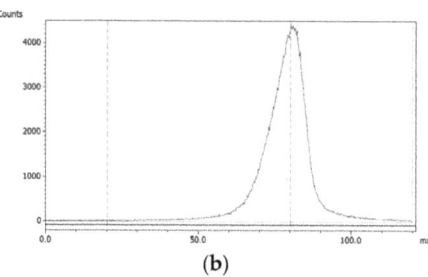

(a) (b)

Figure 4. Chromatograms of (a) $^{68}$Ga-colloid; (b) [$^{68}$Ga]Ga-EDTA.

Validation, with respect to the specificity (successful) and accuracy (attempted) of the paper chromatography method for the determination of the radiochemical purity of [$^{68}$Ga]Ga-EDTA, was performed according to EANM guidelines for the validation of analytical methods for radiopharmaceuticals [17].

To determine the method's specificity, individual chromatograms were produced of $^{68}$Ga-colloid and [$^{68}$Ga]Ga-EDTA. The resolution factor (Rs) was determined to be 3.70 from [17]:

$$Rs = \frac{1.18a(R_2 - R_1)}{W_{h2} - W_{h1}}$$

where $R_{1,2}$ are the retention factors, $W_{h(1,2)}$ are the peak widths at half-height and $\alpha$ is the migration distance of the solvent front. Indices 1 and 2 stand for $^{68}$Ga-colloid and [$^{68}$Ga]Ga-EDTA, respectively. The requirement for Rs is typically higher than 1.5.

To evaluate the accuracy of the method, a standard solution of [$^{68}$Ga]Ga-EDTA spiked with a known activity of $^{68}$Ga-colloid (3.0%) was analyzed to determine the amount of impurity in the product using the analytical method. Only 1.8% of activity of the $^{68}$Ga-colloid was detected (Figure 5a). By mixing an EDTA kit with $^{68}$Ga-colloid, [$^{68}$Ga]Ga-EDTA was formed, indicating that $^{68}$Ga-colloid can be unstable in the presence of a strong chelator EDTA (Figure 5b). Due to this, it was not possible to design a method to determine the accuracy.

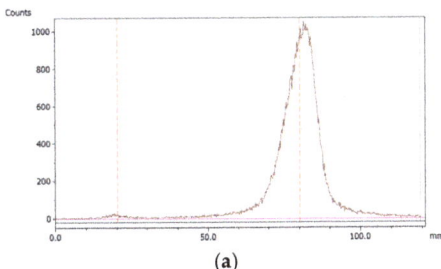
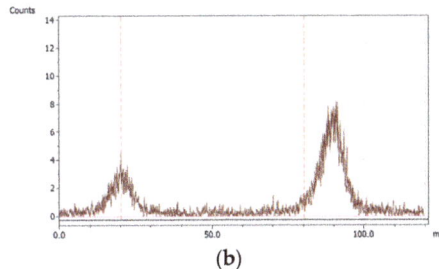

**Figure 5.** Chromatograms of (**a**) [$^{68}$Ga]Ga-EDTA spiked with 3% of $^{68}$Ga-colloid.; (**b**) EDTA kit mixed with $^{68}$Ga-colloid. In both chromatograms, Region 1 corresponds to $^{68}$Ga-colloid and Region 2 corresponds to [$^{68}$Ga]Ga-EDTA.

The analysis results were independent of whether the sample run was performed immediately after application of the spot or the spot was allowed to dry first. Thus, the method was robust. It was also precise, as shown from repeated measurements and comparison of the individual results (Table 3).

See the Discussion section for a further description of the development of the paper chromatography method used for this validation.

## 3. Discussion

Positron Emission Tomography has become a widespread diagnostic technique, which provides the possibility of a both accurate quantitative and qualitative assessment of physiological processes. $^{68}$Ga is one of the most common radionuclides used in PET-imaging. $^{68}$Ga conjugated with EDTA is a physiologically stable metal chelate that can be used for glomerular filtration rate (GFR) estimation and is reported to be suitable for renal function assessments. Gündel et al. have investigated and demonstrated the suitability of [$^{68}$Ga]Ga-EDTA as a tracer for GFR calculation from PET-imaging in small animals, which is shown to conform well to the gold standard of inulin-based GFR-measurement. He also found that [$^{68}$Ga]Ga-EDTA had no protein binding, whereas [$^{68}$Ga]Ga-DTPA had a high level of protein binding, which resulted in the underestimation of GFR [18]. Others have also demonstrated the potential of this radiotracer for split GFR calculations in animals and expect it to have clinical application in human patients in the coming years [10,11].

At present, however, there is no commercially available cold kit for the preparation of [$^{68}$Ga]Ga-EDTA that allows for diagnostic use directly after labeling, and hence has easy applications in clinical practice. The aim of this work was, therefore, to develop and validate a simple cold-kit, stored as a solution in a freezer, to enable the easy production of [$^{68}$Ga]Ga-EDTA for clinical use, which conforms to the regulations set by the Danish regulatory authorities (DMA).

### 3.1. Determination of EDTA Kit Composition

Determination of the simple cold-kit composition provided a sterile and soluble product with the correct pH for both the conjugation reaction and the final labelled [$^{68}$Ga]Ga-EDTA product. The volume of the labelled product was set to 10 mL and the volume of 0.1 N HCl eluent from the $^{68}$Ge/$^{68}$Ga-generator was set at 7 mL. Consequently, the volume of the EDTA kit was 3 mL. Thereafter, the composition of the EDTA kit was designed with the following order.

#### 3.1.1. Amount of Na$_2$EDTA Dihydrate

The concentration of Na$_2$EDTA dihydrate in the radiolabelled product was determined to be 0.0005 M. This corresponded to 0.005 mmol in each EDTA kit and 0.0620 g in 100 mL EDTA kit solution. With a maximum injected volume of 10 mL, i.e., a maximum injected amount of Na$_2$EDTA of 1.86 mg (which was also the total amount in an EDTA-kit), this corresponded to the EDTA kit containing 3 mL, with a concentration of Na$_2$EDTA dihydrate of 0.62 mg/mL.

For higher concentrations, a better radiochemical yield of the complexation reaction is to be expected; however, as a trade-off, more toxicologic considerations need to be applied. Our aim was to maintain a low concentration of Na$_2$EDTA dihydrate. Na$_2$EDTA is used in therapy doses with a maximum of 3 g over 24 hours for the emergency treatment of hypercalcemia and the control of ventricular arrhythmias associated with digitalis toxicity [19]. Other concentrations of Na$_2$EDTA have been reported in the literature, e.g., [$^{51}$Cr]Cr-EDTA, which was previously used to measure GFR, with EDTA doses of up to 50 mg [20], and similarly for [$^{68}$Ga]Ga-EDTA administered by i.v. injection, with doses of 0.05 M (18.6 mg/mL) with a maximum injected volume of 10 mL (i.e., a maximum injected Na$_2$EDTA of 186 mg [9]). In this context, the 0.0005 M concentration of Na$_2$EDTA dihydrate in our radiolabelled product was low.

#### 3.1.2. Amount of NaOAc Trihydrate

The amount of NaOAc trihydrate in the radiolabelled product was set to 0.1 M, as per Hofman et al. [9], corresponding to 1.00 mmol in each EDTA kit and 4.54 g in 100 mL EDTA kit solution.

#### 3.1.3. Amount of NaOH

The amount of 3 M NaOH was adjusted according to the following principles: half of the sodium acetate (0.5 mmol in each EDTA kit) should be protonated to offer a good acetate buffer capacity. This was achieved during the addition of 0.1 N HCl from the generator, where a total amount of 0.7 mmol HCl was added. The excess of 0.2 mmol of HCl should be neutralized by NaOH contained in the EDTA kit, resulting in 2.22 mL of 3 M NaOH in the 100 mL EDTA kit solution.

The kit was designed for use with a GalliaPharm $^{68}$Ge/$^{68}$Ga-generator (Eckert&Ziegler) using 0.1 N HCl for elution. If other generators are considered, with other eluents or other volumes of eluent, the design can be adjusted accordingly, following the description and rationales provided above. However, this will require a separate validation.

### 3.2. Na$_2$EDTA Dihydrate Concentration Determination as a Quality Control of EDTA Kit

The HPLC method used to determine the concentration of Na$_2$EDTA dihydrate was implemented as a quality control test of the EDTA kit. Since EDTA itself does not absorb UV light, a Fe$^{3+}$ complex was formed, which could be measured by UV detection on a HPLC system [14–16]. The method described in the following determined the concentration of Na$_2$EDTA dihydrate with a certainty of ±10%, which was acceptable as we only required a rough estimate of the content to ensure no larger error was introduced during production. Furthermore, the amount given to the patient depends on the radioactivity concentration at the time of injection. This precision could be enhanced by introducing an internal standard in the chromatographic method; however, this was not the scope of this work.

*3.3. Development of Labeling Method*

An automatic, preprogrammed production method using the ModularLab Pharm-Tracer was used to ensure the radiation safety of the personnel. The method needs to be rapid, reproducible and yield a high radiochemically pure product. Labeling reactions were carried out in the disposable cassettes, providing an easy and fast method that routinely achieved very high radiochemical yield and purity >99%. Elution of the generator into the kit takes 3.5 minutes and the complexation reaction between $^{68}$Ga$^{3+}$ and EDTA takes place immediately. Thus, labeling of the product can be performed immediately prior to diagnostic examinations, ensuring minimum loss of radioactivity between production and patient administration.

*3.4. Development of Chromatographic Method for [$^{68}$Ga]Ga-EDTA*

Thin-layer chromatography (TLC) and paper chromatography are commonly utilized, easy methods for the determination of impurities in radiopharmaceuticals. We aimed to develop a method to determine the formation of the $^{68}$Ga-colloid impurity in the [$^{68}$Ga]Ga-EDTA product. Ga$^{3+}$ ions are prone to hydrolysis in aqueous solutions and form different mono- and polynuclear hydroxide species depending on pH, temperature and ionic strength conditions [12]. Free $^{68}$Ga-ions are not expected in the product, since the strong chelator EDTA will ensure that all $^{68}$Ga-ions are coordinated to EDTA (details on experimental proof are explained at the end of this section). Technically, it should be possible to separate [$^{68}$Ga]Ga-EDTA and $^{68}$Ga-colloid, so we investigated the chromatography systems described in the literature for $^{68}$Ga-labelled peptides and [$^{68}$Ga]Ga-EDTA to determine their applicability to our system.

For this investigation, we wanted to prepare the $^{68}$Ga-colloid impurities. During our studies, the method for its preparation, as described in the Ph.Eur. [21], was replaced by the Bench titration method [22], which provides a more precise pH adjustment and which, in our opinion, is a superior method. The $^{68}$Ga-colloid formation was influenced by pH, and unstable oxides or hydroxides were able to complex with EDTA, leading to [$^{68}$Ga]Ga-EDTA [23].

The applied methods using iTLC-SG as the stationary phase are summarized in Table 4. One examined method was based on the analysis of the radiochemical purity of [$^{68}$Ga]Ga-PSMA-HBED-CC in the Ph.Eur. monograph [21], which provides a nice sharp peak at Rf = 0.0 for the $^{68}$Ga-colloid. However, this method was not suitable for [$^{68}$Ga]Ga-EDTA, since this complex provided a broad tailing peak (Table 4, entry 1). The three other applied methods (Table 4, entries 2–4) had the common feature that the $^{68}$Ga-colloid peak did not stay at Rf = 0, which, according to the literature, was to be expected [24–26]. A plausible reason for this is that the $^{68}$Ga-colloids used as reference samples in ours and published studies may not have the same stability. In our studies, it could be argued that the eluents containing either EDTA or TFA can lead to transchelation from $^{68}$Ga-colloid, or rather $^{68}$Ga-oxides or $^{68}$Ga-hydroxides, to complexes with EDTA or trifluoroacetate as ligands.

Another stationary phase was examined using Whatman Grade 1 Chr as the stationary phase and the eluent was a mixture of water:ethanol:pyridine (4:2:1) [23,27]. Here, [$^{68}$Ga]Ga-EDTA produced a clean peak, whereas the $^{68}$Ga-colloid peak was not sharp (Table 4, entry 5).

The results presented here indicate that iTLC-SG strips do not provide a perfect stationary phase for the development of [$^{68}$Ga]Ga-EDTA chromatograms, since the obtained peaks were wide, asymmetric and tailed. Due to this observation, it was decided to proceed with Whatman Grade 1 Chr strips for further trials. Of all the combinations tested in this study, only one system resulted in $^{68}$Ga-colloid peaks, with Rf = 0 (Table 4, entry 1). The others showed some transchelation, resulting in Rf > 0.

Table 4. Results from TLC-studies with reference to the literature. SP: stationary phase, MP: mobile phase.

| Entry | TLC System | $^{68}$Ga-Colloid | [$^{68}$Ga]Ga-EDTA |
|---|---|---|---|
| 1 | SP: iTLC-SG<br>MP: 77 g/L NH$_4$OAc$_{(aq)}$/MeOH (1/1)<br>$^{68}$Ga-colloid: Ph.Eur.<br>Reference: [21] | | |
| 2 | SP: iTLC-SG<br>MP: 0.1M EDTA in 0.25 M NH$_4$Ac, pH 5.5<br>$^{68}$Ga-colloid: Ph.Eur.<br>Reference: [24] | | |
| 3 | SP: iTLC-SG<br>MP: TFA 4%<br>$^{68}$Ga-colloid: Ph.Eur.<br>Reference: [25] | | |
| 4 | SP: iTLC-SG<br>MP: 0.9% NaCl/MeCN (1/1) + 0.08% TFA<br>$^{68}$Ga-colloid: Bench titration<br>Reference: [26] | | |
| 5 | SP: Whatman Grade 1CHR<br>MP: Water/ethanol/pyridine (4/2/1)<br>$^{68}$Ga-colloid: Ph.Eur.<br>Reference: [23,27] | | |

To address this issue, we also investigated HCl with pH = 5.6 as an eluent, i.e. the same pH as the method used to prepare $^{68}$Ga-colloid. The simple mobile phase of HCl, adjusted with NaOH to pH = 5.6, gave the best and most well-defined $^{68}$Ga-colloid peak using both iTLC-SG and Whatman Grade 1 Chr as stationary phases (Table 5, entries 1 and 2). Thus, it was concluded that a chromatography system consisting of the combination of Whatman Grade 1 Chr and HCl (pH 5.6) was the optimum method for quality control of the labelled product (Table 5, entry 2). Additionally, since this method required the adjustment of HCl to pH = 5.6 with NaOH, we also investigated whether the use of saline as a mobile phase instead of HCl was useful. Table 5, entry 3, demonstrates this alternative method of using a system consisting of Whatman Grade 1 Chr as a stationary phase and saline as a mobile phase as a routine quality control for [$^{68}$Ga]Ga-EDTA. The advantage of this method is that it is simple, fast, cheap and reproducible.

Table 5. Results from TLC- and paper chromatograpy studies using local combinations of stationary and mobile phases with inspiration taken from the results presented in Table 4. SP: stationary phase, MP: mobile phase.

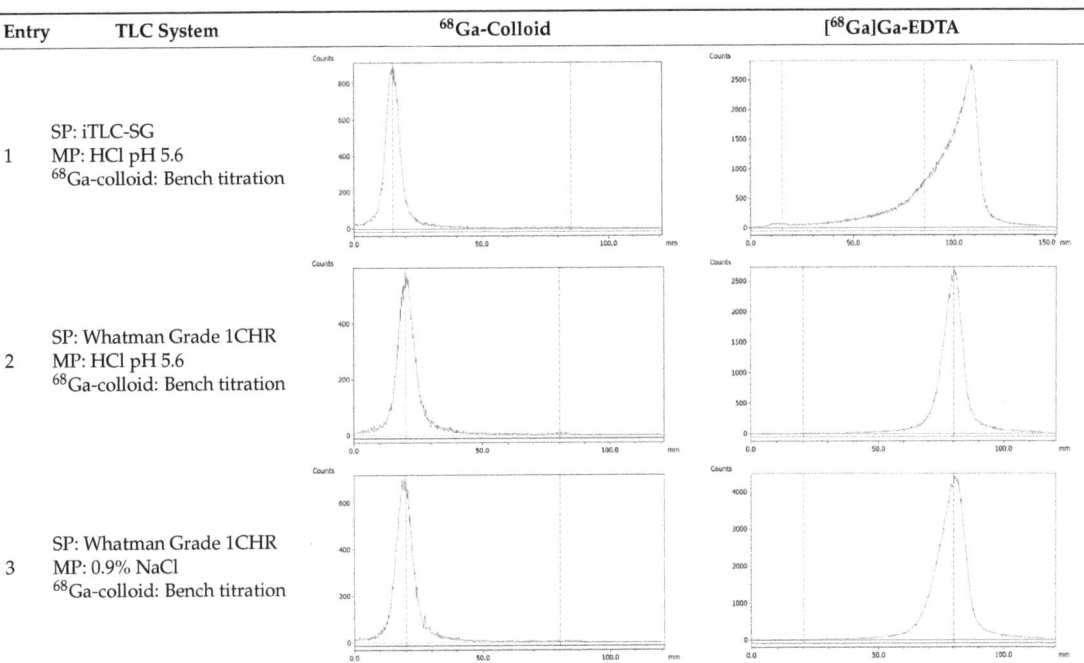

| Entry | TLC System | $^{68}$Ga-Colloid | [$^{68}$Ga]Ga-EDTA |
|---|---|---|---|
| 1 | SP: iTLC-SG<br>MP: HCl pH 5.6<br>$^{68}$Ga-colloid: Bench titration | | |
| 2 | SP: Whatman Grade 1CHR<br>MP: HCl pH 5.6<br>$^{68}$Ga-colloid: Bench titration | | |
| 3 | SP: Whatman Grade 1CHR<br>MP: 0.9% NaCl<br>$^{68}$Ga-colloid: Bench titration | | |

In this study, we do not expect the presence of free [$^{68}$Ga]Ga$^{3+}$ in the product, with the hypothesis being that any free ions would either coordinate to EDTA or form $^{68}$Ga-colloid immediately under the production conditions. To provide supporting evidence for this assumption, we created three solutions, ((i) generator eluate, (ii) kit without EDTA and (iii) [$^{68}$Ga]Ga-EDTA product), on which a) paper chromatography using the optimized paper chromatography method and b) HPLC were performed to determine EDTA content, as described in Sections 2.1.3 and 2.2.2, respectively.

i  Generator eluate consisting of [$^{68}$Ga]GaCl$_3$ was analyzed using Whatman paper, showing that radioactivity developed to the eluent front only as [$^{68}$Ga]Ga-EDTA. Therefore, if free [$^{68}$Ga]Ga$^{3+}$ exists, it cannot be separated from the intended product.

ii The $^{68}$Ge/$^{68}$Ga-generator was eluted into a kit prepared without EDTA (analogous to the production of [$^{68}$Ga]Ga-EDTA). On the Whatman paper, the product stayed at Rf = 0 showing that, if EDTA is not present, free [$^{68}$Ga]Ga$^{3+}$ does not exist in the solution.

iii The intended [$^{68}$Ga]Ga-EDTA product analyzed using HPLC provided a single clear peak at Rt = 5.9 min. However, the eluate solution and the kit without EDTA did not produce signals on the HPLC chromatograms, thus indicating that $^{68}$Ga was trapped on the HPLC column in both cases.

These observations confirm that there is no considerable free [$^{68}$Ga]Ga$^{3+}$ present in the product solution and, as such, it is not necessary to analyze this under general production.

### 4. Materials and Methods

*4.1. EDTA kit Production*

The raw materials used to prepare 100 mL of EDTA kit solution were: TRIPLEX III (ethylenedinithrilotetraacetic acid disodium salt dihydrate = Na$_2$EDTA·2H$_2$O; Merck,

1.37004.1000, VWR), sodium acetate trihydrate (Merck, 1.06235.1000, VWR), sodium hydroxide: sterile 3 M solution 10 × 10 mL (Hospital Pharmacy) and sterile water (SAD, solvent for parenteral use, 100 mL bottles).

EDTA kit production was aseptically performed in a Laminar Air Flow cabinet (GMP grade A) situated in a clean room (grade C), with microbiological monitoring using settle plates and particle monitoring (MET One 3415 particle counter) under the complete duration of critical processing.

In the clean room, reagents were weighed in sterile weight bottles with their lids and transferred to the LAF-cabinet. In the LAF-cabinet, reagents were transferred to the volumetric flask (100 mL) and dissolved in sterile water. After complete dissolution using a magnet stirrer, the product was sterile-filtered (filter unit Cathivex GV 0.22 µm Merck Millipore), portioned manually with a Finnpipette with a sterile tip and sealed in sterile 10 mL glass vials with 3 mL of product in each vial. Vials were frozen at −18 °C.

The ingredients for 100 mL EDTA kit solution are shown in Table 6.

Table 6. Amount of reagents in 100 mL EDTA kit solution.

| Reagent | Molar Weight (g/mol) | Amount of Substance (mol) | Mass (g) | Volume (mL) |
|---|---|---|---|---|
| Disodium EDTA dihydrate | 372.24 | 0.000167 | 0.0620 | - |
| Sodium acetate trihydrate | 136.08 | 0.0333 | 4.54 | - |
| NaOH (3 M) | - | 0.00666 | - | 2.22 |
| Sterile water | - | - | - | Up to 100 |

### 4.2. EDTA Kit Quality Control

The sterile filter used for the bulk production was tested manually for integrity. For this purpose, 3 mL of sterile water was drawn into a 10 mL syringe, followed by air. The filter was then attached to the syringe and the syringe's content rapidly expelled. The filter was intact if the syringe's piston returned to its starting position. The produced EDTA kits were individually tested visually for appearance and volume, determined by comparison to a standard volume. The product pH was measured on a single sample using a calibrated ISO 9001 certified pH-meter (HACH HQ411d with provided PHC705 electrode). Sterility testing of a single sample of the product was carried out by the hospital microbiology department to determine the amount of CFU in the product.

The content of Disodium EDTA dihydrate in the EDTA kit was measured using High-Pressure Liquid Chromatography analysis (HPLC), based on a complexation reaction between EDTA and $Fe^{3+}$ (formation constant for Fe-EDTA, $K_f = 1.3 \times 10^{25}$) [14], reversed phase column and ion pair reagent, as described in [14–16]. The HPLC system consisted of LC-20AD UFLC Shimadzu pump, SPD-20A HPLC UV-VIS detector (wavelength 254 nm), chromatographic column (Kinetex 5 µm XB-C18 100a 150 × 4.6 mm) and associated LabSolutions software. The following HPLC parameters were used: flow: 1 mL/min; injection volume: 30 µL; eluent: 4.5 g sodium acetate trihydrate with 800 mL of water added, pH adjusted to 4.0. Thereafter, 4.0 g tetrabutylammonium bromide was added and filled up to 1 L with water. To create a calibration curve, 10 reference samples of $Na_2$EDTA dihydrate in water were produced with concentrations ranging from 0.01 mg/mL to 0.1 mg/mL in steps of 0.01 mg/mL. A solution of $FeCl_3 \cdot 6H_2O$ in 30% acetic acid/water (V/V) with a concentration of 0.175 mg/mL was produced. For HPLC analysis, 10 samples were prepared by mixing the $Na_2$EDTA dihydrate reference sample and $FeCl_3 \cdot 6H_2O$ solution 1/1. Samples from EDTA kits were diluted 1/10 with water and mixed with $FeCl_3 \cdot 6H_2O$ solution 1/1 prior to HPLC analysis.

Finally, a complete labelling reaction of an EDTA kit with $^{68}$Ga, producing [$^{68}$Ga]Ga-EDTA according to the procedure described below, was required for approval of each EDTA kit batch.

### 4.3. [$^{68}$Ga]Ga-EDTA Production

The production of [$^{68}$Ga]Ga-EDTA was performed in a grade C clean room, with all critical processes conducted in a GMP-grade A Laminar Air Flow cabinet. Microbiological monitoring with settle plates, glove print and particle monitoring (MET One 3415 particle counter) was performed for the complete duration of critical processing. Synthesis was performed automatically using a PharmTracer ModularLab.

The critical sterile procedure in which the thawed, sterile, sealed vial with 3 mL of on-site, pre-produced EDTA kit, was equipped with a vent needle and a needle with a filter unit for sterile filtering (Cathivex GV 0.22 mm Merck Millipore), which was performed in a GMP-grade A LAF cabinet. The product vial was then connected to the elution cassette's outlet and placed into the shielded container, after which PharmTracer's elution software was executed.

The GMP-compliant (production and test compliance with Ph.Eur. and DMA regulations) GalliaPharm $^{68}$Ge/$^{68}$Ga-generator (Eckert and Ziegler 1.85 GBq), was eluted with 7 mL of sterile ultrapure 0.1 N HCl for the direct elution of GalliaPharm into the thawed sterile, sealed vial containing the on-site pre-produced EDTA kit. The process was fully automated using the elution cassette for Modular-Lab PharmTracer (Eckert and Ziegler Eurotop GmbH), Figure 6.

**Figure 6.** Synthesis setup. A: 0.1 N HCl; B: GalliaPharm $^{68}$Ge/$^{68}$Ga-generator; C: dispensing syringe; D: Stopcock manifold; E: lead shield with product vial.

The use of Modular-Lab PharmTracer system, together with Software Modular-Lab and elution/synthesis cassettes (Eckert and Ziegler Eurotope GmbH), provides an efficient, routine GMP production of radiopharmaceuticals and prevents cross-contamination issues. The system was fully qualified and validated to perform [$^{68}$Ga]Ga-EDTA synthesis based on the use of 0.1 N HCl for the elution of the $^{68}$Ge/$^{68}$Ga-generator. Additionally, the automatic synthesis was reproducible and provided the benefit of reductions in the radiation dose to the staff. The elution cassette was inserted into the PharmTracer's stopcock manifold, and the generator connected to the inlet port of the cassette.

The shielded GalliaPharm system is an approved radionuclide generator for medical use, allowing for the elution of [$^{68}$Ga]GaCl$_3$ from a titanium dioxide column, onto which the parent radionuclide $^{68}$Ge is adsorbed. $^{68}$Ga is eluted using sterile ultrapure 0.1 N HCl. The sterile ultrapure HCl solution was connected to the generator's inlet port and the eluate collected at the outlet port with intended use in medicinal production. The generator was eluted a minimum of 24 h prior to labeling in order to avoid the accumulation of free long-life $^{68}$Ge ions and metal impurities, e.g., zinc ions (Zn$^{2+}$) arising as the decay product from $^{68}$Ga, which can interfere with the labeling reaction [2,28].

The $^{68}$Ga eluate was regularly investigated for sterility as well as for $^{68}$Ge breakthrough by a gamma spectrum test in the laboratory. Other potential metal ion impurities, such as Fe and Zn ions, were defined by the manufacturer (Eckert and Ziegler, Berlin, Germany) as being lower than the levels allowed by the European Pharmacopeia.

*4.4. [$^{68}$Ga]Ga-EDTA Quality Control*

Routine quality control of the labelled product included pH verification with indicator papers and the visual determination of appearance and volume, as assessed by comparison with a standard. Testing of the filter integrity was performed as described earlier for EDTA kits and radionuclide purity was tested by measuring the half-life and $^{68}$Ge-breakthrough in a product sample. The product's radionuclide identity was confirmed by half-life determination according to the Ph.Eur. monograph for $^{68}$Ge/$^{68}$Ga generators: three measurements of radioactivity within 15 min in a dose calibrator (Capintec CRC-55TR), which was routinely checked for stability and accuracy. Results were plotted logarithmically as a function of time. $^{68}$Ge breakthrough was determined at a minimum 48 h after the [$^{68}$Ga]Ga-EDTA end of synthesis (EOS) in an automatic gamma-counter (Perkin Elmer, Wizard 2480). In accordance with the Ph.Eur., the results were expressed as a percentage of total eluted $^{68}$Ga.

Standard paper chromatography method was used to determine the radiochemical purity (RCP) and identity of [$^{68}$Ga]Ga-EDTA. The product was applied 2 cm from the bottom of a Whatman Grade 1 Chr (GE Healthcare) chromatography paper strip (12 cm × 2 cm) and directly transferred to a chromatography tank with 10 mL of 0.9% sodium chloride (NaCl). When the solvent front reached 8 cm from the bottom, the strip was removed from the tank and analyzed using a LabLogic TLC-scanner Scan-RAM (Laura software and PS Plastic/PMT radio-detector; 120 mm; speed 1 mm/s). To validate this paper chromatography method, 0.9% NaCl was used as an eluent and Whatman Grade 1 Chr plates as the stationary phase; a standard solution of [$^{68}$Ga]Ga-EDTA (product) and a reference solution of $^{68}$Ga-colloid (impurities) were used. Since the presence of $^{68}$Ga-ions was not expected in a product with large excess of EDTA, $^{68}$Ga-ions were not considered. [$^{68}$Ga]Ga-EDTA was produced using the method described in the previous method section. $^{68}$Ga-colloid was prepared using the bench titration of $^{68}$Ga-eluate with sodium hydroxide solutions to pH 5.6 ± 0.2 [22].

Quantitative endotoxin detection in the product was measured by an EndoSafe Nexgen PTS (Charles River) spectrophotometer, which utilized disposable cartridges with Limulus amebocyte lysate (LAL) reagents (product number PTS201F).

Additional extended quality control procedures (XQC), carried out at regular intervals, included: (1) gamma spectrum measurement by a high purity germanium detector; (2) sterility testing analyzed by the Department of Microbiology, as described above, for

EDTA kits and (3) stability studies (repetition of pH and RCP/ID measurements), no less than 2 h post-EOS.

## 5. Conclusions

This paper describes the development and validation of methods for the production and quality control of a simple EDTA kit and its labeling with $^{68}$Ga. Both production procedures are performed under aseptic conditions compliant with GMP regulations.

Quality control of both the simple kit and labelled product consisted of a visual assessment, volume designation, pH measuring, filter integrity test and test for sterility. The EDTA kit was controlled using the HPLC method to measure precursor (EDTA) content. The final product, [$^{68}$Ga]Ga-EDTA, was controlled for radiochemical purity using the developed, validated and established paper chromatography method with Whatman Grade 1 Chr strips as the solid phase and 0.9% NaCl as an eluent, together with measurements of the radionuclidic purity from the determination of the half-life, gamma spectrum and $^{68}$Ge-breakthrough.

It was shown that the developed processes are reliable, highly reproducible, and easily implemented for local clinical use.

**Author Contributions:** Conceptualization, M.S. and L.F.; methodology and validation, M.S. and L.F.; writing—original draft preparation, M.S.; writing—review and editing, L.F.; supervision, L.F.; funding acquisition, L.F. All authors have read and agreed to the published version of the manuscript.

**Funding:** This research received no external funding except from the contribution mentioned under Acknowledgments.

**Institutional Review Board Statement:** Not applicable.

**Informed Consent Statement:** Not applicable.

**Data Availability Statement:** The data presented in this study are available on request from the corresponding author.

**Acknowledgments:** Toyota-Fonden, Denmark is greatly acknowledged for its financial contribution to the TLC scanner used for chromatograpic studies. Claire Fynbo is acknowledged for proofreading the manuscript.

**Conflicts of Interest:** The authors declare no conflict of interest.

**Sample Availability:** Not applicable.

## References

1. Nudat. NUDAT. Available online: https://www.nndc.bnl.gov/nudat2/reCenter.jsp?z=31&n=37 (accessed on 9 October 2022).
2. Velikyan, I. 68Ga-Based Radiopharmaceuticals: Production and Application Relationship. *Molecules* **2015**, *20*, 12913–12943. [CrossRef] [PubMed]
3. Pozzilli, C.; Bernardi, S.; Mansi, L.; Picozzi, P.; Iannotti, F.; Alfano, B.; Bozzao, L.; Lenzi, G.L.; Salvatore, M.; Conforti, P. Quantitative assessment of blood-brain barrier permeability in multiple sclerosis using 68-Ga-EDTA and positron emission tomography. *J. Neurol. Neurosurg. Psychiatry* **1988**, *51*, 1058–1062. [CrossRef] [PubMed]
4. Gottschalk, A.; Anger, H. The sensitivity of the positron scintillation camera for detecting simulated brain tumors with gallium 68-EDTA. *Am. J. Roentgenol. Radium Therapy Nucl. Med.* **1964**, *92*, 174–176.
5. Roesch, F.; Riss, P.J. The Renaissance of the 68Ge/68Ga Radionuclide Generator Initiates New Developments in 68Ga Radiopharmaceutical Chemistry. *Curr. Top. Med. Chem.* **2010**, *10*, 1633–1668. [CrossRef] [PubMed]
6. Banerjee, S.R.; Pomper, M.G. Clinical applications of Gallium-68. *Appl. Radiat. Isot.* **2013**, *76*, 2–13. [CrossRef] [PubMed]
7. Van Leeuwen, P.; Emmett, L.; Hruby, G.; Kneebone, A.; Stricker, P. 548 68Ga-PSMA has high detection rate of prostate cancer recurrence outside the prostatic fossa in patients being considered for salvage radiation treatment. *Eur. Urol. Suppl.* **2016**, *15*, e548. [CrossRef]
8. Yamashita, M.; Inaba, T.; Kawase, Y.; Iiorii, H.; Wakita, K.; Fujii, R.; Nakahashi, H. Quantitative measurement of renal function using Ga-68-EDTA. *Tohoku J. Exp. Med.* **1988**, *155*, 207–208. [CrossRef] [PubMed]
9. Hofman, M.S.; Hicks, R.J. Gallium-68 EDTA PET/CT for Renal Imaging. *68ga-PET Curr. Status* **2016**, *46*, 448–461. [CrossRef]
10. Ding, Y.; Liu, Y.; Zhang, L.; Deng, Y.; Chen, H.; Lan, X.; Jiang, D.; Cao, W. Quantitative assessment of renal functions using 68Ga-EDTA dynamic PET imaging in renal injury in mice of different origins. *Front. Med.* **2023**, *10*, 1143473. [CrossRef]

11. Fontana, A.O.; Melo, M.G.; Allenbach, G.; Georgantas, C.; Wang, R.; Braissant, O.; Barbey, F.; Prior, J.O.; Ballhausen, D.; Viertl, D. The use of 68Ga-EDTA PET allows detecting progressive decline of renal function in rats. *Am. J. Nucl. Med. Mol. Imaging* **2021**, *11*, 519–528.
12. Hacht, B. Gallium(III) Ion Hydrolysis under Physiological Conditions. *Bull. Korean Chem. Soc.* **2008**, *29*, 372.
13. Bauwens, M.; Chekol, R.; Vanbilloen, H.; Bormans, G.; Verbruggen, A. Optimal buffer choice of the radiosynthesis of 68Ga–Dotatoc for clinical application. *Nucl. Med. Commun.* **2010**, *31*, 753–758. [CrossRef] [PubMed]
14. Heydari, R.; Shamsipur, M.; Naleini, N. Simultaneous Determination of EDTA, Sorbic Acid, and Diclofenac Sodium in Pharmaceutical Preparations Using High-Performance Liquid Chromatography. *AAPS PharmSciTech* **2013**, *14*, 764–769. [CrossRef] [PubMed]
15. Narola, B.; Singh, A.; Mitra, M.; Santhakumar, P.; Chandrashekhar, T. A Validated Reverse Phase HPLC Method for the Determination of Disodium EDTA in Meropenem Drug Substance with UV-Detection using Precolumn Derivatization Technique. *Anal. Chem. Insights* **2011**, *6*, 7–14. [CrossRef] [PubMed]
16. Bergers, P.J.M.; De Groot, A.C. The Analysis of EDTA in Water by HPLC. *Water Res.* **1994**, *28*, 639–642. [CrossRef]
17. Gillings, N.; Todde, S.; Behe, M.; Decristoforo, C.; Elsinga, P.; Ferrari, V.; Hjelstuen, O.; Peitl, P.K.; Koziorowski, J.; Laverman, P.; et al. EANM guideline on the validation of analytical methods for radiopharmaceuticals. *EJNMMI Radiopharm. Chem.* **2020**, *5*, 7–29. [CrossRef] [PubMed]
18. Gündel, D.; Pohle, U.; Prell, E.; Odparlik, A.; Thews, O. Assessing Glomerular Filtration in Small Animals Using [68Ga]DTPA and [68Ga]EDTA with PET Imaging. *Mol. Imaging Biol.* **2017**, *20*, 457–464. [CrossRef]
19. ENDRATE Package Insert. Available online: https://dailymed.nlm.nih.gov/dailymed/fda/fdaDrugXsl.cfm?setid=290c3e9c-c0 c6-440a-1a9c-46e3e2b07a77&type=display (accessed on 10 August 2022).
20. ANSTRO (Chromium [51Cr]CrEDTA package inlet). Available online: https://www.ansto.gov.au/sites/default/files/2020-05/ Chromium-51Cr%20Edetate%20Injection%20BP%202020.pdf (accessed on 10 August 2022).
21. European Pharmacopeia. European Pharmacopeia. European Pharmacopeia Monograph "Gallium (68Ga) PSMA-11 Innjection" (04/2021:3044). In *European Pharmacopeia*, 10th ed.; European Pharmacopoeia: Strasbourg, France, 2020.
22. Ali, M.; Hsieh, W.; Tsopelas, C. An improved assay for68Ga-hydroxide in68Ga-DOTATATE formulations intended for neuroendocrine tumour imaging. *J. Label. Compd. Radiopharm.* **2015**, *58*, 383–389. [CrossRef]
23. Hnatowich, D.J. A method for the preparation and quality control of 68Ga radiopharmaceuticals. *J. Nucl. Med.* **1975**, *16*, 764–768.
24. Brom, M.; Franssen, G.M.; Joosten, L.; Gotthardt, M.; Boerman, O.C. The effect of purification of Ga-68-labeled exendin on in vivo distribution. *EJNMMI Res.* **2016**, *6*, 65. [CrossRef]
25. Larenkov, A.A.; Ya Maruk, A. Radiochemical Purity of 68Ga-BCA-Peptides: Separation of All 68Ga species with a Single iTLC Strip. *Int. J. Chem. Mol. Nucl. Mater. Metall. Eng.* **2016**, *10*, 1212.
26. Larenkov, A.; Rakhimov, M.; Lunyova, K.; Klementyeva, O.; Maruk, A.; Machulkin, A. Pharmacokinetic Properties of [68]Ga-Labelled Folic Acid Conjugates: Improvement Using HEHE Tag. *Molecules* **2020**, *25*, 2712. [CrossRef]
27. Yamashita, M.; Horii, H.; Hashiba, M.; Imahori, Y.; Mizukawa, N. The use of computed radiography for the determination of the impurities in 68Ga-EDTA by paper chromatography. *Radioisotopes* **1986**, *35*, 478–481. [CrossRef] [PubMed]
28. Summary Product Characteristics for Galliapharm, Radionuclide Generator. Available online: https://produktresume.dk/ AppBuilder/serach?q=Galliapharm%2C+radionuclide+generator+0.74-1.85+GBq.docx (accessed on 10 August 2022).

**Disclaimer/Publisher's Note:** The statements, opinions and data contained in all publications are solely those of the individual author(s) and contributor(s) and not of MDPI and/or the editor(s). MDPI and/or the editor(s) disclaim responsibility for any injury to people or property resulting from any ideas, methods, instructions or products referred to in the content.

Article

# Optimization of Deuteron Irradiation of $^{176}$Yb for Producing $^{177}$Lu of High Specific Activity Exceeding 3000 GBq/mg

Lin Shao

Department of Nuclear Engineering, Texas A&M University, College Station, TX 77843, USA; lshao@tamu.edu

**Abstract:** The irradiation of $^{176}$Yb with deuterons offers a promising pathway for the production of the theranostic radionuclide $^{177}$Lu. To optimize this process, calculations integrating deuteron transport, isotope production, and decay have been performed. In pure $^{176}$Yb, the undesired production of $^{174g+m}$Lu occurs at higher deuteron energies, corresponding to a distribution slightly shallower than that of $^{177}$Lu. Hence, $^{174g+m}$Lu can be effectively filtered out by employing either a low-energy deuteron beam or stacked foils. The utilization of stacked foils enables the production of $^{177}$Lu using a high-energy linear accelerator. Another unwanted isotope, $^{176m}$Lu, is produced roughly at the same depth as $^{177}$Lu, but its concentration can be significantly reduced by selecting an appropriate post-irradiation processing time, owing to its relatively short half-life. The modeling approach extended to the mapping of yields as a function of irradiation time and post-irradiation processing time. An optimized processing time window was identified. The study demonstrates that a high-energy deuteron beam can be employed to produce $^{177}$Lu with high specific activity exceeding 3000 GBq/mg. The effect of different purity levels (ranging from 98% to 100%) was also discussed. The impurity levels have a slight impact. The modeling demonstrates the feasibility of obtaining $^{177}$Lu with a specific activity > 3000 GBq/mg and radionuclidic purity > 99.5% when using a commercially available $^{176}$Yb target of 99.6% purity.

**Keywords:** isotope production; theranostic radionuclide; accelerator; lutetium; cancer treatment; medical isotope

Citation: Shao, L. Optimization of Deuteron Irradiation of $^{176}$Yb for Producing $^{177}$Lu of High Specific Activity Exceeding 3000 GBq/mg. *Molecules* 2023, 28, 6053. https://doi.org/10.3390/molecules28166053

Academic Editor: Svend Borup Jensen

Received: 10 June 2023
Revised: 8 August 2023
Accepted: 9 August 2023
Published: 14 August 2023

Copyright: © 2023 by the author. Licensee MDPI, Basel, Switzerland. This article is an open access article distributed under the terms and conditions of the Creative Commons Attribution (CC BY) license (https://creativecommons.org/licenses/by/4.0/).

## 1. Introduction

Theranostics is an emerging approach that combines therapeutic and diagnostic elements for effective cancer treatment [1]. In this approach, radionuclides emitting low-energy gamma rays are used for diagnostic purposes, while those emitting charged particles such as beta rays and alpha particles are utilized for therapy. This fusion of diagnostics and therapy represents a significant advancement in personalized cancer treatment [2,3]. $^{177}$Lu has garnered considerable interest as a theranostic radionuclide due to its emission of beta rays with an energy of 134 keV and low-energy gamma rays at 208 keV [4,5]. Combining $^{177}$Lu with other therapeutic radionuclides, such as $^{90}$Y/$^{177}$Lu and $^{67}$Cu/$^{177}$Lu, has shown great promise in cancer treatment [6,7]. The efficacy of $^{177}$Lu in neuroendocrine tumors has been acknowledged by the US Food and Drug Administration (approved in 2018) and the European Medicines Agency (approved in 2017) [8].

The term "carrier-free" is used to describe radionuclides with the highest specific activity. This means that the final product has 100% isotopic abundance and is not contaminated with stable isotopes. $^{177}$Lu, which has a high specific activity, is particularly important in certain radiation therapies, although it may not be necessary for all types. For instance, in peptide receptor radionuclide therapy, the limited concentration of different cellular cognate receptors expressed on the tumor cell surface necessitates the use of $^{177}$Lu with high specific activity.

Currently, the production of $^{177}$Lu relies primarily on reactors using two production routes [9,10]: the "direct" route and the "indirect" route. In the direct route, the $^{176}$Lu target is subjected to neutron irradiation via $^{176}$Lu(n,γ)$^{177}$Lu reactions. In comparison, the indirect route involves the use of a $^{176}$Yb target via $^{176}$Yb(n,γ)$^{177}$Yb → $^{177}$Lu reactions. Each route has its advantages and disadvantages.

The direct route benefits from the high thermal neutron capture cross-sections of $^{176}$Lu, which is as high as 2090 barn [11,12]. However, a drawback of the direct route is the production of $^{177m}$Lu as an impurity of concern. $^{177m}$Lu has a relatively long half-life of 160.5 days, leading to an increasing ratio of $^{177m}$Lu/$^{177}$Lu over time. This poses challenges in hospital preparations, as the presence of $^{177m}$Lu triggers concerns regarding radioactive waste management. Currently, the average specific activity of the direct route is approximately 740 to 1100 GBq/mg, which needs further improvement.

The indirect production route is based on the $^{176}$Yb(n,γ)$^{177}$Yb → $^{177}$Lu reactions. This route offers the advantage of producing high specific activity, approximately 2960 GBq/mg, as it does not generate $^{177m}$Lu as an impurity. Furthermore, the product obtained via the indirect route can be carrier-free. However, there are some disadvantages associated with the indirect production route. It has low production yields due to the low cross-section of $^{176}$Yb, which is only 2.5 barns. Additionally, the chemical properties of Yb and Lu are very similar, posing challenges in the separation of Yb and Lu [13]. For more information on the various methods under development for Yb/Lu separation, a comprehensive review can be found in reference [13].

Considering the anticipated high market demand, alternative approaches utilizing accelerators have been investigated. These methods involve the use of protons [14], deuterons [15–18], alpha particles [19], and electron beams [20]. Among various accelerator-based techniques, the most efficient method for producing $^{177}$Lu is via the irradiation of a pure Yb target with deuterium [21]. This method has the highest yield in comparison with other possible choices, including $^{nat}$Yb(d,x)$^{177}$Lu, $^{nat}$Hf(p,x)$^{177}$Lu, $^{nat}$Hf(d,x)$^{177}$Lu, $^{nat}$Lu(p,x)$^{177}$Lu, $^{nat}$Lu(d,x)$^{177}$Lu, and $^{nat}$Yb(α,x)$^{177}$Lu.

Previous studies have modeled the utilization of deuteron irradiation on a Yb target for $^{177}$Lu production [22,23]. Kambali compared the production yields between (d,n) and (d,p) reactions [22]. Nagai et al. systematically modeled the activities and specific activities of $^{177}$Lu using deuteron beams of different energies and Yb targets of varying purities [23]. Both studies suggested the feasibility of the overall processes. The present study aims to accomplish two objectives: firstly, to assess the feasibility of producing $^{177}$Lu using a high-energy linear accelerator (LINAC); secondly, to optimize both the irradiation time and the post-irradiation processing time to attain the highest achievable specific activity.

LINAC is a unique type of accelerator that can achieve very high beam energy at a relatively low cost. However, both the ion species and beam energies are fixed characteristics determined by the beam design. In other words, if it is designed at high energy, it cannot be operated at low energies. A high-energy LINAC provides opportunities for isotope production that requires high threshold beam energies. Conversely, a high-energy beam may be less suitable for isotope production that necessitates low threshold energies. An example of such a high-energy accelerator is the LINAC, located in Denton, Texas. Originally manufactured as part of the superconducting supercollider project, it was designed to operate at a beam energy close to 70 MeV for isotope production. This LINAC utilized a design that originated from Los Alamos National Laboratory in the 1980s, specifically tailored for an energy range of 70 to 90 MeV for the nuclear medicine program at that time [24,25]. Therefore, exploring the applications of high-energy LINACs for isotope production holds commercial value. It is worthwhile to investigate the feasibility of utilizing high-energy accelerators as versatile instruments capable of accommodating a wide range of isotopes requiring thermal energy at different energy levels, both low and high.

The current study utilized 80 MeV as an illustrative example of a high-energy LINAC. However, the proposed methodology is applicable to various high-energy LINACs, regardless of whether they operate at 80 MeV or not. The multiple foil target configurations, as proposed in the present study, can be adjusted based on the specific beam energy of any LINAC, making the approach versatile and not limited to Denton LINAC. For the same reason, the beam current was chosen to be typical of the Denton LINAC. Nevertheless, the obtained yields can be readily converted for other LINACs. Hence, the impact of this study extends to general high-energy LINACs and is not specific to the Denton LINAC.

## 2. Modeling Procedure

The modeling approach employed in this study consists of the following steps: (1) a Monte Carlo simulation code was utilized to determine the energy of deuterons at different penetration depths; (2) the localized energy is converted into localized isotope production using energy-dependent activation functions; (3) the effects of continuous ion bombardment (gain) and decay (loss) are calculated as functions of irradiation time and post-irradiation processing time; and (4) the amount of produced isotopes is integrated over the region of interest to calculate activity and specific activity.

The Stopping and Range of Ions in Matter (SRIM) code has been widely used in materials science for irradiation studies [26]. However, SRIM does not directly provide detailed information about the local beam energy at different penetration depths as an output. Nevertheless, this information can be estimated reasonably well by analyzing the projected range of ions at various incident energies. Figure 1 plots the projected ranges of deuterons as a function of incident energy. For instance, at an incident energy of 80 MeV, the projected range of deuterium in Yb is approximately 9.2 mm. Conversely, at a lower energy of 60 MeV, the projected range reduces to 8 mm. This difference of 1 mm in range suggests that, in order for an 80 MeV beam to stop at 9.2 mm, its energy at a depth of 1 mm must be around 60 MeV. In other words, if we denote the projected range curve as $R(E)$, where $E$ is the incident energy as a variable, then for a selected incident energy $E_0$, the energy at a depth of $R(E_0) - R(E)$ from the surface is $E$. Figure 2 plots the local energy as a function of depth for different incident energies, calculated using the procedure described.

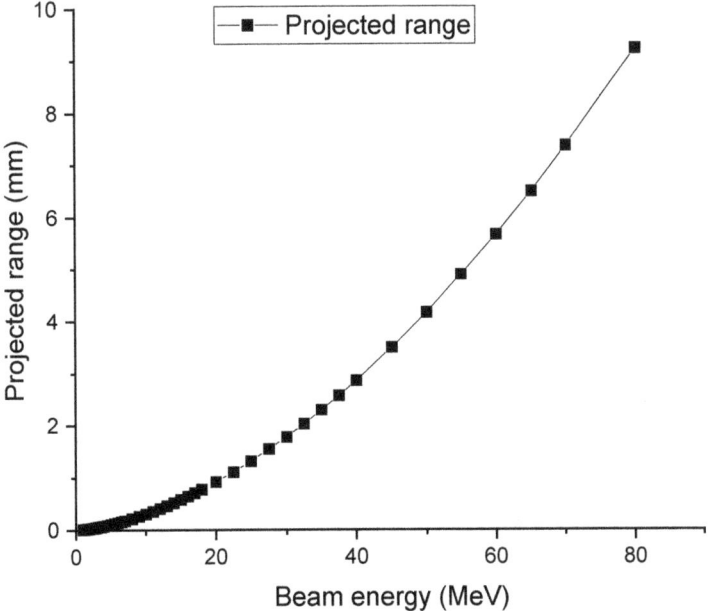

**Figure 1.** Projected ranges of deuteron as a function of deuteron incident energy in Yb.

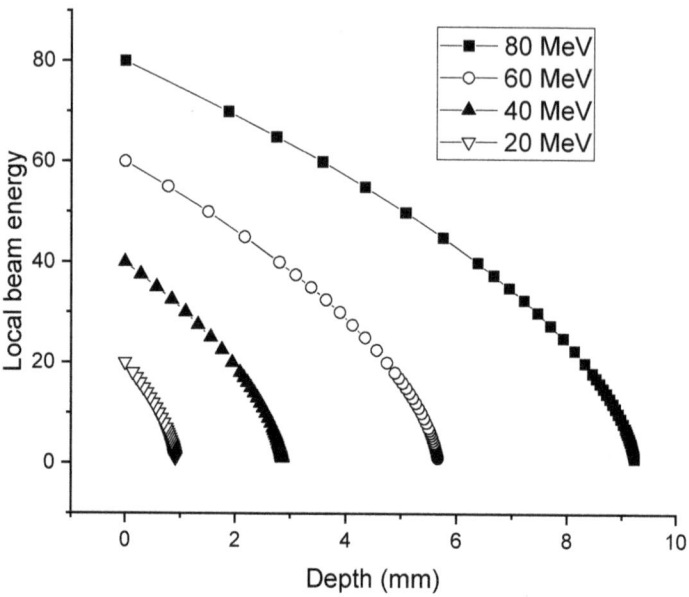

**Figure 2.** Local deuteron energy as a function of depth in Yb for incident energies of 20, 40, 60, and 80 MeV.

The utilization of Figure 2 to convert local beam energy into isotope production, based on energy-dependent cross-sections, assumes that straggling can be neglected. As a result, for a given incident energy, the energy values at each depth point are precise and have minimal uncertainty. This approximation holds true for protons and deuterons because their collisions with target atoms are primarily influenced by glancing angle collisions. Consequently, the ion trajectory remains a straight line for most of the penetration, except towards the very end, where low-energy collisions favor the creation of small damage cascades.

The available cross-section data for deuterium bombardment of $^{176}$Yb and other Yb isotopes are quite limited. Khandaker et al. conducted measurements on natural Yb and reported cross-section data for (d,x) reactions up to 24 MeV [27]. Nagai et al. fitted these cross-section data and experimentally validated the integrated yields [23]. Figure 3 plots the experimental activation functions for producing $^{177}$Lu in pure $^{176}$Yb [27], along with the fitted functions [23]. Two reactions result in the production of $^{177}$Lu. The first reaction is $^{176}$Yb(d,n)$^{177}$Lu, and the second reaction is $^{176}$Yb(d,n)$^{177g+m}$Yb $\rightarrow$ $^{177}$Lu. The cross-section of the $^{176}$Yb(d,n)$^{177g+m}$Yb $\rightarrow$ $^{177}$Lu reaction is higher than that of the $^{176}$Yb(d,n)$^{177}$Lu reaction. $^{177}$Yb undergoes β emission followed by gamma transitions in $^{177}$Lu, with a relatively short half-life of 1.88 h. Therefore, $^{177}$Lu can be approximated as the direct product of deuteron irradiation. It is important to note that the available experimental data is limited to energies up to 24 MeV. Further validation is required to assess the accuracy of the data at higher energies.

To maximize the specific activity of $^{177}$Lu, minimizing the production of other isotopes is crucial. This is why purified $^{176}$Yb is preferred over natural Yb. Natural Yb consists of seven stable isotopes: $^{176}$Yb, $^{174}$Yb, $^{173}$Yb, $^{172}$Yb, $^{171}$Yb, $^{170}$Yb, and $^{168}$Yb, with $^{174}$Yb being the most abundant at 31.8% of natural abundance. Figure 4a–f presents the fitted cross-sections for producing isotopes other than $^{177}$Lu in $^{176}$Yb, $^{174}$Yb, $^{173}$Yb, $^{172}$Yb, $^{171}$Yb, and $^{170}$Yb, respectively. $^{168}$Yb is not included due to its very low abundance (0.126%). It is observed that almost all cross-sections of (d,x) reactions are higher than that for producing $^{177}$Lu in $^{176}$Yb (Figure 3). Therefore, utilizing purified $^{176}$Yb instead of natural Yb is the most effective approach to avoid the generation of unwanted isotopes.

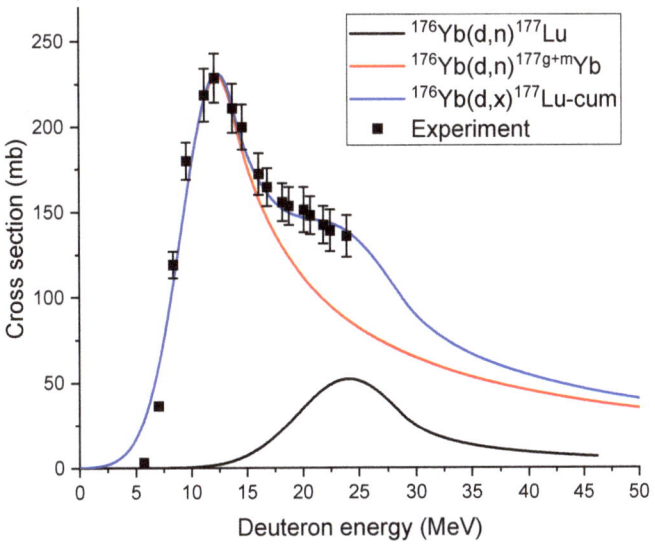

**Figure 3.** Cross-sections of producing $^{177}$Lu as a function of deuteron energy, from experiments [26] and fitting [23].

Table 1 summarizes Lu isotopes (only relevant ones are selected) and their half-lives and decay modes [28,29]. As for specific isotope products in Figure 4, $^{173}$Lu produced from the bombardment of $^{174}$Yb or $^{173}$Yb decays into $^{173}$Yb via electron capture (EC) with a half-life of 1.37 years. For $^{172g+m}$Lu produced from both $^{173}$Yb and $^{172}$Yb, $^{172}$Lu decays into $^{172}$Yb via β+ decay with a half-life of 6.7 days. $^{172m}$Lu decays into $^{172}$Lu via isomeric transition (IT) with a half-life of 3.7 min. For $^{171g+m}$Lu produced from $^{172}$Yb and $^{171}$Yb, $^{171}$Lu decays to $^{171}$Yb via b β+ decay with a half-life of 8.24 days. $^{171m}$Yb decays to $^{171}$Lu via IT decay with a half-life of 79 s. For $^{170g+m}$Lu produced from $^{171}$Yb and $^{170}$Yb, $^{170}$Lu undergoes β+ decay into $^{170}$Yb with a half-life of 2 days. $^{170m}$Lu decays rapidly into $^{170}$Lu via IT decay with a half-life of 670 milliseconds. For $^{169}$Lu produced from bombarding $^{170}$Yb, it decays into $^{169}$Yb with a half-life of 34 h.

**Table 1.** Selected Lu isotopes and their decay characteristics [28,29].

| Nuclide | Half-Life | Decay | Daughter Isotope |
|---|---|---|---|
| $^{169}$Lu | 34.06(5) h | β+ | $^{169}$Yb |
| $^{170}$Lu | 2.012(20) d | β+ | $^{170}$Yb |
| $^{170m}$Lu | 670(100) ms | IT | $^{170}$Lu |
| $^{171}$Lu | 8.24(3) d | β+ | $^{171}$Yb |
| $^{171m}$Lu | 79(2) s | IT | $^{171}$Lu |
| $^{172}$Lu | 6.70(3) d | β+ | $^{172}$Yb |
| $^{172m}$Lu | 3.7(5) min | IT | $^{172}$Lu |
| $^{173}$Lu | 1.37(1) y | EC | $^{173}$Yb |
| $^{174}$Lu | 3.31(5) y | β+ | $^{174}$Yb |
| $^{174m}$Lu | 142(2) d | IT (99.38%) | $^{174}$Lu |
|  |  | EC (0.62%) | $^{174}$Yb |
| $^{176}$Lu | 38.5(7) × 10$^9$ y | β− | $^{176}$Hf |
| $^{176m}$Lu | 3.664(19) h | β− (99.9%) | $^{176}$Hf |
|  |  | EC (0.095%) | $^{176}$Yb |
| $^{177}$Lu | 6.6475(20) d | β− | $^{177}$Hf |

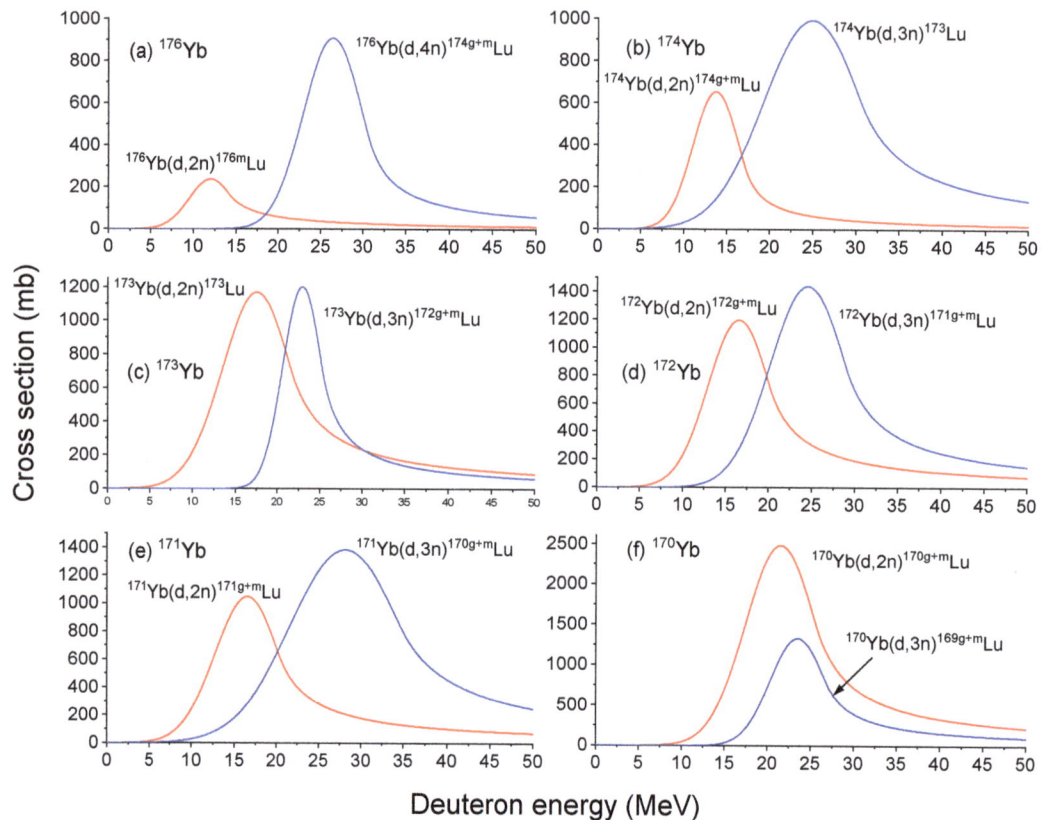

**Figure 4.** Cross-sections of various (d,x) reactions in deuteron-irradiated (**a**) $^{176}$Yb, (**b**) $^{174}$Yb, (**c**) $^{173}$Yb, (**d**) $^{172}$Yb, (**e**) $^{171}$Yb, and (**f**) $^{170}$Yb, as a function of deuteron energy [23].

Figure 4a is important in minimizing the presence of unwanted isotopes when using a purified $^{176}$Yb target. $^{176m}$Lu, in particular, undergoes $\beta^-$ decay into $^{176}$Hf (99.9%) and EC decay into $^{176}$Yb (0.095%). With a relatively short half-life of 3.664 h, it is feasible to wait for a sufficient time period for $^{176m}$Lu to decay before initiating the chemical separation process. By optimizing the timing, the specific activity of $^{177}$Lu can be maximized while minimizing the presence of $^{176m}$Lu and its decay products.

$^{174g}$Lu, with a half-life of 3.31 years, decays into $^{174}$Yb via $\beta^+$ decay. The long half-life of $^{174g}$Lu poses a challenge in achieving a high specific activity of $^{177}$Lu. However, the reactions leading to its production start above 15 MeV and become significant above 20 MeV. Therefore, the concentration of $^{174g}$Lu can be minimized by optimizing the beam energy (E < 20 MeV). Regarding $^{174m}$Lu, it undergoes IT decay into $^{174}$Lu (99.38%) and EC decay into $^{174}$Yb (0.62%). Its relatively long half-life of 142 days presents concerns about quality control. However, since $^{174m}$Lu has a higher threshold energy at 20 MeV, its concentration can also be minimized via beam energy optimization (E < 20 MeV).

The distinctive energy dependence of the production of $^{177}$Lu and other isotopes in pure $^{176}$Yb leads to the difference in their isotope distribution profiles, as shown in Figure 5 for the case of 80 MeV deuteron irradiation. The dashed line in Figure 5 represents the local deuteron energy as a function of penetration depth. The symbols represent the cross-section of producing $^{177}$Lu, $^{176m}$Lu, $^{174g}$Lu, and $^{174m}$Lu. At shallow depths, where the deuteron energy is high, the production of isotopes is relatively low. As the deuteron energy decreases to approximately 40 MeV, corresponding to a depth of around 6 mm, the cross-section for producing $^{174g+m}$Lu increases (as shown in Figure 4a). At an energy of approximately 30 MeV, corresponding to a depth of about 7.5 mm, $^{177}$Lu production starts to increase. At an energy of around 20 MeV and a corresponding depth of 8.3 mm, the cross-section for $^{176m}$Lu production begins to rise. The $^{174m}$Lu cross-section reaches its peak at a depth of approximately 7.7 mm and then decreases to almost zero at a depth of 8.3 mm. For $^{174g}$Lu, its cross-section reaches its peak at a depth of around 7.9 mm and drops to almost zero at a depth of 8.5 mm. For both $^{177}$Lu and $^{176m}$Lu, they exhibit increasing cross-sections and peak at a depth of approximately 8.8 mm. The corresponding deuteron energy at this depth is approximately 13 MeV. Both $^{177}$Lu and $^{176}$Lu cross-sections decrease and approach zero at a depth of approximately 9 mm. As to be discussed shortly, the differences in in-depth profiles of different isotopes are important in maximizing the specific activity of $^{177}$Lu.

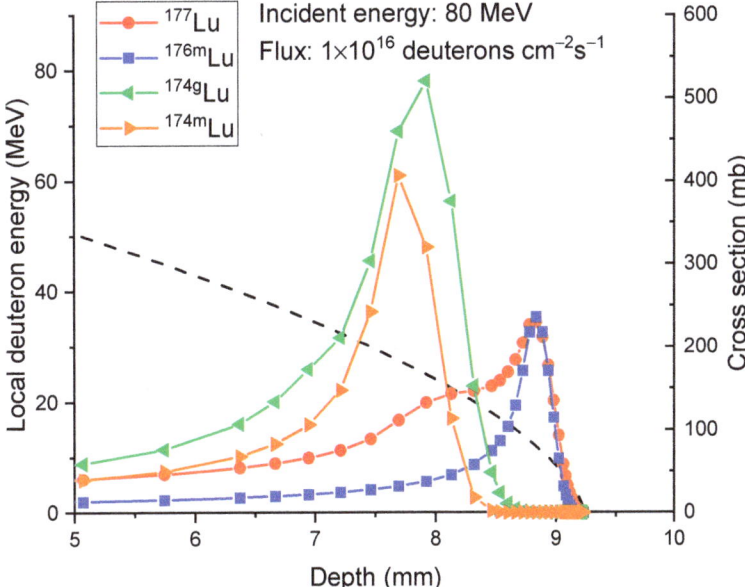

**Figure 5.** Local deuteron energy (dashed line) and cross-sections of producing $^{177}$Lu (red circle), $^{176m}$Lu (blue square), $^{174g}$Lu (green triangle), and $^{174m}$Lu (orange triangle), as a function of depth, for the case of 80 MeV deuteron irradiation of pure $^{176}$Yb.

The depth-dependent cross-sections are used to calculate the quantity of isotope products at different depths for specific bombardment time and processing time. The change of isotopes per unit volume (N) is determined by the following equation [30],

$$dN = \begin{cases} (R - \lambda N)dt & : 0 < t < t_b \\ -\lambda N dt & : t_b < t \end{cases},$$

with the solution

$$N = \frac{R(1 - e^{-\lambda t_b})e^{-\lambda(t - t_b)}}{\lambda} \quad (for\ t > t_b)$$

$$\lambda = 0.693/t_{1/2}$$

$$R = \sigma N_0 \Phi$$

where $t$ is time, $t_b$ is the end of irradiation time, $R$ is the reaction rate per unit volume, $\lambda$ is the decay constant, $N_0$ denotes the atomic density (which is $2.422 \times 10^{22}$ atoms/cm$^3$ for Yb), and $\Phi$ the number of deuterons arriving on the surface per unit area per unit of time. For $\Phi$, a value of $1 \times 10^{16}$ deuterons/cm$^2$/s is used, which is typical for a high-performance plasma source. Note that $1 \times 10^{16}$ deuterons/cm$^2$/s corresponds to 1.6 mA/cm$^2$. The decay constants $\lambda$ used in the calculations are based on the half-life times $t_{1/2}$: 6.65d for $^{177}$Lu, 3.66h for $^{176m}$Lu, 3.31y for $^{174g}$Lu, and 142d for $^{174m}$Lu.

Figure 6 plots the calculated depth profiles of $^{177}$Lu in 80 MeV deuteron-irradiated $^{176}$Yb at three different time points: 0 h, 3 h, and 24 h after the completion of irradiation. The irradiation time is set to 3 days. Three hours after the irradiation, the concentration changes of $^{177}$Lu due to its decay are minimal. In 24 h after the irradiation, concentration decreases become apparent. However, the overall concentration changes are not significant. At the depth of peak concentration, 8.8 mm, the concentration of $^{177}$Lu decreases from $1.24 \times 10^{19}$ atoms/cm$^3$ to $1.21 \times 10^{19}$ atoms/cm$^3$ after three hours and further decreases to $1.12 \times 10^{19}$ atoms/cm$^3$ after 24 h.

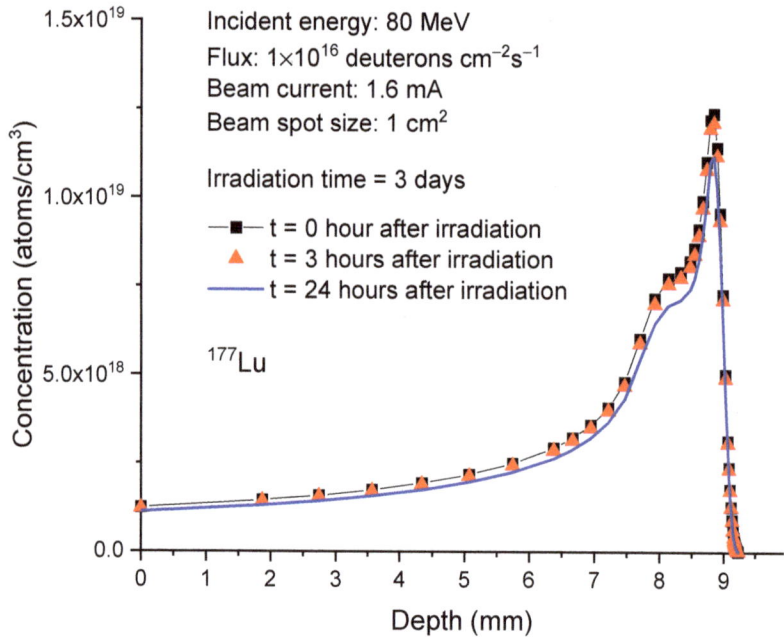

**Figure 6.** Concentrations of $^{177}$Lu as a function of depth in 80 MeV deuteron-irradiated $^{176}$Yb. The irradiation time is 3 days. For comparison, the post-irradiation processing times considered are 0 h, 3 h, and 24 h. The beam current is 1.6 mA. The beam spot size is 1 cm$^2$. This corresponds to $1 \times 10^{16}$ deuterons/cm$^2$/s.

As a result of its much faster decay, the concentration change of $^{176m}$Lu differs significantly from that of $^{177}$Lu. Figure 7 plots the $^{176m}$Lu concentration profiles at the same time points as shown in Figure 6. At the end of irradiation, the peak concentration of $^{176m}$Lu is $1.09 \times 10^{18}$ atoms/cm$^3$. After 3 h, the concentration decreases to $4.22 \times 10^{17}$ atoms/cm$^3$ and further decreases to $1.16 \times 10^{16}$ atoms/cm$^3$ after 24 h. In contrast, both $^{174g}$Lu and $^{174m}$Lu exhibit negligible concentration changes, as shown in Figures 8 and 9. This lack of changes is due to their relatively long half-lives, which are 3.31 years for $^{174g}$Lu and 142 days for $^{174m}$Lu. It is worth noting that the peak concentrations of both isotopes, at $2.5$–$3.5 \times 10^{19}$/cm$^3$, are higher than the peak concentration of $^{177}$Lu (about $1 \times 10^{19}$/cm$^3$).

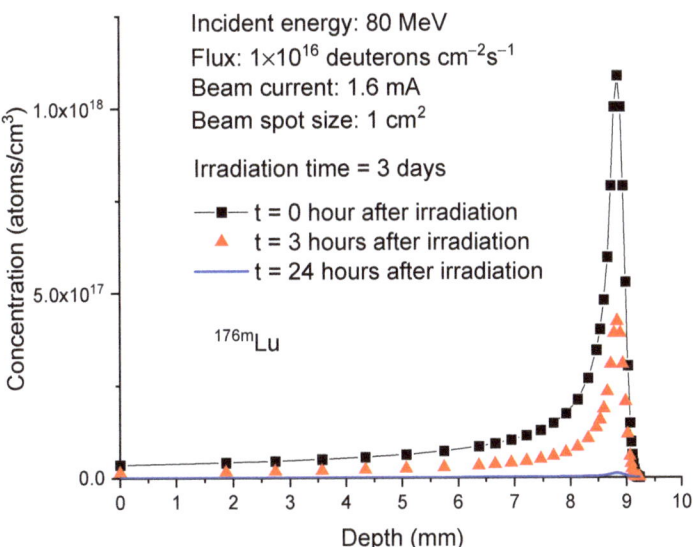

**Figure 7.** Concentrations of $^{176m}$Lu as a function of depth in 80 MeV deuteron-irradiated $^{176}$Yb. The irradiation time is 3 days. The post-irradiation processing times are 0 h, 3 h, and 24 h. The beam current is 1.6 mA. The beam spot size is 1 cm². This corresponds to $1 \times 10^{16}$ deuterons/cm²/s.

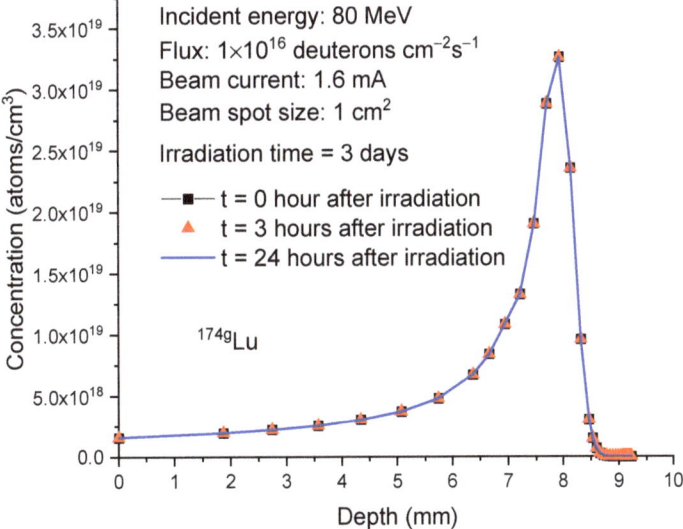

**Figure 8.** Concentrations of $^{174g}$Lu as a function of depth in 80 MeV deuteron-irradiated $^{176}$Yb. The irradiation time is 3 days. The post-irradiation processing times are 0 h, 3 h, and 24 h. The beam current is 1.6 mA. The beam spot size is 1 cm². This corresponds to $1 \times 10^{16}$ deuterons/cm²/s.

Considering the differences in concentration distributions and decay rates, we propose two methods to increase the specific activity of $^{177}$Lu. The first method involves reducing the deuteron beam energy to a value close to 21 MeV. This energy is high enough to activate the production of $^{177}$Lu but not high enough to produce significant amounts of $^{174g}$Lu and $^{174m}$Lu. Figure 10 compares the isotope profiles for an incident energy of 80 MeV and an incident energy of 21 MeV. The calculations correspond to a post-irradiation processing time of 24 h, with the irradiation time set to 3 days. In the case of 21 MeV, $^{177}$Lu reaches its peak concentration at a depth of 0.6 mm and drops to zero at approximately 1 mm depth. The unwanted isotope $^{174g}$Lu is present in small quantities near the surface and rapidly decreases to nearly zero at a depth of around 0.4 mm. The concentration of $^{174m}$Lu is located at even shallower depths and is negligible. Due to its rapid decay, the concentration of $^{176m}$Lu is reduced to a negligible level as well.

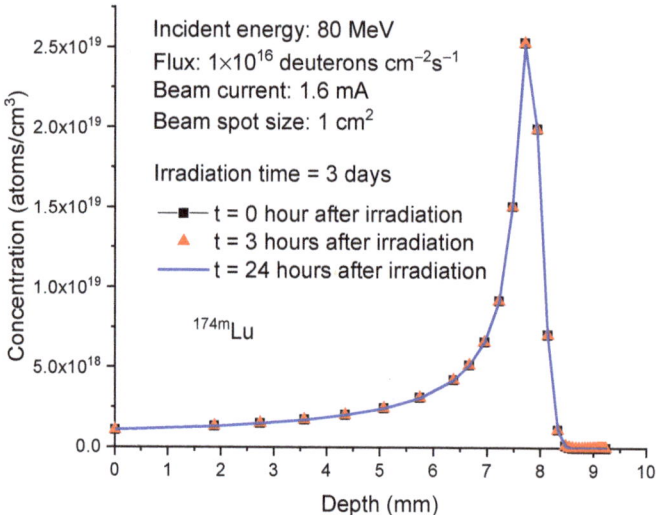

**Figure 9.** Concentrations of $^{174m}$Lu as a function of depth in 80 MeV deuteron-irradiated $^{176}$Yb. The irradiation time is 3 days. The post-irradiation processing times are 0 h, 3 h, and 24 h. The beam current is 1.6 mA. The beam spot size is 1 cm$^2$.

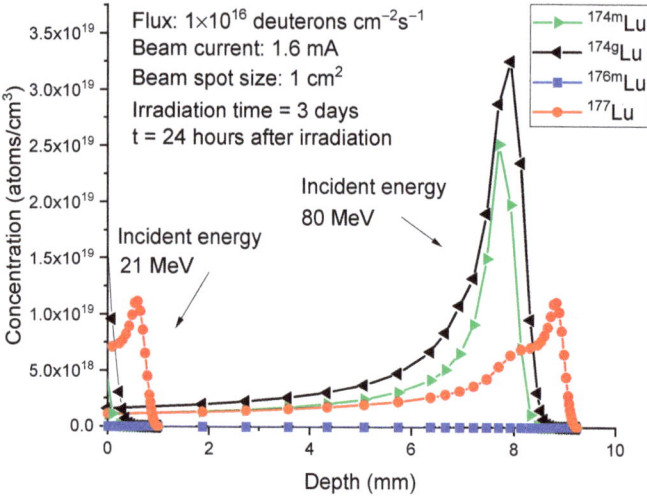

**Figure 10.** Concentration profiles of $^{177}$Lu (red circle), $^{176m}$Lu (blue square), $^{174g}$Lu (black triangle), and $^{174m}$Lu (green triangle) for the incident energies of 80 MeV and 40 MeV. The irradiation time is 3 days. The calculations correspond to 24 h after the irradiation. The beam current is 1.6 mA. The beam spot size is 1 cm$^2$.

The second method involves the use of a stacked foil target. The stacked-foil technique is commonly employed to measure excitation functions of nuclear reactions involving light ions as projectiles [31–33]. In this technique, a target composed of multiple foils is used. In the present study, stacked foils are proposed to harvest isotopes at specific beam energies. Figure 11 illustrates the arrangement where $^{176}$Yb targets with approximately 1 mm thickness are stacked and exposed to deuteron beams. The thickness of the target foils is slightly adjusted to ensure that the last foil corresponds to a depth ranging from 8.22 mm to 9.22 mm. The gray colored box indicates the last foil, where the specific activity of $^{177}$Lu is maximized, and the unwanted isotopes are minimized.

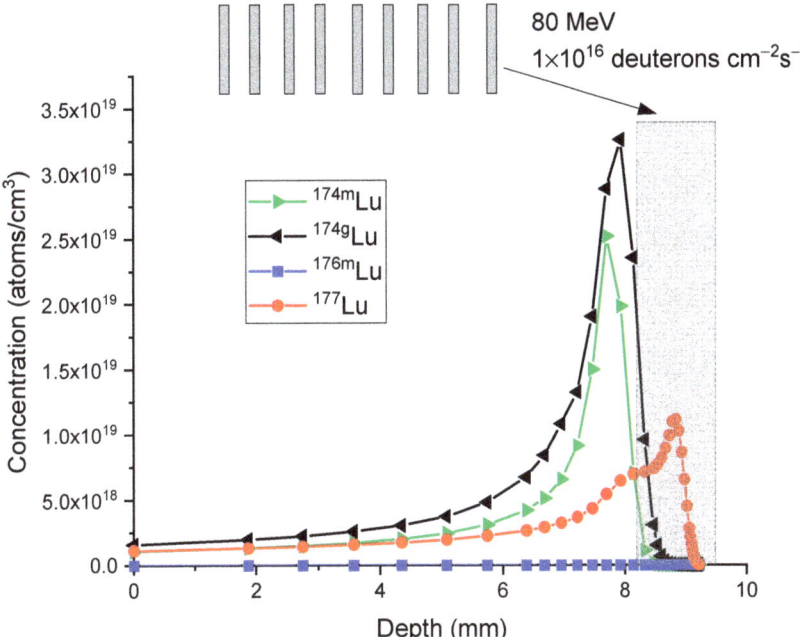

**Figure 11.** Concentration profiles of $^{177}$Lu (red circle), $^{176m}$Lu (blue square), $^{174g}$Lu (black triangle), and $^{174m}$Lu (green triangle) for the incident energies of 80 MeV in a stacked foil target configuration. The irradiation time is 3 days. The calculations correspond to 24 h after the irradiation. The beam current is 1.6 mA. The beam spot size is 1 cm$^2$.

Tables 2 and 3 compare the daily isotope production from different depth regions: 8.22 mm to 9.22 mm (the last foil) and the entire depth range of 0 to 9.22 mm (all foils). The daily production is based on two rounds of irradiation, with each round lasting 12 h. For each round, the post-irradiation chemical separation time is 28 h. This condition arises after optimization to maximize specific activities. It does not represent the conditions to maximize radionuclidic purity nor the conditions optimized to balance both specific activity and radionuclidic purity. The conditions for different optimization needs will be discussed shortly. Table 2 presents the first optimization condition, prioritizing specific activity. The tables provide information on the activity, mass, specific activity, and radionuclidic purity of $^{174m}$Lu, $^{174g}$Lu, $^{176m}$Lu, and $^{177}$Lu.

In Table 2, which pertains to the last foil only, the activity of daily produced $^{177}$Lu is 304 GBq, and its specific activity is 3084 GBq/mg. Although the activity is relatively low, the specific activity is remarkably high. Note that the majority of reactor-based neutron irradiation produces $^{177}$Lu with a specific activity ranging from 740 to 1100 GBq/mg. In Table 3, which covers the entire depth region from 0 to 9.22 mm, the activity of $^{177}$Lu is 1119 GBq, with an average specific activity of 1022 GBq/mg. Even though the specific activity is lower compared to the previous cases, it is still comparable to the quality of reactor-produced isotopes.

**Table 2.** The daily production of Lu isotopes in the depth region of 8.22 mm to 9.22 mm in 80 MeV deuteron-irradiated pure $^{176}$Yb. The beam current is 1.6 mA. The beam spot size is 1 cm$^2$. The irradiation time is 12 h for each batch. The post-irradiation processing time is 28 h. This is the condition optimized to maximize specific activity. The conditions to maximize radionuclide activity and the conditions to balance both specific activity and radionuclide activity are different and will be separately discussed shortly.

|  | Activity (GBq) | Mass (mg) | Specific Activity (GBq/mg) | Radionuclidic Purity |
|---|---|---|---|---|
| $^{174m}$Lu | 0.555 | 0.003 | 5.92 | 0.2% |
| $^{174g}$Lu | 0.481 | 0.021 | 5.18 | 0.1% |
| $^{176m}$Lu | 13.7 | $7.02 \times 10^{-5}$ | 139 | 4.3% |
| $^{177}$Lu | 304 | 0.074 | 3084 | 95.4% |

**Table 3.** The daily production of Lu isotopes in the depth region of 0 mm to 9.22 mm. The irradiation condition is the same as in Table 2.

|  | Activity (GBq) | Mass (mg) | Specific Activity (GBq/mg) | Radionuclidic Purity |
|---|---|---|---|---|
| $^{174m}$Lu | 62.5 | 0.32 | 57 | 5.1% |
| $^{174g}$Lu | 11.5 | 0.50 | 10.4 | 0.9% |
| $^{176m}$Lu | 30.3 | $1.56 \times 10^{-4}$ | 27.8 | 2.5% |
| $^{177}$Lu | 1119 | 0.27 | 1022 | 91.5% |

To maximize $^{177}$Lu production, both in terms of total activity and specific activity, calculations were performed to determine the optimized conditions. The calculations changed both the irradiation time and the post-irradiation processing time incrementally. Figures 12 and 13 compare the isotope quality of the last foil and all foils. These plots emphasize the significance of selecting the appropriate processing time window. In the calculations, irradiation time changes using a step of 12 h, while post-irradiation processing time changes using a step of 4 h. All plots are normalized to one day of operation. If each batch requires a 12 h irradiation period, then one day of operation combines the yields from two batches. The post-irradiation processing start time is not limited to a duration shorter than one day. The calculation takes into account continuous operations, allowing the post-irradiation processing time to extend beyond one day if necessary.

Figure 12 provides a summary of the product quality of the last foil. All figures correspond to one day of operation (two batches per day). In Figure 12a, the total activity of $^{177}$Lu reaches the maximum of approximately 337 GBq/day for an irradiation time of 12 h and a post-irradiation processing time of 4 h. Increasing either the irradiation time or the post-irradiation processing time leads to a decrease in activity. Figure 12b plots the specific activity and exhibits the existence of a peak region. At a given irradiation time, the specific activity initially increases and then decreases with increasing post-irradiation processing time. The peak value is approximately 3071 GBq/mg. For maximizing specific activity, the optimal post-irradiation processing time is around 20 h for short irradiation. The optimal post-irradiation processing time slightly reduces to about 12 h for long irradiation. Prolonged irradiation and post-irradiation processing times result in a drop in specific activity, with the lowest value reaching approximately 2590 GBq/mg, representing the worst scenario in the mapping. However, an overall high specific activity can still be ensured.

**(a) Activity, the last foil only**

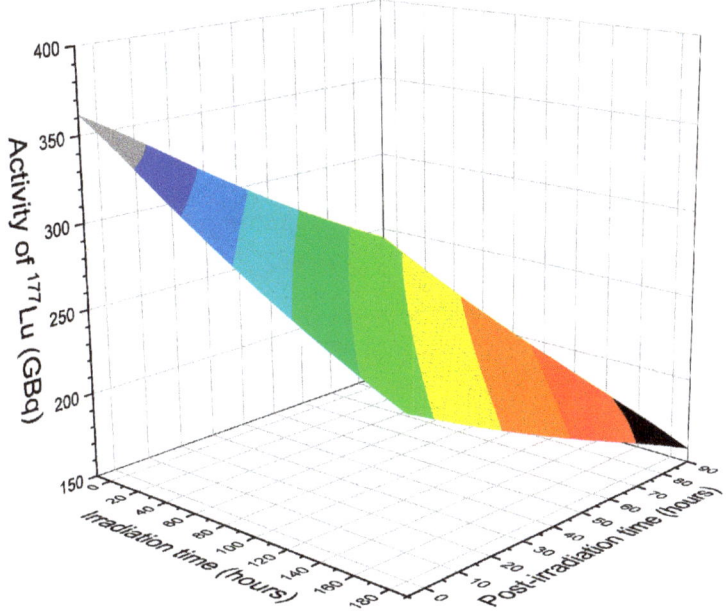

**(b) Specific activity, the last foil only**

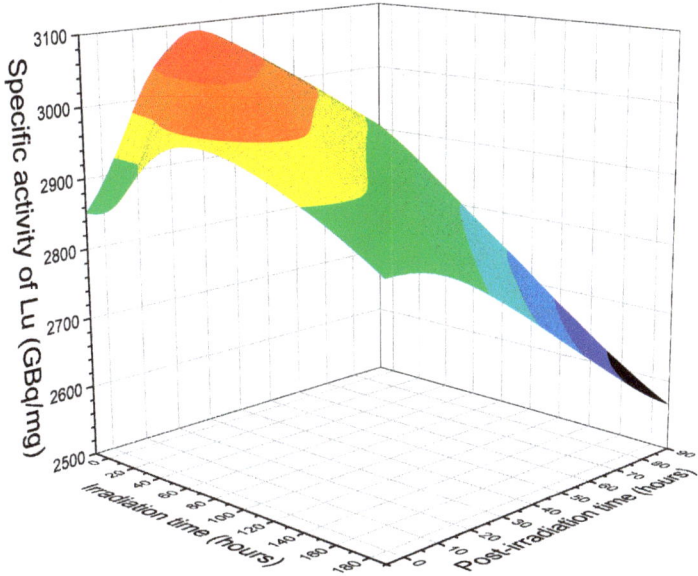

**Figure 12.** *Cont.*

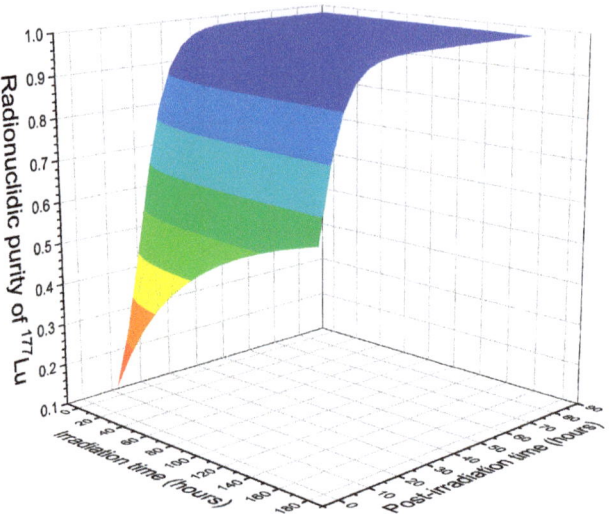

**Figure 12.** (**a**) Activity, (**b**) specific activity, and (**c**) radionuclidic purity of $^{177}$Lu collected from the last foil (corresponding to a depth from 8.22 mm to 9.22 mm) of $^{176}$Yb irradiated using 80 MeV deuterons. The x-axis represents the irradiation time, and the y-axis represents the post-irradiation processing time. All yields are normalized to six days of operation. The flux is $1 \times 10^{16}$ deuterons/cm$^2$/s, which is equivalent to a beam current of 1.6 mA over a beam spot size of 1 cm$^2$.

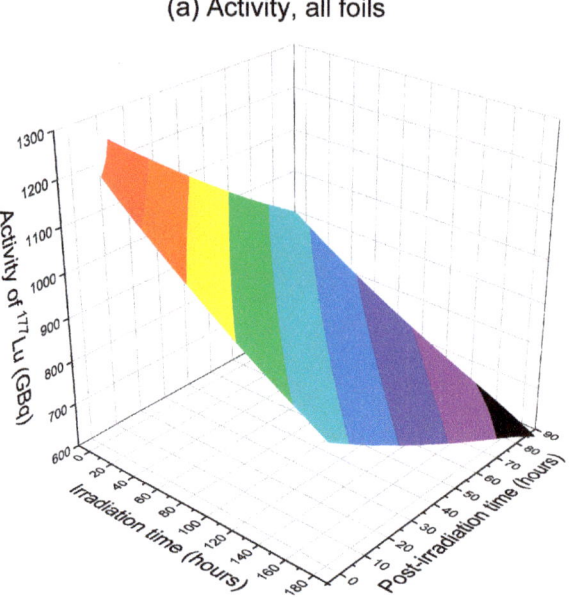

**Figure 13.** *Cont.*

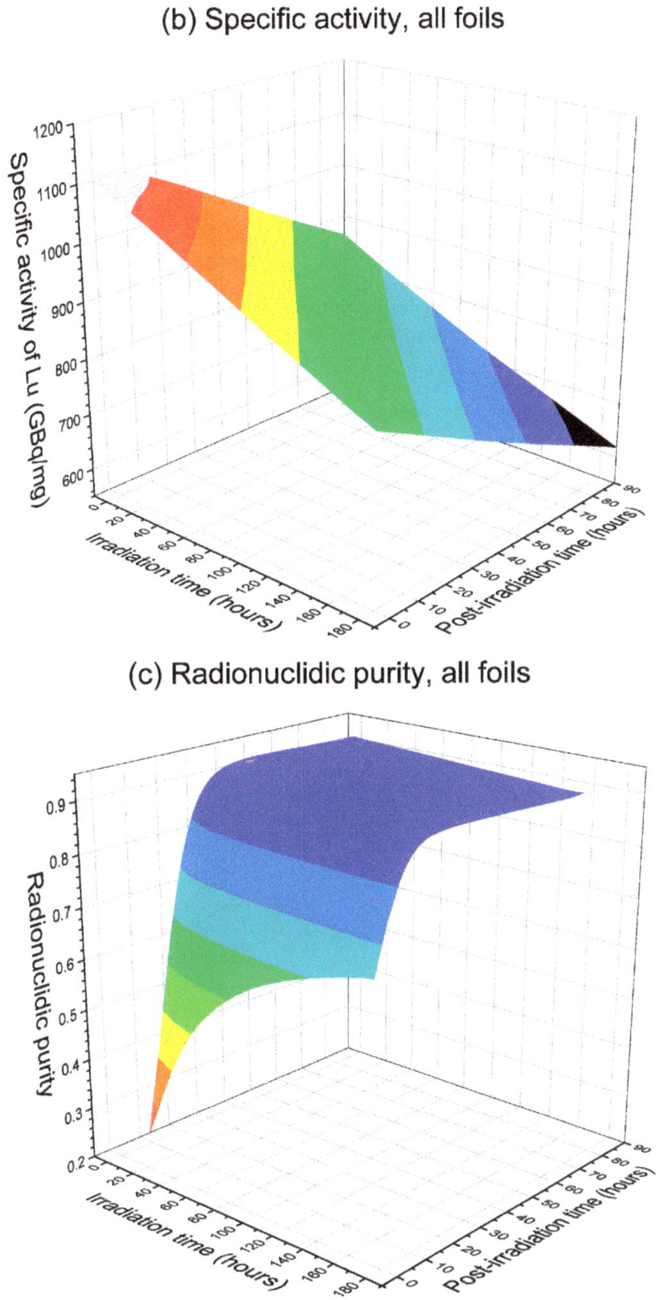

**Figure 13.** (**a**) Activity, (**b**) specific activity, and (**c**) radionuclidic purity of $^{177}$Lu collected from all foils (corresponding to a depth from 0 mm to 9.22 mm) of $^{176}$Yb irradiated using 80 MeV deuterons. The x-axis represents the irradiation time, and the y-axis represents the post-irradiation processing time. All yields are normalized to one day of operation. The flux is $1 \times 10^{16}$ deuterons/cm$^2$/s, which is equivalent to a beam current of 1.6 mA over a beam spot size of 1 cm$^2$.

In Figure 12c, the radionuclidic purity of the last foil is plotted. Unlike activity and specific activity, the radionuclidic purity approaches a saturated maximum of 99.6%. To achieve a high radionuclidic purity, an extended post-irradiation processing time is needed. By combining Figure 12a–c, an optimized time window can be identified. In summary, maximizing activity requires a short irradiation time and a short post-irradiation processing time, maximizing specific activity necessitates a post-irradiation processing time window of approximately 20 h, and maximizing radionuclidic purity calls for a prolonged post-irradiation processing time. A balanced condition can be achieved, such as selecting a 12 h irradiation time and a 28 h post-irradiation processing time, resulting in an activity of 304 GBq/day, a specific activity of 3084 GBq/mg, and a radionuclidic purity of 95.4%. As mentioned earlier, this is not the condition to maximize radionuclidic purity. The condition for maximizing radionuclidic purity will be discussed shortly.

Figure 13a–c plots the activity, specific activity, and radionuclidic purity for all foils (corresponding to a depth region of 0 to 9.22 mm). As mentioned earlier, including these foils containing a high concentration of $^{174g}$Lu and $^{174m}$Lu degrade the overall isotope quality. However, these foils can still be used in certain medical applications where low specific activity is acceptable. In Figure 13a, it is evident that including all foils significantly increases the total activity. The highest activity of ~1260 GBq/day can be achieved by minimizing both the irradiation time (12 h) and the post-irradiation processing time (4 h). Regarding specific activity, as shown in Figure 13b, the highest value of about 1100 GBq/mg can be obtained by minimizing the irradiation and post-irradiation processing time. Even in the worst-case scenario with a prolonged post-irradiation processing time exceeding 60 h, the specific activity remains at approximately 740 GBq/mg or above, which is comparable to typical products obtained via reactor irradiation. For radionuclidic purity, a prolonged post-irradiation processing time is favored, and the maximum value saturates at around 93%.

Different from Figure 12b, which shows a post-irradiation processing time window for achieving maximum specific activity, there is no such window observed from Figure 13b. This discrepancy arises from the high sensitivity of specific activity to the changes of $^{176m}$Lu quantity in the last foil, given that $^{176m}$Lu and $^{177}$Lu are the predominant isotopes present, with $^{176m}$Lu having a rapid decay. In contrast, when all foils are considered, the mass changes contributed by $^{176m}$Lu are relatively insignificant due to the substantial contributions from $^{174g}$Lu and $^{174m}$Lu, which have almost negligible changes due to their relatively long half-lives. Consequently, the specific activity becomes less sensitive to changes in $^{176m}$Lu and more responsive to the activity changes of $^{177}$Lu itself.

Back to the approach of harvesting $^{177}$Lu from the last foil only, the obtained $^{177}$Lu under optimized conditions, as shown in Figure 12b, exhibits a remarkably high specific activity of about 3084 GBq/mg. This makes the process very attractive for hospital applications. This specific activity clearly surpasses the current achieved results using reactors. Table 4 provides a summary of the typical specific activities of $^{177}$Lu from different sources [13]. It is worth noting that 3084 GBq/mg is about 75% of the highest specific activity achievable in carrier-free $^{177}$Lu. The theoretically predicted maximum is 4104 GBq/mg [13].

**Table 4.** Summary of the specific activity ranges of $^{177}$Lu commercially available [13].

| Suppliers | Specific Activity (GBq/mg) |
|---|---|
| Perkin Elmer, USA | ~740 |
| ORNL, USA | 1850–2960 |
| MURR, USA | 925 |
| MDS Nordion, Canada | 1665 |
| ITG, Garching, Germany | 2960 |
| IDB Holland BV | ~740 |

Having all foils made of $^{176}$Yb is not a requirement in the stacked foil approach. Sacrificial materials of less expensive materials can be used in the other foils while reserving the last foil specifically for $^{176}$Yb. Alternatively, for isotope production other than $^{177}$Lu, different target materials can be used to replace the other foils. This flexibility is particularly valuable for isotopes that require a higher threshold energy for production.

The discussion above has assumed a $^{176}$Yb target with a purity level of 100%, which represents an ideal scenario. In reality, targets of a purity level of >99.6% are commercially available (i.e., from ISOFLEX USA, San Francisco, CA, USA). In commercially enriched $^{176}$Yb targets, impurity levels of other Yb isotopes decrease significantly as the mass numbers of Yb decrease [23]. Consequently, $^{174}$Yb is more abundant than other isotope impurities in $^{176}$Yb-enriched targets. The three unwanted products originating from $^{174}$Yb are $^{174g+m}$Lu and $^{173}$Lu. These isotopes have relatively long lifetimes compared to others, resulting in their minimal contribution to the radionuclide impurities of $^{177}$Lu.

To model the effect of purity, calculations were performed for the production and decay of $^{177}$Lu, $^{174m}$Lu, $^{174g}$Lu, $^{176m}$Lu, and $^{173}$Lu. The production calculations utilized cross-section data from Figures 4 and 5, while the decay calculations were based on the half-life times provided in Table 1. Table 5 compares the activity, specific activity, and radionuclidic purity of $^{177}$Lu in $^{176}$Yb targets at purity levels of 100%, 99.6%, 99%, and 98%, respectively. All values correspond to daily production, assuming two rounds of irradiation per day with a 12 h irradiation time for each batch and a 28 h post-irradiation processing time for each batch. The conditions include an 80 MeV beam energy, 1.6 mA beam current, and a 1cm$^2$ beam spot size. The comparison is made for collecting Lu from all foils (0 to 9.22 mm) and from the last foil (8.22 mm to 9.22 mm).

**Table 5.** The daily production of $^{177}$Lu in $^{176}$Yb targets with purities of 100%, 99.6%, 99%, and 98%. Comparison was made for the depth region of 8.22 mm to 9.22 mm (the last foil) and the range of 0 mm to 9.22 mm (all foils) under 80 MeV deuteron irradiation. The beam current is 1.6 mA. The beam spot size is 1 cm$^2$. The irradiation time for each batch is 12 h, and the post-irradiation processing time is 28 h as the optimized conditions. The condition is not optimized to maximize radionuclidic purity.

| Starting $^{176}$Yb Purity | Depth Region | Activity (GBq) per Day | Mass (mg) per Day | Specific Activity (GBq/mg) | Radionuclidic Purity |
|---|---|---|---|---|---|
| 100% | 8.22 mm to 9.22 mm | 304 | 0.074 | 3084 | 95.4% |
| | 0 mm to 9.22 mm | 1119 | 0.273 | 1022 | 91.5% |
| 99.6% | 8.22 mm to 9.22 mm | 302 | 0.074 | 3052 | 95.4% |
| | 0 mm to 9.22 mm | 1115 | 0.272 | 1017 | 91.4% |
| 99% | 8.22 mm to 9.22 mm | 301 | 0.073 | 3004 | 95.3% |
| | 0 mm to 9.22 mm | 1108 | 0.270 | 1008 | 91.4% |
| 98% | 8.22 mm to 9.22 mm | 297 | 0.073 | 2912 | 95.3% |
| | 0 mm to 9.22 mm | 1097 | 0.267 | 994 | 91.3% |

Table 5 shows that there is no significant difference when the purity is reduced from 100% to 99.6%. In the case of the last foil, the activity changes from 304 to 302 GBq and the specific activity changes from 3084 to 3052 GBq/mg. Even in the worst-case scenario of 98% purity, the activity is 335 GBq, and the specific activity is 2912 GBq/mg. These results suggest that a purity level of around 98% and above does not significantly degrade the quality of the final product.

Using 99.6% purity as the example, Table 6 lists the activity, mass, and specific activity of each Lu isotope, giving further details about the impurity effect. Judged by the radionuclidic purity, the largest impurity effect comes from $^{176m}$Lu, at a value of 4.3%. $^{176m}$Lu has a specific activity of 137 GBq/mg. Although its mass is not the largest, its short half-life of 3.66 h makes its contribution large. On the other hand, $^{173}$Lu, as a unique product from $^{174}$Yb, has the smallest contribution in influencing $^{177}$Lu radionuclidic purity.

As one highlight of the present study, irradiation, and processing conditions are identified to provide the quality range that is of particular interest to the isotope production community. The most demanded quality is specific activity > 2960 GBq/mg and radionuclidic purity > 99.5%. Figure 14a–d are for $^{176}$Yb of 100% purity and for the last foil only (depth from 8.22 mm to 9.22 mm). Figure 14a maps the region where the activity exceeds 185 GBq/day, a condition that can be easily achieved based on the modeled conditions. The majority of regions surpass 259 GBq/day, except for a specific corner associated with significantly longer irradiation time and post-irradiation processing times. In Figure 14b, the plot showcases the region that surpasses 99.5% radionuclidic purity, which can be attained with post-irradiation processing times longer than 40 h. This region appears to be less sensitive to variations in irradiation times. Figure 14c presents the region of the specific activity exceeding 2960 GBq/mg. It shows a trade-off between irradiation time and post-irradiation processing times, whereby a longer irradiation time necessitates a shorter post-irradiation processing time. Figure 14d provides a summary of the region that satisfies the combined criteria of activity > 259 GBq/day, specific activity > 2960 GBq/mg, and radionuclidic purity > 99.5%. It represents the optimized conditions obtained by considering all restrictions. This region corresponds to irradiation times shorter than 60 h and post-irradiation processing times between 44 and 64 h.

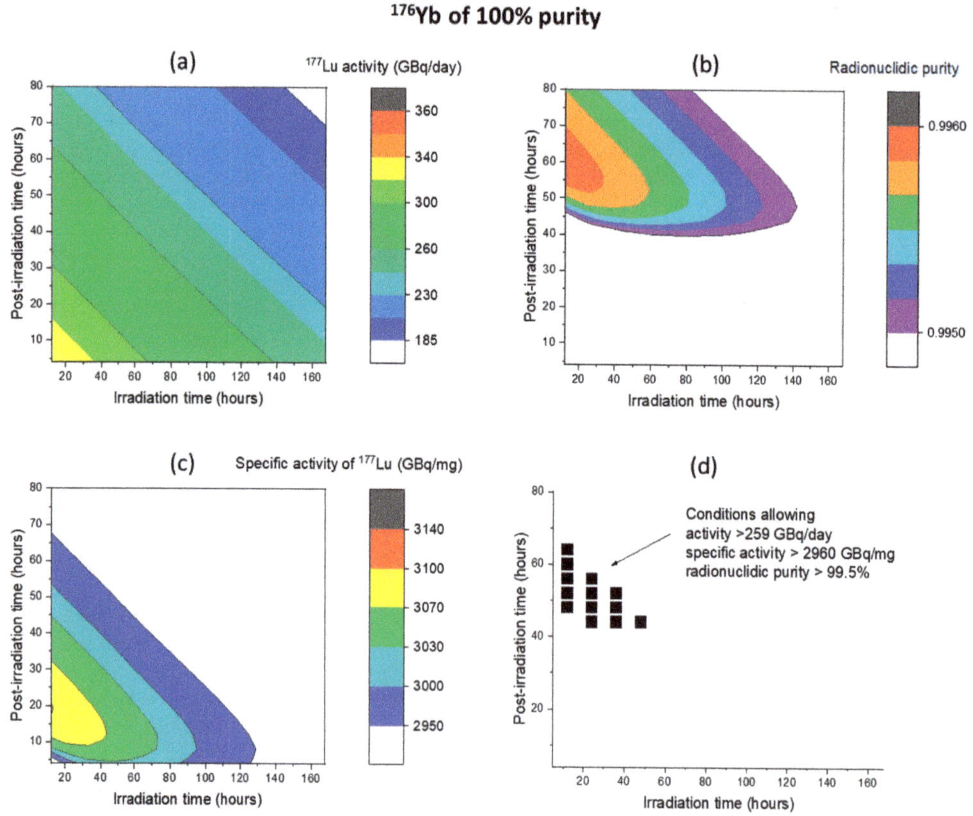

**Figure 14.** (**a**) Activity, (**b**) specific activity, (**c**) radionuclidic purity of $^{177}$Lu obtained from 80 MeV deuteron irradiation of $^{176}$Yb with 100% purity, and (**d**) processing time window capable of achieving activity > 259 GBq/day, specific activity > 2960 GBq/mg, and radionuclidic purity > 99.5%. Only the last foil is processed. The beam current is 1.6 mA, and the beam spot size is 1 cm$^2$.

For a purity of 99.6%, Figure 15a–d plots the corresponding activity, specific activity, radionuclidic purity, and the optimized conditions for achieving activity > 259 GBq/day, specific activity > 2960 GBq/mg, and radionuclidic purity > 99.5%. These maps exhibit similarities to the case of 100% purity, with the expected difference that the regions are slightly smaller. The shrinking specific activity region and specific activity region impose further constraints on the conditions. As shown in Figure 15d, although conditions are narrowed down, a region still exists for attaining the required quality. The necessary irradiation time ranges from 12 to 24 h, while the post-irradiation processing time is around 48 h.

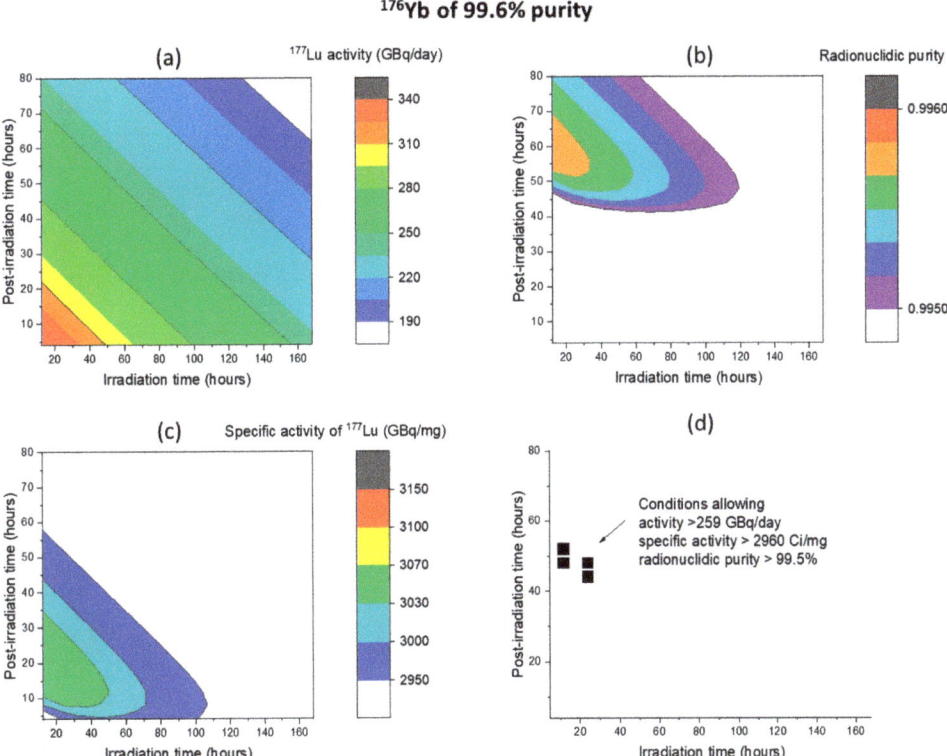

**Figure 15.** (**a**) Activity, (**b**) specific activity, (**c**) radionuclidic purity of $^{177}$Lu obtained from 80 MeV deuteron irradiation of $^{176}$Yb with 99.6% purity, and (**d**) processing time window capable of achieving activity > 259 GBq/day, specific activity > 2960 GBq/mg, and radionuclidic purity > 99.5%. Only the last foil is processed. The beam current is 1.6 mA, and the beam spot size is 1 cm$^2$.

As pointed out in the introduction, the present study used 80 MeV as one example of high-energy LINAC facilities. However, the proposed multiple foil targets can be easily adjusted for other energies. The deuteron ions enter into the last foil at an energy of about 21 MeV. For other beam energies, as long as the foils prior to the last one can reduce beam energies down to 21 MeV, the product of the same quality can be obtained.

As shown in Figure 2, for deuterons reaching 21 MeV, deuterons penetrate about 4.7 mm for an initial beam energy of 60 MeV and about 1.9 mm for an initial beam energy of 40 MeV. As long as the foils ahead of the last one have a combined thickness matching these numbers, the same optimized conditions can be reached for the last foil.

Currently, isotope production is primarily dominated by reactor irradiation and cyclotrons. Most cyclotrons operate at relatively lower beam energies, typically around

40 MeV or lower. In contrast, LINACs are capable of delivering beam energies exceeding 40 MeV. LINACs can serve as versatile devices for isotope production, even for isotopes that favor lower beam energies, utilizing techniques like the one proposed in this study. It is important to note that typical beam currents for cyclotrons are around 100s µA, whereas LINACs can achieve beam currents of a few mA. For a high-energy LINAC, it is not ideal for Lu isotope production since much power is wasted if only the last foil is used. However, the design allows replacing these wasted foils with targets for producing other isotopes that require higher threshold (or optimized) energy.

**Table 6.** The daily production of various Lu isotopes in the $^{176}$Yb target with a purity of 99.6%. Collection is limited to the last foil of the depth region of 8.22 mm to 9.22 mm. The beam current is 1.6 mA, and the beam spot size is 1 cm$^2$. The irradiation time for each batch is 12 h, and the post-irradiation processing time is 28 h as the optimized conditions. The parameters are optimized to maximize specific capacity. If radionuclidic purity needs to be maximized, a different condition is needed.

| Nuclide | Activity (GBq) per Day | Mass (mg) per Day | Specific Activity (GBq/mg) | Radionuclidic Purity |
|---|---|---|---|---|
| $^{177}$Lu | 302 | 0.0737 | 3052 | 95.36% |
| $^{174m}$Lu | 0.555 | 0.003 | 5.74 | 0.17% |
| $^{174g}$Lu | 0.518 | 0.022 | 5.07 | 0.16% |
| $^{176m}$Lu | 13.6 | 0.00007 | 137 | 4.30% |
| $^{173}$Lu | 0.0296 | 0.0005 | 0.296 | 0.01% |

## 3. Conclusions

By employing a simulation approach that combines particle transport and isotope production/decay simulations, the process of producing $^{177}$Lu via deuteron irradiation of $^{176}$Yb has been optimized, with a particular focus on maximizing the specific activity of $^{177}$Lu. Stacked foils have been proposed as the target for a linear accelerator operating at relatively high design energies. The selection of foils can be tailored to minimize the presence of $^{174m}$Lu and $^{174g}$Lu due to their distinct energy-dependent cross-sections. To minimize the production of $^{176m}$Lu, optimization of the post-irradiation processing time times has been investigated. Upon optimization, deuteron irradiation at 80 MeV and at a flux of $1 \times 10^{16}$ deuterons/cm$^2$/s (equivalent to a beam current of 1.6 mA over a beam spot size of 1 cm$^2$) can result in daily production of 304 GBq of $^{177}$Lu, with a high specific activity of 3084 GBq/mg and a high radionuclidic purity of 95.4%. If specific activity and radionuclidic purity need to be balanced, there is an optimized condition that can obtain activity greater than 259 GBq/day, specific activity greater than 2960 GBq/mg, and radionuclidic purity greater than 99.5%. The study suggested the feasibility of utilizing a high energy, high current linear accelerator for the production of $^{177}$Lu at a specific activity level better than a typical reactor produced $^{177}$Lu.

**Funding:** This research received no external funding.

**Informed Consent Statement:** Not applicable.

**Data Availability Statement:** Data are available from the author on reasonable request.

**Conflicts of Interest:** The authors declare no conflict of interest.

**Sample Availability:** Not applicable.

## References

1. Funkhouser, J. Reinventing pharma: The theranostic revolution. *Curr. Drug Discov.* **2002**, *2*, 17.
2. Srivastava, S.C.; Mausner, L.F. *Therapeutic Radionuclides: Production, Physical Characteristics, and Applications in Therapeutic Nuclear Medicine, Medical Radiology*; Baum, R.P., Ed.; Springer: Berlin/Heidelberg, Germany, 2013; p. 11.
3. Cutler, C.S.; Hennkens, H.M.; Sisay, S.; Huclier-Markai, S.; Jurisson, S.S. Radiometals for combined imaging and therapy. *Chem. Rev.* **2013**, *113*, 858–883. [CrossRef] [PubMed]
4. Zaknun, J.J.; Bodei, L.; Mueller-Brand, J.; Pavel, M.E.; Baum, R.P.; Horsch, D.; O'Dorisio, T.M.; Howe, J.R.; Cremonesi, M.; Kwekkeboom, D.J. The joint IAEA, EANM, and SNMMI practical guidance on peptide receptor radionuclide therapy (PRRNT) in neuroendocrine tumours. *Eur. J. Nucl. Med. Mol. Imaging* **2013**, *40*, 800–816. [CrossRef] [PubMed]
5. Bé, M.-M.; Chisté, V.; Dulieu, C.; Browne, E.; Chechev, V.; Kuzmenko, N.; Helmer, R.; Nichols, A.; Schönfeld, E.; Dersch, R. Table of Radionuclides, Monographie BIPM-5 (Vol. 2-A = 151 to 242), Bureau International des Poids et Mesures, Sèvres. 2004. Available online: https://www.bipm.org/documents/20126/53814638/Monographie+BIPM-5+-+Volume+2+%282004%29.pdf/047c963d-1f83-ab5b-7983-744d9f48848a (accessed on 7 August 2023).
6. Seregni, E.; Maccauro, M.; Chiesa, C.; Mariani, L.; Pascali, C.; Mazzaferro, V.; De Braud, F.; Buzzoni, R.; Milione, M.; Lorenzoni, A.; et al. Treatment with tandem [$^{90}$Y]DOTA-TATE and [$^{177}$Lu]DOTA-TATE of neuroendocrine tumours refractory to conventional therapy. *Eur. J. Nucl. Med. Mol. Imaging* **2014**, *41*, 223–230. [CrossRef]
7. Cullinane, C.; Jeffery, C.M.; Roselt, P.D.; van Dam, E.M.; Jackson, S.; Kuan, K.; Jackson, P.; Binns, D.; van Zuylekom, J.; Harris, M.J.; et al. Peptide Receptor Radionuclide Therapy with $^{67}$Cu-CuSarTATE Is Highly Efficacious Against a Somatostatin-Positive Neuroendocrine Tumor Model. *J. Nucl. Med.* **2020**, *61*, 1800–1805. [CrossRef] [PubMed]
8. Abbott, A.; Sakellis, C.G.; Andersen, E.; Kuzuhara, Y.; Gilbert, L.; Boyle, K.; Kulke, M.H.; Chan, J.A.; Jacene, A.; Van den Abbeele, A.D. Guidance on $^{177}$Lu-DOTATATE Peptide Receptor Radionuclide Therapy from the Experience of a Single Nuclear Medicine Division. *J. Nucl. Med. Technol.* **2018**, *46*, 237–244. [CrossRef] [PubMed]
9. Balasubramanian, P.S. Separation of carrier-free lutetium-177 from neutron irradiated natural ytterbium target. *J. Radioanal. Nucl. Chem.* **1994**, *185*, 305–310. [CrossRef]
10. Hashimoto, K.; Matsuoka, H.; Uchida, S. Production of no-carrier-added $^{177}$Lu via the $^{176}$Yb(n,γ)$^{177}$Yb→$^{177}$Lu process. *Radioanal. Nucl. Chem.* **2003**, *255*, 575–579. [CrossRef]
11. Zhernosekov, K.P.; Perego, R.C.; Dvorakova, Z.; Henkelmann, R.; Türler, A. Target burn-up corrected specific activity of $^{177}$Lu produced via $^{176}$Lu(n, gamma) $^{177}$Lu nuclear reactions. *Appl. Radiat. Isot.* **2008**, *66*, 1218–1220. [CrossRef]
12. Chakraborty, S.; Vimalnath, K.V.; Lohar, S.P.; Shetty, P.; Dash, A. On the practical aspects of large-scale production of $^{177}$Lu for peptide receptor radionuclide therapy using direct neutron activation of $^{176}$Lu in a medium flux research reactor: The Indian experience. *J. Radioanal. Nucl. Chem.* **2014**, *302*, 233–243.
13. Dash, A.; Pillai, M.R.A.; Knapp, F.F., Jr. Production of $^{177}$Lu for Targeted Radionuclide Therapy: Available Options. *Nucl. Med. Mol. Imaging* **2015**, *49*, 85–107. [CrossRef] [PubMed]
14. Siiskonen, T.; Huikari, J.; Haavisto, T.; Bergman, J.; Heselius, S.-J.; Lill, J.-O.; Lönnroth, T.; Peräjärvi, K.; Vartti, V.-P. Excitation functions for proton-induced reactions on natural hafnium: Production of Lu-177 for medical use. *Nucl. Instrum. Methods Phys. Res. Sect. B* **2009**, *267*, 3500–3504. [CrossRef]
15. Hermanne, A.; Takács, S.; Goldberg, M.B.; Lavie, E.; Shubin, Y.N.; Kovalev, S. Deuteron-induced cross sections and rationale for production pathways of carrier-free, medically relevant radionuclides. *Nucl. Instrum. Methods Phys. Res. Sect. B* **2006**, *247*, 223. [CrossRef]
16. Manenti, S.; Groppi, F.; Gandini, A.; Gini, L.; Abbas, K.; Holzwarth, U.; Simonelli, F.; Bonardi, M. Excitation function for deuteron induced nuclear reactions on natural ytterbium for production of high specific activity $^{177g}$Lu in no-carrier-added form for metabolic radiotherapy. *Appl. Radiat. Isot.* **2011**, *69*, 37. [CrossRef] [PubMed]
17. Tárkányi, F.; Ditrói, F.; Takács, S.; Hermanne, A.; Yamazaki, H.; Baba, M.; Mohammadi, A.; Ignatyuk, A.V. Activation cross-sections of longer lived products of deuteron induced nuclear reactions on ytterbium up to 40 MeV. *Nucl. Instrum. Methods Phys. Res. Sect. B* **2013**, *304*, 36. [CrossRef]
18. Khandaker, M.U.; Haba, H.; Kassim, H.A. Production of $^{177}$Lu, a potential radionuclide for diagnostic and therapeutic applications. *AIP Conf. Proc.* **2015**, *1657*, 120003.
19. Király, B.; Tárkányi, F.; Takács, S.; Hermanne, A.; Kovalev, S.F.; Ignatyuk, A.V. Excitation functions of alpha-particle induced nuclear reactions on natural ytterbium. *Nucl. Instrum. Methods Phys. Res. Sect. B* **2008**, *266*, 3919. [CrossRef]
20. Kazakov, A.G.; Belyshev, S.S.; Ekatova, T.Y.; Khankin, V.V.; Kuznetsov, A.A.; Aliev, R.A. Production of 177Lu by hafnium irradiation using 55-MeV bremsstrahlung photons. *J. Radioanal. Nucl. Chem.* **2018**, *317*, 1469. [CrossRef]
21. Tárkányi, F.; Takács, S.; Ditrói, F.; Hermanne, A.; Adam, R.; Ignatyuk, A.V. Investigation of the deuteron induced nuclear reaction cross sections on lutetium up to 50 MeV: Review of production routes for 177Lu, 175Hf and 172Hf via charged particle activation. *J. Radioanal. Nucl. Chem.* **2020**, *324*, 1405–1421. [CrossRef]
22. Kambali, I. Production of Lu-177 radionuclide using deuteron beams: Comparison between (d,n) and (d,p) nuclear reactions. *J. Phys. Conf. Ser.* **2018**, *1120*, 012011. [CrossRef]
23. Nagai, Y.; Kawabata, M.; Hashimoto, S.; Tsukada, K.; Hashimoto, K.; Motoishi, S.; Saeki, H.; Motomura, A.; Minato, F.; Itoh, M. Estimated isotopic compositions of Yb in enriched $^{176}$Yb for producing $^{177}$Lu with high radionuclide purity by $^{176}$Yb(d,x)$^{177}$Lu. *J. Phys. Soc. Jpn.* **2022**, *91*, 044201. [CrossRef]

24. Stovall, J.E.; Hansborough, L.D.; O'Brien, H.A., Jr. Radioisotope-production LINAC. In Proceedings of the 1981 Linear Accelerator Conference, Santa Fe, NM, USA, 19–23 October 1981; p. 344.
25. DePalma, A. International isotopes expands production capabilities. *Pharmaceutical Online*, 3 April 1998.
26. Ziegler, J.F.; Ziegler, M.D.; Biersack, J.P. SRIM—The stopping and range of ions in matter. *Nucl. Instrum. Methods Phys. Res. Sect. B Beam Interact. Mater. At.* **2010**, *268*, 1818–1823. [CrossRef]
27. Khandaker, M.U.; Haba, H.; Otuka, N.; Usman, A.R. Investigation of (d,x) nuclear reactions on natural ytterbium up to 24 MeV. *Nucl. Instrum. Methods Phys. Res. B* **2014**, *335*, 8–18. [CrossRef]
28. Kondev, F.G.; Wang, M.; Huang, W.J.; Naimi, S.; Audi, G. The NUBASE2020 evaluation of nuclear properties (PDF). *Chin. Phys. C* **2021**, *45*, 030001. [CrossRef]
29. Belli, P.; Bernabei, R.; Danevich, F.A.; Incicchitti, A.; Tretyak, V.I. Experimental searches for rare alpha and beta decays. *Eur. Phys. J. A* **2019**, *55*, 140. [CrossRef]
30. Lamere, E.; Couder, M.; Beard, M.; Simon, A.; Simonetti, A.; Skulski, M.; Seymour, G.; Huestis, P.; Manukyan, K.; Meisel, Z.; et al. Proton-induced reactions on molybdenum. *Phys. Rev. C* **2019**, *100*, 034614. [CrossRef]
31. Kim, J.H.; Park, H.; Kim, S.; Lee, J.S.; Chun, K.S. Proton beam energy measurement with the stacked Cu foil technique for medical radioisotope production. *J. Korean Phys. Soc.* **2006**, *48*, 755–758.
32. Ryan, K.; Chapman, A.S.; Voyles, N.G.; Lee, A.; Bernstein, J.; Bevins, E. Measurement of the $^{160}$Gd(p,n)$^{160}$Tb excitation function from 4–18 MeV using stacked-target activation. *Appl. Radiat. Isot.* **2021**, *171*, 109647.
33. Vermeulen, C.; Steyn, G.F.; Szelecsényi, F.; Kovács, Z.; Suzuki, K.; Nagatsu, K.; Fukumura, T.; Hohn, A.; van der Walt, T.N. Cross sections of proton-induced reactions on natGd with special emphasis on the production possibilities of $^{152}$Tb and $^{155}$Tb. *Nucl. Instrum. Methods Phys. Res. Sect. B Beam Interact. Mater. At.* **2012**, *275*, 24–32. [CrossRef]

**Disclaimer/Publisher's Note:** The statements, opinions and data contained in all publications are solely those of the individual author(s) and contributor(s) and not of MDPI and/or the editor(s). MDPI and/or the editor(s) disclaim responsibility for any injury to people or property resulting from any ideas, methods, instructions or products referred to in the content.

Article

# A Reliable Production System of Large Quantities of [¹³N]Ammonia for Multiple Human Injections

Luis Michel Alonso Martinez [1], Nabil Naim [1], Alejandro Hernandez Saiz [1], José-Mathieu Simard [1,2], Mehdi Boudjemeline [1,2], Daniel Juneau [1,3] and Jean N. DaSilva [1,3,*]

[1] Radiochemistry and Cyclotron Platform, Centre de Recherche du Centre Hospitalier de l'Université de Montréal (CRCHUM), 900 Rue Saint Denis, Montréal, QC H2X 0A9, Canada
[2] Radiopharmaceutical Science Laboratory, CHU de Québec, 2250 Boul. Henri-Bourassa, Québec, QC G1J 5B3, Canada
[3] Department of Radiology, Radio-Oncology and Nuclear Medicine, UdeM, Pavillon Roger-Gaudry S-716, 2900 Boul. Édouard Montpetit, Montréal, QC H3C 3J7, Canada
* Correspondence: jean.dasilva@umontreal.ca; Tel.: +1-514-890-8000 (ext. 30653)

**Abstract:** [¹³N]Ammonia is one of the most commonly used Positron Emission Tomography (PET) radiotracers in humans to assess myocardial perfusion and measure myocardial blood flow. Here, we report a reliable semi-automated process to manufacture large quantities of [¹³N]ammonia in high purity by proton-irradiation of a 10 mM aqueous ethanol solution using an in-target process under aseptic conditions. Our simplified production system is based on two syringe driver units and an in-line anion-exchange purification for up to three consecutive productions of ~30 GBq (~800 mCi) (radiochemical yield = $69 \pm 3$% n.d.c.) per day. The total manufacturing time, including purification, sterile filtration, reformulation, and quality control (QC) analyses performed before batch release, is approximately 11 min from the End of Bombardment (EOB). The drug product complies with FDA/USP specifications and is supplied in a multidose vial allowing for two doses per patient, two patients per batch (4 doses/batch) on two separate PET scanners simultaneously. After four years of use, this production system has proved to be easy to operate and maintain at low costs. Over the last four years, more than 1000 patients have been imaged using this simplified procedure, demonstrating its reliability for the routine production of large quantities of current Good Manufacturing Practices (cGMP)-compliant [¹³N]ammonia for human use.

**Keywords:** [¹³N]Ammonia; cGMP production; semi-automated production system; PET imaging

## 1. Introduction

[¹³N]Ammonia is one of the most commonly used Positron Emission Tomography (PET) radiotracers to perform myocardial perfusion imaging and to quantitatively measure myocardial blood flow and myocardial flow reserve in patients with suspected coronary artery disease and multiple other cardiac conditions [1,2]. The [¹³N]ammonium cation is extracted from the blood and metabolically trapped in the tissue of interest, including myocardium, mainly as [¹³N]glutamine according to blood flow [3]. While [²⁰¹Tl]thallium chloride, [⁹⁹ᵐTc]technetium sestamibi and [⁹⁹ᵐTc]technetium tetrofosmin provide qualitative images of relative blood flow distribution, [¹³N]ammonium cation allows both qualitative measurements of relative flow and myocardial blood flow quantification [4,5]. Unlike Single Photon Emission Computed Tomography (SPECT) radioisotopes (i.e., thallium-201 and technetium-99m), the positron-emitting radionuclide nitrogen-13 has two 511 keV annihilation gamma rays detected in coincidence by the PET scanner to accurately assess the blood-flow dependent distribution of [¹³N]ammonia in the heart and quantitatively measure flow [6].

[⁸²Rb]RbCl and [¹⁵O]water are also used for the evaluation of myocardial blood flow as positron-emitting radiopharmaceuticals. However, to date, only [¹³N]ammonia and generator-produced [$^{82}$Rb]RbCl are FDA approved as PET myocardial perfusion agents. While a rubidium-82 generator is certainly an interesting option for centers with a high volume of patients, [$^{13}$N]ammonia is an attractive option for centers with an on-site cyclotron, as it exhibits better physical characteristics for PET imaging [7], such as lower positron energy of N-13 1.199 MeV ($\beta^+$ $E_{max}$) versus Rb-82 3.378 MeV ($\beta^+$ $E_{max}$) reducing the positron range significantly before annihilation with an electron and thus improving image resolution [8]. However, the nitrogen-13 half-life of 9.96 min makes the manufacturing and time delivery of [$^{13}$N]ammonia challenging. Hence, different strategies for [$^{13}$N]NH$_3$ production have been adopted by several labs varying from simple in-target production (without further purification) to more complex procedures with in-line purification, dedicated production systems or automated synthesis modules. The in-target production of [$^{13}$N]ammonia with aqueous ethanol (5 mM) as a radical scavenger under pressurized conditions is considered the classic synthetic route [9]. Nonetheless, the absence of purification steps renders the final formulation less than optimal due to the low isotonicity and potential presence of radionuclidic contaminants from the target material. To circumvent these problems, a purification process through an anion-exchange resin has been utilized to remove all anionic impurities with a subsequent reformulation in saline for dispensing $^{13}$[N]NH$_3$ in the form of $^{13}$[N]NH$_4^+$ ions [10].

Complex dedicated production systems have been evaluated to increase product yields and faster manufacturing processes. Frank et al. [11] proposed a dedicated production system consisting of nine parallel Waters CM-Light Sep-Pak cartridges connected to an ISAR chip. The compact microfluidic system allows nine consecutive productions to produce [$^{13}$N]ammonia at >96% radiochemical yield (RCY) per batch, including approximately five minutes for the purification and reformulation processes. Despite all these advantages, this system is yet in the prototype stage and still needs to be thoroughly evaluated. In recent years, the widespread use of commercial synthesis modules was also applied in the automated production of $^{13}$[N]ammonia. Several examples can be found in the literature, with a synthesis time of around five minutes and RCYs of 90% or higher. Notably, Yokell et al. [12] set out a simple method that could also be adapted and validated on commercially available synthesis units. This approach utilizes quaternary methyl ammonium (QMA) and carboxymethyl (CM) cartridges in series to accomplish the purification, formulation, and sterile filtration of [$^{13}$N]ammonia (21 ± 2 GBq, 572 ± 60 mCi) in about five minutes. Considering that [$^{13}$N]NH$_3$ is among the oldest clinically used PET radiopharmaceuticals whose cost have historically been low (due to the low cost of the target material and the process), the price and availability in radiochemistry laboratories of such commercial modules and their often-cumbersome usage to customized productions have limited its widespread utilization in nuclear medicine.

In this paper, we present our institutional experience to reliably produce [$^{13}$N]ammonia in large quantities for clinical use. The proposed [$^{13}$N]ammonia production system has been in operation since 2018 allowing an average of two doses per patient, two patients per batch (four doses/batch) on two separate PET scanners simultaneously. So far, more than 1000 patients have been scanned, demonstrating system reliability for the routine production of large quantities of [$^{13}$N]ammonia for human use.

## 2. Results

The results of six productions (two consecutive batches per day) using the 2XL target at 10 mM aqueous ethanol solution are presented in Table 1. All batches were found to meet the FDA/USP product specifications with slight variations of pH but within the acceptance range (4.5–7.5 at time of injection) after extensive dilution of [$^{13}$N]NH$_3$ with 0.9% sodium chloride in the nuclear medicine PET unit before administration to patients. Analytical High-Performance Liquid Chromatography (HPLC) chromatograms for [$^{nat}$N]NH$_4$Cl standard and [$^{13}$N]ammonia samples are presented in Figure 1a–c. The radiochemical identity was determined by peak comparison of the retention times ($t_R$) between the conductivity peaks of [$^{nat}$N]ammonium chloride and [$^{13}$N]ammonia (both at

$t_R$ = 2.83 min, Figure 1a,b). The radiation signal from [$^{13}$N]ammonia came out after the conductivity peak ($t_R$ = 3.17 min, Figure 1c) since the radiation detector is placed after the conductivity detector. In this HPLC system, the [$^{13}$N]ammonium cation can be adequately separated from the known negatively charged impurities, [$^{18}$F]fluoride and [$^{13}$N]nitrous oxide (NO$_x^-$) eluted with the void volume of IC-Pak cation exchange column. None of these impurities were observed in any [$^{13}$N]ammonia batches. Particular interest was given to the analysis of long-lived radionuclidic impurities by HP(Ge) spectrometry, where no detectable radionuclide contaminants from the cyclotron target body or window were observed in any validation runs. The system described in this article was able to routinely produce 29 ± 2 GBq (793 ± 49 mCi, n = 345) of [$^{13}$N]ammonia per batch with the cyclotron 2XL target at 55 µA for up to 40 min of bombardment. [$^{13}$N]Ammonia was produced in approximately five minutes at yields higher than 90% (d.c.). The whole process, including QC testing until batch release, was carried out in approximately 11 min from EOB. The product was found to be stable at room temperature up to 60 min after EOM, and thus, expiry time was set at 60 min post-EOM. To meet clinical needs, up to three batches can be produced per day.

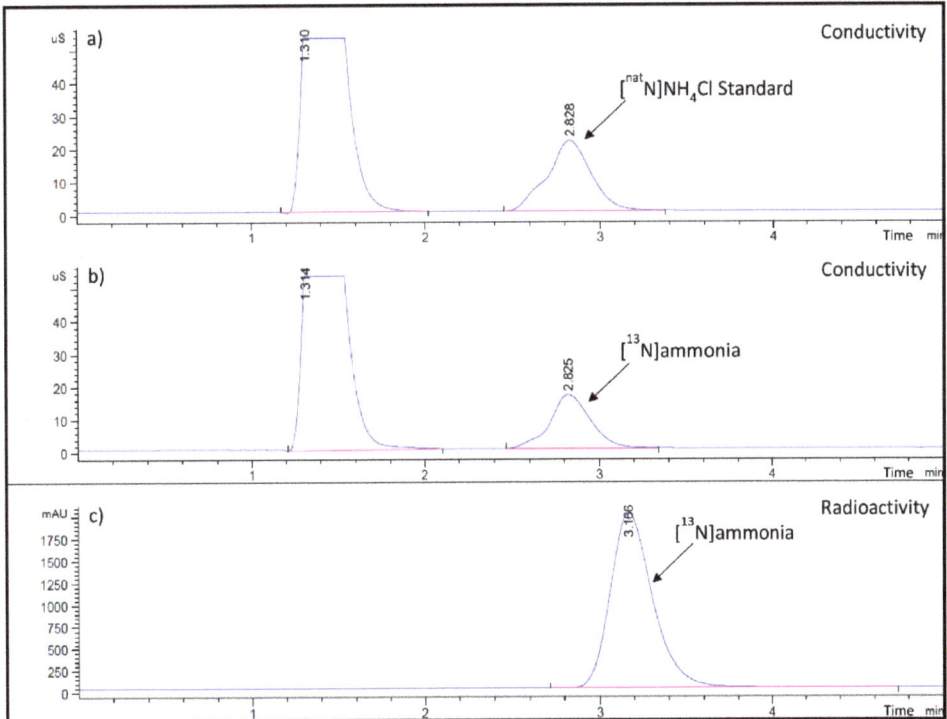

**Figure 1.** Representative analytical HPLC chromatograms for [$^{13}$N]ammonia production: (**a**) [$^{nat}$N]ammonium chloride standard by conductivity detection, $t_R$ = 2.83 min; (**b**) [$^{13}$N]ammonia sample analysis from product vial by conductivity detection, $t_R$ = 2.83 min; (**c**) [$^{13}$N]ammonia sample analysis from product vial by radioactivity detection, $t_R$ = 3.17 min. Differences in retention times are intrinsic to the HPLC system due to the fact that the radiation detector is placed after the conductivity detector.

**Table 1.** Quality control test results of [$^{13}$N]ammonia for three production days (two consecutive batches per day).

| Specification | Acceptance Criteria | Day 1 * | Day 2 * | Day 3 * |
|---|---|---|---|---|
| Activity EOB GBq/mCi | - | 43.03/1163 44.55/1204 | 40.33/1090 42.18/1140 | 42.74/1155 43.07/1164 |
| Activity EOM (GBq/mCi) | - | 28.19/762 28.82/779 | 29.23/790 29.97/810 | 29.49/797 29.49/797 |
| Elapsed time EOB to EOM | - | 5 min 5 min, 15 s | 4 min, 40 s 4 min, 42 s | 4 min, 55 s 5 min, 15 s |
| Yield (d.c. to EOB) | - | 0.93 0.93 | 1.00 0.99 | 0.97 0.99 |
| Appearance | Clear colourless solution with no particulate matter | Pass | Pass | Pass |
| pH | 3.5–8.5 | 4.0 4.0 | 4.7 4.4 | 4.5 4.0 |
| Filter integrity | ≥manufacturer specification of 0.34 MPa (50 psi) | Pass | Pass | Pass |
| Radionuclide identity | Half-life: $t_{1/2} = 10 \pm 1$ min Photopeak: 466 keV < E < 556 keV | 9.03 min 9.84 min Pass | 9.86 min 9.91 min Pass | 10.39 min 9.95 min Pass |
| Radiochemical identity and Purity | HPLC $t_R$ 3.0 to 3.8 min Purity ≥ 95% | 3.17 min 3.27 min 99.99% 99.98% | 3.33 min 3.25 min 99.99% 99.99% | 3.26 min 3.19 min 99.98% 99.99% |
| Radionuclidic purity | Purity ≥ 99.5% | 99.98% 99.92% | 99.99% 99.99% | 99.74% 99.98% |
| Bacterial endotoxins [#] | ≤175 EU/vial | Pass | Pass | Pass |
| Sterility [#] | No growth after 14 days | Pass | Pass | Pass |

* Second lines correspond to the data from the second batch of the day. [#] Test performed with the pooled fraction of all batches of the day. Abbreviations: EOB: End of Bombardment, EOM: End of Manipulation, d.c.: decay corrected, $t_R$: retention time, EU: endotoxin units. Note: "Pass" means all six productions passed QC.

## 3. Discussion

Manufacturing large quantities of [$^{13}$N]ammonia for patient PET studies can be challenging, taking into account the duration of the methodological process and the quality control (QC) assays. For instance, producing [$^{13}$N]NH$_3$ using Devarda's alloy process usually takes 20 min. However, the determination of residual alumina is mandatory, adding an extra QC test before release [13,14]. Despite the fact that the final formulation usually displays values within specifications, this approach has an inevitable drawback given the length of the whole process before batch release and the short half-life of nitrogen-13 (9.96 min) preventing access to large quantities of [$^{13}$N]ammonia for patients. The introduction of ethanol (5–10 mM in water) as a radical scavenger in the target solution has become the current method of choice for the in-target production of [$^{13}$N]ammonia. In this manuscript, manufacturing of large quantities of [$^{13}$N]ammonia in high purity was carried out by proton-irradiation of a 10 mM aqueous ethanol solution followed by QMA cartridge purification and sterile filtration using a dedicated semi-automatic system designed to speed up and simplify the production process. In addition to the simplicity of the method, the fact that it does not require an automated radiosynthesis unit or expensive cassette-based systems sold in sealed sterile containers reduces the costs of the whole process, which adds to the workflow of operations.

The production system comprises two syringe drivers, both connected to the cyclotron 2XL target and the cyclotron helium-unload mechanism to transfer liquids to a class A dispensing hot cell. The rinsing syringe driver was programmed to work separately from the cyclotron software using a minimalist syntax in MS-DOS language. Two dedicated homemade short programs for rinsing and cleaning the cyclotron transfer lines from the target to the dispensing hot cell are carried out by a simple "one-click" procedure. These processes were intended to facilitate the chemist's work during the preparation of the production and for the post-production sanitization with ethanol to achieve manufacturing with current Good Manufacturing Practices (cGMP) compliance. Before the first batch of the day, the target and transfer lines are rinsed with the freshly prepared target solution, followed by the preconditioning and drying of the QMA Sep-Pak cartridge. The design of a serial arrangement of a QMA cartridge and a 0.22 µm filter allowing purification and sterile filtration in-series has a considerable impact on shortening the duration of the production process and on the reduction in manipulation errors. In fact, each batch takes approximately five minutes for completion. Considering the presence of radionuclide impurities that come from the activation of HAVAR foils and niobium material and the natural presence of 0.2% oxygen-18 in target water (that produces fluorine-18) [15–17], most [$^{13}$N]ammonia production processes use in-series anion-cation cartridges aided by valves to purify [$^{13}$N]NH$_3$ [11,12,18]. Hence, an anion exchange resin is mainly utilized to trap negatively charged impurities ([$^{18}$F]F$^-$, [$^{13}$N]NO$_x^-$ or [$^{48}$V]VO$_4^{3-}$), and a cation exchange resin is used to reformulate [$^{13}$N]NH$_4^+$ in 0.9% saline [14]. According to our experience, the fact that our system does not utilize the trap and release of [$^{13}$N]NH$_4^+$ (available in most automated synthesis modules) can be considered an advantage since operators do not need to turn valves with manipulator arms to purify and isolate [$^{13}$N]ammonia for reformulation. However, our approach requires sufficient dilution with saline to guarantee the pH and isotonicity of the final [$^{13}$N]ammonia formulation.

Considering the aforementioned, one might think that a set of two ion exchange (IEX) cartridges is strictly necessary to have the radionuclidic contaminants at their lowest level during [$^{13}$N]ammonia production. However, several authors have demonstrated that using only a CM cartridge (cation exchange) to selectively trap [$^{13}$N]NH$_4^+$ from anionic impurities in the flow-through is sufficient to achieve more than 99.96% radionuclidic purity [10,11,19,20]. Taking this into account, we based our simplified procedure on a single QMA cartridge which gave [$^{13}$N]ammonia with a radionuclidic purity higher than 99.5%. As reported by Bormans et al. [21], the purification of [$^{13}$N]NH$_3$ with an anion exchange cartridge (5 × 10 mm, Dowex AG 1-X8) also revealed the presence of other unidentified radionuclidic impurities but in low concentration (≤0.001% EOB) after two days decay analysis by gamma spectrometry.

Given the duration of PET scans usually performed at rest/stress, batches of around 26–33 GBq (717–885 mCi) are normally produced back-to-back to fulfill patient demands in our PET unit. The production time and yields were similar to those using other prototype systems [11,18,19] or commercial synthesis modules [12,20]. Therefore, our simplified manufacturing method allows the production of large quantities of [$^{13}$N]ammonia with less automation or manual interventions, thus providing an alternative to laboratories without access to automated technology. To date, the number of lots produced since 2018 demonstrates the robustness and reliability of our simplified cGMP-compliant [$^{13}$N]NH$_3$ production system for human use. Importantly, with this production set-up, we continue to guarantee consistent and uninterrupted on-demand [$^{13}$N]ammonia to meet clinical needs.

## 4. Materials and Methods

### 4.1. Chemical and Reagents

All chemicals and reagents were commercially available and used without further purification.

### 4.2. System Description

- Materials:

The system consists of two syringe pumps (model: V6 48K, Norgren, Denver, CO, USA) installed in a panel located outside the vault, a six-position Valco valve (model: C5-2036, VICI, Houston, TX, USA) and the nitrogen-13 target controlled by the IBA C18/9 cyclotron (software version: Bill), a four ports valve (model: C2-3184UMH, VICI, Houston, TX, USA) that allows selecting between helium or target solution, and a pneumatic Valco valve (model: C5-2034UMH, VICI, Houston, TX, USA) to select the liquid transfer between two dispensing hot cells (optional).

- System operation:

The syringe pump is controlled by volume, valve position, number of cycles, and speed with a dedicated laptop located in the production cleanroom (ISO 7/class 10,000). A programmed script file (Batch file) sends the signal to the syringe pump board to perform the required tasks during the production process. The operator selects the execution of specific files depending on where the [$^{13}$N]ammonia manufacturing is. For rinsing the target and transfer lines (Figure 2(a1)), the operator sets the cyclotron valve position to "Unload" to connect the target solution and the transfer line. Then, the corresponding script file is selected, and the pump starts pushing the target solution to fill all lines and the cyclotron target. To rinse the Overflow line (Figure 2(a2)), the operator sets the cyclotron Target valve on the "Load" position and starts several successive fillings to rinse this line. The target overflow goes directly to a bottle installed beside the cyclotron. At the end of these operations, the lines are dried with high-purity helium (grade 5.0) for a few minutes by activating the three-way valve. The target is then loaded with the target solution using the loading system before starting the beam (Figure 2b). Once the target is loaded, the cyclotron valve software closes all ports, and the system is ready to start the beam. After beaming, the activity is transferred with He from the target to the selected dispensing hot cell (Figure 2c) by changing the position of the cyclotron Target valve to unload.

### 4.3. Production of [$^{13}$N]Ammonia Injection

[$^{13}$N]Ammonia is manufactured in two steps, starting with the nitrogen-13 production in the cyclotron target followed by the reformulation and filtration to render [$^{13}$N]ammonia isotonic, sterile and non-pyrogenic. Prior to each production, the transfer line from the cyclotron to the class A dispensing hot cell (Comecer, Castel Bolognese, Italy, ISO 5/class 100) is rinsed with a freshly prepared 60 mL target solution (10 mM ethanol USP in water for injection (WFI) using the rinsing system presented in Figure 2(a1). Then, the anion exchange cartridge (Sep-Pak Accell plus® QMA plus light, Waters, Milford, MA, USA) is connected to the cyclotron line as depicted in Figure 3a and eluted with 50 mL target solution for eight minutes (Figure 2(a1)), followed by drying with He for five minutes. Each month, the production system (target and delivery line) is sanitized with ethanol USP followed by five flushes of 10 mL target solution.

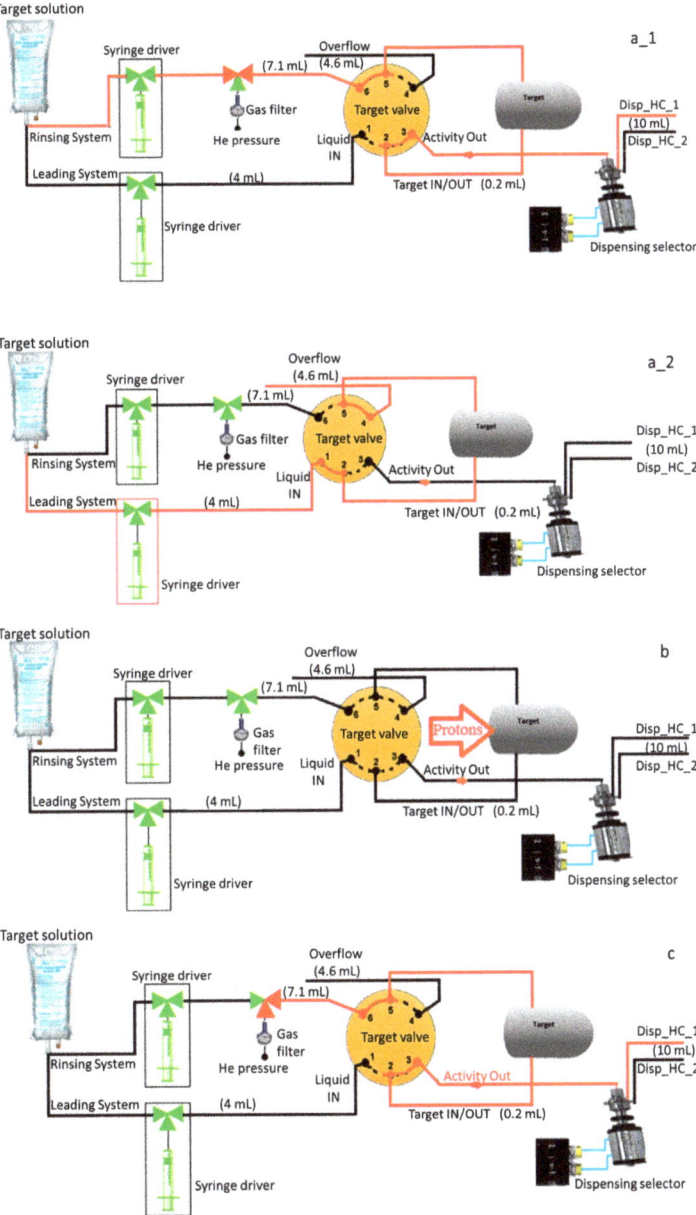

**Figure 2.** Schemes describing the [$^{13}$N]ammonia production system consisting of two syringe drivers connected to the cyclotron target. The line highlighted in red displays the production workflow: (**a1**) For rinsing, a dedicated syringe driver transfers the freshly prepared target solution through the target towards the class A dispensing hot cell (Disp_HC); (**a2**) For loading, a separate syringe driver fills the target prior to the bombardment; (**b**) Beaming sequence; and (**c**) Unloading sequence with helium and transfer to the selected dispensing hot cell.

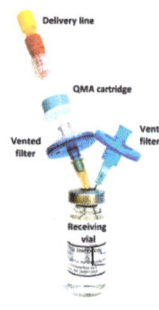
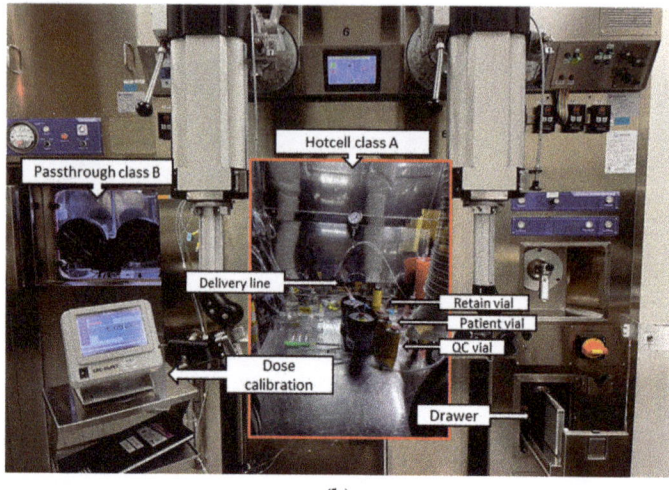

**Figure 3.** Aseptic set-up for [$^{13}$N]ammonia production: (**a**) Receiving formulation vial and connection from cyclotron target; (**b**) Class A dispensing hot cell with a class B pre-chamber (passthrough) for material entrance.

- [$^{13}$N]Ammonia is produced with a Cyclone® 18/9 cyclotron (IBA, Louvain-La-Neuve, Belgium) via the nuclear reaction $^{16}$O(p,α)$^{13}$N with 18-MeV protons degraded to 16 MeV using an aluminum degrader (300–400 μm) installed in the target collimator to prevent the production of oxygen-15 via the $^{16}$O(p,pn)$^{15}$O nuclear reaction. The 2XL target is filled with 3.5 mL of target solution (10 mM ethanol in WFI) (Figure 2(a2)) and bombarded at a current of 55 μA for 20–40 min (Figure 2b). The target body is made of niobium, and the entrance windows are made of a 12.5 μm thick titanium foil (ø 23.5 mm) and a 35 μm thick Havar foil (ø 43 mm).
- The set-up of the Class A dispensing hot cell consists of two vial assemblies containing either four (first batch of the day) or three (second batch) 10 mL vials (Hollister Stier, Montréal, QC, Canada), enough syringes and needles of the appropriate size for the manipulation to be performed, along with a 3 mL syringe containing 2.5 mL of 3% saline. Using the manipulating gloves, before the transfer of the dose, a hydrophilic 0.22 μm vented filter (Millex GV, Millipore, Burlington, MA, USA) and a venting 0.22 μm filter (PharmAssure, Pall, Port Washington, NY, USA) are inserted into the patient (receiving) 10 mL vial. The vented filter is then connected to the rinsed and dried (see above) QMA + delivery line, as shown in Figure 3a. The remaining three vials are used for QC, retain and sterility/endotoxin samples, respectively (Figure 3b). Since the sterility sample of the second batch is pooled with the first one, the vial assembly for the second batch contains only three 10 mL vials. An acrylic plate with three pH strips (0 to 6) and a vented 100 mL vial full of target solution to perform the filter integrity test complete the setup for the [$^{13}$N]ammonia production.
- At the EOB, the irradiated water/ethanol/[$^{13}$N]NH$_3$ solution is flushed out of the target (Figure 2c) with 0.25–0.5 MPa (2.5–5 bar) flow of He through an in-line QMA cartridge to trap any anionic impurities such as [$^{18}$F]fluoride and [$^{13}$N]NO$_x{}^-$, into a dispensing hot cell previously set up for the aseptic preparation (see above). Right after, an additional target rinse with 2.5 mL of target solution is performed to recover the remaining activity.
- Approximately 0.1 mL of [$^{13}$N]NH$_3$ is withdrawn to start the QC, then 2.5 mL of 3% saline is added to the bulk vial to produce isotonic [$^{13}$N]ammonia final product. From this, it is understood that the QC steps start before finishing the sampling and

reformulation of the final product (both stages overlapping). Sterility/endotoxin and retain samples (0.3 mL sterility/endotoxin + 0.1 mL retain) are taken at this step of the process (Figure 3b) while visual inspection, pH and filter integrity tests are performed subsequently. Sterility/endotoxin samples of all batches of the day are pooled, while retained samples from each batch are kept separately in case a sterility and/or endotoxin re-test is required afterwards.

- The final [$^{13}$N]ammonia product is removed from the dispensing hot cell using a dedicated drawer (Figure 3b) and transferred to the Nuclear Medicine department in less than a minute by a built-in pneumatic system while the QC assays are performed in order to save time.

After EOB, the total manufacturing time from the target, purification (via QMA), sterile filtration, reformulation and QC testing until the batch release of [$^{13}$N]ammonia is approximately 11 min. Our validation studies have revealed that the same steps as described above can be repeated for up to three cGMP-compliant back-to-back production batches per day. In that case, the same target solution and QMA are used, but different vial assemblies containing new product vials, filters and syringes are changed after each batch of the day.

### 4.4. Quality Control Testing of [$^{13}$N]Ammonia Injection

The quality control tests of [$^{13}$N] ammonia are performed for each batch in about eight minutes (some QC assays are started before the EOM in order to save time) to ensure the final drug product meets all the specifications for human use. Radiochemical identity and purity are determined by analytical HPLC (column: Waters IC-Pak Cation exchange M/D 150 × 3.9 mm, mobile phase: 3 mM $HNO_3$/0.1 mM EDTA, flow rate: 1.3 mL/min) using a conductivity (Waters 432, Milford, MA, USA) and radioactivity (Raytest Gabi Star, Straubenhardt, Germany) detectors. The retention time of [$^{13}$N]$NH_3$ is compared to that of the [$^{nat}$N]$NH_4$Cl reference standard solution and must be ± 10% (see example in Figure 1). Stability testing by analytical HPLC was conducted at 0, 30, and 60 min after EOM. Radionuclidic identity is assayed for half-life (10 ± 1 min) and annihilation peak (466–556 keV) using a Capintec CRC®-15tW Dose Calibrator containing a NaI(Tl) well-counter coupled to a multichannel analyzer. Radionuclidic purity (>99.5%) is analyzed by HP(Ge) gamma spectrometry as a periodic quality indicator test during the validation runs and re-tested annually afterwards at Nuclear Activation Analysis Lab (Ecole Polytechnique, Montréal, QC, Canada). Bacterial endotoxins (<175 EU per dose) are quantified as a post-release test by LAL EndoSafe Portable Testing System (Charles River, Wilmington, MA, USA) on the pooled samples from each batch of the day. Sterility testing is assayed (on the pooled samples) for no growth after 14 days following USP 71 with a contracted organization (Isologic Innovative Radiopharmaceuticals, Montréal, QC, Canada).

## 5. Conclusions

Our institutional experience of producing large quantities of [$^{13}$N]ammonia using two syringe drivers and in-line anion-exchange resin purification have been thoroughly validated after four years of continuous production. Hence, [$^{13}$N]$NH_3$ injection was found to meet all the product specifications. The simplified production system described in this work has been demonstrated to be reliable and easy to operate and maintain, aiding to keep the [$^{13}$N]ammonia production costs low while satisfying clinical needs.

**Author Contributions:** Conceptualization, N.N., J.-M.S., M.B. and J.N.D.; methodology, N.N., J.-M.S., M.B. and J.N.D.; software, N.N.; validation, J.-M.S., N.N., A.H.S. and M.B.; formal analysis, N.N., J.-M.S., M.B., A.H.S. and J.N.D.; investigation, D.J. and J.N.D.; resources, D.J. and J.N.D.; writing—original draft preparation, L.M.A.M., N.N. and J.N.D.; writing—review and editing, L.M.A.M., M.B., A.H.S., D.J. and J.N.D.; supervision, J.N.D.; project administration, D.J. and J.N.D.; funding acquisition, D.J. and J.N.D. All authors have read and agreed to the published version of the manuscript.

**Funding:** This research received no external funding.

**Institutional Review Board Statement:** The study was conducted in accordance with the Declaration of Helsinki and approved by the Ethics Board of the CHUM (protocol number 2018–7108, initially approved 15 January 2018).

**Informed Consent Statement:** Not applicable.

**Data Availability Statement:** Not applicable.

**Acknowledgments:** The authors acknowledge the valuable contribution of Charles-Olivier Normandeau and Baptiste Fezas. The authors also would like to thank Cornelia Chilian and Darren Hall (Nuclear Activation Analysis Laboratory—Ecole Polytechnique, Montréal) for helpful technical discussions in gamma spectrometry.

**Conflicts of Interest:** The authors declare no conflict of interest. The funders had no role in the design of the study; in the collection, analyses, or interpretation of data; in the writing of the manuscript; or in the decision to publish the results.

**Sample Availability:** Not applicable.

# References

1. Bateman, T.M.; Heller, G.V.; Beanlands, R.; Calnon, D.A.; Case, J.; deKemp, R.; DePuey, E.G.; Di Carli, M.; Guler, E.C.; Murthy, V.L.; et al. Practical guide for interpreting and reporting cardiac PET measurements of myocardial blood flow: An information statement from the american society of nuclear cardiology, and the society of nuclear medicine and molecular imaging. *J. Nucl. Med.* **2021**, *62*, 1599–1615. [CrossRef] [PubMed]
2. Dorbala, S.; Di Carli, M.F.; Delbeke, D.; Abbara, S.; DePuey, E.G.; Dilsizian, V.; Forrester, J.; Janowitz, W.; Kaufmann, P.A.; Mahmarian, J.; et al. SNMMI/ASNC/SCCT guideline for cardiac SPECT/CT and PET/CT 1.0. *J. Nucl. Med.* **2013**, *54*, 1485. [CrossRef] [PubMed]
3. Rosenspire, K.C.; Schwaiger, M.; Mangner, T.J.; Hutchins, G.D.; Sutorik, A.; Kuhl, D.E. Metabolic fate of [$^{13}$N]Ammonia in human and canine blood. *J. Nucl. Med.* **1990**, *31*, 163–167. [PubMed]
4. Herzog, B.A.; Husmann, L.; Valenta, I.; Gaemperli, O.; Siegrist, P.T.; Tay, F.M.; Burkhard, N.; Wyss, C.A.; Kaufmann, P.A. Long-term prognostic value of 13N-ammonia myocardial perfusion positron emission tomography: Added value of coronary flow reserve. *J. Am. Coll. Cardiol.* **2009**, *54*, 150–156. [CrossRef] [PubMed]
5. Fiechter, M.; Ghadri, J.R.; Gebhard, C.; Fuchs, T.A.; Pazhenkottil, A.P.; Nkoulou, R.N.; Herzog, B.A.; Wyss, C.A.; Gaemperli, O.; Kaufmann, P.A. Diagnostic value of 13N-ammonia myocardial perfusion PET: Added value of myocardial flow reserve. *J. Nucl. Med.* **2012**, *53*, 1230. [CrossRef] [PubMed]
6. Muzik, O.; Beanlands, R.S.B.; Hutchins, G.D.; Mangner, T.J.; Nguyen, N.; Schwaiger, M. Validation of nitrogen-13-ammonia tracer kinetic model for quantification of myocardial blood flow using PET. *J. Nucl. Med.* **1993**, *34*, 83–91. [PubMed]
7. Pieper, J.; Patel, V.N.; Escolero, S.; Nelson, J.R.; Poitrasson-Rivière, A.; Shreves, C.K.; Freiburger, N.; Hubers, D.; Rothley, J.; Corbett, J.R.; et al. Initial clinical experience of N13-ammonia myocardial perfusion PET/CT using a compact superconducting production system. *J. Nucl. Cardiol.* **2021**, *28*, 295–299. [CrossRef] [PubMed]
8. Vermeulen, K.; Vandamme, M.; Bormans, G.; Cleeren, F. Design and challenges of radiopharmaceuticals. *Semin. Nucl. Med.* **2019**, *49*, 339–356. [CrossRef] [PubMed]
9. Wieland, B.; Bida, G.; Padgett, H.; Hendry, G.; Zippi, E.; Kabalka, G.; Morelle, J.-L.; Verbruggen, R.; Ghyoot, M. In-target production of [13N]ammonia via proton irradiation of dilute aqueous ethanol and acetic acid mixtures. *Int. J. Rad. Appl. Instrum. A* **1991**, *42*, 1095–1098. [CrossRef] [PubMed]
10. Kumar, R.; Singh, H.; Jacob, M.; Anand, S.P.; Bandopadhyaya, G.P. Production of nitrogen-13-labeled ammonia by using 11MeV medical cyclotron: Our experience. *Hell. J. Nucl. Med.* **2009**, *12*, 248–250. [PubMed]
11. Frank, C.; Winter, G.; Rensei, F.; Samper, V.; Brooks, A.F.; Hockley, B.G.; Henderson, B.D.; Rensch, C.; Scott, P.J.H. Development and implementation of ISAR, a new synthesis platform for radiopharmaceutical production. *EJNMMI Radiopharm. Chem.* **2019**, *4*, 24. [CrossRef] [PubMed]
12. Yokell, D.L.; Rice, P.A.; Neelamegam, R.; El Fakhri, G. Development, validation and regulatory acceptance of improved purification and simplified quality control of [$^{13}$N]Ammonia. *EJNMMI Radiopharm. Chem.* **2020**, *5*, 11. [CrossRef] [PubMed]
13. Vaalburg, W.; Kamphuis, J.A.A.; Beerling-van der Molen, H.D.; Reiffers, S.; Rijskamp, A.; Woldring, M.G. An improved method for the cyclotron production of 13N-labelled ammonia. *Int. J. Appl. Radiat. Isot.* **1975**, *26*, 316–318. [CrossRef] [PubMed]
14. Scott, P.J.H. Synthesis of [$^{13}$N]ammonia ([$^{13}$N]NH3). In *Radiochemical Syntheses*; John Wiley & Sons, Inc.: Hoboken, NJ, USA, 2012; pp. 313–320. ISBN 9781118140345.
15. Ito, S.; Sakane, H.; Deji, S.; Saze, T.; Nishizawa, K. Radioactive byproducts in [18O]H2O used to produce 18F for [18F]FDG synthesis. *Appl. Radiat. Isot.* **2006**, *64*, 298–305. [CrossRef] [PubMed]
16. Marengo, M.; Lodi, F.; Magi, S.; Cicoria, G.; Pancaldi, D.; Boschi, S. Assessment of radionuclidic impurities in 2-[18F]fluoro-2-deoxy-d-glucose ([18F]FDG) routine production. *Appl. Radiat. Isot.* **2008**, *66*, 295–302. [CrossRef] [PubMed]

17. Köhler, M.; Degering, D.; Zessin, J.; Füchtner, F.; Konheiser, J. Radionuclide impurities in [18F]F− and [18F]FDG for positron emission tomography. *Appl. Radiat. Isot.* **2013**, *81*, 268–271. [CrossRef]
18. Kamar, F.; Kovacs, M.S.; Hicks, J.W. Low cost and open source purification apparatus for GMP [$^{13}$N]ammonia production. *Appl. Radiat. Isot.* **2022**, *185*, 110214. [CrossRef] [PubMed]
19. Bars, E.; Sajjad, M. Automated system for the production of multiple doses of nitrogen-13 labeled ammonia. *J. Nucl. Med.* **2012**, *53*, 1459.
20. Akhilesh, S.; Shanker, N.; Subhash, K.; Sanjay, G.; Dixit, M. Fully automated synthesis of nitrogen-13-NH3 by SHIs HM-18 cyclotron and dedicated module for routine clinical studies: Our institutional experiences. *Indian J. Nucl. Med.* **2022**, *37*, 50–53. [CrossRef]
21. Bormans, G.; Langendries, W.; Verbruggen, A.; Mortelmans, L. On-line anion exchange purification of [$^{13}$N]NH3 produced by 10 MeV proton irradiation of dilute aqueous ethanol. *Appl. Radiat. Isot.* **1995**, *46*, 83–86. [CrossRef]

**Disclaimer/Publisher's Note:** The statements, opinions and data contained in all publications are solely those of the individual author(s) and contributor(s) and not of MDPI and/or the editor(s). MDPI and/or the editor(s) disclaim responsibility for any injury to people or property resulting from any ideas, methods, instructions or products referred to in the content.

Article

# One-Pot Radiosynthesis of [$^{18}$F]Anle138b—5-(3-Bromophenyl)-3-(6-[$^{18}$F]fluorobenzo[d][1,3]dioxol-5-yl)-1H-pyrazole—A Potential PET Radiotracer Targeting α-Synuclein Aggregates

Viktoriya V. Orlovskaya [1], Olga S. Fedorova [1], Nikolai B. Viktorov [2], Daria D. Vaulina [1] and Raisa N. Krasikova [1,*]

[1] N.P. Bechtereva Institute of the Human Brain, Russian Academy of Science, 197376 St. Petersburg, Russia
[2] St. Petersburg State Technological Institute, Technical University, 190013 St. Petersburg, Russia
\* Correspondence: raisa@ihb.spb.ru

**Abstract:** Availability of PET imaging radiotracers targeting α-synuclein aggregates is important for early diagnosis of Parkinson's disease and related α-synucleinopathies, as well as for the development of new therapeutics. Derived from a pyrazole backbone, $^{11}$C-labelled derivatives of anle138b (3-(1,3-benzodioxol-5-yl)-5-(3-bromophenyl)-1H-pyrazole)—an inhibitor of α-synuclein and prion protein oligomerization—are currently in active development as the candidates for PET imaging α-syn aggregates. This work outlines the synthesis of a radiotracer based on the original structure of anle138b, labelled with fluorine-18 isotope, eminently suitable for PET imaging due to half-life and decay energy characteristics (97% β+ decay, 109.7 min half-life, and 635 keV positron energy). A three-step radiosynthesis was developed starting from 6-[$^{18}$F]fluoropiperonal (6-[$^{18}$F]FP) that was prepared using (piperonyl)(phenyl)iodonium bromide as a labelling precursor. The obtained 6-[$^{18}$F]FP was used directly in the condensation reaction with tosylhydrazide followed by 1,3-cycloaddition of the intermediate with 3′-bromophenylacetylene eliminating any midway without any intermediate purifications. This one-pot approach allowed the complete synthesis of [$^{18}$F]anle138b within 105 min with RCY of 15 ± 3% ($n$ = 3) and $A_m$ in the range of 32–78 GBq/µmol. The [$^{18}$F]fluoride processing and synthesis were performed in a custom-built semi-automated module, but the method can be implemented in all the modern automated platforms. While there is definitely space for further optimization, the procedure developed is well suited for preclinical studies of this novel radiotracer in animal models and/or cell cultures.

**Keywords:** α-synuclein; fluorine-18; radiofluorination; 6-[$^{18}$F]fluoropiperonal; [$^{18}$F]anle138b; one-pot synthesis

**Citation:** Orlovskaya, V.V.; Fedorova, O.S.; Viktorov, N.B.; Vaulina, D.D.; Krasikova, R.N. One-Pot Radiosynthesis of [$^{18}$F]Anle138b—5-(3-Bromophenyl)-3-(6-[$^{18}$F]fluorobenzo[d][1,3]dioxol-5-yl)-1H-pyrazole—A Potential PET Radiotracer Targeting α-Synuclein Aggregates. *Molecules* 2023, 28, 2732. https://doi.org/10.3390/molecules28062732

Received: 12 February 2023
Revised: 12 March 2023
Accepted: 13 March 2023
Published: 17 March 2023

**Copyright:** © 2023 by the authors. Licensee MDPI, Basel, Switzerland. This article is an open access article distributed under the terms and conditions of the Creative Commons Attribution (CC BY) license (https://creativecommons.org/licenses/by/4.0/).

## 1. Introduction

Positron emission tomography (PET) is a sensitive and versatile imaging modality, often used in conjunction with CT or MRI, offering unique opportunity for a dynamic 3D-visualization of in vivo processes with a mm-scale resolution. The method employs molecular probes labeled with short-lived positron-emitting radionuclides (PET-radiotracers) interacting with specific protein targets of interest. Apart from widespread diagnostic application in oncology, PET has become an essential tool for assessing pathological processes in a variety of neurodegenerative disorders [1]. Parkinson disease (PD) is the second most widespread neurodegenerative condition worldwide, the prevalence of which is increasing as population ages. PD is characterized by progressive degeneration of the dopaminergic system with loss of dopaminergic neurons in the substantia nigra, associated with clinical presentation of motor symptoms. For in vivo assessment of the central dopaminergic function, the 6-L-[$^{18}$F]fluoro-3,4-dihydroxyphenyl-alanine has been seen as the "gold standard" since 1983 [2]. Following metabolic pathway of L-DOPA, the

$^{18}$F-fluorinated analogue crosses the blood-brain barrier and is subsequently decarboxylated to 6-[$^{18}$F]fluorodopamine, which is accumulated within vesicles of dopaminergic neurons [3]. Synaptic dopamine levels are regulated by dopamine reuptake via dopamine transporter (DAT), whereas reduced levels of DAT indirectly reflect degeneration of nigrostriatal neurons [4,5]. PET imaging of DAT with [$^{18}$F]FE-PE2I [6,7] enables the detection of presynaptic dopamine deficiency and provides another biomarker for PD progression assessment.

Together with the loss of midbrain nigrostriatal dopaminergic neurons, the pathologic processes in idiopathic PD and related disorders, such as dementia with Lewy bodies (DLB) and multiple system atrophy (MSA), are characterized by accumulation of α-synuclein aggregates [8–10]. α-Synuclein (α-syn) is an intrinsically disordered protein that is seen as a major component in Lewy bodies (LBs), Lewy neurites, and glial cytoplasmic inclusion type aggregates [11–13]. It has been assumed that a nigrostriatal dopaminergic tract related symptoms are preceded by LBs formation [11,12]. Development of molecular probes that enable in vivo visualisation and quantification of α-syn aggregates may allow for an earlier and more accurate diagnosis of PD and related α-synucleinopathies, and aid in the development of new treatments.

Over the past few years, extensive efforts have been focused on finding α-syn-specific PET radioligands (for the recent rev., see [13–16]). In 2022, the first human study on α-syn imaging using [$^{18}$F]ACI12589 developed by AC Immune was reported [17]. The radiotracer is said to have high affinity and selectivity for α-syn and shows uptake in the brain areas (such as the basal ganglia and cerebellar white matter) known to be affected by pathological processes in patients with MSA—a rare, atypical parkinsonian syndrome. However, it remains unclear whether [$^{18}$F]ACI12589 is able to detect α-syn depositions in the subjects with more common α-synucleinopathies, such as idiopathic PD and LBD [17].

The importance of α-syn as a target continues to stimulate research in the area, and numerous studies are ongoing, with the aim to develop other labelled small molecules allowing for imaging of α-syn aggregates. Derived from pyrazole backbone, labelled analogues of anle138b (3-(1,3-benzodioxol-5-yl)-5-(3-bromophenyl)-1H-pyrazole) (Figure 1)—an inhibitor of α-syn and prion protein oligomerization [18,19]—are in active development as the candidates for α-syn imaging. Anle138b was designed for treatment of the rapidly progressing MSA and PD with potential to be applied to other synucleinopathies, such as DLB. It was demonstrated that anle138b interacts with aggregate forms of α-syn with moderate affinity ($K_d$ of 190 ± 120 nM) [20] and strongly inhibited formation of pathological oligomers and neuronal degeneration in mouse models of α-syn and prion disease with improved survival rates [18–21]. The compound showed a high oral bioavailability and excellent blood-brain barrier penetration. In a range of different mouse models of PD and MSA, anle138b administration reduced protein aggregate deposition in the brain and improved dopamine neuron function and movement, even when treatment began after development of motor symptoms [21]. Recently, anle138b safety, tolerability and pharmacokinetics has been evaluated in healthy volunteers [22]. Phase I clinical trials in PD patients are currently ongoing.

The first PET tracer based on the lead structure from the library of the reported 3,5-diarylpyrazoles [18] has been the carbon-11 labelled anle253b—a molecule with a suitable position for $^{11}$C-labelling [23]. Evaluation of the radiotracer in healthy rats showed low brain uptake and suboptimal pharmacokinetics [23]. Soon after, the same group introduced a derivative of anle253b called MODAG-001 [24], in which one of the phenyl groups was replaced with pyridine. PET imaging in mice showed excellent brain uptake but detected the formation of two brain-penetrating radio-metabolites—something that hampers quantification of [$^{11}$C]MODAG-001 uptake. To inhibit the metabolic demethylation process, a deuterated derivative (d3)-[$^{11}$C]MODAG-001 was developed and was shown to be able to bind to pre-formed α-syn fibrils (α-PFF) in a protein deposition rat model [25]. However, no evidence of binding to aggregated α-syn was observed in human brain sections from DLB patients. Further evaluation of (d3)-[$^{11}$C]MODAG-001 in a porcine brain with intracerebral injection of α-PFF and post-mortem human AD revealed that the radiotracer was not very selective for α-syn and exhibited significant binding in the AD regions [25].

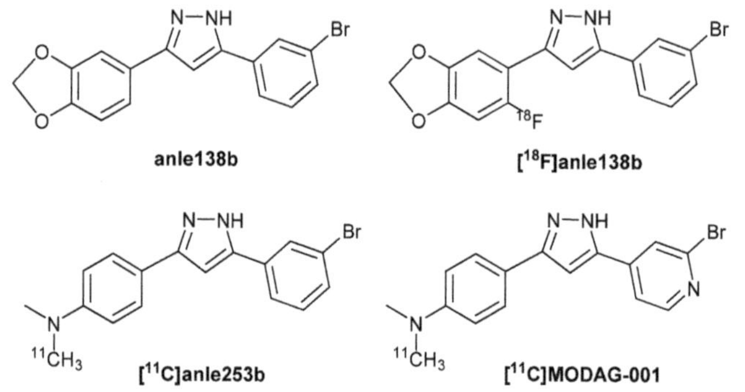

**Figure 1.** Chemical structures of 3,5-diarylpyrazole derivatives: anle138b, [$^{18}$F]anle138b, [$^{11}$C]anle253b and [$^{11}$C]MODAG-001.

Despite mixed results of anle138b/MODAG development when it came to carbon-11 labelling [24,25], we considered it worthwhile to attempt synthesis of a radiotracer based on the original structure of anle138b but labelled with longer-lived fluorine-18 isotope (T$_{1/2}$ 110 min) which displays excellent decay characteristics for PET imaging (97% β+ decay, 635 keV positron energy). The structure–activity (SAR) studies revealed that the placement of bromine in *meta*-position of the 5-phenyl ring led to the highest inhibitory activity of anle138b, whereas further modification of that part of the molecule may result in reduced inhibition [18]. Therefore, in the present work, the 3-substituted aryl moiety was chosen as a suitable labelling position for [$^{18}$F]anle138b (Figure 1). To introduce fluorine-18 into this non-activated position of the aromatic ring, a three-step radiolabelling strategy was attempted starting from 6-[$^{18}$F]fluoropiperonal, obtained through different radiolabelling approaches. As suitability for automation is one of the most important requirements for the labelling method application, a one-pot procedure without intermediate purifications was developed and implemented in a semi-automated module. As a result, [$^{18}$F]anle138b was obtained in 15.1 ± 2.3% (*n* = 3) radiochemical yield (decay-corrected) and $A_m$ in the range of 31.5–79.5 GBq/µmol within a total synthesis time of ca. 105 min.

## 2. Results and Discussion

### 2.1. Radiolabeling Approach for [$^{18}$F]Anle138b

The most common method for introduction of fluorine-18 into majority of PET radiotracers is an aliphatic nucleophilic substitution reaction between no-carrier-added [$^{18}$F]fluoride and a precursor possessing a suitable leaving group in the presence of phase-transfer catalyst (PTC) to enhance [$^{18}$F]fluoride reactivity. For aromatic compounds, the classical S$_N$Ar method requires presence of a leaving group as well as an electron-withdrawing group, preferably in *ortho*- or *para*-position [26]—a requirement difficult to accommodate or engineer in case of complex compounds, such as anle138b. Therefore, the labelling often necessitates multi-steps "built-up" procedures where fluorine-18 is initially introduced into the aromatic ring of a simple and reactive substrate. All nucleophilic fluorinations begin with isolation of [$^{18}$F]fluoride ion from proton-irradiated [$^{18}$O]water; typically, this is achieved by trapping [$^{18}$F]F$^-$ on a quaternary ammonium ion-exchange resin followed by elution with a basic solution of a phase-transfer agent (e.g., Kryptofix K$_{2.2.2}$/K$_2$CO$_3$ mixture in CH$_3$CN/H$_2$O). The [$^{18}$F]fluoride ion complex with the PTC can then be dried to remove water and provide a reactive intermediate. In some instances, bulky counter-ions used for solubilisation of the [$^{18}$F]fluoride, such as R$_4$N$^+$ (R = Et or *n*Bu), are introduced separately as salts (e.g., HCO$_3$, tosylate, triflate); use of such phase-transfer catalysts can simplify trapping/elution procedures [27].

The multi-step labelling strategy for [$^{18}$F]anle138b—(5-(3-bromophenyl)-3-(6-[$^{18}$F] fluorobenzo[*d*][1,3]dioxol-5-yl)-1*H*-pyrazole)—has been drafted by Zarrad et al. [28,29], and consisted of 1,3-cycloaddition between $^{18}$F-fluorinated phenyldiazomethane generated in situ from the corresponding tosylhydrazone with 3′-bromophenyl acetylene as a key synthesis step (Scheme 1). The $^{18}$F label was introduced into the commercially available precursor—4,5-methylendioxy-2-nitrobenzaldehyde (6-nitropiperonal, Step 1, Scheme 1, Figure 2)— possessing carbonyl functional group in *ortho*-position, acting as an electron-withdrawing group. Fluorination was performed with Et$_4$NHCO$_3$ as a PTC and was followed by the isolation of the obtained 6-[$^{18}$F]fluoropiperonal (6-[$^{18}$F]FP) via semi-preparative HPLC and two solid-phase extraction (SPE) purifications. As a result, the 6-[$^{18}$F]FP was obtained in 22% radiochemical yield (RCY, decay-corrected) with high radiochemical and chemical purity. Without this time-consuming and difficult to automate intermediate step aimed at the removal of unreacted 6-nitropiperonal, the condensation reaction of 6-[$^{18}$F]FP with tosylhydrazide (Step 2, Scheme 1) would not proceed to yield [$^{18}$F]anle138b. The suggested three step labelling procedure with intermediate purifications afforded [$^{18}$F]anle138b with RCY of 1% (decay-corrected) which, while being sufficient for in vitro studies, would generally be seen as too low to be practical for most other uses [29].

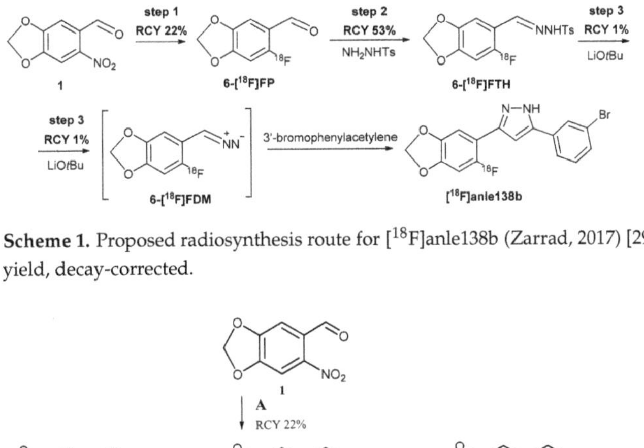

**Scheme 1.** Proposed radiosynthesis route for [$^{18}$F]anle138b (Zarrad, 2017) [29]; RCY—radiochemical yield, decay-corrected.

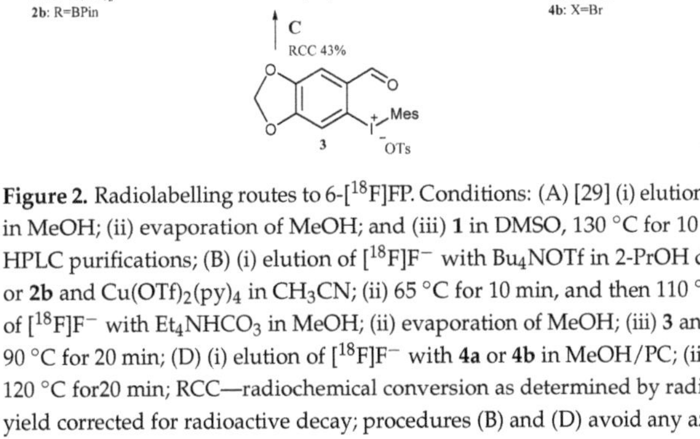

**Figure 2.** Radiolabelling routes to 6-[$^{18}$F]FP. Conditions: (A) [29] (i) elution of [$^{18}$F]F$^−$ with Et$_4$NHCO$_3$ in MeOH; (ii) evaporation of MeOH; and (iii) **1** in DMSO, 130 °C for 10 min with two SPE and one HPLC purifications; (B) (i) elution of [$^{18}$F]F$^−$ with Bu$_4$NOTf in 2-PrOH directly to the solution of **2a** or **2b** and Cu(OTf)$_2$(py)$_4$ in CH$_3$CN; (ii) 65 °C for 10 min, and then 110 °C for 10 min; (C) (i) elution of [$^{18}$F]F$^−$ with Et$_4$NHCO$_3$ in MeOH; (ii) evaporation of MeOH; (iii) **3** and Cu(CH$_3$CN)$_4$OTf in DMF, 90 °C for 20 min; (D) (i) elution of [$^{18}$F]F$^−$ with **4a** or **4b** in MeOH/PC; (ii) 85 °C for 10 min, and then 120 °C for20 min; RCC—radiochemical conversion as determined by radioTLC; RCY—radiochemical yield corrected for radioactive decay; procedures (B) and (D) avoid any azeotropic drying or solvents evaporation steps.

Based on this proof-of-concept study, we have focused our efforts on the development of a simplified one-pot labelling procedure for [$^{18}$F]anle138b with the aim of eliminating most, if not all, intermediate purification steps in the preparation of the 6-[$^{18}$F]FP.

Taking into consideration the recent advances in radiofluorination of electron-rich aromatic strictures [30–33], we explored alternative routes to 6-[$^{18}$F]FP using different radiolabelling precursors and conditions (Figure 2B–D).

## 2.2. Synthesis of 6-[$^{18}$F]FP via Copper-Mediated $^{18}$F-Fluorodeboronation (Method B)

As a starting point, we investigated the feasibility of a copper-mediated approach using the commercially available catalyst Cu(py)$_4$(OTf)$_2$, originally developed for radiofluorination of pinacol arylboronates (ArylBPin) [34]. Its use was further extended to fluorination of organoborons [35] and (hetero)aryl organostannanes [36] and has been found to be useful in the preparation of a wide range of radiotracers [33]. As this methodology has been taken up for the labelling of various precursors, a number of factors have been shown to influence this complex catalytic process. Among them are the sensitivity of the Cu-mediated process to the basic conditions during PTC-controlled solubilisation of [$^{18}$F]fluoride, reaction solvents used, precursor/copper catalyst ratio and others [30,37–39].

The feasibility of $^{18}$F-fluorodeboronation for the synthesis of 6-[$^{18}$F]FP was previously examined using two labeling precursors—commercially available derivatives of boronic acid—6-formylbenzo[*d*][1,3]dioxol-5-yl)boronic acid (**2a**, Figure 2) and pinacolboronate 6-(4,4,5,5-tetramethyl-1,3,2-dioxaborolan-2-yl)benzo[*d*][1,3]dioxole-5-carbaldehyde (**2b**, Figure 2) [29] through application of the so-called "alcohol-enhanced" radiofluorination method [37]. After elution of [$^{18}$F]F$^-$ from the quaternary methyl ammonium (QMA) resin with a solution of Et$_4$NHCO$_3$ in MeOH and evaporation of MeOH, a solution of precursor and Cu(py)$_4$(OTf)$_2$ in DMA/*n*-BuOH was added to [$^{18}$F]Et$_4$NF, followed by heating. The beneficial effect of alcohols as co-solvent in the Cu-mediated process was confirmed for various ArylBPIn substrates [40]. However, according to [29], the attempted Cu-mediated radiofluorination of **2a** and **2b** in DMA/*n*-BuOH 7/3 (100 °C, 10–20 min, Et$_4$NHCO$_3$ as PTC) did not result in the formation of 6-[$^{18}$F]FP.

Using the two labelling precursors—**2a** and **2b**—we applied our previously developed [$^{18}$F]fluoride trapping-elution protocol by replacing Et$_4$NHCO$_3$ [29] with the non-basic Bu$_4$NOTf (10 µmol) in conjunction with 2-PrOH (0.6 mL) as the eluting solvent [41,42]. The fluoride-PTC complex was eluted directly into the reaction vial containing the labelling precursor and Cu(py)$_4$(OTf)$_2$ in a suitable solvent (0.8 mL), avoiding any evaporation steps. Optimisation of radiofluorination parameters in terms of the solvents used and amounts of the reactants was realized using commercially available boronic acid derivative **2a** (Figure 2). From several solvents investigated (Table 1), the highest fluorination (RCC of 96 ± 2%, Table 1, Entry 5) was achieved carrying out radiofluorination in the mixture of 2-PrOH/CH$_3$CN (2/3) at precursor-to-copper catalyst ratio of 20/20 µmol. This protocol worked equally for radiofluorination of ArylBPin precursor **2b**, providing the desired 6-[$^{18}$F]FP with RCC of >97%. For practical reasons, the use of **2a** as a commercially available precursor would of course be preferable. Further reduction in the reactants amounts down to 10/10 µmol has resulted in ca. 50% decrease in the yield of radiofluorination of **2a** (Table 1, Entry 6). No product formation could be observed when radiofluorination of **2a** was carried out in DMA, a solvent typically used in Cu-mediated radiofluorinations, with 2-PrOH as co-solvent (Table 1, Entry 1). The relatively low conversion rates of **2a** (Table 1, Entries 3, 4) were accompanied by substantial losses of radioactivity on the inner surfaces of the reaction vessel (up to 70% of total radioactivity in the case of neat 2-PrOH) and substantial RCC variability. This could be explained by poor solubility of the labelling precursor in the solvents used.

Table 1. Optimization of Cu-mediated radiofluorination of **2a**; elution of [$^{18}$F]F$^-$ by 10 µmol Bu$_4$NOTf in 2-PrOH (0.6 mL); precursor/catalyst in 0.8 mL of 2-PrOH/solvent; reaction mixture volume of 1.4 mL; [a] 110 °C, 20 min; [b] 65 °C, 10 min followed by 110 °C, 10 min; RCC—radiochemical conversion of **2a** into 6-[$^{18}$F]FP as determined from radioTLC data.

| Entry | 2a/Cu(py)$_4$(OTf)$_2$, µmol | Radiofluorination Conditions | RCC, % |
|---|---|---|---|
| 1 | 10/20 | 2-PrOH/DMA [a] | 0; 5 |
| 2 | 10/20 | 2-PrOH/acetone [a] | 13; 26 |
| 3 | 10/20 | 2-PrOH [a] | 29 ± 20 (n = 3) |
| 4 | 20/20 | 2-PrOH [a] | 32 ± 22 (n = 3) |
| 5 | 20/20 | 2-PrOH/CH$_3$CN [b] | 96 ± 2 (n = 6) |
| 6 | 10/10 | 2-PrOH/CH$_3$CN [b] | 48 (n = 1) |
| 7 | 15/15 | 2-PrOH/CH$_3$CN [b] | 86 (n = 1) |
| 8 | 20/15 | 2-PrOH/CH$_3$CN [b] | 64 (n = 1) |

## 2.3. Synthesis of 6-[$^{18}$F]FP via Diaryliodonium Salts Precursors (Methods C and D)

Another effective approach to the incorporation of [$^{18}$F]fluoride into (hetero)aromatic substrates is the use diaryliodonium (DAI) salt precursors—route pioneered by the Pike group [43]. A very practical approach for performing radiofluorination has been introduced for the onium and DAI salts precursors [44]. The advantageous feature of this approach is that the [$^{18}$F]fluoride retained on the anion-exchange resin is eluted directly with the solution of the DAI precursor in MeOH; following quick removal, the solvent is then directly followed by a fluorination reaction. This type of procedure uses neither the PTC/base nor other additives, it reduces the number of operational steps, saves time and is compatible with base-sensitive precursors and products.

For the synthesis of 6-[$^{18}$F]FP, two types of iodonium salts precursors were investigated (Figure 2): (piperonyl)(mesityl)iodonium *p*-toluenesulfonate (**3**), and (piperonyl)(phenyl) iodonium *p*-toluenesulfonate (**4a**) and bromide (**4b**).

A copper-mediated radiofluorination of (mesityl)(aryl)iodonium (MAI) salts using the commercially available (CH$_3$CN)$_4$CuOTf complex and [$^{18}$F]KF/18-crown-6 for the activation of the fluorine-18 was suggested in 2014 [45] as an effective route for radiofluorination of the electron rich arenes. It was shown that when involving the use of a bulky mesityl group as an auxiliary force, the nucleophilic substitution acted towards the less sterically hindered site on the arene ring [45]. A recently developed protocol has been applied for the synthesis radiotracers using MAI salts as labelling precursors [46]. In brief, [$^{18}$F]fluoride is eluted from the anion-exchange matrix with solution of MAI precursor in MeOH/DMF (20% MeOH), followed by Cu-mediated fluorination in the same solvent mixture. Despite the high [$^{18}$F]F$^-$ elution and fluorination efficiency achieved for a series of $^{18}$F-fluorinated aromatic amino acids [46], utilising this approach for the preparation of 6-[$^{18}$F]FP from MAI salt **3** (Figure 2) resulted in elution efficiency from anion-exchange matrix using **3** (20 µmol) in 20% MeOH/DMF (0.72 mL) being low, as was the radiofluorination reaction yield (Table 2, Entry 1). Eluting [$^{18}$F]F$^-$ with solution of **3** (20 µmol) in pure MeOH (1 mL) provided over 90% [$^{18}$F]F$^-$ elution efficiency; however, heating for the purpose of MeOH removal resulted in a complete decomposition of the labelling precursor **3**. As a result, no product was formed (Table 2, Entry 2).

Table 2. Synthesis of 6-[$^{18}$F]FP via radiofluorination of diaryliodinium salts 3, 4a, and 4b using various [$^{18}$F]F$^-$ sorption/elution and radiofluorination protocols; RCC—radiochemical conversion of 3, 4a, and 4b into 6-[$^{18}$F]FP as determined from radioTLC data.

| Entry | Eluent Composition | Evaporation Step | Precursor/μmol | (CH$_3$CN)$_4$CuOTf, μmol | Radiofluorination | | RCC, % |
|---|---|---|---|---|---|---|---|
| | | | | | Solvent, mL | T, °C/t, min | |
| 1 | 3 (20 μmol) in 20% MeOH/DMF (0.72 mL) | - | - | 20 | MeOH/DMF (0.72) | 90/20 | 10 |
| 2 | 3 (20 μmol) in MeOH (1 mL) | + | - | 20 | DMF (0.5) | 90/20 | 0 |
| 3 | Et$_4$NHCO$_3$/MeOH (1 mL) | + | 3/20 | 20 | DMF (0.5) | 90/20 | 35 |
| 4 | Et$_4$NHCO$_3$/MeOH (1 mL) | + | 3/30 | 30 | DMF (0.5) | 90/20 | 43 |
| 5 | Bu$_4$NOTs/MeOH (1 mL) | + | 3/20 | 20 | DMF (0.5) | 90/20 | 27 |
| 6 | Bu$_4$NOTs/MeOH (1 mL) | + | 3/30 | 30 | DMF (0.5) | 90/20 | 52 |
| 7 | 4a (20 μmol) in 20% MeOH/DMF (0.72 mL) | - | - | - | MeOH/DMF (0.72) | 100/20 | 3 |
| 8 | 4a (20 μmol) in 20% MeOH/DMF (0.72 mL) | - | - | - | MeOH/DMF (0.72) | 140/20 | 11 |
| 9 | 4a (20 μmol) in MeOH (1 mL) | - | - | - | MeOH/DMSO (1/0.5) | 160/15 | 5 |
| 10 | 4a (20 μmol) in MeOH (1 mL) | + | - | - | DMSO (0.5) | 160/15 | 0 |
| 11 | Et$_4$NHCO$_3$/MeOH (1 mL) | + | 4a/20 | - | DMSO (0.5) | 130/15 | 4 |
| 12 | Et$_4$NHCO$_3$/MeOH (1 mL) | + | 4a/20 | - | DMSO (0.5) CH$_3$CN (0.5) | 160/15 | 3 |
| 13 | Et$_4$NHCO$_3$/MeOH (1 mL) | + | 4a/20 | - | DMSO (0.5) CH$_3$CN (0.5) | 80/20 | 3 |
| 14 | 4a (20 μmol) in 44% MeOH/PC (0.9 mL) | - | - | - | MeOH/PC (0.9) | (1) 75–85/10 (2) 120/20 | 3 ± 1 (n = 3) |
| 15 | 4b (20 μmol) in MeOH (0.8 mL) | + | - | - | PC (0.6) | 110/15 | 0 |
| 16 | 4b (20 μmol) in 44% MeOH/PC (0.9 mL) | - | - | - | MeOH/PC (0.9) | (1) 75–85/10 (2) 120/20 | 84 ± 5 (n = 6) |

Consequently, we moved on to the more conventional approach for nucleophilic radiofluorination using tetraalkylammonium salts as PTCs for the activation of [$^{18}$F]fluoride. Radionuclide was eluted from the cartridge using 4 μmol of Et$_4$NHCO$_3$ or Bu$_4$NOTs in 1 mL of MeOH followed by the solvent evaporation. The reactive [$^{18}$F]fluoride thus obtained was allowed to react with 3 in the presence of the Cu(I)-catalyst in DMF, affording 6-[$^{18}$F]FP in the RCC of ca. 40–50% when high amounts of the reactants were used (Table 2, Entries 4 and 6). However, despite the improvement in radiofluorination efficiency, the radioactivity yield was not high enough considering the desired 6-[$^{18}$F]FP is an intermediate in a multistep synthesis.

A second strategy which does not involve transition metal catalysts is the radiofluorination of the (phenyl)iodonium salts precursors 4a and 4b with different counter ions (Figure 2). Starting with the tosylate salt 4a, we failed to produce 6-[$^{18}$F]FP consistently using either approach described earlier (Table 2, Entries 7–10) or employing tetralkylammoinum salt for the activation of [$^{18}$F]fluoride (Table 2, Entries 11–14). Gratifyingly, we were able to reach high RCC of 84 ± 5% (n = 4) (Supplementary Materials, Figure S18) in radiofluorination of bromide iodonium salt 4b (Table 2, Entry 16) when performing radiofluorination with two-step heating sequence. Introduction of such a heating procedure was based on our UV-HPLC analysis of samples of the reaction mixture that revealed decomposition of 4b during the MeOH evaporation step (Table 2, Entry 15). To prevent precursor decomposition, MeOH was removed during the first heating step at 75–85 °C under gentle agitation with nitrogen gas flow, while the radiofluorination step was performed at 120 °C for 20 min in a neat propylene carbonate (PC), affording 6-[$^{18}$F]FP in high RCC. However, using the same protocol, only around 3% RCC was observed in the radiofluorination of the tosylate salt 4a (Table 2, Entry 14). Such a low radiofluorination efficiency of 4a, as compared to the bromide salt 4b, is not supported by the literature data demonstrating that tosylate is one of the most reactive salts towards the [$^{18}$F]fluoride ion [47]. We assume that the radiofluorination might be suppressed by the presence of residual silver in **4a** that was prepared from **4b** by reaction with silver tosylate.

Finally, from all of the investigated radiolabeling procedures, highest RCCs for 6-[$^{18}$F]FP were achieved with Cu-mediated radiofluorination of commercially available **2a** using [$^{18}$F]Bu$_4$NOF in 2-PrOH/CH$_3$CN (Table 1, Entry 5). However, the radiofluorination of (phenyl)iodonium salt **4b** in MeOH/PC (Table 2, Entry 15) avoiding Cu-catalyst turned out to be most suitable route for preparing 6-[$^{18}$F]FP with additional benefits for the synthesis automation through the use of a fairly simple and straightforward trapping/elution protocol for [$^{18}$F]F$^-$ (elution efficiency was over 85%).

## 2.4. Radiosynthesis of [$^{18}$F]Anle138b

Radiolabeling of [$^{18}$F]anle138b starting from **4b** as a labelling precursor is depicted in Scheme 2.

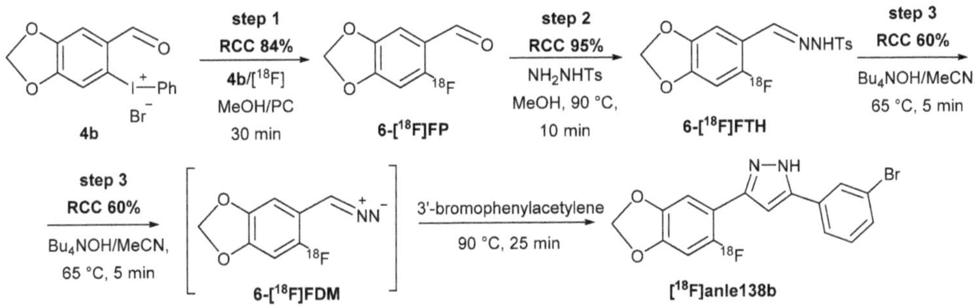

**Scheme 2.** Radiosynthesis scheme for [$^{18}$F]anle138b. Conditions—step 1: Elution of [$^{18}$F]F$^-$ with 20 µmol **4b** in 44% MeOH/PC (0.9 mL). Heating for 85 °C, 10 min, then 120 °C, 20 min; step 2: NH$_2$NHTs (40 µmol) in MeOH (1 mL). Heating for 90 °C, 10 min under stirring by N$_2$ flow; step 3: Bu$_4$NOH (26 µmol in 0.4 mL CH$_3$CN, 65 °C, 5 min), then 3′-bromophenylacetylene (0.4 mmol in 0.4 mL CH$_3$CN, 90 °C, 25 min).

When developing our method for [$^{18}$F]anle138b, we were, in a significant part, guided by the previously outlined synthetic strategy [29], while focusing our attention on making the synthesis as simple as possible by eliminating intermediate purification steps and employing the one-pot approach. The [$^{18}$F]fluoride isolation and synthesis itself were conducted in a custom built synthesis module, described in the experimental part. For monitoring of radiolabelling progress, aliquots of the relevant reaction mixture were taken after each synthesis step (Scheme 2) and analyzed by radio-HPLC (Figure 3).

As described above, the first synthesis step—the radiofluorination of aryliodonium bromide **4b** afforded 6-[$^{18}$F]FP in the RCC of ca. 84%. Non-isolated 6-[$^{18}$F]FP (solution in MeOH/PC) was used directly in the tosyl hydrazide condensation step (40 µmol of hydrazide in 1 mL of MeOH) at 90 °C for 10 min (step 2, Scheme 2). The reaction proceeded with almost quantitative conversion to 6-[$^{18}$F]fluoro-3,4-methylendioxybenzylidene tosylhydrazone (6-[$^{18}$F]FTH) (Figure 3B). The 6-[$^{18}$F]FTH obtained was converted in situ into the corresponding 6-[$^{18}$F]fluorophenyldiazomethane (6-[$^{18}$F]FDM) by addition of Bu$_4$NOH in CH$_3$CN and heating to 65 °C for 5 min. Finally, the cycloaddition reaction of 6-[$^{18}$F]FDM with 3′-bromophenylacetylene afforded crude [$^{18}$F]anle138b. Remarkably, the efficiency of the cycloaddition reaction greatly depends on the nature of base employed in the previous step. From several bases investigated (DBU, NaOH, K$_2$CO$_3$ and LiOtBu) [29], the highest conversion rate of ca. 50% was achieved using LiOtBu. Unfortunately, LiOtBu is insoluble in most organic solvents, including PC that was an essential component of the reaction mixture in our one-pot three-step synthesis procedure. To address this issue, Bu$_4$NOH soluble in both CH$_3$CN and PC was used as a base. Furthermore, according to the data previously reported [48], the efficiency of one-pot cycloaddition process could be improved by sequential introduction of the base and corresponding acetylene into reaction mixture. Taking those considerations into account, we adjusted our procedure to employ sequential addition of reagents in the following order: The solution of Bu$_4$OH in CH$_3$CN was added first with heating at 65 °C for 5 min, followed by 3′-bromophenylacetylene addition and a second round of heating at 90 °C for 25 min. With this approach, we were able to obtain [$^{18}$F]anle138b in the RCC of 60% (Figure 3C) using 0.4 mmol 3′-bromophenylacetylene and 26 µmol Bu$_4$NOH.

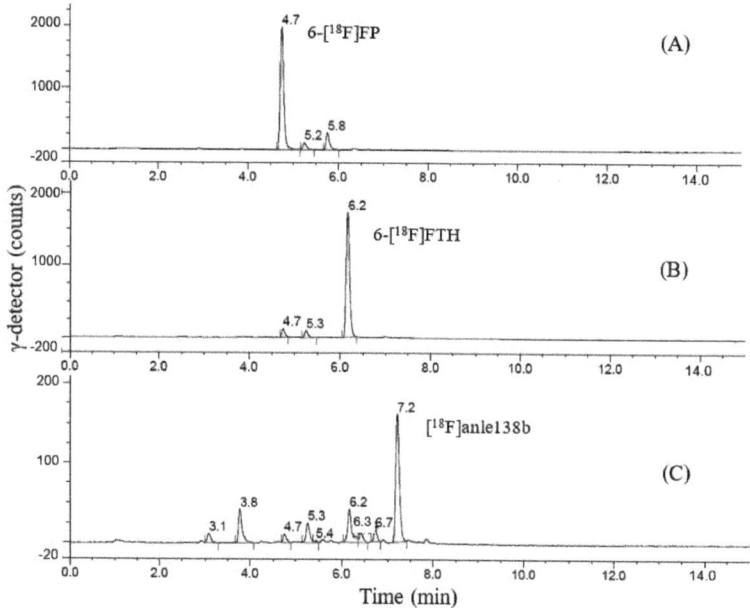

**Figure 3.** Analytical radioHPLC chromatograms of the reaction mixture (**A**,**B**) and final product (**C**) in the synthesis of [$^{18}$F]anle138b. (**A**) 6-[$^{18}$F]FP; (**B**) 6-[$^{18}$F]FTH; and (**C**) [$^{18}$F]anle138b. HPLC column X-Bridge C18, 150 × 4.6 mm (Waters), gradient of 0.1%TFA/CH$_3$CN, and flow rate of 2.0 mL/min.

For the identification of the target product, the authentic reference standard [$^{19}$F]anle138b was prepared (Scheme 3); the synthesis is described in the experimental part.

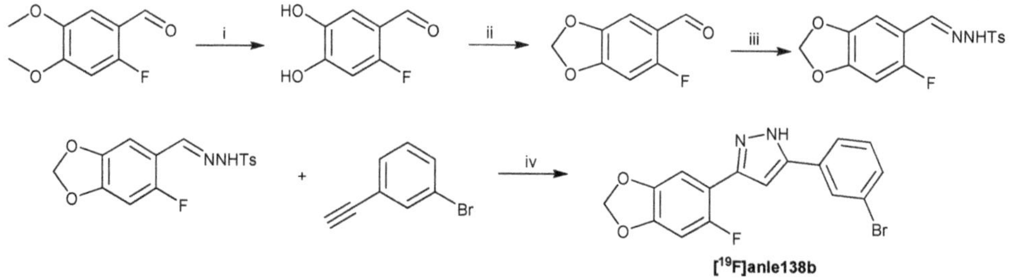

**Scheme 3.** Synthesis of [$^{19}$F]anle138b. Reagents and conditions: (i) BBr$_3$, DCM, −80–20 °C, 30 h; (ii) CH$_2$Br$_2$, K$_2$CO$_3$, DMF, 110 °C, 5 h; (iii) TsNHNH$_2$, CH$_3$OH, 20 °C, 4 h; and (iv) LiO*t*Bu, CH$_3$CN, reflux, 23 h.

## 2.5. Purification and Quality Control of [$^{18}$F]Anle138b

Due to high lipophilicity of anle138b, isolation of the $^{18}$F-fluorinated derivative by conventional semi-preparative HPLC presents a challenge. Given the "one-pot" synthesis approach and the resulting multicomponent reaction mixture, the purification task becomes even more complicated.

The isolation of [$^{18}$F]anle138b from the reaction mixture for two different HPLC systems was evaluated. The first one (System A) is an integral part of the GE Tracerlab FX C Pro module, equipped with UV- and β-radioactivity flow detectors. The HPLC separation was performed on a reverse-phase Ascentis RP-Amide semi-preparative column, 250 × 10 mm (Sigma-Aldrich GmbH, Steinheim, Germany). The content of the reaction vessel (1.2 mL) was diluted with the mobile phase and transferred to the 2 mL injection loop. For practical reasons, biocompatible ethanol-containing eluent would be preferable. However, when using H$_2$O/EtOH gradient system (gradient conditions 1, Materials and Methods) at a flow rate of 3.0 mL/min, only 5% of the injected radioactivity was recovered as [$^{18}$F]anle138b ($R_t$ of 25–26 min) with radiochemical purity (RCP) in order of 80%. Replacing EtOH with CH$_3$CN in the mobile phase and modifying gradient (gradient conditions 2, Materials and Methods) afforded [$^{18}$F]anle138b in 90% RCP with slightly increased recovery of the radioactivity (8%) in the product fraction ($R_t$ of 28–32 min).

Alternatively, we investigated applicability of a different HPLC column, Chromolith SemiPrep RP-18e, 100 × 10 mm equipped with a UV and radioactivity detector, a gradient pump and the Rheodyne-type injector with a 100 µL loop (System B). Under gradient conditions using EtOH-based mobile phase at a flow rate of 4 mL/min, the [$^{18}$F]anle138b (fraction with $R_t$ 9.7–9.9 min, 1 mL volume, 26% recovery of the product) was obtained in more than 98% RCP according to radioHPLC (Figure 4), and ca. 100% according to radioTLC (Supplementary Materials, Figure S19) As the ethanol-containing eluent was applied, cartridge based reformulation of the product was not required.

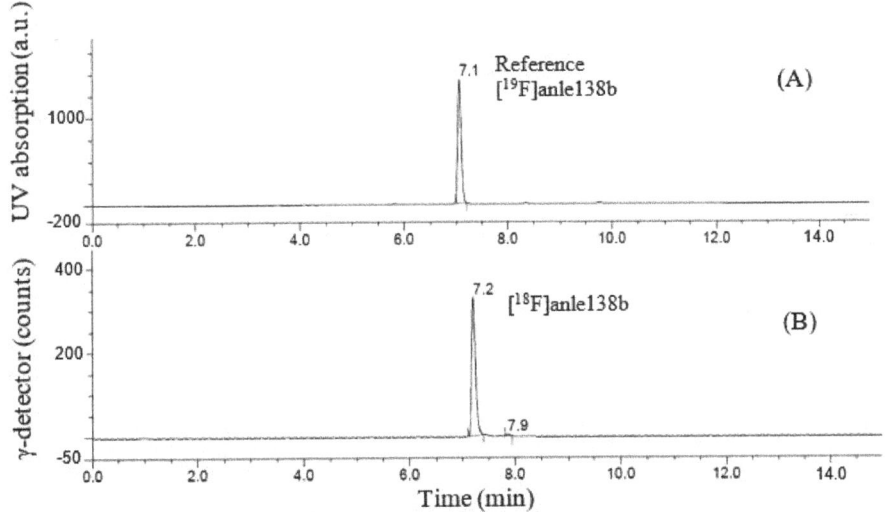

**Figure 4.** HPLC analysis of the formulated [$^{18}$F]anle138b. Conditions: HPLC column X-Bridge C18, 150 × 4.6 mm (Waters Corporation (Millford, CT, USA), gradient of 0.1%TFA/CH$_3$CN with a flow rate 2.0 mL/min, UV is254 nm: (**A**) reference standard [$^{19}$F]anle138b; (**B**) the formulated [$^{18}$F]anle138b.

The main limitation of this method is a small maximum injection volume (100 µL); with higher injected volumes, the column performance would degrade due to overloading. Therefore, multiple injections would be required for purification of all reaction volume produced in the synthesis (1.2 mL total). Nonetheless, the procedure developed is well suited for preclinical studies where small amounts of the radiotracer are typically injected.

Recalculating the radioactivity of the volume injected (100 µL) to total reaction volume (1.2 mL), decay-corrected RCY of [$^{18}$F]anle138b was 15.1 ± 2.3% ($n$ = 3) with synthesis time of ca. 105 min. $A_m$ was in the range of 31.5–79.5 GBq/µmol.

## 3. Materials and Methods

### 3.1. General Chemistry

All commercially available chemicals were used without any further purification. Analytical thin-layer chromatography (TLC) on Merck 60 F254 silica gel plates with UV visualization, 254 nm (Merck KGaA, Darmstadt, Germany) was performed. Column chromatography has been carried out on silica gel 60 (0.035–0.070 mm (Acros Organics, Geel, Belgium) with the indicated eluents. NMR spectra were recorded on a Bruker Avance 400 spectrometer (Bruker Optik GmbH, Ettlingen, Germany) in CDCl$_3$ or DMSO-d6 ($^1$H, $^{13}$C and $^{19}$F at 400.17, 100.62 and 376.54 MHz, respectively). HRMS (ESI) analysis was done on a Bruker micrOTOF mass spectrometer (Bruker Optik GmbH, Ettlingen, Germany).

3.1.1. The 4 Steps Synthesis of 3-(6-Fluoro-1,3-benzodioxol-5-yl)-5-(3-bromophenyl)-1H-pyrazole ([$^{19}$F]Anle138b) (Scheme 3)

2-Fluoro-4,5-dihydroxybenzaldehyde [49]

Under inert atmosphere, 6-fluoroveratraldehyde (1.00 g, 5.43 mmol) was dissolved in anhydrous DCM (10 mL). The mixture was cooled near −80 °C and then boron tribromide (4.4 g, ~17 mmol) was added dropwise for 5 min. The reaction mixture was allowed to warm to room temperature and stirred for 30 h. The red–violet mixture was poured on the ice (150 g) and extracted with ethyl acetate (3 × 20 mL). After drying with MgSO$_4$ and concentrating under reduced pressure, the residue was purified by column chromatography on silica gel (heptane/ethyl acetate: 1/1) to afford 2-fluoro-4,5-dihydroxybenzaldehyde (0.65 g, yield of 77%) as dark solid. NMR spectra have been shown in the Supplementary Materials. $^1$H-NMR (400 MHz, DMSO-$d_6$) δ 6.65 (d, $^3J_{HF}$ = 11.9 Hz, 1H, C$^3$H); 7.10 (d, $^4J_{HF}$ = 6.9 Hz, 1H, C$^6$H); 9.59 (br s, OH); 9.98 (s, 1H, CHO); 10.69 (bs, OH). $^{13}$C-NMR (101 MHz, DMSO-$d_6$) δ 103.05 (d, $J_{CF}$ =24.5 Hz); 112.23; 115.24 (d, $J_{CF}$ =8.9 Hz); 142.64; 153.82; 159.26 (d, $^1J_{CF}$ =249.4 Hz, C$^2$); 185.61 (d, $^3J_{CF}$ =5.7 Hz, CHO).

6-Fluorobenzo-1,3-dioxole-5-carbaldehyde [50]

Under inert atmosphere, the mixture of 2-fluoro-4,5-dihydroxybenzaldehyde (0.65 g, 4.16 mmol), dibromomethane (1.50 g, 8.62 mmol) and K$_2$CO$_3$ (1.38 g, 10.00 mmol) in DMF (10 mL) was stirred at 110 °C for 5 h. After cooling, ethyl acetate (10 mL) was added, and mixture was filtered. The solid was washed with ethyl acetate (20 mL). The liquid was concentrating under reduced pressure to the volume of ~2 mL. The residue was purified by column chromatography on silica gel (heptane/ethyl acetate: 1/1) to afford 6-fluorobenzo-1,3-dioxole-5-carbaldehyde (0.35 g, yield of 50%) as yellow solid. NMR spectra have been shown in the Supplementary Materials. $^1$H-NMR (400 MHz, CDCl$_3$) δ 6.04 (s, 2H, OCH$_2$O); 6.60 (d, $^3J_{HF}$ = 9.7 Hz, 1H, C$^7$H); 7.16 (d, $^4J_{HF}$ = 5.6 Hz, 1H, C$^4$H); 10.12 (s, 1H, CHO). $^{13}$C-NMR (101 MHz, CDCl$_3$) δ 97.97 (d, $J_{CF}$ =28.9 Hz); 102.97 (OCH$_2$O); 104.94 (d, $J_{CF}$ =3.0 Hz); 117.87 (d, $J_{CF}$ =9.0 Hz); 144.87; 154.09 (d, $J_{CF}$ =14.8 Hz); 162.60 (d, $^1J_{CF}$ =254.0 Hz, C$^6$); 185.48 (d, $^3J_{CF}$ =8.5 Hz, CHO). $^{19}$F-NMR (376.5 MHz, CDCl$_3$) δ -126.1.

N'-((6-Fluorobenzo-1,3-dioxole-5-yl)methylene)-4-methylbenzenesulfonohydrazide

A mixture of 6-fluorobenzo-1,3-dioxole-5-carbaldehyde (0.45 g, 2.68 mmol) and p-toluenesulfonyl hydrazide (0.55 g, 2.95 mmol) in methanol (10 mL) was stirred at room temperature for 4 h. Yellow precipitate was filtered and washed with MeOH (2 × 2 mL) to afford the title compound (0.50 g, yield of 56%) as a grey solid. NMR spectra have been shown in the Supplementary Materials. $^1$H-NMR (400 MHz, DMSO-$d_6$) δ 2.36 (s, 3H, CH$_3$);

6.10 (s, 2H, OCH$_2$O); 6.98 (d, $^3J_{HF}$ = 10.0 Hz, 1H, C$^7$H); 7.05 (d, $^4J_{HF}$ = 6.0 Hz, 1H, C$^4$H); 7.41 (d, $J$ = 8.0 Hz, 2H); 7.76 (d, $J$ = 8.0 Hz, 2H); 7.97 (s, 1H); 11.44 (br s, NH). $^{13}$C-NMR (101 MHz, DMSO-$d_6$) δ 21.10; 98.20 (d, $J_{CF}$ = 30.0 Hz); 102.71 (OCH$_2$O); 113.64 (d, $J_{CF}$ = 11.8 Hz); 127.35; 129.81; 135.93; 139.90; 139.95; 143.69; 144.52; 150.14 (d, $J_{CF}$ =15.1 Hz); 156.58 (d, $^1J_{CF}$ =244.0 Hz, C$^6$). $^{19}$F-NMR (376.5 MHz, DMSO-$d_6$) δ -126.4.

(6-Fluoro-1,3-benzodioxol-5-yl)-5-(3-bromophenyl)-1H-pyrazole ([$^{19}$F]Anle138b)

Under inert atmosphere, lithium tert-butoxide (0.22 g, 2.76 mmol) was added into a solution of N'-((6-fluorobenzo-1,3-dioxole-5-yl)methylene)-4-methylbenzenesulfonohydrazide (0.34 g, 1.01 mmol) in dry acetonitrile (10 mL) and stirred at room temperature for 15 min. Then, 3-bromophenylacetylene (0.50 g, 2.76 mmol) was added and the reaction was refluxed for 23 h. The mixture was evaporated in vacuo, and water (10 mL) was added. After extraction with ethyl acetate (2 × 20 mL), the organic phase was concentrated under reduced pressure. The residue was purified by column chromatography on silica gel (heptane/ethyl acetate: 1/1) to afford [$^{19}$F]anle138b (0.10 g, yield of 28%) as a grey solid. $^1$H-NMR (400 MHz, CDCl$_3$) δ 6.05 (s, 2H, OCH$_2$O); 6.72 (d, $J$ = 10.6 Hz, 1H); 6.82 (s, 1H); 7.13 (br d, $J$ = 6.0 Hz, 1H); 7.30 (t, $J$ = 8.0 Hz, 1H); 7.47 (m, 1H); 7.74 (br d, $J$ = 8.0 Hz, 1H); 7.97 (br s, 1H); 10.61 (br s, NH). $^{19}$F-NMR (376.5 MHz, CDCl$_3$) δ -121.4. HRMS (ESI): $m/z$ [M + H]$^+$ calculated for C$_{16}$H$_{10}$BrFN$_2$O$_2$: 360.9982; found: 360.9985. This process involved correct isotopic pattern.

3.1.2. (6-Formylbenzo-1,3-dioxole-5-yl)(phenyl)iodonium Bromide (**4b**) (Figure 2) [44]

Boron trifluoride etherate (0.74 g, 5.21 mmol) was added dropwise to an ice cold solution of 2-formyl-4,5-methylenedioxyphenylboronic acid (0.50 g, 2.58 mmol) in anhydrous DCM (15 mL) under Ar. The resulting yellow suspension was stirred for 30 min at 0 °C. (Diacetoxyiodo)benzene (1.25 g, 3.87 mmol) was added, and the mixture was stirred for 1 h at 20 °C. The solvent was removed under reduced pressure. The residue was taken into saturated NaBr-H$_2$O (15 mL) for 10 min. The resulting yellow suspension was filtered and washed with a H$_2$O/CH$_3$CN mixture (8/1, 10 mL). The solid was refluxed in DCM (10 mL). Insoluble material was filtered off and dried in vacuo to afford **4b** (0.95 g, yield of 85%) as grey solid. $^1$H-NMR (400 MHz, DMSO-$d_6$) δ 6.27 (s, 2H, OCH$_2$O); 7.27 (s, 1H); 7.56 (t, $J$ = 7.8 Hz, 2H); 7.68–7.76 (m, 1H); 7.72 (s, 1H); 8.24 (d, $J$ = 7.8 Hz, 2H); 10.06 (s, 1H, CHO). $^{13}$C-NMR (101 MHz, DMSO-$d_6$) δ 104.20; 112.65; 113.13; 114.04; 116.93; 127.72; 131.76; 132.21; 136.01; 150.27; 154.44; 192.43.

3.1.3. (6-Formylbenzo-1,3-dioxole-5-yl)(phenyl)iodonium Tosylate (**4a**) (Figure 2)

The solution of TsOAg (0.42 g, 1.50 mmol) in MeOH/MeCN mixture (1/3; 8 mL) was added to the suspension of compound **4b** (0.60 g, 1.38 mmol) in MeOH/MeCN mixture (1/1; 10 mL) at 20 °C. After stirring for 1 h, the suspension was filtered, and the resulting solution was evaporated under reduced pressure. The residue was refluxed in DCM (10 mL) and filtered (7 times). Insoluble material was dried in vacuo to afford **4a** (0.23 g, yield of 32%) as grey solid. $^1$H-NMR (400 MHz, DMSO-$d_6$) δ 2.28 (s, 3H); 6.30 (s, 2H, OCH$_2$O); 7.07 (s, 1H); 7.10 (d, $J$ = 7.9 Hz, 2H, TsO); 7.46 (d, $J$ = 7.9 Hz, 2H, TsO); 7.65 (t, $J$ = 7.8 Hz, 2H); 7.81 (m, 1H); 7.83 (s, 1H); 8.28 (dd, $J$ = 8.2; 1.0 Hz, 2H); 10.06 (s, 1H, CHO). $^{13}$C-NMR (101 MHz, DMSO-$d_6$) δ 20.80; 104.54; 107.36; 113.23; 113.76; 114.25; 125.49; 127.53; 128.05; 132.19; 133.07; 136.57; 137.60; 145.75; 150.61; 154.96; 192.70.

3.1.4. (6-Formylbenzo-1,3-dioxole-5-yl)(2,4,6-trimethylphenyl)iodonium Bromide

This compound was prepared analogously to compound **4b** using 2,4,6-trimethyl (diacetoxyiodo)benzene; the crude solid was refluxed in acetone (10 mL) instead of DCM for purification. 6-formylbenzo-1,3-dioxole-5-yl)(2,4,6-trimethylphenyl)iodonium bromide (0.65 g, yield of 66%) was obtained as a grey–yellow solid. $^1$H-NMR (400 MHz, CDCl$_3$) δ 2.38 (s, 3H, CH$_3$); 2.58 (s, 6H, 2CH$_3$); 6.14 (s, 1H); 6.15 (s, 2H, OCH$_2$O); 7.14 (s, 2H); 7.57 (s,

1H); 10.02 (s, 1H, CHO). $^{13}$C-NMR (101 MHz, CDCl$_3$) δ 21.16; 26.91; 103.90; 109.77; 112.17; 115.88; 121.21; 127.89; 129.97; 141.96; 143.84; 149.98; 155.99; 191.45.

3.1.5. (6-Formylbenzo-1,3-dioxole-5-yl)(2,4,6-trimethylphenyl)iodonium Tosylate (3) (Figure 2)

The solution of TsOAg (0.23 g, 0.82 mmol) in MeOH/MeCN mixture (3/1; 4 mL) was added to the solution of compound (6-formylbenzo-1,3-dioxole-5-yl)(2,4,6-trimethylphenyl) iodonium bromide (0.36 g, 0.75 mmol) in a MeOH/MeCN mixture (5/1; 6 mL) at 20 °C. After stirring for 1 h, the suspension was filtered and resulted solution was evaporated under reduced pressure. The residue was refluxed in DCM (10 mL) and filtered. The solution was evaporated under reduced pressure to afford 3 (0.23 g, yield of 54%) as a grey solid. $^1$H-NMR (400 MHz, CDCl$_3$) δ 2.29 (s, 3H); 2.37 (s, 3H); 2.50 (s, 6H, 2CH$_3$); 6.14 (s, 1H); 6.16 (s, 2H, OCH$_2$O); 6.98 (d, $J$ = 7.7 Hz, 2H, TsO); 7.05 (s, 2H); 7.43 (d, $J$ = 7.7 Hz, 2H, TsO); 7.71 (s, 1H); 10.16 (s, 1H, CHO). $^{13}$C-NMR (101 MHz, CDCl$_3$) δ 21.41; 21.42; 26.93; 104.22; 104.73; 109.35; 116.62; 118.38; 125.94; 127.98; 128.37; 130.08; 139.05; 143.03; 143.46; 144.55; 150.78; 156.35; 192.09.3.5.

*3.2. Radiochemistry*

3.2.1. General

Unless otherwise stated, reagents and solvents were commercially available and were used without further purification. Anhydrous acid free MeCN (max 10 ppm H$_2$O) was purchased from Kriochrom, St. Petersburg, Russia. The precursors **3**, **4a**, **4b**, reference standard 6-fluorobenzo-1,3-dioxole-5-carbaldehyde (6-[$^{19}$F]FP) and reference standard [$^{19}$F]anle138b were prepared as described above. The 6-formylbenzo[*d*][1,3]dioxol-5-yl)boronic acid (**2a**) was obtained from Sigma-Aldrich GmbH (Steinheim, Germany). Cu(MeCN)$_4$OTf and Cu(OTf)$_2$(py)$_4$ were obtained from Sigma-Aldrich GmbH (Steinheim, Germany) and stored under argon. Deionized water (18.2 MOhm*cm) from an in-house Millipore Simplicity purification system (Merck KGaA, Darmstadt, Germany) was used for the preparation of all aqueous solutions. [$^{18}$O]H$_2$O (97% enrichment) was purchased from Global Scientific Technologies, Sosnovyi Bor, Russia. Sep-Pak Accell Plus QMA Plus Light Cartridges (130 mg) were acquired from Waters Corporation (Millford, CT, USA) and were conditioned with 10 mL of 0.05 M NaHCO$_3$, followed by 10 mL of H$_2$O before application.

Radio-TLC analyses were carried out on silica gel plates (60 F254, Merck or Sorbfil, Lenchrom, Russia); radioactivity distribution was determined using a Scan-RAM radioTLC scanner controlled by the chromatography software package Laura (v6.0.4.92) for PET (LabLogic, Sheffield, UK). An aliquot (2–3 μL) of the crude reaction mixture diluted with acetonitrile, was applied onto a TLC plate, and the plate was then developed in ethyl acetate. The R$_f$ values for [$^{18}$F]fluoride, 6-[$^{18}$F]FP and [$^{18}$F]anle138b were 0.05, 0.57 and 0.67, correspondingly. The radiochemical conversion (RCC) measured by radioTLC was defined as the ratio of the product peak area to the total peak area on the TLC. RCC values were not corrected for radioactive decay.

Analytical HPLC was performed on a Dionex ISC-5000 system (Dionex, Sunnyvale, CA, USA). It was equipped with a gradient pump, Rheodyne type injector with a 20 μL loop and a UV absorbance detector with variable wavelength (set to 254 nm) connected in series with a radiodetector (Carrol and Ramsey Associates, CA, USA, model 105-S) giving a delay of 0.1 min. The identity, radiochemical and chemical purity of the [$^{18}$F]anle138b and analysis of the reaction mixture at each stage of the synthesis were determined under the following HPLC conditions: X-Bridge C18 HPLC column, 150 × 4.6 mm (Waters Corporation, Millford, CT, USA), eluent with 5–95% gradient (0.1% aq. TFA/MeCN), and a flow rate of 2.0 mL/min. Overall, there was a 0–8.0 min 5–95% MeCN linear increase; 8.0–11.0 min 95% MeCN isocratic; 11.0–11.2 min 95–5% MeCN linear decrease; and an 11.2–15.0 min 5% MeCN isocratic. The R$_t$ values for the precursor, reference and radiolabelled intermediates are presented in Table 3.

Table 3. $R_t$ values for the precursor **4b**, reference [$^{19}$F]anle138b and radiolabelled intermediates.

| Synthesis Step | Compound | Retention Time by γ-Detector, min | Retention Time by UV-254, min |
|---|---|---|---|
| | Precursor **4b** | | 2.9 |
| Step 1 | 6-[$^{18}$F]FP | 4.7 | |
| | 6-[$^{19}$F]FP | | 4.6 |
| Step 2 | 6-[$^{18}$F]FTH | 6.1 | |
| Step 3 | [$^{18}$F]anle138b | 7.1 | |
| Reference | [$^{19}$F]anle138b | | 7.0 |

For the isolation of [$^{18}$F]anle138b, two different HPLC systems were used.

System A employed an HPLC package available on GE Tracerlab FX C Pro module (GE Healthcare, Waukesha, WI, USA), consisting of a SYCAM S1122 pump, UV detector KNF (λ = 254 nm), LAB LABOPORT, 2 mL injection loop and a β-radioactivity detector. Conditions for this system consisted of a column Ascentis RP-AMIDE, 250 × 10 mm, 5 μm (Supelco, Bellefonte, PA, USA). For gradient 1 ($H_2O$/EtOH), the flow rate was 3.0 mL/min with 0–10.0 min 60% EtOH, 10.0–30.0 min 80% EtOH; gradient 2: ($H_2O$/$CH_3CN$), flow rate of 4.5 mL/min with 0–27.0 min 55% $CH_3CN$; 27.0–35.0 min $CH_3CN$.

System B employed Dionex ISC-5000 HPLC system (Dionex, Sunnyvale, CA, USA), described above, with a 100 μL injection loop and Chromolith SemiPrep RP-18e column, 100 × 10 mm (Merck KGaA, Darmstadt, Germany). Conditions: 5 to 95% gradient ($H_2O$/EtOH) with a flow rate of 4.0 mL/min; 0–10.0 min 5–95% EtOH linear increase; 10.0–12.0 min 95% EtOH isocratic; 12.0–12.1 min 95–5% EtOH linear decrease; and 12.1–16.0 min 5% EtOH isocratic.

3.2.2. Production of [$^{18}$F]Fluoride

[$^{18}$F]Fluoride was produced via the $^{18}O(p,n)^{18}F$ nuclear reaction by irradiation of [$^{18}O$]$H_2O$ (97% enrichment, Global Scientific Technologies, Sosnovyj Bor, Russia) in a niobium target (1.4 mL) with 16.4 MeV protons at a PETtrace 4 cyclotron (GE Healthcare, Uppsala, Sweden). The irradiated [$^{18}O$]$H_2O$ was transferred from the target using a flow of helium into the collection vial.

3.2.3. Synthesis of 6-[$^{18}$F]Fluoropiperonal (6-[$^{18}$F]FP)

Method B (Figure 2). Cu-mediated radiofluorination of pinacol arylboronates (**2a**, **2b**).

A solution of [$^{18}$F]F$^-$ (0.5–1.0 GBq) in [$^{18}O$]$H_2O$ was loaded from the male side onto a QMA cartridge. The cartridge was flushed from the male side with 2-PrOH (4 mL) and dried with $N_2$ gas for 2 min. $^{18}F^-$ was eluted from the female side of the cartridge with a solution of Bu$_4$NOTf (4 mg, 10 μmol) in 2-PrOH (0.6 mL) into the 2 mL reaction vessel prefilled with a solution of **2a** or **2b** (20 μmol) and Cu(py)$_4$(OTf)$_2$ (20 μmol) in MeCN (0.8 mL). The reaction mixture was heated at 65 °C for 10 min, followed by a second round of heating 110 °C for 10 min, while the reactor was sealed via valve 16 (Figure 5). Then, the reaction vessel was cooled down to 40 °C.

Method C (Figure 2). Cu-mediated radiofluorination of (piperonyl)(mesityl)iodonium *p*-toluenesulfonate (**3**).

A solution of [$^{18}$F]F$^-$ (0.5–1.0 GBq) in [$^{18}O$]$H_2O$ was loaded from the male side onto a QMA cartridge. The cartridge was flushed from the male side with MeOH (2 mL) and dried with $N_2$ gas for 2 min. $^{18}F^-$ was eluted from the female side of the cartridge with a solution of Et$_4$NHCO$_3$ (0.8 mg, 4.2 μmol) in MeOH (1 mL) into the 2 mL reaction vessel. The solvent was evaporated by heating to 75 °C under gas flow, and the reaction vessel was then cooled to 50 °C. The solution of **3** (30 μmol) and Cu(MeCN)$_4$OTf (30 μmol) in DMF (0.5 mL) was then added to the dried residue and the reaction mixture, and the reaction mixture was heated at 90 °C for 20 min, while the reactor was sealed via valve 16 (Figure 5). Then, the reaction vessel cooled down to 40 °C.

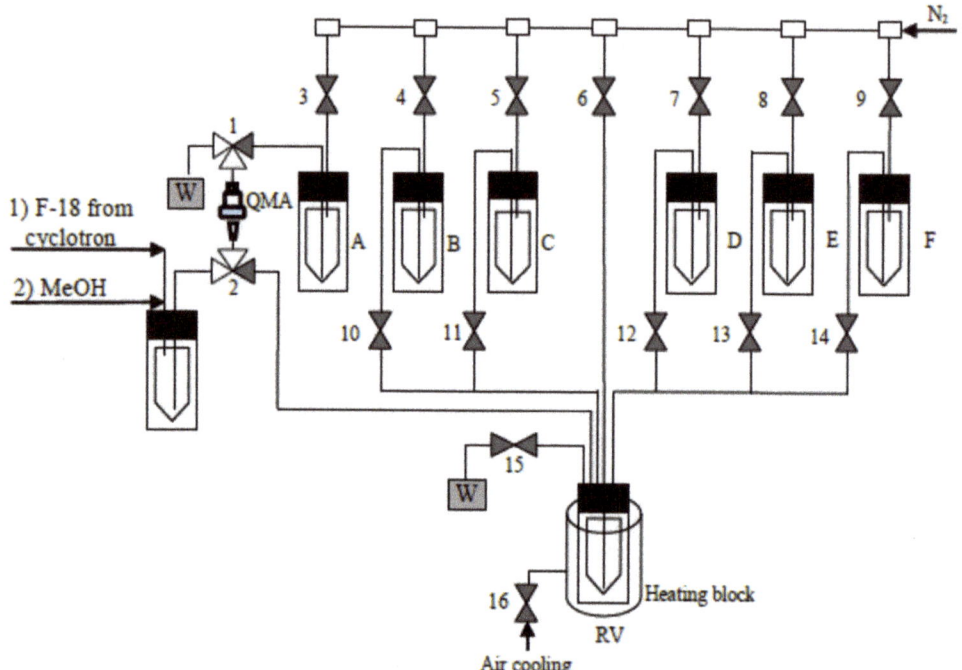

**Figure 5.** Process flow diagram (PFD) for the semi-automated radiosynthesis of [$^{18}$F]anle138b. (A) a solution of precursor **4b** (20 µmol) in 44% MeOH/PC (0.9 mL); (B) a solution of NH$_2$NHTs (40 µmol) in MeOH (1 mL); (C) a solution of Bu$_4$NOH (26 µmol) in MeCN (0.4 mL); and (D) a solution of 3′-bromophenylacetylene (0.4 mmol) in CH$_3$CN (0.4 mL); E and F are not used.

Method D (Figure 2). Radiofluorination of diaryliodonium salts **4a**, **4b**.

A solution of [$^{18}$F]F$^-$ (0.5–1.0 GBq) in [$^{18}$O]H$_2$O was loaded from the male side on a QMA cartridge. The cartridge was flushed from the male side with MeOH (2 mL) and dried with N$_2$ gas for 2 min. $^{18}$F$^-$ was eluted from the female side of the cartridge with a solution of the respective radiolabelling precursor **4a** or **4b** (20 µmol) in 44% MeOH/PC (0.9 mL). The reaction mixture was heated at 85 °C for 10 min with stirring N$_2$ followed by the second round of heating at 120 °C for 20 min, while the reactor was sealed via valve 16 (Figure 5). The reaction vessel was cooled down to 40 °C.

3.2.4. Synthesis of [$^{18}$F]Anle138b from **4b** (Scheme 2)

Step 1. Radiofluorination of diaryliodonium salt **4b** (Method D).

A solution of [$^{18}$F]F$^-$ (9.0–11.0 GBq) in [$^{18}$O]H$_2$O was loaded onto a QMA cartridge from the male side followed by flushing with MeOH (2 mL) and drying with N$_2$ gas for 2 min. $^{18}$F$^-$ was eluted from the female side of the cartridge with a solution of **4b** (20 µmol) in 44% MeOH/PC (0.9 mL). The reaction mixture was heated at 85 °C for 10 min with stirring by nitrogen flow followed by the second round of heating at 120 °C for 20 min, while the reactor was sealed via valve 16 (Figure 5). The reaction vessel was cooled down to 40 °C.

Step 2. Synthesis of 6-[$^{18}$F]fluoro-3,4-methylendioxybenzylidene tosylhydrazone (6-[$^{18}$F]FTH).

The solution of NH$_2$NHTs (40 µmol) in MeOH (1 mL) was added to the reaction mixture obtained at step 1; the content was heated for 10 min at 90 °C under stirring by nitrogen flow. The reaction vessel was cooled down to 65 °C.

Step 3. Synthesis of [$^{18}$F]anle138b.

The solution of Bu$_4$NOH (26 µmol) in MeCN (0.4 mL) was added to the reaction mixture obtained at step 2 and the content was heated for 5 min at 65 °C. Then the solution of 3′-bromophenylacetylene (0.4 mmol) in MeCN (0.4 mL) was added and the reaction mixture was heated for 25 min at 90 °C. The reaction vessel was cooled down to 40 °C.

### 3.2.5. HPLC Purification

System A. The content of the reaction vessel (1.2 mL volume) was diluted with 0.8 mL 50% EtOH or with 0.8 mL of CH$_3$CN. The resulting solution was transferred to 2 mL HPLC loop. The fraction containing [$^{18}$F]anle138b (R$_t$ of 25–26 min under gradient 1 conditions or 28–32 min under gradient 2 conditions) was collected and analysed for radiochemical and chemical purity.

System B. The aliquot (100 µL) of the reaction mixture (from 1.2 mL total volume) was loaded onto 100 µL of HPLC loop. The fraction containing [$^{18}$F]anle138b (R$_t$ of 9.7–9.9 min, 1 mL volume) was collected through a 0.22 µm filter (Millipore, Burlington, MA, USA) attached to a vented sterile vial prefilled with the formulation buffer.

### 3.3. Semi-Automated Synthesis of [$^{18}$F]Anle138b (Figure 5)

The [$^{18}$F]fluoride processing and all the synthesis steps were completed in a custom-built synthesis module described in detail elsewhere [46]. The reactions were performed in a 5 mL V-vial (RV, Figure 5) with a screw cap (Wheaton-vials, Sigma-Aldrich GmbH, Steinheim, Germany). Nitrogen gas was applied for the reagents transfer.

1. Loading of [$^{18}$F]fluoride (9.0–11.0 GBq) onto a QMA anion exchange cartridge.
2. Washing of the cartridge with MeOH (2 mL) and drying with N$_2$ gas for 2 min.
3. Elution of [$^{18}$F]fluoride from the QMA cartridge with a solution of **4b** (20 µmol) in 44% MeOH/PC (0.9 mL) into reaction vessel (RV).
4. Heating reaction mixture in the RV at 85 °C for 10 min with stirring by N$_2$ with valve 16 open; then, valve 16 is closed and the reaction mixture is heated at 120 °C for 20 min.
5. Cooling down RV to 40 °C; valve 16 is open.
6. Addition of NH$_2$NHTs (40 µmol) in MeOH (1 mL).
7. Heating reaction mixture at 90 °C for 10 min with stirring by N$_2$ gas.
8. Cooling down RV to 65 °C.
9. Addition of Bu$_4$NOH (26 µmol) in MeCN (0.4 mL).
10. Heating reaction mixture at 65 °C for 5 min; valve 16 is closed.
11. Addition of the solution of 3′-bromophenylacetylene (0.4 mmol) in MeCN (0.4 mL) (valve 16 is open).
12. Valve 16 is closed, heating reaction mixture in RV at 90 °C for 25 min.
13. Cooling down RV to 40 °C; valve 16 is open.
14. Loading reaction mixture (100 µL) into HPLC loop (HPLC system B).
15. Isolation of the product using EtOH-H$_2$O gradient system with flow rate of 4 mL/min.
16. Manual collection of the product fraction into a vented collection vial.

## 4. Conclusions

In the current paper, we have outlined a novel, but relatively simple, one-pot three-step procedure for the radiosynthesis of [$^{18}$F]anle138b bypassing any intermediate purification steps. Using base- and phase-transfer catalyst-free radiofluorination procedure with diaryliodonium salt precursor (**4b**), 6-[$^{18}$F]fluoropiperonal—the starting building block for the entire [$^{18}$F]anle138b molecule—was obtained with RCC of >85%. Careful optimisation of the conditions for the following condensation and cycloaddition reactions performed without solvent exchange steps allowed to complete synthesis of [$^{18}$F]anle138b within 105 min with RCY of 15 ± 3% (n = 3) and with $A_m$ in the range of 32–80 GBq/µmol, significantly improving upon previously published results in both yield and synthesis time. While there is still space for further optimization, in particular in the area of HPLC purification, the procedure developed is well suited for [$^{18}$F]anle138b production for use in preclinical studies in animal or cell models. In addition, the suggested methodology may find further use in the preparation of other PET imaging agents derived from the pyrazoles backbone.

**Supplementary Materials:** The following supporting information can be downloaded at https://www.mdpi.com/article/10.3390/molecules28062732/s1, Block I: NMR and HRMS (ESI) spectra. Figure S1: $^1$H NMR spectrum for 6-fluorobenzo-1,3-dioxole-5-carbaldehyde, Figure S2: $^{13}$C NMR spectrum for 6-fluorobenzo-1,3-dioxole-5-carbaldehyde, Figure S3: $^{19}$F NMR spectrum of 6-fluorobenzo-1,3-dioxole-5-carbaldehyde, Figure S4: $^1$H NMR spectrum of $N'$-((6-fluorobenzo-1,3-dioxole-5-yl)methylene)-4-methylbenzenesulfonohydrazide, Figure S5: $^{13}$C NMR spectrum of $N'$-((6-fluorobenzo-1,3-dioxole-5-yl)methylene)-4-methylbenzenesulfonohydrazide, Figure S6: $^{19}$F NMR spectrum of $N'$-((6-fluorobenzo-1,3-dioxole-5-yl)methylene)-4-methylbenzenesulfonohydrazide, Figure S7: $^1$H NMR spectrum of [$^{19}$F]anle138b, Figure S8: $^{19}$F NMR spectrum of [$^{19}$F]anle138b, Figure S9: HRMS (ESI) spectrum of [$^{19}$F]anle138b, Figure S10: $^1$H NMR spectrum of compound **4b**, Figure S11: $^{13}$C NMR spectrum of compound **4b**, Figure S12: $^1$H NMR spectrum for compound **4a**, Figure S13: $^{13}$C NMR spectrum for compound **4a**, Figure S14: $^1$H NMR spectrum of (6-formylbenzo-1,3-dioxole-5-yl)(2,4,6-trimethylphenyl)iodonium bromide, Figure S15: $^{13}$C NMR spectrum of (6-formylbenzo-1,3-dioxole-5-yl)(2,4,6-trimethylphenyl)iodonium bromide, Figure S16: $^1$H NMR spectrum of (6-formylbenzo-1,3-dioxole-5-yl)(2,4,6-trimethylphenyl)iodonium tosylate, and Figure S17: $^{13}$C NMR spectrum of (6-formylbenzo-1,3-dioxole-5-yl)(2,4,6-trimethylphenyl)iodonium tosylate. Block II: RadioTLC chromatograms. Figure S18: RadioTLC analysis of 6-[$^{18}$F]FP obtained via radiofluorination of **4b**, Figure S19: radioTLC analysis of the formulated [$^{18}$F]anle138b.

**Author Contributions:** V.V.O., O.S.F. and D.D.V. performed radiosyntheses; V.V.O. and O.S.F. carried out synthesis automation; V.V.O. and R.N.K. conceived and designed experiments; O.S.F. performed HPLC and TLC procedures; N.B.V. carried out cold chemistry; V.V.O. and R.N.K. wrote the manuscript. All authors have read and agreed to the published version of the manuscript.

**Funding:** This work was supported by the Russian Foundation of Basic Research 20-53-12030/22; the research on the synthesis automation at the IHB RAS was supported by the state assignment of the Ministry of Education and Science of Russian Federation.

**Institutional Review Board Statement:** Not applicable.

**Informed Consent Statement:** Not applicable.

**Data Availability Statement:** The data presented in this study are available on request from the corresponding author.

**Acknowledgments:** The authors thank Boris D. Zlatopolskiy (Institute of Radiochemistry and Experimental Molecular Imaging, University Clinic Cologne, Cologne, Germany) for kindly providing pinacolboronate precursor **2a** and iodonium salt precursor **4b**.

**Conflicts of Interest:** The authors declare no conflict of interest.

**Sample Availability:** Samples of the compounds **3**, **4a**, **4b** and [$^{19}$F]anle138b are available from the authors.

# References

1. Uzuegbunam, B.C.; Librizzi, D.; Hooshyar Yousefi, B. PET radiopharmaceuticals for Alzheimer's disease and Parkinson's disease diagnosis, the current and future landscape. *Molecules* **2020**, *25*, 977. [CrossRef]
2. Garnett, E.S.; Firnau, G.; Nahmias, C. Dopamine visualized in the basal ganglia of living man. *Nature* **1983**, *305*, 137–138. [CrossRef]
3. Calabria, F.F.; Calabria, E.; Gangemi, V.; Cascini, G.L. Current status and future challenges of brain imaging with [$^{18}$F]-DOPA PET for movement disorders. *Hell. J. Nucl. Med.* **2016**, *19*, 33–41. [CrossRef]
4. Sioka, C.; Fotopoulos, A.; Kyritsis, A.P. Recent advances in PET imaging for evaluation of Parkinson's disease. *Eur. J. Nucl. Med. Mol. Imaging.* **2010**, *37*, 1594–1603. [CrossRef]
5. Varrone, A.; Halldin, C. New developments of dopaminergic imaging in Parkinson's disease. *Q.J. Nucl. Med. Mol. Imaging.* **2012**, *56*, 68–82. [PubMed]
6. Schou, M.C.; Steiger, C.; Varrone, A.; Guilloteau, D.; Halldin, C. Synthesis, radiolabeling and preliminary in vivo evaluation of [$^{18}$F]FE-PE2I, a new probe for the dopamine transporter. *Bioorg. Med. Chem.* **2009**, *19*, 4843–4845. [CrossRef]

7. Fazio, P.; Svenningsson, P.; Cselényi, Z.; Halldin, C.; Farde, L.; Varrone, A. Nigrostriatal dopamine transporter availability in early Parkinson's disease. *Mov. Disord.* **2018**, *33*, 592–599. [CrossRef]
8. Atik, A.; Stewart, T.; Zhang, J. Alpha-Synuclein as a Biomarker for Parkinson's Disease. *Brain Pathol.* **2016**, *26*, 410–418. [CrossRef]
9. Ingelsson, M. Alpha-Synuclein Oligomers—Neurotoxic Molecules in Parkinson's Disease and Other Lewy Body Disorders. *Front. Neurosci.* **2016**, *10*, 408. [CrossRef]
10. Benskey, M.J.; Perez, R.G.; Manfredsson, F.P. The contribution of alpha synuclein to neuronal survival and function—Implications for Parkinson's disease. *J. Neurochem.* **2016**, *137*, 331–359. [CrossRef]
11. Spillantini, M.G.; Schmidt, M.L.; Lee, V.M.-Y.; Trojanowski, J.Q.; Jakes, R.; Goedert, M. α-Synuclein in Lewy Bodies. *Nature* **1997**, *388*, 839–840. [CrossRef]
12. Spillantini, M.G.; Crowther, R.A.; Jakes, R.; Hasegawa, M.; Goedert, M. α-Synuclein in Filamentous Inclusions of Lewy Bodies from Parkinson's Disease and Dementia with Lewy Bodies. *Proc. Natl. Acad. Sci. USA* **1998**, *95*, 6469–6473. [CrossRef]
13. Korat, Š.; Bidesi, N.S.R.; Bonanno, F.; Di Nanni, A.; Hoàng, A.N.N.; Herfert, K.; Maurer, A.; Battisti, U.M.; Bowden, G.D.; Thonon, D.; et al. Alpha-Synuclein PET tracer development-an overview about current efforts. *Pharmaceuticals* **2021**, *14*, 847. [CrossRef]
14. Henriques, A.; Rouvière, L.; Giorla, E.; Farrugia, C.; El Waly, B.; Poindron, P.; Callizot, N. Alpha-Synuclein: The Spark That Flames Dopaminergic Neurons, In Vitro and In Vivo Evidence. *Int. J. Mol. Sci.* **2022**, *23*, 9864. [CrossRef]
15. Prange, S.; Theis, H.; Banwinkler, M.; van Eimeren, T. Molecular Imaging in Parkinsonian Disorders-What's New and Hot? *Brain Sci.* **2022**, *12*, 1146. [CrossRef]
16. Alzghool, O.M.; van Dongen, G.; van de Giessen, E.; Schoonmade, L.; Beaino, W. α-Synuclein Radiotracer Development and In Vivo Imaging: Recent Advancements and New Perspectives. *Mov. Disord.* **2022**, *37*, 936–948. [CrossRef]
17. Capotosti, F.; Vokali, E.; Molette, J.; Ravache, M.; Delgado, C.; Kocher, J.; Pittet, L.; Vallet, C.; Serra, A.; Piorkowska, K.; et al. Discovery of [$^{18}$F]ACI-12589, a Novel and Promising PET-Tracer for Alpha-Synuclein. *Alzheimer's Dement.* **2022**, *18*, e064680. [CrossRef]
18. Wagner, J.; Ryazanov, S.; Leonov, A.; Levin, J.; Shi, S.; Schmidt, F.; Prix, C.; Pan-Montojo, F.; Bertsch, U.; Mitteregger-Kretzschmar, G.; et al. Anle138b: A novel oligomer modulator for disease-modifying therapy of neurodegenerative diseases such as prion and Parkinson's disease. *Acta Neuropathol.* **2013**, *125*, 795–813. [CrossRef]
19. Levin, J.; Schmidt, F.; Boehm, C.; Prix, C.; Bötzel, K.; Ryazanov, S.; Leonov, A.; Griesinger, C.; Giese, A. The Oligomer Modulator anle138b Inhibits Disease Progression in a Parkinson Mouse Model Even with Treatment Started after Disease Onset. *Acta Neuropathol.* **2014**, *127*, 779–780. [CrossRef]
20. Deeg, A.A.; Reiner, A.M.; Schmidt, F.; Schueder, F.; Ryazanov, S.; Ruf, V.C.; Giller, K.; Becker, S.; Leonov, A.; Griesinger, C.; et al. Anle138b and related compounds are aggregation specific fluorescence markers and reveal high affinity binding to α-synuclein aggregates. *Biochim. Biophys. Acta* **2015**, *1850*, 1884–1890. [CrossRef]
21. Heras-Garvin, A.; Weckbecker, D.; Ryazanov, S.; Leonov, A.; Griesinger, C.; Giese, A.; Wenning, G.K.; Stefanova, N. Anle138b modulates α-synuclein oligomerization and prevents motor decline and neurodegeneration in a mouse model of multiple system atrophy. *Mov. Disord.* **2019**, *34*, 255–263. [CrossRef]
22. Levin, J.; Sing, N.; Melbourne, S.; Morgan, A.; Mariner, C.; Spillantini, M.G.; Wegrzynowicz, M.; Dalley, J.W.; Langer, S.; Ryazanov, S.; et al. Safety, tolerability and pharmacokinetics of the oligomer modulator anle138b with exposure levels sufficient for therapeutic efficacy in a murine Parkinson model: A randomised, double-blind, placebo-controlled phase 1a trial. *EBioMedicine* **2022**, *80*, 104021. [CrossRef]
23. Maurer, A.; Leonov, A.; Ryazanov, S.; Herfert, K.; Kuebler, L.; Buss, S.; Schmidt, F.; Weckbecker, D.; Linder, R.; Bender, D.; et al. $^{11}$C Radiolabeling of anle253b: A Putative PET Tracer for Parkinson's Disease That Binds to α-Synuclein Fibrils in vitro and Crosses the Blood-Brain Barrier. *ChemMedChem.* **2020**, *15*, 411–415. [CrossRef]
24. Kuebler, L.; Buss, S.; Leonov, A.; Ryazanov, S.; Schmidt, F.; Maurer, A.; Weckbecker, D.; Landau, A.M.; Lillethorup, T.P.; Bleher, D.; et al. $^{11}$CMODAG-001-towards a PET tracer targeting α-synuclein aggregates. *Eur. J. Nucl. Med. Mol. Imaging.* **2021**, *48*, 1759–1772. [CrossRef]
25. Raval, N.R.; Madsen, C.A.; Shalgunov, V.; Nasser, A.; Battisti, U.M.; Beaman, E.E.; Juhl, M.; Jørgensen, L.M.; Herth, M.M.; Hansen, H.D.; et al. Evaluation of the α-synuclein PET radiotracer (d$_3$)-[$^{11}$C]MODAG-001 in pigs. *Nucl. Med. Biol.* **2022**, *114–115*, 42–48. [CrossRef]
26. Coenen, H.H.; Ermert, J. $^{18}$F-labelling innovations and their potential for clinical application. *Clin. Transl. Imaging.* **2018**, *6*, 169–193. [CrossRef]
27. Krasikova, R.N.; Orlovskaya, V.V. Phase Transfer Catalysts and Role of Reaction Environment in Nucleophilc Radiofluorinations in Automated Synthesizers. *Appl. Sci.* **2022**, *12*, 321. [CrossRef]
28. Zarrad, F.; Zlatopolskiy, B.D.; Urusova, E.A.; Neumaier, B. First radiosynthesis of F-18-labeled anle138b a potential tracer for imaging of neurodegenerative diseases associated with protein deposition in brain. *J. Label. Compd. Radiopharm.* **2015**, *58*, S241. [CrossRef]
29. Zarrad, F. Efficient Preparation of PET Tracers for Visualization of Age-Related Disorders Using Emerging Methods of Radiofluorination. Ph.D. Dissertation, University of Koln, Koln, Germany, 2017. Available online: https://kups.ub.uni-koeln.de/7640/ (accessed on 12 January 2033).

30. Preshlock, S.; Tredwell, M.; Gouverneur, V. $^{18}$F-Labeling of Arenes and Heteroarenes for Applications in Positron Emission Tomography. *Chem. Rev.* **2016**, *116*, 719–766. [CrossRef]
31. Zarganes-Tzitzikas, T.; Clemente, G.S.; Elsinga, P.H.; Dömling, A. MCR Scaffolds Get Hotter with $^{18}$F-Labeling. *Molecules.* **2019**, *24*, 1327. [CrossRef]
32. Pike, V.W. Hypervalent Aryliodine Compounds as Precursors for Radiofluorination. *J. Label. Compds. Radiopharm.* **2018**, *61*, 196–227. [CrossRef]
33. Wright, J.S.; Kaur, T.; Preshlock, S.; Tanzeyl, S.S.; Winton, W.P.; Sharninghausen, L.S.; Wiesner, N.; Brooks, A.F.; Sanford, M.S.; Scott, P.J.H. Copper-mediated late-stage radiofuorination: Five years of impact on preclinical and clinical PET imaging. *Clin. Transl. Imaging.* **2020**, *8*, 167–206. [CrossRef]
34. Tredwell, M.; Preshlock, S.M.; Taylor, N.J.; Gruber, S.; Huiban, M.; Passchier, J.; Mercier, J.; Genicot, C.; Gouverneur, V. A general copper-mediated nucleophilic $^{18}$F-luorination of arenes. *Angew. Chem. Int. Ed.* **2014**, *53*, 7751–7755. [CrossRef] [PubMed]
35. Mossine, A.V.; Brooks, A.F.; Makaravage, K.J.; Miller, J.M.; Ichiishi, N.; Sanford, M.S.; Scott, P.J. Synthesis of [$^{18}$F] Arenes via the Copper-Mediated [$^{18}$F] Fluorination of Boronic Acids. *Org. Lett.* **2015**, *17*, 5780–5783. [CrossRef]
36. Makaravage, K.J.; Brooks, A.F.; Mossine, A.V.; Sanford, M.S.; Scott, P.J.H. Copper-Mediated Radiofluorination of Arylstannanes with [$^{18}$F] KF. *Org. Lett.* **2016**, *18*, 5440–5443. [CrossRef]
37. Zischler, J.; Kolks, N.; Modemann, D.; Neumaier, B.; Zlatopolskiy, B.D. Alcohol-Enhanced Cu-Mediated Radiofluorination. *Chem. A Eur. J.* **2017**, *23*, 3251–3256. [CrossRef] [PubMed]
38. Antuganov, D.; Zykov, M.; Timofeev, V.; Timofeeva, K.; Antuganova, Y.; Fedorova, O.; Orlovskaya, V.; Krasikova, R. Copper-mediated radiofluorination of aryl pinacolboronate esters: A straightforward protocol using pyridinium sulfonates. *Eur. J. Org. Chem.* **2019**, *2019*, 918–922. [CrossRef]
39. Mossine, A.V.; Brooks, A.F.; Ichiishi, N.; Makaravage, K.J.; Sanford, M.S.; Scott, P.J. Development of Customized [$^{18}$F] Fluoride Elution Techniques for the Enhancement of Copper-Mediated Late-Stage Radiofluorination. *Sci. Rep.* **2017**, *7*, 233. [CrossRef]
40. Zlatopolskiy, B.D.; Zischler, J.; Schäfer, D.; Urusova, E.A.; Guliyev, M.; Bannykh, O.; Endepols, H.; Neumaier, B. Discovery of 7-[$^{18}$F]Fluorotryptophan as a Novel Positron Emission Tomography (PET) Probe for the Visualization of Tryptophan Metabolism in Vivo. *Med. Chem.* **2018**, *61*, 189–206. [CrossRef] [PubMed]
41. Orlovskaya, V.; Fedorova, O.; Kuznetsova, O.; Krasikova, R. Cu-Mediated Radiofluorination of Aryl Pinacolboronate Esters: Alcohols as Solvents with Application to 6-L-[$^{18}$F]FDOPA Synthesis. *Eur. J. Org. Chem.* **2020**, *2020*, 7079–7086. [CrossRef]
42. Orlovskaya, V.V.; Craig, A.S.; Fedorova, O.S.; Kuznetsova, O.F.; Neumaier, B.; Krasikova, R.N.; Zlatopolskiy, B.D. Production of 6-L-[$^{18}$F] Fluoro-m-tyrosine in an Automated Synthesis Module for $^{11}$C-Labeling. *Molecules.* **2021**, *26*, 5550. [CrossRef] [PubMed]
43. Pike, V.W.; Aigbirhio, F.I. Reactions of cyclotron-produced [$^{18}$F]fluoride with diaryliodonium salts—A novel single-step route to no-carrier-added [$^{18}$F]fluoroarenses. *J. Chem. Soc., Chem. Commun.* **1995**, *21*, 2215–2216. [CrossRef]
44. Richarz, R.; Krapf, P.; Zarrad, F.; Urusova, E.A.; Neumaier, B.; Zlatopolskiy, B.D. Neither azeotropic drying, nor base nor other additives: A minimalist approach to $^{18}$F-labeling. *Org. Biomol. Chem.* **2014**, *12*, 8094–8099. [CrossRef] [PubMed]
45. Ichiishi, N.; Brooks, A.F.; Topczewski, J.J.; Rodnick, M.E.; Sanford, M.S.; Scott, P.J.H. Copper-Catalyzed [$^{18}$F]Fluorination of (Mesityl)(aryl)iodonium Salts. *Org. Lett.* **2014**, *16*, 3224–3227. [CrossRef] [PubMed]
46. Orlovskaya, V.V.; Modemann, D.J.; Kuznetsova, O.F.; Fedorova, O.S.; Urusova, E.A.; Kolks, N.; Neumaier, B.; Krasikova, R.N.; Zlatopolskiy, B.D. Alcohol-supported Cu-mediated $^{18}$F-fluorination of iodonium salts under "minimalist" conditions. *Molecules.* **2019**, *24*, 3197. [CrossRef] [PubMed]
47. Chun, J.H.; Lu, S.; Lee, Y.S.; Pike, V.W. Fast and high-yield microreactor syntheses of ortho-substituted [(18)F]fluoroarenes from reactions of [(18)F]fluoride ion with diaryliodonium salts. *J. Org. Chem.* **2010**, *75*, 10–3332. [CrossRef] [PubMed]
48. Aggarwal, V.K.; de Vicente, J.; Bonnert, R.V. A Novel One-Pot Method for the Preparation of Pyrazoles by 1,3-Dipolar Cycloadditions of Diazo Compounds Generated in Situ. *J. Org. Chem.* **2003**, *68*, 5381–5383. [CrossRef]
49. Kirk, L.K.; Cantacuzene, D.; Nimitkitpaisan, Y.; McCulloh, D.; Padgett, W.L.; Daly, J.W.; Creveling, C.R. Synthesis and biological properties of 2-, 5-, and 6-fluoronorepinephrines. *J. Med. Chem.* **1979**, *22*, 1493–1495. [CrossRef]
50. Moreau, A.; Couture, A.; Deniau, E.; Grandclaudon, P.; Lebrun, S. A new approach to isoindoloisoquinolinones. A simple synthesis of nuevamine. *Tetrahedron* **2004**, *60*, 6169–6176. [CrossRef]

**Disclaimer/Publisher's Note:** The statements, opinions and data contained in all publications are solely those of the individual author(s) and contributor(s) and not of MDPI and/or the editor(s). MDPI and/or the editor(s) disclaim responsibility for any injury to people or property resulting from any ideas, methods, instructions or products referred to in the content.

*Article*

# Dissolution of Molybdenum in Hydrogen Peroxide: A Thermodynamic, Kinetic and Microscopic Study of a Green Process for $^{99m}$Tc Production

Flavio Cicconi [1], Alberto Ubaldini [2,*], Angela Fiore [3], Antonietta Rizzo [2], Sebastiano Cataldo [1], Pietro Agostini [1], Antonino Pietropaolo [4], Stefano Salvi [1], Vincenzo Cuzzola [1] and on behalf of the SRF Collaboration [†]

1. ENEA, C.R. Brasimone, 40032 Camugnano, Italy
2. ENEA, Via Martiri di Monte Sole 4, 40129 Bologna, Italy
3. ENEA, S.S.7 "Appia" Km 706, 72100 Brindisi, Italy
4. ENEA, Via E. Fermi 45, 00044 Frascati, Italy
* Correspondence: alberto.ubaldini@enea.it
† Collaborators of the SRF Collaboration are provided in Acknowledgments.

**Abstract:** $^{99m}$Tc-based radiopharmaceuticals are the most commonly used medical radioactive tracers in nuclear medicine for diagnostic imaging. Due to the expected global shortage of $^{99}$Mo, the parent radionuclide from which $^{99m}$Tc is produced, new production methods should be developed. The SORGENTINA-RF (SRF) project aims at developing a prototypical medium-intensity D-T 14-MeV fusion neutron source specifically designed for production of medical radioisotopes with a focus on $^{99}$Mo. The scope of this work was to develop an efficient, cost-effective and green procedure for dissolution of solid molybdenum in hydrogen peroxide solutions compatible for $^{99m}$Tc production via the SRF neutron source. The dissolution process was extensively studied for two different target geometries: pellets and powder. The first showed better characteristics and properties for the dissolution procedure, and up to 100 g of pellets were successfully dissolved in 250–280 min. The dissolution mechanism on the pellets was investigated by means of scanning electron microscopy and energy-dispersive X-ray spectroscopy. After the procedure, sodium molybdate crystals were characterized via X-ray diffraction, Raman and infrared spectroscopy and the high purity of the compound was established by means of inductively coupled plasma mass spectroscopy. The study confirmed the feasibility of the procedure for production of $^{99m}$Tc in SRF as it is very cost-effective, with minimal consumption of peroxide and controlled low temperature.

**Keywords:** molybdenum; dissolution thermodynamics and kinetics; hydrogen peroxide; neutron source; fusion

## 1. Introduction

Technetium-99m ($^{99m}$Tc) is a metastable nuclear isomer of technetium-99 that is used in tens of millions of medical diagnostic procedures annually [1], making it the most commonly used medical radioisotope in the world [2]. Radiopharmaceuticals based on $^{99m}$Tc are used mainly in single-photon emission computed tomography (SPECT), and, for this reason, this isotope is of great importance in nuclear medicine [3]. None of the Tc isotopes are stable, the one with the longest half-life being ($t_{1/2}$) $^{98}$Tc, equal to 4.2 million years [4]. This means that this element can only be found in traces in nature, and, hence, all isotopes must be artificially produced by nuclear reactions, in particular $^{99m}$Tc ($t_{1/2}$ = 6.0 h), which is normally derived from its transient equilibrium parent, $^{99}$Mo ($t_{1/2}$ = 66 h) [1,2].

$^{99}$Mo decays by emitting a beta particle (an electron). About 88% of the decays produce $^{99m}$Tc, which subsequently decays to the ground state, $^{99}$Tc, by emitting a gamma ray. About

12% of the nuclear decays produce $^{99}$Tc directly. In turn, it decays to stable ruthenium-99 ($^{99}$Ru) after emitting a beta particle with a half-life of 211.1 thousand years [3,5].

A technetium-99m generator based on molybdenum-99 is commercially available. The generator is easy to transport and use, which are some of the reasons why it is so widely used in hospitals all over the world [6].

$^{99}$Mo can be produced following different methods, for example, using accelerated charged-particle beams ($\alpha$-particle capture via $^{96}$Zr ($\alpha$, n)$^{99}$Mo reaction or fast proton interaction with $^{100}$Mo via $^{100}$Mo (p,2n) $^{99m}$Tc reaction) or according nuclear reactions in which fast neutrons are involved: neutron photo-production in $^{100}$Mo via $^{100}$Mo ($\gamma$, n)$^{99}$Mo reaction or fast neutron interaction in $^{100}$Mo via $^{100}$Mo(n,2n)$^{99}$Mo inelastic reaction [1,7–11].

However, despite these methods, at present, $^{99}$Mo is almost exclusively obtained from fission of $^{235}$U-containing targets, irradiated in a small number of research nuclear fission reactors in the world [2,12,13].

This fact can lead to a series of non-negligible issues. A global shortage of $^{99}$Mo is a risk, and it happened during the late 2000s because of frequent shutdown due to extended maintenance periods of the main reactors for $^{99}$Mo production, namely the Chalk River National Research Universal (NRU) nuclear fission reactor in Canada and the High Flux Reactor (HFR) in the Netherlands. They are capable of meeting about two-thirds of $^{99}$Mo world demand [14,15]. These events highlighted vulnerabilities in the supply chain of medical radionuclides that relies on nuclear fission reactors.

Indeed, as a fission product, $^{99}$Mo is produced together with many other isotopes of various elements, from which it must be purified [13]. This requires development and implementation of specific and complex radiochemical processes to separate the isotope of interest from all the rest, which, therefore, constitutes a waste product. Therefore, there is a general problem of waste management, and the threat of nuclear proliferation must always be considered [12,15]. It should be kept in mind that $^{99}$Mo accounts for only about 6% of uranium fission products [13,16]. Large volumes of hazardous chemicals, including strong acids, are required for this purpose [13], and, for this reason, $^{99}$Mo production cannot be considered environmentally friendly.

In this context, the SORGENTINA-RF (SRF) project aims at developing a prototypical medium-intensity D-T 14-MeV fusion neutron source mostly dedicated to production of medical radioisotopes, with a special focus on $^{99}$Mo. Indeed, the fusion neutron route is very interesting for a series of reasons, but the lack of a very intense 14 MeV neutron source is a limitation factor for $^{99}$Mo production. SRF will be a prototype plant to assess this production route [1,17].

Theoretically, an alternative route can be followed in order to produce $^{99}$Mo, relying on use of 14 MeV neutrons from a deuterium–tritium fusion reaction:

$$D + T \rightarrow {}^4He + n + 17.6 \text{ MeV}$$

and on inelastic channel $^{100}$Mo(n,2n)$^{99}$Mo [1,8,9,18].

The idea is to exploit the neutrons generated by the fusion process to irradiate a metal target made up of metallic natural molybdenum, where $^{100}$Mo has 10% abundance. The accelerator will operate with deuterons and tritons that will be implanted onto a titanium layer a few microns thick where they interact, in turn producing a neutron field, the main component being that from the D-T reaction mentioned above [19].

First, calculations and projections from a dedicated study indicate that the end of irradiation (EoI) activity of $^{99}$Mo is in the range 2–5 Ci after 24 h continuous irradiation starting from an initial mass of about 10 kg. This yield is more than enough for the daily needs of $^{99m}$Tc of the entire Emilia Romagna, an Italian region, and could be improved by using samples enriched in $^{100}$Mo and higher potencies. The SRF project is, therefore, extremely promising due to the numerous advantages it can offer compared to more traditional methods [1].

In contrast to the traditional production methods, with the SRF method, there are no radiochemical purification issues. The main challenge becomes finding an effective and ecologically acceptable method to transform metallic molybdenum into a stock solution of sodium molybdate, $Na_2MoO_4$, which is used for feeding the Mo/Tc generators [13].

In the case of the SRF prototype, once irradiated, the molybdenum target is transferred into shielded hot cells for dissolution and radiochemical processing.

The stock solution could be prepared by dissolving the metal target using concentrated strong acids or aqua regia, and, in the past, this has usually been completed [13]. However, a greener and more ecologically acceptable approach is to use less aggressive reagents and in the smallest amount possible to achieve the same result. Micrometric Mo powders react vigorously with hydrogen peroxide, even if diluted. In this case, the only by-products are water vapor and oxygen, which is a clear advantage in terms of the sustainability of the process. If coarse pieces of a few centimeters are used, the overall process has slower chemical kinetics (it should also be kept in mind that the molybdenum dissolution process, in the context of SRF, must be completed in a maximum time equal to that necessary for irradiation of the target, i.e., 24 h [20]). Nevertheless, it has been reported that a large amount of highly concentrated hydrogen peroxide can be effective also in the case of samples with the shape of disks [21,22]. In particular, the authors of [22] used hydrogen peroxide for direct production of $^{99m}Tc$ from $^{100}Mo$ by cyclotrons.

The aim of this work is, therefore, to find the most efficient conditions possible to optimize the dissolution process using hydrogen peroxide in a suitable time. Furthermore, a description of the chemical path followed to arrive at formation of the stock solution is also presented.

The chemical behavior of metallic molybdenum towards hydrogen peroxide is studied, from a thermodynamic and kinetic point of view, by means of scanning electron microscopy, infrared and Raman spectroscopy, X-ray diffractometry, pH, temperature and conductivity measurements. Use of metal not exposed to the neutron beam is acceptable in this context because it can be assumed that the physicochemical properties are not substantially modified by irradiation. Two different target geometries have been investigated: pellets and powder. Therefore, once the most suitable conditions for dissolution of the non-irradiated target have been determined, they can also be applied to the target irradiated by SRF.

## 2. Experimental Sections

### 2.1. Materials and Methods

The molybdenum pellets were supplied by Luoyang Combat Tungsten & Molybdenum Material Co., Ltd. (Luoyang, China). They are shaped as small cylinders, approximately 1.7 mm long and 1.5 mm in diameter. Figure 1 shows the optical micrograph images of them in order to show the aspect and morphology. All pieces are very similar and just small variations in volume can be observed. Their surface is usually dark, probably because of the presence of a very thin layer of molybdenum oxide on the surface. Purity of the samples has been analyzed by means of ICP-mass spectrometry (see Section 3.3).

**Figure 1.** Optical micrograph of the metallic molybdenum pieces at different magnifications.

The dissolution process has been studied, performing many experiments with hydrogen peroxide solutions whose concentration ranged from to 3% to 35% $w/w$ and using a mass of metallic molybdenum between 0.5 and 100 g. Different hydrogen peroxide solutions ranging from 30 to 40% $w/w$ were obtained from Carlo Erba Reagents (Milan, Italy), Titolchimica (Pontecchio Polesine, Italy) and Sigma-Aldrich (St. Louis, MO, USA), and they were diluted, if necessary, to the desired concentration for the experiments.

An external ice bath was used in order to control the temperature because the large exothermicity of the reaction can lead to a very fast increase of the system temperature.

Highly concentrated hydrogen peroxide is unstable and decomposes over time. For this reason, fresh batches were used and the solutions were kept in freezer at −18 °C.

The dissolution experiments were performed under fume hood. For each experiment, the procedure has been the following: the hydrogen peroxide solution is prepared with the desired concentration from a stock solution. The solution is then transferred in the flask with an ice-cooled water bath, the molybdenum pellets are weighed with analytical scale and then are added slowly to the hydrogen peroxide solution under magnetic stirring. A circuit with a peristaltic pump connected to an inlet of cold water ensures circulation of the water bath. The low temperature of the solution delays the initial states of the reaction and prevents formation of foam. Furthermore, if the solution is poured directly over the molybdenum pellets, the reaction starts too vigorously and a considerable fraction of the peroxides are consumed via self-decomposition and subsequent formation of a huge amount of foam can overflow from the flask, leading to failure of the experiment. The procedure described above is, therefore, much cleaner and milder and avoids drastic and uncontrollable changes in the temperature reaction.

### 2.2. Characterization Methods

In the present work, Raman spectra of the compounds were acquired by a BWTEK i-Raman plus spectrometer equipped with a 785 nm laser to stimulate Raman scattering, which is measured in the range 100–3500 $cm^{-1}$ with a spectral resolution of 3.5 $cm^{-1}$. The measurement parameters, acquisition time, number of repetitions and laser energy have been selected for each sample in order to maximize the signal to noise ratio. For each spectrum, reference acquisition with the same parameters was previously carried out to subtract the instrumental background. Infrared spectra were acquired with a Thermo Fisher Scientific Nicolet 6700 FT-IR with SMART iTX ATR System (Thermo Fischer Scientific, Waltham, MA, USA) in the mid-, far- and near-IR ranges from 4000 to 400 $cm^{-1}$).

On selected materials, X-ray powder diffraction (XRPD) investigations were performed in order to determine the crystalline phases present using a Philips X'Pert PRO 3040/60 diffractometer operating at 40 kV, 40 mA, Bragg–Brentano geometry, equipped with a Cu Kα source (1.54178 Å), Ni filtered and a curved graphite monochromator. PANalytical High Score software was utilized for data elaboration.

Characterization of the samples, morphology and composition has been performed by scanning electron microscopy (SEM-FEI Inspect-S) coupled with energy-dispersive X-ray spectroscopy (EDX-EDAX Genesis).

The pH and EC of the solutions during the dissolution process were measured using a Crison Basic 20 pH-meter and a Crison Basic 30 EC-Meter (Crison, HachLange, Spain).

A triple quadrupole inductively coupled plasma mass spectrometer (QQQ-ICP-MS, 8800 model, Agilent Technologies, Santa Clara, CA, USA) equipped with two quadrupoles, one (Q1) before and one (Q2) after the octopole reaction system (ORS3), installed in a dedicated Clean Room ISO Class 6, (ISO 14644-1 Clean room) with controlled pressure, temperature and humidity was used for trace analysis of residual metals in the molybdenum.

The instrument was calibrated with reference samples from a multi-element stock standard solution containing 13 elements (100 mg/L, each in 7% $HNO_3$ $v/v$, P/N: CCS-6) supplied by Inorganic Ventures (Christiansburg, VA, USA) as an external calibration. The multi-element standard includes 13 elements that are mainly transition metals that are used in the metallurgic industries, namely V, Cr, Mn, Fe, Co, Ni, Cu, Zn, Ag, Cd, Hg, Tl, Pb. Hg, Tl, Pb, which were also added in the analysis for their toxicological concerns even though they are not commonly present in metallic matrixes.

All data collection and analysis were performed in ICP-MS Tandem MS/MS Helium Mode using Mass Hunter software. Quality control standards (QCs) were added in the batch of analysis in order to control the accuracy of the analytical method. The analyte recoveries were within 90% of the true values.

LODs and LOQs were estimated, respectively, as three and ten times the standard deviation ($\sigma$) of 10 consecutive measurements of the reagent blanks according to EURACHEM recommendation [23].

The collision cell in Helium Mode MS (MS/MS) configuration ensures removal of any spectroscopic interferences are caused by atomic or molecular ions that have the same mass-to-charge as analytes of interest. Regarding those isobaric interferences for the case of molybdenum, some of its oxide's ions can lead to overestimation of Cadmium isotopes; in fact, $^{95}Mo^{16}O^+$ is isobaric to $^{111}Cd$ and $^{98}Mo^{16}O^+$ is isobaric to $^{114}Cd$ [24].

All the reagents used in the experiment were analytical reagent grade. Trace SELECT® grade 69% $HNO_3$ and 37% HCl and ultra-pure grade 30–32% $H_2O_2$ were acquired from Sigma-Aldrich (St. Louis, MO, USA) and Carlo Erba Reagents (Milan, Italy), respectively.

High-purity de-ionized water (resistivity 18.2 MOhm cm$^{-1}$) was obtained from a Milli-Q Advantage A10 water purification system (Millipore, Bedford, MA, USA).

## 3. Discussion

### 3.1. Studies on the Dissolutions Process

Molybdenum has many industrial applications because of its excellent physical properties: for instance, high resistance versus corrosion [25], much slower than steel [26], low linear expansion coefficient, relatively high thermal and electrical conductivity and excellent mechanical characteristics, even at very high temperature, such as high tensile strength and stiffness [27]. Therefore, many of its chemical physical features have been deeply studied.

It is a transition metal with moderate reactivity, which is strongly determined by specific surface, presence of impurities and traces of oxide and, in general, by presence of defects on the surface as they can act as catalytic starting points for any reaction or transformation [28].

This metal belongs to the chromium group and has a rich chemistry because it has oxidation states ranging from $-II$ to $+VI$ and coordination numbers from 0 to 8 [29]. It can form several oxides [30], among which the most important and commonly observed are $MoO_3$ (that has, in turn, some different polymorphs [31]) and $MoO_2$, but at least seven other oxides with molybdenum oxidation state comprised between 4 and 6 exist, for instance, $Mo_8O_{23}$ [32] and $Mo_{17}O_{47}$ [33]. The oxide of trivalent Mo ion $Mo_2O_3$ also exists; indeed, some hydrated compounds have been reported, such as $MoO_3$–$2H_2O$ or $MoO_3$–$H_2O$. In aqueous solutions, higher oxidation states are more relevant, and, in general, compounds of Mo (VI) are more soluble, so the goal of the dissolution process is to form a solution containing molybdate ions ($MoO_4^{2-}$) [29,34].

This metal also has several commonalities with the chemistry of tungsten (W), even if it reacts more easily with strong inorganic acids and oxidants, such as hydrogen peroxide.

In the case of reaction with hydrogen peroxide, schematically, the process can be described according to the following formal steps:

1. Oxidation of metallic molybdenum from the surface toward the center to form a mixture of low- and intermediate-valence molybdenum oxide.
2. Strong oxidation of these ions to the highest valence state by reaction with concentrated hydrogen peroxide to form some soluble oxyhydroxides and peroxides of molybdenum, such $MoO_2(O_2)_2^{2-}$ or $MoO_2HO(O_2)_2^{2-}$ [34].
3. Reaction of these species with NaOH to form $Na_2MoO_4$.

The stoichiometry of the first two steps cannot be perfectly defined, so an overall formal reaction can be written as such:

$$\text{Mo(s)} + 3\,\text{H}_2\text{O}_2\,(l) + 2\,\text{NaOH}\,(l) \rightarrow \text{Na}_2\text{MoO}_4\,(l) + 4\,\text{H}_2\text{O} \tag{1}$$

Figure 2 shows the stages of dissolution: on the left, the initial stage with ice bath and the peristaltic pump for a temperature-controlled reaction; in the center, the orange solution at the end of the dissolution with the intense orange color due to molybdenum-peroxo complexes species; on the right, the transparent solution of sodium molybdate after addition of sodium hydroxide.

**Figure 2.** Images of the dissolution process of the molybdenum pieces.

It has been previously studied by Tkac et al. [35] that the dissolution rates and foam produced by the reaction are highly dependent on the specific brand of peroxide solutions using sintered Mo disk. This is presumably caused by the different peroxide stabilizers used in the final product. The most common stabilizers are metal-based (such as Sn or Al) or phosphate-based; they can both affect the dissolution reaction, and the metals can produce a catalytic effect, increasing the dissolution rate, for example, or the stabilizer simply reduces formation of free peroxide radicals, leading to a different reaction rate. They also reported that dissolution rate is dependent on the specific surface of the sintered disk, as expected. In the preliminary phases of this study, experiments have been conducted regarding how the different peroxides brands can affect reaction rate with a specific ice-cooled bath and peristaltic pump set-up. The reaction rate experiments were not affected by the different peroxide type: Titolchimica, Carlo Erba or Sigma-Aldrich. This is probably because of the controlled temperature that disadvantages collateral reactions, such as parasitic self-decomposition of peroxide itself. For these reasons, for subsequent experiments, a single brand has been chosen in order to enhance repeatability and only the Sigma-Aldrich "For Trace Analysis" grade has been used.

Electrochemical methods have also been investigated. Cieszykowska et al. [36] studied electrochemical dissolution of sintered molybdenum disks in the scale of about 10 g using additional peroxide solution and potassium hydroxide as electrolyte with current density of 365 mA/cm$^2$ and temperature of 55 °C.

In general, the dissolution processes of solids in liquids are faster if the reaction surface is greater. For this reason, molybdenum powders would be preferable and they react with the peroxide much faster than macroscopic pieces due to the fact that, for the same mass, the surface is bigger. Therefore, very large pieces are not suitable for achieving this purpose, which is to have rapid formation of the stock solution, because the process would take too long or require much greater quantities of peroxide.

Use of molybdenum metal powders makes the reaction very fast, to the point that, as has been observed during the present research, a great deal of foam can form during the process even when the reaction is ice-cooled. Associated with this phenomenon, it has also been observed that even small quantities of powders, when they are quickly brought into contact with the liquid, lead to a large increase in system temperature, up to about 380 K. Furthermore, the foam could easily fill the chemical reactor and eventually overflow, with obvious safety problems, and can remove the smallest and lightest metallic particles from the reaction environment, making it difficult to achieve complete dissolution. The strong increase in temperature leads to decomposition of hydrogen peroxide before it has completely reacted, and, consequently, a large excess of solution will be necessary.

However, powders are not suitable within the SRF project for one fundamental reason: once irradiated, the molybdenum target becomes strongly radioactive and cannot be manipulated or transported by a human operator. This means that all the phases following irradiation must be automated. The target will be automatically transferred by means of a pneumatic system from the bunker where the neutron irradiation is carried out to the hot cells for radiochemical processing. For this reason, small particles or powders cannot be used because they can clog the filters, remain attached to the walls of the pipes and be easily and dangerously dispersed into the facility. The ideal geometry has, therefore, been identified as pellets or beads of millimeter size.

Several tests were carried out using different concentrations of hydrogen peroxide ranging from 3% to 30% $w/w$. Taking constant all the other parameters, the time required for the dissolution process decreases almost linearly with peroxide concentration, suggesting that the control steps are phenomena occurring at metallic pellets surface. It is interesting to note that, during this rapid reaction, the color of the solution can change long before reaching the final orange–yellow one because different oxyanions and molybdenum complexes are formed. Even though, from a formal point of view, the reaction is quite simple, in reality, the chemical path leading to formation of the desired products is rather complex [34,37].

It has been observed experimentally in this study that temperature does not increase immediately when the reactants come into contact; there is an initial induction time before which the temperature remains practically constant, and, immediately after, it increases dramatically. This induction time is generally of few seconds and depends on peroxide concentration (as well as temperature and other experimental factors), suggesting that the reaction acts as if it were self-catalytic, as can be seen in Figure 3. It shows the trend of the time required for appearance of the first bubbles and its inverse as a function of concentration of hydrogen peroxide solution, when the metallic pieces are thrown into the liquid. In this case, just a few milligrams of Mo and a few milliliters of peroxide were used.

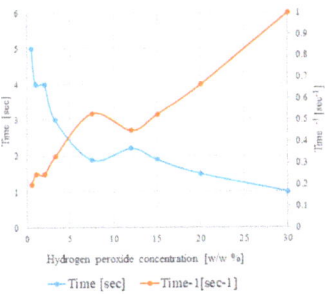

**Figure 3.** Dependence of time of appearance of the first bubbles at surface of metallic pieces and its inverse as a function of concentration.

The trend of the inverse time clearly suggests that the concentration of the peroxide is the fundamental factor and the kinetics of the process are likely controlled by phenomena at solid/liquid interface.

A possible explanation for this phenomenon can be found in the fact that the molybdenum pieces are covered with a very thin layer of $MoO_2$ that acts as a passivation layer, and that is why they appear dark and somehow dusty on the surface. Indeed, because of this layer, generally, in aqueous conditions, molybdenum exhibits considerably lower corrosion rates than many metals, such as iron [34,38]. This oxide forms easily because of favorable redox potentials in these two half-reactions (0.152 V and 1.23 V, respectively) [34,39]:

$$Mo + 2 H_2O \rightarrow MoO_2 + 4 H^+ + 4 e^-$$

$$O_2 + 4 H^+ + 4 e^- \rightarrow 2 H_2O$$

The reaction path must necessarily involve this layer as well. Tyurin [40] suggested that, in alkaline, neutral and weakly acidic solutions, this passive layer is stable.

However, this oxide can be oxidized to the +VI state at higher potentials and low pH, as shown in Figure 4 regarding the E–pH diagram for pure molybdenum [40–42]. Hydrogen peroxide itself is a weak acid, but, as the reaction proceeds, the pH decreases so that the conditions favor dissolution.

Thus, molybdenum dissolves at more positive potentials as an oxyanion of molybdic acid depending upon pH, i.e., as molybdic acid $H_2MoO_4$ at low pH, as hydrogen molybdate anion between pH 3 and 5 and as molybdate anion above about pH 5–6. Some more complex equilibria can exist and other polyoxomolybdates may form [43,44].

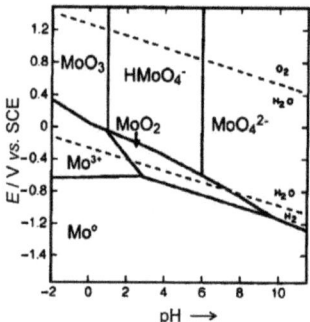

**Figure 4.** E-pH diagram for molybdenum taken from [41].

As the oxidation process proceeds, according to Badawy [45], other oxides form, such as $Mo_2O_5$, with lower protective capacity and formation of a thick oxidation product. This could offer a clue as to whether induction time is observed. As long as the passivating layer of $MoO_2$ exists, the oxidation reaction of the metal cannot take place; on the contrary, the more this passivation layer is destroyed, the more the underlying metal reacts quickly. Furthermore, as the reaction proceeds, pH decreases because the main chemical species with which the molybdenum passes into solution is anion $HMoO_4^-$, making the reaction even faster. For electroneutrality of a system, pH obviously decreases.

Figure 5a shows the trend of pH and electroconductivity as a function of time during the dissolution process of 20 g of molybdenum using a 30% peroxide solution with external ice bath, and Figure 5b shows the trend of the temperature during the experiment. The mass variation (defined as $(Mi - Mf)/Mi \times 100$, where Mi and Mf are the initial and final mass, respectively) of the undissolved metal was also measured in other experiments and shown in the same figure. The final yield of this test after 300 min was over 99%, meaning that dissolution of the metal is complete under these conditions.

It is important to note that, due to the external ice bath, in the first half of the experiment, pH, electrical conductivity of the solution and temperature are almost constant and the mass of metal remains almost constant, indicating that the reaction is very slow at this stage. However, after about 200 min, all these parameters change drastically. The pH of the solution decreases by about 0.5 units and the conductivity increases more than ten times. The temperature, which, until then, had remained roughly stable, increases more than 303 K in a few minutes despite the ice bath, and the mass of the samples begins to decrease dramatically and the inflection points coincide with a drastic increase in temperature.

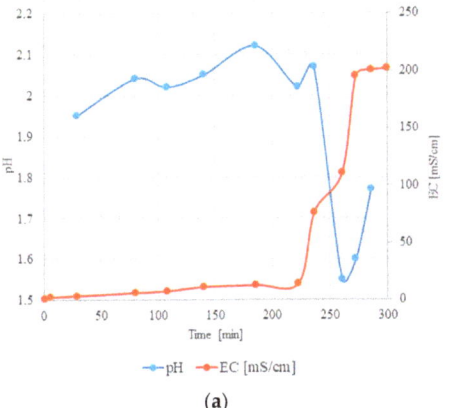
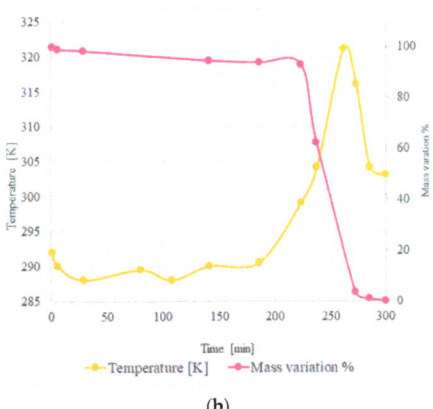

**Figure 5.** (a) pH and electroconductivity of solution during the dissolution process and (b) temperature and mass variation dependence on time during the same process.

The onset of temperature rise coincides with the moment when pH and conductivity begin to vary significantly. The pH fluctuations must be evaluated even though the measurement conditions were not optimal due to the increase in viscosity and the different complexes present in the solution. It is possible that the final rise in pH may be linked to a variation in the anionic species present in the solution and their equilibria. For example, monomeric molybdate ion can easily form polymolybdate anions [43,44] (in most cases, with general formula $Mo_xO_{3x+12}^-$ or $HMo_xO_{3x+1}^-$) and possibly peroxopolymolybdates, in particular when the concentration, as in this case, is high [46]. Indeed, under particularly acidic conditions, anion $MoO_4^{2-}$ can be subject to some protonation reaction, which leads to formation of species such as $MoO_3(OH)^-$ and $MoO_2(OH)_2(OH_2)_2$. All these chemical equilibria can lead to an increase in pH [46]. After complete dissolution of the molybdenum, an orange-colored limpid solution is obtained.

All these observations are in excellent agreement with the proposed model for molybdenum dissolution. In fact, by using the ice bath, it is possible to slow down the first step of the process so that it is easier to observe the drastic variation in the reaction after induction time. As long as the $MoO_2$ passivation layer is almost intact [42,46,47], the process is very slow, whereas, after its removal, it becomes extremely fast and self-catalytic. At this point, the temperature spikes and the electrical conductivity of the system increases dramatically, meaning that the concentration of ions in solution increased significantly. In fact, it is also observed that the mass of the sample begins to decrease only in correspondence with these phenomena and the pH also decreases, in agreement with the hypothesis that, initially, the main molybdenum species in the solution is anion $HMoO_4^-$.

These observations also suggest that, in order to optimize the reaction rate for larger batches, the system temperature should be controlled; it should not be too low, i.e., below about 300 K; otherwise, the process is too slow, and it should not be too high as this would lead to competitive decomposition of the peroxide before complete dissolution, thus requiring larger amounts of hydrogen peroxide, and this is not suitable for an environmentally cleaner process. The optimum process temperature appears to be around 320 K.

It is also observed, in excellent agreement with this chemical path model, that the molybdenum pellets appear with a metallic-silvery surface and no longer black just before sudden changes take place.

Finally, solid sodium hydroxide was added to the intensely orange acid solution until complete discoloration, obtaining a final pH of 14. During the reaction, the NaOH reacts with the peroxides according to the reaction $2\ NaOH + H_2O_2 \rightarrow Na_2O_2 + 2\ H_2O$. The reaction is exothermic and, therefore, favors removal of the residual hydrogen peroxide.

## 3.2. Microscopic Characterization

In order to investigate the dissolution mechanism more deeply, some pieces of molybdenum were observed under electron microscope at different reaction times, keeping all the other parameters constant.

Figure 6 shows the temporal evolution of their surface and the appearance of corrosion signs.

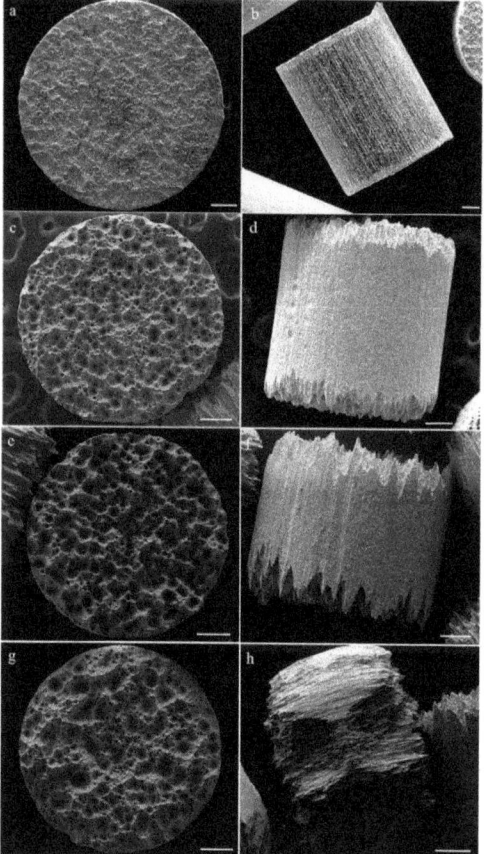

**Figure 6.** SEM images of some molybdenum pieces from top row to bottom row: 0 min ((**a**) basal view, (**b**) lateral view) 30 min ((**c**) basal view, (**d**) lateral view), 60 min ((**e**) basal view, (**f**) lateral view), 120 min ((**g**) basal view, (**h**) lateral view).

In this figure are shown low-magnification secondary electron-SEM pictures of Mo pellets dissolved in peroxide solution at different reaction times, basal (left column) and lateral (right column) overviews of: Figure 6a,b starting Mo pellets; Figure 6c,d Mo pellets after 30 minutes; Figure 6e,f after 60 minutes and Figure 6g,h after 120 minutes of reaction; all scale bars are 200 µm).

Figure 6a,b reports a representative starting Mo pellet showing the basal face without any cavities and holes. After dissolution in peroxide solution, the surface becomes defective and the pieces become smaller on average. Specifically, pristine Mo pellets clearly present dissolution and corrosion signs. The basal faces display presence of numerous cavities and lateral overviews provide direct observation of corrosion lines focused on the edges of the cylinder (Figure 6d,f,h). Upon increasing reaction time, extent of corrosion lines increases such that, after 120 min of reaction time, some pellets achieve a barrel shape (Figure 6h).

Obviously, as the atoms are oxidized to the hexavalent state and pass into solution, the mass of the metallic piece decreases.

However, the size reduction is not uniform in all directions. Although the pieces are not all exactly alike, this observation is absolutely general. In other words, the diameter decreases in percentage much more slowly than length, indicating that, even at the microscopic level, the process follows a particular path. Loss of mass is, therefore, not a simple and progressive release of matter from the external surface of the beads towards the liquid phase until they are completely consumed.

The effects upon corrosion lines of reaction time are reported in Tables 1 and 2; the trend is similar to behavior already observed with regard to pellet evolution: the diameter is almost constant, whereas the length rapidly decreases.

**Table 1.** Mean dimensions of the diameter and the lengths of the pellets at various dissolution times in peroxide solution 30%.

| Reaction Time | Mean Pellet Diameter (µm) | Mean Pellet Length (µm) |
| --- | --- | --- |
| 0 min | 1490 ± 10 | 1710 ± 10 |
| 30 min | 1253 ± 10 | 1113 ± 10 |
| 60 min | 1265 ± 10 | 935 ± 10 |
| 120 min | 1210 ± 10 | 626 ± 10 |

**Table 2.** Mean dimensions of the diameter and the lengths of the pellets and maximum values for diameter and length of corrosion lines as obtained by SEM image analysis at various dissolution times in peroxide solution 30%.

| Reaction Time | Mean Corrosion Lines Diameter (µm) | Mean Corrosion Lines Length (µm) | Max Value of Corrosion Lines Diameter (µm) | Max Value Corrosion Lines Length (µm) |
| --- | --- | --- | --- | --- |
| 30 min | 1253 ± 10 | 1113 ± 10 | 118 | 271 |
| 60 min | 1265 ± 10 | 935 ± 10 | 108 | 408 |
| 120 min | 1210 ± 10 | 626 ± 10 | 126 | 626 |

EDX-SEM investigations (not reported here) were carried out on the entire surface of our four samples to provide compositional information, with a particular focus on understanding the presence of oxygen. Slight oxidation was reported both on flat and lateral surfaces, with uniform distribution of Mo and oxygen content. However, comparing the analysis of the four samples, no significant difference was detected, both in terms of content and distribution of two elements.

Corrosion of a solid surface by liquids can be a very complex phenomenon from a physicochemical point of view and there are many different microscopic mechanisms that can be involved [48].

There are some cases where corrosion occurs in a uniform manner on the whole surface, often called uniform or general corrosion, referring to a process that proceeds at approximately the same rate over the whole exposed surface. However, often, corrosion preferentially starts where the surface energy is greater, i.e., in correspondence with cracks, defects, grain boundaries and presence of impurities [49]. Localized corrosion leads to pitting, which greatly enhances dissolution process kinetics [50]. Regardless of the specific mechanism, localized corrosion begins at specific sites, and, once the process has begun, these initial sites become larger over time and involve the entire surface.

Corrosion can be intragranular or intergranular [51], being the second a form of attack where the boundaries of crystallites of the material are more susceptible to the dissolution process than their insides. Intergranular corrosion is a form of localized corrosion, where the corrosion takes place in a quite narrow path preferentially along the grain boundaries in the metal structure.

It seems realistic to imagine that these small pieces are manufactured by hot extrusion and that, therefore, the grains align themselves preferentially along their axis. Then it can be imagined that they are mechanically cut to the desired size. Their shape can be approximated to cylinders. Often, traces of carbon used as a lubricant for extrusion can remain on the side surfaces, but not on the planar ones. These molybdenum pieces are of high purity, so the planar surfaces are quite uniform. Furthermore, it is possible that the $MoO_2$ layer on the side surface may be thicker because it can be formed during a heat treatment.

Based on the SEM images, it appears that molybdenum dissolution by peroxide is a very complex combination of uniform and localized corrosion. The fact that the sample pieces progressively shorten while their diameter remains roughly the same indicates uniform corrosion, which mainly affects flat surfaces. The metallic pieces progressively lose mass from them, without any particular corrosion points appearing in the first stages at least. On the contrary, the lateral surface shows evident signs of very localized corrosion, with lines that become progressively more evident. In this case, the corrosion is more of the intergranular type since a combination of the two processes cannot be completely ruled out at least initially; once the process has taken hold, it becomes prevalent.

It is, therefore, possible to suppose a different mechanism on the different sides: uniform corrosion on the planar faces and localized corrosion on the lateral surface.

Dissolution on the planar faces starts at the beginning uniformly over the entire surface and on both sides in a similar manner. However very soon, there is formation of some deep corrosion points on these surfaces, with morphology that is quite reminiscent of pitting (Figure 7c). These craters have a conical appearance in the sense that, the deeper they are, the smaller their radius. These corrosion marks, especially the larger ones, are often separated by thin walls that meet at angles of about 120°. In general, pitting is a form of extremely localized corrosion that leads to random creation of small holes and often is due to the de-passivation of a small area. The corrosion rate is, among other factors, determined by the size of the surface that is corroded: the greater, the faster the process. In this case, as the signs of corrosion appear new, surface attackable by the peroxide is created, making the dissolution faster and faster.

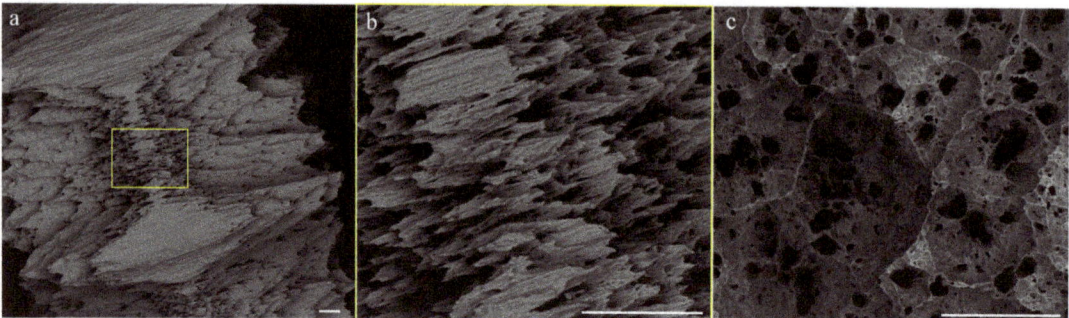

**Figure 7.** Backscattered SEM images of Mo pellet as obtained after 120 min of reaction time. Lateral (**a**,**b**) and basal (**c**) overview of Mo pellet presenting evident localized corrosion. (**b**) Zoom of Figure 7a showing the corrosion lines run along the lateral surface until reaching the middle of the barrel-shaped pellet. All scale bars are of 50 μm.

Since this does not occur on the lateral surface, this can offer an explanation for the fact that the loss of mass occurs mainly from the basal surface rather than from the lateral surface.

Actually, the later surface does not participate to the dissolution process in the very early stages; only its extremities are involved and, in these positions, the signs of corrosion are very evident. After some time, some dissolution lines propagate from these initial points and run along the lateral surface from both sides until they arrive to touch and cover the entire lateral surface (Figure 7a,b). At this point, corrosion proceeds quickly and also inwards. The possible microscopic dissolution mechanism that has been discussed is depicted schematically in Figure 8.

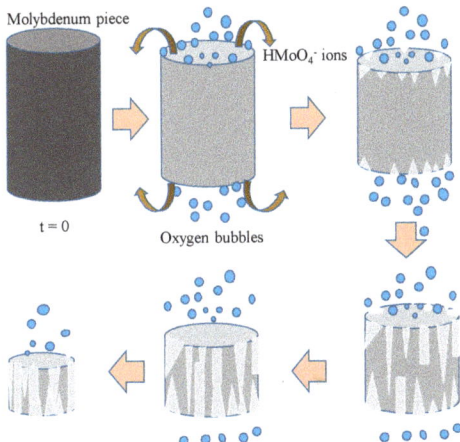

**Figure 8.** A model representation of the proposed microscopic mechanism.

### 3.3. Spectroscopic Characterization

As described in the previous sections, the final aim of the radiochemistry task within the SRF project is to efficiently and quickly prepare a stock solution to obtain $^{99m}$Tc. However, one of the indirect ways to prove the quality of the method used in this work and demonstrate the purity of the solution is to obtain crystals of sodium molybdate, $Na_2MoO_4$, and characterize them spectroscopically. Furthermore, this compound is an interesting material on its own due to its various properties and is worth studying as it stands. For example, it is a promising material for optoelectronics [52].

After preparation of the sodium molybdate solution, the crystals have been slowly evaporated, dried in a laboratory oven and analyzed.

Sodium molybdate at room temperature and pressure crystallizes in the spinel type structure, with Fd-3 m symmetry and cell parameter a = 9.10888 Å [53–55]. It is possible to observe high purity and excellent quality of the sample obtained in this way since almost all the peaks of the diffractogram can be uniquely attributed only to the pure $Na_2MoO_4$ compound and there are only few spurious signals that are due to a very small quantity of sodium carbonate, $Na_2CO_3$, which, in turn, very likely derives from small quantity of impurities in the NaOH and can, therefore, be avoided using this reagent with better quality and higher purity. This also means that the method, when perfectly controlled, is very effective in obtaining the stock solution for feeding the Mo/Tc generators.

Figure 9 below shows the XRD pattern of a sample crystallized from this solution by evaporating the liquid, and Figures 10 and 11 show Raman spectrum and IR spectrum of the same sample, respectively.

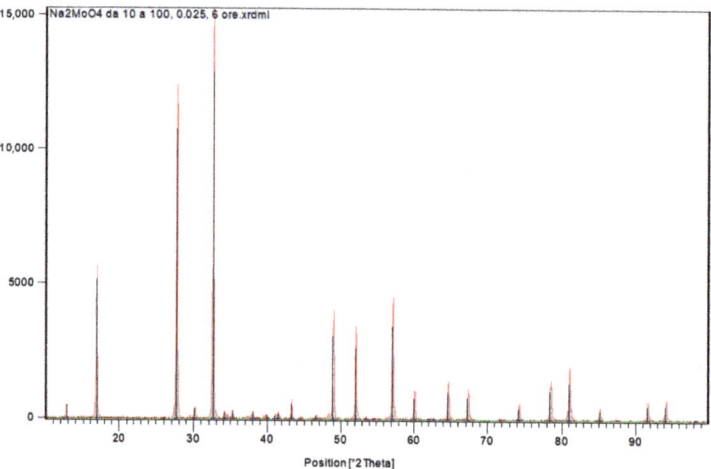

**Figure 9.** XRD pattern of sodium molybdate.

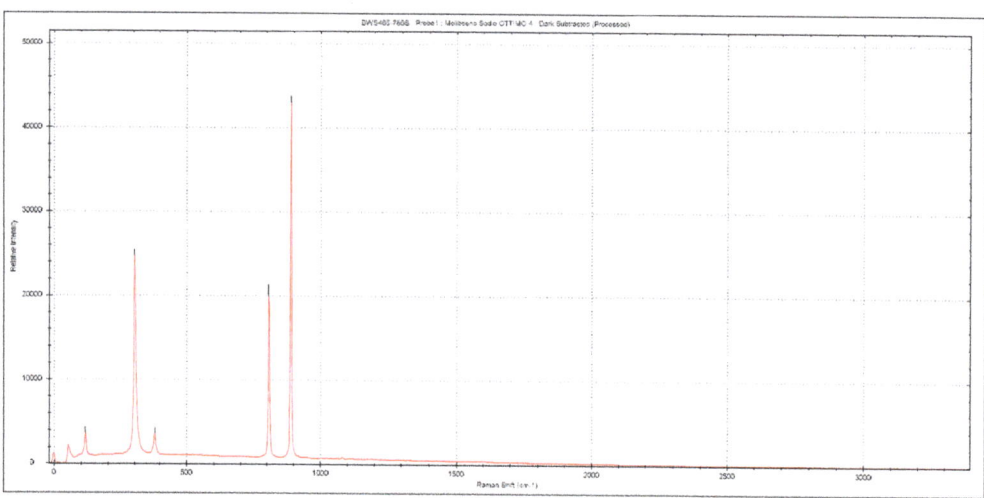

**Figure 10.** Raman spectrum of sodium molybdate.

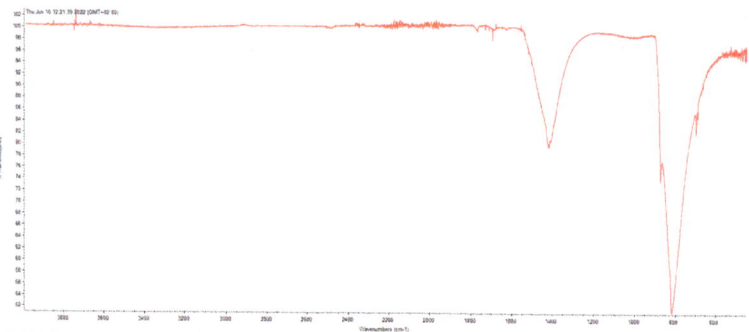

**Figure 11.** IR spectrum of the same sample.

All the Raman bands and signals in IR spectrum can be uniquely attributed only to the pure anhydrous $Na_2MoO_4$ compound and no spurious peaks can be detected.

The crystalline structure consists of isolated $MoO_4$ tetrahedra and the crystals have eight formula units per unit cell [36–38]. In the crystal, $MoO_4$ groups occupy Td sites and Na ions occupy D3d sites. Vibrational modes of these tetrahedra are the main features that can be observed in both types of spectra and the position of observed bands depends on the characteristic vibrations of the O–Mo–O bonds and $MoO_4$ blocks, even if collective modes of the whole crystalline lattice can have a role. According to standard group theory, the irreducible representation $\Gamma$ admits 39 optical modes that are [53–55].

$$\Gamma = A_{1g} + E_g + F_{1g} + 3F_{2g} + 2A_{2u} + 2E_u + 4F_{1u} + 2F_{2u}$$

Selection rules enable $A_{1g}$, $E_g$, and $F_{2g}$ modes to be Raman-active due to symmetric stretching ($\nu 1$), asymmetric stretching ($\nu 3$), symmetric bending ($\delta 2$) and asymmetric bending ($\delta 4$) of the $MoO_4$ units. The measured bands match what has been reported in the literature: 891 cm$^{-1}$, 808 cm$^{-1}$, 380 cm$^{-1}$ and 302 cm$^{-1}$. In addition to these bands, it is possible to note a further signal at about 115 cm$^{-1}$ that can be assigned to translational collective mode of the crystalline lattice [53]. The assignments are reported in Table 3. In the case of IR spectrum instead, the band at about 550 cm$^{-1}$ can been attributed to the bending vibration of the $MoO_4^{2-}$ tetrahedron, while the signals in the range 650–800 cm$^{-1}$ are due to the stretching vibration of the tetrahedra observed also; in particular, the very intense band at about 830 cm$^{-1}$ is due to O–Mo–O stretching [54,55].

It is important to note the absence of signals in around 3000 cm$^{-1}$, which is the region due to hydrogen–oxygen stretching. This indicates significant absence of water, of crystallization nature or absorbed, in the sample.

Table 3. Assignments of the Raman bands of sodium molybdate.

| Wavenumber (cm$^{-1}$) | Attribution |
|---|---|
| 115 | Collective Mode (MoO4) |
| 302 | $\delta$ (Mo–O) |
| 380 | $\delta$ (Mo–O) |
| 808 | $\nu_{as}$ (Mo–O) |
| 891 | $\nu_s$ (Mo–O) |

The purity of starting materials and final product have been verified by ICP-mass spectrometry and resulted in very high values, as can been seen in Table 4.

Table 4. Impurities concentration in the initial molybdenum pieces and in the final product (* below LOQ).

| Element | Concentration (ppb) | | |
|---|---|---|---|
| | Pellets (LCTMM Co.) | Powder (Merck, DE) | Purified Sodium Molybdate |
| V | 0.361 | 1.061 | 0.252 |
| Cr | 23.72 | 12.18 | 8.793 |
| Mn | 2.998 | 1.184 | 1.620 |
| Fe | 161.2 | 75.78 | 76.81 |
| Co | 8.181 | 2.478 | 9.574 |
| Ni | 16.99 | 18.51 | 9.995 |
| Cu | 2.204 | 1.397 | 0.901 |
| Zn | 4.130 | 2.446 | 1.242 |
| Ag | 0 * | 0 * | 0 * |
| Cd | 71.28 | 53.69 | 77.63 |
| Hg | 0.449 | 0.452 | 0 * |
| Tl | 0 * | 0 * | 0 * |
| Pb | 1.653 | 1.026 | 8.556 |

The molybdenum samples were prepared in a clean laboratory for sample treatment and preparation of the ICP-MS analysis with a microwave digestion system, Speedwave Four model (Berghof, Germany), equipped with temperature and pressure control for digestion vessels made of PTFE.

Both the molybdenum pellets, which are the chosen geometry for this study, and the molybdenum powder, which was used in the preliminary phases of this research, were analyzed and compared to the purified crystal of sodium molybdate at the end of the process. In this way, it has been possible to evaluate any unexpected contamination from the beginning to the end of the process.

The values of the analyzed elements were all well beyond the limits of the certificate analysis of the products (<500 ppm for both powder and pellets), and the requirements for trace metal analysis are met, confirming the high purity of the starting material. Importantly, secondary elements concentration did not increase (with the only exceptions of lead and cadmium, whose concentrations remain, in any case, far below the safety limits); on the contrary, it decreased, meaning that the dissolution/recrystallization process increases purity.

RSD for all measurements were below 5%.

## 4. Conclusions

In this experimental study, a green, mild and cost-effective molybdenum dissolution procedure has been investigated. Up to 100 g of solid molybdenum have been successfully dissolved in less than 6 h, demonstrating the feasibility of the process for preparation of the solution used in Mo99/Tcm99 generators. The developed method is rapid and involves mild reaction conditions, minimal reagents consumption and quantitative yields. Furthermore, reaction temperature is easily controlled with ice bath, and, therefore, risks and hazards are minimized.

Chemical and kinetic measurements have also been performed for parameters optimization, such as solution pH and EC, hydrogen peroxide concentration and induction time during the dissolution process.

Sodium molybdate crystals have been successively recrystallized and characterized by means of X-ray diffractometry, Raman spectroscopy, infrared spectroscopy and inductively coupled plasma mass spectroscopy, showing the very high purity of the compound and the extremely low contamination of other metals (<80 ppb).

We further studied the mechanism of dissolution and corrosion of molybdenum pellet via scanning electron microscopy and EDX analysis in order to better understand the dissolution chemistry of the specific solid molybdenum geometry and optimize future larger-scale experiments with specifically designed reaction vessels.

We can, therefore, conclude that the method fits the standards needed for radiochemical processing of irradiated molybdenum by 14 MeV fusion neutron source SORGENTINA-RF.

**Author Contributions:** Conceptualization, F.C. and A.U.; methodology, F.C., A.U. and A.F.; investigation, F.C., A.U., A.F., S.C., S.S. and V.C.; writing—original draft preparation, F.C. and A.U.; writing—review and editing, All Authors; supervision, P.A. and A.R.; project administration, A.P. and P.A. All authors have read and agreed to the published version of the manuscript.

**Funding:** The research has been co-funded in the frame of the Regione Emilia Romagna-ENEA agreement for the project "SORGENTINA RF–Thermomechanical Demonstration".

**Institutional Review Board Statement:** Not applicable.

**Informed Consent Statement:** Not applicable.

**Data Availability Statement:** Not applicable.

**Acknowledgments:** The authors acknowledge the SRF—Collaboration: Massimo Angiolini, Ciro Alberghi, Luigi Candido, Marco Capogni, Mauro Capone, Gian Marco Contessa, Francesco D'Annibale, Marco D'Arienzo, Alessio Del Dotto, Dario Diamanti, Danilo Dongiovanni, Mirko Farini, Paolo Ferrari, Davide Flammini, Manuela Frisoni, Gianni Gadani, Angelo Gentili, Giacomo Grasso, Manuela Guardati, David Guidoni, Marco Lamberti, Luigi Lepore, Andrea Mancini, Andrea Mariani, Ranieri Marinari, Giuseppe A. Marzo, Bruno Mastroianni, Fabio Moro, Agostina Orefice, Valerio Orsetti, Tonio Pinna, Alexander Rydzy, Demis Santoli, Alessia Santucci, Luca Saraceno, Camillo Sartorio, Valerio Sermenghi, Emanuele Serra, Andrea Simonetti, Nicholas Terranova, Silvano Tosti, Marco Utili, Konstantina Voukelatou, Danilo Zola, Giuseppe Zummo and the Region Emilia Romagna-ENEA agreement for the project "SORGENTINA RF—Thermomechanical Demonstration".

**Conflicts of Interest:** The authors declare no conflict of interest.

**Sample Availability:** Samples of the compounds are available from the authors.

# References

1. Agostini, P.; Angiolini, M.; Alberghi, C.; Candido, L.; Capogni, M.; Capone, M.; Cataldo, S.; D'Annibale, F.; D'Arienzo, M.; Diamanti, D.; et al. SORGENTINA-RF project: Fusion neutrons for 99Mo medical radioisotope SORGENTINA-RF. *Eur. Phys. J. Plus* **2021**, *136*, 1140. [CrossRef]
2. Capogni, M.; Pietropaolo, A.; Quintieri, L. $^{99m}$Tc production via 100Mo(n,2n)99Mousing 14 MeV neutrons from a D-T neutron source: Discussion for a scientific case. In *RT/2016/32/ENEA*; ENEA: Rome, Italy, 2016.
3. Gregoire, T. Meeting the Moly-99 Challenge. Nuclear News. 2018. Available online: http://www.shinefusion.com/wp-content/uploads/2018/07/2018.06.20-Meeting-the-moly-99-challenge-Nuclear-News-ANS-article.pdf/ (accessed on 3 March 2022).
4. Kobayashi, T.; Sueki, K.; Ebihara, M.; Nakahara, H.; Imamura, M.; Masuda, A. Half-lives of Technetium 97, 98. *Radiochim. Acta* **1993**, *63*, 29. [CrossRef]
5. Pillai MR, A.; Dash, A.; Knapp, F.F., Jr. Sustained Availability of $^{99m}$Tc: Possible Paths Forward. *J. Nucl. Med.* **2013**, *54*, 313. [CrossRef] [PubMed]
6. Chattopadhyay, S.; Das, S.S.; Barua, L. A simple and rapid technique for recovery of $^{99m}$Tc from low specific activity (n,γ)$^{99}$Mo based on solvent extraction and column chromatography. *Appl. Radiat. Isot.* **2012**, *68*, 1–4. [CrossRef] [PubMed]
7. NEA-OECD. *Report on the Supply of Medical Radioisotopes: An Assessment of Long-Term Global Demand for Technetium-99m*; Nuclear European Agency, Organisation for Economic Co-Operation and Development: Paris, France, 2011.
8. Martini, P.; Boschi, A.; Cicoria, G.; Uccelli, L.; Pasquali, M.; Duatti, A.; Pupillo, G.; Marengo, M.; Loriggiola, M.; Esposito, J. A solvent-extraction module for cyclotron production of high-purity technetium-99m. *Appl. Radiat. Isot.* **2016**, *118*, 302. [CrossRef]
9. Martini, P.; Boschi, A.; Cicoria, G.; Zagni, F.; Corazza, A.; Uccelli, L.; Pasquali, M.; Pupillo, G.; Marengo, M.; Loriggiola, M.; et al. In-house cyclotron production of high-purity Tc-99m and Tc-99m radiopharmaceuticals. *Appl. Radiat. Isot.* **2018**, *139*, 325. [CrossRef]
10. Esposito, J.; Vecchi, G.; Pupillo, G.; Taibi, A.; Uccelli, L.; Boschi, A.; Gambaccini, M. Evaluation of $^{99}$Mo and $^{99m}$Tc Productions Based on a High-Performance Cyclotron. *Sci. Technol. Nucl. Install.* **2013**, *2013*, 972381. [CrossRef]
11. Van der Keur, H. Medical Radioisotopes Production without a Nuclear Reactor. 2010. Available online: WorldWideScience.org (accessed on 12 June 2022).
12. Wang, Y.; Chen, D.; dos Santos Augusto, R.; Liang, J.; Qin, Z.; Liu, J.; Liu, Z. Production Review of Accelerator-Based Medical Isotopes. *Molecules* **2022**, *27*, 5294. [CrossRef]
13. Hasan, S.; Prelas, M.A. Molybdenum-99 production pathways and the sorbents for $^{99}$Mo/$^{99m}$Tc generator systems using (n, γ) $^{99}$Mo: A review. *SN Appl. Sci.* **2020**, *2*, 1782. [CrossRef]
14. NEA, N.E. The Supply of Medical Radioisotopes. 2018 Medical Isotope Demand and Capacity Projection for the 2018–2023 Period. NEA/SEN/HLGMR. 2018. Available online: www.oecd-nea.org (accessed on 1 July 2022).
15. Ferrucci, B.; Ottaviano, G.; Rizzo, A.; Ubaldini, A. Future development of global molybdenum-99 production and saving of atmospheric radioxenon emissions by using nuclear fusion-based approaches. *J. Environ. Radioact.* **2022**, *255*, 107049. [CrossRef]
16. Tonchev, A.P.; Stoyer, M.A.; Becker, J.A.; Macri, R.; Ryan, C.; Sheets, S.A.; Gooden, M.E.; Arnold, C.W.; Bond, E.; Bredeweg, T.; et al. Energy Evolution of the Fission-Product Yields from Neutron-Induced Fission of $^{235}$U, $^{238}$U, and $^{239}$Pu: An Unexpected Observation. In *Fission and Properties of Neutron-Rich Nuclei*; LLNL-PROC: Sanibel Island, FL, USA, 2017; p. 381.
17. Capogni, M.; Pietropaolo, A.; Quintieri, L.; Angelone, M.; Boschi, A.; Capone, M.; Cherubini, N.; De Felice, P.; Dodaro, A.; Duatti, A.; et al. 14 MeV Neutrons for $^{99}$Mo/$^{99m}$Tc Production: Experiments, Simulations and Perspectives. *Molecules* **2018**, *23*, 1872. [CrossRef] [PubMed]
18. Flammini, D.; Bedogni, R.; Moro, F.; Pietropaolo, A. On the Slowing down of 14 MeV Neutrons. *J. Neutr. Res.* **2020**, *22*, 249. [CrossRef]
19. Fonnesu, N.; Scaglione, S.; Spassovsky, I.P.; Pietropaolo, A.; Zito, P.; The SRF Collaboration. On the definition of the deuterium-tritium ion beam parameters for the SORGENTINA-RF fusion neutron source. *Eur. Phys. J. Plus* **2022**, *137*, 1150. [CrossRef]
20. Nawar, M.F.; Türler, A. New strategies for a sustainable $^{99m}$Tc supply to meet increasing medical demands: Promising solutions for current problems. *Front. Chem.* **2022**, *10*, 926258. [CrossRef]

21. Schaffer, P.; Bénard, F.; Bernstein, A.; Buckley, K.; Celler, A.; Cockburn, N.; Corsaut, J.; Dodd, M.; Economou, C.; Eriksson, T.; et al. Direct Production Of $^{99m}$Tc via $^{100}$Mo(p,2n) on Small Medical Cyclotrons. *Phys. Procedia* **2015**, *66*, 383. [CrossRef]
22. Tkac, P.; Rotsch, D.A.; Chemerisov, S.D.; James, P.; Byrnes, J.P.; Bailey, J.L.; Wiedmeyer, S.G. Large-scale dissolution of sintered Mo disks. *J. Radioanal. Nucl. Chem.* **2021**, *327*, 617. [CrossRef]
23. Magnusson, O. Eurachem Guide: The Fitness for Purpose of Analytical Methods—A Laboratory Guide to Method Validation and Related Topics. In *Eurachem Guide*, 2nd ed.; 2014; ISBN 978-91-87461-59-0.
24. Beary, E.S.; Paulsen, P.J. Selective application of chemical separations to isotope dilution inductively coupled plasma mass spectrometric analyses of standard reference materials. *Anal. Chem.* **1993**, *65*, 1602. [CrossRef]
25. Hazza, M.L.; El-Dahshan, M.E. The effect of molybdenum on the corrosion behaviour of some steel alloys. *Desalination* **1994**, *95*, 199. [CrossRef]
26. Cairang, W.; Li, T.; Xue, D.; Yang, H.; Cheng, P.; Chen, C.; Sun, Y.; Zeng, Y.; Ding, X.; Sun, J. Enhancement of the corrosion resistance of Molybdenum by $La_2O_3$ dispersion. *Corros. Sci.* **2021**, *186*, 109469. [CrossRef]
27. Ahmadein, M.; El-Kady, O.A.; Mohammed, M.M.; Essa, F.A.; Alsaleh, N.A.; Djuansjah, J.; Elsheikh, A.H. Improving the mechanical properties and coefficient of thermal expansion of molybdenum-reinforced copper using powder metallurgy. *Mater. Res. Express* **2021**, *8*, 096502. [CrossRef]
28. Barceloux, D.G. Molybdenum. *J. Toxicol. Clin. Toxicol.* **1999**, *37*, 231. [CrossRef] [PubMed]
29. Shabalin, I.L. Molybdenum. In *Ultra-High Temperature Materials, I*; Springer: Dordrecht, The Netherlands, 1999; Volume 231. [CrossRef]
30. Alves de Castro, I.; Shankar Datta, R.; Ou, J.O.; Castellanos-Gomez, A.; Sriram, S.; Daeneke, T.; Kalantar-zade, K. Molybdenum Oxides—From Fundamentals to Functionality. *Adv. Mater.* **2017**, *29*, 1701619. [CrossRef] [PubMed]
31. Zhang, S.; Yajima, T.; Soma, T.; Ohtomo, A. Epitaxial growth of $MoO_3$ polymorphs and impacts of Li-ion electrochemical reactions on their structural and electronic properties. *Appl. Phys. Express* **2022**, *15*, 055505. [CrossRef]
32. Sato, M.; Fujishita, H.; Sato, S.; Hoshino, S. Structural transitions in $Mo_8O_{23}$. *J. Phys. C Solid State Phys.* **1986**, *19*, 3059. [CrossRef]
33. Ekström, T. Formation of ternary phases of $Mo_5O_{14}$ and $Mo_{17}O_{47}$ structure in the molybdenum-wolfram-oxygen system. *Mater. Res. Bull.* **1972**, *7*, 19. [CrossRef]
34. Guro, V.P. Molybdenum dissolution in mixtures of $H_2O_2$ and concentrated $HNO_3$ and $H_2SO_4$ in the presence of tungsten. *Inorg. Mater.* **2008**, *44*, 239. [CrossRef]
35. Tkac, P.; Rotsch, D.A.; Chemerisov, S.D.; Bailey, J.L.; Krebs, J.F.; Vandergrift, G.F. *Optimization of the Dissolution of Molybdenum Disks: FY-16 Results*; Nuclear Engineering Division, Argonne National Laboratories, ANL/NE: Argonne, IL, USA, 2016.
36. Cieszykowska, I.; Jerzyk, K.; Żółtowska, M.; Janiak, T.; Birnbaum, G. Studies on electrochemical dissolution of sintered molybdenum discs as a potential method for targets dissolution in $^{99m}$Tc production. *J. Radioanal. Nucl. Chem.* **2022**, *331*, 1029. [CrossRef]
37. Lima, C.L.; Saraiva, G.D.; Freire PT, C.; Maczka, M.; Paraguassu, W.; de Sousae, F.F.; Mendes Filhoa, J. Temperature-induced phase transformations in $Na_2WO_4$ and $Na_2MoO_4$ crystals. *J. Raman Spectrosc.* **2011**, *42*, 799. [CrossRef]
38. Sydorchuk, V.; Khalameida, S.; Zazhigalov, V.; Zakutevskii, O. Some properties of a vanadium molybdenum oxide composite produced by mechanochemical treatment in various media. *Russ. J. Inorg. Chem.* **2013**, *58*, 1349. [CrossRef]
39. Lyon, S.B. Corrosion of Molybdenum and its Alloys. In *Shreir's Corrosion*; Cottis, B., Graham, M., Lindsay, R., Lyon, S., Richardson, T., Scantlebury, D., Stott, H., Eds.; Elsevier: Oxford, UK, 2010; pp. 2157–2167; ISBN 978-0-444-52787-5.
40. Tyurin, A.G. Estimation of the Effect of Molybdenum on Chemical and Electrochemical Stability of Iron-Based Alloys. *Prot. Met.* **2003**, *39*, 367. [CrossRef]
41. Morris, M.J. Molybdenum 1994. *Coord. Chem. Rev.* **1996**, *152*, 309. [CrossRef]
42. Gough, K.; Holmes, S.; Matosky, N.; Oney, T.; Rutledge, J. *Pressure Oxidation of Molybdenum Concentrates*; Technical Report Kroll Institute for Extractive Metallurgy; Colorado School of Mines: Golden, CO, USA, 2013. [CrossRef]
43. Zhang, N.; Königsberger, E.; Duan, S.; Lin, K.; Yi, H.; Zeng, D.; Zhao, Z.; Hefter, G. Nature of Monomeric Molybdenum (VI) Cations in Acid Solutions Using Theoretical Calculations and Raman Spectroscopy. *J. Phys. Chem. B* **2019**, *123*, 3304. [CrossRef] [PubMed]
44. Zhang, J.-L.; Hua, J.-T.; Zhang, L.-F. Raman Studies on Species in Single and Mixed Solutions of Molybdate and Vanadate. *Chin. J. Chem. Phy.* **2016**, *29*, 425. [CrossRef]
45. Badawy, W.A.; Al-Kharafi, F.M. Corrosion and passivation behaviors of molybdenum in aqueous solutions of different pH. *Electrochim. Acta* **1998**, *44*, 693. [CrossRef]
46. Aveston, J.; Anacker, E.W.; Johnson, J.S. Hydrolysis of Molybdenum (VI). Ultracentrifugation, acidity measurements and Raman spectra of polymolybdates. *Inor. Chem.* **1964**, *3*, 735. [CrossRef]
47. Misirlioglu, Z.; Aksüt, A. Corrosion of Molybdenum in Aqueous Media. *Corrosion* **2002**, *58*, 899. [CrossRef]
48. Harsimran, S.; Santoshc, K.; Rakesh, K. Overview of corrosion and its control: A critical review. *Proc. Eng. Sci.* **2021**, *3*, 13–24. [CrossRef]
49. Page, C.L. Corrosion and protection of reinforcing steel in concrete. Durability of Concrete and Cement Composites. In *Woodhead Publishing Series in Civil and Structural Engineering*; Page, C., Page, M., Eds.; Woodhead Publishing: Sawston, UK, 2007; p. 136.
50. Wang, Z.; Di-Franco, F.; Seyeux, A.; Zanna, S.; Maurice, V.; Marcus, P. Passivation-Induced Physicochemical Alterations of the Native Surface Oxide Film on 316L Austenitic Stainless Steel. *J. Electrochem. Soc.* **2019**, *166*, C3376. [CrossRef]

51. Frankelt, G.S. Pitting Corrosion of Metals: A Review of the Critical Factors. *J. Electrochem. Soc.* **1998**, *145*, 2186. [CrossRef]
52. Abbas, S.A.; Mahmood, I.; Sajjad, M.; Noor, N.A.; Mahmood, Q.; Naeem, M.A.; Mahmood, A.; Ramay, S.M. Spinel-type $Na_2MoO_4$ and $Na_2WO_4$ as promising optoelectronic materials: First-principle DFT calculations. *Chem. Phys.* **2020**, *538*, 110902. [CrossRef]
53. Fortes, A.D. Crystal structures of spinel-type $Na_2MoO_4$ and $Na_2WO_4$ revisited using neutron powder diffraction. *Acta Cryst. E Cryst. Commun.* **2015**, *71*, 59. [CrossRef] [PubMed]
54. Dkhilalli, F.; Megdiche Borchani, S.; Rasheed, M.; Barille, R.; Shihab, S.; Guidara, K.; Megdiche, M. Characterizations and morphology of sodium tungstate particles. *R. Soc. Open Sci.* **2018**, *5*, 172214. [CrossRef] [PubMed]
55. Pope, S.J.; West, Y.D. Use of the FT Raman spectrum of $Na_2MoO_4$ to study sample heating by the laser. *Spectrochim. Acta Part A Mol. Biomol. Spectrosc.* **1995**, *51*, 2011. [CrossRef]

**Disclaimer/Publisher's Note:** The statements, opinions and data contained in all publications are solely those of the individual author(s) and contributor(s) and not of MDPI and/or the editor(s). MDPI and/or the editor(s) disclaim responsibility for any injury to people or property resulting from any ideas, methods, instructions or products referred to in the content.

Review

# Recent Developments in Carbon-11 Chemistry and Applications for First-In-Human PET Studies

Anna Pees [1], Melissa Chassé [1,2], Anton Lindberg [1] and Neil Vasdev [1,2,3,*]

[1] Azrieli Centre for Neuro-Radiochemistry, Brain Health Imaging Centre, Centre for Addiction and Mental Health (CAMH), Toronto, ON M5T 1R8, Canada
[2] Institute of Medical Science, University of Toronto, Toronto, ON M5S 1A8, Canada
[3] Department of Psychiatry, University of Toronto, Toronto, ON M5T 1R8, Canada
* Correspondence: neil.vasdev@utoronto.ca

**Abstract:** Positron emission tomography (PET) is a molecular imaging technique that makes use of radiolabelled molecules for in vivo evaluation. Carbon-11 is a frequently used radionuclide for the labelling of small molecule PET tracers and can be incorporated into organic molecules without changing their physicochemical properties. While the short half-life of carbon-11 ($^{11}$C; $t_{1/2}$ = 20.4 min) offers other advantages for imaging including multiple PET scans in the same subject on the same day, its use is limited to facilities that have an on-site cyclotron, and the radiochemical transformations are consequently more restrictive. Many researchers have embraced this challenge by discovering novel carbon-11 radiolabelling methodologies to broaden the synthetic versatility of this radionuclide. This review presents new carbon-11 building blocks and radiochemical transformations as well as PET tracers that have advanced to first-in-human studies over the past five years.

**Keywords:** Carbon-11; positron emission tomography (PET); radiochemistry; radiotracer; first-in-human

Citation: Pees, A.; Chassé, M.; Lindberg, A.; Vasdev, N. Recent Developments in Carbon-11 Chemistry and Applications for First-In-Human PET Studies. *Molecules* **2023**, *28*, 931. https://doi.org/10.3390/molecules28030931

Academic Editor: Svend Borup Jensen

Received: 2 December 2022
Revised: 9 January 2023
Accepted: 10 January 2023
Published: 17 January 2023

**Copyright:** © 2023 by the authors. Licensee MDPI, Basel, Switzerland. This article is an open access article distributed under the terms and conditions of the Creative Commons Attribution (CC BY) license (https://creativecommons.org/licenses/by/4.0/).

## 1. Introduction

Positron emission tomography (PET) is a molecular imaging technique that utilizes radiotracers for in vivo studies. The radionuclides fluorine-18 and carbon-11 are the most commonly used for labelling PET tracers because of the growing use of organofluorine drugs and as carbon is ubiquitous in nearly every drug or biomolecule. Additionally, their suitable decay characteristics and half-lives match the in vivo pharmacokinetics of small molecules. Consequently, developing new radiochemistry methods for the introduction of these short-lived radionuclides into organic molecules has emerged as one of the greatest challenges in PET radiopharmaceutical chemistry. Our ultimate goal is to radiolabel any molecule for medical imaging—a concept analogous to total synthesis that we introduced as "total radiosynthesis" [1]. Because the radiochemistry of fluorine-18 has been extensively reviewed in recent years [for example see: [2–11]], the focus of this review is on recent radiochemistry methodologies with carbon-11 and translation of $^{11}$C-labelled PET tracers to first-in-human (FIH) PET imaging studies.

Carbon-11 ($^{11}$C) has a half-life of 20.4 min and is produced in a cyclotron by proton bombardment of nitrogen gas in presence of trace amounts of oxygen (0.1–2%) or hydrogen (5–10%) where it is obtained as [$^{11}$C]CO$_2$ or [$^{11}$C]CH$_4$, respectively [12,13]. [$^{11}$C]CO$_2$ and [$^{11}$C]CH$_4$ can either be used directly in radiolabelling reactions or further converted to other $^{11}$C-building blocks (see Scheme 1) [14]. The most common carbon-11 labelling strategy for PET tracers is $^{11}$C-methylation of hydroxy or amino groups using [$^{11}$C]methyl iodide or [$^{11}$C]methyl triflate, which are routinely obtained from [$^{11}$C]CO$_2$ and/or [$^{11}$C]CH$_4$. The advantages of $^{11}$C-methylation are the accessibility of precursors and carbon-11 methylating agents, as well as the general prevalence of methyl groups in pharmaceutical compounds. However, amongst molecules targeting the central nervous system (CNS) the prevalence of such methyl groups is rather low (<35%). Furthermore, metabolic demethylation can lead to cleavage of the radiolabel in vivo [15,16].

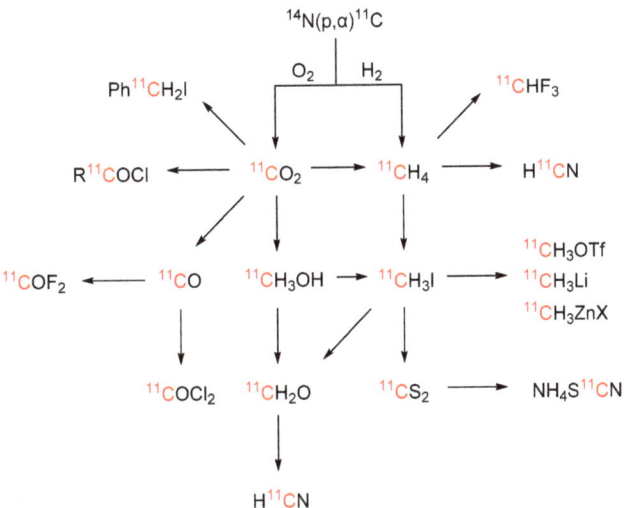

**Scheme 1.** Selected carbon-11 labelled building blocks.

Synthetic efforts have been made in recent years to expand the toolbox for $^{11}$C-chemistry beyond $^{11}$C-methylation (Scheme 1). Particular interest has been paid to the development of [$^{11}$C]CO and [$^{11}$C]CO$_2$ chemistry, in order to gain access to $^{11}$C-labelled carbonyl-based functional groups. These radiochemistry methods open the door to labelling >75% of the compounds in CNS drug pipelines [15]. And other promising building blocks and synthetic strategies have been developed, such as new reactions with [$^{11}$C]methyl iodide and related alkylating reagents, [$^{11}$C]hydrogen cyanide, [$^{11}$C]fluoroform, [$^{11}$C]carbonyl difluoride, [$^{11}$C]carbon disulfide, [$^{11}$C]thiocyanate and [$^{11}$C]formaldehyde (vide infra), further broadening the scope of compounds that can be labelled with carbon-11 and paving the way for our ultimate goal of total radiosynthesis.

It should be noted that all yields are reported as they are stated or defined in their original articles (radiochemical yield (RCY), radiochemical conversion (RCC), radiochemical purity (RCP)) and might not necessarily reflect their definition as reported in the nomenclature guidelines [17,18]. The molar activity (A$_m$) depends on several factors including the starting amount of radioactivity and is therefore difficult to compare.

## 2. Carbon-11 Methodologies

### 2.1. [$^{11}$C]Carbon Dioxide

Historically, [$^{11}$C]CO$_2$ has been a challenging building block for radiochemists to use due to its moderate reactivity and potentially low A$_m$ caused by isotopic dilution with atmospheric CO$_2$. The introduction of bulky organic "fixation" bases such as 1,8-diazabicyclo [5.4.0]undec-7-ene (DBU) and 2-tert-butylimino-2-diethylamino-1,3-dimethylperhydro-1,3,2-diazaphosphorine (BEMP) for trapping of [$^{11}$C]CO$_2$ [19–21] was inspired by green chemistry for capturing atmospheric CO$_2$ and represents a major advance for $^{11}$C-chemistry: the "fixation" bases allow [$^{11}$C]CO$_2$ to be easily trapped in a reaction vessel at room temperature and enable access to high oxidation state functional groups such as carbon-11 labelled carboxylic acids, amides, formamides, ureas, carbamates and other functional groups [22]. This methodology has contributed to the accessibility of [$^{11}$C]CO$_2$ as a building block and, in consequence, a wide array of new [$^{11}$C]CO$_2$ chemistry applications and PET tracers have emerged over the past decade. This review will focus on novel [$^{11}$C]CO$_2$ fixation reactions reported within the last five years.

While [$^{11}$C]CO$_2$ is directly produced in the cyclotron, the irradiated cyclotron target gas contains many undesired chemical and radiochemical entities. To purify [$^{11}$C]CO$_2$ from carrier gases and other by-products, it is typically trapped using liquid nitrogen or by physical adsorption on porous polymers, such as carbon molecular sieves or polydivinylbenzene copolymers. A new method for purifying [$^{11}$C]CO$_2$, also inspired by green chemistry literature, has recently been reported by our laboratories which employs chemisorption by solid polyamine-based adsorbents. This method uses small amounts of silica-grafted polyethyleneimine to trap [$^{11}$C]CO$_2$ at room temperature and quantitatively release it under mild heating (85 °C). Trapping efficiencies (TEs) as high as 79 ± 12% were observed but decreased over multiple cycles, indicating a limited reusability of the capture material. This technology was applied to synthesize a PET tracer by [$^{11}$C]CO$_2$ fixation reactions, and could potentially be applied for solid phase reactions as well as enable the transportation of carbon isotopes [23].

Traditionally, direct use of [$^{11}$C]CO$_2$ is achieved by use of Grignard reagents, organolithiums or silanamines to yield $^{11}$C-labelled amides or carboxylic acids. However, these reagents are challenging to implement in automated PET tracer production due to their hygroscopic nature, tendency to absorb atmospheric CO$_2$, and corrosiveness [22]. As such, many new methodologies for the preparation of [$^{11}$C]carboxylic acids have been developed over the past five years by novel [$^{11}$C]CO$_2$ fixation reactions that employ the aforementioned "fixation" bases. Our laboratory reported the use of aryl and heteroaryl stannanes as precursors which were carboxylated in a copper(I)-mediated reaction with [$^{11}$C]CO$_2$ (see Scheme 2A) [24]. The method was fully automated and applied for an alternative synthesis of [$^{11}$C]bexarotene (previously synthesized by reaction of [$^{11}$C]CO$_2$ with a boronic ester precursor mediated by a copper(I) source [25,26]), and was obtained with a RCY of 32 ± 5% (decay-corrected (dc)) and a A$_m$ of 38 ± 23 GBq/µmol. The strategy was also applied by García-Vázquez et al. to the synthesis of $^{11}$C-carboxylated tetrazines for the labelling of trans-cyclooctene-functionalized PeptoBrushes [27]. After optimization of the original reaction conditions (CuI instead of CuTC, NMP instead of DMF and addition of TBAT as fluoride ion source), two tetrazines were successfully $^{11}$C-carboxylated with RCYs of 10–15% and "clicked" to the TCO-PeptoBrushes. It is noteworthy that Goudou et al. reported the copper-catalyzed radiosynthesis of [$^{11}$C]carboxylic acids by reaction of [$^{11}$C]CO$_2$ with terminal alkynes in presence of DBU (see Scheme 2B) [28]. A small library of [$^{11}$C]propiolic acids was obtained with RCYs between 7 and 28%. A different approach using trimethyl and trialkoxy silanes as precursor has been described by Bongarzone et al. (see Scheme 2C) [29]. In this desilylative carboxylation reaction, aromatic silane precursors were activated by fluoride, forming a pentavalent silicate which was then reacted in a copper-catalyzed reaction with [$^{11}$C]CO$_2$. [$^{11}$C]Carboxylic acids were obtained with RCYs of 19–93% and TEs of 21–89%. A more general approach for the synthesis of [$^{11}$C]carboxylic acids was introduced by Qu et al. (see Scheme 2D) [30]. Sp-, sp$^2$- and sp$^3$-hybridized carbon-attached trimethylsilanes were $^{11}$C-carboxylated in a fluoride-mediated desilylation (FMDS) reaction, resulting in a broad substrate scope and high RCYs (up to 98%). The applicability of the method was demonstrated by synthesizing two carboxylic acid PET tracers via the FMDS approach.

[$^{11}$C]Carboxylic acids can also be synthesized by isotopic exchange reactions. Destro et al. reported the isotopic exchange reaction of cesium salt precursors with $^{13}$C, $^{14}$C, and a few selected examples of $^{11}$C (see Scheme 3A) [31]. While good yields were obtained for [$^{13}$C]CO$_2$ and [$^{14}$C]CO$_2$, yields were low for [$^{11}$C]CO$_2$ (3–50%) due to low TEs. Another take on this strategy was demonstrated by Kong et al., who employed photoredox catalysis and obtained similar results (see Scheme 3B) [32]. In both cases, A$_m$ was low, as expected (<0.2 GBq/µmol). A very recent addition to the portfolio of carboxylic acid labelling strategies by isotopic exchange was presented by Bsharat et al. [33]. These authors developed an aldehyde-catalyzed carboxylate exchange reaction in α-amino acids (see Scheme 3C) with $^{13}$C and $^{11}$C. For the $^{11}$C-reactions, imine carboxylates were pre-formed by condensation of α-amino acids with aryl aldehydes and subsequently subjected to the carboxylate exchange reaction with [$^{11}$C]CO$_2$. An array of α-amino acids was labelled with RCYs of 4–24%, and the modest yields were also attributed to low TEs of the [$^{11}$C]CO$_2$. Phenylalanine was isolated by this reaction with a A$_m$ of 8.4 GBq/mmol.

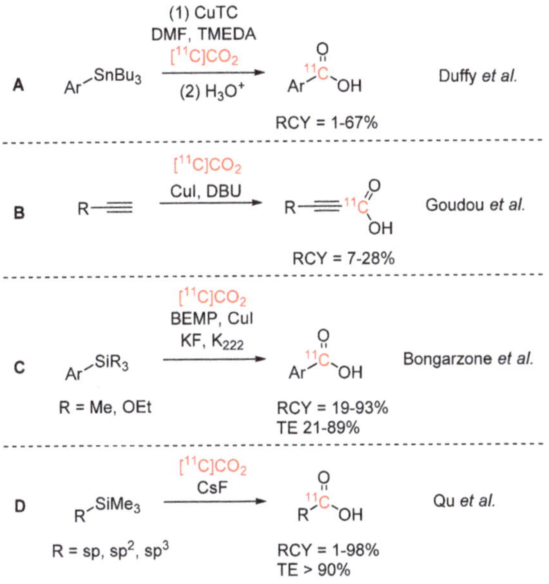

**Scheme 2.** New synthetic strategies for [$^{11}$C]carboxylic acids [24,28–30].

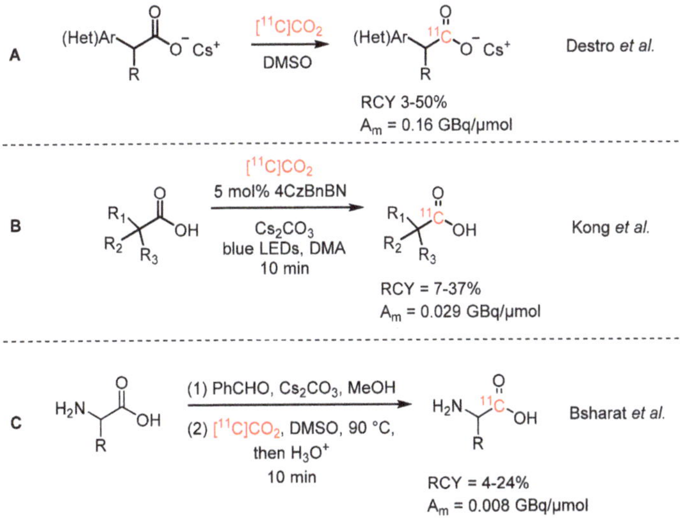

**Scheme 3.** Synthesis of [$^{11}$C]carboxylic acids via isotopic exchange [31–33].

Scheme 4 gives an overview of the proposed mechanism of [$^{11}$C]CO$_2$ fixation with fixation bases such as BEMP and DBU and formation of the [$^{11}$C]isocyanate, as well as the $^{11}$C-labelled functional groups that can be obtained via this pathway [18]. While early works focused on the synthesis of carbamates, the scope of $^{11}$C-labelled functional groups has broadened immensely over time.

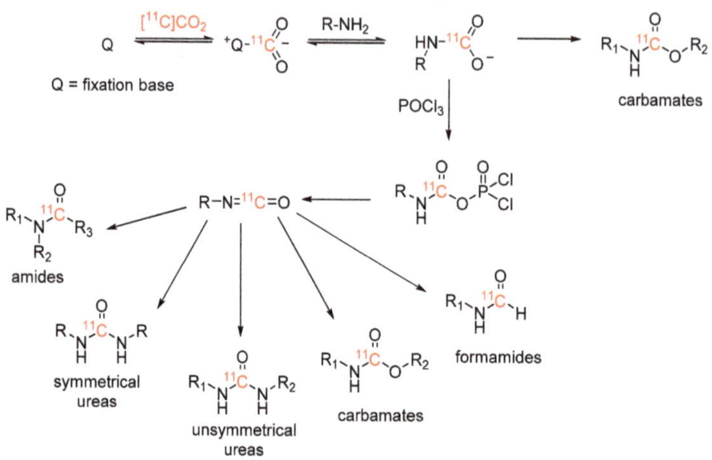

**Scheme 4.** [$^{11}$C]CO$_2$ fixation, [$^{11}$C]isocyanate formation and $^{11}$C-products.

The efficient syntheses of carbon-11 labelled amides, ureas, and formamides have been a longstanding goal in PET radiochemistry and have seen an emergence of interest in recent years. Bongarzone et al. reported a rapid one-pot synthesis of amides via a Mitsunobu reaction (see Scheme 5A) [34]. [$^{11}$C]CO$_2$ was trapped with DBU, converted to [$^{11}$C]isocyanate (or an [$^{11}$C]oxyphosphonium intermediate) using Mitsunobu reagents and subsequently reacted with a Grignard reagent to form the respective amide. RCYs of up to 50% were obtained. The substrate scope was not investigated for [$^{11}$C]CO$_2$, but [$^{11}$C]melatonin was synthesized to demonstrate the applicability of this method to biologically relevant compounds. Mair et al. used organozinc iodides as alternatives to Grignard reagents in a rhodium-catalyzed addition to [$^{11}$C]isocyanates (see Scheme 5B) [35]. The isocyanates were generated similarly to the previous method and reacted with the organozinc iodides in presence of a rhodium catalyst with RCYs of 5–99%. One model compound was isolated with a RCY of 12% and $A_m$ of 267 GBq/μmol to demonstrate suitability for PET tracer production. In order to develop an efficient synthesis strategy for the benzimidazolone PET tracer (S)-[$^{11}$C]CGP12177, Horkka et al. reported a BEMP/Mitsunobu-based strategy for the synthesis of cyclic aromatic ureas: *ortho*-Phenylenediamines were reacted with [$^{11}$C]CO$_2$ in presence of BEMP as fixation base. Mitsunobu reagents (DBAD, nBu$_3$P) were added to form the [$^{11}$C]isocyanate intermediates which then reacted intramolecularly to yield the respective $^{11}$C-labelled urea (see Scheme 5C) [36]. The strategy was also applied to cyclic carbamates and thiocarbamates, as well as the tracer (S)-[$^{11}$C]CGP12177, which was obtained in 23% RCY (dc) with a $A_m$ of 14 GBq/μmol. Luzi et al. reported the synthesis of [$^{11}$C]formamides (see Scheme 5D) [37]. [$^{11}$C]CO$_2$ was trapped with BEMP in diglyme and was reacted with aromatic and aliphatic primary amines to form the respective [$^{11}$C]isocyanates, which were subsequently reduced to the [$^{11}$C]formamides with sodium borohydride. The method performed better for aliphatic amines compared to aromatic amines.

In an attempt to make [$^{11}$C]CO$_2$ fixation with BEMP and DBU more widely accessible and amenable to automation, two strategies of "in-loop" [$^{11}$C]CO$_2$ fixation have been developed. While our laboratory developed this method using a standard stainless-steel HPLC loop for [$^{11}$C]CO$_2$ fixation, Downey et al. applied a disposable ethylene tetrafluoroethylene loop [38,39]. In both cases, [$^{11}$C]CO$_2$ was captured in the loop in the presence of an amine precursor and fixation base, prior to reaction with a model substrate. The "in-loop" fixation has been applied to synthesize $^{11}$C-labelled carbamates, unsymmetrical, and symmetrical ureas.

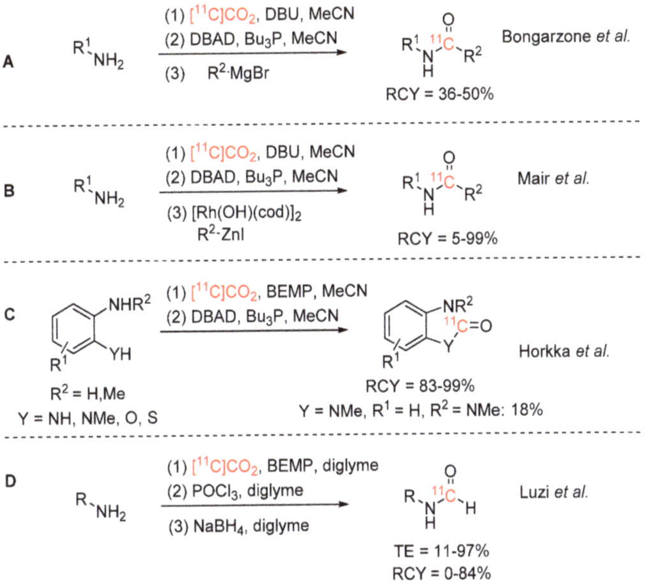

**Scheme 5.** Synthesis of [$^{11}$C]amides and [$^{11}$C]formamides via [$^{11}$C]isocyanates [34–37].

A different approach to access ureas and carbamates via [$^{11}$C]isocyanate was presented by Audisio and co-workers (see Scheme 6). The [$^{11}$C]isocyanate intermediates were generated through a Staudinger aza-Wittig reaction from the respective azide, then reacted either intramolecularly to form cyclic [$^{11}$C]ureas [40] and [$^{11}$C]carbamates [41] or intermolecularly with an amine to form linear ureas [42]. All three strategies were applied for the synthesis of $^{13}$C-, $^{14}$C- and $^{11}$C-labelled compounds. RCYs of the isolated $^{11}$C-compounds generally ranged between 20 to 50%.

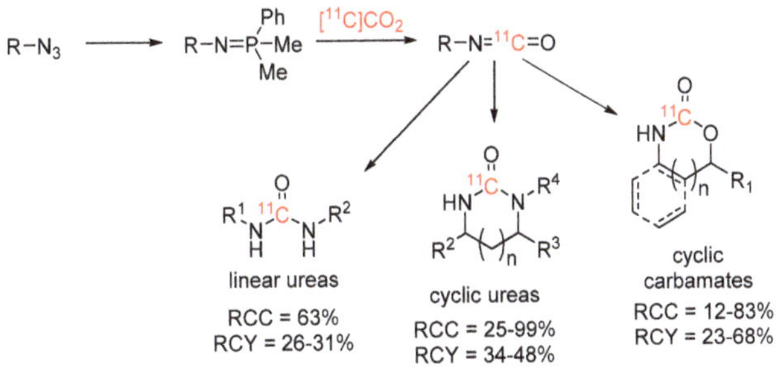

**Scheme 6.** $^{11}$C-labelled ureas and carbamates via the Staudinger aza-Wittig reaction [40–42].

To avoid the multi-step syntheses and limited substrate scope of previously reported methods, Liger et al. reported a novel radiolabelling strategy for benzimidazoles and benzothiazoles (see Scheme 7). In this work, [$^{11}$C]CO$_2$ was reacted with aromatic diamines and aminobenzenethiols in presence of 1,3-bis(2,6-diisopropylphenyl)imidazol-2-ylidene (IPr), zinc chloride, and phenylsilane as reducing reagent to obtain various benzimidazoles and benzothiazoles [43].

Previously, the synthesis of [$^{11}$C]carbonates could only be achieved using the esoteric building block [$^{11}$C]phosgene, which is technically challenging to prepare and requires specialized apparatus. To access this functional group directly from [$^{11}$C]CO$_2$, Dheere et al. developed a procedure involving an alkyl chloride, an alcohol, TBAI and base in DMF (see

Scheme 8) [44]. The procedure was used for the synthesis of one model compound, and resulted in either moderate RCY (31 ± 2%) and higher $A_m$ (10–20 GBq/μmol; low amounts of $^{11}$C), or high RCY (up to 82%) and lower $A_m$ (2 GBq/μmol), depending on the base.

**Scheme 7.** Synthesis of carbon-11 labelled benzimidazoles and benzothiazoles [43].

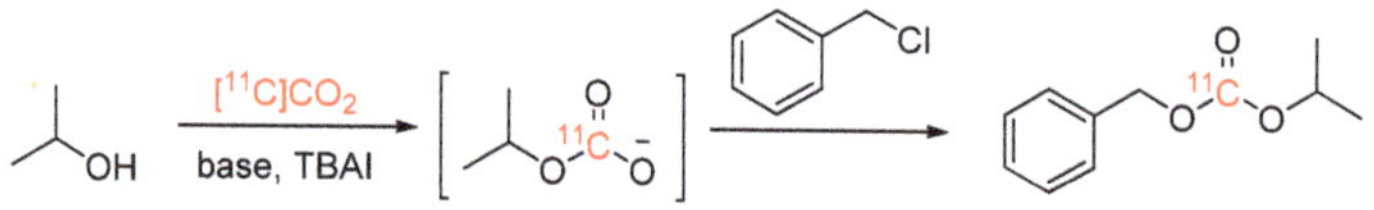

**Scheme 8.** Synthesis of a [$^{11}$C]carbonate from [$^{11}$C]CO$_2$ (RCY and $A_m$ are base-dependent) [44].

A novel method for ring-opening non-activated aziridines with [$^{11}$C]CO$_2$ using DBU/DBN halide ionic liquids was developed by our laboratory (see Scheme 9) [45]. [$^{11}$C]CO$_2$ was introduced to a pre-activated mixture of benzyl aziridine and the ionic liquid giving 4-benzyl [$^{11}$C]oxazolidine-2-one with 77% radiochemical conversion (RCC) and 78% TE. The method was applied to radiolabel an array of [$^{11}$C]oxazolidinones (RCCs 5–95%) as well as a MAO-B inhibitor, [$^{11}$C]toloxatone, as a proof of concept.

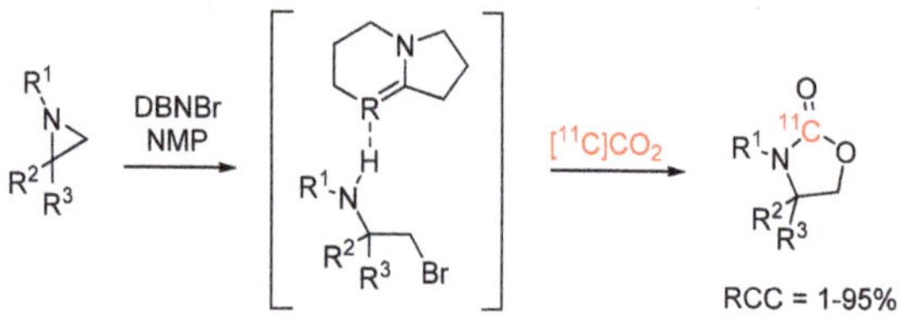

**Scheme 9.** Ring-opening of non-activated aziridines with [$^{11}$C]CO$_2$ [45].

### 2.2. [$^{11}$C]Carbon Monoxide

[$^{11}$C]Carbon monoxide has gained much interest in recent years. Novel $^{11}$CO-chemistry will not be covered within this review but we refer to recent comprehensive reviews of [$^{11}$C]CO production methods and $^{11}$C-carbonylation chemistry [46–50]. Although many straightforward routes for [$^{11}$C]CO production have been established, and a diverse portfolio of [$^{11}$C]carbonylation reactions has been developed, this branch of carbon-11 chemistry is still heavily underrepresented in PET tracer synthesis. In fact, of the 100+ labelled compounds synthesized from [$^{11}$C]CO, only four are reported for human use to our knowledge [51]. One likely reason for the hampered translation of [$^{11}$C]CO radiochemistry to the clinic can be attributed to the historic lack of commercially available automated synthesis units for [$^{11}$C]CO. This has now been overcome with systems such as the TracerMaker$^{TM}$ which is used by our laboratories for the syntheses of $N$-[$^{11}$C]acrylamide PET tracers for imaging Bruton's tyrosine kinase via a palladium-NiXantphos-mediated carbonylation using [$^{11}$C]CO [51,52]. The synthesis of the same class of compounds has also recently been automated as "in-loop" procedure using the GE TracerLab synthesis modules [53]. Prior to this recent work, $^{11}$C-labelled $N$-acrylamides were synthesized from [$^{11}$C]acrylic acid or [$^{11}$C]acryloyl chloride (formed by carboxylation of Grignard or organolithium reagents with [$^{11}$C]CO$_2$) and were not suitable for human translation.

### 2.3. [$^{11}$C]Methyl Iodide and Other $^{11}$C-Alkylation Agents

[$^{11}$C]Methyl iodide and [$^{11}$C]methyl triflate have been known for many decades [54–57] and are by far the most commonly used $^{11}$C-labelling agents. Their widespread use is attributed to their routine radiosyntheses and high reactivity. Both [$^{11}$C]methyl iodide and [$^{11}$C]methyl triflate can be easily synthesized from the primary cyclotron products (i.e., [$^{11}$C]CH$_4$ or [$^{11}$C]CO$_2$) using the classical wet-chemistry approach with lithium aluminium hydride and HI or the gas-phase method involving I$_2$, and dedicated synthesis devices with fully automated procedures are commercially available [58]. Mostly, [$^{11}$C]methyl iodide and [$^{11}$C]methyl triflate are employed in $^{11}$C-methylation reactions of hydroxyl, amine or thiol precursors, but also many different $^{11}$C-C coupling reactions have been established, including Suzuki, Stille, and Negishi couplings. For an overview of $^{11}$C-C cross-coupling strategies, we refer the reader to a comprehensive review from H. Doi [59]. Recent progress in the field has been made by Rokka et al., who systematically studied the reaction of various organoborane precursors with [$^{11}$C]methyl iodide in two different reaction media, DMF(/water) and THF/water, to determine the best precursor and solvent for Suzuki-type cross coupling reactions in $^{11}$C-chemistry [60]. These authors found that for their model compound (1-[$^{11}$C]methylnaphthalene), the boronic acid and pinacol ester precursors gave the highest yields, while the solvent mixture THF/water was equal or superior in any tested reaction. Recent work focused on broadening the substrate scope to diversify $^{11}$C-methylation chemistry. Pipal et al. reported the $^{11}$C-methylation of aromatic and aliphatic bromides via metallaphotoredox catalysis (see Scheme 10A) [61]. The applicability of this labelling strategy was demonstrated by synthesizing 11 $^{11}$C-labelled biologically active compounds, including the PET tracers [$^{11}$C]UCB-J and [$^{11}$C]PHNO, in RCYs of 13–72% for proof of concept. Qu et al. extended their fluoride-mediated desilylation of organosilanes, initially developed for [$^{11}$C]CO$_2$ fixation (vide supra), to [$^{11}$C]methyl iodide and succeeded in labelling a diverse library of silane substrates with RCYs of up to 93% (see Scheme 10B) [25].

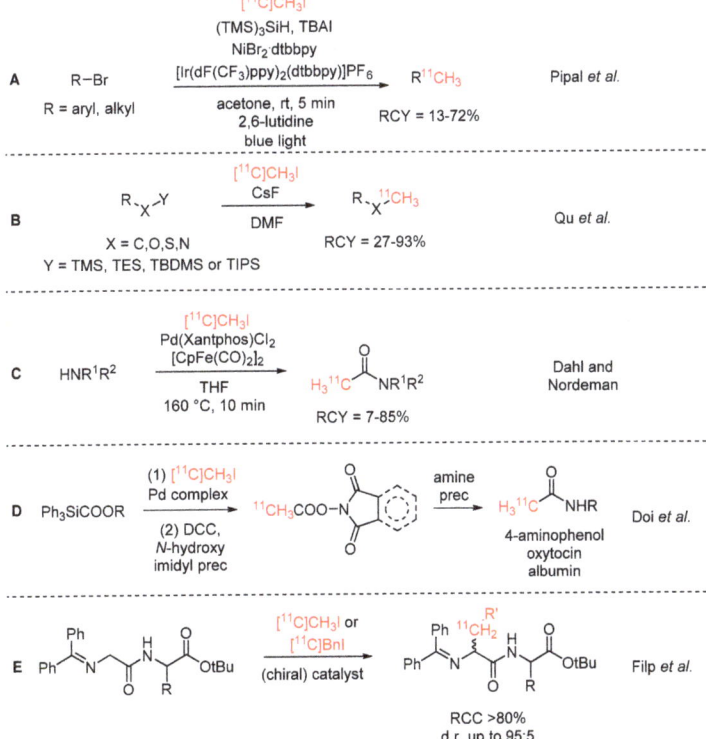

**Scheme 10.** Recent progress in [$^{11}$C]CH$_3$I chemistry [25,61–64].

As an alternative to [$^{11}$C]CO or [$^{11}$C]acetyl chloride chemistry, Dahl and Nordeman developed a procedure for $^{11}$C-acetylation of amines with [$^{11}$C]methyl iodide (see Scheme 10C) [62]. Bis(cyclopentadienyldicarbonyliron) was used as the CO source in the Pd-mediated reaction. The reaction was established for a range of primary amine precursors, including three biologically relevant compounds, and a few examples of secondary amines. A different approach to the same functional group was presented by Doi et al., whereby [$^{11}$C]acetic acid was synthesized in a palladium-mediated cross-coupling reaction from [$^{11}$C]methyl iodide and carboxytriphenylsilane, then converted to the [$^{11}$C]acetic acid phthalimidyl ester or succinimidyl ester (see Scheme 10D) [63]. The imidyl esters were subsequently employed in a $^{11}$C-acetylation reaction with small, medium-sized, and large molecules.

In an effort to develop a stereoselective $^{11}$C-alkylation procedure for diastereomerically enriched dipeptides, Filp et al. investigated the use of various quaternary ammonium salts as chiral phase-transfer catalysts in the $^{11}$C-alkylation of N-terminal glycine Schiff bases (see Scheme 10E) [64]. Next to [$^{11}$C]methyl iodide, the procedure was also applied to [$^{11}$C]benzyl iodide. RCCs of >80% and high diastereomeric ratios (d.r.) of up to 95:5 were obtained. A similar strategy has been used by Pekošak et al. for the stereoselective $^{11}$C-labelling of the tetrapeptide Phe-D-Trp-Lys-Thr with [$^{11}$C]benzyl iodide [65]. [$^{11}$C]Phe-D-Trp-Lys-Thr was synthesized over five steps starting from [$^{11}$C]CO$_2$ and isolated with high stereoselectivity (94% de), RCYs of 9–10% (dc), and A$_m$ of 15–35 GBq/μmol.

To address the shortcomings of current cross-coupling strategies with [$^{11}$C]methyl iodide, Helbert et al. developed a new cross-coupling procedure with [$^{11}$C]methyllithium. [$^{11}$C]Methyllithium was developed as a more reactive alternative for [$^{11}$C]methyl iodide and can be synthesized by reaction of [$^{11}$C]methyl iodide with n-butyllithium [66,67]. In the procedure of Herbert et al., [$^{11}$C]methyllithium was added without intermediate purification to the aryl bromide precursors and a selection of relevant PET tracers was labelled by palladium-mediated $^{11}$C-C cross-coupling with RCYs of 33–57% (see Scheme 11) [68].

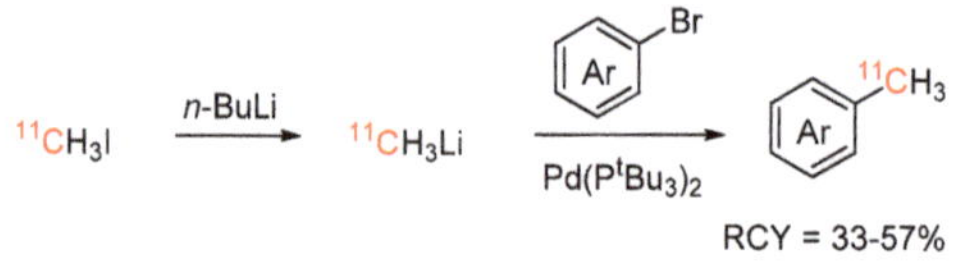

**Scheme 11.** $^{11}$C-C cross-coupling with [$^{11}$C]methyllithium [68].

### 2.4. [$^{11}$C]Hydrogen Cyanide

Since its inception in the 1960s [69], [$^{11}$C]HCN has developed into a versatile building block for the $^{11}$C-labelling of neurotransmitters, amino acids, and other molecules. This is mainly due to its versatility: It can function as nucleophile as well as electrophile, and [$^{11}$C]cyanide incorporation generates many different functionalities, such as nitriles, hydantoins, (thio)cyanates and, through subsequent reaction, carboxylic acids, aldehydes, amides and amines. Two extensive reviews on [$^{11}$C]hydrogen cyanide have been recently published, therefore this $^{11}$C-building block will not be discussed in detail herein [70,71]. Since [$^{11}$C]hydrogen cyanide is one of the few $^{11}$C-building blocks used for FIH PET tracers in recent years (vide infra), we will provide a brief summary of recent work that has not been covered by other reviews.

[$^{11}$C]Hydrogen cyanide is typically produced by reacting [$^{11}$C]CH$_4$ with NH$_3$ gas on a platinum catalyst at 1000 °C. While fully automated production systems are commercially available, [$^{11}$C]HCN is not widely used. In an effort to make [$^{11}$C]hydrogen cyanide more accessible, Kikuchi et al. developed a novel synthesis strategy from widely available [$^{11}$C]methyl iodide (see Scheme 12) [72]. This method involves passing [$^{11}$C]methyl iodide over a heated reaction column, in which it is first converted to [$^{11}$C]formaldehyde and subsequently to [$^{11}$C]hydrogen cyanide. The [$^{11}$C]hydrogen cyanide is obtained fast and with RCYs comparable to the traditional method (50–60% at EOB), without the need for specialized equipment.

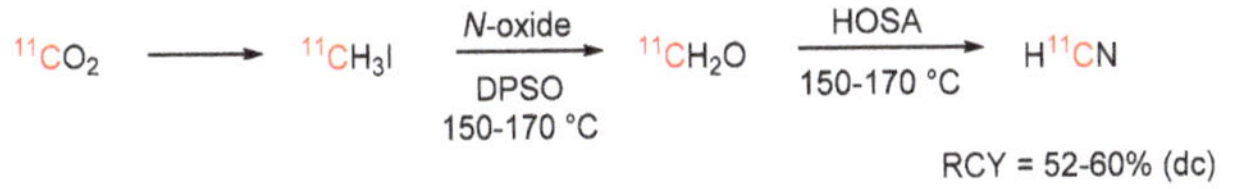

**Scheme 12.** Production of [$^{11}$C]hydrogen cyanide from [$^{11}$C]methyl iodide [72].

### 2.5. [$^{11}$C]Fluoroform

Due to the prevalence of CF$_3$ groups in drugs and other biologically active compounds, there has been much interest in labelling this group with carbon-11 and fluorine-18. Haskali et al. published a synthesis procedure for carbon-11 labelled fluoroform in 2017, where cyclotron-produced [$^{11}$C]methane was fluorinated by passing it over a CoF$_3$ column at elevated temperatures (270 °C) [73]. [$^{11}$C]Fluoroform was obtained with RCYs of ~60%. The process was not only fast and reproducible, but the developed system also required very little maintenance. [$^{11}$C]Fluoroform was reacted with various model compounds (see Scheme 13), in addition to three biologically active compounds.

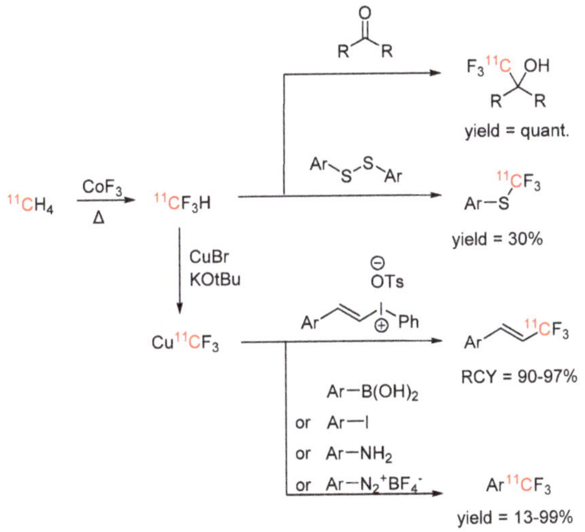

**Scheme 13.** [$^{11}$C]Fluoroform chemistry [73–75].

Whereas flourine-18 labelled fluoroform generally suffers from low molar activities ($\leq$1 GBq/µmol) and only few examples of higher molar activities are known, high molar activities of >200 GBq/µmol were easily obtained with carbon-11 labelled fluoroform. In later works, the substrate scope of reactions with [$^{11}$C]fluoroform was broadened from aryl boronates, aryl iodides, ketones, diazonium salts, and diarylsulfanes to aryl amines and arylvinyl iodonium tosylates (see Scheme 13) [74,75].

### 2.6. [$^{11}$C]Carbonyl Difluoride

As an alternative strategy to access carbon-11 labelled ureas, carbamates, and thiocarbamates, Jakobsson et al. presented the [$^{11}$C]carbonyl group transfer agent [$^{11}$C]carbonyl difluoride [76]. [$^{11}$C]Carbonyl difluoride was synthesized quantitatively by passing [$^{11}$C]CO over a AgF$_2$ column at room temperature. The building block was subsequently reacted with diamines, aminoalcohols, and aminothiols to form the corresponding cyclic azolidin-2-ones (see Scheme 14) under mild conditions with very low precursor quantities, and even in presence of water. The same laboratory expanded their procedure to linear unsymmetrical ureas and established reaction conditions for a broad scope of aryl and aliphatic amines [77]. For the aryl amines, [$^{11}$C]carbonyl fluoride was trapped in a solution with the aryl amine precursor and subsequently reacted with another amine. For the aliphatic amines, alkylammonium tosylate precursors were used in the first step to lower the reactivity of the amine and prevent symmetrical urea formation. Pyridine was used to improve [$^{11}$C]carbonyl fluoride trapping. Suitability for PET tracer synthesis was demonstrated by labelling the epoxide hydrolase inhibitor [$^{11}$C]AR-9281, which was obtained after optimization in high RCYs of 80%.

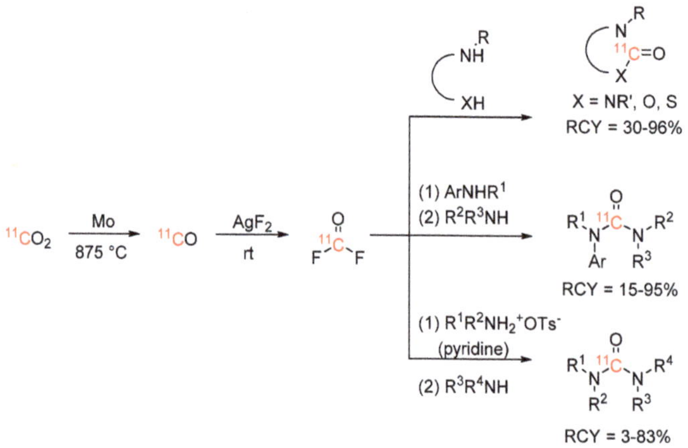

**Scheme 14.** [$^{11}$C]Carbonyl difluoride synthesis and subsequent reaction to cyclic products and linear unsymmetrical ureas [76,77].

## 2.7. [$^{11}$C]Carbon Disulfide

[$^{11}$C]Carbon disulfide, the sulfur analog of [$^{11}$C]carbon dioxide, is an interesting $^{11}$C-building block for the synthesis of organosulfur compounds. It has first been described in 1984, and had limited utility until a decade ago when Miller and Bender proposed a new synthesis strategy, which was further improved by Haywood et al. [78–80]. It can now be readily obtained through the reaction of [$^{11}$C]CH$_3$I with elemental sulfur and has been used to synthesize [$^{11}$C]thioureas, thiocarbamates and related structures. Cesarec et al. recently published a procedure for the synthesis of late transition metal complexes with [$^{11}$C]dithiocarbamate ligands [81]. To this end, [$^{11}$C]carbon disulfide was reacted with diethyl amine or dibenzyl amine to form the respective ammonium [$^{11}$C]dithiocarbamate salt and subsequently reacted with Au(I), Au(III), Pd(II) or Pt(II) complexes to form the respective complexes in RCYs > 70% (see Scheme 15).

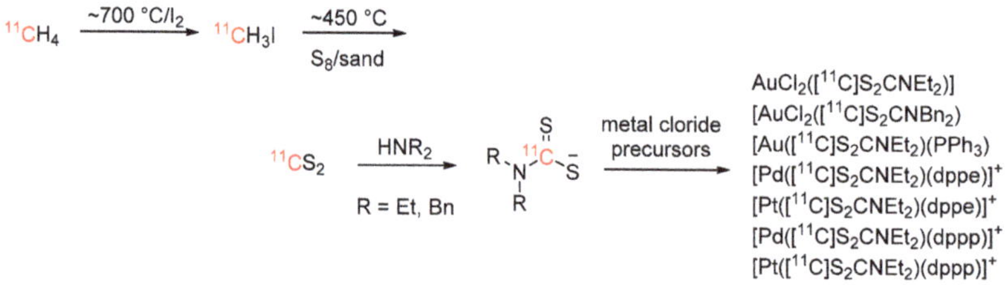

**Scheme 15.** Synthesis of [$^{11}$C]carbon disulfide and formation of [$^{11}$C]dithiocarbamate transition metal complexes [81].

## 2.8. [$^{11}$C]Thiocyanate

[$^{11}$C]Thiocyanate is an interesting $^{11}$C-building block because of its reactivity and potential to give access to a wide range of organosulfur derivatives. Up until recently, its production relied on the use of [$^{11}$C]HCN [82]. Haywood et al. presented a new way to synthesize this $^{11}$C-building block by reacting [$^{11}$C]carbon disulfide with ammonia at 90 °C to form ammonium [$^{11}$C]thiocyanate in near quantitative RCC [83]. The ammonium [$^{11}$C]thiocyanate was subsequently reacted with benzyl bromide, a range of α-ketobromides, and mannose triflate in high RCYs of ≥75% (see Scheme 16). The α-[$^{11}$C]thiocyanatophenones could also be cyclized in the presence of sulfuric and acetic acid to $^{11}$C-thiazolones.

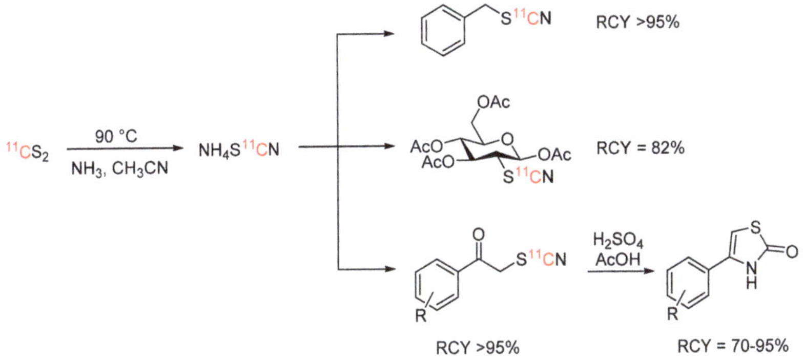

Scheme 16. Synthesis of and reactions with ammonium [$^{11}$C]thiocyanate [83].

## 2.9. [$^{11}$C]Formaldehyde

[$^{11}$C]Formaldehyde is an established and versatile building block for carbon-11 chemistry (see [84] and references therein). Many different synthetic strategies have been proposed, traditionally involving reduction in cyclotron-produced [$^{11}$C]CO$_2$ to [$^{11}$C]CH$_3$OH and subsequent oxidation to [$^{11}$C]formaldehyde. Nader et al. recently proposed the use of XeF$_2$ as an oxidizing agent (see Scheme 17) [85]. [$^{11}$C]Formaldehyde was obtained in non-decay corrected RCYs of 54 ± 5% starting from [$^{11}$C]CO$_2$ and was used in a proof-of principle synthesis to form α-(N-[$^{11}$C]methylamino)isobutyric acid via reductive $^{11}$C-methylation.

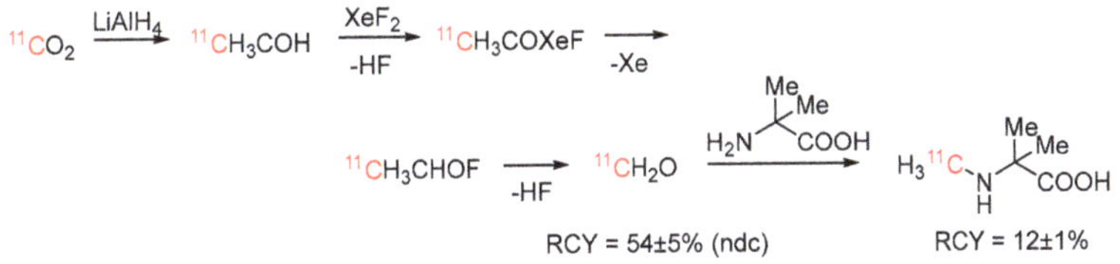

Scheme 17. Synthesis of [$^{11}$C]formaldehyde using XeF$_2$ as oxidizing agent and reaction to [$^{11}$C]Me-AIB [85].

## 3. First-in-Human Translation

Despite the short physical half-life and the need for an on-site cyclotron, $^{11}$C continues to be a favoured radionuclide for small molecule PET tracers. As discussed in this review, innovations continue in $^{11}$C-radiolabelling strategies for applications in $^{11}$C-tracer development. Within the past five years, to our knowledge at least 27 novel $^{11}$C-labelled PET tracers have been translated for FIH PET studies (see Figure 1) [86–109]. Unsurprisingly, the vast majority of these PET tracers were designed to image targets within the CNS (see Table 1). Carbon-11 is ideal for CNS PET because the substitution of naturally occurring $^{12}$C with $^{11}$C does not change the physicochemical properties of the compound, thereby enabling imaging with isotopologues of the molecules of interest for accurate determination of brain penetrance, target affinity, pharmacokinetics, or pharmacodynamics of the molecule, and multiple scans can be performed in the same subject in the same day. Interestingly, many of the ¹¹C-labelled PET tracers for FIH use focused on imaging markers of neuroinflammation, a critical component in the etiology and pathology of several neurodegenerative diseases, including Alzheimer's disease (AD), Parkinson's disease (PD), and amyotrophic lateral sclerosis (ALS) [110–114]. The remaining PET tracers translated for FIH studies that were reported in the past five years strived to image non-CNS targets, including bacterial infection and lung inflammation.

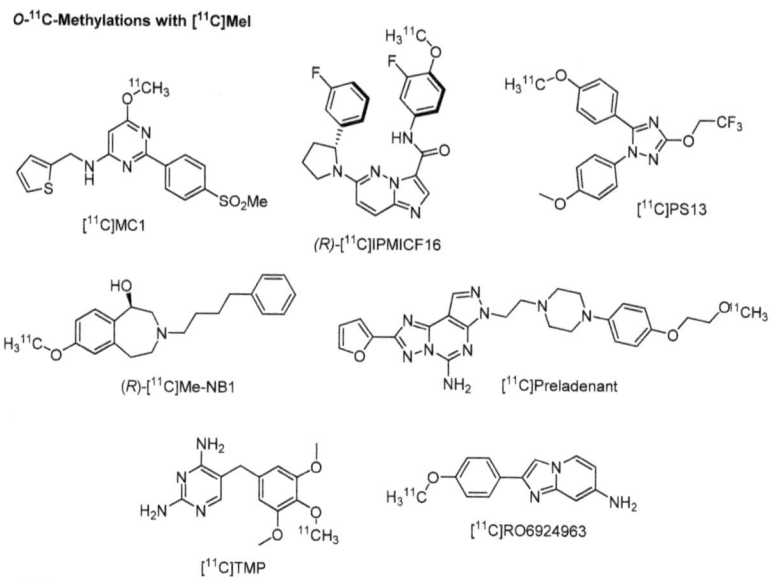

Figure 1. Cont.

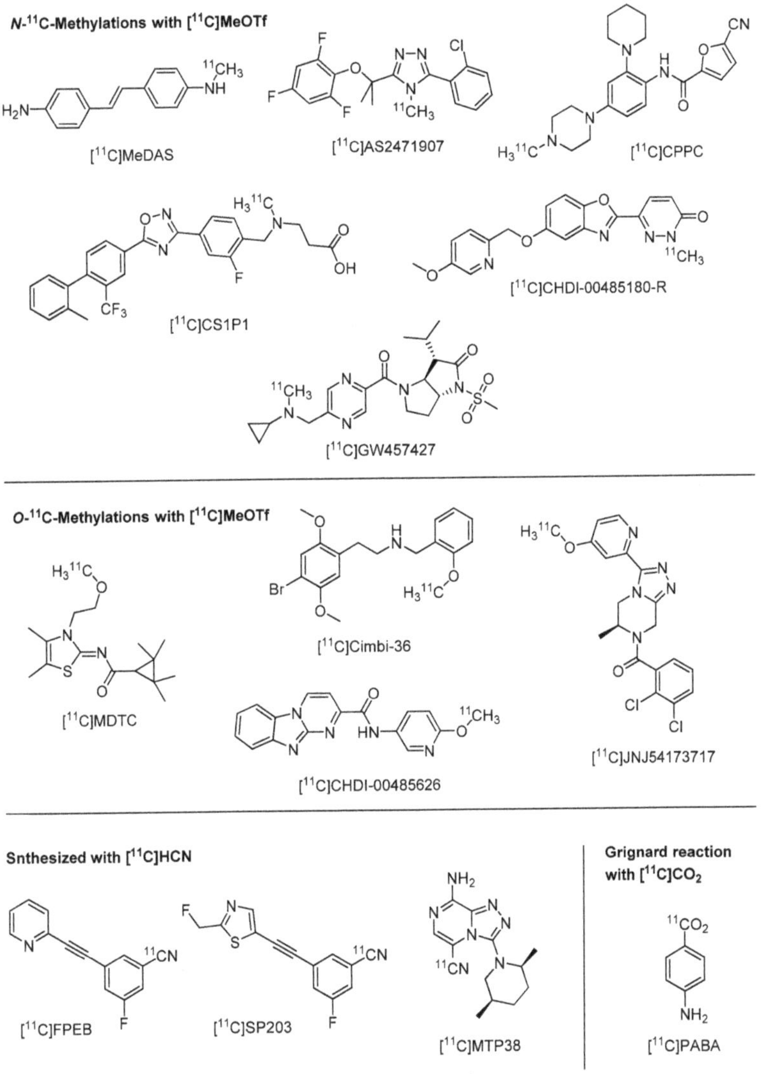

**Figure 1.** Chemical structures of the majority of first-in-human PET tracers labelled with $^{11}$C since 2017. Targets, publication years and references of the tracers are listed in Table 1.

When sorting the tracers according to the labelling method, it becomes immediately apparent that the predominant synthetic strategy remains $^{11}$C-methylation of hydroxy or amino precursors: more than $\frac{3}{4}$ of all tracers were synthesized via this strategy, either using [$^{11}$C]methyl iodide or [$^{11}$C]methyl triflate as the $^{11}$C-buidling block (see Figure 2). This can be attributed to the accessibility of these building blocks from cyclotron-produced [$^{11}$C]CO$_2$ or [$^{11}$C]CH$_4$ and the availability of commercial synthesis devices (vide supra) [58]. Other tracers have been synthesized by alternate $^{11}$C-labelling strategies, using [$^{11}$C]HCN or Grignard reactions with [$^{11}$C]CO$_2$. The latest developments in $^{11}$C-chemistry are not represented among the FIH tracers, which is not surprising since it usually takes time for a new method to be implemented by the broader community. However, many of the existing $^{11}$C-building blocks have not been introduced in the past few years but have been around for decades and should, therefore, be available for clinical application. As indicated in some cases (e.g., [$^{11}$C]CO chemistry), it may be the historic lack of specialized or commercially available radiosynthesis equipment that hampers FIH translation of new PET tracers. Other reasons could be that new $^{11}$C-methodologies are often only developed up to the point of proof-of-principle and not optimized for automated tracer production. Rather than further broadening the scope of $^{11}$C-chemistry, future efforts should focus on closing the gap between new method development and clinical translation.

Table 1. First-in-human $^{11}$C PET tracers and their targets reported since 2017.

| Tracer | Target | $^{11}$C-Building Block | Year | Ref. |
| --- | --- | --- | --- | --- |
| [$^{11}$C]ER176 | TSPO | [$^{11}$C]CH$_3$I | 2017 | [92] |
| [$^{11}$C]K-2 | AMPA receptors | [$^{11}$C]CH$_3$I | 2020 | [96] |
| [$^{11}$C]rifampin | Tuberculosis meningitis | [$^{11}$C]CH$_3$I | 2018 | [105] |
| [$^{11}$C]MC1 | COX-2 | [$^{11}$C]CH$_3$I | 2020 | [98] |
| (R)-[$^{11}$C]IPMICF16 | TrkB/C receptors | [$^{11}$C]CH$_3$I | 2017 | [87] |
| [$^{11}$C]RO6924963 | Tau | [$^{11}$C]CH$_3$I | 2018 | [109] |
| [$^{11}$C]RO6931643 | Tau | [$^{11}$C]CH$_3$I | 2018 | [109] |
| (R)-[$^{11}$C]Me-NB1 | GluN2B-containing NMDA receptors | [$^{11}$C]CH$_3$I | 2022 | [86] |
| [$^{11}$C]Preladenant | Adenosine A$_{2A}$ receptors | [$^{11}$C]CH$_3$I | 2017 | [103] |
| [$^{11}$C]TMP | Bacterial infection | [$^{11}$C]CH$_3$I | 2021 | [107] |
| [$^{11}$C]PS13 | COX-1 | [$^{11}$C]CH$_3$I | 2020 | [104] |
| [$^{11}$C]TTFD | Drug pharmacokinetics | [$^{11}$C]CH$_3$I | 2021 | [108] |
| [$^{11}$C]LSN3172176 | M1 muscarinic acetylcholine receptors | [$^{11}$C]CH$_3$I | 2020 | [97] |
| [$^{11}$C]MeDAS | Myelin | [$^{11}$C]MeOTf | 2022 | [100] |
| [$^{11}$C]AS2471907 | 11ß-hydroxysteroid dehydrogenase type 1 | [$^{11}$C]MeOTf | 2019 | [88] |
| [$^{11}$C]CPPC | CSF1 receptor | [$^{11}$C]MeOTf | 2022 | [91] |
| [$^{11}$C]CS1P1 | Sphingosine-1-phoshate receptor 1 | [$^{11}$C]MeOTf | 2022 | [93] |
| [$^{11}$C]CHDI-00485180-R | mHTT | [$^{11}$C]MeOTf | 2022 | [89] |
| [$^{11}$C]GW457427 | Neutrophil elastase | [$^{11}$C]MeOTf | 2022 | [94] |
| [$^{11}$C]MDTC | CB2 receptor | [$^{11}$C]MeOTf | 2022 | [99] |
| [$^{11}$C]Cimbi-36 | 5-HT2A receptor | [$^{11}$C]MeOTf | 2019 | [90] |
| [$^{11}$C]CHDI-00485626 | mHTT | [$^{11}$C]MeOTf | 2022 | [89] |
| [$^{11}$C]JNJ54173717 | P2X7 receptor | [$^{11}$C]MeOTf | 2019 | [95] |
| [$^{11}$C]FPEB | mGluR5 | [$^{11}$C]HCN | 2017 | [106] |
| [$^{11}$C]SP203 | mGluR5 | [$^{11}$C]HCN | 2017 | [106] |
| [$^{11}$C]MTP38 | PDE7 | [$^{11}$C]HCN | 2021 | [101] |
| [$^{11}$C]PABA | Renal imaging | [$^{11}$C]CO$_2$ | 2020 | [102] |

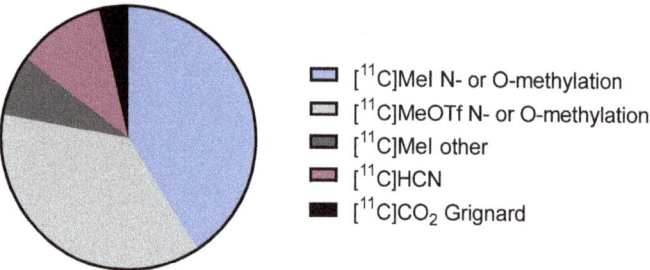

- [$^{11}$C]MeI N- or O-methylation
- [$^{11}$C]MeOTf N- or O-methylation
- [$^{11}$C]MeI other
- [$^{11}$C]HCN
- [$^{11}$C]CO$_2$ Grignard

Figure 2. Graphical representation of the proportions of $^{11}$C-labelling strategies used for FIH PET tracers since 2017.

**Author Contributions:** Conceptualization, A.P. and N.V.; writing—original draft preparation, A.P.; writing—review and editing, M.C., A.L. and N.V.; visualization, A.P.; supervision, N.V. All authors have read and agreed to the published version of the manuscript.

**Funding:** A.P. is supported by the CAMH Discovery Fund. M.C. was supported by the Canadian Institute for Health Research with a Canadian Graduate Scholarship (CGS-M), as well as a Mitacs Accelerate Internship award provided by the Structural Genomics Consortium (SGC). N.V. is supported by the Azrieli Foundation, the Canada Research Chairs Program, Canada Foundation for Innovation and the Ontario Research Fund.

**Institutional Review Board Statement:** Not applicable.

**Informed Consent Statement:** Not applicable.

**Data Availability Statement:** Not applicable.

**Conflicts of Interest:** The authors declare no relevant conflict of interest.

## Abbreviations

| | |
|---|---|
| 4CzBnBN | (2,3,4,6)-3-Benzyl-2,4,5,6-tetra(9H-carbazol-9-yl)benzonitrile |
| 5-HT$_{2A}$ receptor | 5-Hydroxy-tryptamine 2A receptor |
| AD | Alzheimer's disease |
| ALS | Amyotrophic lateral sclerosis |
| A$_m$ | Molar activity |
| AMPA receptor | α-amino-3-hydroxy-5-methyl-4-isoxazolepropionic acid receptor |
| BEMP | 2-*tert*-butylimino-2-diethylamino-1,3-dimethylperhydro-1,3,2-diazaphosphorine |
| CB2 receptor | Cannabinoid receptor type 2 |
| CNS | Central nervous system |
| COX-1, -2 | Cyclooxygenase-1, -2 |
| CSF1 | colony stimulating factor 1 |
| CuTC | Copper(I) thiophene-2-carboxylate |
| DBAD | Di-*tert*-butyl azodicarboxylate |
| DBN | 1,5-Diazabicyclo [4.3.0]non-5-ene |
| DBU | 1,8-diazabicyclo [5.4.0]undec-7-ene |
| dc | Decay-corrected |
| DCC | N,N′-Dicyclohexylcarbodiimide |
| de | Diastereomeric excess |
| DMA | Dimethylacetamide |
| DMF | Dimethylformamide |
| DMSO | Dimethyl sulfoxide |
| DPSO | Diphenyl sulfoxide |
| d.r. | Diastereomeric ratio |
| dtbbpy | 4,4′-Di-*tert*-butyl-2,2′-bipyridine |
| FIH | First-in-human |
| FMDS | Fluoride-mediated desilylation |
| HOSA | Hydroxylamine-*O*-sulfonic acid |
| HPLC | High-performance liquid chromatography |
| IPr | 1,3-bis(2,6-diisopropylphenyl)imidazol-2-ylidene) |
| K$_{222}$ | 4,7,13,16,21,24-Hexaoxa-1,10-diazabicyclo [8.8.8]hexacosane |
| MAO-B | Monoamine oxidase B |
| mGluR5 | Metabotropic glutamate receptor 5 |
| mHTT | Mutant huntingtin protein |
| NMDA | *N*-methyl-D-aspartate |
| NMP | *N*-Methyl-2-pyrrolidone |
| PD | Parkinson's disease |
| PDE7 | Phosphodiesterase 7 |
| PET | Positron emission tomography |
| ppy | 2-Phenylpyridine |
| prec | precursor |

| | |
|---|---|
| quant. | quantitative |
| RCC | Radiochemical conversion |
| RCP | Radiochemical purity |
| RCY | Radiochemical yield |
| Ref. | Reference |
| rt | Room temperature |
| $t_{1/2}$ | Half-life |
| TBAT | Tetrabutylammonium difluorotriphenylsilicate |
| TBAI | Tetra-*n*-butylammonium iodide |
| TE | Trapping efficiency |
| THF | Tetrahydrofuran |
| TMEDA | Tetramethylethylenediamine |
| TMS | Trimetylsilyl |
| TrkB/C | Tropomyosin receptor kinase B/C |
| TSPO | Translocator protein |

## References

1. Liang, S.H.; Vasdev, N. Total Radiosynthesis: Thinking Outside 'the Box'. *Aust. J. Chem.* **2015**, *68*, 1319–1328. [CrossRef] [PubMed]
2. Deng, X.; Rong, J.; Wang, L.; Vasdev, N.; Zhang, L.; Josephson, L.; Liang, S.H. Chemistry for Positron Emission Tomography: Recent Advances in $^{11}$C-, $^{18}$F-, $^{13}$N-, and $^{15}$O-Labeling Reactions. *Angew. Chem. Int. Ed.* **2019**, *58*, 2580–2605. [CrossRef]
3. Ajenjo, J.; Destro, G.; Cornelissen, B.; Gouverneur, V. Closing the Gap between $^{19}$F and $^{18}$F Chemistry. *EJNMMI Radiopharm. Chem.* **2021**, *6*, 33. [CrossRef]
4. Van der Born, D.; Pees, A.; Poot, A.J.; Orru, R.V.A.; Windhorst, A.D.; Vugts, D.J. Fluorine-18 Labelled Building Blocks for PET Tracer Synthesis. *Chem. Soc. Rev.* **2017**, *46*, 4709–4773. [CrossRef] [PubMed]
5. Bernard-Gauthier, V.; Lepage, M.L.; Waengler, B.; Bailey, J.J.; Liang, S.H.; Perrin, D.M.; Vasdev, N.; Schirrmacher, R. Recent Advances in $^{18}$F Radiochemistry: A Focus on B-$^{18}$F, Si-$^{18}$F, Al-$^{18}$F, and C-$^{18}$F Radiofluorination via Spirocyclic Iodonium Ylides. *J. Nucl. Med.* **2018**, *59*, 568–572. [CrossRef]
6. Goud, N.S.; Joshi, R.K.; Bharath, R.D.; Kumar, P. Fluorine-18: A Radionuclide with Diverse Range of Radiochemistry and Synthesis Strategies for Target Based PET Diagnosis. *Eur. J. Med. Chem.* **2020**, *187*, 111979. [CrossRef] [PubMed]
7. Wang, Y.; Lin, Q.; Shi, H.; Cheng, D. Fluorine-18: Radiochemistry and Target-Specific PET Molecular Probes Design. *Front. Chem.* **2022**, *10*, 884517. [CrossRef]
8. Bratteby, K.; Shalgunov, V.; Herth, M.M. Aliphatic $^{18}$F-Radiofluorination: Recent Advances in the Labeling of Base-Sensitive Substrates. *ChemMedChem* **2021**, *16*, 2612–2622. [CrossRef]
9. Francis, F.; Wuest, F. Advances in [$^{18}$F]Trifluoromethylation Chemistry for PET Imaging. *Molecules* **2021**, *26*, 6478. [CrossRef]
10. Wright, J.S.; Kaur, T.; Preshlock, S.; Tanzey, S.S.; Winton, W.P.; Sharninghausen, L.S.; Wiesner, N.; Brooks, A.F.; Sanford, M.S.; Scott, P.J.H. Copper-Mediated Late-Stage Radiofluorination: Five Years of Impact on Preclinical and Clinical PET Imaging. *Clin. Transl. Imaging* **2020**, *8*, 167–206. [CrossRef]
11. Bui, T.T.; Kim, H. Recent Advances in Photo-mediated Radiofluorination. *Chem. Asian J.* **2021**, *16*, 2155–2167. [CrossRef]
12. Goud, N.S.; Bhattacharya, A.; Joshi, R.K.; Nagaraj, C.; Bharath, R.D.; Kumar, P. Carbon-11: Radiochemistry and Target-Based PET Molecular Imaging Applications in Oncology, Cardiology, and Neurology. *J. Med. Chem.* **2021**, *64*, 1223–1259. [CrossRef] [PubMed]
13. Boscutti, G.; Huiban, M.; Passchier, J. Use of Carbon-11 Labelled Tool Compounds in Support of Drug Development. *Drug Discov. Today Technol.* **2017**, *25*, 3–10. [CrossRef] [PubMed]
14. Dahl, K.; Halldin, C.; Schou, M. New Methodologies for the Preparation of Carbon-11 Labeled Radiopharmaceuticals. *Clin. Transl. Imaging* **2017**, *5*, 275–289. [CrossRef] [PubMed]
15. Rotstein, B.H.; Liang, S.H.; Placzek, M.S.; Hooker, J.M.; Gee, A.D.; Dollé, F.; Wilson, A.A.; Vasdev, N. $^{11}$C=O Bonds Made Easily for Positron Emission Tomography Radiopharmaceuticals. *Chem. Soc. Rev.* **2016**, *45*, 4708–4726. [CrossRef]
16. Pike, V.W. PET Radiotracers: Crossing the Blood–Brain Barrier and Surviving Metabolism. *Trends Pharmacol. Sci.* **2009**, *30*, 431–440. [CrossRef]
17. Coenen, H.H.; Gee, A.D.; Adam, M.; Antoni, G.; Cutler, C.S.; Fujibayashi, Y.; Jeong, J.M.; Mach, R.H.; Mindt, T.L.; Pike, V.W.; et al. Consensus Nomenclature Rules for Radiopharmaceutical Chemistry—Setting the Record Straight. *Nucl. Med. Biol.* **2017**, *55*, v–xi. [CrossRef]
18. Herth, M.M.; Ametamey, S.; Antuganov, D.; Bauman, A.; Berndt, M.; Brooks, A.F.; Bormans, G.; Choe, Y.S.; Gillings, N.; Häfeli, U.O.; et al. On the Consensus Nomenclature Rules for Radiopharmaceutical Chemistry—Reconsideration of Radiochemical Conversion. *Nucl. Med. Biol.* **2021**, *93*, 19–21. [CrossRef]
19. Hooker, J.M.; Reibel, A.T.; Hill, S.M.; Schueller, M.J.; Fowler, J.S. One-Pot, Direct Incorporation of [$^{11}$C]CO$_2$ into Carbamates. *Angew. Chem. Int. Ed.* **2009**, *48*, 3482–3485. [CrossRef]
20. Wilson, A.A.; Garcia, A.; Houle, S.; Sadovski, O.; Vasdev, N. Synthesis and Application of Isocyanates Radiolabeled with Carbon-11. *Chem. Eur. J.* **2011**, *17*, 259–264. [CrossRef]

21. Wilson, A.A.; Garcia, A.; Houle, S.; Vasdev, N. Direct Fixation of [$^{11}$C]CO$_2$ by Amines: Formation of [$^{11}$C-Carbonyl]-Methylcarbamates. *Org. Biomol. Chem.* **2010**, *8*, 428–432. [CrossRef] [PubMed]
22. Rotstein, B.H.; Liang, S.H.; Holland, J.P.; Collier, T.L.; Hooker, J.M.; Wilson, A.A.; Vasdev, N. $^{11}$CO$_2$ Fixation: A Renaissance in PET Radiochemistry. *Chem. Commun.* **2013**, *49*, 5621–5629. [CrossRef]
23. Chassé, M.; Sen, R.; Goeppert, A.; Prakash, G.K.S.; Vasdev, N. Polyamine Based Solid CO$_2$ Adsorbents for [$^{11}$C]CO$_2$ Purification and Radiosynthesis. *J. CO2 Util.* **2022**, *64*, 102137. [CrossRef]
24. Duffy, I.R.; Vasdev, N.; Dahl, K. Copper(I)-Mediated $^{11}$C-Carboxylation of (Hetero)Arylstannanes. *ACS Omega* **2020**, *5*, 8242–8250. [CrossRef]
25. Riss, P.J.; Lu, S.; Telu, S.; Aigbirhio, F.I.; Pike, V.W. CuI-Catalyzed $^{11}$C Carboxylation of Boronic Acid Esters: A Rapid and Convenient Entry to $^{11}$C-Labeled Carboxylic Acids, Esters, and Amides. *Angew. Chem. Int. Ed.* **2012**, *51*, 2698–2702. [CrossRef]
26. Rotstein, B.H.; Hooker, J.M.; Woo, J.; Collier, T.L.; Brady, T.J.; Liang, S.H.; Vasdev, N. Synthesis of [$^{11}$C]Bexarotene by Cu-Mediated [$^{11}$C]Carbon Dioxide Fixation and Preliminary PET Imaging. *ACS Med. Chem. Lett.* **2014**, *5*, 668–672. [CrossRef]
27. García-Vázquez, R.; Battisti, U.M.; Shalgunov, V.; Schäfer, G.; Barz, M.; Herth, M.M. [$^{11}$C]Carboxylated Tetrazines for Facile Labeling of Trans-Cyclooctene-Functionalized PeptoBrushes. *Macromol. Rapid Commun.* **2022**, *43*, 2100655. [CrossRef]
28. Goudou, F.; Gee, A.D.; Bongarzone, S. Carbon-11 Carboxylation of Terminal Alkynes with [$^{11}$C]CO$_2$. *J. Label. Compd. Radiopharm.* **2021**, *64*, 237–242. [CrossRef]
29. Bongarzone, S.; Raucci, N.; Fontana, I.C.; Luzi, F.; Gee, A.D. Carbon-11 Carboxylation of Trialkoxysilane and Trimethylsilane Derivatives Using [$^{11}$C]CO$_2$. *Chem. Commun.* **2020**, *56*, 4668–4671. [CrossRef]
30. Qu, W.; Hu, B.; Babich, J.W.; Waterhouse, N.; Dooley, M.; Ponnala, S.; Urgiles, J. A General $^{11}$C-Labeling Approach Enabled by Fluoride-Mediated Desilylation of Organosilanes. *Nat. Commun.* **2020**, *11*, 1736. [CrossRef]
31. Destro, G.; Horkka, K.; Loreau, O.; Buisson, D.; Kingston, L.; Del Vecchio, A.; Schou, M.; Elmore, C.S.; Taran, F.; Cantat, T.; et al. Transition-Metal-Free Carbon Isotope Exchange of Phenyl Acetic Acids. *Angew. Chem.* **2020**, *132*, 13592–13597. [CrossRef]
32. Kong, D.; Munch, M.; Qiqige, Q.; Cooze, C.J.C.; Rotstein, B.H.; Lundgren, R.J. Fast Carbon Isotope Exchange of Carboxylic Acids Enabled by Organic Photoredox Catalysis. *J. Am. Chem. Soc.* **2021**, *143*, 2200–2206. [CrossRef] [PubMed]
33. Bsharat, O.; Doyle, M.G.J.; Munch, M.; Mair, B.A.; Cooze, C.J.C.; Derdau, V.; Bauer, A.; Kong, D.; Rotstein, B.H.; Lundgren, R.J. Aldehyde-Catalysed Carboxylate Exchange in α-Amino Acids with Isotopically Labelled CO$_2$. *Nat. Chem.* **2022**, *14*, 1367–1374. [CrossRef]
34. Bongarzone, S.; Runser, A.; Taddei, C.; Dheere, A.K.H.; Gee, A.D. From [$^{11}$C]CO$_2$ to [$^{11}$C]Amides: A Rapid One-Pot Synthesis via the Mitsunobu Reaction. *Chem. Commun.* **2017**, *53*, 5334–5337. [CrossRef]
35. Mair, B.A.; Fouad, M.H.; Ismailani, U.S.; Munch, M.; Rotstein, B.H. Rhodium-Catalyzed Addition of Organozinc Iodides to Carbon-11 Isocyanates. *Org. Lett.* **2020**, *22*, 2746–2750. [CrossRef]
36. Horkka, K.; Dahl, K.; Bergare, J.; Elmore, C.S.; Halldin, C.; Schou, M. Rapid and Efficient Synthesis of $^{11}$C-Labeled Benzimidazolones Using [$^{11}$C]Carbon Dioxide. *ChemistrySelect* **2019**, *4*, 1846–1849. [CrossRef]
37. Luzi, F.; Gee, A.D.; Bongarzone, S. Rapid One-Pot Radiosynthesis of [Carbonyl-$^{11}$C]Formamides from Primary Amines and [$^{11}$C]CO$_2$. *EJNMMI Radiopharm. Chem.* **2020**, *5*, 20. [CrossRef]
38. Dahl, K.; Collier, T.L.; Cheng, R.; Zhang, X.; Sadovski, O.; Liang, S.H.; Vasdev, N. "In-Loop" [$^{11}$C]CO$_2$ Fixation: Prototype and Proof of Concept. *J. Label. Compd. Radiopharm.* **2018**, *61*, 252–262. [CrossRef]
39. Downey, J.; Bongarzone, S.; Hader, S.; Gee, A.D. In-Loop Flow [$^{11}$C]CO$_2$ Fixation and Radiosynthesis of $N,N'$-[$^{11}$C]Dibenzylurea. *J. Label. Compd. Radiopharm.* **2018**, *61*, 263–271. [CrossRef]
40. Del Vecchio, A.; Caillé, F.; Chevalier, A.; Loreau, O.; Horkka, K.; Halldin, C.; Schou, M.; Camus, N.; Kessler, P.; Kuhnast, B.; et al. Late-Stage Isotopic Carbon Labeling of Pharmaceutically Relevant Cyclic Ureas Directly from CO$_2$. *Angew. Chem.* **2018**, *130*, 9892–9896. [CrossRef]
41. Del Vecchio, A.; Talbot, A.; Caillé, F.; Chevalier, A.; Sallustrau, A.; Loreau, O.; Destro, G.; Taran, F.; Audisio, D. Carbon Isotope Labeling of Carbamates by Late-Stage [$^{11}$C], [$^{13}$C] and [$^{14}$C]Carbon Dioxide Incorporation. *Chem. Commun.* **2020**, *56*, 11677–11680. [CrossRef]
42. Babin, V.; Sallustrau, A.; Loreau, O.; Caillé, F.; Goudet, A.; Cahuzac, H.; Del Vecchio, A.; Taran, F.; Audisio, D. A General Procedure for Carbon Isotope Labeling of Linear Urea Derivatives with Carbon Dioxide. *Chem. Commun.* **2021**, *57*, 6680–6683. [CrossRef] [PubMed]
43. Liger, F.; Cadarossanesaib, F.; Iecker, T.; Tourvieille, C.; Le Bars, D.; Billard, T. $^{11}$C-Labeling: Intracyclic Incorporation of Carbon-11 into Heterocycles: $^{11}$C-Labeling: Intracyclic Incorporation of Carbon-11 into Heterocycles. *Eur. J. Org. Chem.* **2019**, *2019*, 6968–6972. [CrossRef]
44. Haji Dheere, A.K.; Bongarzone, S.; Shakir, D.; Gee, A. Direct Incorporation of [$^{11}$C]CO$_2$ into Asymmetric [$^{11}$C]Carbonates. *J. Chem.* **2018**, *2018*, 7641304. [CrossRef]
45. Lindberg, A.; Vasdev, N. Ring-Opening of Non-Activated Aziridines with [$^{11}$C]CO$_2$ via Novel Ionic Liquids. *RSC Adv.* **2022**, *12*, 21417–21421. [CrossRef] [PubMed]
46. Taddei, C.; Pike, V.W. [$^{11}$C]Carbon Monoxide: Advances in Production and Application to PET Radiotracer Development over the Past 15 Years. *EJNMMI Radiopharm. Chem.* **2019**, *4*, 25. [CrossRef] [PubMed]
47. Eriksson, J.; Antoni, G.; Långström, B.; Itsenko, O. The Development of $^{11}$C-Carbonylation Chemistry: A Systematic View. *Nucl. Med. Biol.* **2021**, *92*, 115–137. [CrossRef]

48. Nielsen, D.U.; Neumann, K.T.; Lindhardt, A.T.; Skrydstrup, T. Recent Developments in Carbonylation Chemistry Using [$^{13}$C]CO, [$^{11}$C]CO, and [$^{14}$C]CO. *J. Label. Compd. Radiopharm.* **2018**, *61*, 949–987. [CrossRef]
49. Taddei, C.; Gee, A.D. Recent Progress in [$^{11}$C]Carbon Dioxide ([$^{11}$C]CO$_2$) and [$^{11}$C]Carbon Monoxide ([$^{11}$C]CO) Chemistry. *J. Label. Compd. Radiopharm.* **2018**, *61*, 237–251. [CrossRef]
50. Shegani, A.; Kealey, S.; Luzi, F.; Basagni, F.; Machado, J.D.M.; Ekici, S.D.; Ferocino, A.; Gee, A.D.; Bongarzone, S. Radiosynthesis, Preclinical, and Clinical Positron Emission Tomography Studies of Carbon-11 Labeled Endogenous and Natural Exogenous Compounds. *Chem. Rev.* **2023**, *123*, 105–229. [CrossRef] [PubMed]
51. Dahl, K.; Turner, T.; Vasdev, N. Radiosynthesis of a Bruton's Tyrosine Kinase Inhibitor, [$^{11}$C]Tolebrutinib, via Palladium-NiXantphos-mediated Carbonylation. *J. Label. Compd. Radiopharm.* **2020**, *63*, 482–487. [CrossRef] [PubMed]
52. Lindberg, A.; Boyle, A.J.; Tong, J.; Harkness, M.B.; Garcia, A.; Tran, T.; Zhai, D.; Liu, F.; Donnelly, D.J.; Vasdev, N. Radiosynthesis of [$^{11}$C]Ibrutinib via Pd-Mediated [$^{11}$C]CO Carbonylation: Preliminary PET Imaging in Experimental Autoimmune Encephalomyelitis Mice. *Front. Nucl. Med.* **2021**, *1*, 772289. [CrossRef]
53. Donnelly, D.J.; Preshlock, S.; Kaur, T.; Tran, T.; Wilson, T.C.; Mhanna, K.; Henderson, B.D.; Batalla, D.; Scott, P.J.H.; Shao, X. Synthesis of Radiopharmaceuticals via "In-Loop" $^{11}$C-Carbonylation as Exemplified by the Radiolabeling of Inhibitors of Bruton's Tyrosine Kinase. *Front. Nucl. Med.* **2022**, *1*, 820235. [CrossRef]
54. Langstrom, B.; Lundqvist, H. The Preparation of $^{11}$C-Methyl Iodide and Its Use in the Synthesis of $^{11}$C-Methyl-Methionine. *Int. J. Appl. Radiat. Isot.* **1976**, *27*, 357–363. [CrossRef] [PubMed]
55. Comar, D.; Cartron, J.-C.; Maziere, M.; Marazano, C. Labelling and Metabolism of Methionine-Methyl-$^{11}$C. *Eur. J. Nucl. Med.* **1976**, *1*, 11–14. [CrossRef]
56. Marazano, C.; Maziere, M.; Berger, G.; Comar, D. Synthesis of Methyl Iodide-$^{11}$C and Formaldehyde-$^{11}$C. *Int. J. Appl. Radiat. Isot.* **1977**, *28*, 49–52. [CrossRef]
57. Jewett, D.M. A Simple Synthesis of [$^{11}$C]Methyl Triflate. *Appl. Radiat. Isot.* **1992**, *43*, 1383–1385. [CrossRef]
58. Mock, B. Automated C-11 Methyl Iodide/Triflate Production: Current State of the Art. *Curr. Org. Chem.* **2013**, *17*, 2119–2126. [CrossRef]
59. Doi, H. Pd-Mediated Rapid Cross-Couplings Using [$^{11}$C]Methyl Iodide: Groundbreaking Labeling Methods in $^{11}$C Radiochemistry: Development of Pd-Mediated Rapid C-[$^{11}$C]Methylations. *J. Label. Compd. Radiopharm.* **2015**, *58*, 73–85. [CrossRef]
60. Rokka, J.; Nordeman, P.; Roslin, S.; Eriksson, J. A Comparative Study on Suzuki-type $^{11}$C-methylation of Aromatic Organoboranes Performed in Two Reaction Media. *J. Label. Compd. Radiopharm.* **2021**, *64*, 447–455. [CrossRef]
61. Pipal, R.W.; Stout, K.T.; Musacchio, P.Z.; Ren, S.; Graham, T.J.A.; Verhoog, S.; Gantert, L.; Lohith, T.G.; Schmitz, A.; Lee, H.S.; et al. Metallaphotoredox Aryl and Alkyl Radiomethylation for PET Ligand Discovery. *Nature* **2021**, *589*, 542–547. [CrossRef]
62. Dahl, K.; Nordeman, P. $^{11}$C-Acetylation of Amines with [$^{11}$C]Methyl Iodide with Bis(Cyclopentadienyldicarbonyliron) as the CO Source: $^{11}$C-Acetylation of Amines with [$^{11}$C]Methyl Iodide with Bis(Cyclopentadienyldicarbonyliron) as the CO Source. *Eur. J. Org. Chem.* **2017**, *2017*, 5785–5788. [CrossRef]
63. Doi, H.; Goto, M.; Sato, Y. Pd$^0$-Mediated Cross-Coupling of [$^{11}$C]Methyl Iodide with Carboxysilane for Synthesis of [$^{11}$C]Acetic Acid and Its Active Esters: $^{11}$C-Acetylation of Small, Medium, and Large Molecules. *Eur. J. Org. Chem.* **2021**, *2021*, 3970–3979. [CrossRef]
64. Filp, U.; Pekošak, A.; Poot, A.J.; Windhorst, A.D. Stereocontrolled [$^{11}$C]Alkylation of N-Terminal Glycine Schiff Bases To Obtain Dipeptides: Stereocontrolled [$^{11}$C]Alkylation of N-Terminal Glycine Schiff Bases To Obtain Dipeptides. *Eur. J. Org. Chem.* **2017**, *2017*, 5592–5596. [CrossRef]
65. Pekošak, A.; Rotstein, B.H.; Collier, T.L.; Windhorst, A.D.; Vasdev, N.; Poot, A.J. Stereoselective $^{11}$C Labeling of a "Native" Tetrapeptide by Using Asymmetric Phase-Transfer Catalyzed Alkylation Reactions. *Eur. J. Org. Chem.* **2017**, *2017*, 1019–1024. [CrossRef]
66. Reiffers, S.; Vaalburg, W.; Wiegman, T.; Wynberg, H.; Woldring, M.G. Carbon-11 Labelled Methyllithium as Methyl Donating Agent: The Addition to 17-Keto Steroids. *Int. J. Appl. Radiat. Isot.* **1980**, *31*, 535–539. [CrossRef]
67. Berger, G.; Maziere, M.; Prenant, C.; Sastre, J.; Comar, D. Synthesis of High Specific Activity $^{11}$C 17alpha Methyltestosterone. *Int. J. Appl. Radiat. Isot.* **1981**, *32*, 811–815. [CrossRef]
68. Helbert, H.; Antunes, I.F.; Luurtsema, G.; Szymanski, W.; Feringa, B.L.; Elsinga, P.H. Cross-Coupling of [$^{11}$C]Methyllithium for $^{11}$C-Labelled PET Tracer Synthesis. *Chem. Commun.* **2021**, *57*, 203–206. [CrossRef]
69. Dubrin, J.; MacKay, C.; Pandow, M.L.; Wolfgang, R. Reactions of Atomic Carbon with Pi-Bonded Inorganic Molecules. *J. Inorg. Nucl. Chem.* **1964**, *26*, 2113–2122. [CrossRef]
70. Xu, Y.; Qu, W. [$^{11}$C]HCN Radiochemistry: Recent Progress and Future Perspectives. *Eur. J. Org. Chem.* **2021**, *2021*, 4653–4682. [CrossRef]
71. Zhou, Y.-P.; Makaravage, K.J.; Brugarolas, P. Radiolabeling with [$^{11}$C]HCN for Positron Emission Tomography. *Nucl. Med. Biol.* **2021**, *102–103*, 56–86. [CrossRef] [PubMed]
72. Kikuchi, T.; Ogawa, M.; Okamura, T.; Gee, A.D.; Zhang, M.-R. Rapid 'on-Column' Preparation of Hydrogen [$^{11}$C]Cyanide from [$^{11}$C]Methyl Iodide via [$^{11}$C]Formaldehyde. *Chem. Sci.* **2022**, *13*, 3556–3562. [CrossRef] [PubMed]
73. Haskali, M.B.; Pike, V.W. [$^{11}$C]Fluoroform, a Breakthrough for Versatile Labeling of PET Radiotracer Trifluoromethyl Groups in High Molar Activity. *Chem. Eur. J.* **2017**, *23*, 8156–8160. [CrossRef] [PubMed]

74. Jana, S.; Telu, S.; Yang, B.Y.; Haskali, M.B.; Jakobsson, J.E.; Pike, V.W. Rapid Syntheses of [$^{11}$C]Arylvinyltrifluoromethanes through Treatment of (E)-Arylvinyl(Phenyl)Iodonium Tosylates with [$^{11}$C]Trifluoromethylcopper(I). *Org. Lett.* **2020**, *22*, 4574–4578. [CrossRef] [PubMed]
75. Young, N.J.; Pike, V.W.; Taddei, C. Rapid and Efficient Synthesis of [$^{11}$C]Trifluoromethylarenes from Primary Aromatic Amines and [$^{11}$C]CuCF$_3$. *ACS Omega* **2020**, *5*, 19557–19564. [CrossRef]
76. Jakobsson, J.E.; Lu, S.; Telu, S.; Pike, V.W. [$^{11}$C]Carbonyl Difluoride—A New and Highly Efficient [$^{11}$C]Carbonyl Group Transfer Agent. *Angew. Chem. Int. Ed.* **2020**, *59*, 7256–7260. [CrossRef]
77. Jakobsson, J.E.; Telu, S.; Lu, S.; Jana, S.; Pike, V.W. Broad Scope and High-Yield Access to Unsymmetrical Acyclic [$^{11}$C]Ureas for Biomedical Imaging from [$^{11}$C]Carbonyl Difluoride. *Chem. Eur. J.* **2021**, *27*, 10369–10376. [CrossRef]
78. Niisawa, K.; Ogawa, K.; Saito, J.; Taki, K.; Karasawa, T.; Nozaki, T. Production of No-Carrier-Added $^{11}$C-Carbon Disulfide and $^{11}$C-Hydrogen Cyanide by Microwave Discharge. *Int. J. Appl. Radiat. Isot.* **1984**, *35*, 29–33. [CrossRef]
79. Miller, P.W.; Bender, D. [$^{11}$C]Carbon Disulfide: A Versatile Reagent for PET Radiolabelling. *Chem. Eur. J.* **2012**, *18*, 433–436. [CrossRef]
80. Haywood, T.; Kealey, S.; Sánchez-Cabezas, S.; Hall, J.J.; Allott, L.; Smith, G.; Plisson, C.; Miller, P.W. Carbon-11 Radiolabelling of Organosulfur Compounds: $^{11}$C Synthesis of the Progesterone Receptor Agonist Tanaproget. *Chem. Eur. J.* **2015**, *21*, 9034–9038. [CrossRef]
81. Cesarec, S.; Edgar, F.; Lai, T.; Plisson, C.; White, A.J.P.; Miller, P.W. Synthesis of Carbon-11 Radiolabelled Transition Metal Complexes Using $^{11}$C-Dithiocarbamates. *Dalton Trans.* **2022**, *51*, 5004–5008. [CrossRef]
82. Stone-Elander, S.; Roland, P.; Halldin, C.; Hassan, M.; Seitz, R. Synthesis of [$^{11}$C]Sodium Thiocyanate for In Vivo Studies of Anion Kinetics Using Positron Emission Tomography (PET). *Nucl. Med. Biol.* **1989**, *16*, 741–746. [CrossRef]
83. Haywood, T.; Cesarec, S.; Kealey, S.; Plisson, C.; Miller, P.W. Ammonium [$^{11}$C]Thiocyanate: Revised Preparation and Reactivity Studies of a Versatile Nucleophile for Carbon-11 Radiolabelling. *Med. Chem. Commun.* **2018**, *9*, 1311–1314. [CrossRef] [PubMed]
84. Hooker, J.M.; Schönberger, M.; Schieferstein, H.; Fowler, J.S. A Simple, Rapid Method for the Preparation of [$^{11}$C]Formaldehyde. *Angew. Chem. Int. Ed.* **2008**, *47*, 5989–5992. [CrossRef] [PubMed]
85. Nader, M.; Oberdorfer, F.; Herrmann, K. Production of [$^{11}$C]Formaldehyde by the XeF$_2$ Mediated Oxidation of [$^{11}$C]Methanol and Its Application in the Labeling of α-(N-[$^{11}$C]Methylamino)Isobutyric Acid. *Appl. Radiat. Isot.* **2019**, *148*, 178–183. [CrossRef] [PubMed]
86. Rischka, L.; Vraka, C.; Pichler, V.; Rasul, S.; Nics, L.; Gryglewski, G.; Handschuh, P.; Murgaš, M.; Godbersen, G.M.; Silberbauer, L.R.; et al. First-in-Humans Brain PET Imaging of the GluN2B-Containing N-Methyl-d-Aspartate Receptor with (R)-$^{11}$C-Me-NB1. *J. Nucl. Med.* **2022**, *63*, 936–941. [CrossRef]
87. Bernard-Gauthier, V.; Bailey, J.J.; Mossine, A.V.; Lindner, S.; Vomacka, L.; Aliaga, A.; Shao, X.; Quesada, C.A.; Sherman, P.; Mahringer, A.; et al. A Kinome-Wide Selective Radiolabeled TrkB/C Inhibitor for in Vitro and in Vivo Neuroimaging: Synthesis, Preclinical Evaluation, and First-in-Human. *J. Med. Chem.* **2017**, *60*, 6897–6910. [CrossRef]
88. Gallezot, J.-D.; Nabulsi, N.; Henry, S.; Pracitto, R.; Planeta, B.; Ropchan, J.; Lin, S.-F.; Labaree, D.; Kapinos, M.; Shirali, A.; et al. Imaging the Enzyme 11β-Hydroxysteroid Dehydrogenase Type 1 with PET: Evaluation of the Novel Radiotracer $^{11}$C-AS2471907 in Human Brain. *J. Nucl. Med.* **2019**, *60*, 1140–1146. [CrossRef]
89. Delva, A.; Koole, M.; Serdons, K.; Bormans, G.; Liu, L.; Bard, J.; Khetarpal, V.; Dominguez, C.; Munoz-Sanjuan, I.; Wood, A.; et al. Biodistribution and Dosimetry in Human Healthy Volunteers of the PET Radioligands [$^{11}$C]CHDI-00485180-R and [$^{11}$C]CHDI-00485626, Designed for Quantification of Cerebral Aggregated Mutant Huntingtin. *Eur. J. Nucl. Med. Mol. Imaging* **2022**, *50*, 48–60. [CrossRef]
90. Johansen, A.; Holm, S.; Dall, B.; Keller, S.; Kristensen, J.L.; Knudsen, G.M.; Hansen, H.D. Human Biodistribution and Radiation Dosimetry of the 5-HT2A Receptor Agonist Cimbi-36 Labeled with Carbon-11 in Two Positions. *EJNMMI Res.* **2019**, *9*, 71. [CrossRef]
91. Coughlin, J.M.; Du, Y.; Lesniak, W.G.; Harrington, C.K.; Brosnan, M.K.; O'Toole, R.; Zandi, A.; Sweeney, S.E.; Abdallah, R.; Wu, Y.; et al. First-in-Human Use of $^{11}$C-CPPC with Positron Emission Tomography for Imaging the Macrophage Colony-Stimulating Factor 1 Receptor. *EJNMMI Res.* **2022**, *12*, 64. [CrossRef] [PubMed]
92. Ikawa, M.; Lohith, T.G.; Shrestha, S.; Telu, S.; Zoghbi, S.S.; Castellano, S.; Taliani, S.; Da Settimo, F.; Fujita, M.; Pike, V.W.; et al. $^{11}$C-ER176, a Radioligand for 18-KDa Translocator Protein, Has Adequate Sensitivity to Robustly Image All Three Affinity Genotypes in Human Brain. *J. Nucl. Med.* **2017**, *58*, 320–325. [CrossRef] [PubMed]
93. Brier, M. Phase 1 Evaluation of $^{11}$C-CS1P1 to Assess Safety and Dosimetry in Human Participants. *J. Nucl. Med.* **2022**, *64*, 1775–1782. [CrossRef]
94. Antoni, G.; Lubberink, M.; Sörensen, J.; Lindström, E.; Elgland, M.; Eriksson, O.; Hultström, M.; Frithiof, R.; Wanhainen, A.; Sigfridsson, J.; et al. In Vivo Visualization and Quantification of Neutrophil Elastase in Lungs of COVID-19 Patients—A First-In-Human Positron Emission Tomography Study with $^{11}$C-GW457427. *J. Nucl. Med.* **2022**, *64*, 263974. [CrossRef]
95. Van Weehaeghe, D.; Koole, M.; Schmidt, M.E.; Deman, S.; Jacobs, A.H.; Souche, E.; Serdons, K.; Sunaert, S.; Bormans, G.; Vandenberghe, W.; et al. [$^{11}$C]JNJ54173717, a Novel P2X7 Receptor Radioligand as Marker for Neuroinflammation: Human Biodistribution, Dosimetry, Brain Kinetic Modelling and Quantification of Brain P2X7 Receptors in Patients with Parkinson's Disease and Healthy Volunteers. *Eur. J. Nucl. Med. Mol. Imaging* **2019**, *46*, 2051–2064. [CrossRef] [PubMed]

96. Miyazaki, T.; Nakajima, W.; Hatano, M.; Shibata, Y.; Kuroki, Y.; Arisawa, T.; Serizawa, A.; Sano, A.; Kogami, S.; Yamanoue, T.; et al. Visualization of AMPA Receptors in Living Human Brain with Positron Emission Tomography. *Nat. Med.* 2020, 26, 281–288. [CrossRef]
97. Naganawa, M.; Nabulsi, N.; Henry, S.; Matuskey, D.; Lin, S.-F.; Slieker, L.; Schwarz, A.J.; Kant, N.; Jesudason, C.; Ruley, K.; et al. First-in-Human Assessment of $^{11}$C-LSN3172176, an M1 Muscarinic Acetylcholine Receptor PET Radiotracer. *J. Nucl. Med.* 2021, 62, 553–560. [CrossRef]
98. Shrestha, S.; Kim, M.-J.; Eldridge, M.; Lehmann, M.L.; Frankland, M.; Liow, J.-S.; Yu, Z.-X.; Cortes-Salva, M.; Telu, S.; Henter, I.D.; et al. PET Measurement of Cyclooxygenase-2 Using a Novel Radioligand: Upregulation in Primate Neuroinflammation and First-in-Human Study. *J. Neuroinflammation* 2020, 17, 140. [CrossRef]
99. Du, Y.; Coughlin, J.M.; Brosnan, M.K.; Chen, A.; Shinehouse, L.K.; Abdallah, R.; Lodge, M.A.; Mathews, W.B.; Liu, C.; Wu, Y.; et al. PET Imaging of the Cannabinoid Receptor Type 2 in Humans Using [$^{11}$C]MDTC. *Res. Sq.* 2022. [CrossRef]
100. Van der Weijden, C.W.J.; Meilof, J.F.; van der Hoorn, A.; Zhu, J.; Wu, C.; Wang, Y.; Willemsen, A.T.M.; Dierckx, R.A.J.O.; Lammertsma, A.A.; de Vries, E.F.J. Quantitative Assessment of Myelin Density Using [$^{11}$C]MeDAS PET in Patients with Multiple Sclerosis: A First-in-Human Study. *Eur. J. Nucl. Med. Mol. Imaging* 2022, 49, 3492–3507. [CrossRef]
101. Kubota, M.; Seki, C.; Kimura, Y.; Takahata, K.; Shimada, H.; Takado, Y.; Matsuoka, K.; Tagai, K.; Sano, Y.; Yamamoto, Y.; et al. A First-in-Human Study of $^{11}$C-MTP38, a Novel PET Ligand for Phosphodiesterase 7. *Eur. J. Nucl. Med. Mol. Imaging* 2021, 48, 2846–2855. [CrossRef] [PubMed]
102. Ruiz-Bedoya, C.A.; Ordonez, A.A.; Werner, R.A.; Plyku, D.; Klunk, M.H.; Leal, J.; Lesniak, W.G.; Holt, D.P.; Dannals, R.F.; Higuchi, T.; et al. $^{11}$C-PABA as a PET Radiotracer for Functional Renal Imaging: Preclinical and First-in-Human Study. *J. Nucl. Med.* 2020, 61, 1665–1671. [CrossRef] [PubMed]
103. Sakata, M.; Ishibashi, K.; Imai, M.; Wagatsuma, K.; Ishii, K.; Zhou, X.; de Vries, E.F.J.; Elsinga, P.H.; Ishiwata, K.; Toyohara, J. Initial Evaluation of an Adenosine A$_{2A}$ Receptor Ligand, $^{11}$C-Preladenant, in Healthy Human Subjects. *J. Nucl. Med.* 2017, 58, 1464–1470. [CrossRef] [PubMed]
104. Kim, M.-J.; Lee, J.-H.; Juarez Anaya, F.; Hong, J.; Miller, W.; Telu, S.; Singh, P.; Cortes, M.Y.; Henry, K.; Tye, G.L.; et al. First-in-Human Evaluation of [$^{11}$C]PS13, a Novel PET Radioligand, to Quantify Cyclooxygenase-1 in the Brain. *Eur. J. Nucl. Med. Mol. Imaging* 2020, 47, 3143–3151. [CrossRef] [PubMed]
105. Tucker, E.W.; Guglieri-Lopez, B.; Ordonez, A.A.; Ritchie, B.; Klunk, M.H.; Sharma, R.; Chang, Y.S.; Sanchez-Bautista, J.; Frey, S.; Lodge, M.A.; et al. Noninvasive $^{11}$C-Rifampin Positron Emission Tomography Reveals Drug Biodistribution in Tuberculous Meningitis. *Sci. Transl. Med.* 2018, 10, 145. [CrossRef]
106. Lohith, T.G.; Tsujikawa, T.; Siméon, F.G.; Veronese, M.; Zoghbi, S.S.; Lyoo, C.H.; Kimura, Y.; Morse, C.L.; Pike, V.W.; Fujita, M.; et al. Comparison of Two PET Radioligands, [$^{11}$C]FPEB and [$^{11}$C]SP203, for Quantification of Metabotropic Glutamate Receptor 5 in Human Brain. *J. Cereb. Blood Flow Metab.* 2017, 37, 2458–2470. [CrossRef]
107. Lee, I.K.; Jacome, D.A.; Cho, J.K.; Tu, V.; Young, A.; Dominguez, T.; Northrup, J.D.; Etersque, J.M.; Lee, H.S.; Ruff, A.; et al. Imaging Sensitive and Drug-Resistant Bacterial Infection with [$^{11}$C]-Trimethoprim. *J. Clin. Investig.* 2022, 132, e156679. [CrossRef]
108. Watanabe, Y.; Mawatari, A.; Aita, K.; Sato, Y.; Wada, Y.; Nakaoka, T.; Onoe, K.; Yamano, E.; Akamatsu, G.; Ohnishi, A.; et al. PET Imaging of $^{11}$C-Labeled Thiamine Tetrahydrofurfuryl Disulfide, Vitamin B1 Derivative: First-in-Human Study. *Biochem. Biophys. Res. Commun.* 2021, 555, 7–12. [CrossRef]
109. Wong, D.F.; Comley, R.A.; Kuwabara, H.; Rosenberg, P.B.; Resnick, S.M.; Ostrowitzki, S.; Vozzi, C.; Boess, F.; Oh, E.; Lyketsos, C.G.; et al. Characterization of 3 Novel Tau Radiopharmaceuticals, $^{11}$C-RO-963, $^{11}$C-RO-643, and $^{18}$F-RO-948, in Healthy Controls and in Alzheimer Subjects. *J. Nucl. Med.* 2018, 59, 1869–1876. [CrossRef]
110. Masdeu, J.C.; Pascual, B.; Fujita, M. Imaging Neuroinflammation in Neurodegenerative Disorders. *J. Nucl. Med.* 2022, 63, 45S–52S. [CrossRef]
111. Chen, Z.; Haider, A.; Chen, J.; Xiao, Z.; Gobbi, L.; Honer, M.; Grether, U.; Arnold, S.E.; Josephson, L.; Liang, S.H. The Repertoire of Small-Molecule PET Probes for Neuroinflammation Imaging: Challenges and Opportunities beyond TSPO. *J. Med. Chem.* 2021, 64, 17656–17689. [CrossRef] [PubMed]
112. Janssen, B.; Vugts, D.; Windhorst, A.; Mach, R. PET Imaging of Microglial Activation—Beyond Targeting TSPO. *Molecules* 2018, 23, 607. [CrossRef] [PubMed]
113. Jain, P.; Chaney, A.M.; Carlson, M.L.; Jackson, I.M.; Rao, A.; James, M.L. Neuroinflammation PET Imaging: Current Opinion and Future Directions. *J. Nucl. Med.* 2020, 61, 1107–1112. [CrossRef] [PubMed]
114. Narayanaswami, V.; Dahl, K.; Bernard-Gauthier, V.; Josephson, L.; Cumming, P.; Vasdev, N. Emerging PET Radiotracers and Targets for Imaging of Neuroinflammation in Neurodegenerative Diseases: Outlook Beyond TSPO. *Mol. Imaging* 2018, 17, 1536012118792317. [CrossRef]

**Disclaimer/Publisher's Note:** The statements, opinions and data contained in all publications are solely those of the individual author(s) and contributor(s) and not of MDPI and/or the editor(s). MDPI and/or the editor(s) disclaim responsibility for any injury to people or property resulting from any ideas, methods, instructions or products referred to in the content.

Article

# Development of in-House Synthesis and Quality Control of [$^{99m}$Tc]Tc-PSMA-I&S

Elisabeth Plhak [1,2,*], Christopher Pichler [2], Edith Gößnitzer [2], Reingard M. Aigner [1] and Herbert Kvaternik [1]

[1] Division of Nuclear Medicine, Department of Radiology, Medical University of Graz, Auenbruggerplatz 9, 8036 Graz, Austria
[2] Department of Pharmaceutical Chemistry, Institute of Pharmaceutical Sciences, University of Graz, Schubertstraße 1/EG/0122, 8010 Graz, Austria
\* Correspondence: elisabeth.plhak@medunigraz.at; Tel.: +43-316-385-30696

**Abstract:** Many radioactive PSMA inhibitory substances have already been developed for PET diagnostics and therapy of prostate cancer. Because PET radionuclides and instrumentation may not be available, technetium-99 m labelled tracers can be considered as a diagnostic alternative. A suitable tracer is [$^{99m}$Tc]Tc-PSMA-I&S, primarily developed for radio-guided surgery, which has been identified for diagnostics of prostate cancer. However, there is no commercial kit approved for the preparation of [$^{99m}$Tc]Tc-PSMA-I&S on the market. This work presents an automated process for the synthesis of [$^{99m}$Tc]Tc-PSMA-I&S concerning good manufacturing practice (GMP). We used a Scintomics GRP 4 V module, with the SCC software package for programming sequences for this development. The optimum reaction conditions were evaluated in preliminary experiments. The pH of the reaction solution was found to be crucial for the radiochemical yield and radiochemical purity. The validation of [$^{99m}$Tc]Tc-PSMA-I&S ($n$ = 3) achieved a stable radiochemical yield of 58.7 ± 1.5% and stable radiochemical purities of 93.0 ± 0.3%. The amount of free [$^{99m}$Tc]TcO$_4^-$ in the solution and reduced hydrolysed [$^{99m}$Tc]TcO$_2$ was <2%. Our automated preparation of [$^{99m}$Tc]Tc-PSMA-I&S has shown reliability and applicability in the clinical setting.

**Keywords:** [$^{99m}$Tc]Tc-PSMA-I&S; automated preparation; technetium-99m

## 1. Introduction

The prostate-specific membrane antigen (PSMA) is an essential target for the diagnosis and therapy of prostate cancer. It belongs to the membrane-type zinc peptidase family and has two functions: as a receptor, and as a zinc-protease enzyme. It is overexpressed in prostate cancer and its metastatic lesions, which makes it an exciting target for imaging and therapy of prostate cancer [1]. [$^{68}$Ga]Ga-PSMA-11 is currently the most used radiopharmaceutical for the diagnosis of prostate cancer. However, the instrumentation and radionuclides for positron emission tomography (PET) applications are not only expensive but also of limited availability in many countries. Therefore, a lot of effort has been employed in developing PSMA targeting tracers for single-photon emission computed tomography (SPECT) [2]. [$^{99m}$Tc]Tc-PSMA-I&S was first introduced by Robu et al., including the preclinical evaluation and first patient application. The abbreviation I&S (imaging and surgery) means that the tracer represents a dual function: It can be used for diagnostic imaging and the surgical resection of PSMA-positive lesions by using a gamma probe. The bifunctional ligand consists of a mercaptoacetyl-triserin (MAS3) chelator binding the [Tc≡O]3 + core, coupled to the PSMA-targeting peptide Lys-urea-Glu (Figure 1). [$^{99m}$Tc]Tc-PSMA-I&S has a high in vivo stability and blood clearance is relatively slow. The best tissue-to-background ratios are reached at later time points, over 5 h after administration, and steadily increase over time due to the long availability of the stable tracer in the blood [3].

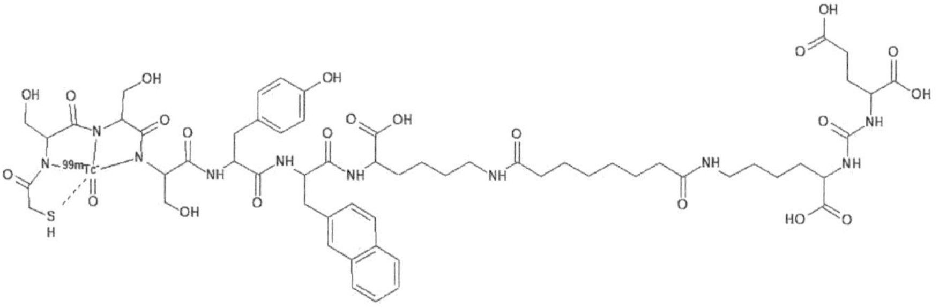

**Figure 1.** Chemical structure of [$^{99m}$Tc]Tc-PSMA-I&S.

Due to its tracer kinetics, [$^{99m}$Tc]Tc-PSMA-I&S is highly suitable for radio-guided surgery [4,5]. According to dosimetry calculations based on the half-life of technetium-99m, the activity in tumour lesions remains in a detectable range for commercially available γ-probes for up to 48 h [6]. In the current literature, the tracer application was constantly performed between 16 and 24 h before surgical treatment [3–5,7,8]. [$^{99m}$Tc]Tc-PSMA-I&S has been tested in robot-assisted radio-guided surgery using a drop-in gamma probe [7–9]. Currently, the benefits of this procedure are being evaluated in a clinical trial. The first interim analysis of the study showed that [$^{99m}$Tc]Tc-PSMA-I&S can help surgeons to identify and remove affected lymphnodes, but is not sensitive enough to identify micrometastatic tissue [8].

In addition the diagnostic performance of [$^{99m}$Tc]Tc-PSMA-I&S is promising. The first study on the diagnostic use of [$^{99m}$Tc]Tc-PSMA-I&S-SPECT/CT showed that the tracer is applicable for evaluating of biochemical recurrence, primary staging, and restaging of prostate cancer. Imaging was performed at 5 h post injection. Although significant tracer accumulation was observed in the liver, the gastrointestinal tract and urinary bladder at this time, additional low dose CT allowed good discrimination between physiological uptake and pathologic lesions. However, at low PSA levels (<4 ng/mL) the detection rate of [$^{99m}$Tc]Tc-PSMA-I&S is inferior to [$^{68}$Ga]Ga-PSMA-11 PET/CT, so it requires a careful patient selection if PET/CT imaging is available [10]. A dosimetry study after administration of 700 MBq [$^{99m}$Tc]Tc-PSMA-I&S, similar to other $^{99m}$Tc-tracers, resulted in an average effective body dose of 3.64 mSv to healthy volunteers [6].

Due to the increasing number of patients, the Division of Nuclear Medicine Graz decided to introduce [$^{99m}$Tc]Tc-PSMA-I&S as a possible partial substitute for [$^{68}$Ga]Ga-PSMA-11. Furthermore, Aalbersberg et al. presented a method of producing [$^{99m}$Tc]Tc-PSMA-I&S on a Scintomics GRP synthesizer using commercially available single-use kits for $^{68}$Ga-peptides [11]. Our goal was to use the free programmable GRP developer software to configure and optimise the kit setup and the automated labelling process. For instance, an additional tubing line to enter technetium-99m pertechnetate ( [$^{99m}$Tc]TcO$_4^-$ ) was introduced, and the composition of the reaction mixture was optimised. In addition, the quality control was carried out similarly to gallium-68 labelled tracers and underwent a full validation.

## 2. Results

### 2.1. Automated Radiolabelling

The automated synthesis was developed on a Scintomics GRP Synthesis module assembled with modified single use kits for the labelling of $^{68}$Ga-peptides. The configuration is shown in Figure 2. The process sequences were created with the Scintomics developer software.

In preparation for the automated process, the PSMA-I&S precursor (40 µg) was diluted in HEPES (4-(2-hydroxyethyl) piperazine-1-ethansulfonic acid) buffer and transferred into the reaction vessel. Then, a tin (II) chloride (SnCl$_2$)/ascorbic acid solution, as well as sodium hydroxide (NaOH), was added to the precursor solution. Furthermore, [$^{99m}$Tc]TcO$_4^-$ was placed in a V-shaped vial with a lead shielding, which was connected with tubes and needles to the apparatus via valves 6 and 7.

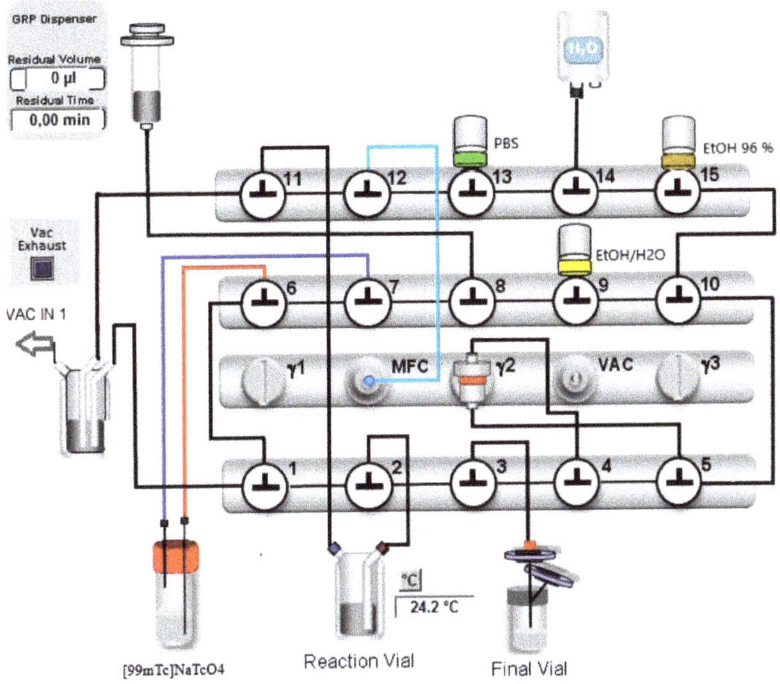

**Figure 2.** Configuration of the Scintomics GRP module for the labelling of [$^{99m}$Tc]Tc-PSMA-I&S.

The synthesis started with the preconditioning of the Sep-pak® Light C18 cartridge with ethanol and water. During the automated process, [$^{99m}$Tc]TcO$_4^-$ was flushed from the V-Vial into the reaction vessel with nitrogen gas (N$_2$). As a result, 99.2 ± 0.2% of the starting activity was successfully transferred to the reaction solution. The labelling was performed at 105 °C within 20 min. After that, the process was likewise carried out to $^{68}$Ga-labelled peptides through purification of the compound over a Sep-Pak® Light C18 cartridge, followed by elution with 50% ethanol and dilution with phosphate-buffered saline (PBS) via a sterile filter. The total volume of the final product was 17.0 ± 1.0 mL. The total processing time was 40 min.

## 2.2. Optimising the Reaction Conditions

The commercially available kit including the reagents and labelling cassettes was originally designed to label $^{68}$Ga-peptides at a pH of about 3.5–5. For the labelling of [$^{99m}$Tc]Tc-PSMA-I&S, preliminary labelling experiments were conducted to determine the optimum reaction conditions. The pH adjustment was done through the addition of 10 M NaOH to the reaction solution. A pH of 5.5 formed a high amount of reduced hydrolysed technetium-99m ( [$^{99m}$Tc]TcO$_2$), retarded in the reaction vessel and onto the Sep-Pak® Light C18 cartridge. The radiochemical yield of [$^{99m}$Tc]Tc-PSMA-I&S was only 0.5%. A reaction pH of 7.2 raised the yield to 46%. The highest radiochemical yields were achieved with reaction conditions between pH 7.8 and 8.2 (Table 1).

Radiochemical purity specifications with HPLC were adopted from the monographs for gallium-68 labelled peptides of the European Pharmacopoeia [12,13]. In the radiochromatogram, two minor regions next to the principal peak of [$^{99m}$Tc]Tc-PSMA-I&S were observed (Figure 3). Experiments showed that the pH value of the reaction mixture could influence the percentage of region 2. For example, in the first batch prepared with 40 µL of 10 M NaOH, the percentage of the principal peak was 70.7%. In the three batches where 80 µL of NaOH were added, the percentage of the principal peak was 87.9 ± 0.5%. However, these batches did not meet the radiochemical purity specification of ≥91%. In three batches prepared with 120 µL of NaOH at pH 8.2, the percentage of the principal peak increased to 93.0 ± 0.3%, which corresponds to the radiochemical purity requirements. The results of this findings are summarized in Table 1.

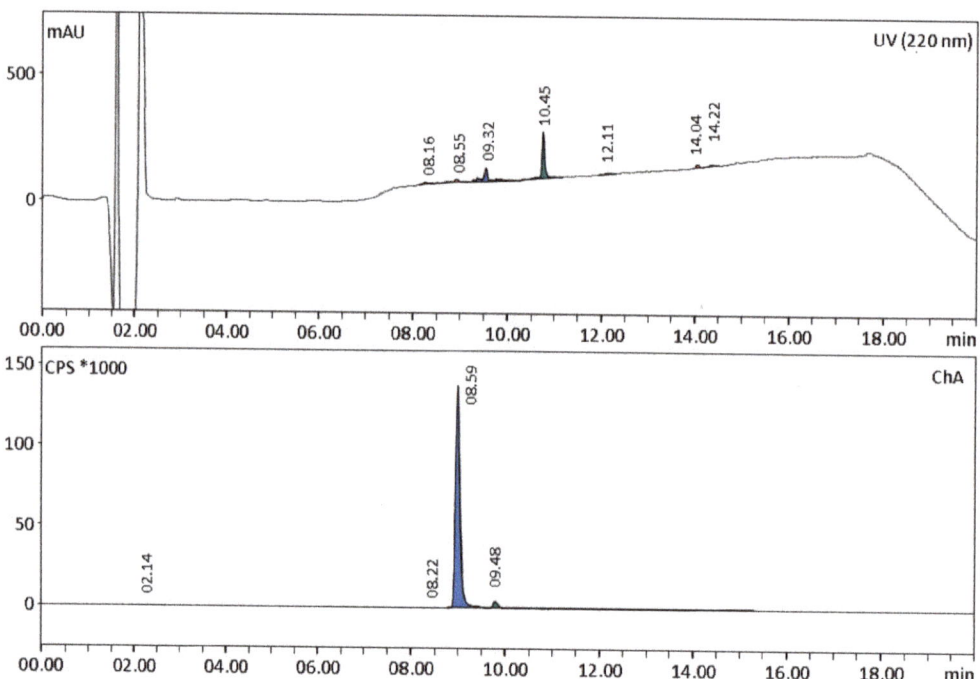

**Figure 3.** Representative radio–HPLC chromatogram of a [$^{99m}$Tc]Tc-PSMA-I&S product solution. In the radio-trace, the principal peak at 8.59 min (93.2%) and the impurity at 9.48 min (5.5%) are visible. At the UV-trace, the peak at 9.32 min was assigned to cold PSMA I&S. The peak at 10.45 min was assigned to the formed dimer.

**Table 1.** Reaction conditions and radiochemical purity evaluation of [$^{99m}$Tc]Tc-PSMA-I&S.

|  | $n = 1$ | $n = 2$ | $n = 3$ | $n = 3$ |
| --- | --- | --- | --- | --- |
| NaOH [mmol] | 0 | 0.4 | 0.8 | 1.2 |
| pH value of the reaction solution | 5.5 | 7.2 | 7.8 | 8.2 |
| Starting activity [MBq] | 1954 (100%) | 2383 (100%) | 2457 ± 309 (100%) | 2378 ± 450 (100%) |
| [$^{99m}$Tc]Tc-PSMA-I&S (EOS) [MBq] | 10 (0.5%) | 1098 (46.1%) | 1453 ± 193 (59.2 ± 4.1%) | 1396 ± 270 (58.7 ± 1.5%) |
| Retained on Sep-Pak® [MBq] | 713 (36.5%) | 151 (6.4%) | 60 ± 27 (2.4 ± 0.9%) | 23 ± 3 (1.0 ± 0.3%) |
| Residue in reaction vial [MBq] | 327 (16.7%) | 130 (5.5%) | 55 ± 31 (2.2 ± 1.1%) | 22 ± 2 (0.9 ± 0.1%) |
| Proportions of the peaks evaluated by HPLC | | | | |

Table 1. Cont.

|  | n = 1 | n = 2 | n = 3 | n = 3 |
|---|---|---|---|---|
| [$^{99m}$Tc]TcO$_4^-$ [%] | n.d. | 0.09 | 0.1 ± 0.03 | 0.1 ± 0.01 |
| Region 1 (impurity) [%] | n.d. | 1.2 | 1.6 ± 0.3 | 1.2 ± 0.03 |
| [$^{99m}$Tc]Tc-PSMA-I&S [%] | n.d. | 70.7 | 87.9 ± 0.5 | 93.0 ± 0.3 |
| Region 2 (impurity) [%] | n.d. | 28.0 | 10.5 ± 0.6 | 5.7 ± 0.3 |
| Amount of reduced hydrolysed technetium-99m (TLC) |  |  |  |  |
| [$^{99m}$Tc]TcO$_2$ [%] | n.d. | 0.3 | 0.2 ± 0.1 | 0.3 ± 0.1 |

## 2.3. Validation of the Automated Labelling of [$^{99m}$Tc]Tc-PSMA-I&S

After exploring the reaction conditions, we did a full validation with the three batches where 120 µL of NaOH were added to the precursor solution. [$^{99m}$Tc]Tc-PSMA-I&S was produced with a mean total activity of 1396 ± 270 MBq. The mean radiochemical yield was calculated based on the starting activity (2378 ± 450 MBq) and was 58.7 ± 1.5%.

The radiochemical purity of the compound was evaluated using HPLC and was 93.0 ± 0.3%. The amount of [$^{99m}$Tc]TcO$_4^-$ evaluated using HPLC was 0.1 ± 0.03%. The colloidal [$^{99m}$Tc]TcO$_2$ was 0.3 ± 0.1% (TLC).

The stability of the [$^{99m}$Tc]Tc-PSMA-I&S was confirmed 6 h prior to preparation using HPLC and TLC (92.8 ± 0.1%) and the amount of free [$^{99m}$Tc]TcO$_4^-$ and [$^{99m}$Tc]TcO2 was ≤0.5%.

The amount of Tc-PSMA-I&S, PSMA-I&S, and related substances in the product solution was evaluated using HPLC by comparing the area under the curve of the peaks found to an external standard of cold PSMA-I&S (5 µg/mL); it was 1.5 ± 0.2 µg/mL. The ligand PSMA-I&S owns a MAS3-group (2-mercaptoacetyl-ser-ser-ser) to specifically bind the technetium-99m (see Figure 1). While working with the unlabelled ligand, we noticed the appearance of a second peak (Rt = 10.5 min) beside PSMA-I&S (Rt = 9.3 min). It is probably the dimer formed through oxidation of the mercaptoacetyl group. Therefore, the dimer peak was also assigned to PSMA-I&S at the HPLC UV trace. Figure 3 shows a representative radio–HPLC chromatogram of a [$^{99m}$Tc]Tc-PSMA-I&S product solution.

Post-release tests included the determination of the ethanol content and HEPES content as well as bacterial endotoxins and sterility testing. Table 2 summarizes quality criteria and the results of the 3 masterbatches of [$^{99m}$Tc]Tc-PSMA-I&S.

Table 2. Results of the quality control of 3 masterbatches of [$^{99m}$Tc]Tc-PSMA-I&S.

| Quality Control | Method | Criteria | Result (n = 3) |
|---|---|---|---|
| Appearance | visual inspection | clear and colourless | conforms |
| pH value | pH indicator strips | 4–8 | 6.6 |
| Radioactivity concentration | dose calibrator |  | 82 ± 16 MBq/mL |
| Identity of [$^{99m}$Tc]Tc-PSMA-I&S (comparison with reference) | HPLC | Rt = 8–12 min | conforms |
| Impurity reduced hydrolysed Technetium-99m | Radio–iTLC | ≤3.0% | 0.3 ± 0.1% |
| Free [$^{99m}$Tc]TcO$_4^-$ | Radio–HPLC | ≤2.0% | 0.1 ± 0.03% |
| Radiochemical purity of [$^{99m}$Tc]Tc-PSMA-I&S | Radio–HPLC | ≥91.0% | 93.0 ± 0.3% |
| Tc-PSMA-I&S, PSMA-I&S and related substances | HPLC | ≤2.4 µg/mL | 1.5 ± 0.2 µg/mL |
| Unspecific impurities | HPLC | ≤2.4 µg/mL | ≤1 µg/mL |
| Ethanol content | gas chromatography | ≤10.0% (v/v) | conforms |
| HEPES content | HPLC | ≤40 µg/mL | 4.4 ± 3.1 µg/mL |

Table 2. *Cont.*

| Quality Control | Method | Criteria | Result ($n = 3$) |
|---|---|---|---|
| Bacterial endotoxins | LAL test | $\leq 175$ IU/V | conforms |
| Sterility | Ph. Eur. | sterile | conforms |

## 3. Discussion

[$^{99m}$Tc]Tc-PSMA-I&S was planned to be introduced at the Division of Nuclear Medicine in Graz as a potential diagnostic alternative to [$^{68}$Ga]Ga-PSMA-11 PET. Unfortunately, no approved kit is available to prepare this tracer; therefore, we decided to synthesize the compound on the Scintomics GRP module according to relevant monographs of the European Pharmacopoeia [14,15] and current good radiopharmacy practice (cGRPP) guidelines of the EANM [16]. With the SCC developer software, we created an automated process, including the transfer of [$^{99m}$Tc]TcO$_4^-$ from a V-shaped vial to the reactor and the purification of the compound via solid phase extraction (SPE).

The precursor was dissolved in 1 mL of HEPES buffer, which is an original part of the ABX reagent and hardware kit. We added a freshly prepared solution of the reducing agent SnCl$_2$ and ascorbic acid to the precursor solution and used a 10 M NaOH solution for pH adjustment. We explored the optimum composition of the reaction solution in preliminary experiments. The pH value of the reaction solution turned out to be crucial for the radiochemical yield and radiochemical purity.

After optimizing the reaction conditions, we validated the labelling process by producing three consecutive master batches of [$^{99m}$Tc]Tc-PSMA-I&S. The purification of the compound was successfully carried out with a Sep-Pak® Light C18 cartridge. Only a minimal amount of activity remained on the cartridge after elution of the compound with 50% ethanol. Free [$^{99m}$Tc]TcO$_4^-$ was almost completely removed, and less than 1% of colloidal [$^{99m}$Tc]TcO$_2$ was found in the product solution.

Within 40 min runtime of the automated process, we prepared up to 1.6 GBq [$^{99m}$Tc]Tc-PSMA-I&S. A radiochemical yield of over 55% related to the starting activity and a radiochemical purity of > 91% was achieved. These results qualify this labelling process for the clinical application.

We adapted an HPLC method to analyse $^{68}$Ga-labelled peptides and validated it for this new compound. For TLC, the standard solvent for the quality control of $^{68}$Ga-peptides was used to evaluate the amount of reduced hydrolysed technetium-99m. The complete quality control of [$^{99m}$Tc]Tc-PSMA-I&S is similar to the routine quality control of $^{68}$Ga-labelled peptides, and can be easily carried out by experienced personnel.

## 4. Materials and Methods

*4.1. Radiolabelling and Purification of [$^{99m}$Tc]Tc-PSMA-I&S*

The radiolabeling was carried out on a Scintomics GRP 4 V module (Fürstenfeldbruck, Bavaria, Germany). The labelling sequence was programmed with the Scintomics developer software. The dedicated reagent and hardware kit (SC-01-H) and the cassettes for synthesis of $^{68}$Ga- peptides (SC-01) were purchased from ABX (Radeberg, Saxony, Germany). The configuration of the labelling cassettes included four modifications: A V-Vial with a perforable seal (DWK, Mainz, Germany) was assembled with two Sterican needles (B. Braun Melsungen AG, Melsungen, Germany). Using silicone tubing lines, the long needle (Ø 0.90 × 70 mm) was connected to valve 7 and the short needle (Ø 0.60 × 30 mm) was connected to valve 6. The position of the silicone tubing to the ventilation port of the reaction vessel was connected to valve 11. The connection to the N$_2$ outlet was changed to valve 12. A detailed description of these changes is shown in Table 3.

Table 3. Summary of modifications of the $^{68}$Ga-. V = Vertical port.

| | Position | Materials | Details |
|---|---|---|---|
| Modification 1 | 6 V | Silicone tubing to V-Vial (purge tube) | 40 cm, 2 blue Luer male fittings, short needle (Ø 0.60 × 30 mm) |
| Modification 2 | 7 V | Silicone tubing to V-Vial (transfer tube) | 40 cm, 2 white Luer male fittings, long needle (Ø 0.90 × 70 mm) |
| Modification 3 | 11 V | Silicone tubing to reaction vial (ventilation port) | Original part of the cassette |
| Modification 4 | 12 V | Silicone tubing to MFC | Original part of the cassette |

The GMP grade precursor PSMA-I&S (40 µg, lyoprotected with 4 mg mannitol in 2 mL vials) was purchased from piChem (Raaba-Grambach, Austria). The specifications of the reagents used were in accordance with the European Pharmacopoeia. Tin (II) chloride dehydrate ($SnCl_2 \times 2 H_2O$), ascorbic acid, and NaOH 10 M in $H_2O$ were purchased from Sigma-Aldrich (Saint Louis, MO, USA) and used without further purification. Hydrochloric acid 1 M (HCl) was purchased from Merck (Darmstadt, Germany) and diluted with water for injection (Fresenius Kabi AG, Graz, Austria) at 0.1 mol/L. A $SnCl_2$/ascorbic acid solution was prepared by dissolving 20 mg $SnCl_2 \times 2 H_2O$ and 20 mg ascorbic acid in 10 mL of 0.1 M HCl. The precursor (40 µg of PSMA-I&S in 4 mg mannitol) was dissolved in 1 mL HEPES buffer (1.5 M, original part of the ABX reagent kit). Then, we added 50 µL of the $SnCl_2$/ascorbic acid solution (2 mg/mL) and adjusted the pH of the precursor solution with 10 M NaOH. The reaction mixture was transferred with a 3 mL syringe into the reaction vial.

Sodium pertechnetate for injection ([$^{99m}$Tc]TcO$_4^-$) was eluted from a Poltechnet $^{99}$Mo/$^{99m}$Tc- Radionuclide generator purchased from POLATOM (Otwok, Poland). For the synthesis, the starting activity was transferred into the V-shaped vial.

### 4.2. Quality Control by HPLC

HPLC was performed on an Agilent 1260 series (Waldbronn, Baden-Wuerttemberg, Germany) equipped with a DAD UV detector (UV-VIS at λ = 220 nm) and a GABI star radiometric detector (Raytest, Straubenhardt, Germany). An ACE®3 C18 column (150 × 3.0 mm, Advanced Chromatography Technologies, Aberdeen, UK) was eluted by gradient elution (0.42 mL/min) with water/TFA 0.1% (solvent A) and ACN/TFA 0.1% (solvent B): Start 24% B; 3–12 min 40% B, 14–16 min 24% B. The total runtime was 30 min. The stock solution was prepared with 40 µg of the lyoprotected precursor in the 2 mL vial by diluting it with 1 mL of 0.01 M NaOH. The calibration standards were prepared by further diluting the stock solution with PBS. To analyse the chemical and radiochemical purity of PSMA-I&S, we validated this HPLC method according to the ICH Q2 (R1) guideline in the operating range of 1.0–10.0 µg/mL [17]. We evaluated a limit of quantification (LOQ) of 1 µg/mL within a linearity with a coefficient of correlation of 0.9996.

### 4.3. Quality Control by TLC

ITLC SG plates (Agilent, Waldbronn, Baden-Wuerttemberg, Germany) and the standard solvent for the quality control for $^{68}$Ga labelled peptides consist of a 1:1 solution of methanol and 1 M ammonium acetate. The radiolabelled compound [$^{99m}$Tc]Tc-PSMA-I&S, as well as [$^{99m}$Tc]TcO$_4^-$, moved to the front, while reduced hydrolysed [$^{99m}$Tc]TcO$_2$ remained at the start.

### 4.4. Evaluation of the pH Value

We determined the pH value of the reaction solutions and the product solutions by using PEHANON pH indicator strips (Macherey-Nagel, Düren, Germany).

### 4.5. Post Release Tests

Instead of a TLC limit test for the HEPES content according to the European pharmacopoeia [12,13], we used our validated HPLC method, modified from Antunes et al. [18]. A 150 × 4.6 mm XBridge C18 5 μm column (Waters, Milford, MA, USA) was used as a stationary phase, and 20 mM solution of ammonium formate (pH 8) as a mobile phase. The flow rate was 0.7 mL/min. The HEPES reference solution (40 μg/mL), the system suitability test with 2, 5-dihydroxybenzoic acid, and the samples were determined at UV $\lambda$ = 195 nm.

For bacterial endotoxin testing, we used a quantitative kinetic chromogenic Limulus Amebocyte Lysate assay (Kinetic-QCL test kit, Lonza, Walkersville, MD, USA), and apyrogenic 96-well microplates (Corning, NY, USA). The measurements were performed at UV $\lambda$ = 405 nm with a FLUOstar Omega microplate reader (BMG Labtech, Ortenberg, Germany) [19].

Accredited laboratory sites tested the ethanol content and sterility according to the European Pharmacopoeia.

## 5. Conclusions

An automated process for the preparation of [$^{99m}$Tc]Tc-PSMA-I&S was developed on a Scintomics GRP synthesizer with respect to good manufacturing practice (GMP) and cGRPP. An advantage is the use of commercially available hardware and reagent kits intended for $^{68}$Ga-labelled peptides with only minor modifications. For successful preparations of [$^{99m}$Tc]Tc-PSMA-I&S, the pH of the reaction mixture must be adjusted with the addition of NaOH. In summary, the presented synthesis, as well as the quality control, can be easily integrated into everyday clinical practice.

**Author Contributions:** Conceptualization, H.K. and E.P.; methodology, E.P. and C.P.; software, E.P.; validation, H.K., E.P. and C.P.; investigation, E.P. and C.P.; resources, R.M.A.; writing—original draft preparation, E.P.; writing—review and editing, E.P.,C.P., E.G., R.M.A. and H.K.; visualization, E.P.; supervision, H.K., R.M.A. and E.G. All authors have read and agreed to the published version of the manuscript.

**Funding:** The authors acknowledge the financial support by the University of Graz.

**Institutional Review Board Statement:** Not applicable.

**Informed Consent Statement:** Not applicable.

**Data Availability Statement:** Data are contained within the article.

**Acknowledgments:** E.P., C.P. and H.K. would like to thank Bernhard Rumpf and Daniel Paul for their practical support during the experiments in the radiopharmaceutical laboratory of the Division of Nuclear Medicine.

**Conflicts of Interest:** The authors declare no conflict of interest.

## References

1. Duatti, A. Review on $^{99m}$Tc radiopharmaceuticals with emphasis on new advancements. *Nucl. Med. Biol.* **2021**, *92*, 202–216. [CrossRef] [PubMed]
2. Schmidkonz, C.; Kuwert, T.; Cordes, M. $^{99m}$Tc-PSMA-SPECT/CT zur Diagnostik des Prostatakarzinoms. *Der Nukl.* **2020**, *43*, 303–308. [CrossRef]
3. Robu, S.; Schottelius, M.; Eiber, M.; Maurer, T.; Gschwend, J.; Schwaiger, M.; Wester, H.J. Preclinical evaluation and first patient application of $^{99m}$Tc-PSMA-I&S for SPECT imaging and radioguided surgery in prostate cancer. *J. Nucl. Med.* **2017**, *58*, 235–242. [CrossRef] [PubMed]
4. Maurer, T.; Robu, S.; Schottelius, M.; Schwamborn, K.; Rauscher, I.; van den Berg, N.S.; van Leeuwen, F.W.B.; Haller, B.; Horn, T.; Heck, M.M.; et al. $^{99m}$Technetium-based prostate-specific membrane antigen–radioguided surgery in recurrent prostate cancer. *Eur. Urol.* **2019**, *75*, 659–666. [CrossRef] [PubMed]

5. Horn, T.; Krönke, M.; Rauscher, I.; Haller, B.; Robu, S.; Wester, H.J.; Schottelius, M.; van Leeuwen, F.W.B.; van der Poel, H.G.; Heck, M.; et al. Single lesion on prostate-specific membrane antigen-ligand positron emission tomography and low prostate-specific antigen are prognostic factors for a favorable biochemical response to prostate-specific membrane antigen-targeted radioguided surgery in recurrent prostate cancer. *Eur. Urol.* **2019**, *76*, 517–523. [CrossRef] [PubMed]
6. Urban, S.; Meyer, C.; Dahlbom, M.; Farkas, I.; Sipka, G.; Besenyi, Z.; Czernin, J.; Calais, J.; Pavics, L. Radiation Dosimetry of 99mTc-PSMA-I&S: A single-center prospective study. *J. Nucl. Med.* **2021**, *62*, 1075–1081. [CrossRef] [PubMed]
7. de Barros, H.A.; van Oosterom, M.N.; Donswijk, M.L.; Hendrikx, J.J.M.A.; Vis, A.N.; Maurer, T.; van Leeuwen, F.W.B.; van der Poel, H.G.; van Leeuwen, P.J. Robot-assisted prostate-specific membrane antigen–radioguided salvage surgery in recurrent prostate cancer using a DROP-IN gamma probe: The first prospective feasibility study. *Eur. Urol.* **2022**, *82*, 97–105. [CrossRef] [PubMed]
8. Gandaglia, G.; Mazzone, E.; Stabile, A.; Pellegrino, A.; Cucchiara, V.; Barletta, F.; Scuderi, S.; Robesti, D.; Leni, R.; Gajate, A.M.S.; et al. Prostate-specific membrane antigen radioguided surgery to detect nodal metastases in primary prostate cancer patients undergoing robot-assisted radical prostatectomy and extended pelvic lymphnode dissection: Results of a planned interim analysis of a prospective phase 2 study. *Eur. Urol.* **2022**, *82*, 411–418. [CrossRef] [PubMed]
9. van Leeuwen, F.W.B.; van Oosterom, M.N.; Meershoek, P.; van Leeuwen, P.J.; Berliner, C.; van der Poel, H.G.; Graefen, M.; Maurer, T. Minimal-invasive robot-assisted image-guided resection of prostate-specific membrane antigen–positive lymph nodes in recurrent prostate cancer. *Clin. Nucl. Med.* **2019**, *44*, 580–581. [CrossRef] [PubMed]
10. Werner, P.; Neumann, C.; Eiber, M.; Wester, H.J.; Schottelius, M. [$^{99\text{cm}}$Tc]Tc-PSMA-I&S-SPECT/CT: Experience in prostate cancer imaging in an outpatient center. *EJNMMI Res.* **2020**, *10*, 45. [CrossRef] [PubMed]
11. Aalbersberg, E.A.; van Andel, L.; Geluk-Jonker, M.M.; Beijnen, J.H.; Stokkel, M.P.M.; Hendrikx, J.J.M.A. Automated synthesis and quality control of [$^{99\text{m}}$Tc]Tc-PSMA for radioguided surgery (in a [$^{68}$Ga]Ga-PSMA workflow). *EJNMMI Radiopharm. Chem.* **2020**, *5*, 10. [CrossRef] [PubMed]
12. European Pharmacopoeia. Monograph Gallium (68Ga) edotreotide injection. 01/2013:2482. In *European Pharmacopeia*, 10th ed.; European Pharmacopoeia: Strasbourg, France, 2020.
13. European Pharmacopoeia. Monograph Gallium (68Ga) PSMA-11 injection. 04/2021:3044. In *European Pharmacopeia*, 10th ed.; European Pharmacopoeia: Strasbourg, France, 2020.
14. European Pharmacopoeia. Monograph Extemporaneous preparation of radiopharmaceuticals. 04/2016:51900. In *European Pharmacopeia*, 10th ed.; European Pharmacopoeia: Strasbourg, France, 2020.
15. European Pharmacopoeia. Monograph Radiopharmaceutical Preparations. 07/2016:0125. In *European Pharmacopeia*, 10th ed.; European Pharmacopoeia: Strasbourg, France, 2020.
16. Gillings, N.; Hjelstuen, O.; Ballinger, J.; Behe, M.; Decristoforo, C.; Elsinga, P.; Ferrari, V.; Kolenc Peitl, P.; Koziorowski, J.; Laverman, P.; et al. Guideline on current good radiopharmacy practice (cGRPP) for the small-scale preparation of radiopharmaceuticals. *EJNMMI Radiopharm. Chem.* **2021**, *6*, 8. [CrossRef] [PubMed]
17. CPMP/ICH/381/95_ICH Harmonized Tripartite Guideline—Validation of Analytical Procedures: Text and Methodology Q2(R1); 2006. Available online: https://www.ema.europa.eu/en/ich-q2r2-validation-analytical-procedures-scientific-guideline (accessed on 5 October 2022).
18. Antunes, I.F.; Franssen, G.M.; Zijlma, R.; van der Woude, G.L.K.; Yim, C.B.; Laverman, P.; Boersma, H.H.; Elsinga, P.H. New sensitive method for HEPES quantification in [Ga-68]-radiopharmaceuticals. *Eur. J. Nucl. Med. Mol. Imaging* **2017**, *44*, 407–408. [CrossRef]
19. Kvaternik, H.; Plhak, E.; Rumpf, B.; Hausberger, D.; Aigner, R.M. Assay of bacterial endotoxins in radiopharmaceuticals by microplate reader. *EJNMMI Radiopharm. Chem.* **2018**, *3* (Suppl. S1), 11.

**Disclaimer/Publisher's Note:** The statements, opinions and data contained in all publications are solely those of the individual author(s) and contributor(s) and not of MDPI and/or the editor(s). MDPI and/or the editor(s) disclaim responsibility for any injury to people or property resulting from any ideas, methods, instructions or products referred to in the content.

*Communication*

# PET Imaging of Fructose Metabolism in a Rodent Model of Neuroinflammation with 6-[$^{18}$F]fluoro-6-deoxy-D-fructose

Amanda J. Boyle [1,2,*], Emily Murrell [1], Junchao Tong [1], Christin Schifani [1], Andrea Narvaez [1], Melinda Wuest [3], Frederick West [3,4], Frank Wuest [3,4] and Neil Vasdev [1,2,*]

[1] Azrieli Centre for Neuro-Radiochemistry, Brain Health Imaging Centre, Centre for Addiction and Mental Health, 250 College St., Toronto, ON M5T 1R8, Canada
[2] Department of Psychiatry, University of Toronto, 250 College St., Toronto, ON M5T 1R8, Canada
[3] Department of Chemistry, University of Alberta, Edmonton, AB T6G 2N4, Canada
[4] Department of Oncology, University of Alberta, Edmonton, AB T6G 1Z2, Canada
* Correspondence: amy.boyle@camh.ca (A.J.B.); neil.vasdev@utoronto.ca (N.V.); Tel.: +1-416-535-8501 (ext. 30884) (A.J.B.); +1-416-535-8501 (ext. 30988) (N.V.)

**Abstract:** Fluorine-18 labeled 6-fluoro-6-deoxy-D-fructose (6-[$^{18}$F]FDF) targets the fructose-preferred facilitative hexose transporter GLUT5, which is expressed predominantly in brain microglia and activated in response to inflammatory stimuli. We hypothesize that 6-[$^{18}$F]FDF will specifically image microglia following neuroinflammatory insult. 6-[$^{18}$F]FDF and, for comparison, [$^{18}$F]FDG were evaluated in unilateral intra-striatal lipopolysaccharide (LPS)-injected male and female rats (50 μg/animal) by longitudinal dynamic PET imaging in vivo. In LPS-injected rats, increased accumulation of 6-[$^{18}$F]FDF was observed at 48 h post-LPS injection, with plateaued uptake (60–120 min) that was significantly higher in the ipsilateral vs. contralateral striatum (0.985 ± 0.047 and 0.819 ± 0.033 SUV, respectively; $p$ = 0.002, n = 4M/3F). The ipsilateral–contralateral difference in striatal 6-[$^{18}$F]FDF uptake expressed as binding potential ($BP_{SRTM}$) peaked at 48 h (0.19 ± 0.11) and was significantly decreased at one and two weeks. In contrast, increased [$^{18}$F]FDG uptake in the ipsilateral striatum was highest at one week post-LPS injection ($BP_{SRTM}$ = 0.25 ± 0.06, n = 4M). Iba-1 and GFAP immunohistochemistry confirmed LPS-induced activation of microglia and astrocytes, respectively, in ipsilateral striatum. This proof-of-concept study revealed an early response of 6-[$^{18}$F]FDF to neuroinflammatory stimuli in rat brain. 6-[$^{18}$F]FDF represents a potential PET radiotracer for imaging microglial GLUT5 density in brain with applications in neuroinflammatory and neurodegenerative diseases.

**Keywords:** fructose; neuroinflammation; PET; fluorine-18; GLUT5; microglia; FDG; FDF

## 1. Introduction

Neuroinflammation occurs in response to viral or bacterial infections, toxins, as well as injury to the central nervous system and involves the activation of innate immune glial cells. Prolonged neuroinflammation is a common feature linked to neurodegenerative diseases such as Alzheimer's disease (AD) that has been corroborated by molecular imaging studies using positron emission tomography (PET) [1–3]. The most common PET imaging biomarker of neuroinflammation is the 18 kDa translocator protein (TSPO) [4], which is not specifically expressed on microglia, but is also found on astrocytes [5–7]. Nonetheless, PET imaging studies targeting TSPO, both in animal models and in humans, have shown neuroinflammation to be an early event in AD pathogenesis [8,9]. A novel PET radiotracer capable of specifically imaging microglia would be critical to further our mechanistic understanding of the link between neuroinflammation and neurodegenerative diseases in the living human brain [10–14].

More specific molecular targets for microglial imaging are highly sought after. Macrophage colony stimulating factor-1 receptor (CSF-1R) is one such target, which is predominantly found on microglia in the brain, with low-level expression occurring in neurons [15,16]. Several efforts to develop potent and selective CSF-1R targeted PET radiotracers for neuroimaging in preclinical and human studies are underway [17–25]. Two other targets of particular interest for PET imaging microglia are the purinergic receptors $P2X_7$ and $P2Y_{12}$ because they are found on M1 and M2 microglial phenotypes, respectively [26]. These receptors represent targets that could elucidate the pro-inflammatory (M1 phenotype) and anti-inflammatory (M2 phenotype) roles of microglia in neuroinflammation. Several PET radiotracers that target the $P2X_7$ receptor have been investigated in preclinical studies [27–32], and recent efforts have also advanced $P2Y_{12}$ receptor-targeted PET radiotracers to preclinical evaluations [26,33–35]. A promising neuroinflammation target is the glucose transporter (GLUT) 5, a high-affinity fructose-specific facilitative hexose transporter which represents the principal fructose transporter in the body [36]. In the brain, GLUT5 is predominantly expressed on microglia [37–39], and cerebral fructose metabolism has been identified as a potential driving mechanism in AD pathology [40]. Thus, GLUT5 represents a novel biomarker for PET imaging of neuroinflammation in neurodegenerative diseases [41]. A fluorine-18 labeled fructose derivative, 6-deoxy-6-fluoro-D-fructose (6-[$^{18}$F]FDF), was developed for PET imaging of fructose metabolism in breast cancer via GLUT5 [42]. PET imaging studies with 6-[$^{18}$F]FDF in breast cancer models also demonstrated the involvement of GLUT2, a low affinity transporter, in the uptake of 6-[$^{18}$F]FDF [43,44], but the relative abundance of GLUT2 versus GLUT5 in brain is unknown. The present study seeks to determine if 6-[$^{18}$F]FDF can be used to specifically image microglia in rodent models of neuroinflammation.

The most common radiotracer for PET imaging is 2-[$^{18}$F]fluoro-2-deoxy-D-glucose ([$^{18}$F]FDG) which is also a hexose. [$^{18}$F]FDG-PET imaging generally captures inflammation as well as changes and differences in glucose metabolism in the brain, and is frequently used for imaging AD and related dementias for neuronal loss and neuroinflammation [45–47]. In this study, we compare PET imaging using 6-[$^{18}$F]FDF with that using [$^{18}$F]FDG in lipopolysaccharide (LPS) rat models of neuroinflammation in the context of our laboratory's previously published results of imaging in this model with the 2nd generation TSPO PET radiopharmaceutical, N-acetyl-N-(2-[$^{18}$F]fluoroethoxybenzyl)-2-phenoxy-5-pyridinamine ([$^{18}$F]FEPPA) [48].

## 2. Results

### 2.1. Early Increase in 6-[$^{18}$F]FDF Uptake in LPS-Injected Striatum

Following injection in rats, 6-[$^{18}$F]FDF accumulated slowly in brain parenchyma after the initial vasculature signal. Due to radiodefluorination of 6-[$^{18}$F]FDF [42,49], bone accumulation of the radioactivity (see Figure S1 for an example) also increased with time, resulting in significant spillover of radioactivity from the skull to adjacent cerebral and cerebellar cortices, with time–activity curves (TACs) in the cortical areas showing increased radioactivity uptake throughout the 120 min acquisition similar to that in the skull. Nevertheless, in subcortical areas such as striatum, thalamus and hippocampus, the TACs plateaued 60 min after the bolus injection and remained stable at 0.8–1.0 standardized uptake values (SUV) for the remainder of the acquisition (see Figures 1 and 2). In this proof-of-concept study, we focused on the striatum and hippocampus which had minimal skull spillover of radioactivity.

As shown in Figure 1, increased 6-[$^{18}$F]FDF uptake was observed in the LPS-injected right striatum vs. the left side at 48 h post-surgery in both male (Figure 1A) and female (Figure 1D) rats. At later time points of one week (Figure 1B,E) and two weeks (Figure 1C,F), increased 6-[$^{18}$F]FDF binding in the ipsilateral vs. contralateral striatum was largely diminished. Therefore, the early increase in 6-[$^{18}$F]FDF uptake following LPS was in sharp contrast to that reported previously for the TSPO ligand [$^{18}$F]FEPPA, which peaked at approximately one week post-LPS injection, and to that of the MAO-B ligand [$^{11}$C]L-deprenyl, which developed after two weeks of LPS injection [48].

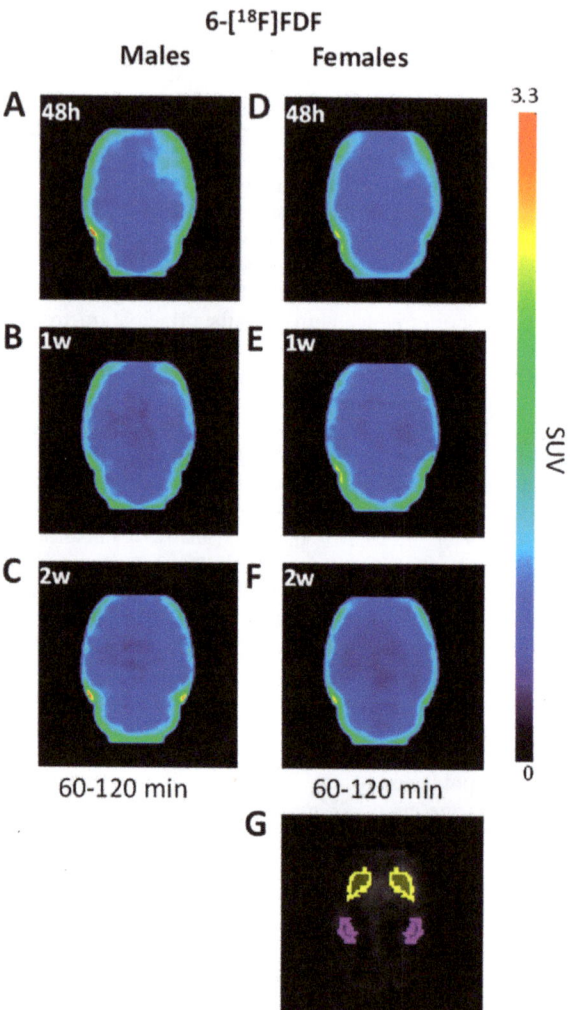

**Figure 1.** Representative static iterative transverse PET images of 6-[$^{18}$F]FDF binding in unilateral lipopolysaccharide (LPS)–injected rat striatum in male (**A–C**) and female rats (**D–F**) at 48 h, 1 week, and 2 weeks post–LPS, and (**G**) MRI indicating the regions of interest of striatum (yellow) and hippocampus (purple).

The results shown in Figure 1 were supported by TAC analyses depicted in Figure 2. Increased 6-[$^{18}$F]FDF uptake (SUV, 60–120 min) in the ipsilateral striatum vs. contralateral side was observed at 48 h after surgery (0.985 ± 0.047 vs. 0.819 ± 0.033, n = 4M/3F, $p = 0.0023$; Figure 2A). Repeated measures ANOVA across the time course revealed a significant effect of LPS treatment (i.e., right vs. left striatum; $F_{1,6} = 22.9$, $p = 0.003$), time ($F_{38,228} = 55.7$, $p < 0.0001$) and time × brain region interaction ($F_{38,228} = 2.81$, $p < 0.0001$), suggesting significantly increased 6-[$^{18}$F]FDF retention with time induced by LPS. No significant difference in TACs between left and right striatum was observed (n = 2M/3F) at one week (treatment: $F_{1,4} = 6.17$, $p = 0.07$ or time × treatment interaction: $F_{38,152} = 0.96$, $p = 0.55$; Figure 2B) or two weeks (treatment: $F_{1,4} = 1.20$, $p = 0.34$ or time × treatment interaction: $F_{38,152} = 1.07$, $p = 0.38$; Figure 2C) after LPS. As a control brain area, hippocampus did not show any significant left and right side difference in TACs at any time point studied ($p > 0.05$; Figure 2D,F), indicating the specificity of LPS-induced local response (see also Figures S2 and S3 for separate TACs of male and female rats).

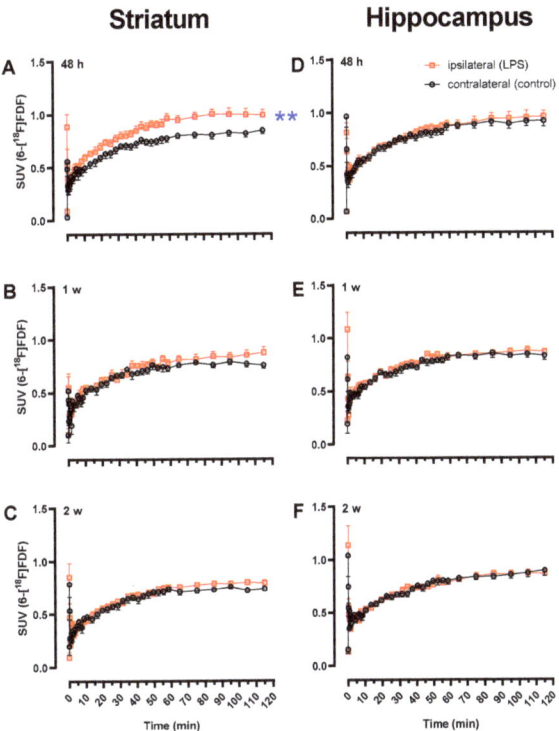

**Figure 2.** Time–activity curves (TACs) of 6-[$^{18}$F]FDF in striatum and hippocampus of rats injected unilaterally with lipopolysaccharide (LPS) in the right striatum. Average TACs (±SEM) of 6-[$^{18}$F]FDF are shown in combined male and female rats in the ipsilateral and contralateral side of striatum (**A–C**) and hippocampus (**D–F**) at 48 h (n = 4M/3F), 1 week (n = 2M/3F), or 2 weeks (n = 2M/3F), post-LPS injection. ** $p = 0.003$, right vs. left striatum at 48 h (repeated measures ANOVA).

With the left-brain region as the reference, the binding potential (BP) on the right side was estimated with simplified reference tissue model (SRTM) [50]. As shown in Figure 3A, one-way ANOVA showed significant change of BP with time in LPS injected striatum ($F_{2,14} = 7.68$, $p = 0.006$), with significantly higher BP at 48 h (0.19 ± 0.11) compared to later time points. In preliminary analysis comparing male and female rats, a two-way ANOVA showed a significant effect of sex ($F_{1,11} = 7.9$, $p = 0.017$) and time ($F_{2,11} = 10.3$, $p = 0.003$) but not sex x time interaction ($F_{2,11} = 1.57$, $p = 0.25$), suggesting that male rats (0.25 ± 0.03; n = 4) had significantly higher BP in the ipsilateral striatum than the females (0.11 ± 0.03; n = 3) at 48 h after LPS injection but the response was later diminished with time in both sexes (Figure 3A, inset).

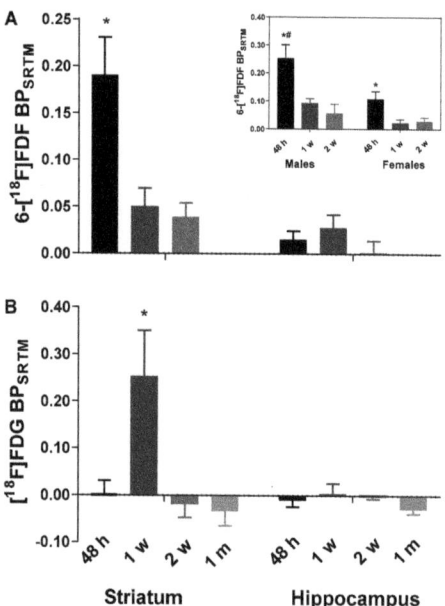

**Figure 3.** Binding potential ($BP_{SRTM}$, ±SEM) of 6-[$^{18}$F]FDF (**A**) and [$^{18}$F]FDG (**B**) in unilateral lipopolysaccharide-injected right striatum of rats. $BP_{SRTM}$ was derived from simplified reference tissue model (SRTM) using contralateral side as the reference tissue. Inset in (**A**) shows separate striatal 6-[$^{18}$F]FDF data for male and female rats. * $p < 0.05$, 6-[$^{18}$F]FDF at 48 h vs. other time points and [$^{18}$F]FDG at one week vs. other time points in the striatum; # $p < 0.05$, male vs. female rats in 6-[$^{18}$F]FDF uptake at 48 h (one-way ANOVA followed by Bonferroni corrections).

## 2.2. Increased [$^{18}$F]FDG Uptake in LPS-Injected Striatum after One Week

We also performed dynamic PET imaging of [$^{18}$F]FDG in the LPS rat model of neuroinflammation for comparison. As expected, [$^{18}$F]FDG was rapidly taken up and retained in the rodent brain throughout the 120 min acquisition. PET scans in male rats following LPS-injection showed increased radioactivity accumulation at one week post-LPS (Figure 4B vs. 4A and 4C). At one week (Figure 5B), but not at other time points (Figure 5A,C,D), TAC analyses demonstrated a significant difference in the right vs. left striatum ($n = 4$; $F_{1,3} = 10.0$, $p = 0.05$) and time x brain region interaction ($F_{38,114} = 1.58$, $p = 0.035$), which is consistent with increased [$^{18}$F]FDG retention local to the LPS-injected striatum. The control brain region hippocampus did not show any significant difference in the right vs. left striatum as indicated by the TACs at any time point ($p > 0.05$; Figure 5E–H). Accordingly, the BP for [$^{18}$F]FDG in the ipsilateral striatum vs. contralateral side (Figure 3B) peaked at one week post-LPS injection (0.25 ± 0.06; $n = 4$), which was significantly higher than at other time points (one-way ANOVA $F_{3,11} = 5.67$, $p = 0.014$). This pattern of BP following LPS-injection into the right striatum is drastically different from that observed following 6-[$^{18}$F]FDF injection, indicating that fructose and glucose metabolism are not occurring in the same population of cells.

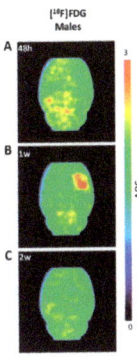

**Figure 4.** Representative static iterative transverse PET images of [$^{18}$F]FDG binding in unilateral lipopolysaccharide (LPS)–injected rat striatum. Shown are PET images in male rats at (**A**) 48 h, (**B**) 1 week, and (**C**) 2 weeks post–LPS injection.

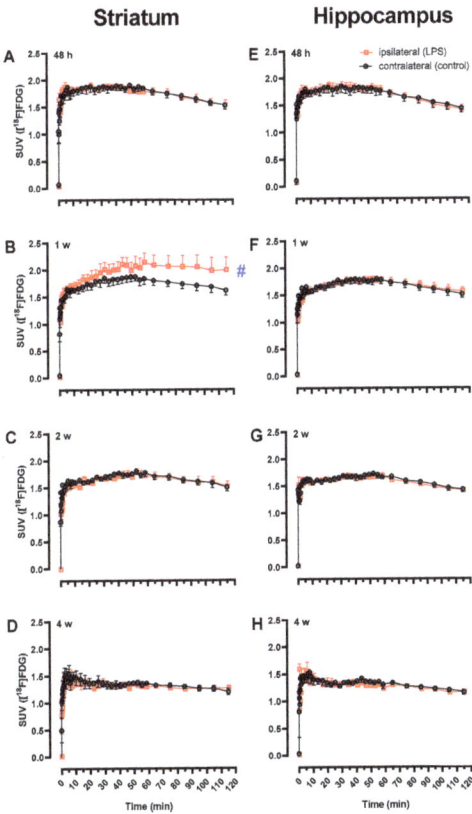

**Figure 5.** Time–activity curves (TACs) of [$^{18}$F]FDG in striatum and hippocampus of male rats injected unilaterally with lipopolysaccharide (LPS) in the right striatum. Average TACs (±SEM) of [$^{18}$F]FDG in ipsilateral and contralateral side of striatum (**A–D**) and hippocampus (**E–H**) at 48 h ($n$ = 4), 1 week ($n$ = 4), 2 weeks ($n$ = 4), or 4 weeks ($n$ = 3) post-LPS injection. $^{\#}$ $p$ = 0.05, right vs. left striatum at 1 week (repeated measures ANOVA).

## 2.3. Immunohistochemistry

Immunohistochemistry shown in Figure 6 demonstrates that an immune response was induced in our neuroinflammation model with activated microglia/macrophages and astrocytes being present on the ipsilateral side one week after LPS injection.

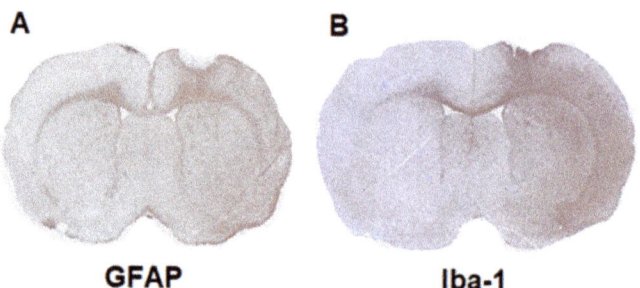

**Figure 6.** Immunohistochemical staining for the presence of microglia and astrocytes in unilateral lipopolysaccharide (LPS)-injected male rat brains. (**A**) Glial fibrillary acidic protein (GFAP, activated astrocytes) and (**B**) ionized calcium binding adaptor molecule 1 (Iba-1, microglia/macrophages) of an LPS-injected male rat brain at 7 days post-LPS injection in the right striatum.

## 3. Discussion

The most common PET biomarker of neuroinflammation is TSPO; however, this target is not exclusive to microglia cells and many TSPO-targeted radiopharmaceuticals for clinical research imaging have been confounded by genetic polymorphisms [51], which complicates the interpretation of TSPO imaging in human brain. Novel PET radiotracers with the ability to specifically image activated microglial cells at different stages are needed to improve our understanding of the role of microglia in neuroinflammation and neurodegenerative diseases [10–14]. In the brain, GLUT5 is predominantly present on microglia [37,39,52], and represents, to our knowledge, an unexplored PET imaging biomarker of neuroinflammation. In this proof-of-concept study, we evaluated the suitability of 6-[$^{18}$F]FDF, a substrate of microglia-located GLUT5, for PET imaging of neuroinflammation in rats injected with LPS into the right striatum and revealed increased radioactivity accumulation in the ipsilateral side compared to the contralateral side (Figure 1). Immunohistochemistry studies have shown that in widely used LPS-induced rodent models of neuroinflammation, microglia activation begins within hours then peaks within 1 to 2 weeks, depending on the biomarker selected for immunohistochemical staining, and then gradually dissipates [53–56]. Indeed, our longitudinal in vivo PET imaging studies found that radioactivity accumulation in the right striatum following 6-[$^{18}$F]FDF administration was highest at 48 h and returned to baseline by 2 weeks post-LPS injection (Figure 2). Therefore, 6-[$^{18}$F]FDF uptake peaks earlier and returns to baseline sooner than that of the TSPO tracer, [$^{18}$F]FEPPA, in male LPS-injected rats [48]. This supports our hypothesis that 6-[$^{18}$F]FDF is likely imaging an early stage of microglial activation, since microgliosis has been shown to start earlier than astrogliosis in response to LPS insults [57].

Our preliminary study showed a trend for sex differences in 6-[$^{18}$F]FDF accumulation in the LPS-injected right striatum over time, with male rats having a greater response than the females. These preliminary results might indicate a difference in microglial response to neuroinflammatory insult with LPS between males and females. Our findings are supported by preclinical studies that have shown sex-related differences in microglial function, microglial expression levels during development, and microglial immune response to LPS injection [58–61]. PET imaging studies have also revealed sex differences in microglia. Another TSPO PET radiotracer, [$^{18}$F]GE-180, revealed higher binding in female mice in a neurodegenerative mouse model for β-amyloid (AppNL-G-F) [62]. Consistently, in human studies, PET imaging with [$^{11}$C]PBR28 showed higher TSPO binding in female healthy control subjects compared to males [63]. The mechanism underlying the sex difference is unknown but could be related to estrous cycle and hormonal status. Taken together, further studies are warranted to confirm the sex differences in male and female cohorts and to examine the effects of hormones on LPS-induced 6-[$^{18}$F]FDF uptake.

[$^{18}$F]FDG is the most widely used PET radiopharmaceutical and is employed for imaging glucose metabolism in neuroinflammation states, including in traumatic brain injuries [64], AD and related dementias [46,47]; however, [$^{18}$F]FDG is not a specific radiotracer for studying inflammation since glucose metabolism of inflammatory microglial and astroglial cells is confounded by glucose metabolism in neurons and other cells in the brain tissue [65]. Interestingly, our studies show that [$^{18}$F]FDG uptake at the site of LPS-injection peaked at 1 week in response to LPS insult (Figure 3B, Figure 5), which coincides with increased [$^{18}$F]FEPPA binding [48]. We postulate that the early stage of neuroinflammation including proliferation of microglia and astrocytes is accompanied by increased energy demand thus glucose metabolism. However, [$^{18}$F]FDG has high background uptake throughout the brain, and therefore subtle changes in glucose metabolism are not likely to be detected by PET imaging. Only male rats were employed in our [$^{18}$F]FDG study as female rats are known to show variable uptake and metabolism with this radiotracer in the brain [66]. Given the sex difference in LPS-induced 6-[$^{18}$F]FDF uptake, future work could consider if female rats respond differently than males to LPS challenge with [$^{18}$F]FDG.

Limitations of this study include a relatively small number of animals examined and defluorination and/or known residual [$^{18}$F]fluoride in the formulation of 6-[$^{18}$F]FDF, as consistently reported in previous studies [49] (see also Section 4.1 below and Figure S4), which could reduce the suitability for translation for human brain PET studies in the present formulation due to proximity to the skull. However, despite bone uptake of [$^{18}$F]fluoride in the skull potentially leading to spill over to adjacent brain regions (e.g., cortices), the partial volume effect was minimal in deep nuclei (e.g., striatum and thalamus) as judged by plateaued rather than continuously increasing TACs during the 120 min acquisition. Overall, 6-[$^{18}$F]FDF was shown to be a promising PET radiotracer for specifically imaging fructose metabolism in microglia via GLUT5 in the brain, and has potential applications for PET imaging of neuroinflammatory and neurodegenerative diseases. We demonstrated the ability of 6-[$^{18}$F]FDF to specifically image early microglial activation in a rodent model of neuroinflammation. Further studies could include more detailed histochemical examination of the expression of GLUT5 and other brain glucose/hexose transporters (e.g., GLUT2), including their cellular localization and relationship to other markers of glial activation (e.g., ionized calcium binding adaptor molecule 1 [Iba1] and glial fibrillary acidic protein [GFAP]) in models of neuroinflammation to correlate with 6-[$^{18}$F]FDF imaging findings. Future studies with 6-[$^{18}$F]FDF could examine PET imaging of neurodegenerative disorders that involve microglia including AD and Parkinson's disease. Another area for future study includes drug addiction as GLUT5 expression and the density of resting microglia have been reported to be increased in brains of methamphetamine users [52]. Further studies could also examine whether differences in 6-[$^{18}$F]FDF uptake among M1 and M2 microglial phenotypes exist. Pro-inflammatory M1 macrophages increase their glucose metabolism, while anti-inflammatory M2 phenotypes have significantly lower glucose consumption than M1 [67]. However, given the inherent heterogeneity of microglia under a pathological condition, in particular in vivo, the M1/M2 dichotomy of microglia might not capture the full picture of microglial status [68–70] and should be considered when developing new biomarkers for neuroinflammation.

## 4. Materials and Methods

### 4.1. Radiochemical Synthesis

Radiochemical synthesis of 6-[$^{18}$F]FDF was performed as previously described [49]. Briefly, the methyl 1,3,4-tri-O-acetyl-6-O-(methylbenzene-sulfonyl)-α/β-D-fructofuranoside precursor (provided by Dr. Frank Wuest of University of Alberta, Edmonton, AB, Canada) was labeled using a [$^{18}$F]KF/Kryptofix (K$_{222}$) complex, followed by acid hydrolysis, and isolation by semi-preparative HPLC (Phenomenex LUNA C18(2) 10 μm 250 mm × 10 mm, 0.1 M sodium acetate buffer, pH = 5, at 2 mL/min). The collected peak was used directly for injection. Radiochemical identity was verified by HPLC and by co-spotting and staining by radio-TLC, and radiochemical purity was determined to be >90% (95:5 MeCN:$H_2O$). The collected 6-[$^{18}$F]FDF peak can be contaminated by a residual [$^{18}$F]fluoride because of similar retention times on the established semi-preparative HPLC conditions (see Figure S4). [$^{18}$F]FDG was purchased from Isologic Innovative Radiopharmaceuticals (ISOLOGIC, Toronto, Canada).

*4.2. Lipopolysaccharide Rat Models of Neuroinflammation*

Rat models of neuroinflammation were prepared by injecting LPS (L2630, serotype O111:B4; Sigma-Aldrich, St. Louis, MO, USA) unilaterally into the right striatum (caudate putamen) as previously described at our laboratory [48]. Briefly, adult male (6-[$^{18}$F]FDF: 357–435 g, n = 4; [$^{18}$F]FDG: 274–337 g, *n* = 4) or female (6-[$^{18}$F]FDF: 207–219 g, n = 3) Sprague Dawley rats were anesthetized by isoflurane in $O_2$ (5%, 2 L/min induction; 3%, 1 L/min maintenance) and positioned in a stereotactic head frame (David Kopf Instruments, Tujunga, CA, USA). Coordinates for the right striatum in relation to bregma were 0.5 mm anteroposterior, 3 mm lateral, and 5.5/4.5 dorsoventral [71]. The arm on stereotactic frame was maneuvered to the appropriate coordinates and a small hole was drilled at this location. A solution of LPS was injected at a rate of 0.5 μL/min via a microinjection pump with the microinjector placed at the appropriate coordinates for injection into the right striatum at a depth of 5.5 mm then 4.5 mm for a total of 50 μg in 4 μL injected.

*4.3. Dynamic PET/MR and PET/CT Acquisition*

PET/ magnetic resonance imaging (MR) or PET/ computed tomography (CT) was performed with 6-[$^{18}$F]FDF in LPS rat models of neuroinflammation at 48 h (*n* = 4M/3F), 1 week (*n* = 2M/3F), and 2 weeks (*n* = 2M/3F) post-LPS injection, as well as with [$^{18}$F]FDG in another cohort of male rats at 48 h (*n* = 4M), 1 week (*n* = 4M), 2 weeks (*n* = 4M) and 4 weeks (*n* = 3M) post-LPS injection. Two male rats in the 6-[$^{18}$F]FDF study were sacrificed at one week post-LPS injection for immunohistochemistry and one male rat in the [$^{18}$F]FDG study died after two weeks post-LPS. PET image acquisition following injection with the radiotracers was performed as previously described [48]. Rats were anesthetized by isoflurane in $O_2$ (4%, 2 L/min induction; 1–2%, 1 L/min maintenance) for lateral tail-vein catheterization then transferred to a nanoScan™ PET/MR 3T or a PET/CT scanner (Mediso, Budapest, Hungary). Anesthesia was maintained throughout the imaging session while body temperature and respiration parameters were monitored closely. A scout MR or CT was acquired for PET field-of-view (FOV) positioning, then MR (gradient echo [GRE] multi-FOV and fast spin echo [FSE] 2D) or CT images were acquired for PET corrections of attenuation and scatter with the segmented material map and for PET/MR or PET/CT co-registration to define anatomic brain regions of interest. Rats were administered a bolus injection of 6-[$^{18}$F]FDF (11.83–27.29 MBq) or [$^{18}$F]FDG (15.39–23.15 MBq) through the tail-vein catheter and a 120 min scan was acquired.

*4.4. PET Data Analysis*

Acquired list mode data were sorted into thirty-nine three-dimensional (3D) (3 × 5 s, 3 × 15 s, 3 × 20 s, 7 × 60 s, 17 × 180 s, and 6 × 600 s), true sinograms (ring difference 84). The 3D sinograms were converted in 2D sinograms using Fourier rebinning [72] with corrections for detector geometry, efficiencies, attenuation, and scatter before image reconstruction using 2D filtered back-projection with a Hann filter at a cut-off of 0.50 $cm^{-1}$. Static images of the complete emission acquisition (0–120 min) and in the time frame of 60–120 min were reconstructed with the manufacturer's proprietary iterative 3D algorithm (6 subsets, 4 iterations). The static iterative images were used for PET and MR or CT co-registration (0–120 min images) and for presentation in figures (60–120 min images; Figures 1 and 4). All data were corrected for dead time and were decay-corrected to the start of acquisition. Dynamic filtered back-projection images were used to extract regional brain TACs using a stereotactic MR atlas [73] following co-registration with subject's MR (T2 weighted 2D FSE, TR 3971 ms, TE 87.5 ms) or CT image implemented in VivoQuant® 2021 software (Invicro, Needham, MA, USA). SUV were calculated by normalizing regional radioactivity for injected radioactivity and body weight. Radiotracer BP in the right striatum was estimated with SRTM, using left striatum as the reference tissue, implemented in PMOD4.203 (PMOD Technologies, Zurich, Switzerland) [50]. TACs of the left and right hippocampus, which were not affected by LPS injection in the striatum, were analyzed as control.

*4.5. Immunohistochemistry*

Immunohistochemistry was performed to examine the expression of Iba-1 and GFAP in brains of LPS-injected male rats ($n = 2$) at 1 week post-LPS injection to confirm the presence of activated microglia and astrocytes, respectively, and also the successful injection of LPS in the right striatum. Brain tissue was fixed in 10% formalin for 48 h then embedded in paraffin and prepared in 4 µm sections onto microscope slides. Slides were dewaxed through changes of xylene, followed by hydration through decreasing grades of alcohol in water (100%, 95%, and 70%). Slides were blocked with 3% hydrogen peroxide, then antigen retrieval was performed with slides being heated at 98 °C in a microwave for 30 min for those being stained for Iba-1. Serum block was applied as directed by the MACH-4 Universal HRP-Polymer kit (Intermedico, BC-M4U534L), followed by incubation with a rabbit anti-Iba1 primary antibody or a rabbit anti-GFAP primary antibody (Abcam, Boston, USA) at room temperature for 1 h. Color was developed using DAB (Agilent Dako, K3468; Carpinteria, CA, USA) and counter stained with hematoxylin. Slides were dehydrated by reversing the rehydration procedure and sections were mounted with mounting medium (Leica, 3801120; Concord, ON, Canada). Slides were scanned with a slide scanner (Olympus, Slideview VS200; Tokyo, Japan).

*4.6. Statistical Analysis*

Data are represented as the mean ± SEM. Statistical analyses were performed by using StatSoft STATISTICA 7.1 (Tulsa, OK, USA). Differences in average SUV 60–120 min between left and right side were examined by paired Student's $t$-test. Differences in 39-frame TACs between left and right side of the brain regions were examined by repeated measures ANOVA. Differences in BP across the time points following LPS injection and between sexes were examined by one-way or two-way ANOVA.

## 5. Conclusions

The major finding of this study is an increased response of 6-[$^{18}$F]FDF to a local bacteria endotoxin insult in rat brain at 48 h post-surgery, suggesting that 6-[$^{18}$F]FDF imaging of fructose metabolism via GLUT5 in microglial cells could be an early biomarker of neuroinflammatory reactions. The preliminary observation of sex differences in 6-[$^{18}$F]FDF response in rats warrants further studies of hormonal influences on microglial reaction.

**Supplementary Materials:** The following are available online at https://www.mdpi.com/article/10.3390/molecules27238529/s1, Figure S1: Unstripped static 6-[18F]FDF image of Figure 1A (48 h post-LPS injection in a male rat), showing radioactivity accumulation in the skull; but also increased uptake in the LPS-injected right striatum vs. the left side. Figure S2: TACs of 6-[18F]FDF in striatum of male and female rats injected unilaterally with LPS in the right striatum. Average TACs (±SEM) of 6-[18F]FDF are shown in the right and left striatum of male (A–C) rats at 48 h (n = 4), 1 week (n = 2), and 2 weeks (n = 2), re-spectively, post-LPS injection and of female rats (D–F) at 48 h (n = 3), 1 week (n = 3), and 2 weeks (n = 3), respectively, post-LPS injection. * $p < 0.05$, right vs left striatum at 48 h (repeated measures ANOVA). Figure S3. TACs of 6-[18F]FDF in hippocampus of male and female rats in-jected unilaterally with LPS in the right striatum. Average TACs (±SEM) of 6-[18F]FDF are shown in the right and left hippocampus of male (AC) rats at 48 h (n = 4), 1 week (n = 2), and 2 weeks (n = 2), respectively, post-LPS injection and of female rats

(D–F) at 48 h (n = 3), 1 week (n = 3), and 2 weeks (n = 3), respectively, post-LPS injection. Figure S4. Results of quality control of a 6-[18F]FDF production with residual [18F]fluoride. (A) Radio-HPLC chromatogram shows 6-[18F]FDF at 3.2 min and residual [18F]fluoride at 2.4 min; (B) Radio-TLC also shows 6-[18F]FDF (93.3%) at 78 mm and residual [18F]fluoride (6.7%) at 53 mm.

**Author Contributions:** A.J.B., E.M., J.T., C.S. and A.N. performed research, analyzed the data and prepared the figures; A.J.B., J.T. and N.V. wrote the main manuscript. A.J.B., E.M., J.T., M.W., F.W. (Frederick West), F.W. (Frank Wuest) and N.V. reviewed the data and all authors reviewed the manuscript. All authors have read and agreed to the published version of the manuscript.

**Funding:** A.J.B. acknowledges support from the CAMH Foundation (Discovery Fund). N.V. thanks the Azrieli Foundation, the Canada Research Chairs Program, Canada Foundation for Innovation, and the Ontario Research Fund for support. A.N. is supported by Enigma Biomedical Group. C.S. acknowledges support from The Brain and Behavior Research Foundation (BBRF; NARSAD Young Investigator Award). F.W. acknowledges support of this work from the Dianne and Irving Kipnes Foundation.

**Institutional Review Board Statement:** Animal studies were conducted under a protocol (#851) approved by the Animal Care Committee at the Centre for Addition and Mental Health, following the Canadian Council on Animal Care guidelines.

**Informed Consent Statement:** Not applicable.

**Data Availability Statement:** The datasets during and/or analyzed during the current study will be made available from the corresponding authors on reasonable request.

**Acknowledgments:** We thank our colleagues at the CAMH Brain Health Imaging Centre for support with the cyclotron, radiochemistry, and preclinical research.

**Conflicts of Interest:** A.N. is employed by Enigma Biomedical Group. N.V. is a co-founder of MedChem Imaging, Inc. All authors declare that the research was conducted in the absence of any commercial or financial relationship that could be construed as a potential conflict of interest.

**Sample Availability:** Samples of the compounds are not available from the authors.

# References

1. Guzman-Martinez, L.; Maccioni, R.B.; Andrade, V.; Navarrete, L.P.; Pastor, M.G.; Ramos-Escobar, N. Neuroinflammation as a Common Feature of Neurodegenerative Disorders. *Front. Pharmacol.* **2019**, *10*, 1008. [CrossRef] [PubMed]
2. Zhou, R.; Ji, B.; Kong, Y.; Qin, L.; Ren, W.; Guan, Y.; Ni, R. PET Imaging of Neuroinflammation in Alzheimer's Disease. *Front. Immunol.* **2021**, *12*, 739130. [CrossRef] [PubMed]
3. Schain, M.; Kreisl, W.C. Neuroinflammation in Neurodegenerative Disorders-a Review. *Curr. Neurol. Neurosci. Rep.* **2017**, *17*, 25. [CrossRef] [PubMed]
4. Beaino, W.; Janssen, B.; Vugts, D.J.; de Vries, H.E.; Windhorst, A.D. Towards PET imaging of the dynamic phenotypes of microglia. *Clin. Exp. Immunol.* **2021**, *206*, 282–300. [CrossRef] [PubMed]
5. Guilarte, T.R.; Kuhlmann, A.C.; O'Callaghan, J.P.; Miceli, R.C. Enhanced expression of peripheral benzodiazepine receptors in trimethyltin-exposed rat brain: A biomarker of neurotoxicity. *Neurotoxicology* **1995**, *16*, 441–450.
6. Nguyen, D.L.; Wimberley, C.; Truillet, C.; Jego, B.; Caillé, F.; Pottier, G.; Boisgard, R.; Buvat, I.; Bouilleret, V. Longitudinal positron emission tomography imaging of glial cell activation in a mouse model of mesial temporal lobe epilepsy: Toward identification of optimal treatment windows. *Epilepsia* **2018**, *59*, 1234–1244. [CrossRef] [PubMed]
7. Pannell, M.; Economopoulos, V.; Wilson, T.C.; Kersemans, V.; Isenegger, P.G.; Larkin, J.R.; Smart, S.; Gilchrist, S.; Gouverneur, V.; Sibson, N.R. Imaging of translocator protein upregulation is selective for pro-inflammatory polarized astrocytes and microglia. *Glia* **2020**, *68*, 280–297. [CrossRef]
8. Hanzel, C.E.; Pichet-Binette, A.; Pimentel, L.S.; Iulita, M.F.; Allard, S.; Ducatenzeiler, A.; Do Carmo, S.; Cuello, A.C. Neuronal driven pre-plaque inflammation in a transgenic rat model of Alzheimer's disease. *Neurobiol. Aging* **2014**, *35*, 2249–2262. [CrossRef]
9. Okello, A.; Edison, P.; Archer, H.A.; Turkheimer, F.E.; Kennedy, J.; Bullock, R.; Walker, Z.; Kennedy, A.; Fox, N.; Rossor, M.; et al. Microglial activation and amyloid deposition in mild cognitive impairment: A PET study. *Neurology* **2009**, *72*, 56–62. [CrossRef]
10. Narayanaswami, V.; Dahl, K.; Bernard-Gauthier, V.; Josephson, L.; Cumming, P.; Vasdev, N. Emerging PET Radiotracers and Targets for Imaging of Neuroinflammation in Neurodegenerative Diseases: Outlook Beyond TSPO. *Mol. Imaging* **2018**, *17*, 1536012118792317. [CrossRef]
11. Jain, P.; Chaney, A.M.; Carlson, M.L.; Jackson, I.M.; Rao, A.; James, M.L. Neuroinflammation PET Imaging: Current Opinion and Future Directions. *J. Nucl. Med.* **2020**, *61*, 1107–1112. [CrossRef] [PubMed]

12. Janssen, B.; Mach, R.H. Development of brain PET imaging agents: Strategies for imaging neuroinflammation in Alzheimer's disease. *Prog. Mol. Biol. Transl. Sci.* **2019**, *165*, 371–399. [PubMed]
13. Janssen, B.; Vugts, D.J.; Windhorst, A.D.; Mach, R.H. PET Imaging of Microglial Activation-Beyond Targeting TSPO. *Molecules* **2018**, *23*, 607. [CrossRef]
14. Chen, Z.; Haider, A.; Chen, J.; Xiao, Z.; Gobbi, L.; Honer, M.; Grether, U.; Arnold, S.E.; Josephson, L.; Liang, S.H. The Repertoire of Small-Molecule PET Probes for Neuroinflammation Imaging: Challenges and Opportunities beyond TSPO. *J. Med. Chem.* **2021**, *64*, 17656–17689. [CrossRef] [PubMed]
15. Akiyama, H.; Nishimura, T.; Kondo, H.; Ikeda, K.; Hayashi, Y.; McGeer, P.L. Expression of the receptor for macrophage colony stimulating factor by brain microglia and its upregulation in brains of patients with Alzheimer's disease and amyotrophic lateral sclerosis. *Brain Res.* **1994**, *639*, 171–174. [CrossRef] [PubMed]
16. Zhang, Y.; Chen, K.; Sloan, S.A.; Bennett, M.L.; Scholze, A.R.; O'Keeffe, S.; Phatnani, H.P.; Guarnieri, P.; Caneda, C.; Ruderisch, N.; et al. An RNA-sequencing transcriptome and splicing database of glia, neurons, and vascular cells of the cerebral cortex. *J. Neurosci.* **2014**, *34*, 11929–11947. [CrossRef] [PubMed]
17. Bernard-Gauthier, V.; Schirrmacher, R. 5-(4-((4-[$^{18}$F]Fluorobenzyl)oxy)-3-methoxybenzyl)pyrimidine-2,4-diamine: A selective dual inhibitor for potential PET imaging of Trk/CSF-1R. *Bioorg. Med. Chem. Lett.* **2014**, *24*, 4784–4790. [CrossRef]
18. Horti, A.G.; Naik, R.; Foss, C.A.; Minn, I.; Misheneva, V.; Du, Y.; Wang, Y.; Mathews, W.B.; Wu, Y.; Hall, A.; et al. PET imaging of microglia by targeting macrophage colony-stimulating factor 1 receptor (CSF1R). *Proc. Natl. Acad. Sci. USA* **2019**, *116*, 1686–1691. [CrossRef] [PubMed]
19. Knight, A.C.; Varlow, C.; Zi, T.; Liang, S.H.; Josephson, L.; Schmidt, K.; Patel, S.; Vasdev, N. In Vitro Evaluation of [$^{3}$H]CPPC as a Tool Radioligand for CSF-1R. *ACS Chem. Neurosci.* **2021**, *12*, 998–1006. [CrossRef]
20. Lee, H.; Park, J.H.; Kim, H.; Woo, S.K.; Choi, J.Y.; Lee, K.H.; Choe, Y.S. Synthesis and Evaluation of a $^{18}$F-Labeled Ligand for PET Imaging of Colony-Stimulating Factor 1 Receptor. *Pharmaceuticals* **2022**, *15*, 276. [CrossRef]
21. Naik, R.; Misheneva, V.; Minn, I.L.; Melnikova, T.; Mathews, W.; Dannals, R.; Pomper, M.; Savonenko, A.; Pletnikov, M.; Horti, A. PET tracer for imaging the macrophage colony stimulating factor receptor (CSF1R) in rodent brain. *J. Nucl. Med.* **2018**, *59*, 547.
22. Tanzey, S.S.; Shao, X.; Stauff, J.; Arteaga, J.; Sherman, P.; Scott, P.J.H.; Mossine, A.V. Synthesis and Initial In Vivo Evaluation of [$^{11}$C]AZ683-A Novel PET Radiotracer for Colony Stimulating Factor 1 Receptor (CSF1R). *Pharmaceuticals* **2018**, *11*, 136. [CrossRef] [PubMed]
23. van der Wildt, B.; Nezam, M.; Kooijman, E.J.M.; Reyes, S.T.; Shen, B.; Windhorst, A.D.; Chin, F.T. Evaluation of carbon-11 labeled 5-(1-methyl-1H-pyrazol-4-yl)-N-(2-methyl-5-(3-(trifluoromethyl)benzamido)phenyl)nicotinamide as PET tracer for imaging of CSF-1R expression in the brain. *Bioorg. Med. Chem.* **2021**, *42*, 116245. [CrossRef]
24. Zhou, X.; Ji, B.; Seki, C.; Nagai, Y.; Minamimoto, T.; Fujinaga, M.; Zhang, M.R.; Saito, T.; Saido, T.C.; Suhara, T.; et al. PET imaging of colony-stimulating factor 1 receptor: A head-to-head comparison of a novel radioligand, $^{11}$C-GW2580, and $^{11}$C-CPPC, in mouse models of acute and chronic neuroinflammation and a rhesus monkey. *J. Cereb. Blood Flow Metab.* **2021**, *41*, 2410–2422. [CrossRef] [PubMed]
25. Coughlin, J.M.; Du, Y.; Lesniak, W.G.; Harrington, C.K.; Brosnan, M.K.; O'Toole, R.; Zandi, A.; Sweeney, S.E.; Abdallah, R.; Wu, Y.; et al. First-in-human use of (11)C-CPPC with positron emission tomography for imaging the macrophage colony-stimulating factor 1 receptor. *EJNMMI Res.* **2022**, *12*, 64. [CrossRef]
26. Beaino, W.; Janssen, B.; Kooij, G.; van der Pol, S.M.A.; van Het Hof, B.; van Horssen, J.; Windhorst, A.D.; de Vries, H.E. Purinergic receptors P2Y12R and P2X7R: Potential targets for PET imaging of microglia phenotypes in multiple sclerosis. *J. Neuroinflammation* **2017**, *14*, 259. [CrossRef]
27. Berdyyeva, T.; Xia, C.; Taylor, N.; He, Y.; Chen, G.; Huang, C.; Zhang, W.; Kolb, H.; Letavic, M.; Bhattacharya, A.; et al. PET Imaging of the P2X7 Ion Channel with a Novel Tracer [$^{18}$F]JNJ-64413739 in a Rat Model of Neuroinflammation. *Mol. Imaging Biol.* **2019**, *21*, 871–878. [CrossRef]
28. Han, J.; Liu, H.; Liu, C.; Jin, H.; Perlmutter, J.S.; Egan, T.M.; Tu, Z. Pharmacologic characterizations of a P2X7 receptor-specific radioligand, [$^{11}$C]GSK1482160 for neuroinflammatory response. *Nucl. Med. Commun.* **2017**, *38*, 372–382. [CrossRef]
29. Janssen, B.; Vugts, D.J.; Wilkinson, S.M.; Ory, D.; Chalon, S.; Hoozemans, J.J.M.; Schuit, R.C.; Beaino, W.; Kooijman, E.J.M.; van den Hoek, J.; et al. Identification of the allosteric P2X7 receptor antagonist [$^{11}$C]SMW139 as a PET tracer of microglial activation. *Sci. Rep.* **2018**, *8*, 6580. [CrossRef]
30. Koole, M.; Schmidt, M.E.; Hijzen, A.; Ravenstijn, P.; Vandermeulen, C.; Van Weehaeghe, D.; Serdons, K.; Celen, S.; Bormans, G.; Ceusters, M.; et al. $^{18}$F-JNJ-64413739, a Novel PET Ligand for the P2X7 Ion Channel: Radiation Dosimetry, Kinetic Modeling, Test-Retest Variability, and Occupancy of the P2X7 Antagonist JNJ-54175446. *J. Nucl. Med.* **2019**, *60*, 683–690. [CrossRef]
31. Territo, P.R.; Meyer, J.A.; Peters, J.S.; Riley, A.A.; McCarthy, B.P.; Gao, M.; Wang, M.; Green, M.A.; Zheng, Q.H.; Hutchins, G.D. Characterization of $^{11}$C-GSK1482160 for Targeting the P2X7 Receptor as a Biomarker for Neuroinflammation. *J. Nucl. Med.* **2017**, *58*, 458–465. [CrossRef] [PubMed]
32. Van Weehaeghe, D.; Van Schoor, E.; De Vocht, J.; Koole, M.; Attili, B.; Celen, S.; Declercq, L.; Thal, D.R.; Van Damme, P.; Bormans, G.; et al. TSPO Versus P2X7 as a Target for Neuroinflammation: An In Vitro and In Vivo Study. *J. Nucl. Med.* **2020**, *61*, 604–607. [CrossRef]

33. Maeda, J.; Minamihisamatsu, T.; Shimojo, M.; Zhou, X.; Ono, M.; Matsuba, Y.; Ji, B.; Ishii, H.; Ogawa, M.; Akatsu, H.; et al. Distinct microglial response against Alzheimer's amyloid and tau pathologies characterized by P2Y12 receptor. *Brain Commun.* **2021**, *3*, fcab011. [CrossRef] [PubMed]
34. van der Wildt, B.; Janssen, B.; Pekošak, A.; Stéen, E.J.L.; Schuit, R.C.; Kooijman, E.J.M.; Beaino, W.; Vugts, D.J.; Windhorst, A.D. Novel Thienopyrimidine-Based PET Tracers for $P_2Y_{12}$ Receptor Imaging in the Brain. *ACS Chem. Neurosci.* **2021**, *12*, 4465–4474. [CrossRef] [PubMed]
35. Villa, A.; Klein, B.; Janssen, B.; Pedragosa, J.; Pepe, G.; Zinnhardt, B.; Vugts, D.J.; Gelosa, P.; Sironi, L.; Beaino, W.; et al. Identification of new molecular targets for PET imaging of the microglial anti-inflammatory activation state. *Theranostics* **2018**, *8*, 5400–5418. [CrossRef] [PubMed]
36. Manolescu, A.R.; Witkowska, K.; Kinnaird, A.; Cessford, T.; Cheeseman, C. Facilitated hexose transporters: New perspectives on form and function. *Physiology (Bethesda)* **2007**, *22*, 234–240. [CrossRef] [PubMed]
37. Horikoshi, Y.; Sasaki, A.; Taguchi, N.; Maeda, M.; Tsukagoshi, H.; Sato, K.; Yamaguchi, H. Human GLUT5 immunolabeling is useful for evaluating microglial status in neuropathological study using paraffin sections. *Acta Neuropathol.* **2003**, *105*, 157–162. [CrossRef]
38. Izumi, Y.; Zorumski, C.F. Glial-neuronal interactions underlying fructose utilization in rat hippocampal slices. *Neuroscience* **2009**, *161*, 847–854. [CrossRef]
39. Payne, J.; Maher, F.; Simpson, I.; Mattice, L.; Davies, P. Glucose transporter Glut 5 expression in microglial cells. *Glia* **1997**, *21*, 327–331. [CrossRef]
40. Johnson, R.J.; Gomez-Pinilla, F.; Nagel, M.; Nakagawa, T.; Rodriguez-Iturbe, B.; Sanchez-Lozada, L.G.; Tolan, D.R.; Lanaspa, M.A. Cerebral Fructose Metabolism as a Potential Mechanism Driving Alzheimer's Disease. *Front. Aging Neurosci.* **2020**, *12*, 560865. [CrossRef]
41. Oppelt, S.A.; Zhang, W.; Tolan, D.R. Specific regions of the brain are capable of fructose metabolism. *Brain Res.* **2017**, *1657*, 312–322. [CrossRef]
42. Wuest, M.; Trayner, B.J.; Grant, T.N.; Jans, H.S.; Mercer, J.R.; Murray, D.; West, F.G.; McEwan, A.J.; Wuest, F.; Cheeseman, C.I. Radiopharmacological evaluation of 6-deoxy-6-[$^{18}$F]fluoro-D-fructose as a radiotracer for PET imaging of GLUT5 in breast cancer. *Nucl. Med. Biol.* **2011**, *38*, 461–475. [CrossRef] [PubMed]
43. Hamann, I.; Krys, D.; Glubrecht, D.; Bouvet, V.; Marshall, A.; Vos, L.; Mackey, J.R.; Wuest, M.; Wuest, F. Expression and function of hexose transporters GLUT1, GLUT2, and GLUT5 in breast cancer-effects of hypoxia. *FASEB J.* **2018**, *32*, 5104–5118. [CrossRef] [PubMed]
44. Wuest, M.; Hamann, I.; Bouvet, V.; Glubrecht, D.; Marshall, A.; Trayner, B.; Soueidan, O.M.; Krys, D.; Wagner, M.; Cheeseman, C.; et al. Molecular Imaging of GLUT1 and GLUT5 in Breast Cancer: A Multitracer Positron Emission Tomography Imaging Study in Mice. *Mol. Pharmacol.* **2018**, *93*, 79–89. [CrossRef] [PubMed]
45. Bouter, C.; Henniges, P.; Franke, T.N.; Irwin, C.; Sahlmann, C.O.; Sichler, M.E.; Beindorff, N.; Bayer, T.A.; Bouter, Y. $^{18}$F-FDG-PET Detects Drastic Changes in Brain Metabolism in the Tg4–42 Model of Alzheimer's Disease. *Front. Aging Neurosci.* **2019**, *10*, 425. [CrossRef] [PubMed]
46. Kato, T.; Inui, Y.; Nakamura, A.; Ito, K. Brain fluorodeoxyglucose (FDG) PET in dementia. *Ageing Res. Rev.* **2016**, *30*, 73–84. [CrossRef] [PubMed]
47. Marcus, C.; Mena, E.; Subramaniam, R.M. Brain PET in the diagnosis of Alzheimer's disease. *Clin. Nucl. Med.* **2014**, *39*, e413–e426. [CrossRef]
48. Narayanaswami, V.; Tong, J.; Schifani, C.; Bloomfield, P.M.; Dahl, K.; Vasdev, N. Preclinical Evaluation of TSPO and MAO-B PET Radiotracers in an LPS Model of Neuroinflammation. *PET Clin.* **2021**, *16*, 233–247. [CrossRef]
49. Bouvet, V.; Jans, H.S.; Wuest, M.; Soueidan, O.M.; Mercer, J.; McEwan, A.J.; West, F.G.; Cheeseman, C.I.; Wuest, F. Automated synthesis and dosimetry of 6-deoxy-6-[$^{18}$F]fluoro-D-fructose (6-[$^{18}$F]FDF): A radiotracer for imaging of GLUT5 in breast cancer. *Am. J. Nucl. Med. Mol. Imaging* **2014**, *4*, 248–259.
50. Lammertsma, A.A.; Hume, S.P. Simplified reference tissue model for PET receptor studies. *Neuroimage* **1996**, *4*, 153–158. [CrossRef]
51. Owen, D.R.; Yeo, A.J.; Gunn, R.N.; Song, K.; Wadsworth, G.; Lewis, A.; Rhodes, C.; Pulford, D.J.; Bennacef, I.; Parker, C.A.; et al. An 18-kDa translocator protein (TSPO) polymorphism explains differences in binding affinity of the PET radioligand PBR28. *J. Cereb. Blood Flow Metab.* **2012**, *32*, 1–5. [CrossRef] [PubMed]
52. Kitamura, O.; Takeichi, T.; Wang, E.L.; Tokunaga, I.; Ishigami, A.; Kubo, S. Microglial and astrocytic changes in the striatum of methamphetamine abusers. *Leg. Med. (Tokyo)* **2010**, *12*, 57–62. [CrossRef] [PubMed]
53. Choi, D.Y.; Liu, M.; Hunter, R.L.; Cass, W.A.; Pandya, J.D.; Sullivan, P.G.; Shin, E.J.; Kim, H.C.; Gash, D.M.; Bing, G. Striatal neuroinflammation promotes Parkinsonism in rats. *PLoS ONE* **2009**, *4*, e5482. [CrossRef]
54. Concannon, R.M.; Okine, B.N.; Finn, D.P.; Dowd, E. Differential upregulation of the cannabinoid $CB_2$ receptor in neurotoxic and inflammation-driven rat models of Parkinson's disease. *Exp. Neurol.* **2015**, *269*, 133–141. [CrossRef] [PubMed]
55. Herrera, A.J.; Castaño, A.; Venero, J.L.; Cano, J.; Machado, A. The single intranigral injection of LPS as a new model for studying the selective effects of inflammatory reactions on dopaminergic system. *Neurobiol. Dis.* **2000**, *7*, 429–447. [CrossRef] [PubMed]
56. Stern, E.L.; Quan, N.; Proescholdt, M.G.; Herkenham, M. Spatiotemporal induction patterns of cytokine and related immune signal molecule mRNAs in response to intrastriatal injection of lipopolysaccharide. *J. Neuroimmunol.* **2000**, *109*, 245–260. [CrossRef] [PubMed]

57. Liddelow, S.A.; Guttenplan, K.A.; Clarke, L.E.; Bennett, F.C.; Bohlen, C.J.; Schirmer, L.; Bennett, M.L.; Münch, A.E.; Chung, W.S.; Peterson, T.C.; et al. Neurotoxic reactive astrocytes are induced by activated microglia. *Nature* **2017**, *541*, 481–487. [CrossRef] [PubMed]
58. Han, J.; Fan, Y.; Zhou, K.; Blomgren, K.; Harris, R.A. Uncovering sex differences of rodent microglia. *J. Neuroinflammation* **2021**, *18*, 74. [CrossRef]
59. Lenz, K.M.; McCarthy, M.M. A starring role for microglia in brain sex differences. *Neuroscientist* **2015**, *21*, 306–321. [CrossRef]
60. Loram, L.C.; Sholar, P.W.; Taylor, F.R.; Wiesler, J.L.; Babb, J.A.; Strand, K.A.; Berkelhammer, D.; Day, H.E.W.; Maier, S.F.; Watkins, L.R. Sex and estradiol influence glial pro-inflammatory responses to lipopolysaccharide in rats. *Psychoneuroendocrinology* **2012**, *37*, 1688–1699. [CrossRef]
61. Schwarz, J.M.; Sholar, P.W.; Bilbo, S.D. Sex differences in microglial colonization of the developing rat brain. *J. Neurochem.* **2012**, *120*, 948–963. [CrossRef] [PubMed]
62. Biechele, G.; Franzmeier, N.; Blume, T.; Ewers, M.; Luque, J.M.; Eckenweber, F.; Sacher, C.; Beyer, L.; Ruch-Rubinstein, F.; Lindner, S.; et al. Glial activation is moderated by sex in response to amyloidosis but not to tau pathology in mouse models of neurodegenerative diseases. *J. Neuroinflammation* **2020**, *17*, 374. [CrossRef] [PubMed]
63. Tuisku, J.; Plavén-Sigray, P.; Gaiser, E.C.; Airas, L.; Al-Abdulrasul, H.; Brück, A.; Carson, R.E.; Chen, M.K.; Cosgrove, K.P.; Ekblad, L.; et al. Effects of age, BMI and sex on the glial cell marker TSPO—A multicentre [$^{11}$C]PBR28 HRRT PET study. *Eur. J. Nucl. Med. Mol. Imaging* **2019**, *46*, 2329–2338. [CrossRef] [PubMed]
64. Ayubcha, C.; Revheim, M.E.; Newberg, A.; Moghbel, M.; Rojulpote, C.; Werner, T.J.; Alavi, A. A critical review of radiotracers in the positron emission tomography imaging of traumatic brain injury: FDG, tau, and amyloid imaging in mild traumatic brain injury and chronic traumatic encephalopathy. *Eur. J. Nucl. Med. Mol. Imaging* **2021**, *48*, 623–641. [CrossRef] [PubMed]
65. Backes, H.; Walberer, M.; Ladwig, A.; Rueger, M.A.; Neumaier, B.; Endepols, H.; Hoehn, M.; Fink, G.R.; Schroeter, M.; Graf, R. Glucose consumption of inflammatory cells masks metabolic deficits in the brain. *Neuroimage* **2016**, *128*, 54–62. [CrossRef]
66. Sijbesma, J.W.A.; van Waarde, A.; Vallez Garcia, D.; Boersma, H.H.; Slart, R.; Dierckx, R.; Doorduin, J. Test-Retest Stability of Cerebral 2-Deoxy-2-[(18)F]Fluoro-D-Glucose ([(18)F]FDG) Positron Emission Tomography (PET) in Male and Female Rats. *Mol. Imaging Biol.* **2019**, *21*, 240–248. [CrossRef]
67. Orihuela, R.; McPherson, C.A.; Harry, G.J. Microglial M$_1$/M$_2$ polarization and metabolic states. *Br. J. Pharmacol.* **2016**, *173*, 649–665. [CrossRef]
68. Jassam, Y.N.; Izzy, S.; Whalen, M.; McGavern, D.B.; El Khoury, J. Neuroimmunology of Traumatic Brain Injury: Time for a Paradigm Shift. *Neuron* **2017**, *95*, 1246–1265. [CrossRef]
69. Ransohoff, R.M. A polarizing question: Do M1 and M2 microglia exist? *Nat. Neurosci.* **2016**, *19*, 987–991. [CrossRef]
70. Hernandez-Baltazar, D.; Nadella, R.; Barrientos Bonilla, A.; Flores Martinez, Y.; Olguin, A.; Heman Bozadas, P.; Rovirosa Hernandez, M.; Cibrian Llanderal, I. Does lipopolysaccharide-based neuroinflammation induce microglia polarization? *Folia Neuropathol.* **2020**, *58*, 113–122. [CrossRef]
71. BÜTtner-Ennever, J. The Rat Brain in Stereotaxic Coordinates, 3rd edn. *J. Anat.* **1997**, *191*, 315–317. [CrossRef]
72. Defrise, M.; Kinahan, P.E.; Townsend, D.W.; Michel, C.; Sibomana, M.; Newport, D.F. Exact and approximate rebinning algorithms for 3-D PET data. *IEEE Trans. Med. Imaging* **1997**, *16*, 145–158. [CrossRef] [PubMed]
73. Schwarz, A.J.; Danckaert, A.; Reese, T.; Gozzi, A.; Paxinos, G.; Watson, C.; Merlo-Pich, E.V.; Bifone, A. A stereotaxic MRI template set for the rat brain with tissue class distribution maps and co-registered anatomical atlas: Application to pharmacological MRI. *Neuroimage* **2006**, *32*, 538–550. [CrossRef] [PubMed]

 molecules

Article

# Recovery of Gallium-68 and Zinc from HNO$_3$-Based Solution by Liquid–Liquid Extraction with Arylamino Phosphonates

Fedor Zhuravlev [1,*], Arif Gulzar [1] and Lise Falborg [2]

[1] Department of Health Technology, Section for Biotherapeutic Engineering and Drug Targeting, Technical University of Denmark, Frederiksborgvej 399, 4000 Roskilde, Denmark
[2] Regionshospitalet Gødstrup, Hospitalsparken 15, 7400 Herning, Denmark
* Correspondence: fezh@dtu.dk; Tel.: +45-4677-5337

**Abstract:** The cyclotron production of gallium-68 via the $^{68}$Zn(p,n)$^{68}$Ga nuclear reaction in liquid targets is gaining significant traction in clinics. This work describes (1) the synthesis of new arylamino phosphonates via the Kabachnik–Fields reaction, (2) their use for liquid–liquid extraction of $^{68}$Ga from 1 M Zn(NO$_3$)$_2$/0.01 M HNO$_3$ in batch and continuous flow, and (3) the use of Raman spectroscopy as a process analytical technology (PAT) tool for in-line measurement of $^{68}$Zn. The highest extraction efficiencies were obtained with the extractants functionalized with trifluoromethyl substituents and ethylene glycol ponytails, which were able to extract up to 90% of gallium-68 in batch and 80% in flow. Only ppm amounts of zinc were co-extracted. The extraction efficiency was a function of pKa and the aqueous solubility of the extractant and showed marked concentration, solvent, and temperature dependence. Raman spectroscopy was found to be a promising PAT tool for the continuous production of gallium-68.

**Keywords:** gallium-68; liquid–liquid extraction; liquid target; arylamino phosphonate; cyclotron production of gallium-68

## 1. Introduction

Gallium-68 radiopharmaceuticals remain one of the cornerstones of positron emission tomography (PET). New $^{68}$Ga PET tracers significantly improve patients' clinical outcomes, and the number of clinical trials and publications involving gallium-68 continues to grow. $^{68}$Ga-PSMA and $^{68}$Ga-NETSPOT™ ($^{68}$Ga-SOMAkit-TOC) are becoming the gold standard for prostate cancer and neuroendocrine tumor diagnostics, [1] and Ga-FAPI is emerging as a new general PET tracer [2].

The clinical success of $^{68}$Ga radiotracers drives the soaring demand for $^{68}$Ga radionuclide. Most $^{68}$Ga is currently supplied by germanium-68 generators, which are convenient to use but expensive and in limited supply. The short life of germanium/gallium generators and their decreasing elution yield due to the decay of parent isotopes cause the total cost of generator ownership to be very high compared to the amount of $^{68}$Ga used for scans. An alternative technology that can potentially solve the $^{68}$Ga shortage has emerged. It is based on irradiation of the stable $^{68}$Zn isotope using medical cyclotrons. The $^{68}$Zn(p,n)$^{68}$Ga nuclear reaction has a large cross-section and can provide $^{68}$Ga with high radionuclidic purity [3]. Traditionally, proton bombardment of solid targets has been used as a primary method of producing PET radiometals. The production of multi-curie amounts of $^{68}$Ga has been reported by using electrodeposited [4], pressed [5], and fused [6,7] $^{68}$Zn solid targets. However, this requires specialized and costly equipment for target preparation, irradiation, cooling, transportation, and post-irradiation target dissolution [8]. In 2011, Jensen and Clark showed that irradiation of a $^{18}$F liquid target charged with a concentrated solution of $^{68}$ZnCl$_2$ can produce clinically-relevant yields of $^{68}$Ga [9]. It was subsequently recognized that liquid targets have a number of advantages over solid targets. Most importantly, liquid target-based production could be readily deployed within the existing clinical infrastructure designed and built for $^{18}$F radiochemistry. The production workflow benefited from easy target preparation, as well as from fast and automated delivery of the irradiated target solution to the hot cells for recovery and purification of $^{68}$Ga. All of this could be performed using the standard $^{18}$F targets and automated radiosynthesis modules. The initial challenges associated with significant outgassing during proton bombardment were overcome by the use of nitric acid solutions of metal nitrate salts [10]. Such solutions are commercially available from Fluidomica (www.fluidomica.pt; accessed on 29 November 2022). The Pandey [11–13] and Alves [14–16] research groups, as well as others [17,18], reported consistent and reliable production of $^{68}$Ga by irradiating liquid targets containing a zinc-68 nitrate solution made by dissolving $^{68}$Zn(NO$_3$)$_2$ (0.6–1.7 M) in nitric acid (0.01–0.3 M) with 30–45 µA proton beams for 30–60 min, yielding as much as 5 GBq at the end of the bombardment. Large regional clinics such as the Mayo Clinic (US) and ICNAS (Portugal) now routinely produce several batches of $^{68}$Ga per day using liquid targets. The European Pharmacopoeia monograph, which regulates the quality of cyclotron-produced $^{68}$Ga, has recently become available [19].

All current solid and liquid target $^{68}$Ga production methods use two-column solid phase extraction (SPE) as the means of purification. The first column, loaded with either hydroxamate [12,18,20] or a strong cation resin [4,5,15] (Dowex 50W-X8, AG 50W-X8), takes advantage of the stronger Lewis acidity of Ga$^{3+}$ retained on the column, while Zn$^{2+}$ is eluted with a HCl, HNO$_3$, or acetone/HBr mixture. The second column, which may contain a phosphine oxide (TK200) [18,20], a cation exchange (AG1) [13,15], or a phosphonate (UTEVA) [4,5] resin, serves mainly to concentrate activity for elution in a formulation-friendly media, such as water or 0.1 M HCl. Since the cost of $^{68}$Zn has a relatively high impact on the overall price of cyclotron-produced $^{68}$Ga, reuse of $^{68}$Zn is desirable. The main limitation of the SPE methodology, at least at the current level of development, is that $^{68}$Zn cannot be immediately reused after SPE purification. The sorption on the first column and the subsequent elution changes both the zinc-68 concentration and the chemical composition of the solution, making it unsuitable for direct use in the cyclotron liquid target.

We have recently reported a proof-of-principle study describing efficient liquid–liquid extraction of radiogallium ($^{66,67,68}$Ga) from ZnCl$_2$/HCl solutions in batch and in flow using a membrane-based separator [21]. We argued that, compared to SPE, liquid–liquid extraction in flow (LLEF) has the advantages of scalability, speed, low cost, and easy recovery of the material through solvent evaporation or back-extraction. Importantly, LLEF does not change the chemical composition of the cyclotron liquid target, which can potentially be reused in-line. The ability to reuse the cyclotron target solution of $^{68}$Zn, coupled with the fluidics-compatible design of solution targets, makes the continuous flow approach an attractive alternative to conventional batch processing. The continuous process can be envisioned as follows: the separation module S is placed next to the liquid target T in the cyclotron vault, and the target is irradiated by the cyclotron beam producing $^{68}$Ga via the $^{68}$Zn(p,n)$^{68}$Ga nuclear reaction (Figure 1). The irradiated solution is then transferred into S, where $^{68}$Ga is separated from $^{68}$Zn. $^{68}$Ga is sent into the hot cell for further downstream processing and radiolabeling. The cyclotron target solution containing $^{68}$Zn is returned to T for new irradiation. The process can be performed in a semi-batch mode, where the target is closed during the bombardment and then opened and processed. Alternatively, the cyclotron target solution can be recirculated through S.

There are several advantages to the continuous approach: (1) The process is scalable and can be run on-demand. This flexibility translates into the maximization of PET scanner occupancy at the hospital. (2) $^{68}$Zn is recycled in-line, saving money and securing the hospital's $^{68}$Ga and $^{68}$Zn supply. (3) Radioactive waste remains contained until the completion of the continuous production campaign. (4) Compatibility with in-line process analytical technology tools (PAT) is an opportunity to implement the FDA's Quality by Design approach to radiopharmaceutical production. In terms of process integration and automatization, LLEF is fully compatible with downstream SPE-based processing, if additional steps are required. Setting the stage for future research, the new liquid extractants can also be used for transferring the solution chemistry to an SPE platform by grafting the extractant onto a solid support.

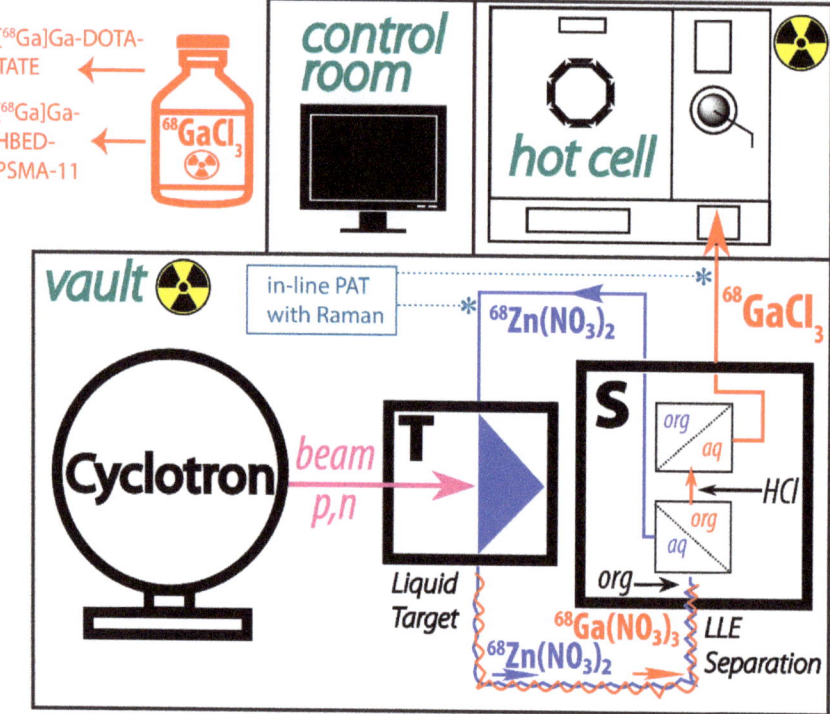

**Figure 1.** Conceptual schematic of the continuous production of $^{68}$Ga using a cyclotron solution target (**T**). Phase separation is performed in module **S** using a membrane separator; the $^{68}$Ga is back-extracted from the organic phase with 0.1 M HCl and is sent directly into the hot cell. The organic phase is discarded.

Central to this conceptual design is a liquid–liquid extraction (LLE)-based separation module. We previously reported near-quantitative LLE of titanium-45 [22,23] and radiogallium [21] from concentrated HCl using a membrane separator with integrated pressure control. The same system could be implemented here with one important caveat: no compound capable of efficient and selective extraction of gallium from nitric acid in the presence of zinc into an organic phase has been reported. Phosphine oxides (Cyanex 923 [24], Cyanex 925 [25]) and phosphoric, phosphonic [26], and phosphinic acids (Cyanex 301 [27]) have been previously tested but extraction was poor.

This study had three objectives. First, we set out to design, synthesize, and test a new family of extractants able to efficiently perform LLE and separation of $^{68}$Ga from zinc nitrate/nitric acid solutions in batch and in flow. Second, we explored the possibility of implementing Raman spectroscopy as a PAT tool and used the Design of Experiment (DoE) technique for LLE optimization. Lastly, we sought to rationalize the results in the context of the extractant's structure and its physical properties.

## 2. Results

*2.1. System Design*

The cyclotron target solutions used in clinics can be prepared in a variety of nitric acid concentrations: from 0.01 M to 1.5 M [15,18,28,29]. From an extraction standpoint, the lowest acidity is preferred. Since 1 M solution of $^{68}$Zn(NO$_3$)$_2$ in 0.01 M HNO$_3$ is also commercially available from Fluidomica for clinical production of $^{68}$Ga, it became our choice of aqueous phase for LLE. In the 1960s, Jagodić reported that arylamino phosphonic acid **1** was able to extract a broad range of metals from various acids [30]. Therefore, we have chosen the arylamino phosphonic acid scaffold for further development, recognizing that it can function as a chelator due to the presence of arylamino moieties (Figure 2). Importantly, the acidity and basicity of NH, and hence the chelation capacity of the arylamino phosphonic acid extractant, can be controlled by the judicious choice of substituents on the aryl rings Ar$_1$ and Ar$_2$. Organic and aqueous solubility can be further tuned by controlling the hydrophilicity of the phosphonic acid monoester moiety.

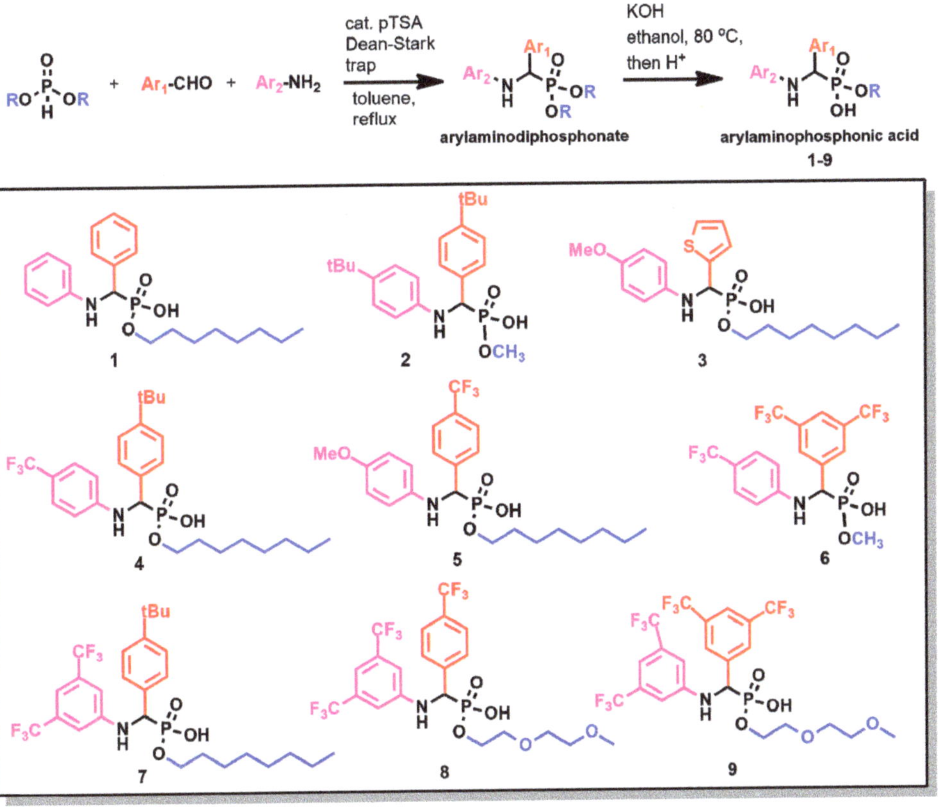

Figure 2. The synthesis of arylamino phosphonic acids **1–9**.

## 2.2. Chemistry

Eight new arylamino phosponic acids **2–9**, together with the previously reported acid **1** (Figure 2), were synthesized. Under the conditions of the Kabachnik–Fields reaction [31], a dialkylphosphite, an aniline, and an aldehyde were refluxed in dry toluene in the presence of the catalytic amount of para-toluene sulfonic acid (pTSA) providing the corresponding aminophosphonate as a single product. The subsequent hydrolysis yielded the requested aminophosphonic monoesters (**1–9**). The synthesis could be conveniently performed as a one-pot, two-step reaction with overall yields in the 40–50% range.

## 2.3. LLE of $^{68}$Ga Using Extractants 1–9 in Batch

The results of LLE of $^{68}$Ga in batch and the calculated pKa and aqueous solubility of extractants 1–9 in different solvents at either room temperature or 50 °C are presented in Table 1.

**Table 1.** Batch LLE of $^{68}$Ga from 1 M zinc nitrate in 0.01 M nitric acid with 10 mM of extractant dissolved in various solvents and performed at RT or 50 °C. The extraction efficiencies (EE), the pKa of the extractants calculated using Cosmotherm, and the experimentally determined aqueous solubilities are listed. Experiments were performed in triplicate; results are presented as means ± standard deviation.

| Entry | Extractant | Solvent | T, °C | $^{68}$Ga EE (%) | pKa$^{Calc}$ | Aqueous Solubility, mM |
|---|---|---|---|---|---|---|
| 1 | | Heptane | RT | 30.6 ± 2.0 | 1.5 | 0.22 |
| 2 | | Anisole | RT | 8.3 ± 3.1 | 1.4 | 0.10 |
|   | |         | 50 °C | 32.9 ± 2.3 | | |
| 3 | | CHCl$_3$/Heptane 3/1 (v/v) | RT | 20.8 ± 2.0 | 1.4 | 0.25 |
| 4 | | Bu$_2$O | RT | 8.1 ± 2.4 | 1.2 | 0.08 |
| 5 | | Bu$_2$O | RT | 34.1 ± 2.0 | 1.2 | 0.04 |
| 6 | | Heptane/TFT 1:1 (v/v) | RT | 49.7 ± 2.6 | 0.7 | 2.60 |
|   | | Bu$_2$O | RT | 24.8 ± 1.9 | | |
| 7 | | Heptane | RT | 16.4 ± 2.0 | 0.9 | 0.08 |
|   | |         | 50 °C | 37.7 ± 2.3 | | |
| 8 | | Bu$_2$O | RT | 41.3 ± 2.5 | 0.7 | 1.60 |
|   | |         | 50 °C | 80.1 ± 3.0 | | |
|   | | TFT | RT | 60.5 ± 1.9 | | |
|   | |     | 50 °C | 88.2 ± 2.3 | | |
| 9 | | TFT | RT | 69.1 ± 3.0 | 0.4 | 1.73 |
|   | |     | 50 °C | 89.6 ± 3.5 | | |

## 2.4. Estimation of pKa Using COSMO-RS

The pKa for compounds 1–9 was computationally estimated using the conductor-like screening model for real solvents (COSMO-RS). This computational technique uses density functional theory to calculate molecular screening charge densities and then applies statistical thermodynamics to yield chemical potentials [32]. pKa can be calculated from the Gibbs free energies of the neutral and ionic compounds [33]. Table 1 shows that the calculated pKa value varies from 1.5 (1) to 0.4 (9). A significant increase in acidity was noted for all extractants functionalized with the CF$_3$ groups (Table 1, entries 6–9).

## 2.5. Batch LLE Optimization Studies

Compounds 1–9 demonstrated significant variability in aqueous solubility, pKa calculations, and extraction efficiency (EE) (Table 1). The compounds where both Ar$_1$ and Ar$_2$ were functionalized with trifluoromethyl groups showed the highest EE. A 7–17% increase in extraction was observed when the concentration of the extractants increased from 10 mM to 30 mM (Figure S1). A marked solvent dependence was also noted. For most extractants, the best results were obtained in TFT, heptane, and Bu$_2$O. Across all tested compounds, an increase in temperature from RT to 50 °C led to a significant increase in EE (Figure 3). In all cases, clear phase separation between the aqueous and organic phases was observed with no detectable extraction of zinc into the organic phase as evidenced by $^{65}$Zn activity measurements. Preliminary experiments indicated that, for an extractant in a given solvent, the concentration of the extractant and the extraction temperature were the main factors affecting extraction efficiency. A traditional approach to optimization is to change one factor (concentration, temperature) at a time. This approach, however, is inefficient as it does not necessarily lead to the optimal experimental conditions, especially when there is interaction between the factors. A better approach is to use a statistically-driven experimental design, called the Design of Experiment (DoE) approach [34]. Guided by a software algorithm, DoE allows one to define a design space and systematically optimize the response (EE) while taking into account the interaction between the factors. To optimize the yield of LLE, we used a central composite face-centered design algorithm implemented in the software package MODDE 9.1.1 (Table 2). Concentration and temperature were varied between 10–30 mM and 25–50 °C, correspondingly. These limits defined the center point (20 mM, 37.5 °C) around which four corner and four median experiments were constructed (see Figure S2 for a graphical representation of the design). The reproducibility of the design was estimated by running the center points in triplicate.

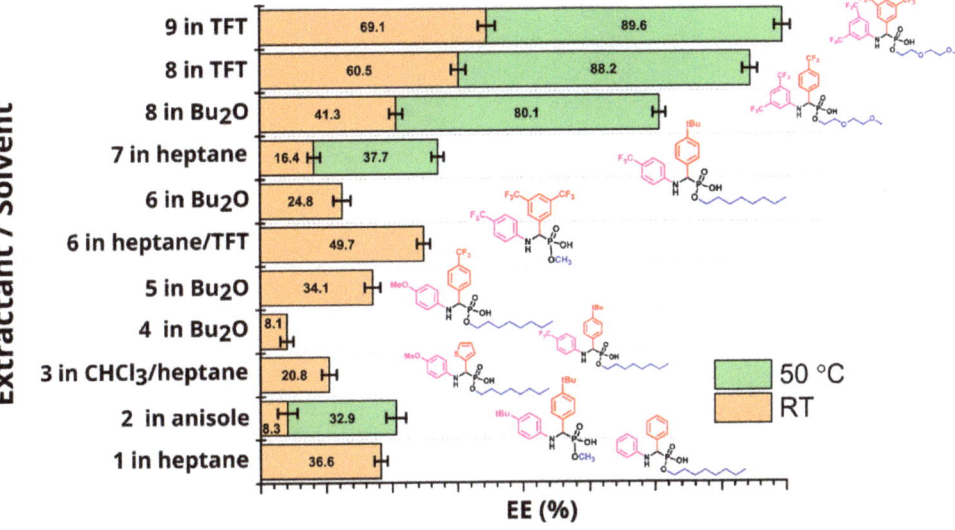

Figure 3. LLE of $^{68}$Ga using extractants 1–9 in batch. In each case, the solvent system was chosen in such a way as to provide the best solubility and phase separation for a given extractant.

The results of 11 runs were fitted with the partial least squares (PLS) algorithm, producing a quadratic model of excellent statistical quality (Figure 4). Both concentration and temperature were positively correlated with LLE efficiency, but the square of concentration was negatively correlated.

Table 2. Batch LLE optimization studies using the central composite face-centered design algorithm implemented in MODDE 9.1.1. Concentration and the temperature were simultaneously varied.

| Exp. No. | Concentration, mM | T, °C | EE (%) |
| --- | --- | --- | --- |
| 1 | 10 | 25 | 69 |
| 2 | 30 | 25 | 71 |
| 3 | 10 | 50 | 83 |
| 4 | 30 | 50 | 91 |
| 5 | 10 | 37.5 | 75 |
| 6 | 30 | 37.5 | 78 |
| 7 | 20 | 25 | 71 |
| 8 | 20 | 50 | 90 |

Table 2. *Cont.*

| Exp. No. | Concentration, mM | T, °C | EE (%) |
|---|---|---|---|
| 9 | 20 | 37.5 | 83 |
| 10 | 20 | 37.5 | 88 |
| 11 | 20 | 37.5 | 91 |

The optimization studies indicated that the extractions performed at and above 20 mM and 37.5 °C consistently produced EE > 83% (Table 2).

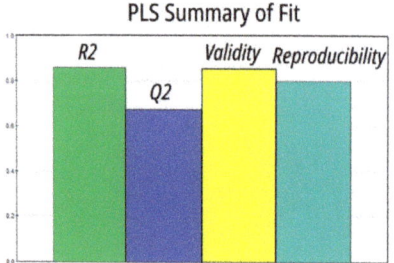
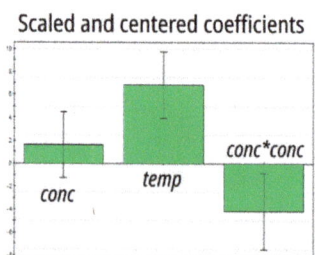

**Figure 4.** Optimization of LLE in batch using extractant **8**: a PLS model based on a central composite face-centered design.

### 2.6. LLE of $^{68}$Ga using Extractants **8** and **9** in Continuous Flow

Having established extractants **8** and **9** as the best performers in batch, we translated batch into flow (Figure 5). The Zaiput membrane separator provided a clean phase separation with no phase breakthrough at both the extraction and stripping stages. At the extraction stage, LLE was on average 10% less efficient in flow. As with the batch experiments, raising the temperature to 50 °C significantly improved EE. Stripping in 2 M HCl was quantitative. ICP analysis of the stripped solution indicated the presence of 11 ppm of zinc.

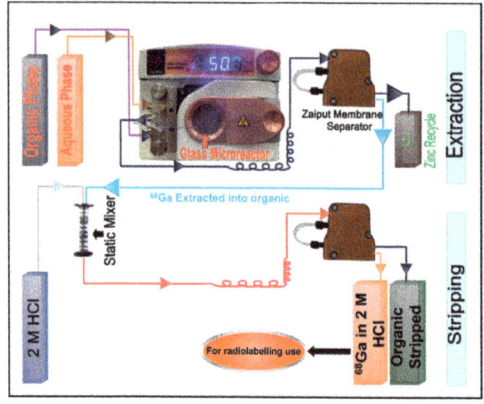
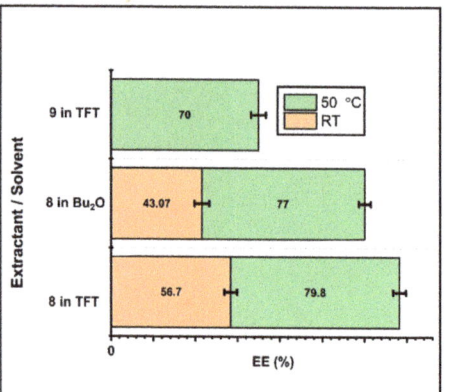

**Figure 5.** (**Left**) A schematic depicting two-stage liquid–liquid extraction in flow. (**Right**) $^{68}$Ga EE (%) obtained from a 1 M Zn(NO$_3$)$_2$ solution in 0.01 M HNO$_3$ using extractants **8** and **9** in TFT and Bu$_2$O at RT and 50 °C.

### 2.7. Zinc Nitrate Quantification Using Raman Spectra

Twelve solutions of zinc nitrate in 0.01 M HNO$_3$ were prepared (with the concentration of zinc varying in the range of 0.3–1.2 M) and the Raman spectra were acquired. A multivariate analysis of the spectra using SIMCA yielded a four-principal component PLS model with $R^2 = 0.99$ and $Q^2 = 0.94$ (Figure 6). The model showed excellent observed vs. predicted linearity ($R^2 = 0.99$) in the whole concentration range and was validated using independently prepared samples. This multivariate calibration was subsequently used to quantify the concentration of zinc before and after LLE.

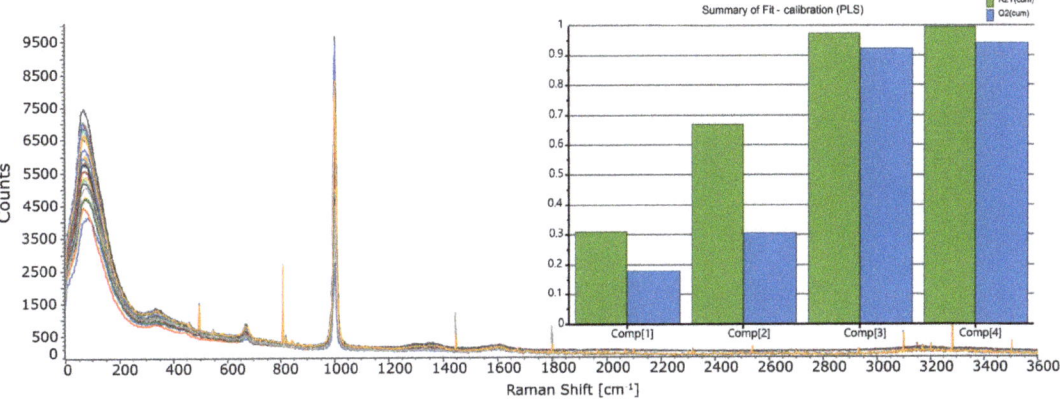

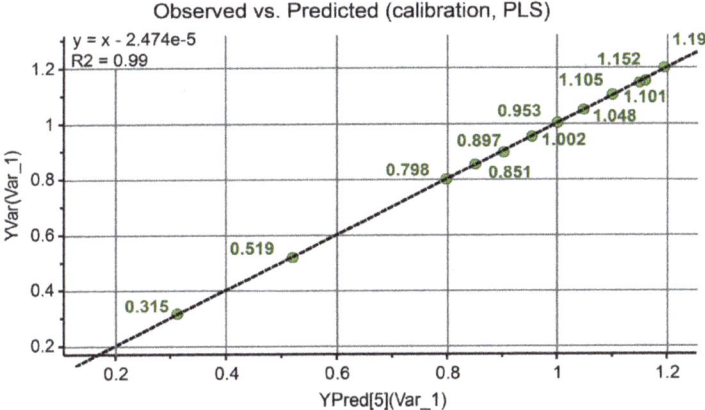

**Figure 6.** (**Top**) The Raman spectra of $Zn(NO_3)_2$ (0.3–1.2 M) prepared in 0.01 M $HNO_3$ solutions. (**Inset**) A four-component PLS calibration model for determination of zinc concentration. (**Bottom**) The performance of the calibration model: observed vs. predicted.

The analysis showed that the overall depletion of zinc in the aqueous phase after LLE was less than 10%. This was independently confirmed by the mass balance measurement and $^{65}Zn$ radiotracing.

## 3. Discussion

The substituents on $Ar_1$, $Ar_2$, and the phosphonic acid affect the properties of compounds **1–9** in two major ways. The electron-withdrawing $CF_3$ groups increase the acidity of the phosphonic acid moiety, whereas electron-releasing tBu and OMe retard dissociation. This is reflected in the calculated pKa value, which varies by more than 1 pKa unit across the series. Experimentally, we found that the increase in the strengths of the phosphonic acid moiety due to $CF_3$ substitution resulted in compounds **8** and **9** being stronger extractants. This is in line with the general observation that the potassium salts of compounds **1–9** proved to be better extractants than the corresponding acids. The effect was particularly pronounced in **4**: EE increased from 8% for the phosphonic acid to 75% for its potassium salt. A similar effect was previously observed by Jagodić and attributed to an increase in the acidity of the solution due to the release of H$^+$ during complexation [30]. Although this explanation is not applicable to the present case due to the picomolar amount of $^{68}$Ga, the importance of phosphonic acid dissociation is further highlighted by the critical role aqueous phase acidity played in extraction. When **1** was used as an extractant, EE dropped from 30% in 0.01 M HNO$_3$ to 4.2% in 0.2 M HNO$_3$, and then to 1.2% in 1 M HNO$_3$. Despite a qualitative correlation between pKa and EE, no statistically significant model emerged from the data. The effect of substituents in Ar$_1$ and Ar$_2$ on the solubility of compounds **1–9** is even more subtle, as both electronic and steric effects play a role. The substitution of an octyl group for a hydrophilic diethylene glycol significantly increased the aqueous solubility of **8** and **9**. Although no model was able to correlate solubility and EE, the combination of COSMO-RS-calculated acidity and solubility yielded a statistically significant PLS model (Figure S3). The combination of higher acidity and higher aqueous solubility favored extraction. A marked solvent, temperature, and concentration dependency can be further rationalized in terms of extractant dimerization, which in the case of alkyl and aryl phosphonic acids has been observed in solvents of low polarity [35]. The negative correlation between EE and the square of the concentration of the extractant we found during batch optimization (Figure 4) suggests that dimerization of extractant **8** competes with extraction. Under this scenario, the rate of dimerization would be proportional to the square of the concentration and would lead to a decrease in EE due to the deactivation of the extractant. Increased temperature is expected to favor the dissociation of the dimer, leading to higher EE. The higher EE of the potassium salts of **1–9** we observed in the preliminary experiments also supports the dimerization hypothesis because the deprotonated species are unable to form aggregates.

Aqueous solutions are essentially transparent to Raman scattering. In our continuous flow design, the laser light was delivered to the flow cell via fiber optics, making Raman spectroscopy an ideal tool for remotely controlled in-line analysis of radioactive mixtures under continuous flow conditions. The Raman spectra of 0.3–1.2 M Zn(NO$_3$)$_2$ prepared in 0.01 M HNO$_3$ solutions were dominated by the symmetric stretching band of nitrate anion centered at 1000 cm$^{-1}$ (Figure 6). The broad peak spanning 305–399 cm$^{-1}$ was assigned to the hexaaquazinc(II) ion [Zn(H$_2$O)$_6$]$^{2+}$ (390 cm$^{-1}$) [36] and the lower-frequency mode corresponding to the nitrate-associated [Zn(H$_2$O)$_x$NO$_3$]$^+$, x < 6 [37]. We found that inclusion of the entire spectral region (3–3600 cm$^{-1}$) and autoscaling the variables to unit variance were essential for obtaining calibrations with the best possible statistical quality.

## 4. Materials and Methods

### 4.1. Materials

All chemicals were reagent grade, purchased from Sigma Aldrich (Merck KGaA, Darmstadt, Germany), and used without additional purification. Dioctyl and diethyleneglycol phosphites were prepared as described previously [38]. The batch and continuous flow extractions were performed using zinc nitrate at natural abundance, spiked with a small amount of zinc-65 for radiotracing purposes. The radionuclide zinc-65 ($^{65}$Zn, t$^1$/$_2$: 244 days) was produced as described previously [21]. The radionuclide gallium-68 ($^{68}$Ga, t$^1$/$_2$: 68 min) was obtained from a $^{68}$Ge/$^{68}$Ga generator produced by Eckert & Ziegler (Berlin, Germany). SEP-10 membrane separators were purchased from Zaiput Flow Technologies (Waltham, MA, USA). Pall PTFE membranes were used for all experiments (47 mm diameter, 0.2 μm pore size, polypropylene support). Perfluoroalkoxy alkane (PFA) diaphragms (0.002″ (00.0508 mm)) were purchased from McMaster Carr (Princeton, NJ, USA). All PFA tubing (1/16″ (1.5875 mm) OD, 0.03″ (0.762 mm) ID) was purchased from Idex Health and Science (West Henrietta, NY, USA). Polytetrafluoroethylene (PTFE) static mixers were purchased from Stamixco (Dinhard, Switzerland). The 15 mL plastic centrifuge tubes with screw caps (SuperClear) were purchased from VWR (Søborg, Denmark).

*4.2. Instrumentation and Methods*

The NMR spectra were recorded on an Agilent 400 MR spectrometer (Agilent, Santa Clara, CA, USA) operating at 400.445 MHz (1H). The Raman spectra were obtained using an Avantes AvaSpec-ULS-RS-TEC spectrometer with 788 nm laser excitation. Zinc-65 and gallium-68 were quantified by gamma spectroscopy using a Princeton Gammatech LGC 5 or Ortec GMX 35195-P germanium detector, calibrated using certified barium-133 and europium-152 sources. Zinc at natural abundance was quantified using a Thermo Scientific iCAP 6000 Series ICP Optical Emission Spectrometer. An Eppendorf 5702 centrifuge was used to assist in phase separation. All experiments used 0.2 μm membrane pore size, a 0.002" (0.051 mm) diaphragm, two 10-element static mixers, and a 108 cm mixing tube. The solutions for the continuous membrane-based separation were pumped using KDS 100 Legacy Syringe pumps. For batch experiments, phase mixing was performed using an IKA ROCKER 3D digital shaker.

*4.3. Batch LLE Extractions*

Initially, 1 M zinc nitrate prepared in 0.01 M $HNO_3$ was used as the aqueous phase for the batch LLE extraction. The organic phase was prepared by dissolving extractants **1–9** in different organic solvents or a mixture of solvents to achieve a final concentration of 10 mM, 20 mM, and 30 mM. A 1 mL aliquot of the aqueous phase was then transferred into a 15 mL centrifuge tube and mixed with 3 mL of the organic phase. The tube was shaken at 80 rpm for 15 min at either 25 °C, 37.5 °C, or 50 °C. After centrifugation for 15 min, the phases where separated and the activity of $^{65}Zn$ and $^{68}Ga$ in the aqueous and organic phases were quantified using gamma spectroscopy. $^{68}Ga$ extraction efficiency was calculated according to the following equation:

$$EE(\%) = (^{68}Ga_{org} \times 100\%)/(^{68}Ga_{org} + {^{68}Ga_{aq}})$$

*4.4. Continuous Membrane-Based LLE*

LLE in flow was performed in two stages. At first, the extraction stage used the same aqueous phase as was used in batch LLE. Extractants **8** and **9**, which showed the best performance in batch extraction, were used as 10 mM TFT solutions for the organic phase. The extraction was performed at 50 °C. The fluidics were driven by two separate syringe pumps equipped with Hamilton glass syringes, which were filled with the organic (9 mL) and the aqueous (3 mL) phases. The aqueous flow rate was 15 mL/h and the organic flow rate was 45 mL/h for all experiments. The aqueous to organic ratio was maintained at 1:3 ($v/v$) at all times. For the extraction process, two phases passed through PFA tubing (1/16" OD, 0.03" ID) and entered the Syriss Chip Climate controller mixer set at 50 °C. The organic and aqueous phases were mixed inside the chip. The two phases were further mixed in a 100 cm long PFA tubing (1/16" OD, 0.03" ID) mixing loop by steady slug flow and then passed into the membrane separator. In the membrane separator, the organic phase permeated the hydrophobic membrane (PTFE/PP, 0.2 μm pore size and 139 μm thickness) and passed through the permeate outlet, while the aqueous phase was retained and passed through the retentate outlet. The 0.002" PFA diaphragm in the membrane separator worked as a form of integrated pressure control, and complete phase separation between the aqueous and the organic phase was obtained with the chosen membrane and diaphragm. In the second stage (stripping), $^{68}Ga$ was back-extracted from the organic phase into the aqueous phase following the experimental protocol described above. This time, however, the organic and aqueous phases were mixed by two PTFE static mixers at room temperature. Again, the aqueous to organic ratio was maintained at 1:3 ($v/v$), where 9 mL of organic phase containing $^{68}Ga$ was stripped with 3 mL of 2 M HCl.

## 4.5. NMR Studies

### 4.5.1. General Comments

NMR spectra were recorded at ambient probe temperatures and referenced as follows (δ, ppm): $^1$H, residual internal DMSO-d5 (2.50); $^{13}$C{$^1$H} internal DMSO-d6 (39.52). The partial structural assignment was performed on a basis of one-dimensional nuclear Overhauser effect spectroscopy (1D NOESY), Heteronuclear Multiple Bond Correlation with adiabatic pulses (HSQCAD), and Gradient-Selected Correlation Spectroscopy (gCOSY) experiments. qNMR studies were performed at ambient temperature using the following acquisition parameters: $^1$H NMR—wet1D pulse sequence with 90° pulse, acquisition time 4 s, relaxation delay 60 s, zero-filling to 256 k, exponential multiplication with 0.3 Hz line broadening, manual phasing, and 5th degree polynomial baseline correction. $^{31}$P NMR—inverse gated decoupling pulse sequence with 90° pulse, acquisition time 1 s, relaxation delay 30 s, zero-filling to 128 k, exponential multiplication with 3 Hz line broadening, manual phasing, and 5th degree polynomial baseline correction.

### 4.5.2. Determination of the Aqueous Solubility of Extractants 1–9

A 1 mL volumetric flask was charged with 10–15 mg of the extractant and 1 mL of $D_2O$. The resulting suspension was sonicated for 30 min and the content was centrifuged, filtered through a 0.45 µm syringe filter, and transferred into an NMR tube. A sealed capillary containing an external calibrant (acetanilide for $^1$H qNMR and triphenylphosphine for $^{31}$P qNMR) was inserted into the NMR tube and the solubility of extractants 1–9 was determined from their $^1$H or $^{31}$P NMR spectra using the formula: $C_x = (I_x/I_{cal}) \times (N_{cal}/N_x) \times C_{cal}$, where I, N, and C are the integral area, number of nuclei, and the concentration of the extractant (x) and the calibrant (cal), respectively.

## 4.6. Multivariate Analysis

The Design of Experiment (DoE) studies were performed using MODDE 9.1.1 (Umetrics AB), using the central composite face-centered design. Multivariate calibration of the zinc nitrate concentrations was performed by acquiring the corresponding Raman spectra using 788 nm laser excitation with 5 s acquisition (10 average) and exporting the spectra as GRAMS SPC files into SIMCA 17.0.0 (Sartorius Stedim Data Analytics AB). The dataset (3–3600 cm$^{-1}$) was scaled to unit variance and processed using PLS. Five independently prepared samples were used for validation of the calibration model.

## 4.7. Computational Methods

All gas-phase and COSMO calculations were performed using the TURBOMOLE 7.5.1 suite of programs using resolution of identity approximation (RI) [39]. The gas-phase structures were optimized at the RI BP/def2-TZVPD level and convergence to the ground state was verified by running analytical frequency calculations. The single-point gas-phase energies were then re-evaluated at the RI MP2/def2-TZVPP level. The COSMO files were obtained at the same theory level in the COSMO phase with a smooth radii-based isosurface cavity, and convergence to the ground state was verified by running numerical frequency calculations. The resulting COSMO files were used to perform solution thermodynamics calculations using COSMOtherm, Version 20.0.0 (Dassault Systèmes), yielding the free energies of solvation, sigma surfaces, and sigma profiles [40].

## 4.8. General Procedure for the Synthesis of Arylaminophosphonic Acids 1–9

A 100 mL round-bottom flask equipped with the Dean–Stark trap was charged with an equimolar (5 mmol) amount of a dialkylphosphite, an aniline, an aldehyde, and a catalytic amount of pTSA dissolved in 40 mL of toluene. The reaction mixture was refluxed overnight. Toluene was removed under reduced pressure, the reaction mixture was redissolved in 50 mL of ethanol, and a solution of 10 mmol of KOH in 4 mL of water was added. The reaction mixture was refluxed overnight, cooled to room temperature, acidified with 3 eq

(15 mmol) of 6 M HCl, and purified on silica (gradient elution: $CHCl_3$ then $CHCl_3/CH_3OH$ 9/1 $v/v$). Analytical purity was confirmed by qNMR as described above.

## 5. Conclusions

A new family of arylaminophosphonic acids (**1–9**) was synthesized. All compounds selectively extracted gallium-68 in the presence of zinc from a solution containing 1 M $Zn(NO_3)_2$ in 0.01 M $HNO_3$. Extractants **8** and **9** were able to extract up to 90% of gallium in batch and up to 80% in flow. Only ppm amounts of zinc were co-extracted. Extraction efficiency correlated with pKa and the aqueous solubility of the extractant. Raman spectroscopy was found to be well suited as a PAT tool for continuous flow production of gallium-68. We are currently evaluating the suitability of cyclotron-produced and LLEF-purified gallium-68 for DOTATATE radiolabeling.

**Supplementary Materials:** The following supporting information can be downloaded at: https://www.mdpi.com/article/10.3390/molecules27238377/s1. Figure S1: The extraction efficiencies for compound **8**, performed at 10 mM and 30 mM in various solvents at RT and 50 °C. The relative increase in EE upon the increase in concentration of compound **8** from 10 to 30 mM is shown on the X axis. Table S1: The distribution ratio Ds = [$^{68}Ga_{tot}$]org/[$^{68}Ga_{tot}$]aq for batch LLE of $^{68}Ga$ from 1 M zinc nitrate in 0.01 M nitric acid with 10 mM of extractant dissolved in various solvents and performed at RT or 50 °C. Figure S2: The design region for the central composite face-centered design shown in Table S1. The numbers in the circles correspond to the experiment numbers in Table S1. The center point was run in triplicate (experiments 9, 10, 11). Figure S3: A PLS model correlating $^{68}Ga$ extraction efficiency with pKa and the aqueous solubility of compounds **1–9**.

**Author Contributions:** Conceptualization, F.Z.; methodology, F.Z. and A.G.; software, F.Z. and A.G.; validation, F.Z. and A.G.; formal analysis, F.Z. and A.G.; investigation, F.Z. and A.G.; resources, F.Z., A.G. and L.F.; data curation, F.Z. and A.G.; writing—original draft preparation, F.Z. and A.G.; writing—review and editing, F.Z., A.G. and L.F.; visualization, F.Z. and A.G.; supervision, F.Z.; project administration, F.Z.; funding acquisition, F.Z. All authors have read and agreed to the published version of the manuscript.

**Funding:** This research was funded by the Novo Nordisk Foundation, grant number NNF21OC0068886.

**Institutional Review Board Statement:** Not applicable.

**Informed Consent Statement:** Not applicable.

**Data Availability Statement:** Not applicable.

**Acknowledgments:** We thank Natan J. W. Straathof for help with HPLC.

**Conflicts of Interest:** The authors declare no conflict of interest.

**Sample Availability:** Samples of the compounds are not available from the authors.

## References

1. Lepareur, N. Cold Kit Labeling: The Future of 68Ga Radiopharmaceuticals? *Front. Med.* **2022**, *9*, 812050. [CrossRef] [PubMed]
2. Kratochwil, C.; Flechsig, P.; Lindner, T.; Abderrahim, L.; Altmann, A.; Mier, W.; Adeberg, S.; Rathke, H.; Röhrich, M.; Winter, H.; et al. 68Ga-FAPI PET/CT: Tracer Uptake in 28 Different Kinds of Cancer. *J. Nucl. Med.* **2019**, *60*, 801–805. [CrossRef] [PubMed]
3. Naik, H.; Suryanarayana, S.V.; Murali, M.S.; Kapote Noy, R. Excitation Function of 68Zn(p,n)68Ga Reaction for the Production of 68Ga. *J. Radioanal. Nucl. Chem.* **2020**, *324*, 285–289. [CrossRef]
4. Lin, M.; Waligorski, G.J.; Lepera, C.G. Production of Curie Quantities of 68Ga with a Medical Cyclotron via the 68Zn(p,n)68Ga Reaction. *Appl. Radiat. Isot.* **2018**, *133*, 1–3. [CrossRef] [PubMed]
5. Nelson, B.J.B.; Wilson, J.; Richter, S.; Duke, M.J.M.; Wuest, M.; Wuest, F. Taking Cyclotron 68Ga Production to the next Level: Expeditious Solid Target Production of 68Ga for Preparation of Radiotracers. *Nucl. Med. Biol.* **2020**, *80–81*, 24–31. [CrossRef]
6. Zeisler, S.; Limoges, A.; Kumlin, J.; Siikanen, J.; Hoehr, C. Fused Zinc Target for the Production of Gallium Radioisotopes. *Instruments* **2019**, *3*, 10. [CrossRef]
7. Thisgaard, H.; Kumlin, J.; Langkjær, N.; Chua, J.; Hook, B.; Jensen, M.; Kassaian, A.; Zeisler, S.; Borjian, S.; Cross, M.; et al. Multi-Curie Production of Gallium-68 on a Biomedical Cyclotron and Automated Radiolabelling of PSMA-11 and DOTATATE. *EJNMMI Radiopharm. Chem.* **2021**, *6*, 1. [CrossRef]

8. Sciacca, G.; Martini, P.; Cisternino, S.; Mou, L.; Amico, J.; Esposito, J.; Gorgoni, G.; Cazzola, E. A Universal Cassette-Based System for the Dissolution of Solid Targets. *Molecules* **2021**, *26*, 6255. [CrossRef]
9. Jensen, M.; Clark, J. Direct Production of Ga-68 from Proton Bombardment of Concentrated Aqueous Solutions of [Zn-68] Zinc Chloride. In Proceedings of the 13th International Workshop on Targetry and Target Chemistry, Roskilde, Denmark, 26–28 July 2010; Danmarks Tekniske Universitet, Risø Nationallaboratoriet for Bæredygtig Energi: Copenhagen, Denmark, 2011; pp. 288–292.
10. Pandey, M.K.; Engelbrecht, H.P.; Byrne, J.F.; Packard, A.B.; DeGrado, T.R. Production of 89Zr via the 89Y(p,n)89Zr Reaction in Aqueous Solution: Effect of Solution Composition on in-Target Chemistry. *Nucl. Med. Biol.* **2014**, *41*, 309–316. [CrossRef]
11. Pandey, M.K.; DeGrado, T.R. Cyclotron Production of PET Radiometals in Liquid Targets: Aspects and Prospects. *Curr. Radiopharm.* **2021**, *14*, 325–339. [CrossRef]
12. Pandey, M.K.; Byrne, J.F.; Schlasner, K.N.; Schmit, N.R.; DeGrado, T.R. Cyclotron Production of 68Ga in a Liquid Target: Effects of Solution Composition and Irradiation Parameters. *Nucl. Med. Biol.* **2019**, *74–75*, 49–55. [CrossRef] [PubMed]
13. Pandey, M.K.; Byrne, J.F.; Jiang, H.; Packard, A.B.; DeGrado, T.R. Cyclotron Production of 68Ga via the 68Zn(p,n)68Ga Reaction in Aqueous Solution. *Am. J. Nucl. Med. Mol. Imaging* **2014**, *4*, 303–310. [PubMed]
14. Alves, F.; Alves, V.H.P.; Do Carmo, S.J.C.; Neves, A.C.B.; Silva, M.; Abrunhosa, A.J. Production of Copper-64 and Gallium-68 with a Medical Cyclotron Using Liquid Targets. *Mod. Phys. Lett. A* **2017**, *32*, 1740013. [CrossRef]
15. Alves, V.; do Carmo, S.; Alves, F.; Abrunhosa, A. Automated Purification of Radiometals Produced by Liquid Targets. *Instruments* **2018**, *2*, 17. [CrossRef]
16. do Carmo, S.J.C.; Scott, P.J.H.; Alves, F. Production of Radiometals in Liquid Targets. *EJNMMI Radiopharm. Chem.* **2020**, *5*, 2. [CrossRef]
17. Nair, M.; Happel, S.; Eriksson, T.; Pandey, M.K.; DeGrado, T.R.; Gagnon, K. Cyclotron Production and Automated New 2-Column Processing of [Ga-68] GaCl3. *Eur. J. Nuclear Med. Mol. Imaging* **2017**, *44*, S275.
18. Riga, S.; Cicoria, G.; Pancaldi, D.; Zagni, F.; Vichi, S.; Dassenno, M.; Mora, L.; Lodi, F.; Morigi, M.P.; Marengo, M. Production of Ga-68 with a General Electric PETtrace Cyclotron by Liquid Target. *Phys. Med.* **2018**, *55*, 116–126. [CrossRef]
19. European Pharmacopoeia Commission, European Directorate for the Quality of Medicines and Healthcare. *European Pharmacopoeia 10 Council of Europe*; Council of Europe: Strasbourg, France, 2020.
20. Rodnick, M.E.; Sollert, C.; Stark, D.; Clark, M.; Katsifis, A.; Hockley, B.G.; Parr, D.C.; Frigell, J.; Henderson, B.D.; Bruton, L.; et al. Synthesis of 68Ga-Radiopharmaceuticals Using Both Generator-Derived and Cyclotron-Produced 68Ga as Exemplified by [68Ga]Ga-PSMA-11 for Prostate Cancer PET Imaging. *Nat. Protoc.* **2022**, *17*, 980–1003. [CrossRef]
21. Pedersen, K.S.; Nielsen, K.M.; Fonslet, J.; Jensen, M.; Zhuravlev, F. Separation of Radiogallium from Zinc Using Membrane-Based Liquid-Liquid Extraction in Flow: Experimental and COSMO-RS Studies. *Solvent Extr. Ion Exch.* **2019**, *37*, 376–391. [CrossRef]
22. Pedersen, K.S.; Imbrogno, J.; Fonslet, J.; Lusardi, M.; Jensen, K.F.; Zhuravlev, F. Liquid–Liquid Extraction in Flow of the Radioisotope Titanium-45 for Positron Emission Tomography Applications. *React. Chem. Eng.* **2018**, *3*, 898–904. [CrossRef]
23. Søborg Pedersen, K.; Baun, C.; Michaelsen Nielsen, K.; Thisgaard, H.; Ingemann Jensen, A.; Zhuravlev, F. Design, Synthesis, Computational, and Preclinical Evaluation of NatTi/45Ti-Labeled Urea-Based Glutamate PSMA Ligand. *Molecules* **2020**, *25*, 1104. [CrossRef]
24. Gupta, B.; Mudhar, N.; Begum, Z.; Singh, I. Extraction and Recovery of Ga(III) from Waste Material Using Cyanex 923. *Hydrometallurgy* **2007**, *87*, 18–26. [CrossRef]
25. Iyer, J.N.; Dhadke, P.M. Liquid-Liquid Extraction and Separation of Gallium (III), Indium (III), and Thallium (III) by Cyanex-925. *Sep. Sci. Technol.* **2001**, *36*, 2773–2784. [CrossRef]
26. Inoue, K.; Baba, Y.; Yoshizuka, K. Solvent Extraction Equilibria of Gallium (III) with Acidic Organophosphorus Compounds from Aqueous Nitrate Media. *Solvent Extr. Ion Exch.* **1988**, *6*, 381–392. [CrossRef]
27. Gupta, B.; Mudhar, N.; Tandon, S.N. Extraction and Separation of Gallium Using Cyanex 301: Its Recovery from Bayer's Liquor. *Ind. Eng. Chem. Res.* **2005**, *44*, 1922–1927. [CrossRef]
28. International Atomic Energy Agency. *Gallium-68 Cyclotron Production*; TECDOC Series; International Atomic Energy Agency: Vienna, Austria, 2019; ISBN 978-92-0-100819-0.
29. DeGrado, T.R.; Pandey, M.K.; Byrne, J.F.; Engelbrecht, H.P.; Jiang, H.; Packard, A.B.; Thomas, K.A.; Jacobson, M.S.; Curran, G.L.; Lowe, V.J. Preparation and Preliminary Evaluation of 63Zn-Zinc Citrate as a Novel PET Imaging Biomarker for Zinc. *J. Nucl. Med.* **2014**, *55*, 1348–1354. [CrossRef]
30. Jagodić, V.; Grdenić, D. Aminophosphonic Acid Mono-Esters as Reagents for Solvent Extraction of Metals. *J. Inorg. Nucl. Chem.* **1964**, *26*, 1103–1109. [CrossRef]
31. Keglevich, G.; Bálint, E. The Kabachnik–Fields Reaction: Mechanism and Synthetic Use. *Molecules* **2012**, *17*, 12821–12835. [CrossRef] [PubMed]
32. Klamt, A. The COSMO and COSMO-RS Solvation Models. *Wiley Interdiscip. Rev. Comput. Mol. Sci.* **2011**, *1*, 699–709. [CrossRef]
33. Klamt, A.; Eckert, F.; Diedenhofen, M.; Beck, M.E. First Principles Calculations of Aqueous PKa Values for Organic and Inorganic Acids Using COSMO−RS Reveal an Inconsistency in the Slope of the PKa Scale. *J. Phys. Chem. A* **2003**, *107*, 9380–9386. [CrossRef]
34. Eriksson, L.; Johansson, E.; Kettaneh-Wold, N.; Wikström, C.; Wold, S. *Design of Experiments: Principles and Applications*, 3rd ed.; Umetrics Academy: Malmo, Sweden, 2008; ISBN 91-973730-4-4.

35. Sasaki, Y.; Oshima, T.; Baba, Y. Mutual Separation of Indium(III), Gallium(III) and Zinc(II) with Alkylated Aminophosphonic Acids with Different Basicities of Amine Moiety. *Sep. Purif. Technol.* **2017**, *173*, 37–43. [CrossRef]
36. Rudolph, W.W.; Pye, C.C. Zinc(II) Hydration in Aqueous Solution. A Raman Spectroscopic Investigation and an Ab-Initio Molecular Orbital Study. *Phys. Chem. Chem. Phys.* **1999**, *1*, 4583–4593. [CrossRef]
37. Ikushima, Y.; Saito, N.; Arai, M. Raman Spectral Studies of Aqueous Zinc Nitrate Solution at High Temperatures and at a High Pressure of 30 MPa. *J. Phys. Chem. B* **1998**, *102*, 3029–3035. [CrossRef]
38. Jagodić, V. Synthesis and Physical Properties of a Novel Aminophosphonic Acid as an Extracting Agent for Metals. *J. Inorg. Nucl. Chem.* **1970**, *32*, 1323. [CrossRef]
39. TURBOMOLE V7.5.1 2020, a Development of University of Karlsruhe and Forschungszentrum Karlsruhe GmbH, 1989–2007, TURBOMOLE GmbH, Since 2007. Available online: http://www.Turbomole.Com (accessed on 29 November 2022).
40. Eckert, F.; Klamt, A. Fast Solvent Screening via Quantum Chemistry: COSMO-RS Approach. *AIChE J.* **2002**, *48*, 369–385. [CrossRef]

Article

# Rapid Purification and Formulation of Radiopharmaceuticals via Thin-Layer Chromatography

Travis S. Laferriere-Holloway [1,2], Alejandra Rios [2,3], Giuseppe Carlucci [1,3,4] and R. Michael van Dam [1,2,3,*]

1. Department of Molecular & Medical Pharmacology, David Geffen School of Medicine, University of California, Los Angeles, CA 90095, USA
2. Crump Institute for Molecular Imaging, University of California, Los Angeles, CA 90095, USA
3. Physics and Biology in Medicine Interdepartmental Graduate Program, David Geffen School of Medicine, University of California, Los Angeles, CA 90095, USA
4. Ahmanson Translational Theranostics Division, Department of Molecular & Medical Pharmacology, University of California, Los Angeles, CA 90095, USA
* Correspondence: mvandam@mednet.ucla.edu

**Citation:** Laferriere-Holloway, T.S.; Rios, A.; Carlucci, G.; van Dam, R.M. Rapid Purification and Formulation of Radiopharmaceuticals via Thin-Layer Chromatography. *Molecules* **2022**, *27*, 8178. https://doi.org/10.3390/molecules27238178

Academic Editor: Svend Borup Jensen

Received: 24 October 2022
Accepted: 18 November 2022
Published: 24 November 2022

**Publisher's Note:** MDPI stays neutral with regard to jurisdictional claims in published maps and institutional affiliations.

**Copyright:** © 2022 by the authors. Licensee MDPI, Basel, Switzerland. This article is an open access article distributed under the terms and conditions of the Creative Commons Attribution (CC BY) license (https://creativecommons.org/licenses/by/4.0/).

**Abstract:** Before formulating radiopharmaceuticals for injection, it is necessary to remove various impurities via purification. Conventional synthesis methods involve relatively large quantities of reagents, requiring high-resolution and high-capacity chromatographic methods (e.g., semi-preparative radio-HPLC) to ensure adequate purity of the radiopharmaceutical. Due to the use of organic solvents during purification, additional processing is needed to reformulate the radiopharmaceutical into an injectable buffer. Recent developments in microscale radiosynthesis have made it possible to synthesize radiopharmaceuticals with vastly reduced reagent masses, minimizing impurities. This enables purification with lower-capacity methods, such as analytical HPLC, with a reduction of purification time and volume (that shortens downstream re-formulation). Still, the need for a bulky and expensive HPLC system undermines many of the advantages of microfluidics. This study demonstrates the feasibility of using radio-TLC for the purification of radiopharmaceuticals. This technique combines high-performance (high-resolution, high-speed separation) with the advantages of a compact and low-cost setup. A further advantage is that no downstream re-formulation step is needed. Production and purification of clinical scale batches of [$^{18}$F]PBR-06 and [$^{18}$F]Fallypride are demonstrated with high yield, purity, and specific activity. Automating this radio-TLC method could provide an attractive solution for the purification step in microscale radiochemistry systems.

**Keywords:** radiopharmaceuticals; microscale radiosynthesis; thin-layer chromatography; miniaturization; preparative TLC

## 1. Introduction

In the last decade, positron-emission tomography (PET) has led to many advances in disease characterization [1,2], drug development [3–5], and monitoring treatment efficacy for various diseases [6,7]. While numerous short-lived radionuclides may be used to label biologically active radiotracers, fluorine-18 remains by far the most common due to its high positron decay ratio (97%), short half-life (109.8 min), low positron energy (635 keV), and wide availability [8–10].

The production of $^{18}$F-labelled radiopharmaceuticals typically involves a late-stage radiofluorination method involving the reaction between [$^{18}$F]fluoride or a prosthetic group labelled with F-18, and a precursor, followed in some instances by the deprotection of functional groups. Subsequently, purification of the crude radiopharmaceutical is required to ensure that all unreacted precursors, reaction by-products, solvents, and other reagents (e.g., phase-transfer catalysts), are removed. The high structural similarity of precursors and byproducts to radiopharmaceuticals, along with the vast precursor excess typically used in radiosyntheses to ensure efficient reaction kinetics, impose significant challenges for the purification process.

Several chromatographic approaches are currently used to purify radiopharmaceuticals, including solid phase extraction (SPE) and high-performance liquid chromatography (HPLC). While these chromatographic methods are both versatile and compatible with a variety of stationary phases (e.g., reverse-phase C18 [11,12], size exclusion (SE) [13,14], and ion exchange (IEX) [15]), they differ in complexity and performance. SPE is generally rapid, but separation resolution is generally regarded as low [16]. A further downside is that developing a suitable SPE-based purification protocol can take considerable time and effort. So far, SPE has only successfully been used to purify a handful of $^{18}$F-labeled radiopharmaceuticals [17–21]. HPLC has high resolution and is used to purify the vast majority of radiotracers. Still, it is time-consuming, bulky, expensive, and often requires a downstream re-formulation process due to bio-incompatible mobile phases [22,23]. In some instances, HPLC can be performed with bio-compatible (e.g., ethanolic) mobile phases, although increased backpressure can become an issue. Another approach that has been used to purify radiopharmaceuticals is molecular imprinting chromatography [24]. However, this technique requires a unique stationary phase for each radiopharmaceutical and is not widely used.

Recently, our group and others have shown that microscale synthesis methods enable efficient reactions while enabling a vast reduction of reagent masses [25–29]. Consequently, the quantity of impurities is drastically reduced, and it appears in some cases that the number of different impurities may also be reduced [28,30]. These factors may allow lower-resolution forms of purification to be employed in the purification of microscale-produced radiopharmaceuticals. For example, purification has been performed using microscale SPE for [$^{18}$F]FDG [31–33] and [$^{18}$F]FLT [34,35]. It has also been attempted for microfluidically-produced [$^{18}$F]Fallypride [36], but sufficient chemical purity was not achieved, suggesting that microscale and conventional SPE may have similar limitations of versatility due to the low separation resolution. Alternatively, our group and others have shown that conventional semi-prep HPLC columns can be replaced with analytical scale columns [37,38], enabling faster purification, higher resolution, and reduced volume of collected pure fraction (enabling faster downstream formulation). However, the continued need for a bulky and expensive instrument to perform purification undermines many of the advantages of microfluidic radiosynthesis.

To overcome these challenges, we propose using thin-layer chromatography (TLC) as a more compact, rapid, and lower-cost way to purify microfluidically-produced radiopharmaceuticals. Purification via TLC is not new and is often used in the pharmaceutical industry for the crude synthesis of candidate molecules [39]. Utilizing preparative TLC plates, crude products are separated, then the product-binding sorbent is removed from the plate and extracted in organic solutions for subsequent processing. Though separations in the pharmaceutical industry usually involve long TLC plates and lengthy separation times, which are incompatible with the production of short-lived radiopharmaceuticals and the goals of miniaturizing the entire radiosynthesis processes, the masses involved in batches of radiopharmaceuticals are far smaller. We hypothesized that short, analytical-scale plates might be suitable for purifying radiopharmaceuticals (Figure 1).

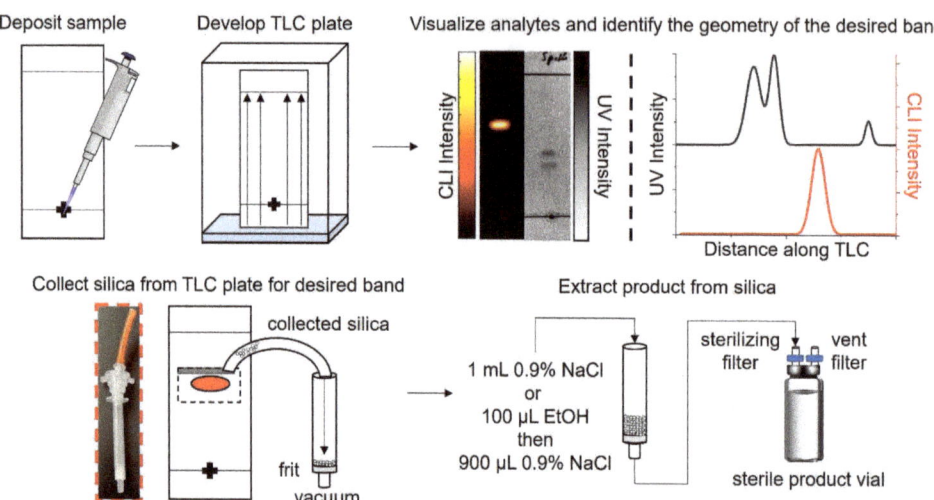

**Figure 1.** Procedure for the purification of microscale-synthesized radiopharmaceuticals using TLC.

Radio-TLC is already widely used in radiochemistry to analyze small samples (e.g., 1 µL) of radiopharmaceuticals. By making use of high-resolution imaging-based readout (e.g., Cerenkov luminescence imaging; CLI and UV imaging), our group has recently shown that separation resolution comparable to radio-HPLC can be achieved on analytical scale TLC plates with very short separation distances (4 cm) and short separation times (<4 min) [40]. In addition to being rapid and having high resolution, TLC is very versatile. We recently adopted the PRISMA algorithm [41] for efficiently optimizing mobile phase compositions to achieve high separation of a wide variety of radiopharmaceuticals from their radioactive and non-radioactive impurities [42]. This paper shows the feasibility of using analytical-scale TLC as a compact, rapid, and high-resolution method for the purification of microfluidically-produced (i.e., low mass scale, low volume) radiopharmaceuticals.

## 2. Results

*2.1. Performance of TLC at the Scale of Crude Reaction Mixtures*

When performing TLC analysis of radiopharmaceuticals, typically, only a small sample volume (0.5 or 1.0 µL) is spotted on the plate via a capillary or pipette. In contrast, the volume of the collected crude product from microscale reactions is on the order of 40–60 µL [43], all of which need to be loaded onto the TLC plate to use this as a purification method.

We have previously used the PRISMA algorithm to establish suitable TLC mobile phases for baseline separation of [$^{18}$F]PBR-06 from radioactive and non-radioactive impurities in crude reaction mixtures (1 µL sample) and for [$^{18}$F]Fallypride from its impurities (1 µL sample) [44]. While the separation resolution is expected to suffer by increasing the sample volume and mass, the degree of resolution reduction needs to be quantified.

To study the effect of sample mass without significantly changing the size of the sample spot, we loaded samples by pipetting crude [$^{18}$F]PBR-06 in 1 µL increments onto the origin while heating the TLC plate with a heat gun (90 °C setting), allowing each droplet to dry (~2 s) before adding the next. Comparison of samples before and after heating showed no changes, including no signs of decomposition. For [$^{18}$F]PBR-06, we discovered that the chromatographic resolution of [$^{18}$F]PBR-06 from its nearest impurity decreased from 2.2 to 1.0 when increasing the total volume of the spotted crude product from 6 to 60 µL. Notably, 60 µL corresponds to an entire batch of crude radiopharmaceutical. When performed manually, sample deposition with this approach took ~5 min to load a 60 µL sample.

We next tried applying the 60 µL volume in a streak rather than a single spot. Each 20 µL portion was deposited as a ~20 mm long line along the origin and then dried at 90 °C (~5 s) before applying the next streak in the same location. By spreading out the mass amount of product over a greater width of the separation medium, the chromatographic load is decreased, achieving nearly the same resolution (2.0) as the plate spotted with only 6 µL. The 60 µL sample could be deposited in <2 min by streaking.

We also tried spotting a 60 µL sample to TLC plates containing a concentrating zone. A large volume can be deposited as a single spot within the concentrating zone. During development, it will be concentrated into a thin line at the boundary of the concentrating zone before its migration and separation within the separation zone. Using this method, the resolution was 1.7. While we expected the resolution to be similar to the streaking approach, the observed resolution may be slightly lower because the plate was a HPTLC plate, which has a thinner sorbent layer (150 µm) than the analytical plates used for other samples (250 µm).

These results are summarized in Figure 2 and Table 1. Due to the high performance of the sample streaking method, in conjunction with analytical TLC plates, they were used for the remainder of the study.

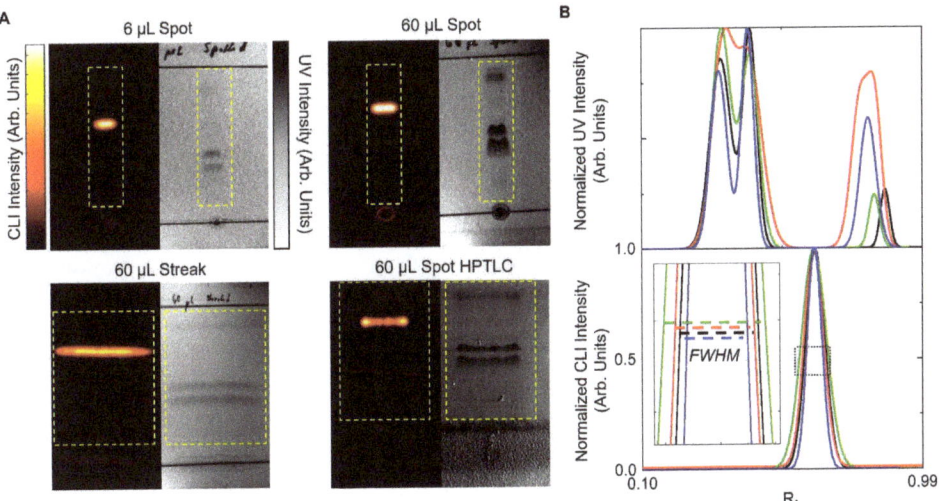

**Figure 2.** Effect of sample deposition parameters on separating [$^{18}$F]PBR-06 samples. (**A**) Images (left: CLI; right: UV) of crude [$^{18}$F]PBR-06 deposited on TLC plates using different volumes and application methods. Yellow lines denote the area of the image used to compute the line profiles shown in panel B, excluding the origin and solvent front lines with a strong signal in the UV images. (**B**) TLC chromatograms generated from the CLI and UV images. Legend: black—6 µL spot, red—60 µL spot, green—60 µL streak, and blue—60 µL spot (HPTLC plate). The inset shows a magnified view of the dashed region to highlight the full-width half maximum (FWHM).

**Table 1.** Effect of sample deposition parameters on chromatographic resolution between [$^{18}$F]PBR-06 and the nearest impurity.

| Sample Volume (µL) | Deposition Method | TLC Plate | Resolution |
| --- | --- | --- | --- |
| 6 | Spot | Analytical | 2.2 |
| 60 | Spot | Analytical | 1.0 |
| 60 | Streak | Analytical | 2.0 |
| 60 | Spot | HPTLC (with concentrating zone) | 1.7 |

In addition to evaluating separation resolution, losses during the sample application process were evaluated. Measurements using a calibrated ion chamber (CRC 25-PET, Capintec, Florham Park, NJ, USA) revealed a ~10–20% loss of initial sample activity on the pipette tip and Eppendorf tube originally containing the crude radiopharmaceutical. The loss could be reduced to <1% if the Eppendorf and pipette tip were rinsed with 20 µL of 9:1 MeOH:H$_2$O ($v/v$). When applied during the sample streaking method, the additional rinse volume did not affect the chromatographic resolution.

## 2.2. Efficiency of Radiopharmaceutical Collection from the TLC Plate

After separation, the product must be collected from the plate. We elected to use a process of scraping the silica stationary phase from the plate in the region of the desired band, followed by extraction of the product into a buffer.

The effectiveness of collecting the sorbent-bound radiopharmaceutical for [$^{18}$F]PBR-06 and [$^{18}$F]Fallypride is demonstrated in Figure 3. Each TLC plate was streaked with 60 µL of crude product, developed, and then measured by a dose calibrator. A CLI image of the plate was then obtained, and the relative abundance of radiochemical species was determined using region of interest (ROI) analysis, as previously described [40,45]. Combining these two measurements, we could estimate the initial quantity of radioactivity corresponding to the radiopharmaceutical product on the TLC plate. Initially, scraping of the sorbent at the position of the radiopharmaceutical band was performed via a small spatula. The sorbent (a fine powder) was collected onto weighing paper and then transferred into a SPE tube. However, using this method, >20% of the radiopharmaceutical activity (and sorbent) could be lost. Instead, we used a piece of plastic tubing with a beveled tip as the scraper. We connected the other end of the tubing through an empty SPE tube fitted with a 0.2 µm frit (Figure 1) to a vacuum source to capture the removed sorbent more efficiently. The entirety of the scraping process took <2 min to complete. The use of vacuum minimized the chance for the dispersing the radioactive powder into the air. Comparison of the collected sorbent activity of the product from the TLC plate (measured via dose calibrator) to the estimate of initial radioactivity of the radiopharmaceutical on the plate indicated that the sorbent-bound product was collected with >97% efficiency for both [$^{18}$F]PBR-06 and [$^{18}$F]Fallypride (Table 2, rows 1 and 3). Additional CLI images of the TLC plates were obtained after the scraping process. ROI analysis showed that the region of the plate corresponding to the product contained ~0% of the initial radioactivity (Figure 3), confirming that the silica removal process is quantitative.

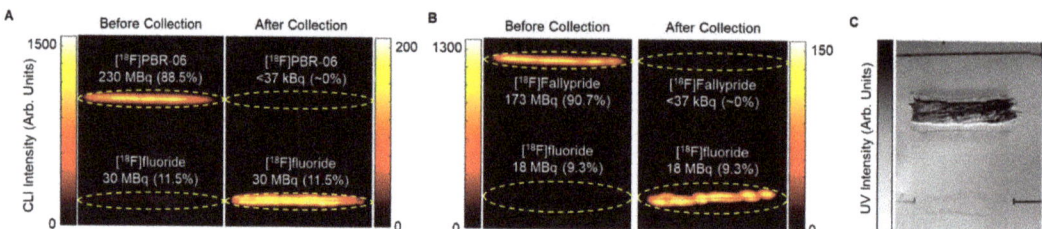

**Figure 3.** CLI images of TLC plates show the effectiveness of the stationary-phase removal step during TLC-based purification. (**A**) Images of analytical TLC plate streaked with crude [$^{18}$F]PBR-06 before and after collection. (**B**) Images of analytical TLC plate streaked with crude [$^{18}$F]Fallypride before and after collection. Yellow bands denote ROIs used in quantifying the proportion of different radiochemical species. (**C**) UV image of an analytical TLC plate after stationary phase removal for recovery of [$^{18}$F]PBR-06.

**Table 2.** Performance of microscale droplet radiosyntheses coupled with the TLC-based purification and formulation. In the extraction step, Method 1 uses 1.0 mL of saline alone, and Method 2 uses 100 µL EtOH, followed by 900 µL saline. The overall collection and extraction efficiency is calculated by multiplying the silica collection efficiency by the extraction efficiency for individual runs and then averaging across replicates. The overall RCY is calculated by multiplying the crude RCY of the droplet synthesis by the silica collection efficiency and the extraction efficiency for individual runs and then averaging across replicates.

| Radiotracer | Activity Level (MBq) | Crude RCY of Droplet Synthesis (%) (n = 8) | Silica Collection Efficiency (%) (n = 8) | Extraction Efficiency (%) Method 1 (n = 4) | Extraction Efficiency (%) Method 2 (n = 4) | Overall Collection and Extraction Efficiency (%) Method 1 (n = 4) | Overall Collection and Extraction Efficiency (%) Method 2 (n = 4) | Overall RCY (%) Method 1 (n = 4) | Overall RCY (%) Method 2 (n = 4) |
|---|---|---|---|---|---|---|---|---|---|
| [$^{18}$F]PBR-06 | 11 | 94.4 ± 1.2 | 98.7 ± 1.3 | 96.4 ± 3.4 | 97.9 ± 1.6 | 95.4 ± 4.6 | 96.3 ± 1.7 | 89.6 ± 3.9 | 91.3 ± 1.9 |
| | 1110–1480 | 91.9 ± 1.8 | 98.1 ± 1.1 | 95.6 ± 2.9 | 98.2 ± 0.3 | 94.2 ± 2.6 | 95.9 ± 0.9 | 86.7 ± 3.7 | 87.9 ± 1.8 |
| [$^{18}$F]Fallypride | 7.5 | 96.5 ± 1.6 | 97.5 ± 1.6 | 95.4 ± 1.1 | 98.4 ± 0.3 | 92.6 ± 2.6 | 96.2 ± 1.3 | 89.4 ± 3.7 | 92.9 ± 2.6 |
| | 740–1480 | 93.2 ± 2.5 | 97.5 ± 1.2 | 97.1 ± 1.0 | 97.8 ± 1.4 | 94.5 ± 1.9 | 95.6 ± 2.8 | 88.1 ± 3.8 | 89.2 ± 4.7 |

### 2.3. Efficiency of Radiopharmaceutical Extraction from the Collected Sorbent

Finally, the purified radiopharmaceutical needs to be separated from the sorbent. This is accomplished by flowing liquid through the sorbent and capturing the eluted liquid while the particles remain trapped by the frit. For this step, the output of the SPE tube (containing the sorbent-bound product) is connected through a sterilizing filter (0.2 µm) to a sterile septum-capped product vial. Vacuum is applied to a sterile filter connected to the vent port of the product vial. The end of the tubing used for scraping the sorbent is then dipped into an Eppendorf tube filled with extractant solution, effectively rinsing the sorbent collection path and eluting the radiopharmaceutical from the collected sorbent.

To avoid needing a later downstream reformulation step, we evaluated the ability to extract the product from the sorbent into biocompatible solutions. Initially, saline was used to extract [$^{18}$F]PBR-06 and [$^{18}$F]Fallypride from the sorbent. Using 1 mL of saline, the extraction efficiency was >95% for both tracers (Table 2, rows 1 and 3). While extraction efficiency with the model radiopharmaceuticals was high, some radiopharmaceuticals require additives, such as EtOH, to improve solubility. For this reason, we also explored the use of other bio-compatible solvents for extraction. Using 100 µL of EtOH, followed by 900 µL of saline, it was possible to extract >97% of the product from the sorbent for both tracers ([$^{18}$F]PBR-06 and [$^{18}$F]Fallypride) (Table 2, rows 1 and 3). Flowing the additional 900 µL of saline through the sorbent provided a final formulated product with <10% EtOH (v/v).

We achieved very high overall radiochemical yield (RCY) for both radiopharmaceuticals with the combination of droplet radiosynthesis and TLC-based purification/formulation (Table 2). Compared to our prior reports of droplet radiosyntheses that used analytical-scale HPLC purification (with purification efficiency of ~80%), the efficiency of the TLC purification and formulation process was significantly higher (nearly quantitative), leading to higher overall radiochemical yield. In particular, a prior report of droplet-based [$^{18}$F]PBR-06 production showed high crude RCY (94 ± 2%, n = 4), but due to losses during HPLC purification, the isolated RCY was only 76% (n = 1) [27], and further losses would have been expected during downstream formulation, which was not performed in that study. Similarly, a prior report of droplet-based [$^{18}$F]Fallypride production exhibited high crude RCY (96 ± 2%, n = 4), but, due to losses during HPLC purification, the isolated yield was 78% (n = 1) [26].

Notably, the entire purification and formulation process with the TLC method was very fast and took <10 min to complete (2 min for sample spotting, >4 min for TLC plate development, 2 min for silica removal, and 2 min for radiopharmaceutical extraction and filtration).

### 2.4. Scale-Up to Clinical Quantities

The ability of the TLC method to purify radiopharmaceuticals at clinically relevant levels were explored. For droplet-based radiosynthesis, scale-up is achieved by simply increasing the amount of radioactivity in the synthesis and does not require increasing the reaction mass scale [38]. For this reason, the chromatographic resolution of the TLC method should not be impaired when utilizing greater activity scales. Indeed, scaling up the amount of radioactivity led to the efficient purification of clinically relevant activity levels of [$^{18}$F]PBR-06 and [$^{18}$F]Fallypride (Table 2, lines 2 and 4). Automating the TLC purification procedure may allow more activity scales to be purified.

### 2.5. Quality Control Testing of Purified [$^{18}$F]PBR-06

A series of selected key quality control (QC) tests were performed to assess the safety and purity of the radiopharmaceuticals purified (and formulated) using the TLC method. Tests performed include appearance (color, clarity), pH, residual phase transfer catalyst, residual solvents, radiochemical purity, chemical purity, and radiochemical identity.

Radiochemical and chemical analyses were performed using HPLC (Figure 4). When we initially analyzed TLC-purified [$^{18}$F]Fallypride (Supplementary Figure S1), we noticed some impurities at early retention times in the UV channel and confirmed that these peaks came from the TLC plate itself. By pre-cleaning the TLC plates, these impurity peaks could be removed (Supplementary Figures S2 and S3). When using pre-cleaned TLC plates, radiochemical and chemical purity standards suitable for injection were achievable.

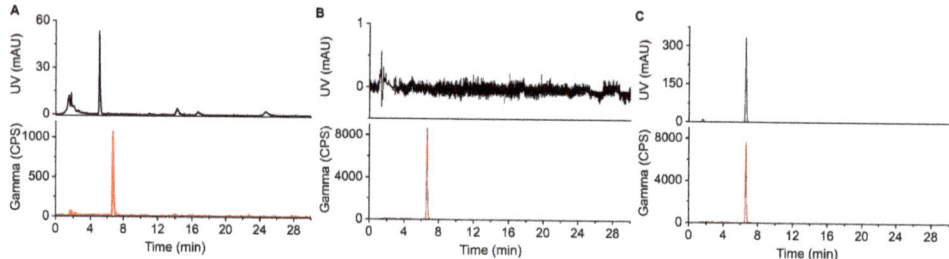

**Figure 4.** HPLC chromatograms of [$^{18}$F]PBR-06. (**A**) Crude reaction mixture. (**B**) TLC-purified (and formulated) product. (**C**) Co-injection of TLC-purified product with the [$^{19}$F]PBR-06 reference standard.

The results of these and additional tests (described in the Supplementary Information) for three consecutive batches of TLC purified [$^{18}$F]PBR-06 are summarized in Table 3. The results suggest that this method could potentially be used to produce tracers for clinical use.

Due to the silica sorbent's integral role in the TLC-purification process, we were concerned that some silica could end up in the final formulation, either as small nanoparticles that pass through the frit and filter or through solubility of silica in aqueous solutions [46]. To determine levels of residual silica, we used ICP-MS to measure Si content of samples that were first digested in $HNO_3$ to ensure any particulate silica was captured into the solution (see Supplementary Information). Si was not detected for the formulated tracer samples (limit of detection 0.83 ng/mL). While the complete elimination of silica in the final radiopharmaceutical formulation cannot be confirmed, it can be concluded that the residual amount is extremely low.

Table 3. Performance and quality control testing results for three consecutive batches of [$^{18}$F]PBR-06.

| Test | Criteria | Batch 1 | Batch 2 | Batch 3 |
|---|---|---|---|---|
| Radioactivity | - | 821 MBq [22.2 mCi] | 744 MBq [20.1 mCi] | 829 MBq [22.4 mCi] |
| Molar Activity | - | 342 GBq/μmol | 315 GBq/μmol | 327 GBq/μmol |
| Appearance | Clear, colorless, and particulate-free | ✓ | ✓ | ✓ |
| Radiochemical Identity | Retention time ratio of radio peak vs. reference standard (0.90–1.10) | 1.01 | 1.01 | 1.01 |
| Residual TBAHCO$_3$ | <104 mg/L | <45 mg/L | <45 mg/L | <45 mg/L |
| Residual Solvents | MeCN < 410 ppm<br>MeOH < 3000 ppm<br>Hexanes < 290 ppm<br>CHCl3 < 60 ppm<br>Et2O < 5000 ppm<br>EtOAc < 5000 ppm<br>AcOH < 5000 ppm<br>Thexyl alcohol < 5000 ppm | <1<br>24<br>6<br><1<br>104<br>21<br>7<br><1 | <1<br>21<br>2<br><1<br>47<br>10<br>5<br><1 | <1<br>24<br>5<br><1<br>102<br>20<br>7<br><1 |
| Radiochemical Purity | >95% | >99% | >99% | >99% |
| Radionuclide Identity (half-life) | 105–115 min | 110.4 | 111.7 | 113.8 |
| pH | 4.5–7.5 | 5.5 | 5.5 | 5.5 |
| Shelf life | Pass appearance, pH, and radiochemical purity after 120 min | ✓ | ✓ | ✓ |

## 3. Discussion

A significant advantage of the TLC-based purification approach described is its high-speed operation. In addition to the rapid separation via TLC, for the radiopharmaceuticals tested, the purified tracer could be recovered in saline (or a mixture with <10% EtOH), eliminating the need for a downstream reformulation step. Current microscale radiopharmaceutical production protocols generally rely on HPLC purification, followed by a separation formulation step performed via solid-phase extraction on a reversed-phase cartridge or via solvent evaporation followed by resuspension in an injectable buffer, requiring 30–60 min to complete [47–49]. In contrast, for the TLC purification (and formulation) method, these steps were completed in <10 min for both [$^{18}$F]PBR-06 and [$^{18}$F]Fallypride. Based on the half-life of fluoride-18, an additional 20–50 min of overall synthesis time would lead to a 12–27% loss of product. Furthermore, during HPLC and cartridge- or evaporation-based reformulation, activity losses are typically substantially higher than the 3–5% loss observed here.

We found the product band retention factors and band heights to be remarkably consistent from run to run for both the [$^{18}$F]PBR-06 and [$^{18}$F]Fallypride product bands (i.e., ($R_f$ = 0.66 ± 0.01, band height = 0.22 ± 0.05 cm, n = 7), ($R_f$ = 0.91 ± 0.01, band height = 0.31 ± 0.05 cm, n = 4), respectively). This allowed us to mark the TLC plate in advance with the expected position of the product band, allowing the sorbent collection without imaging the TLC plate. Batches processed in this fashion had high efficiency (low loss of product) and high chemical and radiochemical purity, equivalent to batches that relied on imaging. This observation suggests that, for well-developed methods, the TLC plate imaging step can potentially be skipped, simplifying the apparatus and procedure. To use this technique reliably requires adequate separation of the desired radiopharmaceutical band from impurity bands (both radioactive and non-radioactive impurities). The mobile phases used for the separation of [$^{18}$F]PBR-06 and [$^{18}$F]Fallypride from impurities were optimized using a recently-reported methodology (PRISMA) to maximize the resolution between the radiopharmaceutical and nearest impurity [44]. This optimization algorithm provides a systematic and resource-efficient way to discover suitable mobile phases for radiopharmaceuticals and appears to have high versatility for a broad range of radiopharmaceuticals [44], suggesting that it will be possible to develop high-resolution TLC-based methods to purify other radiotracers. Despite the presence of various organic solvents in the TLC mobile phases, GC-MS analysis revealed the amounts to be minimal and far below permitted amounts (Table 3). The low values are likely due to (i) the low initial volume of mobile phase "contained" within the silica in the region of the product band, (ii) the application of heat (90 °C for 30 s) to dry the TLC plate after separation, and (iii) the use of vacuum during the sorbent collection step that may further assist in the removal of any residual solvents.

An additional requirement for more widespread use would be to increase the degree of automation to simplify the process and reduce radiation exposure, especially for producing clinical scale or multi-patient batches. Simplifications could be made in the process to reduce exposure, e.g., connecting the SPE tube to the sterilizing filter at the start of the experiment and pulling the vacuum through the sterile vent filter both for collecting the scraped silica into the SPE tube, as well pulling the extraction buffer through the silica. Further automation of each of the processes (sample deposition, TLC separation, and extraction of product) are also needed. While commercially-available systems exist for automated sample deposition in spots, lines, or other patterns (e.g., CAMAG automatic TLC sampler 4 [50]), transfer of the crude radiopharmaceutical to the device, operation time, and system footprint are concerns for use in radiochemistry applications. A more practical approach may be to simply use TLC plates with concentrating zones, which would allow the sample to be dripped at a controlled flow rate onto a single location onto a heated TLC plate rather than the more complicated process of depositing the sample in a streak pattern. The resolution obtained for [$^{18}$F]PBR-06 samples spotted onto concentrating-zone HPTLC plates was nearly as good as for samples streaked onto normal analytical plates and could perhaps be further optimized by comparing different types of concentrating-zone plates. Concentrating zone plates are also likely to reduce the potential dispersion effects if the streak pattern is not perfectly straight. The need for manual handling in the development process can likely be eliminated by integrating the above sample deposition approaches with commercial or custom horizontal TLC setups [51–54]. Commercially available online extraction systems also exist for the collection of identified product bands directly from TLC plates without the need for scraping (e.g., CAMAG TLC-MS Interface 2 [55], Advion Plate Express [56]), using methods such as liquid extraction. However, the manual steps for installation and alignment of TLC plates, system size, and limitations on the band geometry (that will prevent complete collection of the product species) may not be well matched to preparative applications in the radiopharmaceutical field. A more practical approach to automation may be to develop a custom apparatus with adjustable or movable flow cell placed across the product band to extract the species of interest [57,58].

Another strategy for automation may be to leverage the PRISMA procedure to develop mobile phase systems that could provide high separation resolution using other chromatography methods (e.g., silica flash chromatography), which may be easier to automate, or perhaps miniaturize, using microfluidic-based systems with integrated purification media [32,59]. However, it is not clear if the resolution achieved in the column format would match that achieved in the planar TLC format or whether a similar fast operation speed and high recovery efficiencies would be observed. Furthermore, the use of highly-UV-absorbing organic solvents could limit the ability to monitor non-radioactive impurities and obtain a pure product, and the collected radiopharmaceutical would require extensive reformulation to remove relatively large amounts of solvents, making the process more time-consuming and complicated compared to the TLC-based approach.

## 4. Materials and Methods

### 4.1. Reagents and Materials

All reagents and solvents were obtained from commercial suppliers and used without further purification. 2,3-dimethyl-2-butanol (thexyl alcohol; anhydrous, 98%), acetic acid (AcOH;

glacial, >99.9%), acetone (suitable for HPLC, >99.9%), acetonitrile (MeCN, anhydrous, 99.8%), ammonium formate (NH$_4$HCO$_2$, 97%), chloroform (>99.5%, contains 100–200 ppm amylenes as a stabilizer), dichloromethane (DCM; anhydrous, >99.8% contains 40–150 ppm amylene as a stabilizer), diethyl ether (Et$_2$O; >99.9% inhibitor free), ethyl acetate (EtOAc; anhydrous, 99.8%), ethyl alcohol (EtOH; 200 proof, anhydrous, >99.5%), methyl alcohol (MeOH; anhydrous, 99.8%), n-hexanes (98%), Polypropylene SPE tube with PE frits (1 mL, 20 um porosity), Silica with concentrating zone (Silica 60 with diatomaceous earth zone) HPTLC plates, tetrahydrofuran (THF; anhydrous, >99.9% inhibitor free), water (H$_2$O; suitable for ion chromatography) and Whatman Anotop 10 syringe filters (sterile, 0.2 um) were purchased from Sigma-Aldrich (St. Louis, MO, USA). (S)-2,3-dimethoxy-5-[3-[[(4-methylphenyl)-sulfonyl]oxy]-propyl]-N-[[1-(2-propenyl)-2-pyrrolidinyl]methyl]-benzamide ([$^{18}$F]Fallypride precursor, >95%), 5-(3-fluoropropyl)-2,3-dimethoxy-N-(((2S)-1-(2-propenyl)-2-pyrrolidinyl)methyl)benzamide (Fallypride reference standard, >95%), 2-((2,5-dimethoxybenzyl)(2-phenoxyphenyl)amino)-2-oxoethyl 4-methylbenzenesulfonate ([$^{18}$F]PBR-06 precursor, >95%), 2-fluoro-N-(2-methoxy-5-methoxybenzyl)-N-(2-phenoxyphenyl)acetamide (PBR-06 reference standard, >95%), and tetrabutylammonium bicarbonate (TBAHCO$_3$; 75 mM in ethanol) were purchased from ABX Advanced Biochemical Compounds (Radeberg, Germany).

Silica gel 60 F$_{254}$ sheets (aluminum backing, 5 cm × 20 cm) were purchased from Merck KGaA (Darmstadt, Germany). Glass microscope slides (76.2 mm × 50.8 mm, 1 mm thick) were obtained from C&A Scientific (Manassas, VA, USA). Saline (0.9% sodium chloride injection, USP) was obtained from Hospira Inc. (Lake Forest, IL, USA). Sodium phosphate dibasic (Na$_2$HPO$_4$-7H$_2$O) and sodium phosphate monobasic (NaH$_2$PO$_4$-H$_2$O) were purchased from Fisher Scientific (Thermo Fisher Scientific, Waltham, MA, USA).

No-carrier-added [$^{18}$F]fluoride was produced by the (p, n) reaction of [$^{18}$O]H$_2$O (98% isotopic purity, Huayi Isotopes Co., Changshu, China) in a RDS-111 cyclotron (Siemens, Knoxville, TN, USA) at 11 MeV, using a 1.2-mL silver target with Havar foil.

*4.2. Preparation of Radiopharmaceuticals and Reference Standards*

[$^{18}$F]PBR-06 and [$^{18}$F]Fallypride were prepared using droplet radiochemistry methods on Teflon-coated silicon surface tension trap chips [26]. Detailed protocols for preparing these radiotracers have been previously reported [27]. Stock solutions of reference standards were prepared at 20 mM concentrations: 5 mg of Fallypride was added to 685 µL of MeOH, and 5 mg of PBR-06 was added to 632 µL of MeOH.

*4.3. Preparation of TLC Plates*

TLC plates were cut (W × H, 3 × 6 cm), then marked with horizontal pencil lines at 1 cm (origin line) and 5 cm (development line) from the bottom edge.

To eliminate impurities in the TLC plate that can contaminate the radiopharmaceutical, plates were pre-cleaned with solvent, as previously described [60]. Briefly, TLC plates were submerged to the origin line in a mixture of 2:1 EtOAc: MeOH ($v/v$), allowed to develop for 20 min, and then heated for 1 min (at a 120 °C setting) using a heat gun (Furno 500, Wagner).

*4.4. Sample Spotting and Separation*

60 µL of the relevant crude radiopharmaceutical sample was applied to the plate by various methods (e.g., sequential spotting or streaking) by a micro-pipette. Spotting on analytical scale TLC plates was performed by adding 1 µL of the sample and heating with a heat gun at 90 °C (~2 s). Spotting of samples on HPTLC plates occurred with the addition of 10 µL of sample to the concentrating zone, followed by drying at 90 °C (~5 s). Streaking of samples on analytical scale TLC plates were performed by deposition of 20 µL of sample in a thin streak covering ~30 mm, followed by heating at 90 °C (~5 s).

Plates were then developed in the mobile phase up to the development line. The mobile phases for [$^{18}$F]PBR-06 and [$^{18}$F]Fallypride were 29.8:26.9:20.4:22.85:0.05 ($v/v$) Et$_2$O:DCM:CHCl$_3$:n-hexanes:AcOH and 31.3:24.5:34.3:10.0 ($v/v$) THF:acetone:n-hexanes:TEA, respectively. After development, the plates were dried by a heat gun for 30 s at 90 °C.

## 4.5. Readout and Analysis of TLC Plates

The developed TLC plate was covered with a glass plate and visualized, as previously reported [35], to obtain a Cerenkov luminescence image (CLI) (1 min exposure), followed by a UV image (7 ms exposure).

Images were analyzed to determine chromatographic resolution using a custom MATLAB program (Mathworks, Natick, MA, USA) with a graphical user interface (GUI), as previously described [44]. Briefly, the user is guided by the program to create chromatograms from the CLI and UV images, from which peak positions, widths, and resolution are calculated [44]. In the analysis, the lines drawn (for origin and solvent front) are omitted from the selected lanes, since the pencil markings show up as false peaks in the UV chromatogram. The TLC chromatograms were plotted by exporting the data from the Matlab program and processing using OriginPro (OriginLab, Northampton, MA, USA).

## 4.6. TLC Purification of Radiopharmaceuticals

### 4.6.1. Collection of Sorbent from TLC

When performing purification, the CLI and UV images of the TLC plate were used to identify the location of the product band and nearest impurity bands. During the preparation of the TLC plate, a pencil was used to outline the expected position and size of the radiopharmaceutical band (as determined from averaging images of multiple separations from crude batches of the same radiopharmaceutical and identifying the midpoint between the radiopharmaceutical band and its nearest impurities). To scrape the sorbent from the plate, the opening of a piece of plastic tubing cut at a ~45° angle (polyurethane, 1/4″ ID, IDEX) was used. The tubing was connected to the inlet of an empty SPE tube (polypropylene, 1 mL, Sigma Aldrich, St. Louis, MO, USA) that was fitted at the output end with a 10 µm frit (polyethylene, Sigma Aldrich), and the output end was further connected to vacuum. While the desired region was scraped in a series of horizontal lines (raster motion), the sorbent was collected into the SPE tube. The visualization step could be omitted through the pre-calibration step of determining the margins of radiopharmaceutical collection.

### 4.6.2. Extraction of the Radiopharmaceutical from Sorbent

Before extraction, the sterile product vial was fitted with 2 sterile filters (Anotop, 0.2 µm), one prewetted with saline and then connected to the output of the SPE tube and one left dry (vent). The radiopharmaceutical was then eluted from the collected sorbent with biocompatible solvents (1 mL saline, or 100 µL EtOH followed by 900 µL saline) by applying vacuum to the vent filter of the product vial and by moving the tubing 'scraper' into an Eppendorf tube filled with the desired extraction solvent. No separate re-formulation of the collected purified product was required.

## 4.7. HPLC Analyses

Radio-HPLC was used to analyze crude radiopharmaceuticals and to perform tests for radiochemical and chemical purity and radiochemical identity of TLC-purified batches of radiopharmaceuticals. The radio-HPLC system setup comprised a Smartline HPLC system (Knauer, Berlin, Germany) equipped with a degasser (Model 5050), pump (Model 1000), UV detector (254 nm; Eckert & Ziegler, Berlin, Germany), gamma-radiation detector (BFC-4100, Bioscan, Inc., Poway, CA, USA), and counter (BFC-1000; Bioscan, Inc., Poway, CA, USA). A C18 Gemini column was used for separations (250 × 4.6 mm, 5 µm, Phenomenex, Torrance, CA, USA). [$^{18}$F]PBR-06 samples were separated with a mobile phase of 60/40 ($v/v$) MeCN:20 mM sodium phosphate buffer (pH = 5.8) at a flow rate of 1.5 mL/min resulting in a retention time for [$^{18}$F]PBR-06 of 6.5 min. [$^{18}$F]Fallypride samples were separated with a mobile phase of 60% MeCN in 25 mM NH$_4$HCO$_2$ with 1% TEA ($v/v$) at a flow rate of 1.5 mL/min resulting in a retention time for [$^{18}$F]Fallypride of 5.8 min.

### 4.8. Quality Control Testing

Quality control (QC) tests were performed on 3 consecutive batches of [$^{18}$F]PBR-06 produced via a droplet microreactor and purified with the TLC approach described here. Testing focused primarily on color and clarity, radiochemical and chemical purity, molar activity, and residual solvent content to highlight the performance of this novel purification method. A full summary of tests and results can be found in the Supplementary Information Section S1.

### 4.9. ICP-MS Analysis for Silicon Content

To estimate silica content in the final formulation, the amount of silicon was determined in a series of replicate samples, in which the spotting, separation, silica collection and extraction steps (using 1 mL saline) were performed starting with blank TLC plates. Silicon determination was performed via inductively-coupled plasma mass spectrometry (ICP-MS) using a NexION 2000 (Perkin Elmer, Hong Kong, China). For each sample, an area (2.0 × 1.5 cm, W × H) was scraped from a cleaned TLC plate into an SPE tube, and 1 mL of saline was flowed through the silica and a sterile filter and collected into an Eppendorf tube for analysis. Each sample was transferred to a clean Teflon vessel for acid digestion in concentrated $HNO_3$ (65–70%, Trace Metal Grade, Fisher Scientific) with a supplement of $H_2O_2$ (30%, Certified ACS, Fisher Scientific) at 200 °C for 50 min in a microwave digestion system (Titan MPS, Perkin Elmer). Once the sample was cooled to room temperature, it was subsequently diluted to make a final volume of 10 mL by adding filtered DI $H_2O$ for analysis. The calibration curve was established using a standard solution, while the dwell time was 50 ms with thirty sweeps and three replicates with background correction. The detection limit using this procedure was 0.82 ng/mL.

## 5. Conclusions

In this feasibility study, high-resolution radio-TLC was leveraged as a means to perform rapid purification of two clinically-relevant radiopharmaceuticals ([$^{18}$F]PBR-06 and [$^{18}$F]Fallypride) produced via droplet radiochemistry methods. Due to the high chemical and radiochemical purity and the high efficiency of product collection and formulation achieved, it is conceivable that the TLC purification method could serve as a versatile approach for the purification of microscale-produced radiopharmaceuticals. The combination of droplet radiosynthesis with TLC-based purification/formulation for the production of [$^{18}$F]PBR-06 led to high molar activities ($\gtrsim$300 GBq/µmol), comparing favorably to the literature reports (37–222 GBq/µmol [61,62]).

Even with the higher mass loading and volume of the crude radiopharmaceutical (60 µL) compared to typical samples (0.5–1 µL), high separation resolution of the radiopharmaceutical product from radioactive and non-radioactivity impurities was achieved on the TLC plates, as visualized via CLI and UV imaging. The product collection (via sorbent collection from the plate followed by extraction) was nearly quantitative. Notably, by using injectable buffers (saline or EtOH diluted to <10% $v/v$ in saline), the need for subsequent re-formulation is eliminated. Consequently, radio-TLC purification (and formulation) could be completed in under 10 min. Furthermore, due to the low cost of TLC plates, one can consider the purification and formulation system to be disposable (in stark contrast to HPLC-based systems), further simplifying microscale radiosynthesis instruments and eliminating the need for developing and validating cleaning protocols

As a proof-of-concept, several batches of [$^{18}$F]PBR-06 and [$^{18}$F]Fallypride were produced and purified at scales sufficient for clinical imaging. Critical QC tests were performed on multiple batches (e.g., color and clarity, chemical and radiochemical purity, molar activity, and residual solvents) and suggested the potential suitability for clinical production of the TLC purification method.

**Supplementary Materials:** The following supporting information can be downloaded at: https://www.mdpi.com/article/10.3390/molecules27238178/s1, Figure S1: HPLC chromatograms of

[$^{18}$F]Fallypride samples; Figure S2: Images of crude [$^{18}$F]Fallypride on TLC plates using the streaking deposition methods; Figure S3: UV images of non-cleaned and pre-cleaned TLC plates after spotting with the PBR-06 reference standard and developing in the mobile phase for [$^{18}$F]PBR-06; Figure S4: HPLC analysis (using the mobile phase for PBR-06) of mock samples obtained by silica collection and subsequent product extraction; Figure S5: Calibration curve for PBR-06; Figure S6: Calibration curve for Fallypride; Figure S7: Images of iodine-stained plates to test for residual TBAHCO$_3$; Figure S8: Images of Dragendorf-stained plates to test for residual TBAHCO$_3$; Figure S9: pH testing results; Figure S10: HPLC chromatogram of formulated [$^{18}$F]PBR-06 injected 120 min after the end of synthesis. Refs. [63,64] are cited in the Supplementary Materials.

**Author Contributions:** Conceptualization, T.S.L.-H. and R.M.v.D.; methodology, T.S.L.-H. and R.M.v.D.; software, T.S.L.-H.; validation, T.S.L.-H., A.R. and R.M.v.D.; formal analysis, T.S.L.-H.; investigation, T.S.L.-H. and A.R.; resources, R.M.v.D. and G.C.; data curation, T.S.L.-H.; writing—original draft preparation, T.S.L.-H. and R.M.v.D.; writing—review and editing, T.S.L.-H. and R.M.v.D.; visualization, T.S.L.-H.; supervision, R.M.v.D.; project administration, R.M.v.D.; funding acquisition, R.M.v.D. All authors have read and agreed to the published version of the manuscript.

**Funding:** This research was funded in part by the National Cancer Institute (R33 CA240201) and the National Institute of Biomedical Imaging and Bioengineering (R21 EB024243, R01 EB032264, and T32 EB002101).

**Institutional Review Board Statement:** Not applicable.

**Informed Consent Statement:** Not applicable.

**Data Availability Statement:** The data presented in this study are available on request from the corresponding author. The data are not publicly available due to the length of the datasets.

**Acknowledgments:** The authors thank Jeffrey Collins for providing [$^{18}$F]fluoride for these studies. The authors also thank Chris Bobinski at the UCLA Biomedical Research Cyclotron facility for completing some of the QC analyses for these studies. The authors acknowledge the use of the ICP-MS facility within the Nano and Pico Characterization Lab in the California NanoSystems Institute (CNSI) at UCLA and thank Chong Hyun Chang for his assistance with ICP-MS analysis. The authors thank Gregory Khitrov of the UCLA Molecular Instrumentation center for assistance with residual solvent analysis. Microfluidic droplet reaction chips were produced in the Nanofabrication Laboratory (NanoLab) in the CNSI at UCLA.

**Conflicts of Interest:** van Dam is a co-founder of Sofie, Inc. (Dulles, VA, USA) and of DropletPharm, Inc. (Sherman Oaks, CA, USA).

## References

1. Berti, V.; Osorio, R.S.; Mosconi, L.; Li, Y.; Santi, S.D.; de Leon, M.J. Early Detection of Alzheimer's Disease with PET Imaging. *NDD* **2010**, *7*, 131–135. [CrossRef]
2. Dendl, K.; Koerber, S.A.; Kratochwil, C.; Cardinale, J.; Finck, R.; Dabir, M.; Novruzov, E.; Watabe, T.; Kramer, V.; Choyke, P.L.; et al. FAP and FAPI-PET/CT in Malignant and Non-Malignant Diseases: A Perfect Symbiosis? *Cancers* **2021**, *13*, 4946. [CrossRef]
3. Aboagye, E.O.; Price, P.M.; Jones, T. In Vivo Pharmacokinetics and Pharmacodynamics in Drug Development Using Positron-Emission Tomography. *Drug Discov. Today* **2001**, *6*, 293–302. [CrossRef]
4. Bhattacharyya, S. Application of Positron Emission Tomography in Drug Development. *Biochem. Pharm.* **2012**, *1*, 1000e128. [CrossRef] [PubMed]
5. Eckelman, W.C. The Use of Positron Emission Tomography in Drug Discovery and Development. In *Positron Emission Tomography: Basic Sciences*; Bailey, D.L., Townsend, D.W., Valk, P.E., Maisey, M.N., Eds.; Springer: London, UK, 2005; pp. 327–341, ISBN 978-1-84628-007-8.
6. Avril, S.; Muzic, R.F.; Plecha, D.; Traughber, B.J.; Vinayak, S.; Avril, N. 18F-FDG PET/CT for Monitoring of Treatment Response in Breast Cancer. *J. Nucl. Med.* **2016**, *57*, 34S–39S. [CrossRef]
7. Weber, W.A.; Figlin, R. Monitoring Cancer Treatment with PET/CT: Does It Make a Difference? *J. Nucl. Med.* **2007**, *48*, 36S–44S.
8. Banister, S.; Roeda, D.; Dolle, F.; Kassiou, M. Fluorine-18 Chemistry for PET: A Concise Introduction. *Curr. Radiopharm.* **2010**, *3*, 68–80. [CrossRef]
9. Jacobson, O.; Kiesewetter, D.O.; Chen, X. Fluorine-18 Radiochemistry, Labeling Strategies and Synthetic Routes. *Bioconjugate Chem.* **2015**, *26*, 1–18. [CrossRef] [PubMed]
10. Brooks, A.F.; Makaravage, K.J.; Wright, J.; Sanford, M.S.; Scott, P.J.H. Fluorine-18 Radiochemistry. In *Handbook of Radiopharmaceuticals*; John Wiley & Sons, Ltd.: Hoboken, NJ, USA, 2020; pp. 251–289, ISBN 978-1-119-50057-5.

11. Füchtner, F.; Angelberger, P.; Kvaternik, H.; Hammerschmidt, F.; Simovc, B.P.; Steinbach, J. Aspects of 6-[18F]Fluoro-L-DOPA Preparation: Precursor Synthesis, Preparative HPLC Purification and Determination of Radiochemical Purity. *Nucl. Med. Biol.* **2002**, *29*, 477–481. [CrossRef] [PubMed]
12. Boothe, T.E.; Emran, A.M. The Role of High Performance Liquid Chromatography in Radiochemical/Radiopharmaceutical Synthesis and Quality Assurance. In *New Trends in Radiopharmaceutical Synthesis, Quality Assurance, and Regulatory Control*; Emran, A.M., Ed.; Springer: Boston, MA, USA, 1991; pp. 409–422, ISBN 978-1-4899-0626-7.
13. Wester, H.J.; Schottelius, M. Fluorine-18 Labeling of Peptides and Proteins. In *Proceedings of the PET Chemistry*; Schubiger, P.A., Lehmann, L., Friebe, M., Eds.; Springer: Berlin/Heidelberg, Germany, 2007; pp. 79–111.
14. *Size Exclusion Chromatography: Principles and Methods*; GE Healthcare: Chicago, IL, USA, 2014.
15. Waldmann, C.M.; Gomez, A.; Marchis, P.; Bailey, S.T.; Momcilovic, M.; Jones, A.E.; Shackelford, D.B.; Sadeghi, S. An Automated Multidose Synthesis of the Potentiometric PET Probe 4-[$^{18}$F]Fluorobenzyl-Triphenylphosphonium ([$^{18}$F]FBnTP). *Mol. Imaging Biol.* **2018**, *20*, 205–212. [CrossRef]
16. Simpson, N.J.K. *Solid-Phase Extraction: Principles, Techniques, and Applications*; CRC Press: Boca Raton, FL, USA, 2000; ISBN 978-1-4200-5624-2.
17. Nandy, S.K.; Rajan, M.G.R. Simple, Column Purification Technique for the Fully Automated Radiosynthesis of [18F]Fluoroazomycinarabinoside ([18F]FAZA). *Appl. Radiat. Isot.* **2010**, *68*, 1944–1949. [CrossRef]
18. Lee, S.J.; Hyun, J.S.; Oh, S.J.; Yu, K.H.; Ryu, J.S. Development of a New Precursor-Minimizing Base Control Method and Its Application for the Automated Synthesis and SPE Purification of [18F]Fluoromisonidazole ([18F]FMISO). *J. Label. Compd. Radiopharm.* **2013**, *56*, 731–735. [CrossRef] [PubMed]
19. Bogni, A.; Laera, L.; Cucchi, C.; Iwata, R.; Seregni, E.; Pascali, C. An Improved Automated One-Pot Synthesis of O-(2-[18F]Fluoroethyl)-L-Tyrosine ([18F]FET) Based on a Purification by Cartridges. *Nucl. Med. Biol.* **2019**, *72–73*, 11–19. [CrossRef] [PubMed]
20. Cheung, Y.-Y.; Nickels, M.L.; McKinley, E.T.; Buck, J.R.; Manning, H.C. High-Yielding, Automated Production of 3′-Deoxy-3′-[18F]Fluorothymidine Using a Modified Bioscan Coincidence FDG Reaction Module. *Appl. Radiat. Isot.* **2015**, *97*, 47–51. [CrossRef]
21. Lazari, M.; Quinn, K.M.; Claggett, S.B.; Collins, J.; Shah, G.J.; Herman, H.E.; Maraglia, B.; Phelps, M.E.; Moore, M.D.; Dam, R.M. van ELIXYS—A Fully Automated, Three-Reactor High-Pressure Radiosynthesizer for Development and Routine Production of Diverse PET Tracers. *EJNMMI Res.* **2013**, *3*, 52. [CrossRef]
22. Lazarus, C.R. Formulation of Radiopharmaceuticals. In *Radionuclide Imaging in Drug Research*; Wilson, C.G., Hardy, J.G., Frier, M., Davis, S.S., Eds.; Springer: Dordrecht, The Netherlands, 1982; pp. 61–73, ISBN 978-94-011-9728-1.
23. Lau, J.; Rousseau, E.; Kwon, D.; Lin, K.-S.; Bénard, F.; Chen, X. Insight into the Development of PET Radiopharmaceuticals for Oncology. *Cancers* **2020**, *12*, 1312. [CrossRef]
24. Turiel, E.; Martin-Esteban, A. Molecularly Imprinted Polymers: Towards Highly Selective Stationary Phases in Liquid Chromatography and Capillary Electrophoresis. *Anal. Bioanal. Chem.* **2004**, *378*, 1876–1886. [CrossRef]
25. Wang, J.; Chao, P.H.; Hanet, S.; van Dam, R.M. Performing Multi-Step Chemical Reactions in Microliter-Sized Droplets by Leveraging a Simple Passive Transport Mechanism. *Lab Chip* **2017**, *17*, 4342–4355. [CrossRef] [PubMed]
26. Wang, J.; Chao, P.H.; van Dam, R.M. Ultra-Compact, Automated Microdroplet Radiosynthesizer. *Lab Chip* **2019**, *19*, 2415–2424. [CrossRef] [PubMed]
27. Rios, A.; Holloway, T.S.; Chao, P.H.; De Caro, C.; Okoro, C.C.; van Dam, R.M. Microliter-Scale Reaction Arrays for Economical High-Throughput Experimentation in Radiochemistry. *Sci. Rep.* **2022**, *12*, 10263. [CrossRef] [PubMed]
28. Lisova, K.; Wang, J.; Chao, P.H.; van Dam, R.M. A Simple and Efficient Automated Microvolume Radiosynthesis of [18F]Florbetaben. *EJNMMI Radiopharm. Chem.* **2020**, *5*, 30. [CrossRef] [PubMed]
29. Wang, J.; Holloway, T.; Lisova, K.; van Dam, R.M. Green and Efficient Synthesis of the Radiopharmaceutical [18F]FDOPA Using a Microdroplet Reactor. *React. Chem. Eng.* **2020**, *5*, 320–329. [CrossRef] [PubMed]
30. Wang, J.; van Dam, R.M. High-Efficiency Production of Radiopharmaceuticals via Droplet Radiochemistry: A Review of Recent Progress. *Mol. Imaging* **2020**, *19*, 1–21. [CrossRef]
31. Koag, M.C.; Kim, H.-K.; Kim, A.S. Efficient Microscale Synthesis of [18F]-2-Fluoro-2-Deoxy-d-Glucose. *Chem. Eng. J.* **2014**, *258*, 62–68. [CrossRef]
32. Tarn, M.D.; Pascali, G.; De Leonardis, F.; Watts, P.; Salvadori, P.A.; Pamme, N. Purification of 2-[18F]Fluoro-2-Deoxy-d-Glucose by on-Chip Solid-Phase Extraction. *J. Chromatogr. A* **2013**, *1280*, 117–121. [CrossRef]
33. Keng, P.Y.; Chen, S.; Ding, H.; Sadeghi, S.; Shah, G.J.; Dooraghi, A.; Phelps, M.E.; Satyamurthy, N.; Chatziioannou, A.F.; Kim, C.-J. "CJ"; et al. Micro-Chemical Synthesis of Molecular Probes on an Electronic Microfluidic Device. *Proc. Natl. Acad. Sci. USA* **2012**, *109*, 690–695. [CrossRef] [PubMed]
34. Koag, M.C.; Kim, H.-K.; Kim, A.S. Fast and Efficient Microscale Radiosynthesis of 3′-Deoxy-3′-[18F]Fluorothymidine. *J. Fluor. Chem.* **2014**, *166*, 104–109. [CrossRef]
35. Javed, M.R.; Chen, S.; Kim, H.-K.; Wei, L.; Czernin, J.; Kim, C.-J. "CJ"; Dam, R.M. van; Keng, P.Y. Efficient Radiosynthesis of 3′-Deoxy-3′-18F-Fluorothymidine Using Electrowetting-on-Dielectric Digital Microfluidic Chip. *J. Nucl. Med.* **2014**, *55*, 321–328. [CrossRef] [PubMed]

36. Zhang, X.; Liu, F.; Knapp, K.-A.; Nickels, M.L.; Manning, H.C.; Bellan, L.M. A Simple Microfluidic Platform for Rapid and Efficient Production of the Radiotracer [18F]Fallypride. *Lab Chip* **2018**, *18*, 1369–1377. [CrossRef]
37. Lisova, K.; Wang, J.; Rios, A.; van Dam, R.M. Adaptation and Optimization of [F-18] Florbetaben ([F-18] FBB) Radiosynthesis to a Microdroplet Reactor. *J. Label. Compd. Radiopharm.* **2019**, *62*, S353–S354.
38. Wang, J.; Chao, P.H.; Slavik, R.; van Dam, R.M. Multi-GBq Production of the Radiotracer [18F]Fallypride in a Droplet Microreactor. *RSC Adv.* **2020**, *10*, 7828–7838. [CrossRef]
39. Wood, J.L.; Steiner, R.R. Purification of Pharmaceutical Preparations Using Thin-Layer Chromatography to Obtain Mass Spectra with Direct Analysis in Real Time and Accurate Mass Spectrometry. *Drug Test. Anal.* **2011**, *3*, 345–351. [CrossRef]
40. Wang, J.; Rios, A.; Lisova, K.; Slavik, R.; Chatziioannou, A.F.; van Dam, R.M. High-Throughput Radio-TLC Analysis. *Nucl. Med. Biol.* **2020**, *82–83*, 41–48. [CrossRef] [PubMed]
41. Nyiredy, S. Planar Chromatographic Method Development Using the PRISMA Optimization System and Flow Charts. *J. Chromatogr. Sci.* **2002**, *40*, 553–563. [CrossRef] [PubMed]
42. Laferriere-Holloway, T.S.; Rios, A.; Lu, Y.; Okoro, C.C.; van Dam, R.M. A Rapid and Systematic Approach for the Optimization of Radio Thin-Layer Chromatography Resolution. *J. Chromatogr. A* **2022**, *463656*. [CrossRef]
43. Wang, J.; van Dam, R.M. Economical Production of Radiopharmaceuticals for Preclinical Imaging Using Microdroplet Radiochemistry. In *Biomedical Engineering Technologies: Volume 1*; Ossandon, M.R., Baker, H., Rasooly, A., Eds.; Methods in Molecular Biology; Springer: New York, NY, USA, 2022; pp. 813–828, ISBN 978-1-07-161803-5.
44. Holloway, T.; Rios, A.; Okoro, C.; van Dam, R.M. Replacing High-Performance Liquid Chromatography (HPLC) with High-Resolution Thin Layer Chromatography (TLC) for Rapid Radiopharmaceutical Analysis [ABSTRACT]. *Nucl. Med. Biol.* **2021**, *96–97*, S63. [CrossRef]
45. Dooraghi, A.A.; Keng, P.Y.; Chen, S.; Javed, M.R.; Kim, C.-J. "CJ"; Chatziioannou, A.F.; van Dam, R.M. Optimization of Microfluidic PET Tracer Synthesis with Cerenkov Imaging. *Analyst* **2013**, *138*, 5654–5664. [CrossRef]
46. King, E.J. The solubility of silica. *Lancet* **1938**, *231*, 1236–1238. [CrossRef]
47. Lisova, K.; Wang, J.; Hajagos, T.J.; Lu, Y.; Hsiao, A.; Elizarov, A.; van Dam, R.M. Economical Droplet-Based Microfluidic Production of [18F]FET and [18F]Florbetaben Suitable for Human Use. *Sci. Rep.* **2021**, *11*, 20636. [CrossRef] [PubMed]
48. Patt, M.; Schildan, A.; Barthel, H.; Becker, G.; Schultze-Mosgau, M.H.; Rohde, B.; Reininger, C.; Sabri, O. Metabolite Analysis of [18F]Florbetaben (BAY 94-9172) in Human Subjects: A Substudy within a Proof of Mechanism Clinical Trial. *J. Radioanal. Nucl. Chem.* **2010**, *284*, 557–562. [CrossRef]
49. Rominger, A.; Brendel, M.; Burgold, S.; Keppler, K.; Baumann, K.; Xiong, G.; Mille, E.; Gildehaus, F.-J.; Carlsen, J.; Schlichtiger, J.; et al. Longitudinal Assessment of Cerebral β-Amyloid Deposition in Mice Overexpressing Swedish Mutant β-Amyloid Precursor Protein Using 18F-Florbetaben PET. *J. Nucl. Med.* **2013**, *54*, 1127–1134. [CrossRef] [PubMed]
50. CAMAG®Automatic TLC Sampler 4 (ATS 4). Available online: https://www.camag.com/product/camag-automatic-tlc-sampler-4-ats-4 (accessed on 11 November 2022).
51. Arup, U.; Ekman, S.; Lindblom, L.; Mattsson, J.-E. High Performance Thin Layer Chromatography (HPTLC), an Improved Technique for Screening Lichen Substances. *Lichenologist* **1993**, *25*, 61–71. [CrossRef]
52. Tuzimski, T. Basic Principles of Planar Chromatography and Its Potential for Hyphenated Techniques. In *High-Performance Thin-Layer Chromatography (HPTLC)*; Srivastava, M., Ed.; Springer: Berlin/Heidelberg, Germany, 2011; pp. 247–310, ISBN 978-3-642-14025-9.
53. CAMAG® Horizontal Developing Chamber. Available online: https://www.camag.com/product/camag-horizontal-developing-chamber (accessed on 12 November 2022).
54. Hałka-Grysińska, A.; Dzido, T.H.; Sitarczyk, E.; Klimek-Turek, A.; Chomicki, A. A New Semiautomatic Device with Horizontal Developing Chamber for Gradient Thin-Layer Chromatography. *J. Liq. Chromatogr. Relat. Technol.* **2016**, *39*, 257–263. [CrossRef]
55. CAMAG® TLC-MS Interface 2. Available online: https://www.camag.com/product/camag-tlc-ms-interface-2 (accessed on 12 June 2020).
56. Plate Express Automated TLC Plate Reader-Advion X Interchim. Available online: https://www.advion.com/products/plate-express/ (accessed on 11 November 2022).
57. Läufer, K.; Lehmann, J.; Petry, S.; Scheuring, M.; Schmidt-Schuchardt, M. Simple, Inexpensive System for Using Thin-Layer Chromatography for Micro-Preparative Purposes. *J. Chromatogr. A* **1994**, *684*, 370–373. [CrossRef]
58. Pasilis, S.P.; Van Berkel, G.J. Atmospheric Pressure Surface Sampling/Ionization Techniques for Direct Coupling of Planar Separations with Mass Spectrometry. *J. Chromatogr. A* **2010**, *1217*, 3955–3965. [CrossRef]
59. Gerhardt, R.F.; Peretzki, A.J.; Piendl, S.K.; Belder, D. Seamless Combination of High-Pressure Chip-HPLC and Droplet Microfluidics on an Integrated Microfluidic Glass Chip. *Anal. Chem.* **2017**, *89*, 13030–13037. [CrossRef]
60. Kagan, I.A.; Flythe, M.D. Thin-Layer Chromatographic (TLC) Separations and Bioassays of Plant Extracts to Identify Antimicrobial Compounds. *J. Vis. Exp.* **2014**, 51411. [CrossRef]
61. Wang, M.; Gao, M.; Miller, K.D.; Zheng, Q.-H. Synthesis of [11C]PBR06 and [18F]PBR06 as Agents for Positron Emission Tomographic (PET) Imaging of the Translocator Protein (TSPO). *Steroids* **2011**, *76*, 1331–1340. [CrossRef]
62. Lartey, F.M.; Ahn, G.-O.; Shen, B.; Cord, K.-T.; Smith, T.; Chua, J.Y.; Rosenblum, S.; Liu, H.; James, M.L.; Chernikova, S.; et al. PET Imaging of Stroke-Induced Neuroinflammation in Mice Using [18F]PBR06. *Mol. Imaging Biol.* **2014**, *16*, 109–117. [CrossRef]

63. Kuntzsch, M.; Lamparter, D.; Bruggener, N.; Muller, M.; Kienzle, G.J.; Reischl, G. Development and Successful Validation of Simple and Fast TLC Spot Tests for Determination of Kryptofix® 2.2.2 and Tetrabutylammonium in 18F-Labeled Radiopharmaceuticals. *Pharmaceuticals* **2014**, *7*, 621–633. [CrossRef] [PubMed]
64. Tanzey, S.S.; Mossine, A.V.; Sowa, A.R.; Torres, J.; Brooks, A.F.; Sanford, M.S.; Scott, P.J.H. A Spot Test for Determination of Residual TBA Levels in 18F-Radiotracers for Human Use Using Dragendorff Reagent. *Anal. Methods* **2020**, *12*, 5004–5009. [CrossRef] [PubMed]

Article

# Preparation, Optimisation, and In Vitro Evaluation of [$^{18}$F]AlF-NOTA-Pamidronic Acid for Bone Imaging PET

Hishar Hassan [1,\*], Muhamad Faiz Othman [2,\*], Hairil Rashmizal Abdul Razak [3], Zainul Amiruddin Zakaria [4], Fathinul Fikri Ahmad Saad [1], Mohd Azuraidi Osman [5], Loh Hui Yi [5], Zarif Ashhar [6], Jaleezah Idris [6], Mohd Hamdi Noor Abdul Hamid [6] and Zaitulhusna M. Safee [6]

1 Centre for Diagnostic Nuclear Imaging, Universiti Putra Malaysia (UPM), Serdang 43400, Malaysia
2 Department of Pharmacy Practice, Faculty of Pharmacy, Universiti Teknologi MARA, Bandar Puncak Alam 42300, Malaysia
3 Medical Imaging Program, Department of Health and Care Professions, Faculty of Health and Life Sciences, St Luke's Campus, University of Exeter, Devon EX1 2LU, UK
4 Borneo Research on Algesia, Inflammation and Neurodegeneration (BRAIN) Group, Department of Biomedicine, Faculty of Medicine and Health Sciences, Universiti Malaysia Sabah, Kota Kinabalu 88400, Malaysia
5 Department of Cell and Molecular Biology, Faculty of Biotechnology and Biomolecular Sciences, Universiti Putra Malaysia (UPM), Serdang 43400, Malaysia
6 Department of Nuclear Medicine, National Cancer Institute, Putrajaya 62250, Malaysia
\* Correspondence: muhdhishar@upm.edu.my (H.H.); faiz371@uitm.edu.my (M.F.O.); Tel.: +60-3-9769-2498 (H.H.)

**Abstract:** [$^{18}$F]sodium fluoride ([$^{18}$F]NaF) is recognised to be superior to [$^{99m}$Tc]-methyl diphosphate ([$^{99m}$Tc]Tc-MDP) and *2-deoxy-2-[$^{18}$F]fluoro-D-glucose* ([$^{18}$F]FDG) in bone imaging. However, there is concern that [$^{18}$F]NaF uptake is not cancer-specific, leading to a higher number of false-positive interpretations. Therefore, in this work, [$^{18}$F]AlF-NOTA-pamidronic acid was prepared, optimised, and tested for its in vitro uptake. NOTA-pamidronic acid was prepared by an *N*-Hydroxysuccinimide (NHS) ester strategy and validated by liquid chromatography-mass spectrometry analysis (LC-MS/MS). Radiolabeling of [$^{18}$F]AlF-NOTA-pamidronic acid was optimised, and it was ensured that all quality control analysis requirements for the radiopharmaceuticals were met prior to the in vitro cell uptake studies. NOTA-pamidronic acid was successfully prepared and radiolabeled with $^{18}$F. The radiolabel was prepared in a 1:1 molar ratio of aluminium chloride (AlCl$_3$) to NOTA-pamidronic acid and heated at 100 °C for 15 min in the presence of 50% ethanol ($v/v$), which proved to be optimal. The preliminary in vitro results of the binding of the hydroxyapatite showed that [$^{18}$F]AlF-NOTA-pamidronic acid was as sensitive as [$^{18}$F]sodium fluoride ([$^{18}$F]NaF). Normal human osteoblast cell lines (hFOB 1.19) and human osteosarcoma cell lines (Saos-2) were used for the in vitro cellular uptake studies. It was found that [$^{18}$F]NaF was higher in both cell lines, but [$^{18}$F]AlF-NOTA-pamidronic acid showed promising cellular uptake in Saos-2. The preliminary results suggest that further preclinical studies of [$^{18}$F]AlF-NOTA-pamidronic acid are needed before it is transferred to clinical research.

**Keywords:** $^{18}$F; aluminium fluoride (Al-F); pamidronic acid; radiopharmaceuticals; radiochemistry; positron emission tomography; bone imaging

## 1. Introduction

Primary bone cancer is defined as cancer that originates in the bone itself. It can be benign (non-cancerous) or malignant (cancerous), the latter being less common than benign primary bone cancer [1]. However, both types of primary bone cancer are capable of growing and compressing healthy bone tissue. Bone metastases occur when cancer cells from primary cancer migrate into the bone. Osteosarcoma is a classic primary bone cancer characterised by the presence of malignant mesenchymal cells in the bone stroma [2]. The cancer is primary when it is localised, and there is no evidence that the malignant cells have spread outside the bone, and it is considered secondary (metastatic) when it has spread to distant parts of the body [3]. Cancer Research UK reported a survival rate of 40% for people with osteosarcoma who survived their cancer for 5 years among the population in England, while the American Cancer Society reported an average survival rate of 70% [4,5].

On the other hand, bone metastases are a common feature in patients with advanced prostate cancer, breast cancer, and lung cancer and remain the leading cause of death in advanced prostate cancer [6,7]. Bone metastases lead to complications such as severe pain, bone fractures, spinal cord compression, and bone marrow suppression [8,9]. Under these circumstances, the early detection of primary bone cancer and bone metastases is crucial for the prevention of skeletal-related events. For bone diagnostics, in particular, several imaging techniques have been investigated and compared in terms of their sensitivity and specificity [10]. European guidelines recommend cost-effective single-photon emission computed tomography (SPECT) using [$^{99m}$Tc]-methyl diphosphonate ([$^{99m}$Tc]Tc-MDP) for bone diagnosis. However, SPECT imaging has some weaknesses, particularly in quantifying the response to treatment [11]. A major drawback worth highlighting is the slow distribution and excretion of [$^{99m}$Tc]Tc-MDP, possibly due to the direct complexation of the radiometal [$^{99m}$Tc] and methyl diphosphonate (MDP) in the [$^{99m}$Tc]Tc-MDP complex. Furthermore, its specificity is limited as uptake is also observed in non-cancer cells [12,13]. This phenomenon could represent an increased risk for the under-staging and under-treatment of the disease [14]. Meanwhile, positron emission tomography (PET) has relied on *2-deoxy-2-[$^{18}$F]fluoro-D-glucose* (2-[$^{18}$F]FDG) for most oncological cases. However, 2-[$^{18}$F]FDG uptake was variable in blastic lesions, and cranial bone involvement was reportedly overlooked due to physiological brain metabolism [15]. By contrast, a new marker targeting C-X-C chemokine receptor type 4 (CXCR-4) expression, [$^{68}$Ga]Ga-Pentixafor, appears to be suitable only for the diagnosis of the chronic infection of the bone [16].

A multicentre study concluded that [$^{18}$F]sodium fluoride ([$^{18}$F]NaF) is superior to [$^{99m}$Tc]Tc-MDP and 2-[$^{18}$F]FDG in the diagnosis of bone metastases [15,17,18]. In this study, all [$^{18}$F]NaF PET/CT and [$^{99m}$Tc]Tc-MDP scans were positive for bone metastases, while 2-[$^{18}$F]FDG results were negative in some cases. Further cross-sectional imaging proved that [$^{18}$F]NaF PET/CT was advantageous over [$^{99m}$Tc]Tc-MDP scans, as some lesions that were missed on [$^{99m}$Tc]Tc-MDP scans were detected by [$^{18}$F]NaF PET/CT [15]. Nevertheless, there is concern that [$^{18}$F]NaF uptake is not cancer-specific, leading to a higher number of false-positive interpretations [19]. Currently, there are several efforts to identify early markers for bone imaging and new drug targets to improve the quality of life for patients with skeletal-related events caused by bone metastases. Therefore, there is a need for accurate imaging, proper staging, and assessment of the response to treatment and long-term oncological management, as this is directly related to patient morbidity and healthcare costs [19].

Bisphosphonates (BPs) are a group of drugs that were discovered back in the 1960s. They bind strongly to bone minerals, which gives them the unique property of selective uptake [20,21]. BPs bind strongly to bone minerals via calcium coordination in the hydroxyapatite lattice, which differs from the binding of fluoride in that it displaces the hydroxide in the hydroxyapatite lattice and converts it to fluorapatite [22,23]. Over the years, modifications to the structure of BPs have been studied to improve their pharmacodynamic behaviour, mainly to increase their bone-binding affinity. BPs are divided into two main groups: nitrogen-containing BPs (N-BPs) and non-nitrogen-containing BPs. Comparative studies have shown that the N-BPs form additional hydrogen bonds and have a higher bone-binding affinity [21]. The N-BPs include pamidronic acid, zoledronic acid, risedronic acid, alendronic acid, and ibandronic acid. Previous studies concluded that pamidronic acid has the highest bone-binding affinity, followed by alendronic acid, zoledronic acid, risedronic acid, and finally, ibandronic acid [24,25].

In recent years, the number of targeted radiopharmaceutical markers for bone imaging has increased. For example, recent developments using $^{68}$Ga-labeled bisphosphonates have been investigated for PET bone imaging [26–28]. However, there has been insufficient innovation in $^{18}$F radiopharmaceutical derivatives for bone imaging. Therefore, there is a need to develop an $^{18}$F radiopharmaceutical derivative that can be used more specifically for PET bone imaging. $^{18}$F can be produced in high yield in a modern medical cyclotron compared to $^{68}$Ga produced by generators with a limited activity per elution. Another important aspect is that $^{18}$F has a longer half-life, so it can be transported to remote hospitals without a cyclotron on site. One of the most important issues is that approved $^{68}$Ge/$^{68}$Ga generators have become very expensive and have long delivery times, which further weakens the rationale for $^{68}$Ga radiopharmaceuticals [29]. Therefore, replacing $^{68}$Ga with $^{18}$F can lead to a reliable supply with a significant cost reduction [30]. The formation of $^{18}$F linked to the bifunctional chelating agent was laborious until McBride et al., in 2009, developed an excellent method that exploited the fluorophilic nature of aluminium to allow its direct complexation with [$^{18}$F]F$^-$ and to form stable aluminium fluoride complexes ([$^{18}$F]AlF$^{2+}$) [31]. A bifunctional chelating agent can be covalently attached to a peptide, small protein, or compound of interest and seamlessly coordinated with [$^{18}$F]AlF$^{2+}$ complexes using this method. This novel approach leads to shorter reaction times, more efficient radiochemistry, and a better economic approach. It enables radiolabeling in aqueous media through a one-pot reaction [30,32,33]. Furthermore, the method also solves the problem associated with $^{68}$Ga complexation and classical carbon-$^{18}$F radiochemistry [34]. Therefore, it is now possible to label various targeting vectors that were previously labelled with the radiometal $^{68}$Ga or other radionuclides with $^{18}$F in a high yield without having to invest in an expensive $^{68}$Ge/$^{68}$Ga generator which a radiopharmacy centre that does not have one [35].

To find out whether $^{18}$F-labeled bisphosphonate has the same problem as [$^{18}$F]NaF in terms of non-cancer uptake, the present work aims to produce a new targeting vector by first forming the [$^{18}$F]AlF$^{2+}$, followed by coordinating the positively charged [$^{18}$F]AlF$^{2+}$ to the NOTA-pamidronic acid moiety and optimising the radiolabeling of [$^{18}$F]AlF-NOTA-pamidronic acid. Pamidronic acid was selected as a targeting vector for bone imaging because it has the highest bone-binding affinity compared to other bisphosphonates, as reported by Jahnke et al. [25]. Ultimately, this study also aims to demonstrate the cellular uptake of [$^{18}$F]AlF-NOTA-pamidronic acid through an in vitro study using normal human osteoblast cell lines (hFOB 1.19) and human osteosarcoma cell lines (Saos-2) to demonstrate the concept that [$^{18}$F]AlF-NOTA-pamidronic acid can potentially be used for PET bone imaging.

## 2. Results and Discussion

*2.1. Preparation, Validation, and Isolation of NOTA-Pamidronic Acid*

2.1.1. Preparation of NOTA-Pamidronic Acid

The preparation of the NOTA-pamidronic acid (Figure 1) involved the conjugation of NOTA chelator with pamidronic acid, which acted as a vector molecule, using the *N*-hydroxysuccinimide (NHS) ester strategy. The NOTA-NHS ester-activate reacted with the primary amines of the pamidronic acid to form stable amide bonds while NHS was released. The primary amine group (NH$_2$) was known to be an easy target for conjugation [36]. The conjugation was carried out at room temperature for 4 h by adding a NOTA-NHS ester to a pamidronic acid solution adjusted to pH 8 with TEA [37,38].

Although NOTA-NHS was prepared by dissolving in dimethylformamide (DMF), there was a possibility that hydrolysis might occur during the reaction due to the presence of water from the pamidronic acid solution. Consequently, the unconjugated NOTA-NHS was hydrolysed to form free NOTA. Therefore, the mass spectrometric analysis also detected the free NOTA (fragment) with an exact molecular weight of 302.136 g mol$^{-1}$ (Figure 2).

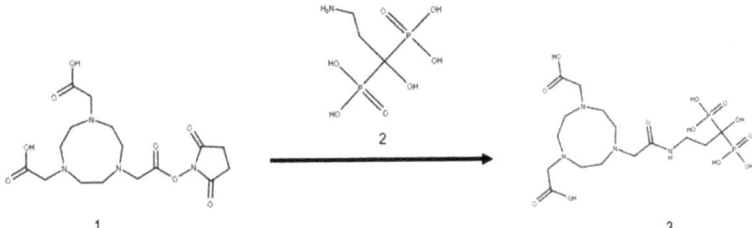

**Figure 1.** Conjugation of NOTA-NHS (1) and pamidronic acid (2) for the formation of NOTA-pamidronic acid (3).

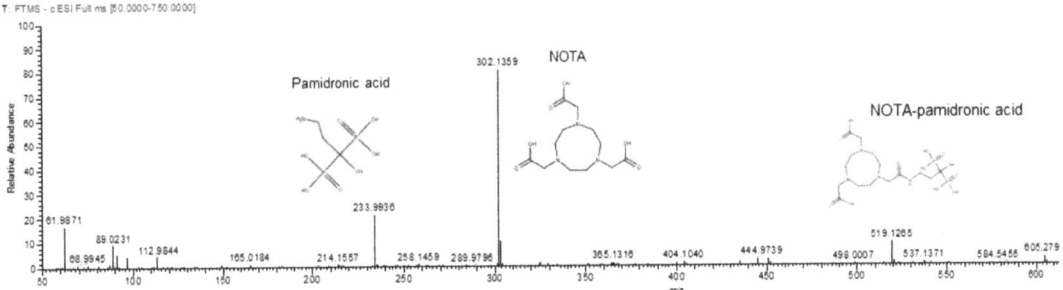

**Figure 2.** The mass spectrum of pamidronic acid, NOTA-pamidronic acid, and free NOTA producing m/z [M-H] of 234, 302, and 519 respectively.

A major reason for dissolving the NOTA-NHS ester chelator in an organic solvent, such as DMF in this case, is that the NHS esters in the NOTA-chelator are relatively insoluble in water and must first be dissolved in an organic solvent [39]. Furthermore, when a compound containing an NHS ester is dissolved in water, it immediately begins to hydrolyse, which can reduce the yield of the NOTA-pamidronic acid [40]. The conditions (pH 8 and room temperature) for this reaction appeared to be sufficient to allow for conjugation. In general, a reasonable physiological to basic pH was sufficient for the reaction to take place.

2.1.2. Validation of NOTA-Pamidronic Acid Using LC-MS Analysis

The validation of NOTA-pamidronic acid from conjugation was performed using liquid chromatography-mass spectrometry (LC-MS) analysis. Liquid chromatography conditions for mass spectrometry analysis were performed according to the ion suppression reversed-phase chromatography. Figure 3 shows that pamidronic acid (m.w. 233.9936), NOTA-pamidronic acid (m.w. 519.1269), and free NOTA (m.w. 302.1359) eluted at the retention times (Rt) of 0.36, 0.58, and 0.71 min, respectively. The result of the chromatogram in Figure 2 confirms that the difference in polarity between these three compounds, with pamidronic acid being the most polar and free NOTA being the least polar, results in the pamidronic acid eluting first, followed by NOTA-pamidronic acid and free NOTA last.

With a molar ratio of 5:1 (Pamidronic acid: NOTA-NHS), a yield of 24.13% was obtained. The figure explains how about 24.13% of the pamidronic acid with an initial weight of 5.875 mg was theoretically converted into 1.418 mg of NOTA-pamidronic acid.

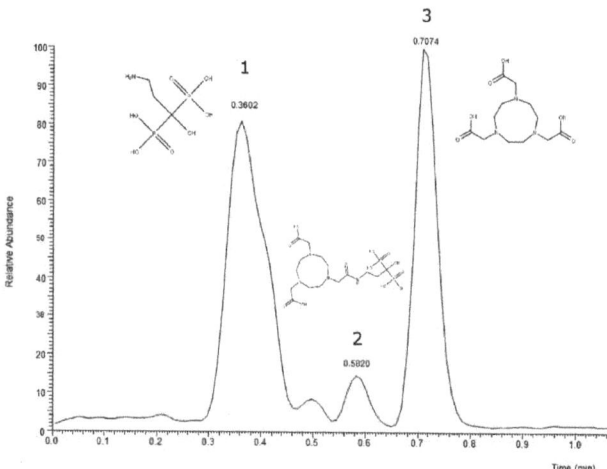

**Figure 3.** LC-MS chromatogram of crude NOTA-pamidronic acid sample. (1) pamidronic acid; (2) NOTA-pamidronic acid; (3) free NOTA.

Table 1 shows how as the concentration of pamidronic acid increased, the amount of NOTA-pamidronic acid also increased. This shows that there was a strong relationship between the percentage of NOTA-pamidronic acid yield and the ratio of pamidronic acid: NOTA-NHS. This supports Hermanson et al., who identified an optimised product yield when the ratio of target molecules to the NHS crosslinker was increased [39]. When the concentration of pamidronic acid (the target molecule) was increased, the amount of NOTA-NHS coupled to the pamidronic acid certainly increased. Unfortunately, the amount of non-conjugated pamidronic acid (free pamidronic acid) also increased, as observed in the peak area for non-conjugated pamidronic acid. This resulted in a decrease in the percentage yield of NOTA-pamidronic when pamidronic acid was included in the calculation of the percentage yield of NOTA-pamidronic acid.

**Table 1.** Percentage yield of NOTA-pamidronic acid.

| Pamidronic Acid: NOTA Molar Ratio | Peak Area (Average, $n = 3$) | | % Yield |
|---|---|---|---|
| | Free Pamidronic Acid | NOTA-Pamidronic Acid | |
| 5:1 | 3,409,680,392 | 1,084,594,404 | 24.13 |
| 10:1 | 5,242,116,420 | 1,155,377,273 | 18.06 |
| 15:1 | 8,343,559,233 | 1,243,259,017 | 12.97 |

A statistical analysis of one-way ANOVA was performed and confirmed that there was a statistical difference between the percentage yield of NOTA-pamidronic acid when the molar ratio of pamidronic acid and NOTA was varied ($p < 0.05$). Furthermore, Bonferroni posthoc analysis was performed to determine the molar ratio of the pamidronic acid to NOTA-NHS, which demonstrated a significant difference in the percentage yield of NOTA-pamidronic acid. The Bonferroni post hoc analysis revealed that changing the molar ratio of pamidronic acid to NOTA-NHS produced a significant difference in the percentage yield of NOTA-pamidronic acid across the group.

Another viable alternative to maximise the amount of NOTA-pamidronic acid produced in the reaction was to increase the molar excess of NOTA-NHS over the pamidronic acid. However, an excess of the NOTA compound (free NOTA) would remain, potentially competing with the NOTA-pamidronic acid for [$^{18}$F]AlF$^{2+}$ coordination during radiolabeling. Therefore, the purification of the NOTA-pamidronic acid crude sample is essential to remove the free NOTA as much as possible from the crude sample.

The possibility of increasing the amount of NOTA-pamidronic acid precursor produced in the reaction was also explored by using dimethyl sulfoxide (DMSO) as an organic solvent, which is more polar than DMF. The experiment was repeated with DMSO to dissolve the NOTA-NHS before pamidronic acid was added at a molar ratio of 5:1. However, the yield of the NOTA-pamidronic acid preparation was only highest at 12.95% when DMSO was used. A statistical analysis of the independent samples $t$-test was performed and confirmed that there was a statistical difference between the percentage yield of NOTA-pamidronic acid when using DMF and DMSO as an organic solvent ($p < 0.05$). An interesting point about using DMSO was that it was more difficult to eliminate DMSO during the post-reaction than DMF. This is consistent with the findings of other researchers who have also used DMSO as an organic solvent, as DMSO takes longer to dry and is more difficult to eliminate [40]. The study, therefore, recommends using DMF as an organic solvent during conjugation.

2.1.3. Mass Spectrometry Analysis of NOTA-Pamidronic Acid

The important piece of information from the mass spectrometry analysis was the mass-to-charge ratio ($m/z$) value of the conjugated product. The $m/z$ value of the conjugated product is crucial to determine whether the preparation of the NOTA-pamidronic acid precursor was successful or not. Most importantly, it can be difficult to independently characterise the conjugated product using analytical liquid chromatography alone, as there is no non-commercially available NOTA-pamidronic acid reference standard.

Since the ionisation mode of electrospray ionisation (ESI) in this experiment was set to a negative mode, the $m/z$ value of the NOTA-pamidronic acid ion was one proton lower due to the abstraction of a proton [M-H]$^-$. The negative ion mode was preferred over the positive ion mode in this experiment because of the presence of carboxyl groups [41]. Therefore, the precursor ion was expected to be [M-H]$^-$ due to deprotonation from the molecular formula.

Table 2 shows that the [M-H]$^-$ $m/z$ values for pamidronic acid (A), NOTA-pamidronic acid (B), and free NOTA (C) were 233.9934, 519.1265, and 302.1358, respectively. The relative error of the obtained $m/z$ values was less than 1 ppm. The relative error for all three compounds was below the acceptable limit of between 2 and 5 ppm, especially for an Orbitrap mass analyser [42].

Table 2. The calculated $m/z$ and obtained $m/z$ of pamidronic acid, NOTA-pamidronic acid, and NOTA by product; ESI negative mode $m/z$ vale [M-H]$^-$ ($n = 3$).

| Compound | [M-H]$^-$ Calculated $m/z$ | [M-H]$^-$ Obtained $m/z$ | Relative Error (ppm) |
| --- | --- | --- | --- |
| Pamidronic acid | 233.9932 | 233.9934 | 0.9972 |
| NOTA-pamidronic acid | 519.1263 | 519.1265 | 0.4495 |
| NOTA | 302.1358 | 302.1358 | 0.0000 |

The evidence confirmed that the samples prepared from the conjugation of pamidronic acid and NOTA-NHS, yielded a NOTA-pamidronic acid precursor, as the obtained $m/z$ (m.w. 519.1265) was similar to the calculated $m/z$ (m.w. 519.1263) (Figure 2).

2.1.4. Fragmentation Analysis of NOTA-Pamidronic Acid

Another important piece of information obtained from the MS–MS analysis was the fragment ions. In the MS–MS analysis, the selected precursor ions of NOTA-pamidronic acid (m.w. 519.1265) were broken down into fragments (product ions). In this study, the MS–MS analysis was used to ensure that the conjugated sample was NOTA-pamidronic acid. For this purpose, the spectral analysis of the NOTA-pamidronic acid fragment (product) ions was compared with the predicted spectra of the fragment ions obtained from the competitive fragmentation modelling for metabolite identification (CFM-ID) (http://cfmid.wishartlab.com, accessed on 25 July 2022) (Table 3).

Table 3. The relative error (ppm) and RDBE for each fragment produced from the MS–MS analysis.

| Obtained m/z | Exact m/z (Predicted) | Relative Error (ppm) | RDBE | Molecular Formula |
|---|---|---|---|---|
| 519.1265 | 519.1263 | 0.3853 | 4.5 | $C_{15}H_{29}N_4O_{12}P_2$ |
| 501.1159 | 501.1157 | 0.3991 | 5.5 | $C_{15}H_{27}N_4O_{11}P_2$ |
| 437.1442 | 437.1443 | 0.2288 | 5.5 | $C_{15}H_{26}N_4O_9P_1$ |
| 393.1546 | 393.1545 | 0.2544 | 4.5 | $C_{14}H_{26}N_4O_7P_1$ |
| 283.1772 | 283.1776 | 1.4125 | 4.5 | $C_{13}H_{23}N_4O_3$ |
| 152.0108 | 152.0118 | 6.5784 | 1.5 | $C_3H_7NO_4P_1$ |
| 142.9294 | 142.9299 | 3.4982 | 1.5 | $H_1O_5P_2$ |
| 134.9841 | 134.9847 | 4.4449 | 2.5 | $C_3H_4O_4P_1$ |

The elemental composition for 519 was determined to be 15 carbon (C), 29 hydrogens (H), 4 nitrogens (N), 12 oxygens (O), and 2 phosphorus (P). Table 3 shows the ring plus double bond equivalence (RDBE) of the NOTA-pamidronic acid precursor ion ($C_{15}H_{29}N_4O_{12}P_2$) was 4.5. This shows that the structure of NOTA-pamidronic acid contains one ring and three double bonds (carbonyl). The value of the RDBE of NOTA-pamidronic acid shows that this precursor ion is an even-electron ion (EE), which is consistent with the nitrogen rule (N Rule) that an odd-numbered precursor ion (RDBE value of 4.5) would have an even number of nitrogens (4 nitrogens) for an EE.

The neutral loss observed in the MS–MS analysis was 18, 64, 82, and 44, derived from $H_2O$, $HPO_2$, $H_3PO_3$, and $CO_2$, respectively. Based on the fragments generated (Figure 4), the primary fragment ions were $m/z$ 501 [M-H-$H_2O$]$^-$, 437 [M-H-$H_2O$-$HPO_2$]$^-$/[M-H-$H_3PO_3$]$^-$, and 393 [M-H-$CO_2$]$^-$/[M-H-$H_3PO_3$-$CO_2$]$^-$. The ESI-MS-MS produced two fragment ions, $m/z$ 501 and 437, through the neutral loss of $H_2O$ and $H_3PO_3$, respectively. The $m/z$ 501 was produced by the -OH dehydration of two phosphorus groups forming a four-membered ring. The fragment ion of $m/z$ 437 resembled [M-H-$H_3PO_3$]$^-$. The ESI-MS-MS of $m/z$ 437 produced three fragment ions, observed at $m/z$ 393, 152, and 135. The $m/z$ 393 was produced by the neutral loss of carbon dioxide [M-H-$H_3PO_3$-$CO_2$]$^-$. The ESI-MS-MS of $m/z$ 501 produced a fragment ion of 143. The $m/z$ 143 in the lower mass series was found to be an identical fragment ion for the compounds with a bisphosphonate group (Figure 5). Figure 4 shows that the base peak was at an $m/z$ value of 437.

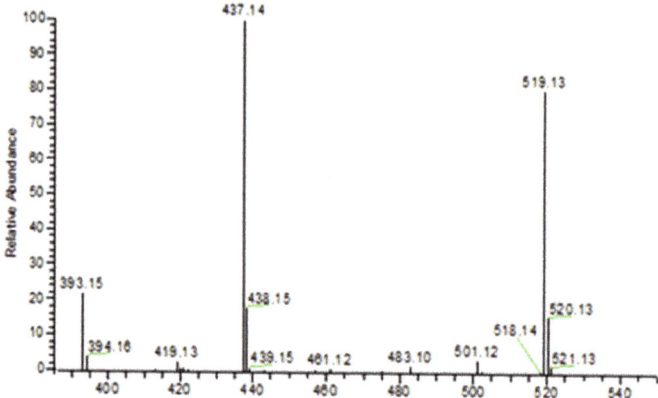

Figure 4. ESI (- negative) mass spectrum of NOTA-pamidronic acid producing 519 [M-H]$^-$ precursor ion and 437 [M-H-$H_3PO_3$]$^-$ base peak.

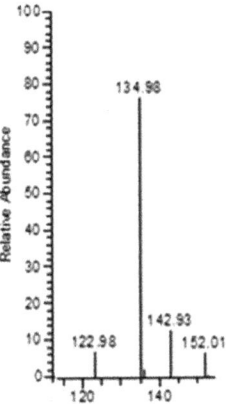

**Figure 5.** Diagnostic ions for BPs were observed at $m/z$ 135, 143, and 152. The ion observed at $m/z$ 143 indicates the presence of a BP group, such as pamidronic acid in this case.

The observed base peak with the $m/z$ value of 437, corresponds to two possible pathways (Figure 6) for the production of the fragment ion $m/z$ 437: [M-H-H$_2$O-HPO$_2$]$^-$ /[M-H-H$_3$PO$_3$]$^-$ [43]. The abundance of the fragment ion $m/z$ 437 was due to the stability and low proton affinity of the neutral loss (Field's rule). The increase in the unsaturation of $m/z$ 437 based on an RDBE value of 5.5 reflects this increased stability.

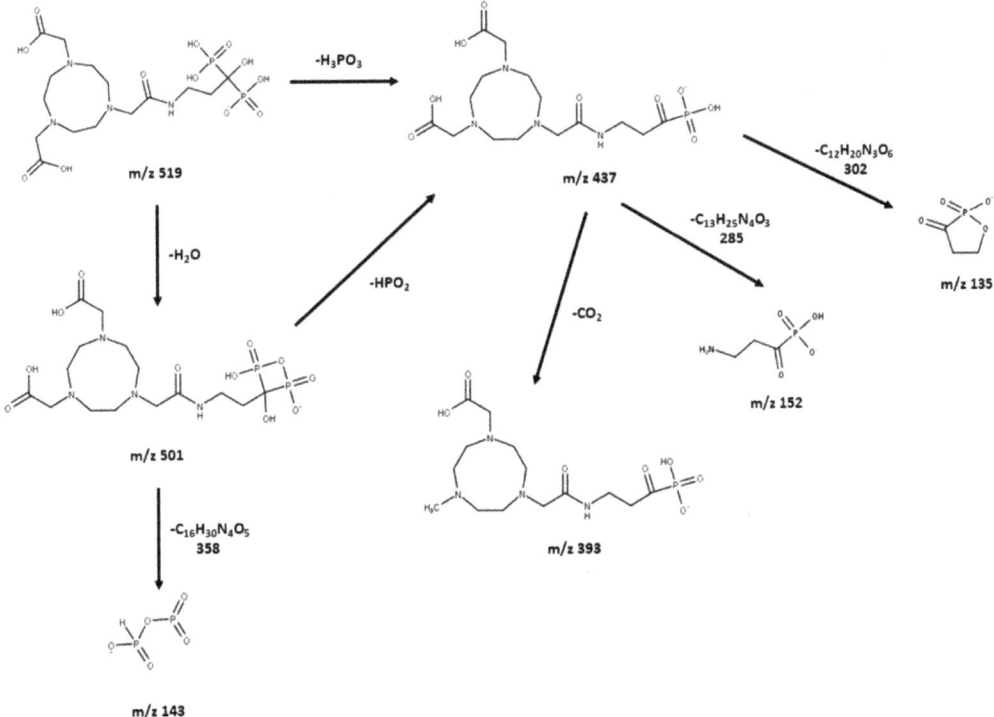

**Figure 6.** Proposed (−) ESI-MS fragmentation of NOTE-pamidronic acid.

The $m/z$ 501 and 437 fragment ions produced characteristic ions in a lower mass series. Three fragment ions with $m/z$ values of 135, 143, and 152 were observed in the lower mass series (Figure 5). The characteristic ion observed, in particular, at an $m/z$ of 143 indicates the presence of a BP group, such as pamidronic acid, in this case, which is consistent with other findings [43]. The fragment ions produced are consistent with the EE rule, which favours the heterolytic process via the charge retention fragmentation (CRF) pathway [44].

2.1.5. Isolation of the NOTA-Pamidronic Acid Fraction from the Crude Sample

The peak corresponding to NOTA-pamidronic acid (Rt: 5.28 min, $SD$ = 0.24 min) was collected using reverse-phase high-performance liquid chromatography (RP-HPLC) (Agilent 1200, USA) equipped with a fraction collector. The isolation of the NOTA-pamidronic acid fractions was carried out under optimal chromatographic conditions as described in Section 3.2.3. The collected NOTA-pamidronic acid fractions were re-analysed using the analytical method RP-HPLC. The purity of the NOTA-pamidronic acid was determined by the peak area of the isolated NOTA-pamidronic acid fractions compared to other peaks of the RP-HPLC chromatogram. Figure 7 below shows the chromatogram of the NOTA-PAM fractions in which NOTA-pamidronic acid was detected at the highest peak of number 4 (Rt: 5.15 min). This shows that the collected NOTA-pamidronic acid fractions had the highest purity. Peak numbers 1, 2, and 3 were detected as unknown impurities. Free NOTA was not detectable in the isolated NOTA-pamidronic acid fractions because the peak at an Rt of 7.56 min ($SD$ = 0.08), corresponding to free NOTA, was not present.

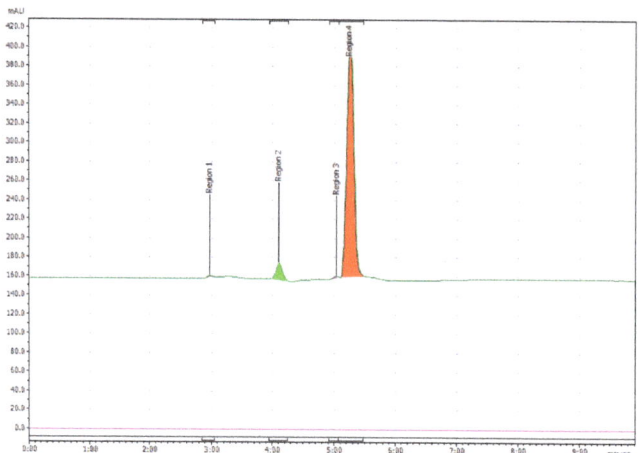

**Figure 7.** Chromatogram of UV 220 nm of pure NOTA-pamidronic acid post isolation.

The purity of the isolated NOTA-pamidronic acid fractions was 92.2% ($SD$ = 1.9, $n$ = 3) with an observed molecular mass of 519.1265 ± 0.0004 (theoretical molecular mass: 519.1263). The collected NOTA-pamidronic acid fractions were then sent for freeze-drying to improve their stability and prevent the possible degradation of the final product by removing water from the final product. The freeze-dried NOTA-pamidronic acid was then stored in the freezer at −20 °C.

2.2. *Optimisation of [$^{18}$F]AlF-NOTA-Pamidronic Acid Radiolabeling Conditions*

The optimisation of the [$^{18}$F]AlF-NOTA-pamidronic acid radiolabeling conditions was carried out in two stages. In the first stage, the optimisation of the radiolabeling conditions for [$^{18}$F]AlF$^{2+}$ complexation was evaluated by examining the AlCl$_3$ concentration (3.2.4.1. the preparation of [$^{18}$F]AlF$^{2+}$ complexes). The RCY of the [$^{18}$F]AlF$^{2+}$ complexes was determined from an aliquot of the reaction solution and evaluated with a Sep-Pak cartridge combination. In the second stage, the radiolabeling conditions for the [$^{18}$F]AlF-NOTA complexation were then optimised using a non-conjugated NOTA-NHS chelator before using a NOTA-pamidronic acid precursor. In attempting to optimise the radiolabeling conditions for the [$^{18}$F]AlF-NOTA complexation, four variables were identified: (1) the molar ratio of AlCl$_3$ to the NOTA-pamidronic acid, (2) the reaction time, (3) the reaction temperature, and (4) the co-solvent that would potentially affect the formation of [$^{18}$F]AlF-NOTA-pamidronic acid complexes.

### 2.2.1. [$^{18}$F]F$^-$ Activity

About 83–95% of the $^{18}$F radioactivity was eluted from fractions two to three. The fractionation technique was able to concentrate 113 MBq of [$^{18}$F]F$^-$ in a volume of 200 µL and minimise contamination (with metallic impurities) due to the smaller volume of the eluate. Some short-lived radionuclides, such as nitrogen-13 ($^{13}$N) and oxygen-15 ($^{15}$O), were among the most likely non-metallic radionuclide contaminants in the aqueous $^{18}$F solution. However, both have extremely short half-lives, of about 10 min and 2 min for $^{13}$N and $^{15}$O, respectively, and would decay naturally during transport (in our case) or before radiolabeling began.

The potential $^{18}$F contaminants varied depending on the target system used in the cyclotron. A niobium target with a Havar foil released mainly manganese (Mn), cobalt (Co), and technetium (Tc) species, which are normally trapped in the Sep-Pak QMA cartridge [45]. This shows that impurities in an aqueous $^{18}$F solution freshly produced from a cyclotron can be removed by solid phase extraction and fractionation before using the aluminium-fluoride (Al-F) technique [46].

Preliminary results showed that the formation of [$^{18}$F]AlF$^{2+}$ was significantly lower when the solid phase extraction and fractional elution were not performed. It was suggested that this was due to other ionic impurities in the aqueous $^{18}$F solutions, such as iron (II) ion (Fe$^{2+}$), copper (II) ion (Cu$^{2+}$), zinc (II) ion (Zn$^{2+}$), ammonium (NH$^{4+}$), which could compete with Al$^{3+}$ in the formation of [$^{18}$F]AlF$^{2+}$. However, the evaluation of radionuclide $^{18}$F in the aqueous solution was not considered in this study. Therefore, it was recommended to perform the solid phase extraction (SPE) and fractional elution and to use the highest fractional [$^{18}$F]F$^-$ activity for the formation of [$^{18}$F]AlF$^{2+}$.

### 2.2.2. Effect of AlCl$_3$ Concentration on the Radiochemical Yield (RCY) of [$^{18}$F]AlF$^{2+}$ Complexes

[$^{18}$F]AlF$^{2+}$ complexes were formed by the reaction of AlCl$_3$ with [$^{18}$F]F$^-$ anions in an aqueous solution between pH 4 and 5. Although 20 mM AlCl$_3$ resulted in the highest formation of [$^{18}$F]AlF$^{2+}$ complexes, 99.94% ($SD = 0.1$, $n = 3$), further statistical analysis of the independent $t$-test revealed that increasing the AlCl$_3$ concentration did not significantly increase the RCY of the [$^{18}$F]AlF$^{2+}$ complexes ($p > 0.05$). The difference in RCY of the [$^{18}$F]AlF$^{2+}$ complexes was too small. While 5 and 20 mM AlCl$_3$ resulted in higher [$^{18}$F]AlF$^{2+}$ complexes ($M = 99.92\%$, $SD = 0.1$ for 5 mM), higher amounts of AlCl$_3$ were required to achieve the exact complexation yield with 2 mM AlCl$_3$ ($M = 99.88\%$, $SD = 0.2$, $n = 3$). Since the results show that a lower concentration of AlCl$_3$ (2 mM) can produce comparable [$^{18}$F]AlF$^{2+}$ complexes to a higher concentration of AlCl$_3$ (5 mM and 20 mM), the use of AlCl$_3$ at a lower concentration is, therefore, more optimal. Therefore, 2 mM AlCl$_3$ was used for the [$^{18}$F]AlF$^{2+}$ complexation.

### 2.2.3. Effect of Reaction Temperature and Time on the Formation of the [$^{18}$F]AlF-NOTA-NHS Complex

The radiolabeling conditions were first optimised using a complexation assay with the inexpensive NOTA NHS chelator. Radiolabeling conditions were set at three different temperatures (60, 80, and 100 °C) and reaction times (5, 10, and 15 min) to investigate the effects of the reaction temperature and time on the RCY of the [$^{18}$F]AlF-NOTA-NHS complex. The RCY of the [$^{18}$F]AlF-NOTA NHS complex was proportional to the increase in the reaction temperature. The [$^{18}$F]AlF-NOTA NHS complex had the highest RCY, 94.6%

(SD 0.9) when the reaction temperature and time was 100 °C for 15 min and the lowest when the reaction temperature was 60 °C ($M = 48.5\%$, $SD = 36.9$). These results were consistent as the NOTA chelator needed to be heated to 100–120 °C [47]. Since the NOTA chelator is cyclic, the activation energies for chelating the metal ions are significantly higher than for the linear chelator [48]. To overcome these considerable kinetic barriers in the radiolabeling of NOTA conjugates with [$^{18}$F]AlF$^{2+}$, reaction solutions were heated to 100–120 °C [48].

Nevertheless, it has been pointed out that these reaction conditions are problematic when the chelator is conjugated with heat-sensitive biomolecules [47,49]. Further statistical analysis of the one-way ANOVA revealed that the reaction time between 5 and 15 min had no significant effect on the RCY of the [$^{18}$F]AlF-NOTA NHS complex when heated at 100 °C ($p > 0.05$). The RCY of the [$^{18}$F]AlF-NOTA NHS complex was above 90% in all experiments with reaction times. Therefore, a viable strategy for a chelator conjugated to heat-sensitive biomolecules is a short reaction time: between 5 and 10 min at 100 °C.

2.2.4. Effect of Organic Solvent and Percentage of Organic Solvent

Previous results showed that the formation of the [$^{18}$F]AlF-NOTA NHS complex was first and second highest when acetonitrile 70% ($v/v$) or ethanol 50% ($v/v$) were added as organic solvents [50]. To ensure reproducibility, the radiolabeling conditions of the NOTA-pamidronic acid precursor were repeated using acetonitrile 70% ($v/v$) and ethanol 50% ($v/v$) as organic solvents. The formation of [$^{18}$F]AlF-NOTA-pamidronic acid from both organic solvents was determined by the radio-thin layer chromatography technique (r-TLC). In contrast to the previous results, the formation of [$^{18}$F]AlF-NOTA-pamidronic acid was highest ($M = 95.50\%$, $SD = 5.34$) when ethanol 50% ($v/v$) was used as an organic solvent ($n = 6$) (Figure 8). The formation of [$^{18}$F]AlF-NOTA-pamidronic acid, determined by r-TLC, was only 75.55% ($SD = 2.21$, $n = 6$) when acetonitrile 70% ($v/v$) was added to the reaction mixture (Figure 8).

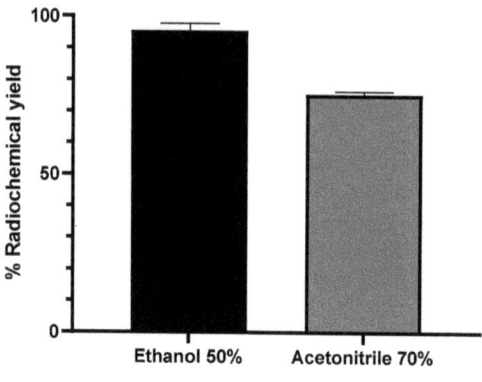

**Figure 8.** RCY of [$^{18}$F]AlF-NOTA-pamidronic acid when different organic solvents were added to the reaction mixture ($n = 6$); mean (SEM).

Further statistical analysis of the independent samples $t$-test showed that the difference in the formation of [$^{18}$F]AlF-NOTA-pamidronic acid was significant when ethanol 50% ($v/v$) was added to the reaction mixture ($p < 0.05$). There is a possible explanation that could justify the higher formation of [$^{18}$F]AlF-NOTA-pamidronic acid when ethanol was used compared to acetonitrile. The presence of ethanol facilitates the interaction of metal cations, [$^{18}$F]AlF$^{2+}$, with donors in the chelate structure, such as, in this case, NOTA-pamidronic acid moiety and, therefore, leads to a higher forms of [$^{18}$F]AlF-NOTA-pamidronic acid [38].

The results showed that the use of acetonitrile was only effective in coordinating [$^{18}$F]F$^-$ with the Al$^{3+}$ to form [$^{18}$F]AlF$^{2+}$ in the previous suggestion. One could speculate that the difference in the structure of NOTA-pamidronic acid compared to NOTA-NHS could contribute to the higher formation of [$^{18}$F]AlF-NOTA-pamidronic acid when ethanol 50% (v/v) was used. Since the formation of [$^{18}$F]AlF-NOTA-pamidronic acid was higher when ethanol 50% (v/v) was used, ethanol was chosen as the organic solvent. Furthermore, ethanol was the most biocompatible of all the solvents [51].

2.2.5. The Optimal Ratio between AlCl$_3$ and NOTA-Pamidronic Acid

Based on the preliminary results [50], the formation of the [$^{18}$F]AlF-NOTA NHS complex was above 80% for all AlCl$_3$-to-NOTA molar ratios prepared. Nevertheless, a statistical analysis of one-way ANOVA revealed that the difference in the percentage of the RCY (formation) of the [$^{18}$F]AlF-NOTA NHS complex between the prepared molar ratios of AlCl$_3$-to-NOTA was insignificant ($p > 0.05$). The results also showed that the RCY of the [$^{18}$F]AlF-NOTA NHS complex decreased when the molar ratio exceeded 1:5. In view of this, the molar ratios of 1:1, 1:3, and 1:5 AlCl$_3$:NOTA-pamidronic acid were chosen. The RCY of [$^{18}$F]AlF-NOTA-pamidronic acid, determined by the radio-TLC scanner using ITLC-SG strips as adsorbents and a mobile phase of 1M ammonium acetate and acetonitrile (1:1), was above 60% for all the molar ratios prepared (Figure 9) and, thus, met the requirement to prepare [$^{18}$F]AlF-NOTA-pamidronic acid with an acceptable RCY between 40 and 60%. The RCY of [$^{18}$F]AlF-NOTA-pamidronic acid was highest ($M = 95.50\%$, $SD = 5.34$) when the NOTA-pamidronic acid was prepared at a molar ratio of 1:1 with AlCl$_3$ ($n = 6$). By contrast, increasing the molar ratio of the NOTA-pamidronic acid precursor to 1:3 and 1:5 did not further increase the formation of [$^{18}$F]AlF-NOTA-pamidronic acid. This result was expected as the presence of excess chelator reduces RCY, as, in this case, it did with the NOTA-pamidronic acid [38].

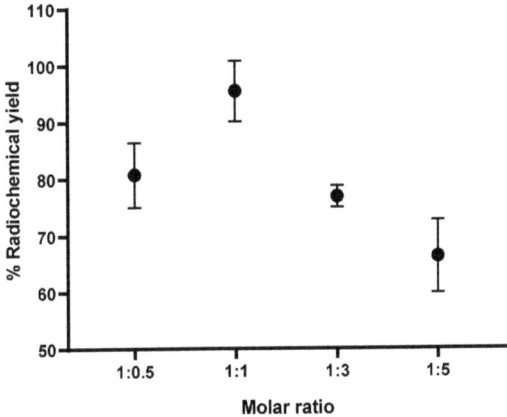

Figure 9. RCY of [$^{18}$F]AlF-NOTA-pamidronic acid ($n = 6$).

Interestingly, the formation of [$^{18}$F]AlF-NOTA-pamidronic acid was the second highest ($M = 80.74\%$, $SD = 5.73$) when the molar ratio of the NOTA-pamidronic acid precursor was reduced to half ($n = 6$). This result suggests that even at a lower molar ratio of AlCl$_3$-to-NOTA-pamidronic acid (2 μmol AlCl$_3$: 1 μmol NOTA-pamidronic acid), an RCY of more than 60% can be successfully achieved. This could bring economic advantages later on when upscaling the radiolabeling of NOTA-pamidronic acid, as production costs could be reduced due to the lower amount of NOTA-pamidronic acid precursor.

Further statistical analysis of the one-way ANOVA revealed that the difference in the percentage of the RCY (formation) of [$^{18}$F]AlF-NOTA-pamidronic acid between the prepared molar ratios of AlCl$_3$-to-NOTA-pamidronic acid was significant ($p < 0.05$). The Games–Howell post hoc analysis of the percentage of RCY of [$^{18}$F]AlF-NOTA-pamidronic acid revealed that increasing the molar ratio of AlCl$_3$-to-NOTA-pamidronic acid beyond

1:1 significantly reduced the percentage of RCY ($p < 0.05$). The difference in the percentage RCY of [$^{18}$F]AlF-NOTA-pamidronic acid was insignificant when the molar ratio of 1:0.5 was compared to the molar ratio of 1:3, indicating that although 1 µmol (molar ratio of 1:0.5) and 6 µmol (molar ratio of 1:3) of the NOTA-pamidronic acid precursor could achieve a yield of more than 60%, a higher amount of the NOTA-pamidronic acid precursor was required (6 µmol) to approach the exact [$^{18}$F]AlF-NOTA-pamidronic acid complexation with a molar ratio of 1:0.5 (1 µmol). The optimal radiolabeling conditions (Figure 10) that could produce [$^{18}$F]AlF-NOTA-pamidronic acid complexation in more than 90% RCY were as follows (Table 4).

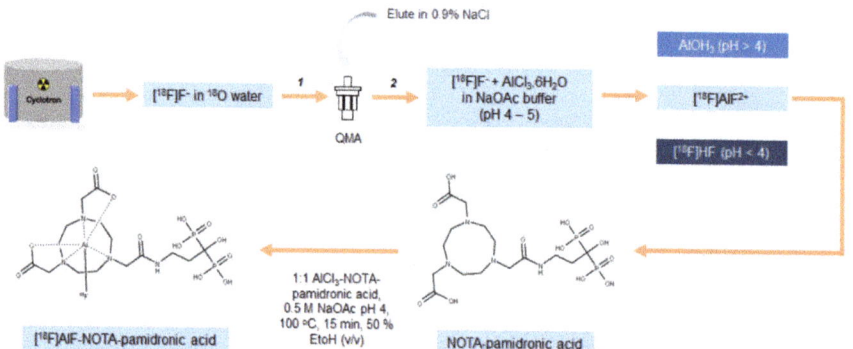

**Figure 10.** Schematic representation of the process for optimal radiolabeling conditions for [$^{18}$F]AlF-NOTA-pamidronic acid (1: purification; 2: fractionation).

**Table 4.** Optimal radiolabeling conditions.

| Variables | Optimal Conditions |
| --- | --- |
| AlCl$_3$ concentration | 2 mM |
| AlCl$_3$-to-NOTA-pamidronic acid molar ratio | 1:1 (2 µmol NOTA-pamidronic acid) |
| Reaction temperature | 100 °C |
| Reaction time | 15 min |
| Organic solvent | Ethanol 50% ($v/v$) |

Our experimental results also confirmed that QMA-bound [$^{18}$F]F$^-$ could be eluted with 0.9% saline without the need for pH adjustment by eluting with 0.4 M KHCO$_3$ followed by acetic acid [52,53]. In the early phase of the radiolabeling studies, the QMA-bound [$^{18}$F]F$^-$ was eluted with 0.4 M KHCO$_3$, and the KHCO$_3$ was subsequently neutralised with acetic acid [52]. Although a one-step labeling approach (one-pot approach) was used in most previous applications, a two-step labeling approach was used in the present study: preparing the NOTA-pamidronic acid precursor in its pure form and using the NOTA-pamidronic acid as a precursor for the labeling step with aqueous [$^{18}$F]AlF$^{2+}$. The two-step labeling approach used in the experimental design differed slightly from that of Vogg et al. and D'Souza et al., as their approach first involved the formation of the purified [Al(OH)(NODA)] complex or peptide-aluminium complex and then the ligand exchange of [OH]$^-$ for [$^{18}$F]F$^-$ [30,52]. Vogg emphasised that even with the correct pH and ethanol in the aqueous reaction medium, they could not achieve an RCY above 80% [30]. They pointed out that heating to 100 °C may have led to the thermal stability of the [Al(OH)(NODA)] complex in the presence of [$^{18}$F]F$^-$ during labeling, releasing the Al$^{3+}$. Al$^{3+}$ then readily combined with the [$^{18}$F]F$^-$ to form [$^{18}$F]AlF$^{2+}$, which, in turn, formed a low concentration of Al-free NODA complexes for reaction. Therefore, they assumed that the addition of the metal-free NODA, right from the beginning of the second labeling step, could have increased the yield to over 80% [30]. This could be the reason why an RCY of over 90% was achieved for the [$^{18}$F]AlF-NOTA-pamidronic acid in this experiment.

In our experiment, the choice of a two-step labeling approach proved to be very applicable, although pamidronic acid could survive at higher temperatures during labeling. The metal-free NOTA-pamidronic acid was prepared in its purest form before the labeling step, with the aqueous [$^{18}$F]AlF$^{2+}$ formed in the first labeling step. The prepared NOTA-pamidronic acid was isolated from the crude sample, leaving the free NOTA. Free NOTA with an $N_3O_2$ donor proved to be the most stable [$^{18}$F]AlF efficiently at higher temperatures (100–120 °C) and could interfere with the radiolabeling of NOTA-pamidronic acid if present. Therefore, instead of [$^{18}$F]AlF-pamidronic acid, [$^{18}$F]AlF-free NOTA moieties could also be formed.

In summary, the radiolabeling of NOTA-pamidronic acid using an aluminium-fluoride technique was straightforward and could be completed within 30 min without time-consuming drying steps and eliminating the need for high-performance liquid chromatography (HPLC) or SPE for purification. In contrast to the [$^{18}$F]AlF labeling strategy, most $^{18}$F labeling strategies are tedious to perform and require multiple purifications of intermediates, resulting in a low RCY. The [$^{18}$F]AlF labeling strategy enables the significant redesign of existing radiopharmaceuticals as replacements for $^{68}$Ga, as demonstrated by the many examples where [$^{18}$F]AlF-derivatives of $^{68}$Ga-peptides have been developed to overcome the limitations of $^{68}$Ga [34]. The preliminary data suggest that $^{68}$Ga and [$^{18}$F]AlF radiopharmaceuticals have similar pharmacokinetic profiles, although differences have been observed in some cases, particularly in biodistribution [54].

2.2.6. Molar Activity ($A_m$)

Under these conditions, [$^{18}$F]AlF-NOTA-pamidronic acid (Figure 11) was obtained with molar activities ($A_m$) of 0.024 GBq µmol$^{-1}$ (SD = 0.002) at the end of the syntheses ($n = 6$)

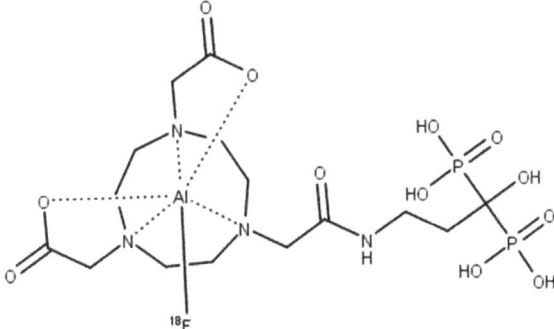

Figure 11. The structure of [$^{18}$F]AlF-NOTA-pamidronic acid.

2.3. *Quality Control Analysis of [$^{18}$F]AlF-NOTA-Pamidronic acid*

2.3.1. Radiochemical Purity (RCP) Analysis of [$^{18}$F]AlF-NOTA-Pamidronic Acid Using RP-HPLC

The RCP of the [$^{18}$F]AlF-NOTA-pamidronic acid was 100%, based on the RP-HPLC analysis ($n = 12$) (Figure 12). The Rt of the [$^{18}$F]AlF-NOTA-pamidronic acid at 5.46 min (SD = 0.05) was similar to the Rt of the NOTA-pamidronic acid precursor. The relative standard deviation (RSD) percentage was within 2% ($n = 6$). No free $^{18}$F or [$^{18}$F]AlF$^{2+}$ was detected in the radiochromatogram, indicating that the NOTA-pamidronic acid precursor was successfully radiolabeled with [$^{18}$F]AlF$^{2+}$ complexes. The unbound [$^{18}$F]F$^{-}$ or [$^{18}$F]AlF$^{2+}$ peak appeared at 2.09 min (SD = 0.02) under this chromatographic condition.

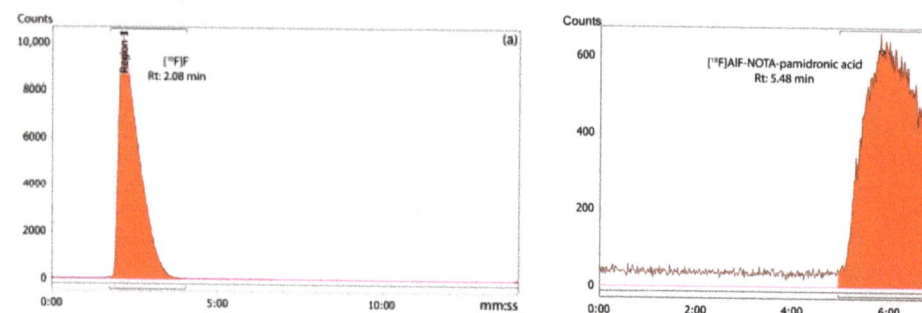

**Figure 12.** (a) Peak corresponding to the $[^{18}F]F^-$ and (b) $[^{18}F]AlF$-NOTA-pamidronic acid from the radiochromatogram of RP-HPLC.

RCY Analysis of the $[^{18}F]AlF$-NOTA-Pamidronic Acid Using r-TLC

The RCY of the $[^{18}F]AlF$-NOTA-pamidronic acid was 100%, based on an r-TLC analysis of the determined $[^{18}F]AlF$-NOTA-pamidronic acid sample obtained when NOTA-pamidronic acid was prepared in a molar ratio of 1:1 with $AlCl_3$ ($n = 6$) (Figure 13). $[^{18}F]AlF$-NOTA-pamidronic acid was spotted onto the ITLC-SG strip at the origin, and the strip was developed to the solvent front in a solvent mixture of 1M of ammonium acetate and acetonitrile (1:1). The retention factor (Rf) of the $[^{18}F]AlF$-NOTA-pamidronic acid, 0.91 (SD = 0.005), was within the acceptance criteria of Rf 0.6 to 1.0. The %RSD was within 2% ($n = 6$). No free $[^{18}F]F^-$ or $[^{18}F]AlF^{2+}$ was detected in the respective radiochromatogram, which is normally retained at an Rf of 0 to 0.4. Based on these findings, the Rf of $[^{18}F]F^-$ or $[^{18}F]AlF^{2+}$ was 0.06 (SD = 0.002) when unbound $[^{18}F]F^-$ or $[^{18}F]AlF^{2+}$ was spotted onto the ITLC-SG strip and developed under the same chromatographic conditions.

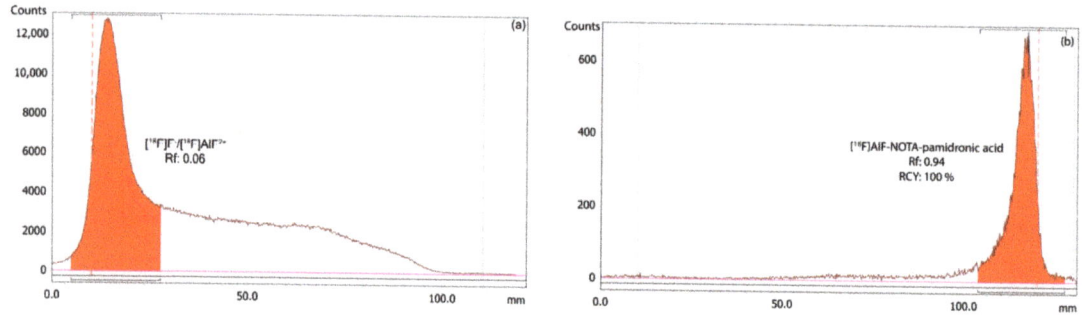

**Figure 13.** (a) Peak corresponding to the $[^{18}F]F^-/[^{18}F]AlF^{2+}$, and (b) $[^{18}F]AlF$-NOTA-pamidronic acid from the radiochromatogram of r-TLC.

2.3.2. Residual Solvents Analysis

Since only ethanol 50% ($v/v$) was added in the radiolabeling of the $[^{18}F]AlF$-NOTA-pamidronic acid, only ethanol was detected in the final formulation by gas chromatography (GC). The residue of ethanol was detected in the $[^{18}F]AlF$-NOTA-pamidronic acid sample at an Rt of 2.61 min (SD = 0.01) ($n = 6$) (Figure 14), which was similar to the previously determined Rt of the ethanol standard solution. The %RSD was also within 2%. Using a previously prepared calibration curve, the concentration of the ethanol in the $[^{18}F]AlF$-NOTA-pamidronic acid sample was determined to be 1.353 mg mL$^{-1}$, which was below the threshold of 5 mg mL$^{-1}$.

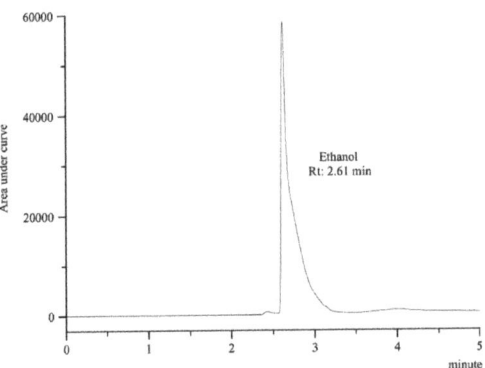

**Figure 14.** GC-FID chromatogram of ethanol peak from [$^{18}$F]AlF-NOTA-pamidronic acid sample.

2.3.3. Stability Study of [$^{18}$F]AlF-NOTA-Pamidronic Acid

The stability of the [$^{18}$F]AlF-NOTA-pamidronic acid was determined at 1, 2, 3, and 4 h after radiolabeling (Figure 15), verifying that the radiochemical purity remained higher than 90% for 4 h in the final formulation vial even without the addition of ascorbic acid as a radioprotective agent. The result was consistent with the fact that the NOTA ligand, in this case, NOTA-pamidronic acid, is known to form stable complexes with the Al$^{3+}$ (AlCl$_3$) [51]. The RCP of [$^{18}$F]AlF-NOTA-pamidronic acid was determined to be 90.15% ($SD$ = 0.06) ($n$ = 6) after 4 h of radiolabeling.

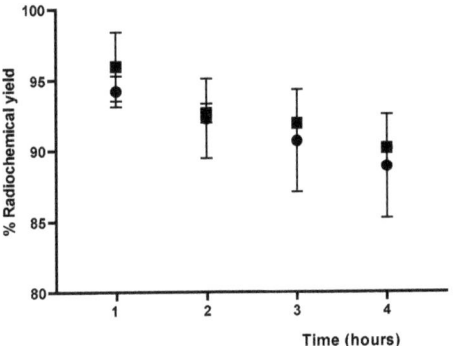

**Figure 15.** Stability of the [$^{18}$F]AlF-NOTA-pamidronic acid in the final formulation and in human plasma up to 4 h [mean (SEM)].

The RCP of the [$^{18}$F]AlF-NOTA-pamidronic acid was 90.67% ($SD$ = 3.62) ($n$ = 6) after 3 h of radiolabeling, indicating that the requirement of more than 90% of the radioactivity was in the form of the [$^{18}$F]AlF-NOTA-pamidronic acid was met in human plasma at 37 °C (3 h). In contrast to other studies that also used NOTA as a chelator or the Al-F technique for radiopharmaceutical development, the corresponding radiopharmaceutical was no longer stable in human plasma from 60 to 90 min onwards [35,55].

A possible explanation could be that the radioactivity of [$^{18}$F]F$^-$ used in this experiment was lower (less than 30 MBq) than in other studies where a higher radioactivity (more than 400 MBq) was used, even with the same NOTA chelator, which could have contributed to the instability of this radiopharmaceutical in the human plasma [35]. For this reason, special care should be taken in the future when upscaling production to prevent the possible instability of [$^{18}$F]AlF-NOTA-pamidronic acid in human plasma. Based on this information, it should be considered that images of PET should be acquired within 30–60 min of the intravenous injection if used for preclinical or clinical imaging in the future.

In summary, [$^{18}$F]AlF-NOTA-pamidronic acid, which was prepared in two consecutive runs, met all the quality control analysis requirements for the appearance, pH, radiochemical purity, and organic solvent analysis for the radiopharmaceutical. [$^{18}$F]AlF-NOTA-pamidronic acid was also stable in the final formulation and in human plasma at 37 °C (4 h). Table 5 below summarises the quality control analyses performed on the [$^{18}$F]AlF-NOTA-pamidronic acid sample with the respective acceptance criteria and observed value.

**Table 5.** Quality control analysis of [$^{18}$F]AlF-NOTA-pamidronic acid ($n = 6$).

| Quality Control Analysis | Acceptance Criteria | [$^{18}$F]AlF-NOTA-Pamidronic Acid |
|---|---|---|
| Appearance | Clear, colourless and free of particles | Verified |
| pH | 4 to 8 | 7 |
| RCP (HPLC) | ≥90% | 100% |
| RCY (ITLC-SG) | 90% | 95% |
| Organic solvent: ethanol (GC) | ≤5 mg mL$^{-1}$ | 1.353 mL$^{-1}$ |

*2.4. In Vitro Binding Studies of [$^{18}$F]AlF-NOTA-Pamidronic Acid*

2.4.1. In Vitro Bone Binding Assay Using Hydroxyapatite (HA)

The in vitro studies in this experiment were performed with synthetic HA using the method described by Meckel et al. [27]. The HA binding assay (Figure 16) showed a higher binding of [$^{18}$F]NaF ($M = 94.31\%$, $SD = 3.28$), followed by [$^{18}$F]AlF-NOTA-pamidronic acid ($M = 93.68\%$, $SD = 4.35$) and [$^{18}$F]AlF-NOTA ($M = 90.04\%$, $SD = 5.77$) ($n = 3$). Statistical analysis of the one-way ANOVA revealed that the difference in the percentage binding of HA was insignificant for all three agents ($p > 0.05$). The results showed that a higher HA binding assay of more than 90% was achieved with the $^{18}$F derivative than when the $^{68}$Ga derivatives were used [23,27,56–58]. The in vitro HA binding assay recorded the highest value of 91% only with $^{68}$Ga-DOTA-pamidronic acid, while the majority ranged from 70 to 85%.

This result was to some extent expected, as both [$^{18}$F]AlF-NOTA-pamidronic acid and [$^{18}$F]NaF show a high binding affinity to HA. The difference in % of HA binding was only 0.63% between the two. The result was similar to that of Keeling et al., who compared the in vitro bone binding between [$^{68}$Ga]Ga-THP-Pamidronate and [$^{18}$F]NaF [23]. [$^{18}$F]NaF binds to HA, displacing the hydroxyl groups within the HA lattice for [$^{18}$F]F$^-$ [23]. This could be the reason why the binding of HA was also observed in [$^{18}$F]AlF-NOTA, as it possibly behaves similar to [$^{18}$F]F$^-$ when NOTA is hydrolysed in 0.9% NaCl solutions and becomes an [$^{18}$F]AlF$^{2+}$ complex that can be incorporated into the HA surface [59].

The results showed that [$^{18}$F]AlF-NOTA-pamidronic acid was also sensitive to the presence of HA and was able to bind to HA similar to [$^{18}$F]NaF. However, the experiment did not provide further evidence for the specificity of [$^{18}$F]AlF-NOTA-pamidronic acid and [$^{18}$F]NaF. Therefore, the in vitro cellular uptake studies conducted in the following section demonstrate the specificity of the uptake of [$^{18}$F]AlF-NOTA-pamidronic acid, [$^{18}$F]NaF, and [$^{18}$F]AlF-NOTA for the cell lines used.

2.4.2. In Vitro Cellular Uptake Studies

In this section, the specificity for the uptake of [$^{18}$F]AlF-NOTA-pamidronic acid, [$^{18}$F]NaF, and [$^{18}$F]AlF-NOTA for the normal human osteoblast cell line (hFOB 1.19) and human osteosarcoma cell lines (Saos-2) is investigated. The radioactivity that accumulated on the surface in both cell lines after 30 min of incubation with [$^{18}$F]AlF-NOTA-pamidronic acid, [$^{18}$F]NaF, and [$^{18}$F]AlF-NOTA was counted using a gamma counter ($n = 3$). The result showed that the uptake of [$^{18}$F]NaF was higher in both cell lines, followed by [$^{18}$F]AlF-

NOTA-pamidronic acid, the compound of interest for this study (Figure 17). This mirrors the results from the in vitro HA binding assay.

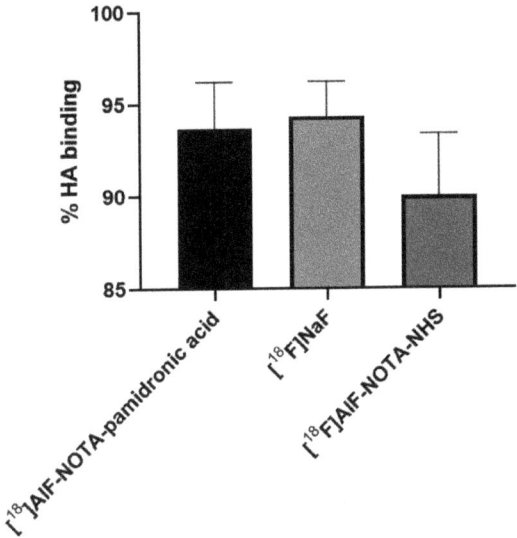

**Figure 16.** The % bone binding assay experimented using HA for [$^{18}$F]AlF-NOTA-pamidronic, [$^{18}$F]NaF and [$^{18}$F]AlF-NOTA ($n$ = 3) [mean (SEM)].

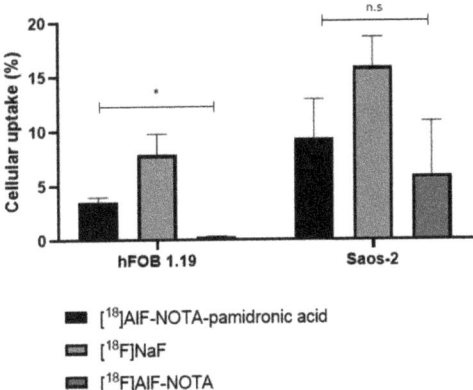

**Figure 17.** Cellular uptake of the [$^{18}$F]AlF-NOTA-pamidronic acid, [$^{18}$F]NaF and [$^{18}$F]AlF-NOTA in hFOB 1.19 and Saos-2 cells ($n$ = 3) at time point 30 min (* $p$ < 0.05, n.s: non-significant) [mean (SEM)].

In the hFOB 1.19 cell lines, the [$^{18}$F]NaF uptake was 7.84% (SD = 3.31), followed by an [$^{18}$F]AlF-NOTA-pamidronic acid uptake of 3.52% (SD = 0.76). The observed cellular uptake of [$^{18}$F]AlF-NOTA for hFOB 1.19 was almost negligible at 0.28% (SD = 0.07). The cellular uptake of [$^{18}$F]NaF was 2-fold higher than the uptake of [$^{18}$F]AlF-NOTA-pamidronic acid in the hFOB 1.19 cell lines. The cellular uptake of [$^{18}$F]AlF-NOTA was very low, possibly due to the presence of a bifunctional chelator of NOTA that limited the surface binding of [$^{18}$F]AlF-NOTA to the hFOB 1.19 cell lines. On the other hand, this shows that NOTA formed stable complexes with [$^{18}$F]AlF$^{2+}$, and the [$^{18}$F]AlF-NOTA complex in this experiment was not further transchelated to [$^{18}$F]AlF$^{2+}$ when performed in vitro [51].

Further statistical analysis of the one-way ANOVA revealed that the difference in cellular uptake was significant for all three radiopharmaceuticals ($p < 0.05$). This result was somewhat expected, as [$^{18}$F]NaF has been used primarily for the early diagnosis and monitoring of bones in PET [59]. Although the result shows that the cellular uptake of [$^{18}$F]AlF-NOTA-pamidronic acid in hFOB 1.19 was 2-fold lower compared to [$^{18}$F]NaF, it was nevertheless demonstrated that [$^{18}$F]AlF-NOTA-pamidronic acid can bind to normal osteoblast cells. The evidence supports the concept that [$^{18}$F]AlF-NOTA-pamidronic acid has the potential to be used for PET bone imaging.

Although [$^{18}$F]NaF showed higher cellular uptake in the Saos-2 cell lines ($M = 15.84\%$, $SD = 4.78$), the difference in the cellular uptake compared to [$^{18}$F]AlF-NOTA-pamidronic acid ($M = 9.28\%$, $SD = 6.25$) was not significant when further statistically analysed using a one-way ANOVA ($p > 0.05$) (Figure 16). The amount of accumulated radioactivity between [$^{18}$F]NaF and [$^{18}$F]AlF-NOTA-pamidronic acid was not 2-fold higher than that previously observed in hFOB 1.19 cell lines.

This finding literally suggests that the compound investigated in this study, [$^{18}$F]AlF-NOTA-pamidronic acid, tends to be more specific than [$^{18}$F]NaF, in the sense that the cellular uptake of [$^{18}$F]AlF-NOTA-pamidronic acid in the Saos-2 cell lines was higher than the cellular uptake of [$^{18}$F]AlF-NOTA-pamidronic acid in the hFOB 1.19 cell lines. Furthermore, the difference in cellular uptake of [$^{18}$F]AlF-NOTA-pamidronic acid and [$^{18}$F]NaF in the Saos-2 cell lines was insignificant, whereas the difference in the cellular uptake of [$^{18}$F]AlF-NOTA-pamidronic acid and [$^{18}$F]NaF in hFOB 1.19 cell lines was significant.

Nevertheless, the study does not attempt to mislead the audience by simply drawing a conclusion about the specificity of [$^{18}$F]AlF-NOTA-pamidronic acid in the Saos-2 cell lines compared to [$^{18}$F]NaF. The focus of this study is on the potential of [$^{18}$F]AlF-NOTA-pamidronic acid in PET bone imaging. There may be an explanation for the insignificant difference in the cellular uptake of [$^{18}$F]AlF-NOTA-pamidronic acid and [$^{18}$F]NaF observed in the Saos-2 cell lines. Although the Saos-2 cell line has several osteoblastic features, further studies have shown that osteosarcomas histologically express two other common features, namely chondroblastic and fibroblastic [3,60,61]. Chondroblasts contribute to the formation of cartilage, while fibroblasts form connective tissues that support and connect other tissues or organs in the body [60]. Since the osteoblastic, chondroblastic, and fibroblastic subtypes are predominantly expressed in osteosarcomas, it is possible that the compound investigated in this study, [$^{18}$F]AlF-NOTA-pamidronic acid, could provide additional information that [$^{18}$F]NaF could not.

Based on this evidence, this study has proved that the concerns of Bastawrous et al. were valid, as the uptake of [$^{18}$F]NaF is non-cancer-specific, and non-malignant cells can also show an uptake, as observed in the case of the significant cellular uptake in normal human osteoblasts (hFOB 1.19). This is consistent with reports of a higher number of false-positive interpretations with [$^{18}$F]NaF [19]. Nevertheless, this study concludes that the investigated compound [$^{18}$F]AlF-NOTA-pamidronic acid has the potential to be used in PET bone imaging and is substantially able to provide additional information that [$^{18}$F]NaF could not provide in the case of osteosarcoma disease.

2.4.3. Limitation of the Study

This study reports only the bone binding affinity of the [$^{18}$F]AlF-NOTA-pamidronic acid on its in vitro binding affinity to HA and in vitro cellular uptake to demonstrate the uptake of [$^{18}$F]AlF-NOTA-pamidronic acid. The in vitro cellular uptake studies were only performed at a one-time point (30 min) due to the distance between the laboratory and the gamma counter facility for counting. The comparison of both studies with [$^{18}$F]NaF and [$^{18}$F]AlF-NOTA have only been discussed superficially and is beyond the scope of this study. The comparison of the in vitro binding affinity studies between [$^{18}$F]AlF-NOTA-pamidronic acid and [$^{18}$F]NaF can be a comprehensive study for future endeavours.

## 3. Materials and Methods

### 3.1. Materials

Pamidronic acid ($C_3H_{11}NO_7P_2$) (Santa Cruz Biotechnology, Dallas, TX, USA), NOTA-NHS ester ($C_6H_{24}N_4O_8$) (CheMatech, Dijon, France), $^{18}F$ radionuclide (National Cancer Institute, Putrajaya, Malaysia), synthetic hydroxyapatite (HA) (Sigma Aldrich, St. Louis, Mo, USA), osteoblast cell line (hFOB 1.19) (ATCC, Manassas, VA, USA), osteosarcoma cell line (Saos-2) (ATCC, USA), human plasma (Department of Medical Microbiology, HPUPM, Serdang, Malaysia), dimethylformamide (DMF), 0.1% trifluoroacetic acid ($C_2HF_3O_2$) >99.9%, aluminium chloride hexahydrate ($AlCl_3.6H_2O$), triethylamine ($C_6H_{15}N$) >99% (Sigma Aldrich, USA), dimethyl sulfoxide (DMSO), ammonium acetate ($NH_4CH_3CO_2$) (Sigma Aldrich, USA), acetic acid ($CH_3COOH$), acetonitrile (Supelco, Bellefonte, USA), formic acid ($CH_2O_2$), ethanol, sodium acetate ($C_2H_3NaO_2$) (Merck, Kenilworth, USA), water for injection (WFI), 0.9% sodium chloride (B. Braun, USA), Geneticin® (G418 sulfate), penicillin streptomycin (ThermoFisher Scientific, Waltham, MA, USA), Foetal bovine serum, Dulbecco's phosphate-buffered saline (PBS), Trypsin/EDTA, sterile DMSO (ATCC, USA), Instant Thin Layer Chromatography-Silica Gel (ITLC-SG) (Agilent, Santa Clara, CA, USA).

### 3.2. Methods

#### 3.2.1. Preparation, Validation, and Isolation of NOTA-Pamidronic Acid Precursor

In general, the preparation of the NOTA-pamidronic acid precursor involves the dilution of the weighed pamidronic acid in 2.5 mL of deionised water. The mixture was vortexed until completely dissolved. A careful adjustment of the pH to 8 followed using pH indicator strips during the addition of 30 µL TEA. Next, NOTA-NHS was weighed and dissolved in 1.5 mL DMF before about 50 µL fractions of the NOTA-NHS solution were added to the dissolved pamidronic acid (650 µL) and prepared in triplicate ($n = 3$).

The pH of the reaction was monitored every hour with a pH indicator strip. The reaction condition should be neutral to slightly basic [37]. After 4 h, and prior to validation, the crude NOTA-pamidronic acid was filtered by SPE using a Sep-Pak C18 Plus Light cartridge (Waters, Milford, MA, USA) and a 0.22 µm nylon syringe filter (PhenexTM-NY, USA) to remove organic impurities. A series of experiments were carried out varying the molar ratio of the pamidronic acid: NOTA-NHS by preparing both materials according to Table 6. Figure 18 shows the workflow for the preparation of the NOTA-pamidronic acid precursor.

**Table 6.** Pamidronic acid and NOTA-NHS molar ratio.

| Pamidronic Acid (mg) | Pamidronic Acid (mM) | NOTA-NHS (mM) | Pamidronic Acid: NOTA-NHS Molar Ratio |
| --- | --- | --- | --- |
| 5.875 | 10 | 2 | 5:1 |
| 11.750 | 20 | 2 | 10:1 |
| 17.625 | 30 | 2 | 15:1 |

#### 3.2.2. Validation of NOTA-Pamidronic Acid Using LC-MS Analysis

LC-MS analysis was performed using an LC Dionex Ultimate 3000 (Thermo Fisher Scientific, Waltham, MA, USA) equipped with an autosampler, quaternary pump, column compartment, UV, and diode-array detector (PDA). Chromatographic analysis was performed using a Synergi Hydro-RP 2.5 µm, 50 × 2.1 mm column (Phenomenex, Torrance, CA, USA). The column temperature was 30 °C throughout the analysis. Since the sample compounds were polar, reversed-phase chromatography with ion suppression was used in this analysis. The mobile phase was 0.1% formic acid in the water, while the flow rate was set at 0.3 mL min$^{-1}$. Approximately 10 µL of the sample was injected for each LC-MS analysis. The ionisation mode was set to a negative ion mode with a collision energy of 30 eV. The $m/z$ scan was performed from $m/z$ 50 to 750. The chromatogram and mass spectrum (precursor ion and isotopic abundance) were analysed using a Thermo Xcalibur

4.2.47. The percentage yield of the NOTA-pamidronic acid precursor and free pamidronic acid was recorded and calculated from the chromatogram.

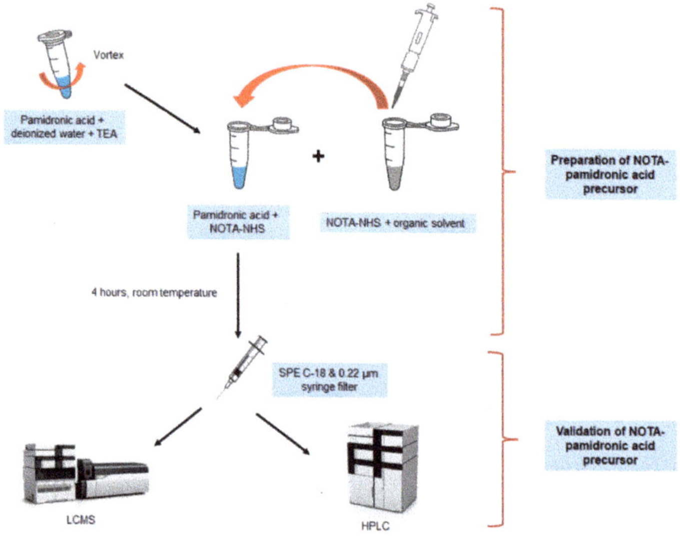

**Figure 18.** Workflow illustration of the NOTA-pamidronic acid preparation process.

### 3.2.3. Isolation of the NOTA-Pamidronic Acid Fraction from the Crude Sample

The NOTA-pamidronic acid fraction was isolated and collected from the crude product using an RP-HPLC (Agilent 1200, USA) equipped with a fraction collector. The isolation of the NOTA-pamidronic acid fractions was carried out under optimal chromatographic conditions using a Synergi 4μ (C18, polar endcapped) column, 0.1% trifluoroacetic acid (TFA) in water as a mobile phase at 0.5 mL min$^{-1}$ with detection at 220 nm. A sample injection volume of 10 μL was manually injected using a microlitre syringe (Glass capillary, Hamilton). The fractions corresponding to each peak in the chromatogram–free pamidronic acid, NOTA-pamidronic acid, and free NOTA were repeatedly collected for 5 mL in separate vials. However, only the collected NOTA-pamidronic acid fractions were re-analysed using analytical RP-HPLC before being sent for freeze-drying.

### 3.2.4. Preparation of [$^{18}$F]AlF$^{2+}$ Complexes

Using the highest fraction of [$^{18}$F]F$^-$ radioactivity previously collected in the (PCR) tube, 10 μL of the [$^{18}$F]F$^-$ with radioactivity of ~5 to 10 MBq was added to a 0.2 mL vial containing 10 μL of 1 mM AlCl$_3$ in 0.5 M sodium acetate buffer solution at pH 4. The reaction mixture was thoroughly mixed and incubated at room temperature for 10 min to allow the formation of [$^{18}$F]AlF$^{2+}$ complexes.

### 3.2.5. Quality Control Analysis for [$^{18}$F]AlF-NOTA-Pamidronic Acid

[$^{18}$F]AlF-NOTA-pamidronic acid was analysed for the appearance of pH, RCP, RCY, residual solvent, and in vitro stability study. Chromatographic separation for the residual solvent analysis was performed in accordance with Hassan et al. [62].

### 3.2.6. In Vitro Stability Study of [$^{18}$F]AlF-NOTA-Pamidronic Acid

The stability study of the [$^{18}$F]AlF-NOTA-pamidronic acid in normal saline (0.9% NaCl) and in vitro human plasma was studied at room temperature after 1, 2, 3, and 4 h using the r-TLC method. The stability of the [$^{18}$F]AlF-NOTA-pamidronic acid (100 μL) was tested by incubating 900 μL of human plasma at 37 °C for up to 4 h. After 1-, 2-, 3-, and 4-h incubation, about 200 µL were precipitated with 200 µL of ethanol and centrifuged at 12,000 rpm for 5 min before being analysed by the r-TLC method [55,63].

3.2.7. In Vitro Binding Studies of [$^{18}$F]AlF-NOTA-Pamidronic Acid
In Vitro Bone Binding Assay Using Hydroxyapatite

A total amount of 50 µL of radiolabeled [$^{18}$F]AlF-NOTA-pamidronic acid was added to the prepared HA assay and incubated for 10 min. After incubation, the supernatant was carefully removed with the Pasteur pipette, leaving HA in the centrifuge tube. Next, 500 µL of normal saline was added and then centrifuged at 3000 rpm. This time, the supernatant was transferred to another centrifuge tube (B), leaving the retained [$^{18}$F]AlF-NOTA-pamidronic acid on HA (A). This experiment was performed simultaneously with [$^{18}$F]NaF and [$^{18}$F]AlF-NOTA. The bone binding assay was determined with a gamma counter and calculated with the following Equation (1):

$$\% \text{ bone binding assay} = \frac{fraction\ A}{fraction\ A + fraction\ B} \times 100\% \qquad (1)$$

In Vitro Cellular Uptake Studies

Cell lines hFOB 1.19 and Saos-2 were seeded in 12-well plates at a density of $1 \times 10^5$ cells/well 2 days prior to the cellular uptake studies. Complete culture media (CCM) was freshly replaced on the day of the experiment. The cells were then incubated with [$^{18}$F]AlF-NOTA-pamidronic acid (90–140 kBq/well), which was adjusted to a final volume of 0.5 mL with CCM. After 30 min, the supernatant was collected, and the cells were washed twice with PBS. The radioactivity that accumulated on the surface of both cell lines was measured with a gamma counter. The radioactive medium and the collected PBS were defined as $C_{out}$. Finally, the cells were harvested with trypsin and washed again twice with PBS. The radioactivity of the harvested cells and PBS was defined as $C_{in}$. The cellular uptake rate was calculated using the following formula (Equation (2)). The above steps were repeated for [$^{18}$F]NaF and [$^{18}$F]AlF-NOTA. Each of the cell lines was prepared in triplicate for each of the radiopharmaceuticals used ($n = 3$).

$$Cellular\ uptake = \frac{Cin}{(Cin + Cout)} \times 100\% \qquad (2)$$

The human osteoblast cell line hFOB 1.19 (ATCC® CRL-11372™) was acquired from the Animal Type Culture Collection (ATCC, Manassas, VA, USA). The human osteosarcoma cell line, Saos-2 (ATCC HTB-85™), was obtained from Dr. Azuraidi Osman at the Department of Cell and Molecular Biology, Faculty of Biotechnology and Biomolecular Sciences, UPM. The in vitro study was performed with the approval of the Institutional Biosafety and Biosecurity Committee (IBBC), Universiti Putra Malaysia (UPM/IBBC/NGMO/2021/R004).

## 4. Conclusions

[$^{18}$F]AlF-NOTA-pamidronic acid was successfully prepared and optimised by exploiting the fluorophilic nature of aluminium to form a stable complex cation [$^{18}$F]AlF$^{2+}$, followed by coordination with the NOTA-pamidronic acid moiety. The preliminary in vitro results demonstrate that the [$^{18}$F]AlF-NOTA-pamidronic acid can potentially be used for PET bone imaging. One of the most intriguing results was to witness its specificity when a higher cellular uptake of [$^{18}$F]AlF-NOTA-pamidronic acid was observed in the Saos-2 cell lines than in the hFOB 1.19 cell lines. These preliminary results suggest that a comprehensive preclinical study of [$^{18}$F]AlF-NOTA-pamidronic acid is required before it can be moved into clinical research.

**Author Contributions:** Conceptualisation, H.H., M.F.O. and H.R.A.R.; Formal analysis, H.H., L.H.Y., J.I., M.H.N.A.H. and Z.M.S.; Funding acquisition, H.H. and M.F.O.; Investigation, H.H., Z.A. and M.H.N.A.H.; Methodology, H.H., M.F.O., H.R.A.R., M.A.O., Z.A. and J.I.; Project administration, H.H.; Supervision, M.F.O., H.R.A.R., Z.A.Z. and F.F.A.S.; Validation, H.H. and M.F.O.; Visualisation, H.H.; Writing—original draft, H.H. and M.F.O.; Writing—review and editing, M.F.O. and H.R.A.R. All authors have read and agreed to the published version of the manuscript.

**Funding:** This work was supported by the Universiti Putra Malaysia (UPM) through Putra Grant (9580700), the Universiti Teknologi MARA (UiTM) through MyRA LPhD grant (600-RMC/GPM LPHD 5/3 (129/2021)) and the Ministry of Higher Education, Malaysia (Hadiah Latihan Persekutuan—SLPP).

**Institutional Review Board Statement:** The in vitro study was performed with the approval of the Institutional Biosafety and Biosecurity Committee (IBBC) of the Universiti Putra Malaysia (UPM/IBBC/NGMO/2021/R004) on 3 January 2022.

**Informed Consent Statement:** Not applicable.

**Data Availability Statement:** The datasets used and/or analysed during the current study are available from the corresponding author upon reasonable request.

**Acknowledgments:** All authors sincerely thank the following funding agencies: the Centre for Diagnostic Nuclear Imaging (CDNI), UPM, the Department of Cell and Molecular Biology, Faculty of Biotechnology and Biomolecular Sciences, UPM, and the Department of Nuclear Medicine, National Cancer Institute, Putrajaya.

**Conflicts of Interest:** The authors declare no conflict of interest regarding this article.

## References

1. Pullan, J.E.; Lotfollahzadeh, S. *Primary Bone Cancer*; StatPearls Publishing: Treasure Island, FL, USA, 2022.
2. de Azevedo, J.W.V.; de Medeiros Fernandes, T.A.A.; Fernandes, J.V.; de Azevedo, J.C.V.; Lanza, D.C.F.; Bezerra, C.M.; Andrade, V.S.; de Araújo, J.M.G.; Fernandes, J.V. Biology and pathogenesis of human osteosarcoma (Review). *Oncol. Lett.* **2020**, *19*, 1099–1116. [CrossRef] [PubMed]
3. Abarrategi, A.; Tornin, J.; Martinez-cruzado, L.; Hamilton, A.; Martinez-campos, E.; Rodrigo, J.P.; González, M.V.; Baldini, N.; Garcia-castro, J.; Rodriguez, R. Osteosarcoma: Cells-of-Origin, Cancer Stem Cells, and Targeted Therapies. *Stem Cells Int.* **2016**, *2016*, 13. [CrossRef] [PubMed]
4. Cancer Research UK Survival for Bone Cancer. Available online: https://www.cancerresearchuk.org/about-cancer/bone-cancer/survival (accessed on 25 July 2022).
5. The American Cancer Society Survival Rates for Osteosarcoma. Available online: https://www.cancer.org/cancer/osteosarcoma/detection-diagnosis-staging/survival-rates.html (accessed on 18 July 2022).
6. Scimeca, M.; Urbano, N.; Rita, B.; Mapelli, S.N.; Catapano, C.V.; Carbone, G.M.; Ciuffa, S.; Tavolozza, M.; Schillaci, O.; Mauriello, A.; et al. Prostate Osteoblast-Like Cells: A Reliable Prognostic Marker of Bone Metastasis in Prostate Cancer Patients. *Contrast Media Mol. Imaging* **2018**, *2018*, 9840962. [CrossRef] [PubMed]
7. Jin, J.K.; Dayyani, F.; Gallick, G.E. Steps in prostate cancer progression that lead to bone metastasis. *Int. J. Cancer* **2011**, *128*, 2545–2561. [CrossRef] [PubMed]
8. Quiroz-Munoz, M.; Izadmehr, S.; Arumugam, D.; Wong, B.; Kirschenbaum, A.; Levine, A.C. Mechanisms of osteoblastic bone metastasis in prostate cancer: Role of prostatic acid phosphatase. *J. Endocr. Soc.* **2019**, *3*, 655–664. [CrossRef]
9. Bayouth, J.E.; Macey, D.J.; Kasi, L.P.; Fossella, F. V Dosimetry and toxicity of samarium-153-EDTMP administered for bone pain due to skeletal metastases. *J. Nucl. Med.* **1994**, *35*, 63–69.
10. Costelloe, C.M.; Chuang, H.H.; Madewell, J.E. FDG PET for the detection of bone metastases: Sensitivity, specificity and comparison with other imaging modalities. *PET Clin.* **2010**, *5*, 281–295. [CrossRef]
11. Azad, G.K.; Cook, G.J. Multi-technique imaging of bone metastases: Spotlight on PET-CT. *Clin. Radiol.* **2016**, *71*, 620–631. [CrossRef]
12. Cuccurullo, V.; Di Stasio, G.D.; Mansi, L. Nuclear Medicine in Prostate Cancer: A New Era for Radiotracers. *World J. Nucl. Med.* **2018**, *17*, 70–78. [CrossRef]
13. Hung, C.-S.; Su, H.-Y.; Liang, H.-H.; Lai, C.-W.; Chang, Y.-C.; Ho, Y.-S.; Wu, C.-H.; Ho, J.-D.; Wei, P.-L.; Chang, Y.-J. High-level expression of CXCR4 in breast cancer is associated with early distant and bone metastases. *Tumor Biol.* **2014**, *35*, 1581–1588. [CrossRef]
14. Calais, J.; Cao, M.; Nickols, N.G. The Utility of PET/CT in the Planning of External Radiation Therapy for Prostate Cancer. *J. Nucl. Med.* **2018**, *59*, 557–567. [CrossRef] [PubMed]
15. Araz, M.; Aras, G.; Küçük, Ö.N. The role of 18F-NaF PET/CT in metastatic bone disease. *J. Bone Oncol.* **2015**, *4*, 92–97. [CrossRef] [PubMed]

16. Bouter, C.; Meller, B.; Sahlmann, C.O.; Staab, W.; Wester, H.J.; Kropf, S.; Meller, J. (68)Ga-Pentixafor PET/CT Imaging of Chemokine Receptor CXCR4 in Chronic Infection of the Bone: First Insights. *J. Nucl. Med.* **2018**, *59*, 320–326. [CrossRef] [PubMed]
17. Schirrmeister, H.; Guhlmann, A.; Elsner, K.; Kotzerke, J.; Glatting, G.; Rentschler, M.; Neumaier, B.; Träger, H.; Nüssle, K.; Reske, S.N. Sensitivity in Detecting Osseous Lesions Depends on Anatomic Localization: Planar Bone Scintigraphy Versus 18F PET. *J. Nucl. Med.* **1999**, *40*, 1623–1629.
18. Schirrmeister, H.; Guhlmann, A.; Kotzerke, J.; Santjohanser, C.; Kühn, T.; Kreienberg, R.; Messer, P.; Nüssle, K.; Elsner, K.; Glatting, G.; et al. Early detection and accurate description of extent of metastatic bone disease in breast cancer with fluoride ion and positron emission tomography. *J. Clin. Oncol. Off. J. Am. Soc. Clin. Oncol.* **1999**, *17*, 2381–2389. [CrossRef]
19. Bastawrous, S.; Bhargava, P.; Behnia, F.; Djang, D.S.W.; Haseley, D.R. Newer PET application with an old tracer: Role of 18F-NaF skeletal PET/CT in oncologic practice. *Radio Graph.* **2014**, *34*, 1295–1316. [CrossRef]
20. Fleisch, H.; Russel, R.; Straumann, F. Effect of Pyrophosphate on Hydroxyapatite and Its Implications in Calcium Homeostasis. *Nature* **1966**, *212*, 901–903. [CrossRef]
21. Russell, R.G.G.; Watts, N.B.; Ebetino, F.H.; Rogers, M.J. Mechanisms of action of bisphosphonates: Similarities and differences and their potential influence on clinical efficacy. *Osteoporos. Int.* **2008**, *19*, 733–759. [CrossRef]
22. Gonzalez-Galofre, Z.N.; Alcaide-Corral, C.J.; Tavares, A.A.S. Effects of administration route on uptake kinetics of 18F-sodium fluoride positron emission tomography in mice. *Sci. Rep.* **2021**, *11*, 5512. [CrossRef]
23. Keeling, G.P.; Sherin, B.; Kim, J.; San Juan, B.; Grus, T.; Eykyn, T.R.; Rösch, F.; Smith, G.E.; Blower, P.J.; Terry, S.Y.A.; et al. [68Ga]Ga-THP-Pam: A Bisphosphonate PET Tracer with Facile Radiolabeling and Broad Calcium Mineral Affinity. *Bioconjug. Chem.* **2020**, *32*, 1276–1289. [CrossRef]
24. Ebetino, E.H.; Barnett, B.L.; Russell, R.G.G. A computational model delineates differences in hydroxyapatite binding affinities of bisphosphonates in clinical use VO—20 RT—Conference Proceedings. *J. Bone Miner. Res.* **2005**, *20*, S259.
25. Jahnke, W.; Henry, C. An in vitro assay to measure targeted drug delivery to bone mineral. *ChemMedChem* **2010**, *5*, 770–776. [CrossRef] [PubMed]
26. Holub, J.; Meckel, M.; Kubíček, V.; Rösch, F.; Hermann, P. Gallium(III) complexes of NOTA-bis (phosphonate) conjugates as PET radiotracers for bone imaging. *Contrast Media Mol. Imaging* **2015**, *10*, 122–134. [CrossRef]
27. Meckel, M.; Bergmann, R.; Miederer, M.; Roesch, F. Bone targeting compounds for radiotherapy and imaging: *Me(III)-DOTA conjugates of bisphosphonic acid, pamidronic acid and zoledronic acid. *EJNMMI Radiopharm. Chem.* **2016**, *1*, 14. [CrossRef]
28. Passah, A.; Tripathi, M.; Ballal, S.; Yadav, M.P.; Kumar, R.; Roesch, F.; Meckel, M.; Sarathi Chakraborty, P.; Bal, C. Evaluation of bone-seeking novel radiotracer (68)Ga-NO2AP-Bisphosphonate for the detection of skeletal metastases in carcinoma breast. *Eur. J. Nucl. Med. Mol. Imaging* **2017**, *44*, 41–49. [CrossRef] [PubMed]
29. Mueller, D.; Fuchs, A.; Leshch, Y.; Proehl, M. The Shortage of Approved $^{68}$Ge/$^{68}$Ga Generators—Incoming Material Inspection and GMP Compliant Use of Non-Approved Generators. *J. Nucl. Med.* **2019**, *60*, 1059.
30. Kang, D.; Simon, U.; Mottaghy, F.M.; Vogg, A.T.J. Labelling via [al18f]2+ using precomplexed al-noda moieties. *Pharmaceuticals* **2021**, *14*, 818. [CrossRef]
31. McBride, W.J.; Sharkey, R.M.; Karacay, H.; D'Souza, C.A.; Rossi, E.A.; Laverman, P.; Chang, C.H.; Boerman, O.C.; Goldenberg, D.M. A novel method of 18F radiolabeling for PET. *J. Nucl. Med.* **2009**, *50*, 991–998. [CrossRef]
32. Huynh, P.T.; Soni, N.; Pal, R.; Sarkar, S.; Jung, J.-M.; Lee, W.; Yoo, J. Direct radiofluorination of heat-sensitive antibody by Al-18F complexation. *New J. Chem.* **2019**, *43*, 15389–15395. [CrossRef]
33. Richter, S.; Wuest, F. 18 F-labeled peptides: The future is bright. *Molecules* **2014**, *19*, 20536–20556. [CrossRef]
34. Archibald, S.J.; Allott, L. The aluminium-[$^{18}$F] fluoride revolution: Simple radiochemistry with a big impact for radiolabelled biomolecules. *EJNMMI Radiopharm. Chem.* **2021**, *6*, 30. [CrossRef] [PubMed]
35. Poschenrieder, A.; Osl, T.; Schottelius, M.; Hoffmann, F.; Wirtz, M.; Schwaiger, M.; Wester, H. First 18 F-Labeled Pentixafor-Based Imaging Agent for PET Imaging of CXCR4 Expression In Vivo. *Tomography* **2016**, *2*, 85–93. [CrossRef] [PubMed]
36. ThermoFisher Scientific Amine-reactive Crosslinker Chemistry. Available online: https://www.thermofisher.com.my/en/home/life-science/protein-biology/protein-biology-learning-center/protein-biology-resource-library/pierce-protein-methods/amine-reactive-crosslinker-chemistry.html (accessed on 23 November 2020).
37. Hermanson, G.T. The Reactions of Bioconjugation. *Bioconjugate Tech.* **2013**, *2013*, 229–258. [CrossRef]
38. Alonso Martinez, L.M.; Harel, F.; Nguyen, Q.T.; Létourneau, M.; D'Oliviera-Sousa, C.; Meloche, B.; Finnerty, V.; Fournier, A.; Dupuis, J.; DaSilva, J.N. Al18F-complexation of DFH17, a NOTA-conjugated adrenomedullin analog, for PET imaging of pulmonary circulation. *Nucl. Med. Biol.* **2018**, *67*, 36–42. [CrossRef]
39. Hermanson, G.T. The Reactions of Bioconjugation. In *Bioconjugate Techniques*; Academic Press: Boston, MA, USA, 2013; pp. 229–258. ISBN 978-0-12-382239-0.
40. Woodman, R.H. Bioconjugation Discussion Reasons for Choosing NHS, TFP, or PFP Esters for Conjugating to Amines. Available online: https://www.researchgate.net/post/Bioconjugation_Discussion_Reasons_for_Choosing_NHS_TFP_or_PFP_esters_for_conjugating_to_amines (accessed on 1 September 2021).
41. Liigand, P.; Kaupmees, K.; Haav, K.; Liigand, J.; Leito, I.; Girod, M.; Antoine, R.; Kruve, A. Think Negative: Finding the Best Electrospray Ionization/MS Mode for Your Analyte. *Anal. Chem.* **2017**, *89*, 5665–5668. [CrossRef] [PubMed]
42. Strupat, K.; Scheibner, O.; Bromirski, M. High-resolution, accurate-mass Orbitrap Mass Spectrometry—Definitions, opportunities, and advantages. *Tech. Note 64287* **2013**, 1–5.

43. Qu, Z.; Chen, X.; Qu, C.; Qu, L.; Yuan, J.; Wei, D.; Li, H.; Huang, X.; Jiang, Y.; Zhao, Y. Fragmentation pathways of eight nitrogen-containing bisphosphonates (BPs) investigated by ESI-MSn in negative ion mode. *Int. J. Mass Spectrom.* **2010**, *295*, 85–93. [CrossRef]
44. Cheng, C.; Gross, M.L. Applications and mechanisms of charge-remote fragmentation. *Mass Spectrom. Rev.* **2000**, *19*, 398–420. [CrossRef]
45. International Atomic Energy Agency. *Cyclotron Produced Radionuclides: Operation and Maintenance of Gas and Liquid Targets*; International Atomic Energy Agency (IAEA): Vienna, Austria, 2012; ISBN 2077-6462.
46. Laverman, P.; McBride, W.J.; Sharkey, R.M.; Goldenberg, D.M.; Boerman, O.C. Al18F labeling of peptides and proteins. *J. Label. Compd. Radiopharm.* **2014**, *57*, 219–223. [CrossRef]
47. Cleeren, F.; Lecina, J.; Billaud, E.M.F.; Ahamed, M.; Verbruggen, A.; Bormans, G.M. New Chelators for Low Temperature Al18F-Labeling of Biomolecules. *Bioconjug. Chem.* **2016**, *27*, 790–798. [CrossRef]
48. Tsionou, M.I.; Knapp, C.E.; Foley, C.A.; Munteanu, C.R.; Cakebread, A.; Imberti, C.; Eykyn, T.R.; Young, J.D.; Paterson, B.M.; Blower, P.J.; et al. Comparison of macrocyclic and acyclic chelators for gallium-68 radiolabelling. *RSC Adv.* **2017**, *7*, 49586–49599. [CrossRef] [PubMed]
49. De Meyer, T.; Muyldermans, S.; Depicker, A. Nanobody-based products as research and diagnostic tools. *Trends Biotechnol.* **2014**, *32*, 263–270. [CrossRef] [PubMed]
50. Hassan, H.; Othman, M.F.; Abdul Razak, H.R. Optimal $^{18}$F-fluorination conditions for the high radiochemical yield of [$^{18}$F]AlF-NOTA-NHS complexes. *Radiochim. Acta* **2021**, *109*, 567–574. [CrossRef]
51. McBride, W.J.; Sharkey, R.M.; Goldenberg, D.M. Radiofluorination using aluminum-fluoride (Al18F). *EJNMMI Res.* **2013**, *3*, 36. [CrossRef] [PubMed]
52. D'Souza, C.A.; McBride, W.J.; Sharkey, R.M.; Todaro, L.J.; Goldenberg, D.M. High-yielding aqueous 18F-labeling of peptides via Al 18F chelation. *Bioconjug. Chem.* **2011**, *22*, 1793–1803. [CrossRef]
53. McBride, W.J.; D'souza, C.A.; Sharkey, R.M.; Karacay, H.; Rossi, E.A.; Chang, C.H.; Goldenberg, D.M. Improved 18F labeling of peptides with a fluoride-aluminum-chelate complex. *Bioconjug. Chem.* **2010**, *21*, 1331–1340. [CrossRef]
54. Hou, J.; Long, T.; Hu, S. Head-to-head Comparison of the 18F-AlF-NOTA-Octreotide and 68Ga-DOTATATE PET/CT within patients with Neuroendocrine Neoplasms. *J. Nucl. Med.* **2020**, *61*, 59.
55. Giglio, J.; Zeni, M.; Savio, E.; Engler, H. Synthesis of an Al18F radiofluorinated GLU-UREA-LYS(AHX)-HBED-CC PSMA ligand in an automated synthesis platform. *EJNMMI Radiopharm. Chem.* **2018**, *3*, 4. [CrossRef]
56. Fellner, M.; Biesalski, B.; Bausbacher, N.; Kubíček, V.; Hermann, P.; Rösch, F.; Thews, O. 68Ga-BPAMD: PET-imaging of bone metastases with a generator based positron emitter. *Nucl. Med. Biol.* **2012**, *39*, 993–999. [CrossRef]
57. Fakhari, A.; Jalilian, A.R.; Johari-daha, F.; Shafiee-ardestani, M.; Khalaj, A. Preparation and Biological Study of 68 Ga-DOTA-alendronate. *Asia Ocean J. Nucl. Med. Biol.* **2016**, *4*, 98–105. [CrossRef]
58. Ashhar, Z.; Yusof, N.A.; Ahmad Saad, F.F.; Mohd Nor, S.M.; Mohammad, F.; Bahrin Wan Kamal, W.H.; Hassan, M.H.; Ahmad Hassali, H.; Al-Lohedan, H.A. Preparation, Characterization, and Radiolabeling of [(68)Ga]Ga-NODAGA-Pamidronic Acid: A Potential PET Bone Imaging Agent. *Molecules* **2020**, *25*, 2668. [CrossRef] [PubMed]
59. Park, P.S.U.; Raynor, W.Y.; Sun, Y.; Werner, T.J.; Rajapakse, C.S.; Alavi, A. 18F-sodium fluoride pet as a diagnostic modality for metabolic, autoimmune, and osteogenic bone disorders: Cellular mechanisms and clinical applications. *Int. J. Mol. Sci.* **2021**, *22*, 6504. [CrossRef] [PubMed]
60. Cutilli, T.; Scarsella, S.; Fabio, D.D.; Oliva, A.; Cargini, P. High-grade chondroblastic and fibroblastic osteosarcoma of the upper jaw. *Ann. Maxillofac. Surg.* **2011**, *1*, 176–180. [CrossRef] [PubMed]
61. Tahmasbi-Arashlow, M.; Barnts, K.L.; Nair, M.K.; Cheng, Y.-S.L.; Reddy, L. V Radiographic manifestations of fibroblastic osteosarcoma: A diagnostic challenge. *Imaging Sci. Dent.* **2019**, *49*, 235–240. [CrossRef]
62. Hassan, H.; Othman, M.F.; Zakaria, Z.A.; Ahmad Saad, F.F.; Abdul Razak, H.R. Multivariate optimisation and validation of the analytical GC-FID for evaluating organic solvents in radiopharmaceutical. *J. King Saud Univ. -Sci.* **2021**, *33*, 101554. [CrossRef]
63. IAEA. *Guidance for Preclinical Studies with Radiopharmaceuticals*; IAEA: Vienna, Austria, 2021.

Article

# Preparation and Evaluation of [¹⁸F]AlF-NOTA-NOC for PET Imaging of Neuroendocrine Tumors: Comparison to [⁶⁸Ga]Ga-DOTA/NOTA-NOC

Johan Hygum Dam [1,2,*], Niels Langkjær [1], Christina Baun [1], Birgitte Brinkmann Olsen [3], Aaraby Yoheswaran Nielsen [1] and Helge Thisgaard [1,2]

[1] Department of Nuclear Medicine, Odense University Hospital, Kløvervænget 47, DK-5000 Odense, Denmark
[2] Department of Clinical Research, University of Southern Denmark, J.B. Winsløws Vej 19, DK-5000 Odense, Denmark
[3] Department of Surgical Pathology, Zealand University Hospital, Sygehusvej 10, DK-4000 Roskilde, Denmark
* Correspondence: johan.dam@rsyd.dk

**Abstract:** Background: The somatostatin receptors 1–5 are overexpressed on neuroendocrine neoplasms and, as such, represent a favorable target for molecular imaging. This study investigates the potential of [¹⁸F]AlF-NOTA-[1-Nal³]-Octreotide and compares it in vivo to DOTA- and NOTA-[1-Nal³]-Octreotide radiolabeled with gallium-68. Methods: DOTA- and NOTA-NOC were radiolabeled with gallium-68 and NOTA-NOC with [¹⁸F]AlF. Biodistributions of the three radioligands were evaluated in AR42J xenografted mice at 1 h p.i and for [¹⁸F]AlF at 3 h p.i. Preclinical PET/CT was applied to confirm the general uptake pattern. Results: Gallium-68 was incorporated into DOTA- and NOTA-NOC in yields and radiochemical purities greater than 96.5%. NOTA-NOC was radiolabeled with [¹⁸F]AlF in yields of 38 ± 8% and radiochemical purity above 99% after purification. The biodistribution in tumor-bearing mice showed a high uptake in tumors of 26.4 ± 10.8 %ID/g for [⁶⁸Ga]Ga-DOTA-NOC and 25.7 ± 5.8 %ID/g for [⁶⁸Ga]Ga-NOTA-NOC. Additionally, [¹⁸F]AlF-NOTA-NOC exhibited a tumor uptake of 37.3 ± 10.5 %ID/g for [¹⁸F]AlF-NOTA-NOC, which further increased to 42.1 ± 5.3 %ID/g at 3 h p.i. Conclusions: The high tumor uptake of all radioligands was observed. However, [¹⁸F]AlF-NOTA-NOC surpassed the other clinically well-established radiotracers in vivo, especially at 3 h p.i. The tumor-to-blood and -liver ratios increased significantly over three hours for [¹⁸F]AlF-NOTA-NOC, making it possible to detect liver metastases. Therefore, [¹⁸F]AlF demonstrates promise as a surrogate pseudo-radiometal to gallium-68.

**Keywords:** octreotide; AlF; neuroendocrine; PET; gallium-68; F-18

## 1. Introduction

Neuroendocrine tumors may occur throughout the body, although they are generally in a gastrointestinal or broncho-pulmonary location, and can release a range of hormones, neurotransmitters, and other compounds with pharmacological effects [1]. For the diagnosis and staging of neuroendocrine tumors, the somatostatin receptor subtypes 1–5, and mainly subtype 2, embody well-known and confirmed targets for molecular imaging and therapy, as exemplified in the NETTER-1 trial applying [¹⁷⁷Lu]Lu-DOTA-TATE for midgut neuroendocrine tumors [2]. The widely accepted and licensed 'kits' SomakitTOC and NETSpot for 68Ga-labeling to prepare [⁶⁸Ga]Ga-DOTATOC and [⁶⁸Ga]Ga-DOTATATE, respectively, for positron emission tomography (PET) has replaced [¹¹¹In]In-DTPA-octreotide (Octreoscan) and [⁹⁹ᵐTc]Tc-EDDA/HYNIC-TOC (Tektrotyd) for scintigraphy [3]. The increasing demand for licensed gallium-68 from commercial [⁶⁸Ge]Ge/[⁶⁸Ga]Ga generators for this application and HBED-PSMA-11 for prostate cancer imaging may outstrip the present-day supply [4]. An emerging technology to meet future demand comprises accelerator-produced gallium-68 on biomedical cyclotrons by proton irradiation on enriched zinc-68 [5–8], but this is not yet widely adopted. At the same time, [$^{18}$F]aluminum fluoride could be regarded as a pseudo-radiometal and a convenient replacement for traditional radiometals. As a development to facilitate radiofluorination without the use of prosthetic groups, McBride et al. expanded the means of direct radiolabeling with fluoride-18 via aluminum chelation [9,10]. This conveys the favorable characteristics of fluoride-18 (availability, longer half-life, short positron range and high positron yield, and fewer gamma emissions) compared to gallium-68, but still via chelation. Other means to directly label with fluoride-18 utilize the affinity of fluoride toward boron or silicon, but this labeling strategy requires the precursor compound to be chemically prepared and does not allow for a change to a therapeutic radioisotope [11,12].

Herein, we describe the radiolabeling and evaluation of the octreotide analog NOTA-[1-Nal$^3$]-Octreotide (NOTA-NOC) with [$^{18}$F]AlF for potential abundant access to a clinically important radiopharmaceutical with an affinity for somatostatin receptor subtype 2, 3, and 5. Additionally, we describe the preclinical comparison of [$^{18}$F]AlF-NOTA-NOC to both [$^{68}$Ga]Ga-DOTA-NOC and [$^{68}$Ga]Ga-NOTA-NOC by biodistribution and PET/CT imaging.

## 2. Results

### 2.1. Radiolabeling and Stability

The radiolabeling of DOTA- and NOTA-NOC with gallium-68 was achieved in yields greater than 97.4% and radiochemical purities (RCP) of 98.5 ± 0.2% ($n$ = 4) and 96.5 ± 0.5% ($n$ = 7), respectively, and was not subjected to further purification. Apparent molar activities of 11.6 ± 0.9 MBq/nmol for [$^{68}$Ga]Ga-DOTA-NOC and 12.1 ± 1.9 for [$^{68}$Ga]Ga-NOTA-NOC were observed.

The radiolabeling comprised of fluorine-18, ethanol, AlCl$_3$, and NOTA-NOC was achieved by mixing all reagents in a sealed polypropylene vial for 15 min at 105 °C. To remove unreacted [$^{18}$F]fluoride and possibly AlCl$_3$ and species thereof, the [$^{18}$F]AlF-NOTA-NOC was purified by solid-phase extraction on an Empore C18 extraction disc to achieve a high RCP of 99.3 ± 0.6% ($n$ = 3). [$^{18}$F]AlF-NOTA-NOC was obtained in yields of 38 ± 8% ($n$ = 3), non-decay corrected (non d.c.), applying up to 6 GBq of fluorine-18. An apparent molar activity of 32 ± 10 MBq/nmol was achieved for [$^{18}$F]AlF-NOTA-NOC at the end of synthesis. The serum stability [$^{18}$F]AlF-NOTA-NOC was found to be high with an initial RCP of 99.2%, decreasing to 98.4% after 3 h.

### 2.2. Determination of LogP and $K_D$

The partition coefficient, LogP, for the three compounds was determined to be −1.20 for [$^{18}$F]AlF-NOTA-NOC, −1.29 for [$^{68}$Ga]Ga-NOTA-NOC, and −1.42 for [$^{68}$Ga]Ga-DOTA-NOC. The apparent dissociation constant, $K_D$, for [$^{18}$F]AlF-NOTA-NOC was determined to be 3.47 nM.

### 2.3. Animal Experiments

In a subcutaneous AR42J mouse model, [$^{18}$F]AlF-NOTA-NOC exhibited a similar uptake pattern by ex vivo biodistribution as that of [$^{68}$Ga]Ga-DOTA/NOTA-NOC with some differences summarized in the discussion. The biodistribution is summarized in Figure 1. Very high uptake in the tumor was observed for all radioconjugates at 1 h post-injection (p.i.), viz., 26.4 ± 10.8% ID/g for [$^{68}$Ga]Ga-DOTA-NOC, 25.7 ± 5.8% ID/g for [$^{68}$Ga]Ga-NOTA-NOC, and 37.3 ± 10.5% ID/g for [$^{18}$F]AlF-NOTA-NOC, which further increased to 42.1 ± 5.3% ID/g at the 3 h time point for [$^{18}$F]AlF-NOTA-NOC. In addition, receptor-positive normal tissues (pancreas, stomach, intestines, and lungs) exhibited an expected moderate to high uptake, whereas the radioactivity seen in the kidneys was primarily due to the kidneys being the major route of excretion.

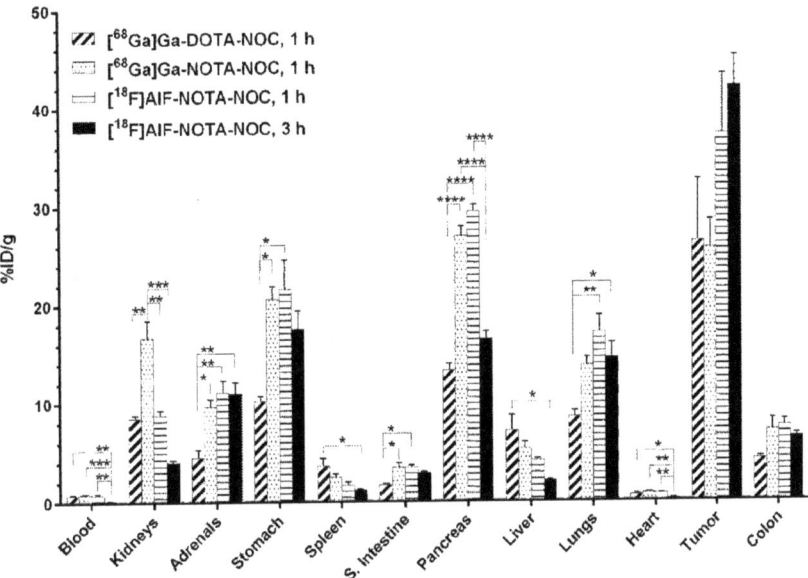

**Figure 1.** Biodistribution of [$^{68}$Ga]Ga-NOTA-NOC (50.2 ± 5 pmol, 627 ± 116 kBq), [$^{68}$Ga]Ga-DOTA-NOC (42.5 ± 6 pmol, 509 ± 24 kBq), and [$^{18}$F]AlF-NOTA-NOC (47 ± 10 pmol, 1017 ± 279 kBq) in AR42J xenograft mice. * $p \leq 0.05$, ** $p \leq 0.01$, *** $p \leq 0.001$, and **** $p \leq 0.0001$.

## 3. Discussion

Initial attempts to radiolabel NOTA-NOC with [$^{18}$F]AlF led to low and irreproducible labeling yields (<5% non d.c.) and required optimization as to the stoichiometric ratio of AlCl$_3$ to NOTA-NOC. Similar to the previous radiolabeling of RGD and bombesin peptides, the best conditions were found to comprise equimolar amounts of peptide and AlCl$_3$ [13,14]. By performing the radiolabeling in 65–70% ethanol, a method developed by Laverman et al. and others [15–17], we achieved the highest radiolabeling yield of up to 46% non d.c. It was, therefore, necessary to remove unlabeled [$^{18}$F]AlF rapidly using the Empore extraction disc to achieve high radiochemical purity. This renders kit-like radiolabeling with [$^{18}$F]AlF less suitable, although precedence exists, but more suitable for automated production [16,18]. The radiolabeling of DOTA- and NOTA-NOC with gallium-68 was performed in nearly quantitative yields and was formulated and applied without further purification.

The ex vivo biodistribution was performed with comparable molar amounts (app. 45 pmol) of the different radioligands. From the biodistribution in Figure 1, it is seen that normal tissues (small intestines, pancreas, and stomach), which are generally somatostatin receptor-positive, display a statistically significant higher uptake of [$^{68}$Ga]Ga-NOTA-NOC and [$^{18}$F]AlF-NOTA-NOC than [$^{68}$Ga]Ga-DOTA-NOC at the 1 h time point. This is less evident in the colon and lungs, which are also somatostatin receptor-positive. The variances in the uptake may be a feature of differences in the conformation of the radiometal chelate complexes, lipophilicity, and their overall charge as the increased uptake is displayed for both NOTA functionalized compounds [12]. On the other hand, both DOTA- and NOTA-NOC tumor uptake remained relatively similar and stable when radiolabeled with gallium-68 ([$^{68}$Ga]Ga-DOTA-NOC: 26.4 ± 10.8% ID/g; [$^{68}$Ga]Ga-NOTA-NOC: 25.7 ± 5.8% ID/g). When applying [$^{18}$F]AlF-NOTA-NOC, the tumor uptake after 1 h was elevated to 37.3 ± 10.5% ID/g and increased further to 42.1 ± 5.3% ID/g at the 3 h time point. Laverman et al. found a comparable tumor uptake of an [$^{18}$F]AlF labeled octreotide analog of 28.3 ± 5.7% ID/g after 2 h of injection of 200 pmol in an AR42J xenograft model, i.e., a 4-fold higher molar injection than in this study [15]. Importantly, at the 3 h time point of biodistribution, at which point the normal tissue has cleared of [$^{18}$F]AlF-NOTA-NOC, the liver displayed a significantly lower uptake of [$^{18}$F]AlF-NOTA-NOC (2.1 ± 0.1% ID/g) than that of [$^{68}$Ga]Ga-DOTA-NOC (7.2 ± 1.6% ID/g) after 1 h ($p$ = 0.014). A surprisingly low uptake in the spleen and a relatively high uptake in the lungs were observed for all ligands, which is very different from the typical uptake pattern in humans. For [$^{68}$Ga]Ga-DOTA-NOC in humans, the spleen is normally the organ with the highest receptor-specific uptake, followed by the kidneys and liver, while the uptake in the lungs is low [19]. However, a tissue uptake pattern similar to ours in tumor-bearing mice in previous studies was seen using radiometal-labeled DOTA-TATE, suggesting inter-species differences [20,21].

The tumor-to-blood and -liver ratios for [$^{68}$Ga]Ga-DOTA-NOC were 35 and 3.7, and for [$^{68}$Ga]Ga-NOTA-NOC, the ratios were determined to be comparable at 31 and 5.2, respectively (Figure 2). However, for [$^{18}$F]AlF-NOTA-NOC, the corresponding ratios were 52 for the tumor-to-blood and 9.0 for the tumor-to-liver ratio at 1 h p.i., increasing to 295 and 21, respectively, at the 3 h time point. Thus, for the liver, which represents a major site of metastasis, an increased tumor-to-organ ratio, and thus image contrast, of a factor of nearly 4-5 was determined for [$^{18}$F]AlF-NOTA-NOC over the other two radioconjugates.

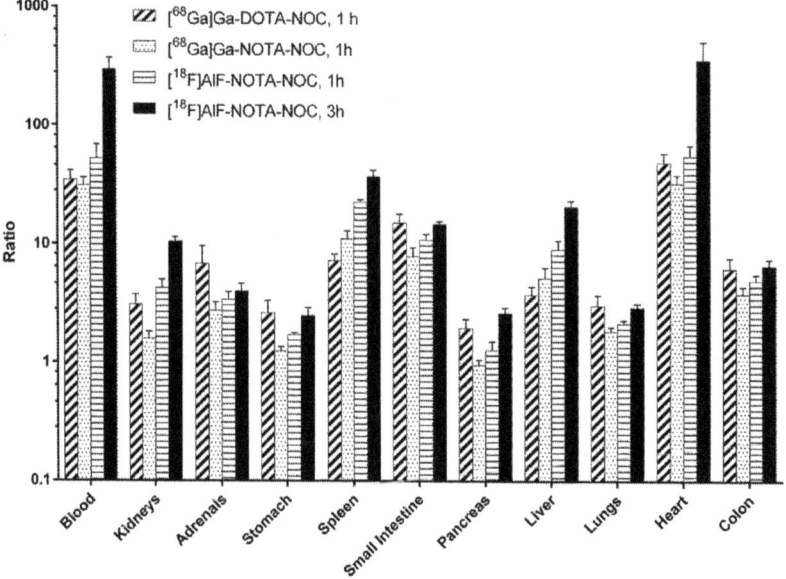

**Figure 2.** Tumor-to-organ ratios for each organ and radiopharmaceutical.

By preclinical PET/CT imaging of the uptake in the xenograft tumor model, Figure 3, the general uptake pattern found in the ex vivo biodistribution was confirmed for all radioconjugates. A pronounced excretion of [$^{18}$F]AlF-NOTA-NOC via the gall bladder, as indicated by PET/CT, could be an effect of increased lipophilicity of this particular radioconjugate (logP −1.20 for [$^{18}$F]AlF-NOTA-NOC vs. −1.29 and −1.42 for [$^{68}$Ga]Ga-NOTA-NOC and [$^{68}$Ga]Ga-DOTA-NOC, respectively), which carries neutral charge with regard to the radiolabeled complex. Additionally, the stable chelation of [$^{18}$F]AlF by the lack of skeletal uptake from unbound fluoride-18 and aluminum species thereof was confirmed and is in agreement with Laverman et al. at both 1 and 3 h p.i. [15].

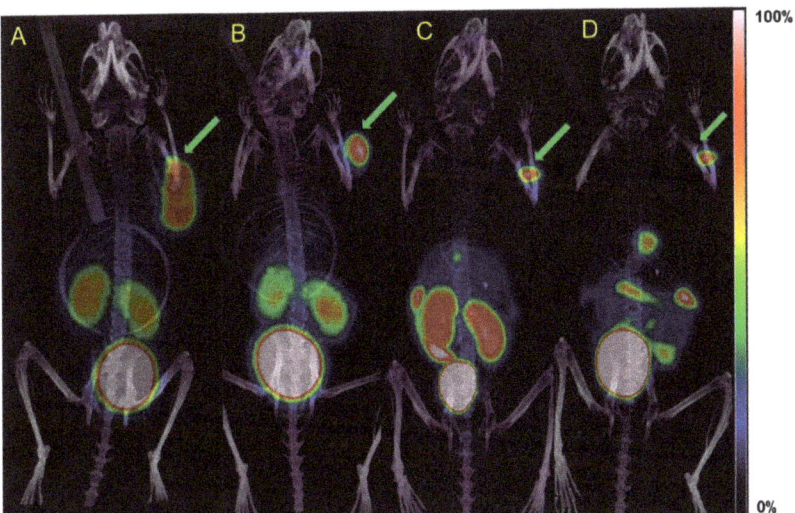

**Figure 3.** PET/CT MIP images of AR42J xenograft mice injected with [$^{68}$Ga]Ga-NOTA-NOC at 1 h (**A**), [$^{68}$Ga]Ga-DOTA-NOC at 1 h (**B**), [$^{18}$F]AlF-NOTA-NOC at 1 h (**C**), and [$^{18}$F]AlF-NOTA-NOC at 3 h (**D**). Tumors are indicated by arrows.

## 4. Materials and Methods

### 4.1. General

The peptidic precursors NOTA-[1-Nal$^3$] octreotide acetate (NOTA-NOC) and DOTA-[1-Nal$^3$] octreotide acetate (DOTA-NOC) were purchased from ABX (Radeberg, Germany). Aluminum chloride (AlCl$_3$·xH$_2$O, 99.999%), sodium acetate buffer for complexometry (pH 4.6), bovine serum albumin, sodium acetate (anhydrous, ≥99.999%), and acetic acid (≥99.0%, traceSELECT) were obtained from Sigma-Aldrich. Deionized (DI) water with a resistivity of 18.2 MΩ·cm was acquired from a MilliQ Direct-Q 3 UV water purification system. AlCl$_3$·xH$_2$O (10 mM) in 0.5 M NaOAc, pH 4.1, was prepared from 408 mg anhydrous sodium acetate with 1.15 mL acetic acid to 50.0 mL DI water. To this solution, 120 mg AlCl$_3$·xH$_2$O (assumed to be fully hydrated, x = 6, Mw 242) was dissolved. Phosphate buffered saline and ethanol (100%) were acquired from the local hospital pharmacy. The fluoride-18 was produced by the standard proton bombardment of [$^{18}$O]H$_2$O on a GE PETtrace cyclotron. The radiochemical yields and purities were determined by reverse phase HPLC (Phenomenex Jupiter C18, 150 × 4.6 mm, H$_2$O + 0.1% TFA:ACN gradient). The radioisotope conjugates were formulated in phosphate buffer containing 0.1% bovine serum albumin and applied for preclinical PET/CT imaging and biodistribution.

### 4.2. In Vitro Experiments

AR42J cells (CLS Cell Lines Service GmbH) were grown in Nutrient Mixture F-12 Ham (Sigma-Aldrich) supplemented with 10% fetal bovine serum, 1% penicillin/streptomycin, and 2 mM L-glutamine (all from Thermo Fisher Scientific) and kept at 37 °C under 5% CO$_2$. The AR42J cells were harvested by trypsinization and prepared in 50 µL medium mixed with extracellular matrix gel (Sigma-Aldrich; ratio 1:1).

For saturation binding analyses, 200,000 AR42J cells were seeded in 24-well plates and allowed to adhere overnight. The next day, the medium was removed and the cells were washed once with binding buffer (F12 containing 1% FBS, 1% P/S, and 0.5% bovine serum albumin) and then incubated for 1 h at 37 °C in fresh binding buffer. Afterward, the plates were placed at 4 °C for 30 min followed by incubation with increasing amounts of [$^{18}$F]AlF-NOTA-NOC (1–75 nM, 0.017–0.97 MBq) for 2 h at 4 °C. Non-specific bindings was determined by co-incubation with 10 µM Octreotide acetate (Sequoia Research Products

Ltd., Pangbourne, UK). The cells were washed twice with cold PBS and solubilized with 1 M NaOH. The cell-associated radioactivity was measured using a 2470 Wizard Automatic gamma counter. The dissociation constant ($K_D$) value was determined by non-linear regression using GraphPad Prism [22].

*4.3. Radiolabelling and Stability*

Radiolabeling with gallium-68 was performed by fractionated elution of an Eckert-Ziegler generator in volumes of 25–30 drops of 0.1 M HCl. DOTA- or NOTA-NOC (3 nmol, 1 nmol/μL), 100 μL eluate, 10 μL ethanol (100%), and 30 μL sodium acetate buffer, pH 4.6, were mixed in a microwave glass vial and sealed. The mixture was heated to 90 °C for 2 min by dynamic microwave irradiation (PETwave).

Fluoride-18 was loaded onto a Chromafix PS-HCO3 cartridge pre-conditioned with deionized water by helium pressure from the cyclotron. The radioactive cartridge was washed with 8 mL DI water, and the fluoride-18 was eluted with 0.9% saline (100 μL) to yield a stock solution of purified fluoride-18. The [$^{18}$F]AlF-NOTA-NOC was prepared in one pot by mixing the fluoride-18 (0.7–6 GBq in 100 μL 0.9% saline) with the peptide (20 nmol in DI water, 27 μL), ethanol (100%, 0.4 mL), $AlCl_3$ (10 mM, 2 μL) in sodium acetate buffer (0.5 M, pH 4.1), and sodium acetate buffer (pH 4.1, 0.5 M, 80 μL), followed by heating in a sealed polypropylene vial for 15 min at 105 °C in a heating block. Then, the reaction mixture was diluted with DI water (8 mL) and loaded onto a pre-conditioned solid-phase extraction disc (3M Empore C18-SD, 35 mg, 6 mL) by the application of an evacuated vial. The disc was then washed with DI water, and the compound was eluted with ethanol (70%, fractionation in app. 5 drops). The stability of [$^{18}$F]AlF-NOTA-NOC was tested as described by Dam et al. by hourly HPLC analysis at 0, 1, 2, and 3 h [23].

*4.4. Determination of LogP*

Radiolabeled peptide (5 μL, 3–5 MBq) was added to Dulbecco's PBS (495 μL) along with 1-Octanol (500 μL) in an Eppendorf tube. The tubes were vortexed for 5 min and centrifuged to separate phases. A volume of 10 μL of each phase was transferred to new Eppendorf tubes. The fractions were counted in an well counter (Atomlab 950) and LogP were calculated [LogP = Counts(1−Octanol)/Counts(PBS)].

*4.5. Animal Experiments*

All animal experiments were planned and performed following the national legislation by the Animal Experiments Inspectorate in Denmark. Male SCID (severe combined immunodeficiency) mice (bred in-house, age 9–12 weeks) had access to water and chow ad libitum. Anesthesia with approximately 2% isoflurane in 100% oxygen was applied and the mice were inoculated subcutaneously in the left shoulder with $1 \times 10^6$ AR42J cells 12–15 days before the experiment (tumor weight 364 ± 58 mg). For biodistribution, the AR42J xenografted mice were injected in the tail vein with [$^{68}$Ga]Ga-NOTA-NOC (50.2 ± 5 pmol, 627 ± 116 kBq, $n$ = 4), [$^{68}$Ga]Ga-DOTA-NOC (42.5 ± 6 pmol, 509 ± 24 kBq, $n$ = 3), or [$^{18}$F]AlF-NOTA-NOC (47.0 ± 10 pmol, 1017 ± 279 kBq, $n$ = 3 for each time point). After 1 or 3 h, the mice were sacrificed and the organs were collected and weighed. The organ radioactivities were quantified in an Atomlab 950 well spectrometer which was cross calibrated to a high-purity germanium detector. The tissue uptake of injected radioactivity was determined as a percentage of the applied total injected activity per gram, %IA/g.

For PET/CT imaging, a Siemens Inveon preclinical scanner (Siemens Healthcare, Knoxville, TN, USA) was applied in docked mode. The animals were anesthetized (app. 2% isoflurane in oxygen) and placed on a dedicated heated scanner bed in the prone position. The scan commenced with a 2-bed CT scan performed with the settings: 360 projections, full rotation, and bin 4 at 80 kV, and 500 μA in 350 mS. The CT scan was followed by a 70 min dynamic PET scan (framed $4 \times 30$ s, $10 \times 60$ s, $1 \times 180$ s, $3 \times 900$ s, and $1 \times 600$ s) initiated immediately before tail-vein injection of [$^{68}$Ga]Ga-DOTA-NOC (3.9 MBq), [$^{68}$Ga]Ga-NOTA-

NOC (4.7 MBq), or [$^{18}$F]AlF-NOTA-NOC (11.3 MBq). For comparable imaging, a 30 min static PET/CT scan was performed at 3 h p.i. of [$^{18}$F]AlF-NOTA-NOC.

CT and PET images were co-registered using a transformation matrix, and the CT-based attenuation corrected PET data were reconstructed using an OSEM3D/MAP algorithm (matrix 128 × 128, with 2 OSEM3D iterations, and 18 MAP iterations, target resolution 1.5 mm) using the image analysis software Inveon Research Workplace (Siemens Healthcare). Three-dimensional regions of interests (ROIs) were drawn on the fused PET/CT images covering the tumor volumes. Maximum intensity projection (MIP) images were obtained after setting the PET signal scale from zero to the maximum of the tumor uptake in the ROIs and adjusted to display a color scale from 0 to the maximum tumor uptake value.

### 4.6. Statistics

GraphPad Prism 6.07 was applied for statistical analysis and data fitting. One-way ANOVA and Bonferroni correction for multiple comparisons were applied for each organ. $P$ values considered statistically significant were * $p \leq 0.05$, ** $p \leq 0.01$, *** $p \leq 0.001$, and **** $p \leq 0.0001$. The Student's two-tailed t-test was applied for exploration statistics within a single organ-to-tumor comparison for one time point. Data are presented as the mean ± SEM.

## 5. Conclusions

In this work, a very high uptake of [$^{18}$F]AlF-NOTA-NOC was demonstrated in a subcutaneous mouse model of AR42J and compared to [$^{68}$Ga]Ga-DOTA/NOTA-NOC. A more pronounced effect on the uptake in receptor-positive normal tissues was observed by changing the chelator from DOTA to NOTA than the effect from changing the radiolabel.

By applying biodistribution and PET imaging at 3 h p.i., the tumor-to-blood and -liver ratios were significantly increased. For the diagnosis and staging of neuroendocrine tumors, [$^{18}$F]AlF shows promise as an excellent surrogate pseudo-radiometal instead of gallium-68 at both 1 and 3 h p.i. A more delicate stoichiometric balance exists for achieving near quantitative radiolabeling yields with [$^{18}$F]AlF on the NOTA chelator than for traditional radiometals. Therefore, further work is needed to automate the production and purification to reach the potential of [$^{18}$F]AlF-NOTA-NOC in the clinic.

**Author Contributions:** J.H.D. and N.L. designed, performed, and analyzed the experiments. C.B., A.Y.N. and B.B.O. performed the in vitro and vivo work. H.T. analyzed the data and interpreted the results. J.H.D., H.T. and N.L. wrote the manuscript. All authors have read and agreed to the published version of the manuscript.

**Funding:** The research project was solely funded by the Department of Nuclear Medicine, Odense University Hospital.

**Institutional Review Board Statement:** The animal study protocol was approved by The Animal Experiments Inspectorate in Denmark: protocol license 2016-15-0201-01027, approved 25 August 2016.

**Informed Consent Statement:** Not applicable.

**Data Availability Statement:** The datasets used and/or analyzed during the current study are available from the corresponding author upon reasonable request.

**Acknowledgments:** The authors acknowledge the Danish Molecular Biomedical Imaging Center (DaMBIC) for the use of the small animal PET/CT facilities at the University of Southern Denmark.

**Conflicts of Interest:** The authors declare no conflict of interest.

## Abbreviations

| | |
|---|---|
| ACN | Acetonitrile |
| AlF | [$^{18}$F]aluminum fluoride |
| PET | Positron emission tomography |
| CT | Computed tomography |
| DI | De-ionized |
| DOTA | 1,4,7,10-Tetraazacyclododecane-$N,N',N'',N'''$-tetraacetic acid |
| HPLC | High pressure liquid chromatography |
| NOC | -[1-Nal3]-Octreotide |
| NOTA | 1,4,7-Triazacyclononane-1,4,7-triyltriacetic acid |
| RCPSCID | Radiochemical puritySevere combined immunodeficiency |
| SEM | Standard error of the mean |
| TATETFA | [Tyr3, Thr8]-octreotideTrifluoroacetic acid |

## References

1. Zandee, W.T.; Kamp, K.; van Adrichem, R.C.; Feelders, R.A.; de Herder, W.W. Effect of hormone secretory syndromes on neuroendocrine tumor prognosis. *Endocr. Relat. Cancer* **2017**, *24*, R261–R274. [CrossRef] [PubMed]
2. Strosberg, J.; El-Haddad, G.; Wolin, E.; Hendifar, A.; Yao, J.; Chasen, B.; Mittra, E.; Kunz, P.L.; Kulke, M.H.; Jacene, H.; et al. Phase 3 Trial of $^{177}$Lu-Dotatate for Midgut Neuroendocrine Tumors. *N. Engl. J. Med.* **2017**, *376*, 125–135. [CrossRef] [PubMed]
3. Sahnoun, S.; Conen, P.; Mottaghy, F.M. The battle on time, money and precision: Da[(18)F] id vs. [(68)Ga]liath. *Eur. J. Nucl. Med. Mol. Imaging* **2020**, *47*, 2944–2946. [CrossRef] [PubMed]
4. Kumar, K. The Current Status of the Production and Supply of Gallium-68. *Cancer Biotherapy Radiopharm.* **2020**, *35*, 163–166. [CrossRef]
5. Synowiecki, M.A.; Perk, L.R.; Nijsen, J.F.W. Production of novel diagnostic radionuclides in small medical cyclotrons. *EJNMMI Radiopharm. Chem.* **2018**, *3*, 3. [CrossRef]
6. Schweinsberg, C.; Johayem, A.; Llamazares, A.; Gagnon, K. The First Curie-Quantity Production of [68Ga]Ga-PSMA-HBED-CC. *J. Label. Comp. Radiopharm.* **2019**, *62*, S276.
7. Kumlin, J.; Dam, J.H.; Chua, C.J.; Borjian, S.; Kassaian, A.; Hook, B.; Zeisler, S.; Schaffer, P.; Thisgaard, H. Multi-Curie Production of Gallium-68 on a Biomedical Cyclotron. *Eur. J. Nuclear Med. Mol. Imaging* **2019**, *46*, S39.
8. Thisgaard, H.; Kumlin, J.; Langkjær, N.; Chua, J.; Hook, B.; Jensen, M.; Kassaian, A.; Zeisler, S.; Borjian, S.; Cross, M.; et al. Multi-curie production of gallium-68 on a biomedical cyclotron and automated radiolabelling of PSMA-11 and DOTATATE. *EJNMMI Radiopharm. Chem.* **2021**, *6*, 1. [CrossRef]
9. McBride, W.J.; Sharkey, R.M.; Karacay, H.; D'Souza, C.A.; Rossi, E.A.; Laverman, P.; Chang, C.H.; Boerman, O.C.; Goldenberg, D.M. A novel method of 18F radiolabeling for PET. *J. Nucl. Med.* **2009**, *50*, 991–998. [CrossRef]
10. McBride, W.J.; Goldenberg, D.M. Methods and Compositions for Improved F-18 Labeling of Proteins, Peptdes and Othermolecules. US Patent 2008/0170989 A1, 17 July 2008.
11. Lisova, K.; Sergeev, M.; Evans-Axelsson, S.; Stuparu, A.D.; Beykan, S.; Collins, J.; Jones, J.; Lassmann, M.; Herrmann, K.; Perrin, D.; et al. Microscale radiosynthesis, preclinical imaging and dosimetry study of [(18)F]AMBF3-TATE: A potential PET tracer for clinical imaging of somatostatin receptors. *Nucl. Med. Biol.* **2018**, *61*, 36–44. [CrossRef]
12. Ilhan, H.; Lindner, S.; Todica, A.; Cyran, C.C.; Tiling, R.; Auernhammer, C.J.; Spitzweg, C.; Boeck, S.; Unterrainer, M.; Gildehaus, F.J.; et al. Biodistribution and first clinical results of (18)F-SiFAlin-TATE PET: A novel (18)F-labeled somatostatin analog for imaging of neuroendocrine tumors. *Eur. J. Nucl. Med. Mol. Imaging* **2020**, *47*, 870–880. [CrossRef] [PubMed]
13. Liu, Y.; Hu, X.; Liu, H.; Bu, L.; Ma, X.; Cheng, K.; Li, J.; Tian, M.; Zhang, H.; Cheng, Z. A Comparative Study of Radiolabeled Bombesin Analogs for the PET Imaging of Prostate Cancer. *J. Nucl. Med.* **2013**, *54*, 2132–2138. [CrossRef] [PubMed]
14. Chatalic, K.L.; Franssen, G.M.; van Weerden, W.M.; McBride, W.J.; Laverman, P.; de Blois, E.; Hajjaj, B.; Brunel, L.; Goldenberg, D.M.; Fehrentz, J.-A.; et al. Preclinical Comparison of Al18F- and 68Ga-Labeled Gastrin-Releasing Peptide Receptor Antagonists for PET Imaging of Prostate Cancer. *J. Nucl. Med.* **2014**, *55*, 2050–2056. [CrossRef]
15. Laverman, P.; McBride, W.J.; Sharkey, R.M.; Eek, A.; Joosten, L.; Oyen, W.J.; Goldenberg, D.M.; Boerman, O.C. A Novel Facile Method of Labeling Octreotide with $^{18}$F-Fluorine. *J. Nucl. Med.* **2010**, *51*, 454–461. [CrossRef] [PubMed]
16. McBride, W.J.; D'Souza, C.A.; Karacay, H.; Sharkey, R.M.; Goldenberg, D.M. New Lyophilized Kit for Rapid Radiofluorination of Peptides. *Bioconjugate Chem.* **2012**, *23*, 538–547. [CrossRef] [PubMed]
17. Persson, M.; Liu, H.; Madsen, J.; Cheng, Z.; Kjaer, A. First 18F-labeled ligand for PET imaging of uPAR: In vivo studies in human prostate cancer xenografts. *Nucl. Med. Biol.* **2013**, *40*, 618–624. [CrossRef]
18. Tshibangu, T.; Cawthorne, C.; Serdons, K.; Pauwels, E.; Gsell, W.; Bormans, G.; Deroose, C.M.; Cleeren, F. Automated GMP compliant production of [(18)F]AlF-NOTA-octreotide. *EJNMMI Radiopharm. Chem.* **2020**, *5*, 4. [CrossRef]
19. Prasad, V.; Baum, R.P. Biodistribution of the Ga-68 labeled somatostatin analogue DOTA-NOC in patients with neuroendocrine tumors: Characterization of uptake in normal organs and tumor lesions. *Q. J. Nucl. Med. Mol. Imaging* **2010**, *54*, 61–67.

20. Thisgaard, H.; Olsen, B.B.; Dam, J.H.; Bollen, P.; Mollenhauer, J.; Høilund-Carlsen, P.F. Evaluation of Cobalt-Labeled Octreotide Analogs for Molecular Imaging and Auger Electron–Based Radionuclide Therapy. *J. Nucl. Med.* **2014**, *55*, 1311–1316. [CrossRef]
21. Schmitt, A.; Bernhardt, P.; Nilsson, O.; Ahlman, H.; Kölby, L.; Schmitt, J. Forssel-Aronsson, E., Biodistribution and dosimetry of 177Lu-labeled [DOTA$^0$, Tyr$^3$]Octreotate in male nude mice with human small cell lung cancer. *Cancer Biother. Radiopharm.* **2003**, *18*, 593–599. [CrossRef]
22. Tolmachev, V.; Orlova, A.; Andersson, K. Methods for Radiolabelling of Monoclonal Antibodies. In *Human Monoclonal Antibodies: Methods and Protocols (Methods in Molecular Biology)*; Steinitz, M., Ed.; Springer Science, LLC: Totowa, NJ, USA, 2014; pp. 309–330. ISBN 978-1-62703-586-6-16.
23. Dam, J.H.; Olsen, B.B.; Baun, C.; Høilund-Carlsen, P.F.; Thisgaard, H. In Vivo Evaluation of a Bombesin Analogue labeled with Ga-68 and Co-55/57. *Mol. Imaging Biol.* **2016**, *18*, 368–376. [CrossRef] [PubMed]

Article

# Effective Preparation of [$^{18}$F]Flumazenil Using Copper-Mediated Late-Stage Radiofluorination of a Stannyl Precursor

Mohammad B. Haskali [1,2,*], Peter D. Roselt [2], Terence J. O'Brien [3], Craig A. Hutton [4,5], Idrish Ali [3], Lucy Vivash [3] and Bianca Jupp [3]

1 Sir Peter MacCallum Department of Oncology, The University of Melbourne, Melbourne, VIC 3010, Australia
2 The Radiopharmaceutical Research Laboratory, The Peter MacCallum Cancer Centre, Melbourne, VIC 3000, Australia
3 Department of Neuroscience, Central Clinical School, Monash University, Melbourne, VIC 3004, Australia
4 School of Chemistry, The University of Melbourne, Melbourne, VIC 3010, Australia
5 Bio21 Molecular Science and Biotechnology Institute, The University of Melbourne, Melbourne, VIC 3010, Australia
* Correspondence: mo.haskali@petermac.org; Tel.: +61-(3)-8559-6913

**Citation:** Haskali, M.B.; Roselt, P.D.; O'Brien, T.J.; Hutton, C.A.; Ali, I.; Vivash, L.; Jupp, B. Effective Preparation of [$^{18}$F]Flumazenil Using Copper-Mediated Late-Stage Radiofluorination of a Stannyl Precursor. *Molecules* **2022**, *27*, 5931. https://doi.org/10.3390/molecules27185931

Academic Editor: Svend Borup Jensen

Received: 1 July 2022
Accepted: 5 September 2022
Published: 13 September 2022

**Publisher's Note:** MDPI stays neutral with regard to jurisdictional claims in published maps and institutional affiliations.

**Copyright:** © 2022 by the authors. Licensee MDPI, Basel, Switzerland. This article is an open access article distributed under the terms and conditions of the Creative Commons Attribution (CC BY) license (https://creativecommons.org/licenses/by/4.0/).

**Abstract:** (1) Background: [$^{18}$F]Flumazenil **1** ([$^{18}$F]FMZ) is an established positron emission tomography (PET) radiotracer for the imaging of the gamma-aminobutyric acid (GABA) receptor subtype, GABA$_A$ in the brain. The production of [$^{18}$F]FMZ **1** for its clinical use has proven to be challenging, requiring harsh radiochemical conditions, while affording low radiochemical yields. Fully characterized, new methods for the improved production of [$^{18}$F]FMZ **1** are needed. (2) Methods: We investigate the use of late-stage copper-mediated radiofluorination of aryl stannanes to improve the production of [$^{18}$F]FMZ **1** that is suitable for clinical use. Mass spectrometry was used to identify the chemical by-products that were produced under the reaction conditions. (3) Results: The radiosynthesis of [$^{18}$F]FMZ **1** was fully automated using the iPhase FlexLab radiochemistry module, affording a 22.2 ± 2.7% (*n* = 5) decay-corrected yield after 80 min. [$^{18}$F]FMZ **1** was obtained with a high radiochemical purity (>98%) and molar activity (247.9 ± 25.9 GBq/µmol). (4) Conclusions: The copper-mediated radiofluorination of the stannyl precursor is an effective strategy for the production of clinically suitable [$^{18}$F]FMZ **1**.

**Keywords:** radiofluorination; PET imaging; GABA$_A$; radiochemistry; Flumazenil 1; benzodiazepine; stannyl; stannane

## 1. Introduction

Flumazenil **1**, marketed as Romazicon, is primarily used clinically for the treatment of a benzodiazepine overdose and for the reversal of the sedative effects of anesthesia [1,2]. Flumazenil **1** is a potent competitive antagonist at the benzodiazepine site of the gamma-aminobutyric acid (GABA) receptor subtype, GABA$_A$. The amino acid GABA is the primary inhibitory neurotransmitter in the central nervous system (CNS) and functions to inhibit neuronal activity through its action at the GABA$_A$ receptors [3]. In addition to binding to GABA, these ligand-gated ion channels also bind benzodiazepines, a class of psychoactive drugs with a core structure encompassing a fused benzene ring and a diazepine ring [4,5] which facilitates the action of GABA at this receptor.

Alterations in GABAergic function including at the GABA$_A$ receptor is associated with a variety of neurological disorders including substance use disorder, schizophrenia, autism spectrum disorder, major depressive disorder and epilepsy [6–11]. Therefore, a quantitative assessment of GABA$_A$ receptors in the brain using positron emission tomography (PET) can provide valuable information concerning the GABAergic function of a broad range of neurological and neuropsychiatric conditions [12,13].

PET represents the most selective and sensitive (pico- to nano-molar range) non-invasive molecular imaging technique for the quantification of receptor density and drug interactions in vivo [14]. PET utilizes biologically active drugs at tracer doses which are radiolabeled with short-lived positron-emitting radionuclides. Fluorine-18 is one of the most useful positron-emitting radionuclides with ideal properties for PET imaging. Therefore, F-18-labelled flumazenil 1 has evolved to be one of the most useful radiopharmaceuticals for the PET imaging of GABA$_A$ receptors [13,15]. [$^{18}$F]Flumazenil 1 ([$^{18}$F]FMZ) is now well established in the clinical management of drug-resistant temporal lobe epilepsy (TLE) with excellent sensitivity and anatomical resolution [13].

[$^{18}$F]FMZ 1 is commonly prepared by the nucleophilic aromatic substitution (S$_N$Ar) of the nitro group on nitromazenil using 'naked' fluoride-18 (Figure 1) [16]. Considering that nucleophilic aromatic substitution by fluoride-18 ion does not proceed well unless it is activated by a strong electron withdrawing group in the ortho- or para-position, the effective production of [$^{18}$F]FMZ 1 by S$_N$Ar has been subjected to considerable optimization to proceed with a satisfactory RCY [17]. The electron withdrawing groups enable the radiofluorination of arenes by reducing the electron density at the target carbon and by resonance stabilization of the arising Meisenheimer complex in an addition-elimination mechanism. Therefore, improved methods for the preparation of [$^{18}$F]FMZ 1 in high yield and under mild chemical conditions, affording a satisfactory chemical purity, has been the subject of over a decade of radiopharmaceutical research.

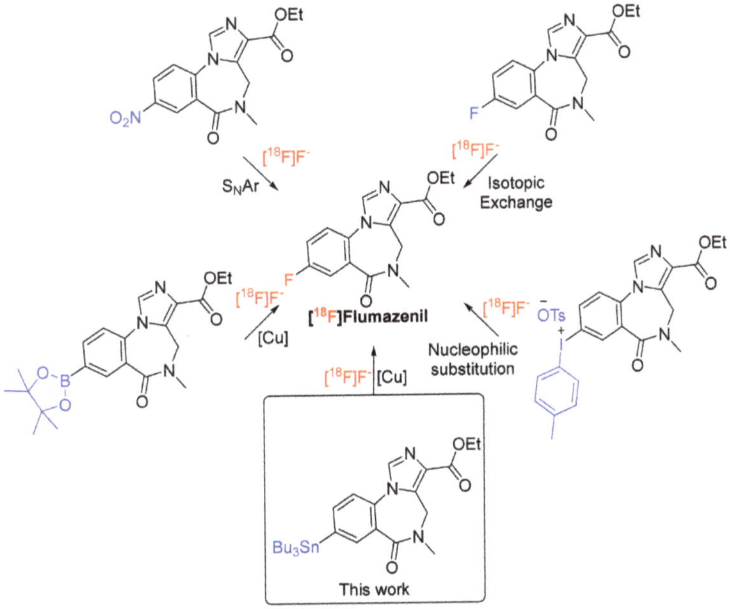

**Figure 1.** Nucleophilic radiofluorination methods for the preparation of [$^{18}$F]flumazenil 1.

Windhorst and co-workers initially reported the radiosynthesis of [$^{18}$F]FMZ 1 by the substitution of the nitro group of nitromazenil or by the isotopic exchange (F-19 substituted by F-18) of flumazenil 1 (Figure 1) [18]. The radiofluorination was performed at an elevated temperature ($\geq$130 °C), a reduced pressure ($\leq$0.8 bar) and with the use of 2,4,6-trimethylpyridine as solvent. These harsh conditions afforded a moderate (44% yield by isotopic exchange) to low yield ($\leq$12% yield by nitro substitution). Furthermore, the method using an isotopic exchange afforded [$^{18}$F]FMZ 1 with a low molar activity (the ratio of the quantity of F-18-labeled molecules to the total quantity of FMZ (i.e., containing either F-18 or F-19)) [19]. The preparation of [$^{18}$F]FMZ 1 by isotopic exchange was further improved by Ryzhikov and co-workers by employing standard radiofluorination conditions, but the final product retained a low molar activity [20]. Halldin and co-workers optimized the radiofluorination conditions for the substitution of the nitro group of nitromazenil to afford a moderate radiochemical yield (30%) and a high molar activity of [$^{18}$F]FMZ **1**, at a high temperature (160 °C for 30 min) [16]. These conditions have since become the method of choice for the preparation of [$^{18}$F]FMZ **1**, with nitromazenil precursor becoming commercially available through specialized suppliers. However, we note that Schirrmacher and co-workers reported only a 2–5% overall radiochemical yield (RCY) when adopting these conditions [21]. Similarly, van Dam and co-workers reported only a 7 ± 2% overall RCY under these conditions, thus highlighting the limitations of this method [22]. Our own experience with this method has also led to 3–4% isolated yields of [$^{18}$F]FMZ **1** [13,23,24]. Some of these variations in RCY could arise from inconsistencies in the measurement of RCY and by not taking into consideration the recovery of radioactivity from the systems that are in use.

A major breakthrough in the radiofluorination of inert aromatic systems was first described by Pike and coworkers, using diaryliodonium salts to facilitate substitution reactions [25]. This method has been applied by Seok Moon and co-workers to prepare [$^{18}$F]FMZ **1** in a high radiochemical yield (67.2 ± 2.7% decay-corrected) [26]. More recent developments in F-18 radiochemistry have paved the way for the preparation of a wide range of radiopharmaceuticals that were previously inaccessible. One key development relates to the copper-mediated late-stage radiofluorination of aryl boronates and aryl stannanes. The application of this method to the preparation of [$^{18}$F]FMZ **1** from a boronic ester precursor has been disclosed (Figure 1), affording a ≤17% decay-corrected radiochemical yield ($n$ = 2) [27]. The usefulness of this method for the preparation of [$^{18}$F]FMZ **1** for clinical use has recently been demonstrated [28]. Importantly, this recent development has optimized the formation of [$^{18}$F]FMZ **1** to over a 30–48% radiochemical yield (non-decay-corrected) when TBA-HCO$_3$ was utilized as a base. One challenging aspect relates to the susceptibility of boronic esters to hydrolysis and protodeboronation under the column chromatography conditions [27]. The copper-mediated radiofluorination of boronic esters requires air for the oxidative transformation of Cu(II) to Cu(III) in the catalytic cycle that facilitates aromatic radiofluorination [17,27,28]. These limitations may reduce the usefulness of this method.

In contrast, the copper-mediated late-stage radiofluorination of aryl stannanes does not require air to proceed, and the aryltrialkylstannanes are generally stable under chromatographic conditions. The preparation of [$^{18}$F]FMZ **1** through the copper-mediated late-stage fluorination of a stannyl precursor has not been reported. This method introduces some potential advantages, including the amenability to prepare large amounts of a stannyl precursor (gram scale) and its simple purification using column chromatography, with potentially high radiochemical yields, high molar activity and simple automation processes. Herein, we report the synthesis of a stannyl precursor to flumazenil and the conditions for its radiofluorination. Furthermore, we translate these conditions using the iPHASE FlexLab radiochemistry module to fully automate the radiosynthesis of [$^{18}$F]FMZ **1**. We also investigate the chemical nature of the side products formed in the reaction using mass spectrometry, and we report on other quality control aspects of the final formulated product and its suitability for clinical use.

## 2. Results

### 2.1. Synthesis of Stannyl Precursor 3

Stannyl precursor **3** was synthesized by the palladium-catalyzed stannylation of bromomazenil **2** (Figure 2). The stannylation reaction proceeded smoothly to produce precursor **3** in good chemical yield (65%) after 6 h in toluene. The stannyl precursor **3** was purified using flash chromatography, affording a pure solid material ready for the radiofluorination reactions.

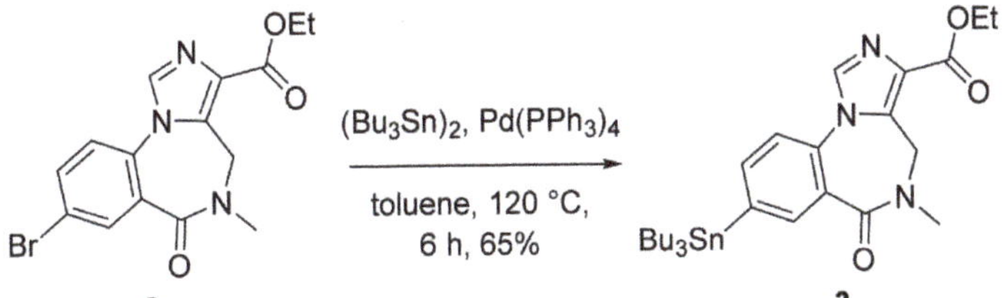

**Figure 2.** Synthesis of stannyl precursor 3.

## 2.2. Radiosynthesis of [$^{18}$F]flumazenil 1

With the pure stannyl precursor 3 in hand, we moved toward the optimization of the copper-catalyzed radiofluorination conditions to obtain [$^{18}$F]FMZ 1 in good radiochemical yield (Figure 3). The optimal radiofluorination conditions using copper (II) triflate, pyridine and DMA were maintained as recommended by Scott and co-workers [29]. We investigated the effects of temperature and time on the yield of [$^{18}$F]FMZ 1 and its formation.

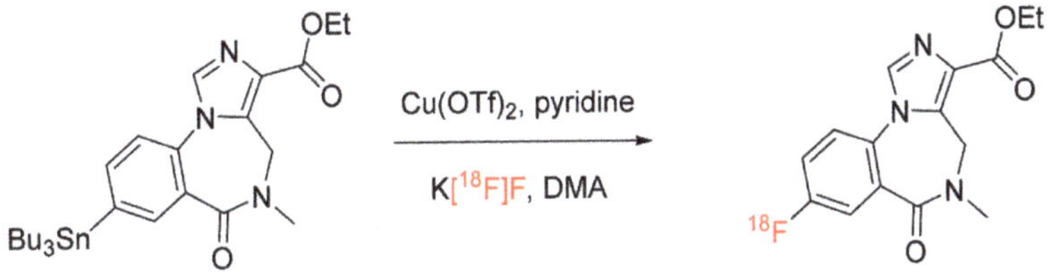

**Figure 3.** Cu-catalyzed radiosynthesis of [$^{18}$F]FMZ 1.

The [$^{18}$F]FMZ 1 % formation was heavily dependent on the reaction temperature. At 80 °C only 10.2% [$^{18}$F]FMZ 1 was observed after 10 min of reaction time, and the total radioactivity that was recovered from the analytical HPLC column was less than 30%. The low recovery from the analytical HPLC column indicates that the radiochemical yield of [$^{18}$F]FMZ 1 may be even lower than was observed (10.2%), as the majority of the free F-18 ions may be retained in the column. Free F-18 is known to have a notorious affinity to silica-based HPLC columns at a low pH [30]. As such, it is essential to consider the % column recovery of the total radioactivity that was injected when determining the radiochemical yield by the integration of a radiochemical HPLC trace.

Increasing the reaction temperature to 100 °C increased the radiochemical yield of [$^{18}$F]FMZ 1 to 51.9%. The % recovery of the radioactivity from the HPLC column also increased to 60% at 100 °C, indicating significant improvements in the radiochemical formation of [$^{18}$F]FMZ 1. Increasing the reaction temperature to 120 °C further increased radiochemical yield to 58.1% and the % recovery of the radioactivity to 63%. However, increasing the reaction temperature further to 140 °C did not increase the radiochemical yield, but it did increase the % recovery of the radioactivity to 80%. This may indicate that the isolated yield of [$^{18}$F]FMZ 1 maybe higher at 140 °C. However, we chose to continue to perform our optimization reactions at 120 °C to avoid the harsher conditions at 140 °C that can damage the reactor apparatus over time.

We then explored the effect of reaction time on the radiochemical yield of [$^{18}$F]FMZ 1. Reducing the reaction time to 5 min reduced the radiochemical yield of [$^{18}$F]FMZ 1 at 120 °C to 51.1%, with similar HPLC column recovery (63.6%). Increasing reaction time to 20 min also reduced the radiochemical yield of [$^{18}$F]FMZ 1 to 51.2%, with a slight increase in HPLC column recovery (71.0%). As such, the optimal conditions for the copper-catalyzed radiofluorination of the stannyl precursor 3 to form [$^{18}$F]FMZ 1 was identified to be at 120 °C for 10 min (Table 1).

**Table 1.** Optimization conditions for the production of [$^{18}$F]FMZ 1.

| Entry [1] | Temperature (°C) | Time (min) | [$^{18}$F]FMZ RCY% [2] | HPLC Column Recovery (%) [3] |
|---|---|---|---|---|
| 1 | 80  | 10 | 10 | 29.7 |
| 2 | 100 | 10 | 52 | 60.0 |
| 3 | 120 | 10 | 58 | 63.0 |
| 4 | 140 | 10 | 57 | 80.0 |
| 5 | 120 | 5  | 51 | 63.6 |
| 6 | 120 | 20 | 51 | 71.0 |

[1] Radiolabeling conditions: 10 µmol stannyl precursor 3, 40 µmol Cu(OTf)$_2$, 150 µmol pyridine in 1 mL DMA. [2] Radiochemical yield determined by integration of the HPLC radiochemical trace obtained from the analytical HPLC analysis of the crude reaction mixture. Reactions were performed once under each of these conditions. HPLC analysis was performed on Kinetex 5 µm XB-C18 4.6 × 150 mm column, 0.1% TFA in 15–90% MeCN:H$_2$O over 7 min. [3] HPLC column recovery (%) obtained by accurately measuring total injected amount of radioactivity and the total collected radioactive eluted from the HPLC column. % Recovery was calculated as following: (collected total radioactivity/injected total radioactivity) * 100. HPLC analysis was performed on Kinetex 5 µm XB-C18 4.6 × 150 mm column, 0.1% TFA in 15–90% MeCN:H$_2$O over 7 min.

The optimal reaction conditions were than translated using the iPHASE Flexlab radiochemistry module to automate the full production process, including the F-18 isolation and workup, radiofluorination, HPLC purification and final formulation of [$^{18}$F]FMZ 1. [$^{18}$F]FMZ 1 was produced using the iPHASE Flexlab module using the configurations that are presented in Table 2. The total automated production time was 80 min to produce the fully formulated [$^{18}$F]FMZ 1 in a 22.2 ± 2.7% (n = 5) isolated yield with was decay-corrected to end of synthesis.

**Table 2.** Preparation details for the automated production of [$^{18}$F]FMZ 1 using the iPHASE Flexlab radiochemistry module.

| Entry | Position | Reagents or Materials | Quantities |
|---|---|---|---|
| 1 | V 13–V 15 | Sep-Pak Light QMA | 1 |
| 2 | V 1 | KOTf/ K$_2$CO$_3$ in H$_2$O | 10 mg KOTf and 50 µg of K$_2$CO$_3$ in 550 µL H$_2$O |
| 3 | V 4 | Stannyl Precursor 3 and Cu(OTf)$_2$/ Pyridine mixture | 5 mg of precursor 3 in 1.0 mL DMA containing 13 mg of Cu(OTf)$_2$ and 14 µL of pyridine |
| 4 | V 6 | 0.1% TFA in MeCN:H$_2$O | 3.0 mL (2.5:0.5) |
| 5 | V 8 | Saline | 9 mL |
| 6 | V 9 | Ethanol | 1 mL |
| 7 | V 11 | Saline | 5 mL |
| 8 | HPLC Flask 1 | Milli-Q Water | 50 mL |
| 9 | V 22–V 46 | C18 SPE Cartridge | 1 |

### 2.3. Chemical Characterization and Quality Control of [$^{18}$F]FMZ 1 Formulation

The [$^{18}$F]FMZ 1 was produced to meet the quality control specifications for F-18-labelled radiopharmaceuticals which are in line with the international standards. Furthermore, special reference was made to the N-[$^{11}$C-methyl]flumazenil injection monograph (BP) when assigning the release criteria in Table 3. [$^{18}$F]FMZ 1 was produced in high radiochemical purity (>98%). The ethanol that was used during the formulation of [$^{18}$F]FMZ 1 was always below 10%, and the other residual solvents that were used throughout the synthesis (MeCN and DMA) remained below their respective limit, as presented in Table 3.

**Table 3.** Quality control specifications of [$^{18}$F]FMZ **1**.

| Parameter | Specification | Observed Results ($n = 3$) |
|---|---|---|
| Appearance | Clear and colorless | Pass |
| pH | 4–8 | 5–6 |
| Residual Solvents (MeCN) (%V/V) | <0.04 | 0.0066% ± 0.0018% (0.006–0.008%)³ |
| Residual Solvents (DMA) (%V/V) | <0.11 | Not detected |
| Ethanol Determination (%V/V) | <10% | 7.16% ± 1.38% (6.1–8.1%)³ |
| Radionuclidic identity (half-life) | 105–115 min | 108–113 |
| Radiochemical identity (HPLC) | Reference standard ± 1.0 min | Reference Std: 8.5 min Product: 8.6 min |
| Radiochemical Purity (HPLC) | ≥95% | >98% |
| Radiochemical Purity (TLC) | ≥98% | >98% |
| Molar Activity | ≥37 GBq/μmol | 247.9 ± 25.9 GBq/μmol (222–274 GBq/μmol) |
| Copper * | ≤34 ppm | 0.0157 ± 0.005 ppm |
| Sterility | Sterile | No growth observed |
| Endotoxin | ≤175 IU/V | Pass |
| Filter Integrity (bubble point test) | ≥50 psi | Pass |

* Copper was quantified by inductively coupled plasma-mass spectrometry (ICP-MS) of the non-radioactive product after complete decay of F-18.

### 2.4. MS Analysis of By-Products Generated during the Radiosynthesis of [$^{18}$F]FMZ **1**

Stannyl precursor **3** was completely consumed after the reaction at 120 °C for 10 min. An MS analysis of the crude reaction mixture indicated the formation of at least three possible by-products as well as a trace amount of the carrier flumazenil **1**. The major by-product that was formed was assigned as hydroxy-mazenil **4** with the minor by-products being assigned as des-fluoro-flumazenil **5** and the dimeric mazenil **6** (Figure 4). The MS analysis of des-fluoro-flumazenil **5** presented the expected $m/z$ of 286 but there was also another $m/z$ peak at 387.2 which we were unable to identify. The formation of by-products **4–6** was further supported by a tandem MS/MS analysis, presenting fragmentation profiles that are consistent with that of an authentic reference standard of flumazenil **1** (MS/MS spectra and observed fragments presented in the Supplementary Materials File).

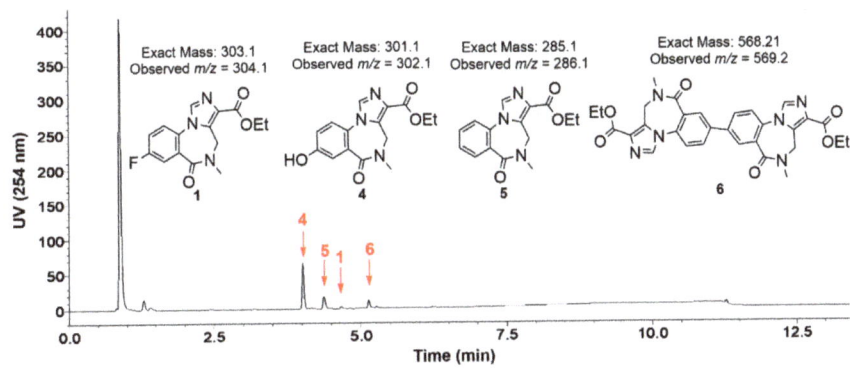

**Figure 4.** LC-MS/MS analysis of major by-products that were formed in the crude reaction mixture of [$^{18}$F]FMZ **1** at 120 °C after 10 min reaction time.

### 2.5. PET Imaging Using [$^{18}$F]flumazenil **1**

PET imaging using [$^{18}$F]FMZ **1** produced by our current method demonstrated uptake and binding that was consistent with that which was previously observed in rodent studies of [$^{18}$F]FMZ **1** PET, with uptake that was primarily concentrated in the cortices and hippocampi, with minimal uptake in the pons. (Figure 5). The average hippocampal $B_{max}$ was 19.45 ± 1.5 pmol/mL, while the $1/K_D$ was 0.25 ± 0.03 pmol/mL (Figure 6), which is in line with the values that were acquired from our previous study in naïve rats [23].

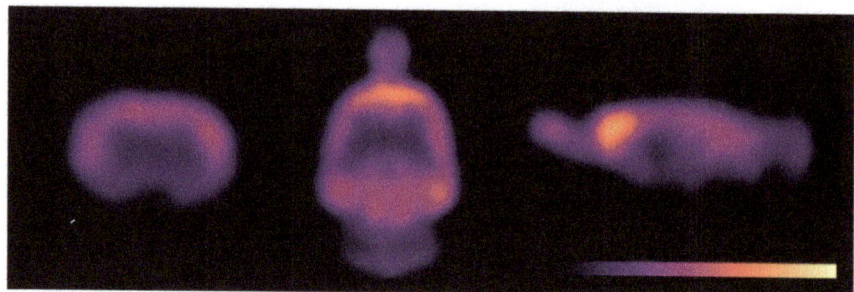

**Figure 5.** Representative image of [$^{18}$F]flumazenil **1** uptake.

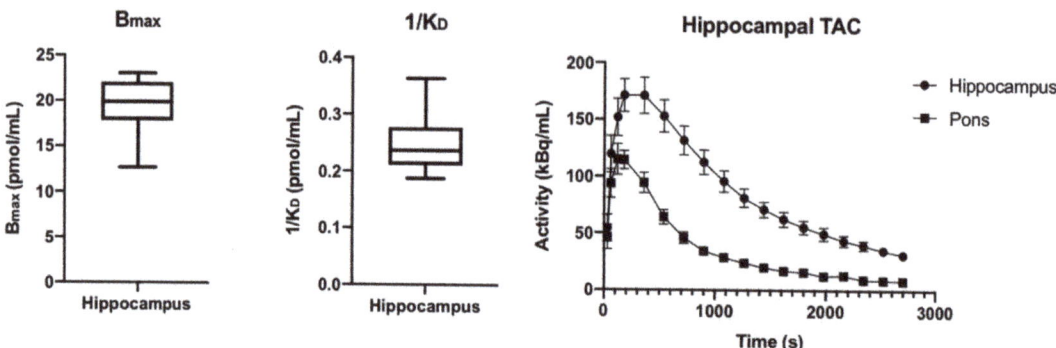

**Figure 6.** Box and whisker plots of hippocampal $B_{max}$ and $1/K_D$ values and average time activity curves (TAC) from hippocampal and pons volumes of interest.

## 3. Discussion

The successful production protocols for a given radiopharmaceutical are determined by many factors, including the amenability of a high-scale precursor synthesis, the efficacy and simplicity of the radiofluorination process and the suitability of the final product for its clinical use as determined by its quality control characteristics. The commonly employed method for the preparation of [$^{18}$F]FMZ **1** through the nucleophilic substitution of a nitro leaving group suffers from a low yield and poor reproducibility. In our laboratory, this method generated [$^{18}$F]FMZ **1** in less than 4% radiochemical yield and with a complex HPLC purification. Alternative methods have been investigated for the preparation of [$^{18}$F]FMZ **1**, including its radiochemical synthesis from diaryliodonium salts and boronic esters [26–28]. The latter method suffers from the instability of the boronic ester precursor under chromatographic conditions, and the need for an aerated reaction vessel [27,28].

The preparation and radiofluorination of the stannyl precursor **3** overcomes many of these complexities. The synthesis of stannyl precursor **3** proceeded in one step from the bromo-precursor **2**, and the synthesis proved to be scalable (≥2 g), affording high purity precursor after a rapid flash column chromatography. Furthermore, precursor **3** has proved stable for over a year when it is stored at 4 °C. The radiofluorination of stannyl precursor **3** proceeded efficiently without the need for aeration or phase transfer catalysts (including kryptofix or tetrabutylammonium bicarbonate). We note that Scott and co-workers have also demonstrated the sensitivity of the Cu-mediated radiofluorination of stannyl precursors to the type of base/phase transfer catalyst that is used, and therefore, potassium triflate that is doped with potassium carbonate was used as the optimal combination [29]. Nonetheless, it is useful that no phase transfer catalyst is required under these conditions, thereby eliminating the need to test the end product for kryptofix or tetrabutylammonium before its clinical administration. Finally, [$^{18}$F]FMZ **1** was obtained in high radiochemical purity after HPLC purification. Nevertheless, small closely eluting UV-active byproducts were formed, and the HPLC conditions needed to be optimized carefully to allow for the adequate separation of the product from the impurities.

The analysis of the crude reaction mixture identified the formation of several chemical by-products which may be addressed in the future to further enhance the yield of [$^{18}$F]FMZ **1**, as well as to improve the purification process. The stannyl precursor **3** was completely consumed in the reaction mixture and the hydroxylated by-product **4** was found to be the major by-product along with by the product of protodestannylation (affording desfluoro-flumazenil **5**). Hydroxylated by-product **4** may arise through the oxidation of the stannane under the reaction conditions [31]. A small amount of biaryl by-product **6** was also formed through homocoupling [32,33]. However, considering that the yield of [$^{18}$F]FMZ **1** was useful for multi-patient dose preparation, and the HPLC purification was successfully optimized, no further optimizations were performed. This new method of [$^{18}$F]FMZ preparation will facilitate the more ready utilization of this highly selective and sensitive radiotracer for the GABA$_A$ receptor in clinical practice, to assist in the localization of the epileptogenic zone in patients with drug resistant TLE [13], as well as other potential clinical applications where the dysfunction GABA$_A$ receptors are believed to play a role such as traumatic brain injury, schizophrenia, addiction and anxiety.

## 4. Materials and Methods

All chemicals obtained commercially were of analytical grade and used without further purification. No-carrier-added fluoride-18 was obtained from a PETtrace 16.5MeV cyclotron (Cyclotek) incorporating a high-pressure niobium target via the $^{18}$O(p,n)$^{18}$F nuclear reaction (98% $^{18}$O isotopic enrichment). Radiochemical synthesis was performed using an iPHASE Flexlab radiochemistry module purchased from iPHASE Technologies Pty. Ltd. Australia. F-18 Separation cartridges (QMA strong anion exchange cartridge, Waters Australia) were preconditioned with 0.5 mL of 0.05M solution of KOTf, followed by 5 mL water. Reversed phase solid phase extraction (SPE) cartridges (33 μm polymeric reversed phase (30 mg/mL), Phenomenex, Lane Cove West, NSW, Australia) were preconditioned with ethanol and rinsed with water before use. Radioactivity measurements were carried out using a CRC-15PET dose calibrator (Capintec, Florham Park, NJ, USA) that was calibrated daily using Cs-137 and Co-57 sources (Isotope Products Laboratories, Valencia, CA, USA). Radiation was detected using a solid-state photodiode scintillator crystal detector (Knauer, Berlin, Germany). Preparative high performance liquid radiochemical chromatography (HPLRC) was performed using a Knauer 1050 pump, 2500 UV detector, and 5050 manager. Radiation was detected using a Knauer solid state photodiode scintillation crystal detector in a TO-5 case. Analytical HPLRC was performed using a Shimadzu HPLC system consisting of a CBM-20A system controller, SIL-20A auto-injector, LC-20AD solvent delivery unit, CTO-20A control valve, DGU-20A degasser and a SPD-M20A detector coupled to a LCMS-8030 triple-quadrupole mass spectrometer. This was coupled to a radiation detector consisting of an Ortec model 276 photomultiplier base with a 925-SCINTACE-mate preamplifier, amplifier, bias supply, and SCA and a Bicron 1M11/2 photomultiplier tube. Gas Chromatography (GC) analysis was performed using a Shimadzu GC-17A instrument coupled with an AOC-20i auto injector. $^1$H NMR spectra of small molecules were obtained using a 400 MHz Agilent DD2 NMR Spectrometer.

**Synthesis of ethyl 5,6-dihydro-5-methyl-6-oxo-8-tributylstannyl-4H-imidazo-[1,5-a][1,4]benzodiazepine-3-carboxylate (3)**: To a solution of ethyl 8-bromo-5,6-dihydro-5-methyl-6-oxo(4H)-imidazo[1,5-a][1,4]benzodiazepine-3-carboxylate **2** (2.2 g, 6.04 mmol) in toluene (20 mL) was added tetrakis-(triphenylphosphine)palladium(0) (0.25 g, 0.22 mmol) and bis(tributyltin) (10.5 g, 18.2 mmol). The reaction was purged with nitrogen in a pressure vessel and then heated to 120 °C for 6 h. The reaction mixture was then cooled and diluted with ethyl acetate (300 mL) and washed with water, dried over anhydrous sodium sulfate, and evaporated to dryness in vacuo. Crude reaction mixture was purified by flash column chromatography starting with hexane (100%) to remove any unreacted bis(tributyltin) ($R_f$ = 0.95), followed by the elution of stannyl precursor 3 using ethyl acetate (100%; $R_f$ = 0.33). The title compound was isolated as an oil (2.25 g, 65%) that slowly crystalized after storage at 4 °C into a sticky white solid. $^1$H-NMR (400 MHz, CDCl$_3$) δ 0.87 (t, $J$ = 7.2 Hz, 9H), 1.1.03–1.19 (m, 6H), 1.28–1.36 (m, 6H), 1.43 (t, $J$ = 7.2 Hz, 3H), 1.47–1.63 (m, 6H), 3.23 (s, 3H), 4.30–4.50 (m, 3H), 5.19 (m, 1H), 7.34 (d, $J$ = 8.0 Hz, 1H), 7.69 (dd, $J$ = 8.0, 1.2 Hz, 1H), 7.87 (s, 1H), 8.11 (d, $J$ = 0.8 Hz, 1H). ESI-MS: $m/z$ 576.2 [M$^+$ + H]. These data match the literature data [26].

**Radiochemistry**: Fluoride-18 (33.3–74.0 GBq) was trapped on QMA cartridge and azeotropically dried according to our previously reported procedures using the iPHASE FlexLab radiochemistry module [34]. Potassium triflate:potassium carbonate (10 mg:0.05 mg; 550 μL) was used to elute F-18 from the QMA cartridge and prepare dried K[$^{18}$F]F. To the dried K[$^{18}$F]F, stannyl precursor 3 (5 mg, 8.7 μmol) in DMA (1 mL) containing copper, triflate:pyridine (14 mg:13 μL; 28.7 μmol:146 μmol) was added. After 10 min at 120 °C, the residue was diluted with 0.1% TFA H$_2$O/MeCN (2.5:0.5, 3 mL). The mixture was then purified by preparative HPLC on a Kinetex 5 μm XB-C18 250 × 150 mm column, 0.1% TFA in 15–80% MeCN:H$_2$O over 40 min. Isolated product was diluted in water (30 mL) and then trapped on a C18 SEP-PAK cartridge. The trapped product was rinsed with saline (5 mL), eluted with ethanol (1 mL) and diluted with saline (10 mL) before sterile filtration (Vented Cathivex GV, 0.22 μm, 25 mm) into a sterile evacuated product vial (FILL-EASE$^{TM}$ Sterile Vials; SVV-15A) was performed to prepare the title compound to be ready for injection (2–8 GBq, 22.2 ± 2.7% isolated yield ($n$ = 5)). The total reaction time was 80 min.

During optimization reactions, a small fraction (10 μL) of the crude reaction mixture was diluted into water:MeCN (75:25, 90 μL) and directly injected onto an analytical HPLC system. The % recovery of radioactivity from the analytical HPLC column was quantified by accurately measuring total injected amount of radioactivity using a dose calibrator. Every radioactive peak was then collected from the waste line after passing the radioactivity detector. % recovery was then calculated as the following: (collected total radioactivity/injected total radioactivity) × 100.

**Sample preparation and extraction protocol for inductively coupled plasma-mass spectrometry (ICP-MS)**

For the quantitative analysis of Copper (Cu), 100 μL of each sample were digested with 50 μL concentrated 65% nitric acid (HNO$_3$,70% Analytical grade from Ajax Finechem) followed by heating at 95 °C for 10 min. The samples were then diluted with Milli-Q water (18.2 MΩ; Milli-Q H$_2$O; Merk Millipore, Australia) (1:10 to a final volume of 1 mL). Samples were briefly vortexed and centrifuged at 15,000 rpm for 25 min and supernatant was immediately transferred to new 1.7 mL microcentrifuge tubes. Sample blanks were prepared in the same manner.

**Inductively coupled plasma-mass spectrometry (ICP-MS)**

Tuning solution containing 1 μg/L of cerium (Ce), cobalt (Co), lithium (Li), thallium (Tl) and Y in 2% ($v/v$) HNO$_3$ (Agilent Technologies, Australia) was used to tune and optimize the Agilent 8900 triple quadrupole ICP-MS (Agilent Technologies, Australia) in a Helium gas analysis mode. A 9-point calibration (0, 1, 5, 10,25, 50, 100, 250 and 500 parts per billion (ppb) in 1% HNO$_3$) standard curve for Cu was prepared using commercially available multi-element standards (Multi-Element Calibration Standard 2A, Agilent Technologies, USA). The $R^2$ value for copper calibration curve was 0.999617. ICP-MS analysis method for Cu detection yielded a limit of detection (LOD) of less than 0.0993 μg/L and a limit of quantitation (LOQ) around 0.33 μg/L. Yittrium ($^{89}$Y) (Agilent Technologies, USA) was used as an internal standard at a concentration of 0.1 μg/mL and used as reference element solution to normalize all measurements. All the samples, calibration standards and reference solution were introduced at the flow rate of 0.4 mL/min using a T-piece and a peristaltic pump. The data was collected in spectrum mode with the average of three technical replicates. The ICP-MS operating parameters were established according to the manufacturer's guidelines and other parameters were optimized for copper in a batch-specific mode prior to each experiment and these are as follows: ICP-MS operating parameters; Scan type: Single Quad, RF Power: 1550W, RF Matching: 1.8V, Nebulizer Gas: 1.05L/min, Extract 1: −12V, Extract 2: 250V, Omega Bias: −120V, Omega Lens: 7.2V, Deflect: −5V, Gas: He Gas, Oct P bias: −18V.

**Small animal PET imaging**: Six adult, male Wistar rats (Monash Animal Resources Centre, $0.500 \pm 0.071$ kg) were anesthetized with isoflurane (induction: 5% in 1 L/min $O_2$, maintenance 1.5–2% in 1 L/min $O_2$) and [$^{18}$F]flumazenil (dose: $23.5 \pm 10$ MBq, mass: $3.6 \pm 1.2$ nmol) was injected as a bolus over 10s via the dorsal penile vein, as previously described [23]. Immediately following tracer injection, dynamic PET scans were acquired for 45 min on a nanoScan-PET/CT (Mediso, Hungary). PET images were reconstructed across the following time frames ($2 \times 30$ s, $2 \times 60$ s, $14 \times 180$ s) using the Tera-tomo 3D algorithm provided by the supplier with correction for scatter, attenuation and dead-time.

PET scans were manually co-registered to a corresponding T2-weighted MRI acquired from each rat acquired using a 9.4T Bruker Avance IIIHD MRI (Bruker, Germany) with actively decoupled 4-channel receive-only surface and volume transmit coils (voxel size $0.1 \times 0.1 \times 0.7$ mm$^3$; matrix $256 \times 256 \times 24$) using ITK-SNAP [35], and resliced using linear interpolation. Volumes of interest (VOI) incorporating both left and right hippocampus and pons were manually delineated on the MR to extract time activity curves from the PET scans in PMOD (PMOD Technologies, Switzerland). $B_{max}$ (receptor number) and $1/K_D$ (receptor affinity) were estimated from each VOI using a nonlinear fit of the bound ligand (B) versus the free ligand (F) (activity in pons): $B_{max} = (F \times B)/(F + K_D)$ in Prism. A representative image is provided in Figure 5.

The study was conducted in accordance with the Australian NH&MRC Code of conduct for use of animals in research and the study protocol was approved by the Alfred Research Alliance Animal Ethics Committee (E/2004/2020/M).

**IUPAC nomenclature:**

*Flumazenil 1*: Ethyl 8-fluoro-5-methyl-6-oxo-5,6-dihydro-4H-benzo[f]imidazo[1,5-a][1,4]diazepine-3-carboxylate

*Stannyl precursor 3*: Ethyl 5-methyl-6-oxo-8-(tributylstannyl)-5,6-dihydro-4H-benzo[f]imidazo[1,5-a][1,4]diazepine-3-carboxylate

*Hydroxylated by-product 4*: Ethyl 8-hydroxy-5-methyl-6-oxo-5,6-dihydro-4H-benzo[f]imidazo[1,5-a][1,4]diazepine-3-carboxylate

*Des-fluoro-flumazenil 5*: Ethyl 5-methyl-6-oxo-5,6-dihydro-4H-benzo[f]imidazo[1,5-a]diazepine-3-carboxylate

*Homodimerization product 6*: 8,8'-Bis[Ethyl 5-methyl-6-oxo-5,6-dihydro-4H-benzo[f]imidazo[1,5-a][1,4]diazepine-3-carboxylate].

**Supplementary Materials:** The following supporting information can be downloaded at: https://www.mdpi.com/article/10.3390/molecules27185931/s1.

**Author Contributions:** M.B.H. developed the chemistry, radiochemistry and performed the chemical analysis described herein. M.B.H. also led the writing of this manuscript. P.D.R. and C.A.H. assisted with the chemistry and radiochemistry developments as well as the write up the manuscript. I.A., B.J., T.J.O. and L.V., funded this work, performed the pharmacological evaluation in using small animal PET imaging and assisted with the write up of the manuscript. All authors have read and agreed to the published version of the manuscript.

**Funding:** This study was funded by a National Health and Medical Research Council Investigator Grant to TJO (#GNT1176426).

**Institutional Review Board Statement:** The study was conducted in accordance with the Australian NH & MRC Code of conduct for use of animals in research and the study protocol was approved by the Alfred Research Alliance Animal Ethics Committee (E/2004/2020/M).

**Informed Consent Statement:** Not applicable.

**Data Availability Statement:** Supplementary Materials File is available.

**Conflicts of Interest:** The authors declare no conflict of interest.

## References

1. Amrein, R.; Leishman, B.; Bentzinger, C.; Roncari, G. Flumazenil in benzodiazepine antagonism. Actions and clinical use in intoxications and anaesthesiology. *Med. Toxicol. Advers. Drug Exp.* **1987**, *2*, 411–429.
2. Benini, A.; Gottardo, R.; Chiamulera, C.; Bertoldi, A.; Zamboni, L.; Lugoboni, F. Continuous Infusion of Flumazenil in the Management of Benzodiazepines Detoxification. *Front. Psychiatry* **2021**, *12*, 646038. [CrossRef] [PubMed]
3. Hoffman, E.J.; Warren, E.W. Flumazenil: A benzodiazepine antagonist. *Clin. Pharm* **1993**, *12*, 699–701.
4. Michel, T.C. CHAPTER 12—Intravenous Anesthetics and Benzodiazepines. In *Anesthesia Secrets*, 4th ed.; Duke, J., Ed.; Mosby: Maryland Heights, MI, USA, 2011; pp. 90–94.
5. Haefely, W. Benzodiazepine interactions with GABA receptors. *Neurosci. Lett.* **1984**, *47*, 201–206. [CrossRef]
6. Charych, E.I.; Liu, F.; Moss, S.J.; Brandon, N.J. GABA(A) receptors and their associated proteins: Implications in the etiology and treatment of schizophrenia and related disorders. *Neuropharmacology* **2009**, *57*, 481–495. [CrossRef]
7. Sarawagi, A.; Soni, N.D.; Patel, A.B. Glutamate and GABA Homeostasis and Neurometabolism in Major Depressive Disorder. *Front. Psychiatry* **2021**, *12*, 637863. [CrossRef]
8. Fogaça, M.V.; Duman, R.S. Cortical GABAergic Dysfunction in Stress and Depression: New Insights for Therapeutic Interventions. *Front. Cell. Neurosci.* **2019**, *13*, 87. [CrossRef]
9. Coghlan, S.; Horder, J.; Inkster, B.; Mendez, M.A.; Murphy, D.G.; Nutt, D.J. GABA system dysfunction in autism and related disorders: From synapse to symptoms. *Neurosci. Biobehav. Rev.* **2012**, *36*, 2044–2055. [CrossRef]
10. Sperk, G.; Furtinger, S.; Schwarzer, C.; Pirker, S. GABA and its receptors in epilepsy. *Adv. Exp. Med. Biol.* **2004**, *548*, 92–103.
11. Barker, J.S.; Hines, R.M. Regulation of GABAA Receptor Subunit Expression in Substance Use Disorders. *Int. J. Mol. Sci.* **2020**, *21*, 4445. [CrossRef]
12. Sarikaya, I. PET studies in epilepsy. *Am. J. Nucl. Med. Mol. Imaging* **2015**, *5*, 416–430. [PubMed]
13. Vivash, L.; Gregoire, M.C.; Lau, E.W.; Ware, R.E.; Binns, D.; Roselt, P.; Bouilleret, V.; Myers, D.E.; Cook, M.J.; Hicks, R.J.; et al. 18F-flumazenil: A γ-aminobutyric acid A-specific PET radiotracer for the localization of drug-resistant temporal lobe epilepsy. *J. Nucl. Med.* **2013**, *54*, 1270–1277. [CrossRef]
14. Heiss, W.-D.; Herholz, K. Brain Receptor Imaging. *J. Nucl. Med.* **2006**, *47*, 302–312.
15. Kim, W.; Park, H.S.; Moon, B.S.; Lee, B.C.; Kim, S.E. PET measurement of "GABA shift" in the rat brain: A preclinical application of bolus plus constant infusion paradigm for [(18)F]flumazenil. *Nucl. Med. Biol.* **2017**, *45*, 30–34. [CrossRef] [PubMed]
16. Ryzhikov, N.N.; Seneca, N.; Krasikova, R.N.; Gomzina, N.A.; Shchukin, E.; Fedorova, O.S.; Vassiliev, D.A.; Gulyás, B.; Hall, H.; Savic, I.; et al. Preparation of highly specific radioactivity [$^{18}$F]flumazenil and its evaluation in cynomolgus monkey by positron emission tomography. *Nucl. Med. Biol.* **2005**, *32*, 109–116. [CrossRef] [PubMed]
17. Preshlock, S.; Tredwell, M.; Gouverneur, V. (18)F-Labeling of Arenes and Heteroarenes for Applications in Positron Emission Tomography. *Chem. Rev.* **2016**, *116*, 719–766. [CrossRef]
18. Windhorst, A.D.; Klok, R.P.; Koolen, C.L.; Visser, G.W.M.; Herscheid, J.D.M. Labeling of [18F]flumazenil via instant fluorination, a new nucleophilic fluorination method. *J. Label. Compd. Radiopharm.* **2001**, *44*, S930–S932. [CrossRef]
19. Krasikova, R.N.; Ryzhikov, N.N.; Gomzina, N.A.; Vassiliev, D.A. Isotopic 18F/19F exchange in the flumazenil molecule using K18F/kryptofix complex. *J. Label. Compd. Radiopharm.* **2003**, *46*, S213.
20. Ryzhikov, N.; Gomzina, N.; Fedorova, O.; Vasil'ev, D.; Kostikov, A.; Krasikova, R. Preparation of [$^{18}$F]Flumazenil, a Potential Radioligand for PET Imaging of Central Benzodiazepine Receptors, by Isotope Exchange. *Radiochemistry* **2004**, *46*, 290–294. [CrossRef]
21. Schirrmacher, R.; Kostikov, A.; Massaweh, G.; Kovacevic, M.; Wängler, C.; Thiel, A. Synthesis of [18F]Flumazenil ([18F]FZ). In *Radiochemical Syntheses*; Wiley: Hoboken, NJ, USA, 2012; pp. 111–123.
22. Collins, J.; Waldmann, C.M.; Drake, C.; Slavik, R.; Ha, N.S.; Sergeev, M.; Lazari, M.; Shen, B.; Chin, F.T.; Moore, M.; et al. Production of diverse PET probes with limited resources: 24 $^{18}$F-labeled compounds prepared with a single radiosynthesizer. *Proc. Natl. Acad. Sci. USA* **2017**, *114*, 11309–11314. [CrossRef]
23. Vivash, L.; Gregoire, M.-C.; Bouilleret, V.; Berard, A.; Wimberley, C.; Binns, D.; Roselt, P.; Katsifis, A.; Myers, D.E.; Hicks, R.J.; et al. In vivo measurement of hippocampal GABAA/cBZR density with [18F]-flumazenil PET for the study of disease progression in an animal model of temporal lobe epilepsy. *PLoS ONE* **2014**, *9*, e86722. [CrossRef]
24. Dedeurwaerdere, S.; Gregoire, M.C.; Vivash, L.; Roselt, P.; Binns, D.; Fookes, C.; Greguric, I.; Pham, T.; Loch, C.; Katsifis, A.; et al. In-vivo imaging characteristics of two fluorinated flumazenil radiotracers in the rat. *Eur. J. Nucl. Med. Mol. Imaging* **2009**, *36*, 958–965. [CrossRef] [PubMed]

25. Pike, V.W.; Aigbirhio, F.I. Reactions of cyclotron-produced [18F]fluoride with diaryliodonium salts—a novel single-step route to no-carrier-added [18]fluoroarenes. *J. Chem. Soc. Chem. Commun.* **1995**, *21*, 2215–2216. [CrossRef]
26. Moon, B.S.; Kil, H.S.; Park, J.H.; Kim, J.S.; Park, J.; Chi, D.Y.; Lee, B.C.; Kim, S.E. Facile aromatic radiofluorination of [18F]flumazenil from diaryliodonium salts with evaluation of their stability and selectivity. *Org. Biomol. Chem.* **2011**, *9*, 8346–8355. [CrossRef] [PubMed]
27. Preshlock, S.; Calderwood, S.; Verhoog, S.; Tredwell, M.; Huiban, M.; Hienzsch, A.; Gruber, S.; Wilson, T.C.; Taylor, N.J.; Cailly, T.; et al. Enhanced copper-mediated 18F-fluorination of aryl boronic esters provides eight radiotracers for PET applications. *Chem. Commun.* **2016**, *52*, 8361–8364. [CrossRef] [PubMed]
28. Gendron, T.; Destro, G.; Straathof, N.J.W.; Sap, J.B.I.; Guibbal, F.; Vriamont, C.; Caygill, C.; Atack, J.R.; Watkins, A.J.; Marshall, C.; et al. Multi-patient dose synthesis of [18F]Flumazenil via a copper-mediated 18F-fluorination. *EJNMMI Radiopharm. Chem.* **2022**, *7*, 5. [CrossRef] [PubMed]
29. Makaravage, K.J.; Brooks, A.F.; Mossine, A.V.; Sanford, M.S.; Scott, P.J.H. Copper-Mediated Radiofluorination of Arylstannanes with [18F]KF. *Org. Lett.* **2016**, *18*, 5440–5443. [CrossRef]
30. Ory, D.; Van den Brande, J.; de Groot, T.; Serdons, K.; Bex, M.; Declercq, L.; Cleeren, F.; Ooms, M.; Van Laere, K.; Verbruggen, A.; et al. Retention of [(18)F]fluoride on reversed phase HPLC columns. *J. Pharm. Biomed. Anal.* **2015**, *111*, 209–214. [CrossRef]
31. Falck, J.R.; Lai, J.-Y.; Ramana, D.V.; Lee, S.-G. Synthesis of alcohols via mild oxidation of perfluoroethylstannanes. *Tetrahedron Lett.* **1999**, *40*, 2715–2718. [CrossRef]
32. Stefani, H.A.; Guarezemini, A.; Cella, R. ChemInform Abstract: Homocoupling Reactions of Alkynes, Alkenes and Alkyl Compounds. *Tetrahedron* **2010**, *66*, 7871–7918. [CrossRef]
33. Espinet, P.; Echavarren, A.M. The Mechanisms of the Stille Reaction. *Angew. Chem. Int. Ed.* **2004**, *43*, 4704–4734.
34. Haskali, M.B.; Farnsworth, A.L.; Roselt, P.D.; Hutton, C.A. 4-Nitrophenyl activated esters are superior synthons for indirect radiofluorination of biomolecules. *RSC Med. Chem.* **2020**, *11*, 919–922. [CrossRef] [PubMed]
35. Yushkevich, P.A.; Yang, G.; Gerig, G. ITK-SNAP: An interactive tool for semi-automatic segmentation of multi-modality biomedical images. In Proceedings of the 2016 38th Annual International Conference of the IEEE Engineering in Medicine and Biology Society (EMBC), Orlando, FL, USA, 16–20 August 2016; pp. 3342–3345.

Article

# Dosimetry of [$^{212}$Pb]VMT01, a MC1R-Targeted Alpha Therapeutic Compound, and Effect of Free $^{208}$Tl on Tissue Absorbed Doses

Kelly D. Orcutt [1], Kelly E. Henry [2], Christine Habjan [2], Keryn Palmer [2], Jack Heimann [2], Julie M. Cupido [2], Vijay Gottumukkala [2], Derek D. Cissell [2], Morgan C. Lyon [2], Amira I. Hussein [2], Dijie Liu [1], Mengshi Li [1], Frances L. Johnson [1] and Michael K. Schultz [1,3,4,5,*]

[1] Viewpoint Molecular Targeting, Inc., Coralville, IA 52241, USA
[2] Invicro, LLC, Needham, MA 02494, USA
[3] Department of Radiology, The University of Iowa, Iowa City, IA 52242, USA
[4] Department of Radiation Oncology, The University of Iowa, Iowa City, IA 52242, USA
[5] Departments of Radiology and Radiation Oncology, The University of Iowa, Iowa City, IA 52242, USA
* Correspondence: michael-schultz@uiowa.edu; Tel.: +1-(319)-335-8017

**Abstract:** [$^{212}$Pb]VMT01 is a melanocortin 1 receptor (MC1R) targeted theranostic ligand in clinical development for alpha particle therapy for melanoma. $^{212}$Pb has an elementally matched gamma-emitting isotope $^{203}$Pb; thus, [$^{203}$Pb]VMT01 can be used as an imaging surrogate for [$^{212}$Pb]VMT01. [$^{212}$Pb]VMT01 human serum stability studies have demonstrated retention of the $^{212}$Bi daughter within the chelator following beta emission of parent $^{212}$Pb. However, the subsequent alpha emission from the decay of $^{212}$Bi into $^{208}$Tl results in the generation of free $^{208}$Tl. Due to the 10.64-hour half-life of $^{212}$Pb, accumulation of free $^{208}$Tl in the injectate will occur. The goal of this work is to estimate the human dosimetry for [$^{212}$Pb]VMT01 and the impact of free $^{208}$Tl in the injectate on human tissue absorbed doses. Human [$^{212}$Pb]VMT01 tissue absorbed doses were estimated from murine [$^{203}$Pb]VMT01 biodistribution data, and human biodistribution values for $^{201}$Tl chloride (a cardiac imaging agent) from published data were used to estimate the dosimetry of free $^{208}$Tl. Results indicate that the dose-limiting tissues for [$^{212}$Pb]VMT01 are the red marrow and the kidneys, with estimated absorbed doses of 1.06 and 8.27 mGy$_{RBE=5}$/MBq. The estimated percent increase in absorbed doses from free $^{208}$Tl in the injectate is 0.03% and 0.09% to the red marrow and the kidneys, respectively. Absorbed doses from free $^{208}$Tl result in a percent increase of no more than 1.2% over [$^{212}$Pb]VMT01 in any organ or tissue. This latter finding indicates that free $^{208}$Tl in the [$^{212}$Pb]VMT01 injectate will not substantially impact estimated tissue absorbed doses in humans.

**Keywords:** $^{212}$Pb; $^{203}$Pb; $^{208}$Tl; MC1R; dosimetry; absorbed dose; melanoma

## 1. Introduction

Melanocortin 1 receptor (MC1R) is a G protein-coupled receptor that is expressed in melanocytes and is implicated in melanogenesis [1]. MC1R is overexpressed on many mouse and human melanoma cells [2,3]. Positron emission tomography (PET) imaging of an MC1R-targeted peptide $^{68}$Ga-DOTA-GGNle-CycNSH$_{hex}$ in melanoma patients has established clinical proof-of-concept of MC1R as a target for imaging and therapy [4].

Targeted alpha-particle therapy (TAT) is a promising therapeutic strategy that is unique in its ability to deliver cytotoxicity circumventing cellular resistance [5] and has demonstrated significant responses in early clinical trials [6–8]. High linear energy transfer (LET) alpha emissions result in clustered DNA double strand breaks [9–16]. In cell culture, alpha emitters have been shown to be more effective in inducing cell death than gamma radiation [17]. Due to short tissue ranges (<100 μm in water, <40 μm in bone), it had previously been believed that TAT may be best suited for the treatment of micrometastases and other disseminated tumors. However, recent TAT studies have demonstrated efficacy in large tumors and there is a growing body of evidence that TAT can activate the immune system and impart both bystander and abscopal effects [18,19]. In the clinical setting, TAT has demonstrated patient benefit even in subjects refractive to beta particle therapy [6].

[$^{203}$Pb]VMT01 is an MC1R-targeted TAT ligand in clinical development (NCT04904120) with elementally matched gamma-emitting [$^{203}$Pb]VMT01 that can be used as an imaging surrogate. [$^{212}$Pb]VMT01 human serum stability and in vivo mouse biodistribution experiments demonstrate robust retention of the $^{212}$Bi daughter within the chelator following beta emission of parent $^{212}$Pb and no evidence of in vivo translocation (Li and collaborators, SNMMI-ACNM Mid-Winter Meeting 2022 Abstract) [20]. In addition, the decay physics for $^{212}$Pb [21,22] (Figure 1) dictates that retention of $^{212}$Bi within the chelator will subsequently lead to alpha decay via the $^{212}$Po or $^{208}$Tl branches at the site of localization due to the short half-lives of $^{212}$Po (0.3 µs) and $^{208}$Tl (3.05 m). Due to recoil energy, the 36% alpha emission from $^{212}$Bi via the $^{208}$Tl branch will result in the accumulation of free $^{208}$Tl in the administered injectate. Here, we calculated [$^{212}$Pb]VMT01 human tissue absorbed doses from murine [$^{203}$Pb]VMT01 biodistribution data and the activity and effect of free $^{208}$Tl in the injectate on tissue absorbed doses.

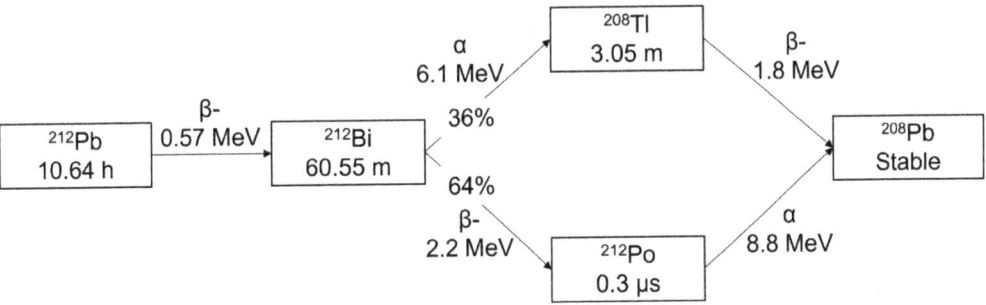

**Figure 1.** $^{212}$Pb decay scheme [21,22].

## 2. Results

### 2.1. Murine [$^{203}$Pb]VMT01 Biodistribution

Murine biodistribution results following intravenous administration of [$^{203}$Pb]VMT01 in female and male CD-1 IGS naïve mice are provided in Supplemental Tables S1 and S2 (Supplementary material). [$^{203}$Pb]VMT01 cleared rapidly through the kidneys with an accumulation of 6.24 ± 0.35% ID/g and 8.30 ± 1.90% ID/g in females and males, respectively at 0.5 h. Kidney activity decreased to 1.09 ± 0.12% ID/g and 0.55 ± 0.10% ID/g in females and males, respectively at 55 h. Accumulation and retention in other organs were minimal.

### 2.2. Dosimetry

[$^{203}$Pb]VMT01 and [$^{212}$Pb]VMT01 TIACs (Table 1) and human tissue absorbed doses (Table 2) are provided for a 2 h bladder voiding model.

Table 1. [$^{203}$Pb]VMT01 and [$^{212}$Pb]VMT01 time-integrated activity coefficients.

| Organ | [$^{203}$Pb]VMT01 TIAC (MBq h/MBq) | | [$^{212}$Pb]VMT01 TIAC (MBq h/MBq) | |
|---|---|---|---|---|
| | Female | Male | Female | Male |
| Adrenal glands | $5.08 \times 10^{-5}$ | $3.72 \times 10^{-4}$ | $4.75 \times 10^{-5}$ | $3.33 \times 10^{-4}$ |
| Brain | $3.53 \times 10^{-4}$ | $1.55 \times 10^{-3}$ | $3.06 \times 10^{-4}$ | $1.24 \times 10^{-3}$ |
| Cortical bone | $1.50 \times 10^{-2}$ | $8.89 \times 10^{-2}$ | $1.20 \times 10^{-2}$ | $6.34 \times 10^{-2}$ |
| Eyes | $4.24 \times 10^{-5}$ | $8.03 \times 10^{-5}$ | $3.71 \times 10^{-5}$ | $6.73 \times 10^{-5}$ |
| Gallbladder | $2.07 \times 10^{-4}$ | $2.22 \times 10^{-4}$ | $1.45 \times 10^{-4}$ | $1.84 \times 10^{-4}$ |
| Heart contents | $2.28 \times 10^{-2}$ | $2.62 \times 10^{-2}$ | $2.22 \times 10^{-2}$ | $2.59 \times 10^{-2}$ |
| Heart wall | $9.93 \times 10^{-4}$ | $1.91 \times 10^{-3}$ | $7.74 \times 10^{-4}$ | $1.43 \times 10^{-3}$ |
| Kidneys | $2.00 \times 10^{-1}$ | $1.48 \times 10^{-1}$ | $9.70 \times 10^{-2}$ | $9.01 \times 10^{-2}$ |
| Left colon | $3.49 \times 10^{-1}$ | $4.26 \times 10^{-1}$ | $1.03 \times 10^{-1}$ | $1.51 \times 10^{-1}$ |
| Liver | $1.09 \times 10^{-1}$ | $8.85 \times 10^{-2}$ | $4.29 \times 10^{-2}$ | $3.68 \times 10^{-2}$ |
| Lungs | $1.79 \times 10^{-2}$ | $2.18 \times 10^{-2}$ | $1.11 \times 10^{-2}$ | $1.66 \times 10^{-2}$ |
| Ovaries | $1.46 \times 10^{-4}$ | - | $4.99 \times 10^{-5}$ | - |
| Pancreas | $3.68 \times 10^{-4}$ | $1.19 \times 10^{-3}$ | $2.95 \times 10^{-4}$ | $9.27 \times 10^{-4}$ |
| Rectum | $2.87 \times 10^{-1}$ | $3.67 \times 10^{-1}$ | $5.04 \times 10^{-2}$ | $8.47 \times 10^{-2}$ |
| Red marrow | $1.58 \times 10^{-1}$ | $2.24 \times 10^{-4}$ | $1.39 \times 10^{-3}$ | $1.79 \times 10^{-4}$ |
| Right colon | $4.23 \times 10^{-1}$ | $4.94 \times 10^{-1}$ | $2.10 \times 10^{-1}$ | $2.69 \times 10^{-1}$ |
| Small intestines | $1.28 \times 10^{-1}$ | $1.91 \times 10^{-1}$ | $1.07 \times 10^{-1}$ | $1.60 \times 10^{-1}$ |
| Spleen | $3.00 \times 10^{-3}$ | $3.51 \times 10^{-3}$ | $1.52 \times 10^{-3}$ | $1.86 \times 10^{-3}$ |
| Stomach contents | $1.09 \times 10^{-2}$ | $1.84 \times 10^{-2}$ | $7.28 \times 10^{-3}$ | $7.34 \times 10^{-3}$ |
| Testes | - | $3.85 \times 10^{-4}$ | - | $2.85 \times 10^{-4}$ |
| Thymus | $6.52 \times 10^{-5}$ | $1.41 \times 10^{-4}$ | $5.29 \times 10^{-5}$ | $1.25 \times 10^{-4}$ |
| Thyroid | $1.38 \times 10^{-4}$ | $4.69 \times 10^{-4}$ | $1.05 \times 10^{-4}$ | $3.41 \times 10^{-4}$ |
| Total body/remainder | $4.55 \times 10^{0}$ | $9.20 \times 10^{-1}$ | $1.18 \times 10^{0}$ | $8.83 \times 10^{-1}$ |
| Trabecular bone | $1.50 \times 10^{-2}$ | $8.89 \times 10^{-2}$ | $1.20 \times 10^{-2}$ | $6.34 \times 10^{-2}$ |
| Urinary bladder | $1.49 \times 10^{0}$ | $1.39 \times 10^{0}$ | $1.40 \times 10^{0}$ | $1.31 \times 10^{0}$ |
| Uterus | $1.05 \times 10^{-3}$ | - | $7.46 \times 10^{-4}$ | - |

Table 2. Human tissue absorbed doses.

| Organ/tissue | $^{203}$Pb Absorbed Dose (mGy/MBq) | | $^{212}$Pb Absorbed Dose (mGy$_{RBE=5}$/MBq) | |
|---|---|---|---|---|
| | Female | Male | Female | Male |
| Adrenals | $1.14 \times 10^{-2}$ | $1.11 \times 10^{-2}$ | $1.06 \times 10^{-1}$ | $5.83 \times 10^{-1}$ |
| Brain | $2.17 \times 10^{-3}$ | $5.86 \times 10^{-4}$ | $7.88 \times 10^{-3}$ | $2.20 \times 10^{-2}$ |
| Breasts | $6.71 \times 10^{-3}$ | - | $4.67 \times 10^{-1}$ | - |
| Oesophagus | $7.55 \times 10^{-3}$ | $3.12 \times 10^{-3}$ | $4.69 \times 10^{-1}$ | $2.90 \times 10^{-1}$ |
| Eyes | $3.78 \times 10^{-3}$ | $1.02 \times 10^{-3}$ | $6.16 \times 10^{-2}$ | $1.08 \times 10^{-1}$ |
| Gallbladder wall | $1.84 \times 10^{-2}$ | $1.14 \times 10^{-2}$ | $4.86 \times 10^{-1}$ | $3.07 \times 10^{-1}$ |
| Left colon | $9.83 \times 10^{-2}$ | $1.16 \times 10^{-1}$ | $8.31 \times 10^{-1}$ | $8.46 \times 10^{-1}$ |
| Small intestine | $2.87 \times 10^{-2}$ | $2.58 \times 10^{-2}$ | $5.96 \times 10^{-1}$ | $4.40 \times 10^{-1}$ |
| Stomach wall | $1.27 \times 10^{-2}$ | $6.97 \times 10^{-3}$ | $4.83 \times 10^{-1}$ | $3.02 \times 10^{-1}$ |
| Right colon | $7.40 \times 10^{-2}$ | $8.24 \times 10^{-2}$ | $8.50 \times 10^{-1}$ | $8.07 \times 10^{-1}$ |
| Rectum | $1.18 \times 10^{-1}$ | $1.17 \times 10^{-1}$ | $7.35 \times 10^{-1}$ | $6.39 \times 10^{-1}$ |
| Heart wall | $7.38 \times 10^{-3}$ | $4.45 \times 10^{-3}$ | $7.86 \times 10^{-1}$ | $7.06 \times 10^{-1}$ |
| Kidneys | $4.23 \times 10^{-2}$ | $2.80 \times 10^{-2}$ | $8.27 \times 10^{0}$ | $6.83 \times 10^{0}$ |
| Liver | $1.15 \times 10^{-2}$ | $7.05 \times 10^{-3}$ | $7.32 \times 10^{-1}$ | $4.91 \times 10^{-1}$ |
| Lungs | $6.59 \times 10^{-3}$ | $2.82 \times 10^{-3}$ | $2.81 \times 10^{-1}$ | $3.30 \times 10^{-1}$ |
| Ovaries | $2.60 \times 10^{-2}$ | - | $1.59 \times 10^{-1}$ | - |
| Pancreas | $1.12 \times 10^{-2}$ | $1.15 \times 10^{-2}$ | $7.45 \times 10^{-2}$ | $1.76 \times 10^{-1}$ |
| Prostate | - | $2.53 \times 10^{-2}$ | - | $3.46 \times 10^{-1}$ |
| Salivary glands | $7.27 \times 10^{-3}$ | $1.54 \times 10^{-3}$ | $4.66 \times 10^{-1}$ | $2.87 \times 10^{-1}$ |
| Red Marrow | $1.06 \times 10^{-3}$ | $5.46 \times 10^{-3}$ | $8.64 \times 10^{-1}$ | $1.06 \times 10^{0}$ |
| Osteogenic Cells | $1.46 \times 10^{-2}$ | $1.23 \times 10^{-2}$ | $3.88 \times 10^{0}$ | $6.95 \times 10^{0}$ |

Table 2. Cont.

| Organ/tissue | $^{203}$Pb Absorbed Dose (mGy/MBq) | | $^{212}$Pb Absorbed Dose (mGy$_{RBE=5}$/MBq) | |
|---|---|---|---|---|
| | Female | Male | Female | Male |
| Spleen | $1.16 \times 10^{-2}$ | $6.11 \times 10^{-3}$ | $2.90 \times 10^{-1}$ | $3.01 \times 10^{-1}$ |
| Testes | - | $5.26 \times 10^{-3}$ | - | $2.07 \times 10^{-1}$ |
| Thymus | $5.47 \times 10^{-3}$ | $1.94 \times 10^{-3}$ | $6.90 \times 10^{-2}$ | $1.23 \times 10^{-1}$ |
| Thyroid | $4.88 \times 10^{-3}$ | $2.16 \times 10^{-3}$ | $1.50 \times 10^{-1}$ | $4.04 \times 10^{-1}$ |
| Urinary bladder wall | $2.29 \times 10^{-1}$ | $1.89 \times 10^{-1}$ | $2.95 \times 10^{0}$ | $2.14 \times 10^{0}$ |
| Uterus | $4.83 \times 10^{-2}$ | - | $3.27 \times 10^{-1}$ | - |

For [$^{203}$Pb]VMT01, the tissue with the highest estimated absorbed dose was the urinary bladder wall (0.23 mGy/MBq for females and 0.19 mGy/MBq for males) and the effective dose was 0.028 mSv/MBq and 0.024 mSv/MBq for females and males, respectively. For [$^{212}$Pb]VMT01, the tissue with the highest estimated absorbed dose was the kidneys (8.27 mGy$_{RBE=5}$/MBq for females and 6.83 mGy$_{RBE=5}$/MBq for males). The anticipated dose limiting tissues for [$^{212}$Pb]VMT01 are the red marrow and kidneys, with estimated absorbed doses of 1.06 and 8.27 mGy$_{RBE=5}$/MBq and maximum tolerated activities of approximately 1.9 GBq and 2.2 GBq, respectively, based on published threshold doses from external beam irradiation data [23,24].

Human biodistribution of $^{201}$Tl chloride published in the literature [25,26] and the calculated activity fraction of free $^{208}$Tl in the injectate at a shelf-life of 6 h was used to estimate human tissue absorbed doses of administered free $^{208}$Tl. The activity fraction of free $^{208}$Tl in the injectate was calculated at a shelf-life of 6 h to be 0.44 MBq $^{208}$Tl per MBq $^{212}$Pb (Table 3).

Table 3. $^{212}$Pb, $^{212}$Bi, and $^{208}$Tl activity and activity fraction in injectate preparation at 0 h and 6 h for nominal 1 MBq $^{212}$Pb-VMT01.

| | 0 h | | 6 h | |
|---|---|---|---|---|
| | Activity (MBq) | Activity Fraction | Activity (MBq) | Activity Fraction |
| $^{212}$Pb | 1.00 | 1.00 | 0.68 | 1.00 |
| $^{212}$Bi | 0.00 | 0.00 | 0.73 | 1.08 |
| $^{208}$Tl | 0.00 | 0.00 | 0.29 | 0.44 |

$^{208}$Tl absorbed tissue doses are provided in Table 4. The estimated percent increase in absorbed tissue doses from free $^{208}$Tl in the injectate was 0.03% and 0.09% in the red marrow and kidneys, respectively. In addition, absorbed doses from free $^{208}$Tl result in a percent increase of less than 1.2% over [$^{212}$Pb]VMT01 in any organ or tissue, and were within the values that would be expected to be the uncertainty in absorbed dose estimates for [$^{212}$Pb]VMT01 alone.

Table 4. Tissue absorbed dose estimates for human adult male for free $^{208}$Tl in the injectate at a shelf-life of 6 h, [$^{212}$Pb]VMT01 human adult male, total absorbed dose, and % increase in absorbed dose from free $^{208}$Tl contribution.

| Organ/Tissue | $^{208}$Tl Absorbed Dose (mGy/MBq) | [$^{212}$Pb]VMT01 Absorbed Dose (mGy$_{RBE=5}$/MBq) | Total Absorbed Dose (mGy$_{RBE=5}$/MBq) | $^{208}$Tl % Increase |
|---|---|---|---|---|
| Adrenals | $3.09 \times 10^{-3}$ | $5.83 \times 10^{-1}$ | $5.84 \times 10^{-1}$ | 0.23 |
| Brain | $5.98 \times 10^{-4}$ | $2.20 \times 10^{-2}$ | $2.23 \times 10^{-2}$ | 1.18 |
| Oesophagus | $9.40 \times 10^{-4}$ | $2.90 \times 10^{-1}$ | $2.90 \times 10^{-1}$ | 0.14 |
| Eyes | $5.09 \times 10^{-4}$ | $1.08 \times 10^{-1}$ | $1.08 \times 10^{-1}$ | 0.21 |
| Gallbladder wall | $1.57 \times 10^{-3}$ | $3.07 \times 10^{-1}$ | $3.08 \times 10^{-1}$ | 0.22 |
| Left colon | $7.38 \times 10^{-3}$ | $8.46 \times 10^{-1}$ | $8.49 \times 10^{-1}$ | 0.38 |
| Small intestine | $7.60 \times 10^{-3}$ | $4.40 \times 10^{-1}$ | $4.43 \times 10^{-1}$ | 0.75 |

Table 4. Cont.

| Organ/Tissue | $^{208}$Tl Absorbed Dose (mGy/MBq) | [$^{212}$Pb]VMT01 Absorbed Dose (mGy$_{RBE=5}$/MBq) | Total Absorbed Dose (mGy$_{RBE=5}$/MBq) | $^{208}$Tl % Increase |
|---|---|---|---|---|
| Stomach wall | $1.14 \times 10^{-3}$ | $3.02 \times 10^{-1}$ | $3.02 \times 10^{-1}$ | 0.16 |
| Right colon | $7.06 \times 10^{-3}$ | $8.07 \times 10^{-1}$ | $8.10 \times 10^{-1}$ | 0.38 |
| Rectum | $6.58 \times 10^{-3}$ | $6.39 \times 10^{-1}$ | $6.42 \times 10^{-1}$ | 0.45 |
| Heart wall | $3.62 \times 10^{-3}$ | $7.06 \times 10^{-1}$ | $7.08 \times 10^{-1}$ | 0.22 |
| Kidneys | $1.41 \times 10^{-2}$ | $6.83 \times 10^{0}$ | $6.84 \times 10^{0}$ | 0.09 |
| Liver | $1.97 \times 10^{-3}$ | $4.91 \times 10^{-1}$ | $4.92 \times 10^{-1}$ | 0.17 |
| Lungs | $8.02 \times 10^{-4}$ | $3.30 \times 10^{-1}$ | $3.30 \times 10^{-1}$ | 0.11 |
| Pancreas | $1.73 \times 10^{-3}$ | $1.76 \times 10^{-1}$ | $1.77 \times 10^{-1}$ | 0.43 |
| Prostate | $1.02 \times 10^{-3}$ | $3.46 \times 10^{-1}$ | $3.46 \times 10^{-1}$ | 0.13 |
| Salivary glands | $5.98 \times 10^{-4}$ | $2.87 \times 10^{-1}$ | $2.87 \times 10^{-1}$ | 0.09 |
| Red marrow | $7.70 \times 10^{-4}$ | $1.06 \times 10^{0}$ | $1.06 \times 10^{0}$ | 0.03 |
| Osteogenic cells | $6.88 \times 10^{-4}$ | $6.95 \times 10^{0}$ | $6.95 \times 10^{0}$ | 0.00 |
| Spleen | $3.24 \times 10^{-3}$ | $3.01 \times 10^{-1}$ | $3.02 \times 10^{-1}$ | 0.47 |
| Testes | $3.59 \times 10^{-3}$ | $2.07 \times 10^{-1}$ | $2.09 \times 10^{-1}$ | 0.75 |
| Thymus | $8.71 \times 10^{-4}$ | $1.23 \times 10^{-1}$ | $1.23 \times 10^{-1}$ | 0.31 |
| Thyroid | $6.39 \times 10^{-4}$ | $4.04 \times 10^{-1}$ | $4.04 \times 10^{-1}$ | 0.07 |
| Urinary bladder wall | $8.85 \times 10^{-4}$ | $2.14 \times 10^{0}$ | $2.14 \times 10^{0}$ | 0.02 |

## 3. Discussion

$^{212}$Pb is a promising alpha-emitting isotope with an elementally matched gamma-emitting isotope $^{203}$Pb that can be used as an imaging surrogate via single photon emission computed tomography (SPECT). $^{212}$Pb physical half-life (10.64 h) is attractive from a clinical translation perspective with regard to patient care and waste management. A recently published phase 1 dose escalation trial of targeted alpha therapy with $^{212}$Pb-DOTAMTATE demonstrated patient safety and promising preliminary efficacy in patients with somatostatin receptor-positive neuroendocrine tumors [27].

From a toxicity standpoint, recoil energy from the emission of an alpha particle decouples the daughter nuclide from any chelator or other chemical bond, and untargeted daughter nuclides are known to accumulate in normal tissues, such as in bone or kidneys [28]. In the work presented here, we calculated estimated human tissue absorbed doses for [$^{212}$Pb]VMT01 from preclinical murine biodistribution data. In addition, we calculated estimated human tissue absorbed doses for free $^{208}$Tl (that will accumulate in the injectate prior to administration).

One limitation in the dosimetry of alpha radiotherapeutics is the unknown RBE value. Here, according to the method published by dos Santos and collaborators [21], an RBE value of 5 was used for $^{212}$Pb alpha emissions and a value of 1 was used for beta and gamma radiation. Recent studies performed in mammary carcinoma NT2.5 cells treated with $^{212}$Pb-labeled anti-HER2 antibody reported an RBE of 8.3 at 37% survival [29]. Notably, the dose contribution of extracellular unbound $^{212}$Pb-labeled antibody to the absorbed dose was about 2 orders of magnitude smaller compared to the bound and internalized $^{212}$Pb, suggesting that extracellular $^{212}$Pb delivers minimal radiation to cells. The authors conclude that these findings suggest that the actual lesion to dose-limiting tissue absorbed dose could be an order of magnitude greater than that predicted by the calculated absorbed dose.

The analysis presented here demonstrates that accumulated $^{208}$Tl in the injectate results in about 1% increase or less in estimated tissue absorbed doses over those projected for [$^{212}$Pb]VMT01. The dosimetry projections for [$^{212}$Pb]VMT01 from [$^{203}$Pb]VMT01 biodistribution data assume that the time-integrated activity coefficient of [$^{212}$Pb]VMT01 applies to all daughter radionuclides. This assumption is valid if there is no in vivo translocation of daughters. Human serum stability and in vivo mouse biodistribution studies demonstrate that $^{212}$Pb and $^{212}$Bi remain stably chelated to VMT01 with no evidence of daughter translocation in vivo (Li and collaborators, SNMMI-ACNM Mid-Winter Meeting 2022

Abstract) [20]. Retention of $^{212}$Pb daughter $^{212}$Bi within the chelator will result in decay of subsequent daughters $^{212}$Po and $^{208}$Tl at the site of localization due to their short half-lives. Prior to administration, accumulation of unchelated $^{208}$Tl will occur in the formulated product due to the recoil energy of the alpha decay from $^{212}$Bi. Accumulation of unchelated $^{212}$Po may also occur prior to administration as a result of beta decay from $^{212}$Bi; this decay has not yet been characterized. However, due to the extremely short half-life of the $^{212}$Po daughter (0.3 µs), decay from any free $^{212}$Po in the intravenously administered product can be assumed to occur in the plasma with negligible radiation to blood cells [29].

## 4. Materials and Methods

### 4.1. Radiolabeling and In Vivo Biodistribution

$^{203}$Pb chloride was obtained from Lantheus Medical Imaging (North Billerica, MA, USA). The structure of VMT01 has been previously published by Li and collaborators [30]. Radiolabeling of VMT01 with $^{203}$Pb was performed as previously described [30]; radio-chemical purity was > 99% as assessed by radio-HPLC. Thirteen-week-old male and female CD-1 IGS mice obtained from Charles River Laboratories (Wilmington, MA, USA) ($n = 28$ per sex, $n = 56$ total) were injected intravenously with [$^{203}$Pb]VMT01 (1.5 ± 0.38 pmol, 74 kBq). Following dosing, animals were sacrificed at 0.5, 1, 2, 4, 6, 24, or 55 h post-injection ($n = 4$ per time point per sex); at each time point whole blood, thymus, thyroid, adrenals, heart, lungs, spleen, bone (femur mid-diaphysis), bone marrow, liver, gallbladder, kidneys (adrenals removed), bladder wall, large intestine (wall and contents), cecum (with contents), small intestines (wall and contents), stomach (wall and contents), pancreas, brain, eyes, skin, muscle (quadriceps), ovaries, testes, uterus, tail, and remaining carcass (at select time points) were resected and assayed for radioactive content by gamma counting. Urine and feces were evaluated for radioactive content using pooled samples from cages.

### 4.2. Ex Vivo Gamma Counting

The activity of each collected tissue was measured on a Wizard 1480 (Perkin Elmer Life and Analytical Sciences, Bridgeport, CT, USA) or Wizard 2470 (Perkin Elmer Life and Analytical Sciences, Bridgeport, CT, USA) with a 279 keV peak position and 68% window coverage in units of counts per minute (CPM). Triplicate aliquots of the radiotracer, pulled from the dose-calibrated bulk injectate prepared fresh on each day of injections, were weighed, and assayed via gamma counting to convert CPM to units of grams of injected material. The uptake (percent of the injected dose, % ID) and concentration (% ID per gram, % ID/g) were calculated for each sample count using the known injected dose mass, corrected for tail uptake. Concentration estimates used the sample weight of the gamma-counted tissue in grams (g).

### 4.3. $^{203}$Pb Dosimetry Analysis

The radioactivity concentration of [$^{203}$Pb]VMT01 in each organ (fraction of injected activity per gram) over time was used to compute time-integrated activity coefficients (TIAC) [31] for each organ. For all organs except the total body and blood, uptake at time zero was assumed to be 0% ID. Total body and blood were assumed to be 100% ID at time zero. Human TIAC values were defined by multiplying individual mouse concentration values by animal body weight and by the human phantom organ weight to body weight ratio. This method is equivalent to the percent kilogram per gram method [32]. The human phantom organ weight to body weight ratios were determined from the ICRP 89 adult male and adult female phantom organ and total body weights from OLINDA/EXM 2.0 (Hermes Medical Solutions, Stockholm, Sweden). Each time point value was computed from the group average of the data.

TIAC through the last experimental time point was generated using trapezoidal integration of the seven data points. The contribution to the TIAC following the last experimental time point was estimated by fitting decay-corrected data to a single or a bi-exponential model to estimate biological clearance or assuming physical decay only following the last time point. The combination of physical decay and biological clearance was then analytically integrated. Human TIAC values were then adjusted for radioactivity leaving the body via the renal and gastrointestinal (GI) systems using the dynamic voiding bladder [33] (2 h void) and human alimentary tract model [34]. Excreted urine activity at each time point was defined as 100%-total body % ID-feces % ID. The fraction of excreted urine activity and the voiding half-life were determined by fitting the data to an exponential function. These coefficients were used with a 2 h human voiding time to calculate the urinary bladder TIAC. The ICRP 100 human alimentary tract (HAT) model [34] was utilized with the assumption that radioactivity enters the GI tract via the small intestine. For all animals in each sex group, the radioactivity (% ID, decay corrected) within the small and large intestine, cecum, and all contents were summed at each time point. The peak sum across time for each sex were then determined and used as input into the HAT model in OLINDA/EXM 2.0 to calculate the small intestines, left colon, right colon, and rectum TIACs. Total body radioactivity was calculated as the sum of all measured tissues except for bladder wall, urine, GI, and feces. Total body % ID human was assumed to be equivalent to total body % ID in mouse. The remainder of body TIAC was calculated by subtracting source organ TIACs except for excreta and those derived from the voiding and HAT models. Cortical and trabecular bone TIACs were calculated based on relative surface densities assuming radioactivity distributed to the bone surface. TIAC values were used to compute tissue absorbed dose values for the human adult male and female using OLINDA/EXM 2.0 with ICRP 89 adult male and female phantoms.

### 4.4. $^{212}$Pb Dosimetry Analysis

[$^{203}$Pb]VMT01 data were extrapolated to [$^{212}$Pb]VMT01 by adjusting the radioactive decay half-life. Assuming transient equilibrium between $^{212}$Pb and its daughters ($^{212}$Bi, $^{212}$Po, and $^{208}$Tl), the same residence times as for $^{212}$Pb were applied to the daughter nuclides as described by dos Santos and collaborators [21] OLINDA/EXM 2.0 calculations were performed for all nuclides manually. For $^{208}$Tl and $^{212}$Po, the relevant branching fraction was applied. A relative biological effectiveness (RBE) value of 5 was used for the alpha emissions from $^{212}$Bi and $^{212}$Po (while an RBE of 1 was used for beta and gamma emissions); absorbed doses are presented in units of Gray (Gy$_{RBE=5}$).

### 4.5. $^{208}$Tl Dosimetry Analysis

Human biodistribution of $^{201}$Tl chloride (a cardiac imaging agent) via scintigraphy imaging published in the literature [25,26] was used to estimate the dosimetry of free $^{208}$Tl. Thallous ion behaves as a potassium analog and tissue uptake is essentially intracellular. Biodistribution of thallium at early times in organs is thus related to regional blood flow. Human % ID values for heart, brain, kidney, liver, intestine, spleen, testes, and the remainder of body were determined from scintigraphy imaging as reported by Svensson and collaborators [26] and Krahwinkel and collaborators [25] for $^{201}$Tl chloride (using the earliest imaging time point from a combination of at rest and after exercise) and conservatively assuming no biological clearance and 100% ID in the total body (Table 5). Radioactive decay of $^{208}$Tl (3.05 m half-life) and resulting TIAC values were used to determine tissue-absorbed doses in the ICRP 89 human adult male using OLINDA/EXM 2.2. The activity fraction of free $^{208}$Tl in the injectate at a shelf-life of 6 h was calculated using the $^{212}$Pb decay scheme and branching fraction of 35.94% for $^{208}$Tl. The $^{208}$Tl activity fraction was used to calculate the $^{208}$Tl mGy$_{RBE=5}$/MBq administered $^{212}$Pb activity.

**Table 5.** $^{208}$Tl human tissue % ID.

| Organ/Tissue | Human % ID |
| --- | --- |
| Heart | 3.2 [26] |
| Brain | 1.5 [25] |
| Kidneys | 12.5 [26] |
| Liver | 5.1 [25] |
| Intestine [a] | 20.1 [25] |
| Spleen | 1.0 [25] |
| Testes | 0.4 [25] |
| Remainder of body | 56.2 [25] |

[a] Activity was split equally between the small intestine, upper large intestine wall, lower large intestine wall, and rectum wall based on ICRP 89 target wall organ masses.

## 5. Conclusions

The critical tissues for [$^{212}$Pb]VMT01 based on human dosimetry estimates from murine [$^{203}$Pb]VMT01 biodistribution data and tissue threshold doses from external beam irradiation data are anticipated to be red marrow and kidneys. Dosimetry analysis indicates that free $^{208}$Tl that will accumulate in the [$^{212}$Pb]VMT01 injectate prior to administration will not substantially impact estimated tissue absorbed doses in humans. The dosimetry estimations support the clinical evaluation of [$^{212}$Pb]VMT01.

**Supplementary Materials:** The following supporting information can be downloaded at https://www.mdpi.com/article/10.3390/molecules27185831/s1, Table S1: Organ/tissue activity concentrations (% ID/g) of [$^{203}$Pb]VMT01 in female CD-1 IGS naïve mice; Table S2: Organ/tissue activity concentrations (% ID/g) of [$^{203}$Pb]VMT01 in male CD-1 IGS naïve mice.

**Author Contributions:** K.D.O. contributed to experimental design, data analysis, and interpretation and preparation of the manuscript. K.E.H. and A.I.H. contributed to experimental design, acquisition, data analysis and interpretation, and manuscript revision. C.H., K.P., J.H. and M.C.L. contributed to data analysis and interpretation. J.M.C. and V.G. contributed to the acquisition and data analysis. D.D.C. contributed to experimental design, acquisition, data analysis, and interpretation. D.L., M.L., F.L.J. and M.K.S. contributed to the experimental design, interpretation, and editing of the manuscript. All authors have read and agreed to the published version of the manuscript.

**Funding:** This work was partially funded by NIH R01CA243014, NIH SBIR program R44CA250872, and NIH SBIR program R44CA254613.

**Institutional Review Board Statement:** The animal study protocol was approved by the Invicro Institutional Animal Care and Use Committee (IACUC). Protocol number: 005-2020.

**Informed Consent Statement:** Not applicable.

**Data Availability Statement:** The data presented in this study may be available on request from the corresponding author.

**Acknowledgments:** The authors gratefully acknowledge Jennifer Tavares, Erin Snay, and the Invicro Discovery Laboratory staff for animal handling and coordination of in vivo experiments, and Amos Hedt, Joseph O'Donoghue, and Edward K Fung for helpful discussions and QC of results. The authors appreciate Fiorenza Ianzini for assistance with the editing and revision of the manuscript.

**Conflicts of Interest:** K.O., M.K.S., F.L.J., M.L. and D.L. are employees of and have a financial interest in Viewpoint Molecular Targeting, Inc.

**Sample Availability:** Samples of the compounds are available on request from the corresponding author.

## References

1. Herraiz, C.; Martínez-Vicente, I.; Maresca, V. The α-melanocyte-stimulating hormone/melanocortin-1 receptor interaction: A driver of pleiotropic effects beyond pigmentation. *Pigment Cell Melanoma Res.* **2021**, *34*, 748–761. [CrossRef] [PubMed]
2. Tatro, J.B.; Wen, Z.; Entwistle, M.L.; Atkins, M.B.; Smith, T.J.; Reichlin, S.; Murphy, J.R. Interaction of an alpha-melanocyte-stimulating hormone-diphtheria toxin fusion protein with melanotropin receptors in human melanoma metastases. *Cancer Res.* **1992**, *52*, 2545–2548. [PubMed]

3. Siegrist, W.; Solca, F.; Stutz, S.; Giuffrè, L.; Carrel, S.; Girard, J.; Eberle, A.N. Characterization of receptors for alpha-melanocyte-stimulating hormone on human melanoma cells. *Cancer Res.* **1989**, *49*, 6352–6358. [PubMed]
4. Yang, J.; Xu, J.; Gonzalez, R.; Lindner, T.; Kratochwil, C.; Miao, Y. $^{68}$Ga-DOTA-GGNle-CycMSH$_{hex}$ targets the melanocortin-1 receptor for melanoma imaging. *Sci. Transl. Med.* **2018**, *10*, eaau4445. [CrossRef]
5. Sgouros, G. α-Particle-Emitter Radiopharmaceutical Therapy: Resistance Is Futile. *Cancer Res.* **2019**, *79*, 5479–5481. [CrossRef] [PubMed]
6. Yadav, M.P.; Ballal, S.; Sahoo, R.K.; Tripathi, M.; Seth, A.; Bal, C. Efficacy and safety of $^{225}$Ac-PSMA-617 targeted alpha therapy in metastatic castration-resistant Prostate Cancer patients. *Theranostics* **2020**, *10*, 9364–9377. [CrossRef] [PubMed]
7. Jadvar, H.; Colletti, P.M. Targeted α-therapy in non-prostate malignancies. *Eur. J. Nucl. Med. Mol. Imaging* **2021**, *49*, 47–53. [CrossRef]
8. Parker, C.; Lewington, V.; Shore, N.; Kratochwil, C.; Levy, M.; Lindén, O.; Noordzij, W.; Park, J.; Saad, F. Targeted Alpha Therapy, an Emerging Class of Cancer Agents: A Review. *JAMA. Oncol.* **2018**, *4*, 1765–1772.
9. Charlton, D.E.; Nikjoo, H.; Humm, J.L. Calculation of initial yields of single- and double-strand breaks in cell nuclei from electrons, protons and alpha particles. *Int. J. Radiat. Biol.* **1989**, *56*, 1–19. [CrossRef]
10. Goodhead, D.T.; Thacker, J.; Cox, R. Weiss Lecture. Effects of radiations of different qualities on cells: Molecular mechanisms of damage and repair. *Int. J. Radiat. Biol.* **1993**, *63*, 543–556. [CrossRef]
11. Goodhead, D.T. Initial events in the cellular effects of ionizing radiations: Clustered damage in DNA. *Int. J. Radiat. Biol.* **1994**, *65*, 7–17. [CrossRef]
12. Barendsen, G.W. The relationships between RBE and LET for different types of lethal damage in mammalian cells: Biophysical and molecular mechanisms. *Radiat. Res.* **1994**, *139*, 257–270. [CrossRef]
13. Goodhead, D.T. Molecular and cell models of biological effects of heavy ion radiation. *Radiat. Environ. Biophys.* **1995**, *34*, 67–72. [CrossRef]
14. Nikjoo, H.; O'Neill, P.; Wilson, W.E.; Goodhead, D.T. Computational approach for determining the spectrum of DNA damage induced by ionizing radiation. *Radiat. Res.* **2001**, *156*, 577–583. [CrossRef]
15. Georgakilas, A.G.; O'Neill, P.; Stewart, R.D. Induction and repair of clustered DNA lesions: What do we know so far? *Radiat. Res.* **2013**, *180*, 100–109. [CrossRef]
16. Nikitaki, Z.; Nikolov, V.; Mavragani, I.V.; Mladenov, E.; Mangelis, A.; Laskaratou, D.A.; Fragkoulis, G.I.; Hellweg, C.E.; Martin, O.A.; Emfietzoglou, D.; et al. Measurement of complex DNA damage induction and repair in human cellular systems after exposure to ionizing radiations of varying linear energy transfer (LET). *Free Radic. Res.* **2016**, *50*, S64–S78. [CrossRef]
17. Yard, B.D.; Gopal, P.; Bannik, K.; Siemeister, G.; Hagemann, U.B.; Abazeed, M.E. Cellular and Genetic Determinants of the Sensitivity of Cancer to α-Particle Irradiation. *Cancer Res.* **2019**, *79*, 5640–5651. [CrossRef]
18. Pouget, J.P.; Constanzo, J. Revisiting the Radiobiology of Targeted Alpha Therapy. *Front. Med.* **2021**, *8*, 692436. [CrossRef]
19. Kratochwil, C.; Bruchertseifer, F.; Giesel, F.L.; Weis, M.; Verburg, F.A.; Mottaghy, F.; Kopka, K.; Apostolidis, C.; Haberkorn, U.; Morgenstern, A. 225Ac-PSMA-617 for PSMA-Targeted α-Radiation Therapy of Metastatic Castration-Resistant Prostate Cancer. *J. Nucl. Med.* **2016**, *57*, 1941–1944. [CrossRef]
20. Li, M.; Baumhover, N.J.; Liu, D.; Boschetti, F.; Lee, D.; Obot, E.R.; Marks, B.M.; Sagastume, E.A.; McAlister, D.; Gabr, M.; et al. Novel chelator modifications to improve in vitro and in vivo stability of $^{212}$Pb/$^{212}$Bi radiopeptide conjugates for alpha-particle radiotherapy. In Proceedings of the SNMMI-ACNM Mid-Winter Meeting, Orlando, FL, USA, 27–29 February 2022.
21. Dos Santos, J.C.; Schäfer, M.; Bauder-Wüst, U.; Lehnert, W.; Leotta, K.; Morgenstern, A.; Kopka, K.; Haberkorn, U.; Mier, W.; Kratochwil, C. Development and dosimetry of $^{203}$Pb/$^{212}$Pb-labelled PSMA ligands: Bringing "the lead" into PSMA-targeted alpha therapy? *Eur. J. Nucl. Med. Mol. Imaging* **2019**, *46*, 1081–1091. [CrossRef]
22. Zaid, N.R.R.; Kletting, P.; Beer, A.J.; Rozgaja Stallons, T.A.; Torgue, J.J.; Glatting, G. Mathematical Modeling of In Vivo Alpha Particle Generators and Chelator Stability. *Cancer Biother. Radiopharm.* **2021**. [CrossRef] [PubMed]
23. Stewart, F.A.; Akleyev, A.V.; Hauer-Jensen, M.; Hendry, J.H.; Kleiman, N.J.; Macvittie, T.J.; Aleman, B.M.; Edgar, A.B.; Mabuchi, K.; Muirhead, C.R.; et al. ICRP publication 118: ICRP statement on tissue reactions and early and late effects of radiation in normal tissues and organs–threshold doses for tissue reactions in a radiation protection context. *Ann. ICRP* **2012**, *41*, 1–322. [CrossRef] [PubMed]
24. Emami, B.; Lyman, J.; Brown, A.; Coia, L.; Goitein, M.; Munzenrider, J.E.; Shank, B.; Solin, L.J.; Wesson, M. Tolerance of normal tissue to therapeutic irradiation. *Int. J. Radiat. Oncol. Biol. Phys.* **1991**, *21*, 109–122. [CrossRef]
25. Krahwinkel, W.; Herzog, H.; Feinendegen, L.E. Pharmacokinetics of thallium-201 in normal individuals after routine myocardial scintigraphy. *J. Nucl. Med.* **1988**, *29*, 1582–1586.
26. Svensson, S.E.; Lomsky, M.; Olsson, L.; Persson, S.; Strauss, H.W.; Westling, H. Non-invasive determination of the distribution of cardiac output in man at rest and during exercise. *Clin. Physiol.* **1982**, *2*, 467–477. [CrossRef]
27. Delpassand, E.S.; Tworowska, I.; Esfandiari, R.; Torgue, J.; Hurt, J.; Shafie, A.; Núñez, R. Targeted Alpha-Emitter Therapy With $^{212}$Pb-DOTAMTATE for the Treatment of Metastatic SSTR-Expressing Neuroendocrine Tumors: First-in-Human, Dose-Escalation Clinical Trial. *J. Nucl. Med.* **2022**, *121*, 263230.
28. Zaid, N.R.R.; Kletting, P.; Winter, G.; Beer, A.J.; Glatting, G. A Whole-Body Physiologically Based Pharmacokinetic Model for Alpha Particle Emitting Bismuth in Rats. *Cancer Biother. Radiopharm.* **2022**, *37*, 41–46. [CrossRef]

29. Liatsou, I.; Yu, J.; Bastiaannet, R.; Li, Z.; Hobbs, R.F.; Torgue, J.; Sgouros, G. $^{212}$Pb-conjugated anti-rat HER2/*neu* antibody against a *neu*-N derived murine mammary carcinoma cell line: Cell kill and RBE in vitro. *Int. J. Radiat. Biol.* **2022**, 1452–1461. [CrossRef]
30. Li, M.; Liu, D.; Lee, D.; Cheng, Y.; Baumhover, N.J.; Marks, B.M.; Sagastume, E.A.; Ballas, Z.K.; Johnson, F.L.; Morris, Z.S.; et al. Targeted Alpha-Particle Radiotherapy and Immune Checkpoint Inhibitors Induces Cooperative Inhibition on Tumor Growth of Malignant Melanoma. *Cancers* **2021**, *13*, 3676. [CrossRef]
31. Bolch, W.E.; Eckerman, K.F.; Sgouros, G.; Thomas, S.R. MIRD pamphlet No. 21: A generalized schema for radiopharmaceutical dosimetry–standardization of nomenclature. *J. Nucl. Med.* **2009**, *50*, 477–484. [CrossRef]
32. Kirschner, A.S.; Ice, R.D.; Beierwaltes, W.H. Radiation Dosimetry of 131-I-19-Iodocholesterol: The Pitfalls of Using Tissue Concentration Data. *J. Nucl. Med.* **1975**, *16*, 247–249.
33. Cloutier, R.J.; Smith, S.A.; Watson, E.E.; Snyder, W.S.; Warner, G.G. Dose to the fetus from radionuclides in the bladder. *Health Phys.* **1973**, *25*, 147–161. [CrossRef]
34. ICRP. Human alimentary tract model for radiological protection. ICRP Publication 100. A report of The International Commission on Radiological Protection. *Ann. ICRP* **2006**, *36*, 25–327.

Review

# Radionuclides for Targeted Therapy: Physical Properties

Caroline Stokke [1,2,*], Monika Kvassheim [1,3] and Johan Blakkisrud [1]

1. Department of Physics and Computational Radiology, Division of Radiology and Nuclear Medicine, Oslo University Hospital, P.O. Box 4959 Nydalen, 0424 Oslo, Norway
2. Department of Physics, University of Oslo, Problemveien 7, 0315 Oslo, Norway
3. Division of Clinical Medicine, University of Oslo, Problemveien 7, 0315 Oslo, Norway
* Correspondence: carsto@ous-hf.no

**Abstract:** A search in PubMed revealed that 72 radionuclides have been considered for molecular or functional targeted radionuclide therapy. As radionuclide therapies increase in number and variations, it is important to understand the role of the radionuclide and the various characteristics that can render it either useful or useless. This review focuses on the physical characteristics of radionuclides that are relevant for radionuclide therapy, such as linear energy transfer, relative biological effectiveness, range, half-life, imaging properties, and radiation protection considerations. All these properties vary considerably between radionuclides and can be optimised for specific targets. Properties that are advantageous for some applications can sometimes be drawbacks for others; for instance, radionuclides that enable easy imaging can introduce more radiation protection concerns than others. Similarly, a long radiation range is beneficial in targets with heterogeneous uptake, but it also increases the radiation dose to tissues surrounding the target, and, hence, a shorter range is likely more beneficial with homogeneous uptake. While one cannot select a collection of characteristics as each radionuclide comes with an unchangeable set, all the 72 radionuclides investigated for therapy—and many more that have not yet been investigated—provide numerous sets to choose between.

**Keywords:** radionuclide; targeted therapy; radionuclide therapy; radioactivity; molecular radiotherapy; beta; alpha; auger

**Citation:** Stokke, C.; Kvassheim, M.; Blakkisrud, J. Radionuclides for Targeted Therapy: Physical Properties. *Molecules* **2022**, *27*, 5429. https://doi.org/10.3390/molecules27175429

Academic Editor: Svend Borup Jensen

Received: 30 June 2022
Accepted: 21 August 2022
Published: 25 August 2022

**Publisher's Note:** MDPI stays neutral with regard to jurisdictional claims in published maps and institutional affiliations.

**Copyright:** © 2022 by the authors. Licensee MDPI, Basel, Switzerland. This article is an open access article distributed under the terms and conditions of the Creative Commons Attribution (CC BY) license (https://creativecommons.org/licenses/by/4.0/).

## 1. Introduction

Therapies with radioactive nuclides have been rapidly increasing in both number and variations over the last few years. Treatments based on beta-minus-emitters (called beta-emitters from hereon) have been in use since the 1930s, and the first two radionuclides used for treatment purpose were sodium-24 ($^{24}$Na) and phosphorus-32 ($^{32}$P), both for the first time in 1936 to treat haematological disease [1]. Iodine-131 ($^{131}$I) was the most commonly investigated beta-emitter for long, also due to the direct targeting properties of $^{131}$I, which allow for accumulation in differentiated thyroid cells [2]. The later emergence of carrier molecules to which radionuclides could be conjugated introduced new radionuclides in targeted therapy, such as yttrium-90 ($^{90}$Y) and lutetium-177 ($^{177}$Lu). While alpha-emitters have been used for decades, they have recently increased in popularity as their short range unlocks potential for tailored treatment of smaller structures [3]. Together with the currently less frequently used auger-emitters, beta- and alpha-emitters constitute the armament for therapies with radionuclides.

The latest collection of radionuclides tabulated by the International Commission on Radiological Protection (ICRP) lists 1252 radionuclides [4]. Besides the general choice of type of radiation emitted (Table 1), there are several important factors to consider when selecting the optimal radionuclide. First, the size of the structures being targeted should be in agreement with the range of the radiation emitted. Potential heterogeneity in the uptake of the carrier molecule could also affect the optimal range; a higher degree of heterogeneity can be evened out by emitters with longer ranges. The half-life of the radionuclide should be selected with care; the pharmacokinetics of the carrier molecule will determine the optimal time for depositing the maximum amount of energy within the target tissue. The abundance of photons emitted is of importance for both radiation protection and imaging approaches. While potential radioactive daughters may add to the total amount of energy released, their possible re-localisation should be addressed.

Table 1. Overview of types of radiation.

| Type of Radiation | Particle | Mass | Typical Energy | Typical Range in Tissue | LET (keV/µm) | RBE |
|---|---|---|---|---|---|---|
| Alpha | 2 protons and 2 neutrons | $6.6 \times 10^{-27}$ kg | Discrete; ~4–10 MeV | ~20–70 µm | ~50–300 | ~5 |
| Beta minus | Electron | $9.1 \times 10^{-31}$ kg | Continuous; maximum of some hundred keV to some MeV | Less than a mm to some mm | ~0.1–2 | ~1 |
| Auger | Electron | $9.1 \times 10^{-31}$ kg | <10 keV | <1 µm | ~4–26 | ~1 or higher |

This review will cover the above-mentioned factors important for selection of radionuclides for targeted therapy, with emphasis on oncological applications. Historical overviews of therapeutic radionuclides both registered in the clinical trials database (https://www.clinicaltrials.gov (accessed on 30 June 2022)) and described in publications listed in PubMed (https://pubmed.ncbi.nlm.nih.gov (accessed on 30 June 2022)) are also included. Besides the physical aspects described here, factors related to radiochemistry and -pharmacy (production, cost, availability, and conjugation) are of vital importance and the reader is referred to other reviews for an overview [5–7].

## 2. Trends

A search to identify trends of publications was conducted using the PubMed database, starting with the 1252 radionuclides listed in ICRP-publication 107 [4]. This yielded a total of 72 radionuclides included in either theoretical consideration for treatment applications, preclinical, or clinical studies. To provide an overview of therapies with molecular or functional targeting mechanisms, brachytherapies with applicators were excluded in the search query, while selective internal radiation therapy (SIRT) was included (search terms are provided in Supplementary File S1). Some publications, for example, the first [131]I studies, have been observed missing as some of them are indexed without the specific radionuclide, but including publications without this term would have resulted in a high number of false positive hits. The timelines in Figure 1 show the publication trends over the last 74 years. A similar search to identify trends of registered clinical trials was conducted using the clinical trials database, starting with the radionuclides identified in the PubMed-search in addition to iodine-125, copper-67, and indium-111 (Figure 2). Some of the applications for the most common radionuclides are listed below.

More than 60 years after its introduction, [131]I is still subject of an extensive number of publications. The radionuclide is commonly used to treat metastatic differentiated thyroid cancer, for ablation purposes, and for benign thyroid diseases [2,8,9]. It is also used in several conjugates; for example, the form of [[131]I]meta-iodobenzylguanidin ([[131]I]MIBG), which, due to its molecular analogy with norepinephrine, can be taken up by neuroendocrine cells. It can, therefore, be used for adult diseases, such as pheochromocytoma, paragangliomas, medullary thyroid carcinoma, and neuroendocrine carcinomas, as well as for neuroblastomas in children [10,11]. [131]I was used in the radioimmunoconjugate [[131]I]I-tositumomab that targets the cluster of differentiation (CD) 20 antigen and was approved by the US Food and Drug Administration (FDA) for non-Hodgkin lymphoma (NHL) [12].

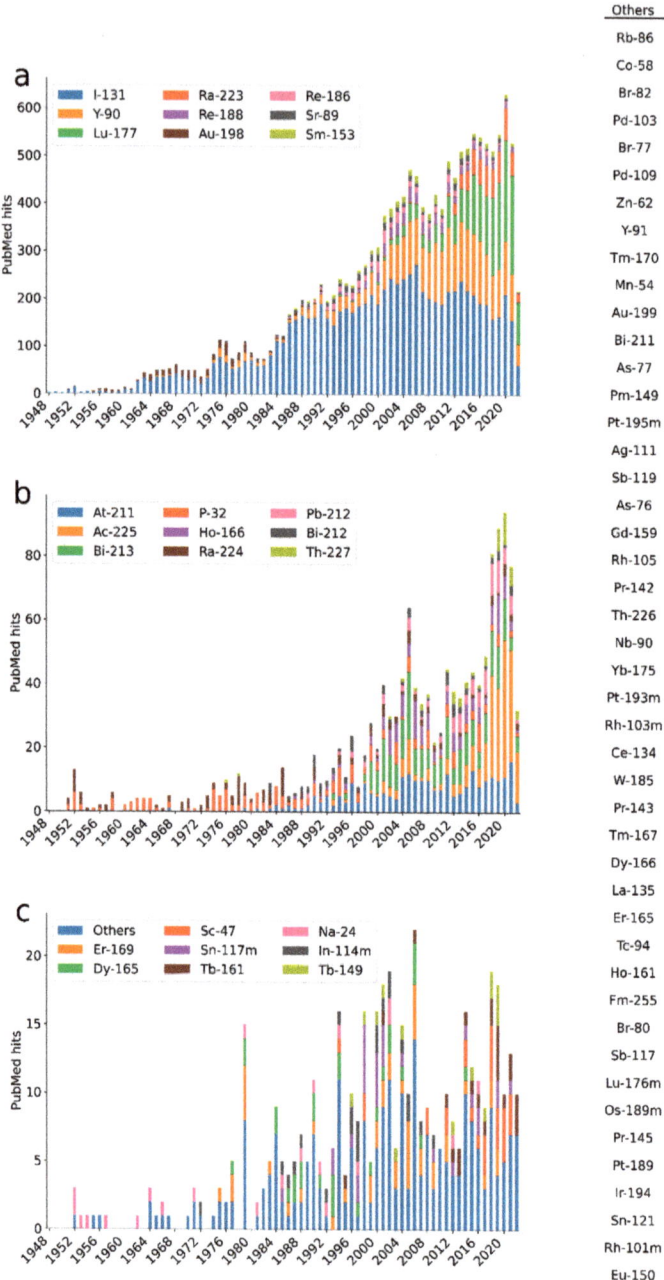

**Figure 1.** The number of hits in PubMed as of 30 June 2022, per radionuclide and year of publication. The radionuclides have been separated according to the aggregated numbers of publications across panels (**a**–**c**). The radionuclides with fewer than 13 hits total have been aggregated into "Others" and are listed to the right in the figure. The search strategy and search strings are described in Supplementary File S1.

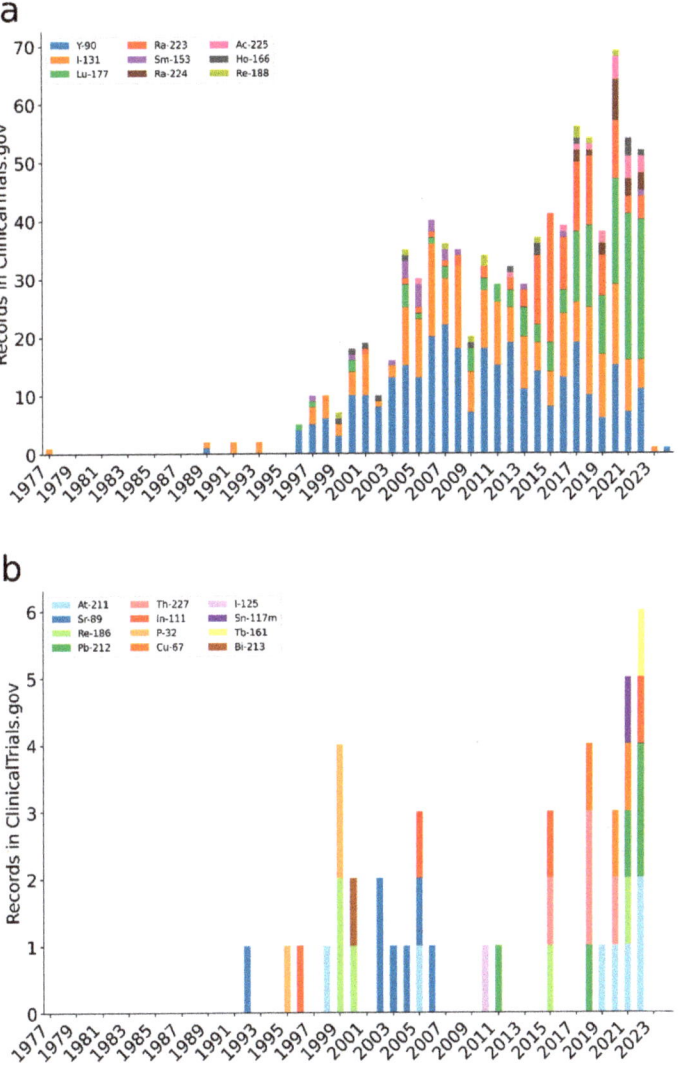

**Figure 2.** The number of records in the clinical trials database as of June 2022 per radionuclide and year. The results are split in panels (**a**,**b**) according to the aggregated number of records. The search strategy and search strings are described in Supplementary File S1.

$^{177}$Lu and $^{90}$Y have been included in more than one thousand aggregated publications each (Figure 1). Both have been explored extensively in somatostatin analogue therapy for neuroendocrine tumours, individually as well as in comparison or combination [13–15]. $^{90}$Y is used in the FDA-approved CD20-targeting radioimmunoconjugate [$^{90}$Y]Y-ibritumomab tiuxetan to treat NHL [16]. In the last few years, SIRT for liver radioembolisation has increased rapidly, and two types of $^{90}$Y-based microspheres are currently approved as medical devices [17]. In addition to the FDA-approved [$^{177}$Lu]Lu-DOTA-0-Tyr$^3$-octreotate ([$^{177}$Lu]Lu-DOTATATE) [18], $^{177}$Lu is part of the recently approved radiopharmaceutical used to target prostate-specific membrane antigen (PSMA) in patients with metastatic castrate-resistant prostate cancer (mCRPC), [$^{177}$Lu]Lu-PSMA-617 [15,19]. Furthermore,

$^{177}$Lu-based treatments are part of several clinical studies, for example, with the promising target fibroblast activation protein (FAP) [20].

The alpha-emitter radium-223 ($^{223}$Ra) was described as one of several potential alpha-emitters for radioimmunotherapy in the late 1990s [21]. The main application has, however, been in the form of [$^{223}$Ra]radiumdichloride, which is currently an FDA-approved treatment of skeletal metastases from castration-resistant prostate cancer [22]. Later explorations of $^{223}$Ra include e.g., nanomicells to increase the efficacy in treatment of osteosarcoma [23].

The beta-emitter $^{32}$P has had many applications since its first use against haematological diseases, including intracavitary applications and treatment of osseous metastases [24,25]. The next radionuclides in terms of aggregated research output were rhenium-186 ($^{186}$Re), rhenium-188 ($^{188}$Re), strontium ($^{89}$Sr), and samarium-153 ($^{153}$Sm), which have—together with other radionuclides—been used in palliative treatment of bone metastases or radiosynovectomy [26,27]. $^{188}$Re has, in recent years, been explored both for melanoma, in a radioimmunoconjugate, and for inoperable hepatocellular carcinoma, with selective administration of [$^{188}$Re]Re-4-hexadecyl-1-2,9,9-tetramethyl-4,7-diaza-1,10-decanethiol/lipiodol [28,29].

The alpha-emitter astatine-211 ($^{211}$At) was produced for the first time as early as in 1940 [30] and has been explored in numerous preclinical models. In its free state, the biodistribution is very similar to iodine's. An up to 12-year follow-up on a cohort with ovarian cancer treated with $^{211}$At conjugated to MX35 F(ab')$_2$, a murine monoclonal antibody, was reported in 2019 [31].

Another alpha-emitter, actinium-225 ($^{225}$Ac), has been investigated for multiple clinical uses, such as in the radioimmunoconjugate [$^{225}$Ac]Ac-lintuzumab in treatment of acute myeloid leukemia [32]. The radionuclide is perhaps currently best known for PSMA directed treatment of mCRPC [33], and a recent meta-analysis pooled safety (225 patients) and efficacy in the form of reported response (263 patients) and outcome (200 patients) [34].

Some radionuclides that are either primarily or exclusively used for diagnostic purposes have been omitted from the publication trend dataset due to difficulties in separating diagnostic and therapeutic reported applications. These include copper-64 ($^{64}$Cu), indium-111 ($^{111}$In), technetium-99m ($^{99m}$Tc), fluorine-18 ($^{18}$F), gallium-68 ($^{68}$Ga), iodine-125 ($^{125}$I), and gallium-67 ($^{67}$Ga). However, some of these have been investigated for explicit therapeutic applications over the years. The beta-plus- and auger-emitter $^{64}$Cu has, for example, been investigated for therapy in the form of a hypoxia-marker [$^{64}$Cu]Cu-diacetyl-bis(N4-methylhiosemicarbazone) ([$^{64}$Cu]Cu-ATSM) [35]. The gamma- and auger-emitter $^{111}$In was initially used to image neuroendocrine disease with somatostatin analogues but was administered with increased activity to function as treatment [36]. Another gamma- and auger-emitter, $^{67}$Ga, has also been reassessed as a potentially viable therapeutic radionuclide [37]. The gamma- and auger-emitter iodine-123 ($^{123}$I) has been investigated in a pre-clinical model for treatment of glioblastoma [38]. The most common emitter in diagnostic nuclear medicine, $^{99m}$Tc, has, in the form of [$^{99m}$Tc]TcO$_4^-$, actually been investigated for therapy in a preclinical breast cancer model because of its auger electron emission and interaction with the human sodium/iodine sympother [39].

### 3. Characteristics

*3.1. Linear Energy Transfer (LET) and Relative Biological Effectiveness (RBE)*

LET is a purely physical quantity, describing charged particles' energy loss per length, measured in keV/µm. This quantity will depend on particle mass, charge, and energy, and may also vary along a particle track as the particle loses energy while traversing the material. For example, alpha-particles are known for their high LET Bragg peak at the end of the track. In general, alpha-particles have higher LET than auger electrons, which again have higher LET than beta-particles (Table 1) [40]. Auger electrons originate from electron capture or internal conversion after an isomeric transition and, hence, differ from beta-particles. They are often emitted in cascades, and will have an extremely short range, resulting in the higher LET [41,42].

Besides its close association with range, the LET is independently important for choice of emitter as it may impact the RBE. The RBE describes the effect of the radiation, for example, in terms of cell killing, and depends on both properties of the radiation and biological factors. It is defined as the biologically iso-effective absorbed doses for a certain radiation in relation to a reference radiation (often 250 keV photons). There is no established one-to-one agreement between LET and RBE, but RBE tends to increase with LET until a peak at around 100–200 keV/μm. Alpha-particles often show RBE values around five, and electrons commonly have RBE values of one (same as for reference photons). However, for auger electrons, higher RBE values have been reported when the emitters are taken up in the nucleus [41,43]. Indirect effects may kill the cells even though the emitters are not directly inserted into the nucleus, but an RBE of one, similar to other electrons, can then be expected [41,42]. While it should be emphasized that RBE values are commonly investigated in in vitro or rodent studies and are challenging to compare clinically when many additional factors contribute, it is clear that a higher efficiency per absorbed dose is to be expected from, e.g., alpha-emitters than beta-emitters. Potential benefits of this are, however, more complicated to generally interpret as both tumours and normal tissues at risk will be subject to the same effects. Differences in dose rate between the target and normal tissues, cell/tissue sensitivity, degree of oxygenation, and other radiobiological factors will impact the RBE. These should ideally be known to decide whether alpha- or beta-emitters are better suited for a specific treatment as, for instance, alpha-radiation-induced damage is independent of oxygenation, while much of the treatment effect with beta radiation requires oxygenation.

### 3.2. Particle Range

The range of a charged particle in tissue is the average distance it is expected to travel before it comes to rest (the expectation value of the path length), and, for therapeutic radiation, range depends on particle energy and particle type [44]. While many of the photons exit a patient injected with a diagnostic radionuclide for imaging, for radionuclide therapies, most of the energy should be deposited within the patient, specifically in the tissue being targeted. Hence, short-range, non-penetrating radiation; beta, alpha, and auger electrons are the ones of interest [45]. Still, the particle range ideal for a target varies greatly between cases. The range of auger electrons is very short (Table 1), and, for therapy with auger electrons to be most efficient, decays should occur in or near to the target cell nucleus [40]. Considerations of cellular uptake and intracellular positioning are, therefore, especially important for auger-emitters. Alpha-particles travel in nearly straight paths as they gradually lose energy to atomic electrons, and the range is measured from the point of decay to where ionisations stop [46]. Alphas have a much longer range than auger electrons, and it is not crucial that the decays occur near the nucleus as the alpha-particles typically traverse a few cell diameters while depositing their energy [40]. For beta-emitters, range is described differently as betas are emitted with a continuous range of energies and follow tortuous paths as they undergo multiple scatterings in a medium. Max range, mean range, and $X_{90}$, the radius of the sphere in which 90% of the beta-emitter's energy is deposited, can all be useful when describing range [46]. Selected properties of the 15 most common beta-emitters identified through the PubMed database search are provided in Table 2. It is important to be aware that, when range is given in literature, it usually describes the total path length travelled by the beta particle, which, in clinical settings, is longer than the depth of penetration. Beta-emitters irradiate a much larger volume than alpha-emitters, and the differences in range between beta-emitters are large enough to produce differences in therapeutic effect.

Table 2. According to our search, the 15 beta-emitters most frequently published on in relation to radionuclide therapy and their properties. The data are taken from ICRP 107 [4].

| Radionuclide | Half-Life | Mean Energy/keV | Max Energy/MeV | $X_{90}$ in Water^/mm | Max CSDA Range in Tissue ᵛ/mm | Photons Intensity > 5% and Energy > 75 keV |
|---|---|---|---|---|---|---|
| I-131 | 8.02 days | 181.9 | 0.81 | 0.9 | 3.3 | 364 keV (82%) 637 keV (7%) 284 keV (6%) |
| Y-90 | 64.10 h | 932.9 | 2.28 | 5.5 | 10.8 | |
| Lu-177 | 6.65 days | 133.3 | 0.50 | 0.6 | 1.7 | 208 keV (11%) 113 keV (6%) |
| P-32 | 14.26 days | 694.8 | 1.71 | 3.7 | 8.0 | |
| Re-188 | 17.00 h | 762.6 | 2.12 | 4.5 | 10.1 | 155 keV (16%) |
| Sr-89 | 50.53 days | 584.5 | 1.50 | 3.3 | 6.9 | |
| Re-186 * | 3.72 days | 346.6 | 1.07 | 1.9 | 4.6 | 137 keV (9%) |
| Sm-153 | 46.50 h | 223.6 | 0.81 | 1.1 | 3.3 | 103 keV (30%) |
| Au-198 | 2.70 days | 312.2 | 1.37 | 1.7 | 6.2 | 412 keV (96%) |
| Ho-166 | 26.80 h | 665.0 | 1.85 | 4.0 | 8.7 | 81 keV (7%) |
| Cu-67 | 61.83 h | 135.9 | 0.56 | 0.6 | 2.0 | 185 keV (49%) 93 keV (16%) 91 keV (7%) |
| Er-169 | 9.40 days | 99.6 | 0.35 | 0.4 | 1.0 | |
| Dy-165 | 2.33 h | 439.7 | 1.29 | 2.5 | 5.8 | |
| Tb-161 | 6.91 days | 154.3 | 0.59 | 0.7 | 2.1 | |
| Sc-47 | 3.35 days | 161.9 | 0.60 | 0.7 | 2.2 | 159 keV (68%) |

* beta-particle yield of ¹⁸⁶Re is 92.5%. ^ Calculated with Geant4 Application for Tomographic Emission (GATE); see Supplementary File S2 for details. ᵛ The continuous-slowing-down approximation (CSDA) range was calculated using the maximum energies with equation A.18 from Prestwich et al. (1989) [47].

In general, one might choose the high energy beta-emitters, such as $^{90}$Y, for targeting large tumours, the lower energy beta-emitters, such as $^{177}$Lu, for smaller tumours, and alpha-emitters for micro metastases. However, various target characteristics affect the ideal range of the therapeutic radiation; examples include size, geometry, and distribution of the molecular target. Sometimes, there can be a variety of tumour characteristics within a patient, and, in such a case, the optimal range could be different for every tumour. Several studies have investigated the impact of particle range for different tumour characteristics, such as studies by O'Donoghue et al. and Bernhardt et al. investigating the relationship between tumour size and range, and Tamborino et al. looking at the influence of radionuclide choice with heterogeneously distributed radiolabelled peptides [48–50].

### 3.2.1. Size of Targets and Tissues at Risk

Ideally, particle range should be optimised to irradiate the entire target volume while minimising radiation to healthy tissue. This ideal range is highly variable as target characteristics vary greatly between diseases, individual patients, and individual targets. To illustrate the effects of size and geometries with different particles and energies, we performed simulations. The method description is found in Supplementary File S2. In Figure 3, we show the simple case of a spherical target volume with homogeneous uptake in the entire sphere. The energy deposited in the sphere as a percentage of total energy is plotted against sphere diameter for alpha sources of 5.5 MeV and for beta sources with the beta emission energy spectra of $^{90}$Y, $^{177}$Lu, and $^{131}$I. As the sphere gets smaller, more of the energy is deposited outside the source volume and the impact of particle type and energy is larger. $^{90}$Y emits beta-particles of high energies, and, already for 10 mm spheres, only approximately 62% of the emitted energy is deposited in the sphere. In contrast, the alpha-emitter deposits more than 90% of its energy within the sphere even when the sphere diameter is only 0.5 mm. Hence, if only range is of interest and there is uptake in an entire spherical tumour volume with a diameter around 0.5 mm, one might want to use an alpha-emitter rather than a high-energy beta-emitter to keep most of the energy in the tumour.

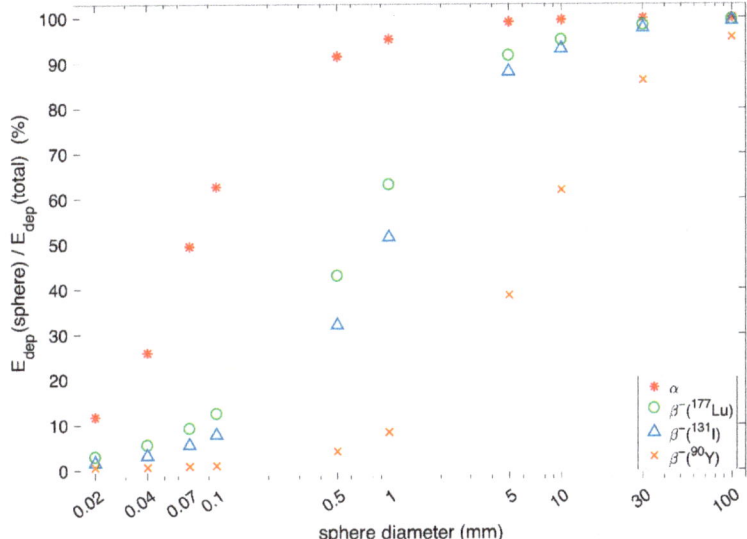

**Figure 3.** The figure shows energy deposited within a spherical source ($E_{dep}$(sphere)) as a percentage of the total energy emitted ($E_{dep}$(total)). The sphere source diameter ranged from 0.02 mm, to approximate a single cell, to 100 mm, to approximate a large tumour. Note that the x-axis is a log scale. Four different sources are shown, three pure beta sources with beta energies following the emission spectra for $^{90}$Y, $^{177}$Lu, and $^{131}$I, and an alpha source with alpha energies of 5.5 MeV. For large spheres, most of the energy will be deposited inside the sphere regardless of type of emitter.

Figure 4 illustrates sources that are spherical shells, a simplified geometry typical for large tumours with varying uptake of the radiotherapeutical due to heterogeneous vascularisation and necrotic areas in the core region [51]. In the simulations, uniform activity distribution is assumed in the shell, without any activity in the core. As the shell thickness increases, the choice of emitter becomes less significant. As an example, with a thin shell of 0.5 mm thickness and a 10 mm outer diameter, more than 95% of the energy from alpha decays is deposited in the shell and only 1.5% is deposited in the core. For the same geometry, beta radiation with the energy spectrum of $^{90}$Y deposits only 20% of the energy in the shell and 26% is deposited in the core. Hence, if one wishes to irradiate a volume without uptake of the radiopharmaceutical, a high-energy beta-emitter is a better choice than an alpha-emitter. How high the beta energy should be will be a trade-off between energy deposited outside the shell being kept reasonably low and sufficiently irradiating the core. If the radius of the core is assumed smaller, one can choose a lower-energy beta-emitter and still deposit sufficient energy in the core to kill the cells while limiting radiation to surrounding healthy tissue. Depending on how vital or radiation-sensitive the surrounding tissue is, in some situations irradiating the core might be less important than protecting surrounding tissues. In cases with active tumour cells without uptake in the core, ranges adequate to reach those cells from areas of uptake will likely be imperative, and, for example, $^{90}$Y might be the favoured radionuclide.

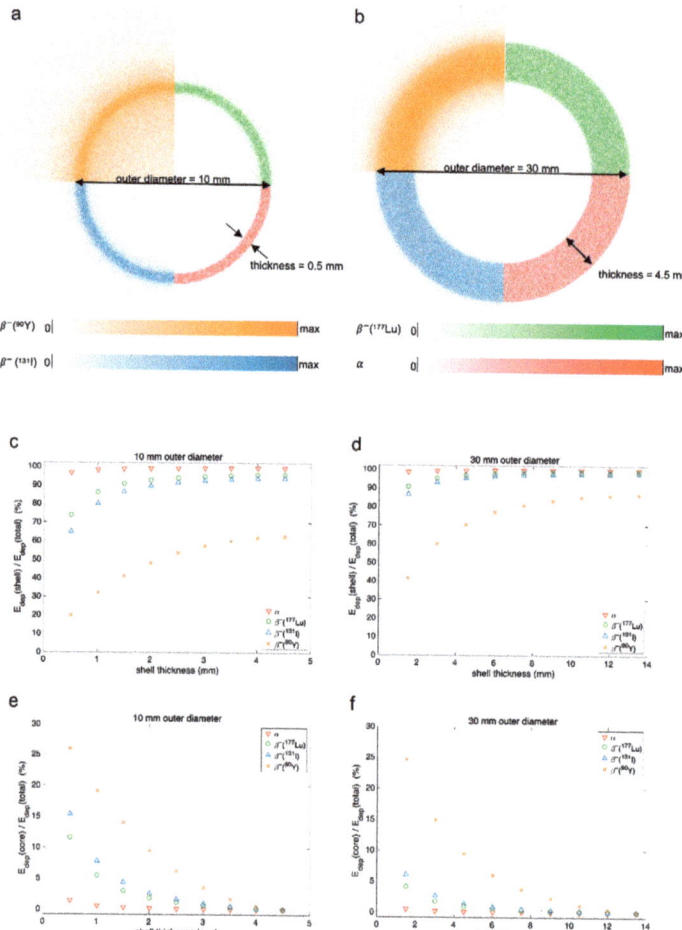

**Figure 4.** The figure shows energy deposited in and around a spherical shell source. This approximates a situation where there is uptake around the outer rim of a core without uptake. Panels (**a,b**) illustrate examples of shell thicknesses for the two outer diameters and show the central slices of images containing the energy deposition maps of the shells, where each quarter shows a different source. In panels (**c,d**), the ratio of energy deposited in the shell source ($E_{dep}$(shell)) to total deposited energy ($E_{dep}$(total)) is plotted against the shell thickness. In panels (**e,f**), the ratio of energy deposited in the core ($E_{dep}$(core)) to total deposited energy is also plotted against shell thickness. Two different outer diameters were used: in (**a,c,e**), shells with an outer diameter of 10 mm are shown, and, in (**b,d,f**), shells with an outer diameter of 30 mm are shown. Four different sources were used, three pure beta sources with beta energies following the emission spectra for $^{90}$Y, $^{177}$Lu, and $^{131}$I, and an alpha source with alpha energies of 5.5 MeV.

Optimal range is not only determined by tumour geometry and heterogeneity; uptake in healthy tissue or surrounding critical organs can be more important. For instance, with PSMA therapy for mCRPC, diagnostic scans prior to therapy reveal areas with uptake. Kratochwil et al. considered diffuse bone marrow infiltration in a patient a contraindication for treatment with beta-emitting [$^{177}$Lu]Lu-PSMA-617 and administered [$^{225}$Ac]Ac-PSMA-617 instead to avoid limiting hematologic toxicity [33]. Hobbs et al. presented a model illustrating the importance of accounting for micron-scale activity distributions and anatomy when calculating doses to bone marrow with alpha-emitters due to the short range of the radiation. They showed that strongly increasing the average absorbed dose to the bone marrow from $^{223}$Ra did not necessarily lead to toxicity since only a smaller fraction of cells were compromised [52]. Hobbs et al. similarly developed a nephron- and cellular-based model for kidneys as average dose to kidney is a poor predictor of biological response for short-range alpha radiation [53]. Hence, while short-range alpha radiation to a smaller degree exposes immediate surrounding tissue to radiation, uptake in smaller structures in healthy tissue can have large consequences due to the high amount of energy released over the short range.

### 3.2.2. Heterogeneity

There are many levels of heterogeneity in radionuclide therapy. There is heterogeneity between patients, between tumours, and within tumours. The most relevant in the context of particle range is heterogeneity within a tumour, specifically heterogeneous distribution of uptake of the radiotherapeutical. This can be caused, for instance, by permeability and extent of tumour vasculature or affinity of the targeting radiopharmaceutical for tumour cells [48]. For the long ranges achieved with beta radiation, much of the energy of the particle is deposited outside the accumulating cell, a mechanism termed cross-fire effect. Cross-fire effect is important for the efficacy of some radiotherapies as it can ensure radiation of tumour cells without uptake of the radiotherapeutical [40]. However, in cases where the range of the radiation is larger than the metastases, the cross-fire effect will deliver the radiation dose to surrounding tissue, causing the ratio of absorbed dose to tumour over healthy tissue to decrease [51]. Figure 4 illustrates a simplified geometry of a tumour, where a large core without uptake is found for the thinner shells. In such a case, the cross-fire effect ensures a more homogeneous absorbed dose distribution. Another typical case of uptake heterogeneity occurs when only single cells or small cell clusters accumulate the radiopharmaceutical and they are positioned at various intervals [51]. Enger et al. showed that the cross-fire effect can be an advantage in these cases as well [51].

### 3.3. Physical Half-Life

The physical half-life, i.e., the time it takes for the activity to reach half its initial value by radioactive decay, is inherent to the radionuclide, unchangeable by the chemical and physical environment and often known to a high degree of precision. The radionuclide with the optimal physical half-life for a certain pharmaceutical and application will depend on several factors, mainly related to the pharmacokinetics in targets and normal tissues at risk. Figure 5 illustrates some examples with normal organ and tumour tissue with different uptake and clearance.

In general, the activity in the different tissues, and, hence, the absorbed dose rate, is continually changing after administration of the radiopharmaceutical. Most radiotherapeuticals, especially those injected intravenously, will display an uptake and a washout phase. The rates and kinetics of the different phases vary between tissues. The absorbed dose deposited from activity within a tissue (the so-called self-dose) in radionuclide therapy is given by the total numbers of disintegrations occurring per mass multiplied by the energy deposited locally. This depends on both the magnitude of the uptake and the length of time the radionuclide resides in the tissue. Contributions from surrounding tissues or other organs can also add to the total absorbed dose, but these are often minor compared to the self-dose for tissues with specific uptake.

Ideally, the initial uptake phase for targets should be short so fewer disintegrations take place en route. While uptake times can be practically instantaneous for selective treatments, for antibody vectors, they can be longer, i.e., hours or days. Schemes to pre-target tumours in order to reduce the uptake time, for example, with biotin–avidin systems, have been proposed [54]. If the uptake phase for targets is sufficiently rapid, the physical half-life of the radionuclide can, in theory, be very short. In practice, production and logistics will then limit the half-life. Radiation protection concerns may also limit a very short half-life as—keeping all other factors constant—a higher amount of activity is needed to achieve the same absorbed dose if the physical half-life is decreased, and this may lead to a higher exposure rate of the surroundings. If the uptake phase for targets is longer than for normal tissues, longer physical half-lives are preferred.

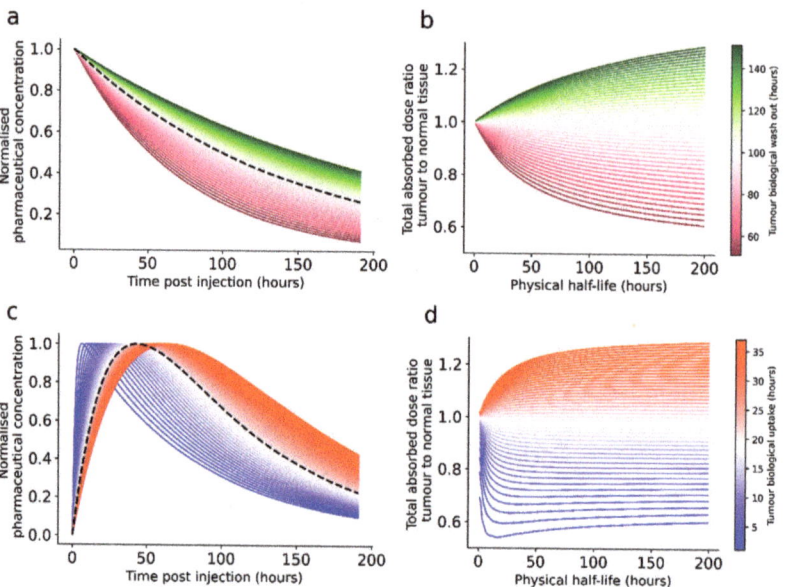

**Figure 5.** The figure illustrates theoretical situations that involve biological uptake and clearance of a radiopharmaceutical in a normal organ and tumours (**a,c**) and the differences in absorbed dose ratios that can be expected by selecting radionuclides with various half-lives for each (**b,d**). Two different types of kinetics are illustrated. In both scenarios, the normal tissue kinetics (illustrated with a dashed black line) are kept fixed and the tumour kinetics are varied. In the first situation, illustrated in panel (**a**), with the corresponding ratios in panel (**b**), an instantaneous uptake and a mono-exponential elimination is assumed for both tumour and normal tissue. The initial amount of radiopharmaceutical per tissue is set identical for both normal organ and tumours. The tumour-curves have been colour-graded according to the biological half-life, where white is equal to the normal tissue elimination (here, 100 h), whereas more saturated green indicates a slower and saturated purple indicates a faster elimination compared to the normal tissue. In panel (**b**), the ratios between the total energy absorption between tumour and normal tissue for the different tumour eliminations have been plotted for a range of physical half-lives. In panel (**c**), a different situation with bi-exponential uptake and washout is illustrated. Here, the rate of wash-out is kept fixed, while the uptake phase is varied. Again, different theoretical tumours are shown in coloured whole lines, where more saturated blue is a faster uptake and more saturated red is a slower uptake, while the normal organ is represented by a black dashed line. The curves here have been normalised to the same maximum amount of radiopharmaceutical per tissue. In panel (**d**), the absorbed dose ratios between the tumours and the normal organ are plotted over a range of physical half-lives for this scenario.

For the wash-out phase as well, the balance between the tissues is vital. If the biological wash-out of the target tissue is slower than for the relevant normal tissues, it suggests a longer physical half-life and vice versa. An additional element to consider is that the physical half-life should not be too long compared to the biological residence time in tumours. This would lead to redistribution of activity outside targets. Radiation protection considerations could also put an upper bond on the radionuclide half-life to avoid the patient from becoming a long-lived (although low-activity) radiation source.

For example, we can consider [$^{177}$Lu]Lu-DOTATATE-treatment of neuroendocrine tumours. Typically, wash-out from tumours and kidneys has been observed to behave exponentially with biological half-lives of the order of 287 h and 77 h, respectively [55,56]. $^{177}$Lu has a physical half-life of 160.8 h, resulting in effective half-lives of 103 and 52 h. If identical uptake per tissue mass is assumed, this results in a tumour to normal tissue absorbed dose ratio of 2.0. If the physical half-life was shorter, say 50 h, this ratio would be lowered to 1.4. Increasing the physical half-life, at least to a certain point, would result in an increased tumour to normal tissue ratio.

### 3.4. Imaging Properties

Post-therapy imaging of the distribution of the radiotherapeutical is essential to validate that the uptake pattern is as intended, and often also to calculate the absorbed doses to target tissue and normal organs at risk [57]. The potential for imaging in general depends on the pharmacokinetics and distribution of the radiotherapeutical, the amount of photons produced by the disintegrations, and the administered activity of the radionuclide. Photon origin, yield, and energy, in addition to camera settings, will also impact the image quality and quantitative properties. Most beta- and auger-emitters will emit some gamma photons, allowing for gamma camera imaging (Table 2). For example, the in vivo distribution of $^{177}$Lu and $^{131}$I can be followed easily due to gammas with appropriate energies and yields [58,59]. Other beta-emitters are more challenging, and, e.g., $^{90}$Y was first believed to be difficult to image accurately. However, bremsstrahlung imaging and later also positron emission tomography (PET) were proven feasible [60], and the approaches have become widely used after SIRT [61]. Still, this treatment is localized, and systemic administered treatments using $^{90}$Y will be more challenging with regard to imaging as for example studies of [$^{90}$Y]Y-ibritumomab tiuxetan and [$^{90}$Y]Y-DOTA-DPhe$^1$-Tyr$^3$-octreotide ([$^{90}$Y]Y-DOTATOC) have demonstrated [62,63]. Companion diagnostics—similar carrier molecules with diagnostic emitters attached—can be used to predict the distribution pattern of a radiotherapeutical, but potential deviations between the radiopharmaceuticals should still be investigated. Additionally, the half-lives of the diagnostic emitters need to approximate those of the therapeutic radionuclides for dosimetric purposes.

Due to high LET, RBE, and decay chains frequently containing multiple alpha- or beta-emitting daughters (Figure 6), the amount of activity used for alpha therapy is typically much lower than for beta therapy. Although some alpha-emitters emit gamma photons suitable for imaging, the low administered activities result in poorer image statistics. In addition, the recorded photon energy spectra commonly include relatively high amounts of bremsstrahlung and X-rays; hence, quantification is often more complex and associated with larger uncertainties. However, imaging has been performed for treatments with radionuclides in the decay chains of radium-224 ($^{224}$Ra) to lead-208 ($^{208}$Pb) [64], thorium-227 ($^{227}$Th) to lead-207 ($^{207}$Pb) [65–68], and radium-225 ($^{225}$Ra) to bismuth-209 ($^{209}$Bi) [69–73]. The fourth decay chain in the same range of atomic numbers, radium-226 ($^{226}$Ra) to lead-210 ($^{210}$Pb) to lead-206 ($^{206}$Pb), has somewhat unsuitable half-lives for most targeted treatments. It should be emphasized that, if imaging is only possible for a single or a few radionuclides in a decay chain, possible redistribution of daughters should be considered. For some alpha-emitters, isotopes with other characteristics can be found and companion diagnostics developed. An example includes lead-203 ($^{203}$Pb) as a surrogate for lead-212 ($^{212}$Pb), which is investigated for PSMA-based treatment of mCRPC [74]. With regard to imaging properties, terbium-149 ($^{149}$Tb) can be highlighted as an alpha-emitter that also emits positrons, allowing for PET acquisition [75].

### 3.5. Radiation Protection Considerations

In general, external exposure from patients will depend on the radionuclides' photon yield and energy, the amount of activity administered, and the pharmacokinetics of the radiopharmaceutical. The exposure limit of the public is set to an effective dose of 1 mSv/year in total by a European directive; however, it is also described that the member states shall ensure that dose constraints are established for the exposure of knowing and willing caretakers or comforters [76]. While thresholds of 1 mSv, 3 mSv, and 15 mSv are often used in this category for children, adults, and adults more than 60 years old, respectively, based on a consensus statement for $^{131}$I treatment [77], some variations in recommended precautions may be found between countries depending on the established limits and calculation methods. $^{131}$I-based treatments are traditionally associated with the highest degree of external exposure due to the high yield of photons (Table 2), and isolation measures and other constraints to limit exposure of the public, professionals, carers, and family members are common [78]. Depending on the characteristics, other beta-emitters may also impinge various radiation protection measures. Treatments such as [$^{177}$Lu]Lu-DOTATATE can be given on both an out-patient and in-patient basis with suitable restrictions [79]. For beta-emitters with low photon yield (such as $^{90}$Y) and most alpha-emitters, external exposure from patients will be less of a concern. However, handling of compounds before patient administration, especially for prolonged durations (for example, for labelling) may still raise concerns [80]. For hospitals, waste management might also be a relevant factor as storing and disposing of radioactive waste can be required.

Fluids from patients may lead to contamination of the surroundings, and excretion in urine, faeces, saliva, blood, and breast milk should be addressed. Stability of the radiopharmaceuticals, release of potential radioactive daughters, and volatile or gaseous radionuclides are important in this context. The biokinetics of a range of individual nuclides are described in a series of ICRP publications [81–84].

## 4. Discussion

While targeted treatment with radionuclides continues to increase in both diversity and the number of therapies performed, the field is currently dominated by three radionuclides: the beta-emitters $^{177}$Lu, $^{131}$I, and $^{90}$Y. This may reflect an inclination to "not change a winning team"; once radionuclides of beneficial properties have been established, the tendency may be that they are often used for novel compounds. The advantages of having well characterised radionuclides are evident with regard to imaging, production, chemistry, etc. The PubMed and clinical trials database searches performed in this work revealed that, over the years, numerous radionuclides have been either suggested or explored as options in treatments. To our knowledge, this is the first search where the starting point has been the full ICRP 107 list of 1252 radionuclides, not simply an author-defined selection.

Of the three most frequently used radionuclides, $^{131}$I is a candidate to be integrated directly into various biomolecules (such as with [$^{131}$I]MIBG), and it has targeting properties of its own. This makes $^{131}$I an ideal radionuclide for some specific applications, such as treatment of metastatic differentiated thyroid cancer [8]. The wide variation in the iodine isotopes, both with respect to half-lives and radiation emitted, also provides excellent opportunities for companion diagnostics. For example, $^{123}$I and iodine-124 ($^{124}$I) are alternatives that can be used for SPECT or PET imaging. However, the high yield of 364 keV photons often renders $^{131}$I less than desirable with regard to radiation protection. Both $^{90}$Y and $^{177}$Lu are radiometals that require chemical linkage to relevant carriers and are, in theory, therefore, more open to be replaced by other radionuclides. Examining the physical properties, it is clear that selection of one or the other may largely depend on the tissues to be targeted, surrounding normal structures, and normal tissue distribution and microstructure. $^{90}$Y emits electrons with longer ranges, which are well suited for heterogeneous structures, such as larger tumours with potentially poorly vascularised cores. $^{177}$Lu emits electrons with shorter ranges and will be preferred if a more homogeneous uptake is expected. An inhomogeneous distribution between different sub-structures in normal tissues, such as observed in kidneys for somatostatin analogues, may also suggest using a short-range beta-emitter [85].

Previous studies have suggested tailored treatments according to the individual patient's tumour signatures and sizes, or even using a cocktail of emitters [86]. For example, peptide receptor radiotherapy (PRRT) with both $^{177}$Lu and $^{90}$Y has been explored using a combination strategy [87,88]. While $^{90}$Y is currently the most frequently used beta-emitter with a long range, alternatives are encouraged due to the lack of imageable signal from $^{90}$Y. As a last consideration on range, it should be noted that, although the physical range of radiation is well defined, it does not itself determine the space in which cells can be affected by radiation. Radiation-induced bystander effects or abscopal effects, which involve cells behaving as though they have been exposed to radiation when they have not, are not fully understood, but they contribute to the efficacy of radionuclide treatment [86,89].

For alpha-emitters, the situation is somewhat different. $^{223}$Ra has been the most frequently used radionuclide since the early 2010s. This is due to [$^{223}$Ra]RaCl$_2$ (Xofigo) treatment for castration-resistant prostate cancer, where the bone-seeking properties of radium as a calcium-analogue are exploited [22]. However, in the last few decades, several alpha-emitters have been investigated, mostly for compounds containing a linker. The choice of radionuclide is then more open. Range can be considered a less relevant factor to decide between different alpha-emitters since the energy will primarily be deposited within a radius of some cell diameters. Production, chemistry, physical half-life, and radioactive daughters will, therefore, be more important for choosing the optimal alpha-emitting radionuclide for a certain pharmaceutical. For example, radionuclides in the decay chain starting with $^{226}$Ra have inconvenient half-lives or chemical properties and are rarely used. Radioactive daughters are an important factor and especially relevant for alpha-emitters (Figure 6). The opportunity to produce additional radioactive nuclides in vivo can be both an advantage and a concern depending on where the surplus energy will be deposited. Especially, the recoil effect can break chemical binding with the carrier, releasing the daughter [90]. Biodistribution, pharmaceutical properties, half-lives, and other characteristics of the daughter will then be among the determining factors. For example, a historical consideration of $^{223}$Ra for therapy deemed the radionuclide as interesting due to the total of four alpha-particles but less than desired due to the radon-219 ($^{219}$Rn) daughter that could redistribute as a noble gas [21]. However, after the clinical introduction, no evidence of re-distribution has been found. While alpha-particles are excellent at ensuring localised deposition of energy (Figure 3), this can also render the treatments less useful for heterogeneous tumours where cross-fire effects are desired. As described, the overall low photon flux from alpha-emitters will often pose challenges for quantitative imaging and dosimetry, which are currently required by a European directive [76].

No treatments based on auger-emitters are currently approved by the FDA or the European Medicines Agency (EMA). Many of the radionuclides familiar from diagnostic tracers (for example, $^{99m}$Tc, $^{123}$I, $^{111}$In) also decay by electron capture and/or undergo internal conversion after an isomeric transition, giving rise to smaller fractions of auger electrons. Over the last two decades, some registered clinical studies with auger-emitters, such as $^{111}$In and $^{125}$I, were found in our search [36,91,92], but none seem to have moved to clinical routine. While the short range of the auger electrons can spare adjacent local tissues at risk [93]—such as for the alpha-emitters—the photon irradiation can be considerable. For example, for PRRT, $^{90}$Y labelled somatostatin analogues gave a lower red marrow absorbed dose than $^{111}$In analogues as a result of the decreased photon contribution from the total body [94].

While it is relatively easy to list the various factors of importance for selection between different radionuclides, overall comparisons with clinical value are more challenging. Absorbed dose is a well-defined parameter used to quantify the total amount of energy deposited per mass; however, corrections for the time dependence, radiation type, and heterogeneity may still be needed to estimate the probabilities for toxicity or response. Especially, different types of emitters are challenging in this regard, and, for some applications, the possibility of having a selection of various radionuclides available may be beneficial. For example, PSMA targeted treatment with $^{225}$Ac is sometimes an option to [$^{177}$Lu]Lu-PSMA treatment for individual mCRPC patients [95].

The concept of theragnostics has gained increasing focus over the last few years. Even with radiotherapeuticals that are possible to image themselves, companion diagnostic tracers are important for stratification purposes, for patient selection, in some cases individual treatment planning, and for response evaluation and follow-up. Selection of optimal therapeutic radionuclides should, therefore, also include considerations of the theragnostic properties and the need and possibility for diagnostic analogies.

In conclusion, selection of radionuclides for targeted therapy depends on a variety of factors. Identifying the optimal radionuclide might not be possible until after the biodistribution, pharmacokinetics, and uptake levels have been accurately described through clinical investigations. However, target sizes, expected intra-target homogeneity and uptake density, characteristics and pharmacokinetics of the radionuclide-bound carrier, administration route, expected normal tissues at risk, and other considerations may serve to guide the selection in terms of physical properties. In our search, 72 radionuclides out of a total of 1252 were found to have been considered for clinical use, and 21 were found to have been included in clinical trials. This illustrates the diversity and opportunities in the field.

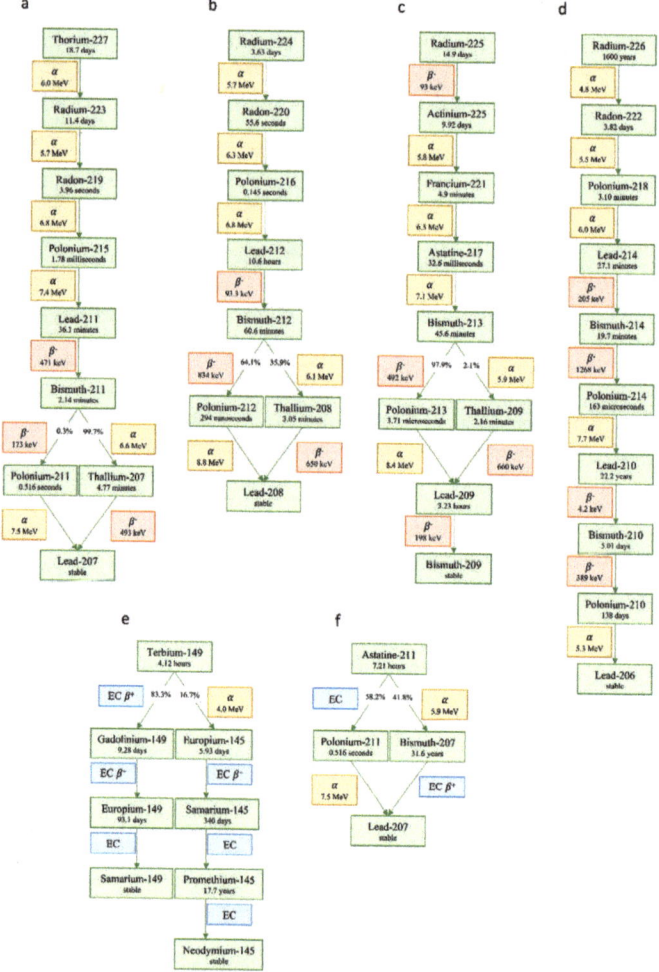

**Figure 6.** Six decay schemes (**a**–**f**), including most alpha-emitters relevant for radionuclide therapy. All branching ratios larger than 0.1% are included. The beta- and alpha-particle energies given are for the highest intensity emission. Data from Ref. [96].

**Supplementary Materials:** The following are available online at https://www.mdpi.com/article/10.3390/molecules27175429/s1, Supplementary Files S1 and S2 [4,97–99].

**Author Contributions:** C.S., M.K. and J.B. contributed to conceptualization, methodology, draft preparation, review, and editing. All authors have read and agreed to the published version of the manuscript.

**Funding:** M.K. was in part supported by South-Eastern Norway Regional Health Authority grant number 2020028.

**Conflicts of Interest:** The authors declare no conflict of interest.

## References

1. Silberstein, E.B. Radionuclide therapy of hematologic disorders. *Semin. Nucl. Med.* **1979**, *9*, 100–107. [CrossRef]
2. Sawin, C.T.; Becker, D.V. Radioiodine and the treatment of hyperthyroidism: The early history. *Thyroid* **1997**, *7*, 163–176. [CrossRef] [PubMed]
3. Sgouros, G.; Bodei, L.; McDevitt, M.R.; Nedrow, J.R. Radiopharmaceutical therapy in cancer: Clinical advances and challenges. *Nat. Rev. Drug. Discov.* **2020**, *19*, 589–608. [CrossRef] [PubMed]
4. Eckerman, K.; Endo, A. ICRP Publication 107. Nuclear decay data for dosimetric calculations. *Ann. ICRP* **2008**, *38*, 7–96. [PubMed]
5. Willowson, K.P. Production of radionuclides for clinical nuclear medicine. *Eur. J. Phys.* **2019**, *40*, 043001. [CrossRef]
6. Van de Voorde, M.; Van Hecke, K.; Cardinaels, T.; Binnemans, K. Radiochemical processing of nuclear-reactor-produced radiolanthanides for medical applications. *Coord. Chem. Rev.* **2019**, *382*, 103–125. [CrossRef]
7. Nelson, B.J.B.; Andersson, J.D.; Wuest, F. Targeted Alpha Therapy: Progress in Radionuclide Production, Radiochemistry, and Applications. *Pharmaceutics* **2021**, *13*, 49. [CrossRef]
8. Luster, M.; Clarke, S.E.; Dietlein, M.; Lassmann, M.; Lind, P.; Oyen, W.J.; Tennvall, J.; Bombardieri, E. Guidelines for radioiodine therapy of differentiated thyroid cancer. *Eur. J. Nucl. Med. Mol. Imaging* **2008**, *35*, 1941–1959. [CrossRef]
9. Silberstein, E.B.; Alavi, A.; Balon, H.R.; Clarke, S.E.; Divgi, C.; Gelfand, M.J.; Goldsmith, S.J.; Jadvar, H.; Marcus, C.S.; Martin, W.H.; et al. The SNMMI practice guideline for therapy of thyroid disease with 131I 3.0. *J. Nucl. Med.* **2012**, *53*, 1633–1651. [CrossRef]
10. Vöö, S.; Bucerius, J.; Mottaghy, F.M. I-131-MIBG therapies. *Methods* **2011**, *55*, 238–245. [CrossRef]
11. Lashford, L.S.; Lewis, I.J.; Fielding, S.L.; Flower, M.A.; Meller, S.; Kemshead, J.T.; Ackery, D. Phase I/II study of iodine 131 metaiodobenzylguanidine in chemoresistant neuroblastoma: A United Kingdom Children's Cancer Study Group investigation. *J. Clin. Oncol.* **1992**, *10*, 1889–1896. [CrossRef]
12. Kaminski, M.S.; Zasadny, K.R.; Francis, I.R.; Milik, A.W.; Ross, C.W.; Moon, S.D.; Crawford, S.M.; Burgess, J.M.; Petry, N.A.; Butchko, G.M.; et al. Radioimmunotherapy of B-cell lymphoma with [131I]anti-B1 (anti-CD20) antibody. *N. Engl. J. Med.* **1993**, *329*, 459–465. [CrossRef]
13. Kwekkeboom, D.J.; Teunissen, J.J.; Bakker, W.H.; Kooij, P.P.; de Herder, W.W.; Feelders, R.A.; van Eijck, C.H.; Esser, J.P.; Kam, B.L.; Krenning, E.P. Radiolabeled somatostatin analog [177Lu-DOTA0,Tyr3] octreotate in patients with endocrine gastroenteropancreatic tumors. *J. Clin. Oncol.* **2005**, *23*, 2754–2762. [CrossRef] [PubMed]
14. Imhof, A.; Brunner, P.; Marincek, N.; Briel, M.; Schindler, C.; Rasch, H.; Mäcke, H.R.; Rochlitz, C.; Müller-Brand, J.; Walter, M.A. Response, survival, and long-term toxicity after therapy with the radiolabeled somatostatin analogue [90Y-DOTA]-TOC in metastasized neuroendocrine cancers. *J. Clin. Oncol.* **2011**, *29*, 2416–2423. [CrossRef]
15. Sartor, O.; de Bono, J.; Chi, K.N.; Fizazi, K.; Herrmann, K.; Rahbar, K.; Tagawa, S.T.; Nordquist, L.T.; Vaishampayan, N.; El-Haddad, G.; et al. Lutetium-177-PSMA-617 for Metastatic Castration-Resistant Prostate Cancer. *N. Engl. J. Med.* **2021**, *385*, 1091–1103. [CrossRef]
16. Witzig, T.E.; Gordon, L.I.; Cabanillas, F.; Czuczman, M.S.; Emmanouilides, C.; Joyce, R.; Pohlman, B.L.; Bartlett, N.L.; Wiseman, G.A.; Padre, N.; et al. Randomized controlled trial of yttrium-90-labeled ibritumomab tiuxetan radioimmunotherapy versus rituximab immunotherapy for patients with relapsed or refractory low-grade, follicular, or transformed B-cell non-Hodgkin's lymphoma. *J. Clin. Oncol.* **2002**, *20*, 2453–2463. [CrossRef] [PubMed]
17. Levillain, H.; Bagni, O.; Deroose, C.M.; Dieudonné, A.; Gnesin, S.; Grosser, O.S.; Kappadath, S.C.; Kennedy, A.; Kokabi, N.; Liu, D.M.; et al. International recommendations for personalised selective internal radiation therapy of primary and metastatic liver diseases with yttrium-90 resin microspheres. *Eur. J. Nucl. Med. Mol. Imaging* **2021**, *48*, 1570–1584. [CrossRef] [PubMed]
18. Strosberg, J.R.; Caplin, M.E.; Kunz, P.L.; Ruszniewski, P.B.; Bodei, L.; Hendifar, A.; Mittra, E.; Wolin, E.M.; Yao, J.C.; Pavel, M.E.; et al. 177Lu-Dotatate plus long-acting octreotide versus high dose long-acting octreotide in patients with midgut neuroendocrine tumours (NETTER-1): Final overall survival and long-term safety results from an open-label, randomised, controlled, phase 3 trial. *Lancet Oncol.* **2021**, *22*, 1752–1763. [CrossRef]
19. FDA Approves Pluvicto/Locametz for Metastatic Castration-Resistant Prostate Cancer. Available online: https://jnm.snmjournals.org/content/63/5/13N.2/tab-article-info (accessed on 30 June 2022).

20. Baum, R.P.; Schuchardt, C.; Singh, A.; Chantadisai, M.; Robiller, F.C.; Zhang, J.; Mueller, D.; Eismant, A.; Almaguel, F.; Zboralski, D.; et al. Feasibility, Biodistribution, and Preliminary Dosimetry in Peptide-Targeted Radionuclide Therapy of Diverse Adenocarcinomas Using (177)Lu-FAP-2286: First-in-Humans Results. *J. Nucl. Med.* **2022**, *63*, 415–423. [CrossRef]
21. McDevitt, M.R.; Sgouros, G.; Finn, R.D.; Humm, J.L.; Jurcic, J.G.; Larson, S.M.; Scheinberg, D.A. Radioimmunotherapy with alpha-emitting nuclides. *Eur. J. Nucl. Med.* **1998**, *25*, 1341–1351. [CrossRef]
22. Parker, C.; Nilsson, S.; Heinrich, D.; Helle, S.I.; O'Sullivan, J.M.; Fosså, S.D.; Chodacki, A.; Wiechno, P.; Logue, J.; Seke, M.; et al. Alpha emitter radium-223 and survival in metastatic prostate cancer. *N. Engl. J. Med.* **2013**, *369*, 213–223. [CrossRef] [PubMed]
23. Souza, B.; Ribeiro, E.; da Silva de Barros, A.O.; Pijeira, M.S.O.; Kenup-Hernandes, H.O.; Ricci-Junior, E.; Diniz Filho, J.F.S.; Dos Santos, C.C.; Alencar, L.M.R.; Attia, M.F.; et al. Nanomicelles of Radium Dichloride [(223)Ra]RaCl(2) Co-Loaded with Radioactive Gold [(198)Au]Au Nanoparticles for Targeted Alpha-Beta Radionuclide Therapy of Osteosarcoma. *Polymers* **2022**, *14*, 1405. [CrossRef]
24. Silberstein, E.B.; Elgazzar, A.H.; Kapilivsky, A. Phosphorus-32 radiopharmaceuticals for the treatment of painful osseous metastases. *Semin. Nucl. Med.* **1992**, *22*, 17–27. [CrossRef]
25. Smith, H.O.; Gaudette, D.E.; Goldberg, G.L.; Milstein, D.M.; DeVictoria, C.L.; Runowicz, C.D. Single-use percutaneous catheters for intraperitoneal P32 therapy. *Cancer* **1994**, *73*, 2633–2637. [CrossRef]
26. Handkiewicz-Junak, D.; Poeppel, T.D.; Bodei, L.; Aktolun, C.; Ezziddin, S.; Giammarile, F.; Delgado-Bolton, R.C.; Gabriel, M. EANM guidelines for radionuclide therapy of bone metastases with beta-emitting radionuclides. *Eur. J. Nucl. Med. Mol. Imaging* **2018**, *45*, 846–859. [CrossRef]
27. Schneider, P.; Farahati, J.; Reiners, C. Radiosynovectomy in rheumatology, orthopedics, and hemophilia. *J. Nucl. Med.* **2005**, *46*, 48s–54s.
28. Shinto, A.S.; Karuppusamy, K.K.; Kurup, R.E.R.; Pandiyan, A.; Jayaraj, A.V. Empirical 188Re-HDD/lipiodol intra-arterial therapy based on tumor volume, in patients with solitary inoperable hepatocellular carcinoma. *Nucl. Med. Commun.* **2021**, *42*, 43–50. [CrossRef]
29. Klein, M.; Lotem, M.; Peretz, T.; Zwas, S.T.; Mizrachi, S.; Liberman, Y.; Chisin, R.; Schachter, J.; Ron, I.G.; Iosilevsky, G.; et al. Safety and efficacy of 188-rhenium-labeled antibody to melanin in patients with metastatic melanoma. *J. Skin Cancer* **2013**, *2013*, 828329. [CrossRef]
30. Corson, D.R.; MacKenzie, K.R.; Segrè, E. Artificially Radioactive Element 85. *Phys. Rev.* **1940**, *58*, 672–678. [CrossRef]
31. Hallqvist, A.; Bergmark, K.; Bäck, T.; Andersson, H.; Dahm-Kähler, P.; Johansson, M.; Lindegren, S.; Jensen, H.; Jacobsson, L.; Hultborn, R.; et al. Intraperitoneal α-Emitting Radioimmunotherapy with (211) At in Relapsed Ovarian Cancer: Long-Term Follow-up with Individual Absorbed Dose Estimations. *J. Nucl. Med.* **2019**, *60*, 1073–1079. [CrossRef]
32. Rosenblat, T.L.; McDevitt, M.R.; Carrasquillo, J.A.; Pandit-Taskar, N.; Frattini, M.G.; Maslak, P.G.; Park, J.H.; Douer, D.; Cicic, D.; Larson, S.M.; et al. Treatment of Patients with Acute Myeloid Leukemia with the Targeted Alpha-Particle Nanogenerator Actinium-225-Lintuzumab. *Clin. Cancer Res.* **2022**, *28*, 2030–2037. [CrossRef] [PubMed]
33. Kratochwil, C.; Bruchertseifer, F.; Giesel, F.L.; Weis, M.; Verburg, F.A.; Mottaghy, F.; Kopka, K.; Apostolidis, C.; Haberkorn, U.; Morgenstern, A. 225Ac-PSMA-617 for PSMA-Targeted α-Radiation Therapy of Metastatic Castration-Resistant Prostate Cancer. *J. Nucl. Med.* **2016**, *57*, 1941–1944. [CrossRef] [PubMed]
34. Ma, J.; Li, L.; Liao, T.; Gong, W.; Zhang, C. Efficacy and Safety of (225)Ac-PSMA-617-Targeted Alpha Therapy in Metastatic Castration-Resistant Prostate Cancer: A Systematic Review and Meta-Analysis. *Front. Oncol.* **2022**, *12*, 796657. [CrossRef] [PubMed]
35. Yoshii, Y.; Furukawa, T.; Kiyono, Y.; Watanabe, R.; Mori, T.; Yoshii, H.; Asai, T.; Okazawa, H.; Welch, M.J.; Fujibayashi, Y. Internal radiotherapy with copper-64-diacetyl-bis (N4-methylthiosemicarbazone) reduces CD133+ highly tumorigenic cells and metastatic ability of mouse colon carcinoma. *Nucl. Med. Biol.* **2011**, *38*, 151–157. [CrossRef]
36. Krenning, E.P.; de Jong, M.; Kooij, P.P.; Breeman, W.A.; Bakker, W.H.; de Herder, W.W.; van Eijck, C.H.; Kwekkeboom, D.J.; Jamar, F.; Pauwels, S.; et al. Radiolabelled somatostatin analogue(s) for peptide receptor scintigraphy and radionuclide therapy. *Ann. Oncol.* **1999**, *10*, S23–S29. [CrossRef]
37. Othman, M.F.; Mitry, N.R.; Lewington, V.J.; Blower, P.J.; Terry, S.Y. Re-assessing gallium-67 as a therapeutic radionuclide. *Nucl. Med. Biol.* **2017**, *46*, 12–18. [CrossRef]
38. Pirovano, G.; Jannetti, S.A.; Carter, L.M.; Sadique, A.; Kossatz, S.; Guru, N.; Demétrio De Souza França, P.; Maeda, M.; Zeglis, B.M.; Lewis, J.S.; et al. Targeted Brain Tumor Radiotherapy Using an Auger Emitter. *Clin. Cancer Res.* **2020**, *26*, 2871–2881. [CrossRef]
39. Costa, I.M.; Siksek, N.; Volpe, A.; Man, F.; Osytek, K.M.; Verger, E.; Schettino, G.; Fruhwirth, G.O.; Terry, S.Y.A. Relationship of In Vitro Toxicity of Technetium-99m to Subcellular Localisation and Absorbed Dose. *Int. J. Mol. Sci.* **2021**, *22*, 13466. [CrossRef]
40. Kassis, A.I. Therapeutic Radionuclides: Biophysical and Radiobiologic Principles. *Semin. Nucl. Med.* **2008**, *38*, 358–366. [CrossRef]
41. Howell, R.W. Advancements in the use of Auger electrons in science and medicine during the period 2015–2019. *Int. J. Radiat. Biol.* **2020**, 1–26. [CrossRef]
42. O'Donoghue, J.A.; Wheldon, T.E. Targeted radiotherapy using Auger electron emitters. *Phys. Med. Biol.* **1996**, *41*, 1973–1992. [CrossRef] [PubMed]
43. Ku, A.; Facca, V.J.; Cai, Z.; Reilly, R.M. Auger electrons for cancer therapy—A review. *EJNMMI Radiopharm. Chem.* **2019**, *4*, 27. [CrossRef] [PubMed]

44. Attix, F.H. Charged-Particle Interactions in Matter. In *Introduction to Radiological Physics and Radiation Dosimetry*; John Wiley & Sons: Hoboken, NJ, USA, 1986; pp. 160–202.
45. Mayles, W.P.M.; Nahum, A.E.; Rosenwald, J.-C. *Handbook of Radiotherapy Physics: Theory and Practice*, 2nd ed.; CRC Press: Boca Raton, FL, USA, 2021.
46. McParland, B.J. Charged Particle Interactions with Matter. In *Nuclear Medicine Radiation Dosimetry: Advanced Theoretical Principles*; Springer: London, UK, 2010; pp. 209–324.
47. Prestwich, W.V.; Nunes, J.; Kwok, C.S. Beta Dose Point Kernels for Radionuclides of Potential Use in Radioimmunotherapy. *J. Nucl. Med.* **1989**, *30*, 1036–1046. [PubMed]
48. O'Donoghue, J.A.; Bardiès, M.; Wheldon, T.E. Relationships between Tumor Size and Curability for Uniformly Targeted Therapy with Beta-Emitting Radionuclides. *J. Nucl. Med.* **1995**, *36*, 1902–1909.
49. Bernhardt, P.; Benjegård, S.A.; Kölby, L.; Johanson, V.; Nilsson, O.; Ahlman, H.; Forssell-Aronsson, E. Dosimetric comparison of radionuclides for therapy of somatostatin receptor-expressing tumors. *Int. J. Radiat. Oncol. Biol. Phys.* **2001**, *51*, 514–524. [CrossRef]
50. Tamborino, G.; Nonnekens, J.; Struelens, L.; De Saint-Hubert, M.; Verburg, F.A.; Konijnenberg, M.W. Therapeutic efficacy of heterogeneously distributed radiolabelled peptides: Influence of radionuclide choice. *Phys. Med.* **2022**, *96*, 90–100. [CrossRef]
51. Enger, S.A.; Hartman, T.; Carlsson, J.; Lundqvist, H. Cross-fire doses from β-emitting radionuclides in targeted radiotherapy. A theoretical study based on experimentally measured tumor characteristics. *Phys. Med. Biol.* **2008**, *53*, 1909–1920. [CrossRef]
52. Hobbs, R.F.; Song, H.; Watchman, C.J.; Bolch, W.E.; Aksnes, A.-K.; Ramdahl, T.; Flux, G.D.; Sgouros, G. A bone marrow toxicity model for 223Ra alpha-emitter radiopharmaceutical therapy. *Phys. Med. Biol.* **2012**, *57*, 3207–3222. [CrossRef] [PubMed]
53. Hobbs, R.F.; Song, H.; Huso, D.L.; Sundel, M.H.; Sgouros, G. A nephron-based model of the kidneys for macro-to-micro α-particle dosimetry. *Phys. Med. Biol.* **2012**, *57*, 4403–4424. [CrossRef]
54. Cremonesi, M.; Ferrari, M.; Chinol, M.; Stabin, M.G.; Grana, C.; Prisco, G.; Robertson, C.; Tosi, G.; Paganelli, G. Three-step radioimmunotherapy with yttrium-90 biotin: Dosimetry and pharmacokinetics in cancer patients. *Eur. J. Nucl. Med.* **1999**, *26*, 110–120. [CrossRef]
55. Sandström, M.; Freedman, N.; Fröss-Baron, K.; Kahn, T.; Sundin, A. Kidney dosimetry in 777 patients during (177)Lu-DOTATATE therapy: Aspects on extrapolations and measurement time points. *EJNMMI Phys.* **2020**, *7*, 73. [CrossRef] [PubMed]
56. Roth, D.; Gustafsson, J.; Warfvinge, C.F.; Sundlöv, A.; Åkesson, A.; Tennvall, J.; Gleisner, K.S. Dosimetric Quantities in Neuroendocrine Tumors over Treatment Cycles with (177)Lu-DOTATATE. *J. Nucl. Med.* **2022**, *63*, 399–405. [CrossRef] [PubMed]
57. Stokke, C.; Gabiña, P.M.; Solný, P.; Cicone, F.; Sandström, M.; Gleisner, K.S.; Chiesa, C.; Spezi, E.; Paphiti, M.; Konijnenberg, M.; et al. Dosimetry-based treatment planning for molecular radiotherapy: A summary of the 2017 report from the Internal Dosimetry Task Force. *EJNMMI Phys.* **2017**, *4*, 27. [CrossRef]
58. Ljungberg, M.; Celler, A.; Konijnenberg, M.W.; Eckerman, K.F.; Dewaraja, Y.K.; Sjögreen-Gleisner, K.; Bolch, W.E.; Brill, A.B.; Fahey, F.; Fisher, D.R.; et al. MIRD Pamphlet No. 26: Joint EANM/MIRD Guidelines for Quantitative 177Lu SPECT Applied for Dosimetry of Radiopharmaceutical Therapy. *J. Nucl. Med.* **2016**, *57*, 151–162. [CrossRef]
59. Dewaraja, Y.K.; Ljungberg, M.; Green, A.J.; Zanzonico, P.B.; Frey, E.C.; Committee, S.M.; Bolch, W.E.; Brill, A.B.; Dunphy, M.; Fisher, D.R.; et al. MIRD Pamphlet No. 24: Guidelines for Quantitative 131I SPECT in Dosimetry Applications. *J. Nucl. Med.* **2013**, *54*, 2182–2188. [CrossRef]
60. Lhommel, R.; van Elmbt, L.; Goffette, P.; Van den Eynde, M.; Jamar, F.; Pauwels, S.; Walrand, S. Feasibility of 90Y TOF PET-based dosimetry in liver metastasis therapy using SIR-Spheres. *Eur. J. Nucl. Med. Mol. Imaging* **2010**, *37*, 1654–1662. [CrossRef]
61. Chiesa, C.; Sjogreen-Gleisner, K.; Walrand, S.; Strigari, L.; Flux, G.; Gear, J.; Stokke, C.; Gabina, P.M.; Bernhardt, P.; Konijnenberg, M. EANM dosimetry committee series on standard operational procedures: A unified methodology for (99m)Tc-MAA pre- and (90)Y peri-therapy dosimetry in liver radioembolization with (90)Y microspheres. *EJNMMI Phys.* **2021**, *8*, 77. [CrossRef] [PubMed]
62. Cremonesi, M.; Ferrari, M.; Zoboli, S.; Chinol, M.; Stabin, M.G.; Orsi, F.; Maecke, H.R.; Jermann, E.; Robertson, C.; Fiorenza, M.; et al. Biokinetics and dosimetry in patients administered with (111)In-DOTA-Tyr(3)-octreotide: Implications for internal radiotherapy with (90)Y-DOTATOC. *Eur. J. Nucl. Med.* **1999**, *26*, 877–886. [CrossRef]
63. Wahl, R.L.; Frey, E.C.; Jacene, H.A.; Kahl, B.S.; Piantadosi, S.; Bianco, J.A.; Hammes, R.J.; Jung, M.; Kasecamp, W.; He, B.; et al. Prospective SPECT-CT Organ Dosimetry-Driven Radiation-Absorbed Dose Escalation Using the In-111 (111In)/Yttrium 90 (90Y) Ibritumomab Tiuxetan (Zevalin®) Theranostic Pair in Patients with Lymphoma at Myeloablative Dose Levels. *Cancers* **2021**, *13*, 2828. [CrossRef]
64. Meredith, R.F.; Torgue, J.; Azure, M.T.; Shen, S.; Saddekni, S.; Banaga, E.; Carlise, R.; Bunch, P.; Yoder, D.; Alvarez, R. Pharmacokinetics and imaging of 212Pb-TCMC-trastuzumab after intraperitoneal administration in ovarian cancer patients. *Cancer Biother. Radiopharm.* **2014**, *29*, 12–17. [CrossRef]
65. Hindorf, C.; Chittenden, S.; Aksnes, A.K.; Parker, C.; Flux, G.D. Quantitative imaging of 223Ra-chloride (Alpharadin) for targeted alpha-emitting radionuclide therapy of bone metastases. *Nucl. Med. Commun.* **2012**, *33*, 726–732. [CrossRef] [PubMed]
66. Larsson, E.; Brolin, G.; Cleton, A.; Ohlsson, T.; Lindén, O.; Hindorf, C. Feasibility of Thorium-227/Radium-223 Gamma-Camera Imaging During Radionuclide Therapy. *Cancer Biother. Radiopharm.* **2020**, *35*, 540–548. [CrossRef] [PubMed]
67. Murray, I.; Rojas, B.; Gear, J.; Callister, R.; Cleton, A.; Flux, G.D. Quantitative Dual-Isotope Planar Imaging of Thorium-227 and Radium-223 Using Defined Energy Windows. *Cancer Biother. Radiopharm.* **2020**, *35*, 530–539. [CrossRef]

68. Pacilio, M.; Ventroni, G.; De Vincentis, G.; Cassano, B.; Pellegrini, R.; Di Castro, E.; Frantellizzi, V.; Follacchio, G.A.; Garkavaya, T.; Lorenzon, L.; et al. Dosimetry of bone metastases in targeted radionuclide therapy with alpha-emitting (223)Ra-dichloride. *Eur. J. Nucl. Med. Mol. Imaging* **2016**, *43*, 21–33. [CrossRef]
69. Gosewisch, A.; Schleske, M.; Gildehaus, F.J.; Berg, I.; Kaiser, L.; Brosch, J.; Bartenstein, P.; Todica, A.; Ilhan, H.; Böning, G. Image-based dosimetry for 225Ac-PSMA-I&T therapy using quantitative SPECT. *Eur. J. Nucl. Med. Mol. Imaging* **2021**, *48*, 1260–1261. [PubMed]
70. Sgouros, G.; Ballangrud, A.M.; Jurcic, J.G.; McDevitt, M.R.; Humm, J.L.; Erdi, Y.E.; Mehta, B.M.; Finn, R.D.; Larson, S.M.; Scheinberg, D.A. Pharmacokinetics and dosimetry of an alpha-particle emitter labeled antibody: 213Bi-HuM195 (anti-CD33) in patients with leukemia. *J. Nucl. Med.* **1999**, *40*, 1935–1946.
71. Kratochwil, C.; Giesel, F.L.; Bruchertseifer, F.; Mier, W.; Apostolidis, C.; Boll, R.; Murphy, K.; Haberkorn, U.; Morgenstern, A. $^{213}$Bi-DOTATOC receptor-targeted alpha-radionuclide therapy induces remission in neuroendocrine tumours refractory to beta radiation: A first-in-human experience. *Eur. J. Nucl. Med. Mol. Imaging* **2014**, *41*, 2106–2119. [CrossRef]
72. Cordier, D.; Forrer, F.; Bruchertseifer, F.; Morgenstern, A.; Apostolidis, C.; Good, S.; Müller-Brand, J.; Mäcke, H.; Reubi, J.C.; Merlo, A. Targeted alpha-radionuclide therapy of functionally critically located gliomas with 213Bi-DOTA-[Thi8,Met(O2)11]-substance P: A pilot trial. *Eur. J. Nucl. Med. Mol. Imaging* **2010**, *37*, 1335–1344. [CrossRef]
73. Usmani, S.; Rasheed, R.; Al Kandari, F.; Marafi, F.; Naqvi, S.A.R. 225Ac Prostate-Specific Membrane Antigen Posttherapy α Imaging: Comparing 2 and 3 Photopeaks. *Clin. Nucl. Med.* **2019**, *44*, 401–403. [CrossRef]
74. Dos Santos, J.C.; Schäfer, M.; Bauder-Wüst, U.; Lehnert, W.; Leotta, K.; Morgenstern, A.; Kopka, K.; Haberkorn, U.; Mier, W.; Kratochwil, C. Development and dosimetry of (203)Pb/(212)Pb-labelled PSMA ligands: Bringing "the lead" into PSMA-targeted alpha therapy? *Eur. J. Nucl. Med. Mol. Imaging* **2019**, *46*, 1081–1091. [CrossRef]
75. Müller, C.; Vermeulen, C.; Köster, U.; Johnston, K.; Türler, A.; Schibli, R.; van der Meulen, N.P. Alpha-PET with terbium-149: Evidence and perspectives for radiotheragnostics. *EJNMMI Radiopharm. Chem.* **2016**, *1*, 5. [CrossRef]
76. European Commission. Council Directive 2013/59/Euratom of 5 December 2013 laying down basic safety standards for protection against the dangers arising from exposure to ionising radiation, and repealing Directives 89/618/Euratom, 90/641/Euratom, 96/29/Euratom, 97/43/Euratom and 2003/122/Euratom. *Official J.* **2014**, *13*, 1–73.
77. Expert Group Ex Art 31. EURATOM: Guidance for Radiation Protection Following Iodine-131 Therapy Concerning Doses Due to Out-Patients or Discharged In-Patients. Available online: https://ec.europa.eu/energy/sites/default/files/opinion_of_article_31_goe_on_the_jrc_report_28_june_2021.pdf (accessed on 30 June 2022).
78. Release of patients after therapy with unsealed radionuclides. *Ann. ICRP* **2004**, *34*, 1–79. Available online: https://pubmed.ncbi.nlm.nih.gov/15571759/ (accessed on 30 June 2022).
79. Levart, D.; Kalogianni, E.; Corcoran, B.; Mulholland, N.; Vivian, G. Radiation precautions for inpatient and outpatient (177)Lu-DOTATATE peptide receptor radionuclide therapy of neuroendocrine tumours. *EJNMMI Phys.* **2019**, *6*, 7. [CrossRef] [PubMed]
80. Cremonesi, M.; Ferrari, M.; Paganelli, G.; Rossi, A.; Chinol, M.; Bartolomei, M.; Prisco, G.; Tosi, G. Radiation protection in radionuclide therapies with (90)Y-conjugates: Risks and safety. *Eur. J. Nucl. Med. Mol. Imaging* **2006**, *33*, 1321–1327. [CrossRef] [PubMed]
81. Paquet, F.; Bailey, M.R.; Leggett, R.W.; Lipsztein, J.; Marsh, J.; Fell, T.P.; Smith, T.; Nosske, D.; Eckerman, K.F.; Berkovski, V.; et al. ICRP Publication 137: Occupational Intakes of Radionuclides: Part 3. *Ann. ICRP* **2017**, *46*, 1–486. [CrossRef] [PubMed]
82. Paquet, F.; Bailey, M.R.; Leggett, R.W.; Etherington, G.; Blanchardon, E.; Smith, T.; Ratia, G.; Melo, D.; Fell, T.P.; Berkovski, V.; et al. ICRP Publication 141: Occupational Intakes of Radionuclides: Part 4. *Ann. ICRP* **2019**, *48*, 9–501. [CrossRef]
83. Paquet, F.; Leggett, R.W.; Blanchardon, E.; Bailey, M.R.; Gregoratto, D.; Smith, T.; Ratia, G.; Davesne, E.; Berkovski, V.; Harrison, J.D. Occupational Intakes of Radionuclides: Part 5. *Ann. ICRP* **2022**, *51*, 11–415. [CrossRef]
84. Paquet, F.; Bailey, M.R.; Leggett, R.W.; Lipsztein, J.; Fell, T.P.; Smith, T.; Nosske, D.; Eckerman, K.F.; Berkovski, V.; Ansoborlo, E.; et al. ICRP Publication 134: Occupational Intakes of Radionuclides: Part 2. *Ann. ICRP* **2016**, *45*, 7–349. [CrossRef]
85. Konijnenberg, M.; Melis, M.; Valkema, R.; Krenning, E.; de Jong, M. Radiation dose distribution in human kidneys by octreotides in peptide receptor radionuclide therapy. *J. Nucl. Med.* **2007**, *48*, 134–142.
86. Haberkorn, U.; Giesel, F.; Morgenstern, A.; Kratochwil, C. The Future of Radioligand Therapy: α, β, or Both? *J. Nucl. Med.* **2017**, *58*, 1017–1018. [CrossRef]
87. Villard, L.; Romer, A.; Marincek, N.; Brunner, P.; Koller, M.T.; Schindler, C.; Ng, Q.K.; Mäcke, H.R.; Müller-Brand, J.; Rochlitz, C.; et al. Cohort study of somatostatin-based radiopeptide therapy with [(90)Y-DOTA]-TOC versus [(90)Y-DOTA]-TOC plus [(177)Lu-DOTA]-TOC in neuroendocrine cancers. *J. Clin. Oncol.* **2012**, *30*, 1100–1106. [CrossRef] [PubMed]
88. Kunikowska, J.; Królicki, L.; Hubalewska-Dydejczyk, A.; Mikołajczak, R.; Sowa-Staszczak, A.; Pawlak, D. Clinical results of radionuclide therapy of neuroendocrine tumours with 90Y-DOTATATE and tandem 90Y/177Lu-DOTATATE: Which is a better therapy option? *Eur. J. Nucl. Med. Mol. Imaging* **2011**, *38*, 1788–1797. [CrossRef] [PubMed]
89. Pouget, J.P.; Georgakilas, A.G.; Ravanat, J.L. Targeted and Off-Target (Bystander and Abscopal) Effects of Radiation Therapy: Redox Mechanisms and Risk/Benefit Analysis. *Antioxid. Redox Signal.* **2018**, *29*, 1447–1487. [CrossRef] [PubMed]
90. de Kruijff, R.M.; Wolterbeek, H.T.; Denkova, A.G. A Critical Review of Alpha Radionuclide Therapy-How to Deal with Recoiling Daughters? *Pharmaceuticals* **2015**, *8*, 321–336. [CrossRef]

91. Vallis, K.A.; Reilly, R.M.; Scollard, D.; Merante, P.; Brade, A.; Velauthapillai, S.; Caldwell, C.; Chan, I.; Freeman, M.; Lockwood, G.; et al. Phase I trial to evaluate the tumor and normal tissue uptake, radiation dosimetry and safety of (111)In-DTPA-human epidermal growth factor in patients with metastatic EGFR-positive breast cancer. *Am. J. Nucl. Med. Mol. Imaging* **2014**, *4*, 181–192.
92. Li, L.; Quang, T.S.; Gracely, E.J.; Kim, J.H.; Emrich, J.G.; Yaeger, T.E.; Jenrette, J.M.; Cohen, S.C.; Black, P.; Brady, L.W. A Phase II study of anti-epidermal growth factor receptor radioimmunotherapy in the treatment of glioblastoma multiforme. *J. Neurosurg.* **2010**, *113*, 192–198. [CrossRef]
93. Konijnenberg, M.W.; Bijster, M.; Krenning, E.P.; De Jong, M. A stylized computational model of the rat for organ dosimetry in support of preclinical evaluations of peptide receptor radionuclide therapy with (90)Y, (111)In, or (177)Lu. *J. Nucl. Med.* **2004**, *45*, 1260–1269.
94. Barone, R.; Walrand, S.; Konijnenberg, M.; Valkema, R.; Kvols, L.K.; Krenning, E.P.; Pauwels, S.; Jamar, F. Therapy using labelled somatostatin analogues: Comparison of the absorbed doses with 111In-DTPA-D-Phe1-octreotide and yttrium-labelled DOTA-D-Phe1-Tyr3-octreotide. *Nucl. Med. Commun.* **2008**, *29*, 283–290. [CrossRef] [PubMed]
95. Feuerecker, B.; Tauber, R.; Knorr, K.; Heck, M.; Beheshti, A.; Seidl, C.; Bruchertseifer, F.; Pickhard, A.; Gafita, A.; Kratochwil, C.; et al. Activity and Adverse Events of Actinium-225-PSMA-617 in Advanced Metastatic Castration-resistant Prostate Cancer After Failure of Lutetium-177-PSMA. *Eur. Urol.* **2021**, *79*, 343–350. [CrossRef]
96. Evaluated Nuclear Structure Data Files. Available online: https://www.iaea.org/resources/databases/evaluated-nuclear-structure-data-file#:~{}:text=ENSDF%20is%20a%20database%20that,Center%20at%20Brookhaven%20National%20Laboratory (accessed on 29 June 2022).
97. Jan, S.; Santin, G.; Strul, D.; Staelens, S.; Assié, K.; Autret, D.; Avner, S.; Barbier, R.; Bardies, M.; Bloomfield, P.M.; et al. GATE: A simulation toolkit for PET and SPECT. *Phys. Med. Biol.* **2004**, *49*, 4543–4561. [CrossRef] [PubMed]
98. Sarrut, D.; Bardiès, M.; Boussion, N.; Freud, N.; Jan, S.; Létang, J.-M.; Loudos, G.; Maigne, L.; Marcatili, S.; Mauxion, T.; et al. A review of the use and potential of the GATE Monte Carlo simulation code for radiation therapy and dosimetry applications. *Med. Phys.* **2014**, *41*, 064301. [CrossRef] [PubMed]
99. RADAR—The RAdiation Dose Assessment Resource. Available online: https://www.doseinfo-radar.com/ (accessed on 18 August 2022).

Review

# Production Review of Accelerator-Based Medical Isotopes

Yiwei Wang [1], Daiyuan Chen [1], Ricardo dos Santos Augusto [2], Jixin Liang [3], Zhi Qin [4], Juntao Liu [1,5,*] and Zhiyi Liu [1,5,*]

[1] School of Nuclear Science and Technology, Lanzhou University, Lanzhou 730000, China
[2] Brookhaven National Laboratory, United States Department of Energy Upton, New York, NY 11973-5000, USA
[3] Department of Nuclear Technology and Application, China Institute of Atomic Energy, Beijing 102413, China
[4] Institute of Modern Physics, Chinese Academy of Sciences, Lanzhou 730000, China
[5] Frontiers Science Center for Rare Isotopes, Lanzhou University, Lanzhou 730000, China
* Correspondence: ljt@lzu.edu.cn (J.L.); zhiyil@lzu.edu.cn (Z.L.)

**Abstract:** The production of reactor-based medical isotopes is fragile, which has meant supply shortages from time to time. This paper reviews alternative production methods in the form of cyclotrons, linear accelerators and neutron generators. Finally, the status of the production of medical isotopes in China is described.

**Keywords:** medical isotope production; accelerator; nuclear medicine; review

## 1. Introduction

### 1.1. Definition of Medical Isotopes

Medical isotopes are radioisotopes that emit positrons or gamma rays for medical diagnosis or particulate radiation, such as alpha or beta particles for medical therapy [1].

### 1.2. Medical Use

The application process for medical isotopes is depicted in Figure 1 and can be summarized in four steps:

(1a) In a reactor, irradiate a suitable target with neutrons to induce a nuclear reaction;
(1b) In an accelerator, irradiate a suitable target with protons, alpha, or deuteron particles to induce a nuclear reaction;
(2) Separate radioisotopes from the irradiated targets;
(3) Combine the ligands with radioisotopes to prepare radiopharmaceuticals;
(4) Employ the radiopharmaceuticals in nuclear medicine.

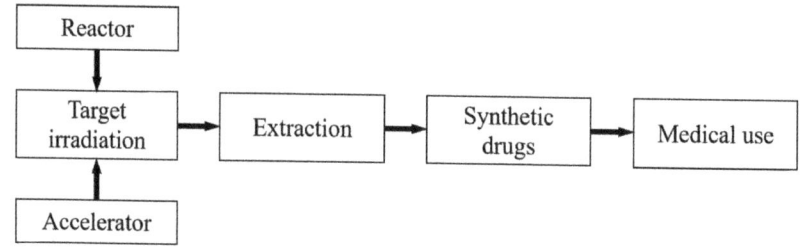

**Figure 1.** Process for the application of medical isotopes.

Depending on the physical characteristics of the isotopes applied, radiopharmaceuticals have different medical uses in diagnosis, therapy, or both (theranostics) [2], leading to a steady increase in the use of medical isotopes in nuclear medicine over time [3,4].

1.2.1. Radiopharmaceuticals for Diagnosis

Radiopharmaceuticals are generally injected intravenously or, in some cases, taken orally [5,6]. They are transported in the blood throughout the body and, due to their high affinities with specific organs, can target different diseases, especially tumors. The γ rays emitted by radiopharmaceuticals are used for imaging. Currently, there are two main imaging applications for diagnosis in nuclear medicine: Single Photon Emission Computed Tomography (SPECT) [7–9] and Positron Emission Tomography (PET) [10–12]. The distribution of radiotracers in vivo can be detected using SPECT and PET cameras.

The main advantage of nuclear medicine diagnosis lies in its ability to find lesions earlier since diseased tissues usually first denote functional changes before later evolving into shape and structural changes [13]. Another major feature of nuclear medicine diagnosis is its ability to specifically show the locations and sizes of tumors, especially when combined with Computed Tomography (CT) or Magnetic Resonance Imaging (MRI) [14,15].

1.2.2. Radiopharmaceuticals for Therapy

Therapeutic radiopharmaceuticals accumulate in diseased tissue after entering the human body. Then, their cumulative radioactive emissions can produce biological effects (e.g., killing tumor cells), which makes radiopharmaceuticals particularly suitable for cancer treatment [16]. The applications of radiopharmaceuticals for therapy include α therapy, β therapy, and Auger therapy. This review focuses on α therapy and β therapy.

1.2.3. Radiopharmaceuticals for Theranostics

In theranostics, radiopharmaceuticals can be used to perform diagnostic imaging and medical treatment [17–19]. Imaging diagnosis is used to determine an optimal treatment modality and can help monitor and evaluate the medical treatment progress [18,20,21]. Currently, radiopharmaceuticals for theranostics use either the same radiopharmaceutical, which emits γ rays for diagnosis and α or β particles for treatment [22,23], or two different radiopharmaceuticals (one for diagnosis and the other for treatment) [24].

Radiopharmaceuticals for theranostics have developed rapidly in recent years with great progress in treating neuroendocrine tumors, thyroid cancer [20,21,25,26], prostate cancer, breast cancer [27,28], and other diseases.

1.3. The Status of Medical Isotope Production

Radioisotopes are divided into natural and artificial radioisotopes. Currently, there are about 200 radioisotopes in use, most of which are produced artificially [29].

With the widespread usage of radiopharmaceuticals, the stable production and supply of medical isotopes is becoming increasingly important.

Medical isotopes are generally produced via either reactors or accelerators. Typically, reactor-based medical isotopes are neutron-rich isotopes commonly characterized by a long half-life, while accelerator-based medical isotopes tend to offer a shorter half-life and usually emit positrons or γ rays [30]. Reactor irradiation is currently the most commonly used method to produce medical isotopes due to their high yield, low cost, and ease of target preparation. However, this supply is sustained by reactors that were built in the 1950–60s (Table 1). The majority of these reactors will gradually shut down before 2030.

Table 1. Information on the world's major reactors producing medical isotopes [31–34].

| Country | Reactor | Power [MW] | Year of First Criticality | Estimated Retirement Time |
|---|---|---|---|---|
| Belgium | BR-2 | 100 | 1961 | 2026 |
| Netherlands | HFR | 45 | 1961 | 2024 |
| Czech Republic | LVR-15 | 10 | 1957 | 2028 |
| Poland | MARIA | 20 | 1974 | 2030 |
| South Africa | SAFARI-1 | 20 | 1965 | 2030 |
| Russia | WWR-TS | 15 | 1964 | 2025 |
| United States | HFIR | 100 | 1965 | 2035 |
| Australia | OPAL | 20 | 2006 | 2057 |
| Germany | FRM-II | 20 | 2004 | 2054 |

Moreover, due to their age, and as part of the decommissioning process, reactors can be expected to have longer periods of down time due to maintenance or unplanned shutdown events for safety or technical reasons [35], increasing the risk of supply interruptions or persistent shortages. Additionally, most irradiated targets for $^{99}$Mo production in a reactor context use highly enriched uranium (HEU) targets that generate considerable amounts of highly radioactive waste and increase the risk of nuclear proliferation [36,37]. These factors strengthen the argument that medical isotopes produced via reactors should be replaced by accelerator-based production [38,39].

The growing interest and recent improvements in accelerator technologies have already led some medical isotopes produced via reactors to be replaced or partly replaced by accelerator-produced isotopes. There are many advantages to using medical isotopes produced by accelerators:

(1) Supervision is easier, and safety is improved [40];
(2) The maintenance and decommissioning costs are lower [29];
(3) The amount of radioactive waste produced is less than 10% of the amount produced by a reactor, and the radiation levels are lower [41];
(4) It has no risk of nuclear proliferation [42].

As shown in Figure 2, the number of cyclotrons producing radioisotopes is increasing, while the number of reactors is slowly decreasing.

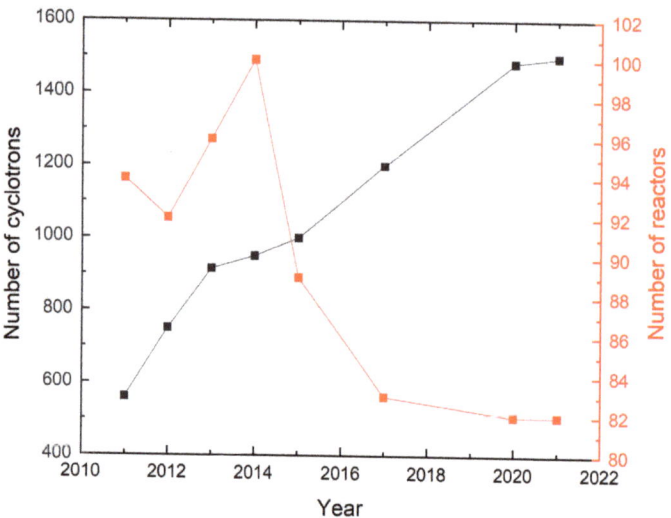

**Figure 2.** A comparison of the number of cyclotrons in the world and the number of reactors reported by the IAEA [43–57].

## 2. Medical Isotopes

This section reviews the medical isotopes produced by cyclotrons, linear accelerators, and neutron generators and lists some of the most commonly used medical isotopes, as well as their characteristics, applications, and production methods.

*2.1. Medical Isotopes Produced by Cyclotrons (1–5: PET Radioisotopes, 6–7: SPECT Radioisotopes, 8–10: Therapeutic Radioisotopes)*

A cyclotron is a particle accelerator that accelerates charged particles and uses an electromagnetic field to get the particles to follow a spiral path to ever-increasing energies until achieving the energy necessary to produce medical isotopes via nuclear interactions [58]. Compared with linear accelerators, the beams from cyclotrons have characteristically lower beam intensity, but their energy can be higher [59]. Cyclotrons are classified according to the energy of the particles they produce. As shown in Table 2, different types of cyclotrons can produce medical isotopes for a wide range of applications.

**Table 2.** Classification of medical cyclotrons [60].

| Type | The Energy of Particles [MeV] | Application |
|---|---|---|
| Small medical cyclotron | <20 | Short-lived radioisotopes for PET |
| Medium-energy cyclotron | 20–35 | Production of SPECT and some PET radioisotopes |
| High-energy cyclotron | >35 | Production of radioisotopes for therapy |

### 2.1.1. $^{18}$F

$^{18}$F ($T_{1/2}$ = 109.8 min) decays and emits positrons with an average energy of 0.25 MeV; hence, the distance traveled until reaching positron annihilation in tissues is short. $^{18}$F is the most commonly used PET radioisotope. At present, the Food and Drug Administration (FDA) has approved $^{18}$F radiopharmaceuticals for use in the diagnosis of a variety of diseases, such as Alzheimer's disease, infections, and many types of cancer, as well as to evaluate treatment outcomes [61,62]. According to clinical data, [$^{18}$F]FDG can distinguish between Parkinson's Disease (PD), MSA with predominant Parkinsonism (MSA-P), and MSA with predominant cerebellar features (MSA-C) [63,64]. PET diagnosis is expensive and can cost over $1000, while doctors can make an early and accurate diagnosis. For that reason, the annual number of PET scans has steadily increased for many years [65]. Most $^{18}$F is produced via cyclotrons by exploiting two nuclear reactions:

(1) $^{18}$O (p, n) $^{18}$F: This reaction requires enriched (and more expensive) $^{18}$O target materials to produce $^{18}$F in a high yield [66]. Technology developments led to improvements in the target system and the production of $^{18}$F up to 34 GBq, as well as specific activities of 350–600 GBq/mmol 30 min after the end of bombardment [67]. Subsequently, it was found that with the irradiation of 11 MeV protons, the yield of $^{18}$F further increased directly with the proton current. However, the impurities also increased such that for a proton current of 20 µA, the yield of $^{56}$Co (4.86 MBq) and $^{110m}$Ag (1.51 MBq) doubled [68]. Many developing countries do not have medical isotope production facilities. If these countries desire to become self-sufficient in the production of medical isotopes, they could start by installing low-energy cyclotrons to produce $^{18}$F [69].

(2) $^{20}$Ne (d, α) $^{18}$F: This is the first production method used to produce $^{18}$F. This reaction is characterized by lower yields and low specific activity, so it is gradually being replaced. However, with production improvements, this method could again become an attractive alternative [70].

### 2.1.2. $^{68}$Ga

$^{68}$Ga ($T_{1/2}$ = 68 min) is a metal PET radioisotope. Currently, there are about 100 ongoing clinical tests with $^{68}$Ga [61], indicating the rapid development of $^{68}$Ga-labelled radiotracers. Radiopharmaceuticals labeled with $^{68}$Ga are used for the diagnosis of neuroendocrine tumors and are highly accurate when used in patients with suspected but yet not localized neuroendocrine tumors [71]. In addition, $^{68}$Ga and $^{177}$Lu ($T_{1/2}$ = 6.7 d) have a similar coordination chemistry, rendering them some of the most promising radiopharmaceuticals for theranostics. For neuroendocrine tumors, both [$^{68}$Ga]Ga-DOTA-TATE and [$^{177}$Lu]Lu-DOTA-TATE have been approved by the FDA for clinical PET diagnosis and medical treatment [72–74]. [$^{68}$Ga]Ga-PSMA-11 is the first radiopharmaceutical approved by the FDA for PET imaging of PSMA-positive prostate cancer, and [$^{177}$Lu]Lu-PSMA-617 has also been used for PSMA-targeted therapy [74–77].

$^{68}$Ga is generally available using a $^{68}$Ge/$^{68}$Ga generator and represents a relatively simple and convenient method [78] that can yield up to 1.85 GBq [79]. With the development of technology, the commercial "ionic" generators have made $^{68}$Ga clinically successful [80,81]. $^{68}$Ga obtained by generators cannot meet the growing demands, however, so the use of accelerators to obtain $^{68}$Ga has aroused scientific interest. Moreover, higher yields of $^{68}$Ga can be obtained with the $^{68}$Zn (p, n) $^{68}$Ga reaction using a small cyclotron [82,83]. The yield when using a solid target was reported as 5.032 GBq/μA·h [83]. After 6 h, impurities such as $^{66}$Ga and $^{67}$Ga only accounted for 0.51% of the total activity [84]. Compared with using a generator, this production method does not require radioactive waste treatment. Although the solid target system is complex, and the separation steps are lengthy, an automated process was developed to separate the solid target and is simpler to operate than alternative methods [85]. This nuclear reaction can also take place in a liquid target, with radiochemical and radionuclidic purities both above 99.9%. However, the yield using a liquid target was found to be significantly lower (192.5 ± 11.0) MBq/μA·h [86]. This production method using the liquid target as an alternative method still needs further optimization to improve the yield.

### 2.1.3. $^{64}$Cu

Upon decay, $^{64}$Cu ($T_{1/2}$ = 12.7 h) emits positrons and electrons that can be utilized for PET diagnosis and have potential applications in β therapy, thus making $^{64}$Cu useful as a radiopharmaceutical for theranostics. Furthermore, $^{64}$Cu and $^{67}$Cu ($T_{1/2}$ = 61.76 h) can be radiopharmaceuticals for theranostics in order to conduct pre-targeted radioimmunotherapy [87]. Presently, the FDA has approved [$^{64}$Cu]Cu-DOTA-TATE to localize somatostatin receptor-positive neuroendocrine tumors in adult patients. In clinical experiments, [$^{64}$Cu]Cu-DOTA-TATE has excellent imaging quality and higher detection rates for lesions [88].

$^{64}$Cu can be produced by small medical cyclotrons via $^{64}$Ni (p, n) $^{64}$Cu reaction with high specific activity. This production method requires an enriched $^{64}$Ni (at least 96%) target to obtain a high yield of 5.89 GBq/μA·h and $^{64}$Cu with radionuclidic purity higher than 99% [89]. The disadvantage is that the $^{64}$Ni target material has a low isotopic abundance (0.926%) in nature [90], meaning that the target material is expensive and must be recycled to improve its cost-effectiveness [91,92]. Alternative methods of $^{64}$Cu production can also be deuteron-zinc reactions such as $^{nat}$Zn (d, x) $^{64}$Cu, and $^{66}$Zn (d, α) $^{64}$Cu. Although they have lower costs, their yields are lower, and high-energy deuterons are required [93]. These factors limit actual production through such reactions.

The $^{64}$Ni (p, n)$^{64}$Cu reaction is the preferred choice for $^{64}$Cu production in clinical applications. During the past decade, more than 20 countries, including the United States, Japan, Finland, and China, have developed $^{64}$Ni (p, n) $^{64}$Cu methods for $^{64}$Cu production [89,91,94], some of which are shown in Table 3.

Table 3. Facilities that have reported the production of $^{64}$Cu [91,94–99].

| Facility/Location | Nuclear Reaction | Irradiation Parameters | Yield |
|---|---|---|---|
| Fukui Medical University | $^{64}$Ni(p, n)$^{64}$Cu | 12 MeV, (50 ± 3) μA | 2-24 GBq in 2 h |
| The University of Sherbrooke PET Imaging Centre | $^{64}$Ni(p, n)$^{64}$Cu | 15 MeV, 18 μA | 3.9 GBq in 4 h |
| IBA | $^{64}$Ni(p, n)$^{64}$Cu | 10 MeV, 12 μA | 5123 MBq in 3 h |
| Paul Scherrer Institute | $^{64}$Ni(p, n)$^{64}$Cu | 11 MeV, 40–50 μA | Max 8.2 GBq in 4–5 h |
| Turku PET Centre | $^{64}$Ni(p, n)$^{64}$Cu | 15.7 MeV, < 100 μA | Max 9.4 GBq after purification |
| Sumitomo HM-20 cyclotron | $^{64}$Ni(p, n)$^{64}$Cu | 12.5 MeV, 20 μA | 7.4 GBq in 5–7 h |
| NIRS AVF-930 cyclotron | $^{64}$Ni(p, n)$^{64}$Cu | 24 MeV HH$^+$, 10 eμA | 5.2-13 GBq in 1–3 h |

### 2.1.4. $^{89}$Zr

$^{89}$Zr ($T_{1/2}$ = 78.4 h) is a positron emitter and a new metal PET radioisotope ideal for immunoimaging [100]. To date, $^{89}$Zr-atezolizumab has been studied in renal cell carcinoma (RCC), but some obstacles were encountered, so further research is needed [101]. $^{89}$Zr is produced by cyclotrons involving the following nuclear reactions:

(1) $^{89}$Y (p, n) $^{89}$Zr: This reaction only requires low-energy protons (5-15 MeV) and targets with natural abundance $^{89}$Y (100%), which reduces the costs significantly. The number of interference nuclear reactions is limited; hence, one can obtain a high specific activity of $^{89}$Zr [102–104]. The yield of this (p, n) reaction can be as high as 44 MBq/μA·h under irradiation of 14 MeV protons [105]. Various methods for the isolation and purification of $^{89}$Zr have been proposed, including solvent extraction, anion exchange chromatography, and weak cation exchange chromatography, which can obtain $^{89}$Zr with high specific activity and radionuclidic purity [106]. The proton energy from small medical cyclotrons installed in hospitals can meet the requirements for bombarding the $^{89}$Y target, which is the main reason why many hospitals have developed $^{89}$Zr production processes.

(2) $^{89}$Y (d, 2n) $^{89}$Zr: This reaction uses low-energy deuterons (also 5–15 MeV) and has the same advantages as the aforementioned production method [102–104], as well as offering a higher yield of 58 MBq/μA·h. However, one must still factor in the availability of the beam of particles and the costs of these two production methods [105]. Thus, more research is needed.

(3) $^{nat}$Sr (α, xn) $^{89}$Zr: Besides requiring α beams, if $^{nat}$Sr targets are used, abundant quantities of impurities such as $^{88}$Zr and $^{86}$Zr can easily be produced. For the moment, this production method is only theoretically feasible [107].

### 2.1.5. $^{124}$I

$^{124}$I ($T_{1/2}$ = 4.176 d) is a PET nuclide that can provide a higher quality diagnostic image [108]. Currently, $^{124}$I is used for the clinical diagnosis of thyroid cancer [109] and neuroblastoma [110]. $^{124}$I and $^{131}$I can also be combined as radiopharmaceuticals for theranostics to treat thyroid cancer [20].

$^{124}$I is produced via cyclotrons through two different production methods:

(1) $^{124}$Te (p, n) $^{124}$I: This is the main production method currently employed. Although this method offers a relatively low production rate, it can achieve high currents and use enriched targets to improve the overall yield [108]. The average yield of this reaction is 16 MBq/μA·h, and at the end of bombardment, the impurity content of $^{123}$I and $^{125}$I only reaches about 1% [111]. Dry distillation is used to extract $^{124}$I [112]. On the downside, the enriched $^{124}$Te target material costs about 10000$/g, which is relatively expensive [113].

(2) $^{124}$Te (d, 2n) $^{124}$I: Has a high production yield of 17.5 MBq/μA·h, however, this reaction requires a beam of deuterons, which may be difficult to obtain and can result in impurities such as $^{125}$I (reaching about 1.7%) [111,114].

### 2.1.6. $^{99}$Mo/$^{99m}$Tc

$^{99m}$Tc ($T_{1/2}$ = 6.02 h) emits single γ rays with 0.141 MeV and is mostly used in SPECT; for the diagnosis of stroke; and to examine bone, myocardium, kidneys, thyroid, salivary glands, and other organs [61,62]. The proportion of nuclear medicine diagnosis applying $^{99m}$Tc accounts for approximately 80% of all nuclear medicine procedures, representing around 40 million examinations worldwide every year [115]. $^{99m}$Tc is mainly produced using a $^{99}$Mo/$^{99m}$Tc generator. Currently, $^{99m}$Tc can be produced by cyclotrons through the following reactions:

(1) $^{100}$Mo (p, 2n) $^{99m}$Tc [116,117]: This is the main production method and is optimal with a proton energy range of 19–24 MeV and a highly enriched $^{100}$Mo target, such that $^{98}$Tc, $^{97}$Tc, and other impurities can be reduced to a minimum. According to

the experimental data, with a proton beam energy of 24 MeV, the yield of $^{99m}$Tc is about 592 GBq/mA·h [118]. A target irradiated with a 24 MeV proton beam at 500 µA for 12 h yielded 2.59 TBq of $^{99m}$Tc [119]. GE PETtrace880 machines have obtained approximately 174 GBq after 6 h [116]. To date, TRIUMF and its partners have successfully verified the feasibility of using a 24 MeV cyclotron to produce $^{99m}$Tc to supply the needs of all applications in Vancouver by developing a complete process based on 16, 19, and 24 MeV cyclotron production and applied the results to relevant patents [120]. Automated modules to separate $^{99m}$Tc from irradiated targets of $^{100}$Mo are under development [121]. However, the shipped distance should be considered based on the direct product and its half-life [122];

(2)  $^{96}$Zr (α, n) $^{99}$Mo→$^{99m}$Tc [123,124]: This production method can produce $^{99m}$Tc with high specific activity. However, it has a low yield, and a beam with a high current is difficult to obtain, which limits the applicability of this production method.

### 2.1.7. $^{123}$I

$^{123}$I ($T_{1/2}$ = 13.2 h) is a γ-ray emitter that can be utilized for SPECT diagnosis. It has especially been used for the diagnosis of Parkinson's disease, primary and metastatic pheochromocytoma, and neuroblastoma. The sensitivity and specificity of this technology are greater than 90% [125]. It also can be used for diagnosis of the thyroid, brain, and myocardium.

Presently, there are three common production routes yielding $^{123}$I:

(1–2) $^{124}$Xe (p, 2n) $^{123}$Cs→$^{123}$Xe→$^{123}$I and $^{124}$Xe (p, pn)$^{123}$Xe→$^{123}$I: These nuclear reactions require a medium-energy cyclotron and can obtain with a high radionuclidic purity. The yield of these reactions simulated by MCNP was 757 MBq/µA·h. Compared with the experimental data, the maximum fluctuation was about 185 MBq/µA·h [126,127]. However, due to the use of enriched $^{124}$Xe targets, these methods are costly [128,129].

(3) $^{123}$Te (p, n) $^{123}$I: This production method can apply a low-energy cyclotron. When enriched targets of $^{123}$Te (enrichment of 99.3%) were used, an ultrapure nuclide was obtained, and the yield increased from nearly 18.5 to 37GBq 30 h after EOB (end of the bombardment) [130–132]. This production method is also costly because of the enriched target of $^{123}$Te. This alternative production method was proven feasible to produce $^{123}$I.

### 2.1.8. $^{225}$Ac

$^{225}$Ac ($T_{1/2}$ = 9.92 d) has a unique decay chain that can emit four α rays, causing it to be more effective in destroying tumor cells than other isotopes. Presently, the first use of [$^{225}$Ac]Ac-PSMA-I&T in a clinical context was successful in treating advanced metastatic castration-resistant prostate cancer [133–135]. Additionally, the research of [$^{225}$Ac]Ac-DOTAGA-SP for the treatment of malignant gliomas is ongoing [136].

$^{225}$Ac can be produced with medium-energy protons via the $^{226}$Ra (p, 2n) $^{225}$Ac reaction. The yield was only about 2.4 MBq after EOB [137], moreover, its radioactive inventory is difficult to handle [137–139]. Production of $^{225}$Ac applying high-energy protons (60–140 MeV) through bombarding a $^{232}$Th target can produce a high yield of 96 GBq, but this yield requires high intensity and energy [140], which are not readily available. Currently, the U.S. Department of Energy Isotope Program produces $^{225}$Ac using a spallation-induced reaction with high-energy protons on natural thorium.

### 2.1.9. $^{211}$At

$^{211}$At ($T_{1/2}$ = 7.2 h) emits α particles that can be utilized in α therapy [141]. Currently, $^{211}$At in the form of [$^{211}$At]At-PA and [$^{211}$At]At-ch81C6 has been studied in glioma and recurrent brain tumors [142,143]. Gothenburg (Sweden) [144] is undergoing a clinical research using [$^{211}$At]At-MX35(Fab)$_2$ to treat ovarian cancer patients, which is an alpha-emitting radionuclide with great clinical potential [145].

$^{211}$At is commonly produced by a medium-energy cyclotron bombarding a $^{209}$Bi target with α particles, causing a $^{209}$Bi (α, 2n) $^{211}$At reaction to take place [146,147]. Purifying the

$^{211}$At from the target material was either done by a wet extraction or a dry distillation. The National Institutes of Health (Bethesda, USA) produced a maximum of 1.71 GBq in one hour, while Sichuan University in China produced a maximum of 200 MBq in 2 h [148]. However, due to the product of toxic impurities such as $^{210}$Po, the energy of the α beam needs to be monitored [148,149].

2.1.10. $^{67}$Cu

$^{67}$Cu ($T_{1/2}$ = 61.76 h) emits γ rays for SPECT diagnosis and β particles that can be used for medical treatment. Thus, $^{67}$Cu can be used individually or with $^{64}$Cu as a radiopharmaceutical for theranostics. Presently, $^{67}$Cu is used for the nuclear medicinal diagnosis of neuroendocrine tumors and lymphomas [150,151] and the medical treatment of lymphoma and colon cancer [152].

$^{67}$Cu is generally produced via the $^{68}$Zn (p, 2p) $^{67}$Cu reaction. This reaction has high recovery and needs both a medium-energy cyclotron and a highly enriched target [153–155]. Due to the need for high-energy protons, there are only a few laboratories in the world that can produce $^{67}$Cu [156]. The yield of the integral physical thick target was calculated and is shown in Figure 3.

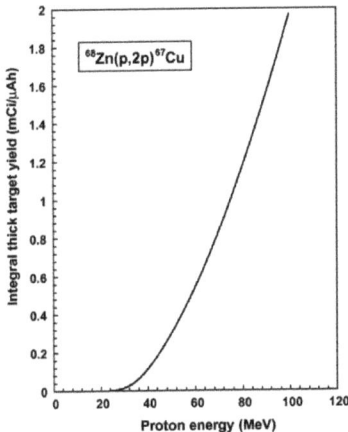

**Figure 3.** Integral physical thick target yields for the $^{68}$Zn (p, 2p) $^{67}$Cu reaction [157].

In addition to the medical isotopes mentioned above, $^{11}$C [158], $^{13}$N [159], $^{15}$O [160], $^{86}$Y [161], $^{44}$Sc [162,163], $^{201}$Tl [164], $^{47}$Sc [165,166], $^{32}$P [167], $^{67}$Ga [168], and other medical isotopes produced by cyclotrons have also been reported.

Cyclotrons are the main accelerator-based drivers of medical isotope production. Their output is constantly improving due to advancements in targets [169,170], research on new nuclear reactions [171–173], and accelerator technology developments [174–176], leading not only to increased yields but also to a reduction in radioactive impurities. Most medical isotopes currently produced by reactors can also alternatively be produced by cyclotrons, and the constant improvements to the medical-isotope-producing abilities of cyclotrons have contributed to the stable supply of medical isotopes.

## 2.2. Medical Isotopes Produced by Linacs

The charged particles accelerated by a linac pass through the focusing magnetic field and the linear acceleration field once without deflection [58]. Once ejected, these particles irradiate their targets to produce medical isotopes. Linac beams are characterized by high beam intensity and lower energy [59].

In terms of linacs currently used to produce medical isotopes, proton linacs can be relatively easily employed in medical isotope production. For example, proton linacs that produce PET nuclides can reduce the weight of cyclotron magnets, and some high-energy and high-fluxes proton linacs can produce therapeutic nuclides [177–179]. While feasibility reports on the ability of electron linacs to produce medical isotopes are common, the pulsed beams and the cross-sections of linacs can create challenges when used in practice [41,180–182]. There are other linacs that accelerate other charged particles; however, these linacs will not be described here.

### 2.2.1. $^{18}$F

PET radioisotopes can be produced with proton linacs. The first compact proton linear accelerator in the United States for the generation of medical isotopes produces $^{18}$F for a local hospital [183]. Additionally, Hitachi, Ltd. and AccSys Technology, Inc. (Hitachi's subsidiary company) also developed a proton linac to produce PET nuclides. After bombardment for one hour, 23.5 GBq $^{18}$F was produced, indicating that batch production of $^{18}$F could be achieved [177].

$^{18}$F ($T_{1/2}$ = 109.8 min) can also be produced by electron linacs through a photonuclear reaction $^{19}$F (γ, n) $^{18}$F, as well as other commonly used PET radioisotopes such as $^{11}$C ($T_{1/2}$ = 20.38 min), $^{13}$N ($T_{1/2}$ = 9.96 min), and $^{15}$O ($T_{1/2}$ = 122 s). When using a photonuclear reaction to produce these PET radioisotopes, the yields are generally lower since the cross-section is 1–2 orders of magnitude lower than that under a proton reaction. However, photonuclear reactions can use a natural target of $^{19}$F, thus providing lower costs compared to proton reactions [177]. Many feasibility reports on producing PET nuclides via photonuclear reactions have been published, but actual production still needs further study.

### 2.2.2. $^{99}$Mo

$^{99}$Mo ($T_{1/2}$ = 66 h) decays into $^{99m}$Tc ($T_{1/2}$ = 6.02 h). An electron linac can be utilized to produce $^{99}$Mo via the photonuclear reaction $^{100}$Mo (γ, n) $^{99}$Mo [184–186]. It was reported that the yield of $^{99}$Mo obtained after 6.5 days of continuous bombardment of a 6 g high-purity $^{100}$Mo target with 36 MeV electrons was 458.8 GBq (average beam power of ~8 kW) [187]. The cost of this production method can be reduced by using a natural target and, although this method will produce the isotopes of Mo, isotopes of Tc will not be produced, making it easy to separate $^{99}$Mo via chemical difference or evaporation temperature difference [188]. NorthStar and its partners have studied this production method and listed it as the main $^{99}$Mo supply option in their long-term plans [187]. Canadian Light Source (CLS) and TRIUMF also conducted feasibility research on this production method and plan to put it into production [189,190].

In addition to the medical isotopes mentioned above, the production of $^{67}$Cu [191–193], $^{64}$Cu [194], $^{225}$Ac [195,196], $^{68}$Ga [197], $^{111}$In [181], $^{177}$Lu [198], $^{47}$Sc [199], and other medical isotopes through linacs have been reported.

Overall, linacs have some disadvantages in terms of their design and yields [41,182, 200,201]. As a backup method for the production of medical isotopes, linacs still require further research.

## 2.3. Medical Isotopes Produced by Neutron Generators

A neutron generator is an accelerator-based neutron source device that is capable of delivering neutrons through nuclear fusion reactions. These neutrons will, in turn, irradiate the target to produce medical isotopes. The nuclear fusion reactions commonly used to produce neutrons are shown in Table 4.

**Table 4.** Fusion reactions that produce neutrons [202–206].

| Reaction | Energy [MeV] | The Suitable Reaction of Isotope Production |
|---|---|---|
| D-D reaction | 2–3 | (n, γ) |
| D-T reaction | 14–15 | (n, 2n) (n, p) |
| D-$^7$Li reaction | 10&13 | (n, 2n) (n, p) |

### 2.3.1. $^{99}$Mo/$^{99m}$Tc

$^{99m}$Tc ($T_{1/2}$ = 6.02h) can be produced by neutron generators [207,208]. After neutron moderation, neutrons with a specific energy can be obtained and then used to produce $^{99m}$Tc via the nuclear reaction of $^{235}$U (n, f) $^{99}$Mo→$^{99m}$Tc. The advantages of this production method include both ease of supervision and overall safety, but the yield will be 1–2 orders of magnitude lower than that produced by a reactor [209]. SHINE and Phoenix Laboratory used a DT neutron generator to bombard UO$_2$SO$_4$ to produce $^{99}$Mo. After irradiation of a 5 L UO$_2$SO$_4$ solution for about 20 h, the yield of $^{99}$Mo was 51.8 GBq [210]. The disadvantage of this production method is that a long-term, stable, and high-intensity beam is difficult to achieve [211].

In addition, $^{99}$Mo can be produced via the nuclear reactions of $^{98}$Mo (n, γ) $^{99}$Mo and $^{100}$Mo (n, 2n) $^{99}$Mo, both of which use Mo targets instead of U targets. Additionally, sufficient activity of $^{99}$Mo can be produced in principle [207,208,212]. The yields of these two nuclear reactions can be increased by improving the fluxes of neutrons and the irradiation time and/or using highly enriched targets, in addition to other methods [213]. However, $^{99}$Mo from an irradiated $^{98}$Mo/$^{100}$Mo target is a carrier-added product with a low specific activity. The biggest challenge for this method is how to develop a new type of $^{99}$Mo/$^{99m}$Tc generator that meets medical requirements.

### 2.3.2. $^{67}$Cu

$^{67}$Cu ($T_{1/2}$ = 61.76 h) is generally produced by cyclotrons. Kin proposed using neutrons to produce $^{67}$Cu [212]. Presently, using neutron generators via the D-T reaction in the form of $^{67}$Zn (n, p) $^{67}$Cu can produce $^{67}$Cu. Due to the developments of neutron generators, $^{67}$Cu can be produced in the hospital without the need to transport the isotope over long distances. This production method does not produce a large number of impurities [156,214], and the activity can reach hundreds to thousands of MBq [212]. However, when dealing with radioactive isotopes with GBq, the radiation facility will result in higher costs [212].

In addition to the medical isotopes mentioned above, $^{89}$Sr [215–217], $^{64}$Cu [218], $^{47}$Sc [219], $^{132}$Xe [220], $^{225}$Ac [212], and other medical isotopes produced by neutron generators have also been reported.

As a neutron source, a neutron generator is essential to produce neutron-rich medical isotopes. Although such generators have the advantages of low cost and target reusability [212,221], providing continuously high fluxes of neutrons and engaging in separation-extraction of the medical isotopes remain challenging topics [221]. Despite these challenges, generators are presently regarded as a viable alternative to the reactor-production method.

## 3. The Status of Medical Isotope Production via Accelerators in China

### 3.1. Available Accelerators for Medical Isotope Production in China

Currently, there are about 160 PET small medical cyclotrons for the routine production of $^{11}$C, $^{18}$F, and other medical isotopes to meet clinical demands in China [222]. Additionally, there are several medium- and high-energy accelerators used for medical isotope production in China.

The Chinese Institute of Atomic Energy (CIAE) and Shanghai Ansheng Kexing Company each have a C-30 cyclotron with adjustable proton energy of 15.5–30 MeV and beam currents up to 350 µA. These can be used to produce medical isotopes such as $^{11}$C, $^{18}$F, $^{64}$Cu, $^{68}$Ge, $^{89}$Zr, $^{123}$I, $^{124}$I, and $^{201}$Tl. CIAE has a 100 MeV proton cyclotron (C-100) with a beam current up to 200 µA capable of producing $^{67}$Cu, $^{225}$Ac, and other medical isotopes of interest.

The Sichuan University owns a cyclotron capable of delivering beams of protons, as well as alpha and deuteron particles (p–26 MeV and α–30 MeV).

The Chinese Academy of Sciences Institute of Modern Physics built a 25 MeV superconducting proton linear accelerator with an intensity in the order of milliamps. At present, the linac can accelerate various beams such as proton beams, $^{3}$He$^{2+}$ beams, and $^{4}$He$^{2+}$ beams. The energy of $^{3}$He$^{2+}$ beams can reach 36 MeV at an intensity of 200 µA, while the energy of $^{4}\text{He}^{2+}$ beams can reach 32 MeV with a current of 100 μA. The accelerator can meet the needs of medical isotope production and produce various radioisotopes such as $^{99}\text{Mo}/^{99m}\text{Tc}$, $^{117m}\text{Sn}$, $^{211}\text{At}$, $^{55}\text{Fe}$, $^{73}\text{As}$, $^{225}\text{Ac}$, $^{109}\text{Cd}$, $^{88}\text{Y}$, and $^{75}\text{Se}$.

Lanzhou University has been instrumental in the development of advanced ion source selection, ion beam extraction, and acceleration system design, as well as target system design. Additionally, the university independently built a series of neutron generators based on D-D and D-T reactions [223].

### 3.2. The Status of Medical Isotope Production via Accelerators

There is a solid research foundation for accelerator-based medical isotope production in China. In the 1980s, Sichuan University and others successfully developed production technology for medical isotopes such as $^{211}\text{At}$, $^{123}\text{I}$, $^{111}\text{In}$, and $^{201}\text{Tl}$ by relying on domestic cyclotrons and a CS-30 cyclotron [224]. Since the 1990s, CIAE has produced medical isotopes such as $^{18}\text{F}$, $^{111}\text{In}$, and $^{201}\text{Tl}$ using a C-30 cyclotron.

In the last two decades, with the popularization and rapid development of domestic nuclear medicine, the amount of PET equipment increased to 427 by 2019. Today, 117 hospitals equipped with small medical cyclotrons routinely produce $^{18}\text{F}$ to meet clinical needs, with an annual consumption of more than 1850 TBq. Additionally, some emerging isotopes such as $^{64}\text{Cu}$, $^{89}\text{Zr}$, and $^{123/124}\text{I}$ have been rapidly developed for medical applications. In 2007, CIAE cooperated with Atom Hitech to carry out research on $^{123}\text{I}$ production using enriched $^{124}\text{Xe}$ gas at 111 GBq for each batch with a C-30 cyclotron. In 2012, Atom Hitech produced carrier-free $^{64}\text{Cu}$ with enriched $^{64}\text{Ni}$ at 37–74 GBq for each batch based on a C-30 cyclotron. In 2016, Sichuan University bombarded an $^{89}\text{Y}$ target with 13 MeV protons and obtained $^{89}\text{Zr}$ with a radionuclidic purity of more than 99% [16]. However, due to the limited availability of high-energy particle accelerators for the production of therapeutic nuclides such as $^{67}\text{Cu}$, $^{225}\text{Ac}$, and $^{223}\text{Ra}$, China is significantly lagging behind the advanced international levels of development. In 2021, for the first time, CIAE obtained around 22.2 MBq of $^{225}\text{Ac}$ with radionuclidic purity greater than 99% using a C-100 cyclotron.

## 4. Summary

Presently, cyclotrons remain the primary facilities for accelerator-based medical isotope production, although linacs and neutron generators are rapidly becoming a viable alternative.

Cyclotrons with adjustable energy ranges or medium energy can produce various kinds of medical isotopes and can cover most radiopharmaceutical production needs in a region [59]. Yield and purity improvements in medical isotopes and the overall cost of cyclotron production have led researchers to explore further possibilities, including proton linacs, which have significant advantages in providing proton beams in the order of tens to hundreds of MeV [179]. These linacs can be developed in research institutes or laboratories conducting scientific experiments and physical research at the same time. For electron linacs, the cross-section of photonuclear interactions is relatively low, which restricts their practical applications. Other factors, such as impurity products and economic costs, also play major roles when evaluating production techniques and methodologies. Attempts to produce medical isotopes through neutron generators are promising and could theoretically yield the medical isotopes that are currently produced by reactors. However, improving the neutron flux rate remains a major consideration.

As medical isotopes produced by reactors often face supply shortages, interest in the use of accelerator-based techniques to produce medical isotopes will increase. We hope to develop an accelerator with the right energy, right beam types, right location, and good shielding facilities, which will play an important role in the supply of medical isotopes.

**Author Contributions:** Conceptualization, Y.W. and D.C.; methodology, D.C.; investigation, Y.W., D.C., R.d.S.A., J.L. (Jixin Liang) and Z.L.; resources, Z.L., R.d.S.A., J.L. (Jixin Liang), Z.Q. and J.L. (Juntao Liu); writing—original draft preparation, Y.W.; writing—review and editing, R.d.S.A. and J.L. (Jixin Liang); supervision, Z.Q. and J.L. (Juntao Liu); project administration, Z.L.; funding acquisition, J.L. (Juntao Liu). All authors have read and agreed to the published version of the manuscript.

**Funding:** This research was funded by the National natural Science Foundation of China, grant number 11975115; Special Projects of the Central Government in Guidance of Local Science and Technology Development (Research and development of three-dimensional prospecting technology based on Cosmic-ray muons), grant number YDZX20216200001297; the Research and Development of Medical Isotopes based on High-current Superconducting Linear Accelerator Project, the Fundamental Research Funds for the Central Universities, grant number lzujbky-2019-54; the Science and Technology Planning Project of Gansu, grant number 20JR10RA645; Lanzhou University Talent Cooperation Research Funds sponsored by Lanzhou City, grant number 561121203 and Gansu provincial science and technology plan projects for talents, grant number 054000029.

**Institutional Review Board Statement:** Not applicable.

**Informed Consent Statement:** Not applicable.

**Data Availability Statement:** No new data were created or analyzed in this study. Data sharing is not applicable to this article.

**Conflicts of Interest:** The authors declare no conflict of interest.

# References

1. Radioisotopes in Medicine [EB/OL]. Available online: https://world-nuclear.org/information-library/non-power-nuclear-applications/radioisotopes-research/radioisotopes-in-medicine.aspx (accessed on 1 June 2022).
2. AMA Manual of Style Committee. *AMA manual of style: A guide for authors and editors*, 10 th ed.; Oxford University Press: New York, NY, USA, 2007; ISBN 978-0-19-517633-9.
3. Reuzé, S.; Schernberg, A.; Orlhac, F.; Sun, R.; Chargari, C.; Dercle, L.; Deutsch, E.; Buvat, I.; Robert, C. Radiomics in nuclear medicine applied to radiation therapy: Methods, pitfalls, and challenges. *Int. J. Radiat. Oncol. Biol. Phys.* **2018**, *102*, 1117–1142. [CrossRef] [PubMed]
4. Langbein, T.; Weber, W.A.; Eiber, M. Future of theranostics: An outlook on precision oncology in nuclear medicine. *J. Nucl. Med.* **2019**, *60*, 13S–19S. [CrossRef] [PubMed]
5. Kar, N.R. Production and applications of radiopharmaceuticals: A review. *Int. J. Pharm. Investig.* **2019**, *9*, 36–42. [CrossRef]
6. Vermeulen, K.; Vandamme, M.; Bormans, G.; Cleeren, F. Design and challenges of radiopharmaceuticals. In *Seminars in Nuclear Medicine*; WB Saunders: Philadelphia, PA, USA, 2019; Volume 49, pp. 339–356.
7. Holly, T.A.; Abbott, B.G.; Al-Mallah, M.; Calnon, D.A.; Cohen, M.C.; DiFilippo, F.P.; Ficaro, E.P.; Freeman, M.R.; Hendel, R.C.; Jain, D.; et al. Single photon-emission computed tomography. *J. Nucl. Cardiol.* **2010**, *17*, 941–973. [CrossRef] [PubMed]
8. Jaszczak, R.J.; Coleman, R.E.; Lim, C.B. SPECT: Single photon emission computed tomography. *IEEE Trans. Nucl. Sci.* **1980**, *27*, 1137–1153. [CrossRef]
9. Jaszczak, R.J.; Coleman, R.E. Single photon emission computed tomography (SPECT). Principles and instrumentation. *Investig. Radiol.* **1985**, *20*, 897–910. [CrossRef]
10. Valk, P.E.; Delbeke, D.; Bailey, D.L.; Townsend, D.W.; Maisey, M.N. *Positron Emission Tomography*; Springer: London, UK, 2005.
11. Kubota, K. From tumor biology to clinical PET: A review of positron emission tomography (PET) in oncology. *Ann. Nucl. Med.* **2001**, *15*, 471–486. [CrossRef]
12. Wagner, H.N., Jr. A brief history of positron emission tomography (PET). In *Seminars in Nuclear Medicine*; WB Saunders: Philadelphia, PA, USA, 1998; Volume 28, pp. 213–220.
13. Wheat, J.M.; Currie, G.M.; Davidson, R.; Kiat, H. An introduction to nuclear medicine. *Radiographer* **2011**, *58*, 38–45. [CrossRef]
14. Mariani, G.; Bruselli, L.; Kuwert, T.; Kim, E.E.; Flotats, A.; Israel, O.; Dondi, M.; Watanabe, N. A review on the clinical uses of SPECT/CT. *Eur. J. Nucl. Med. Mol. Imaging* **2010**, *37*, 1959–1985. [CrossRef]
15. Palmedo, H.; Bucerius, J.; Joe, A.; Strunk, H.; Hortling, N.; Meyka, S.; Roedel, R.; Wolff, M.; Wardelmann, E.; Biersack, H.J.; et al. Integrated PET/CT in differentiated thyroid cancer: Diagnostic accuracy and impact on patient management. *J. Nucl. Med.* **2006**, *47*, 616–624.
16. Liqun, H.; Shufang, L.; Ge, S.; Huan, L.; Jianguo, L.; Quan, A.; Zhongwen, W. Current Applications and Prospects of Radionuclide for Therapy. *J. Isot.* **2021**, *34*, 412.
17. Rösch, F.; Baum, R.P. Generator-based PET radiopharmaceuticals for molecular imaging of tumours: On the way to THERANOSTICS. *Dalton Trans.* **2011**, *40*, 6104–6111. [CrossRef]
18. Notni, J.; Wester, H.J. Re-thinking the role of radiometal isotopes: Towards a future concept for theranostic radiopharmaceuticals. *J. Label. Compd. Radiopharm.* **2018**, *61*, 141–153. [CrossRef]

19. Qaim, S.M.; Scholten, B.; Neumaier, B. New developments in the production of theranostic pairs of radionuclides. *J. Radioanal. Nucl. Chem.* **2018**, *318*, 1493–1509. [CrossRef]
20. Nagarajah, J.; Janssen, M.; Hetkamp, P.; Jentzen, W. Iodine symporter targeting with 124I/131I theranostics. *J. Nucl. Med.* **2017**, *58* (Suppl. S2), 34S–38S. [CrossRef]
21. Eberlein, U.; Cremonesi, M.; Lassmann, M. Individualized dosimetry for theranostics: Necessary, nice to have, or counterproductive? *J. Nucl. Med.* **2017**, *58* (Suppl. S2), 97S–103S. [CrossRef]
22. Braccini, S.; Belver-Aguilar, C.; Carzaniga, T.; Dellepiane, G.; Häffner, P.; Scampoli, P. Novel irradiation methods for theranostic radioisotope production with solid targets at the Bern medical cyclotron. In Proceedings of the International Conference on Cyclotrons and their Applications (CYC), Cape Town, South Africa, 22–27 September 2019; pp. 22–27.
23. Brandt, M.; Cardinale, J.; Aulsebrook, M.L.; Gasser, G.; Mindt, T.L. An overview of PET radiochemistry, part 2: Radiometals. *J. Nucl. Med.* **2018**, *59*, 1500–1506. [CrossRef]
24. Poschenrieder, A.; Schottelius, M.; Schwaiger, M.; Kessler, H.; Wester, H.-J. The influence of different metal-chelate conjugates of pentixafor on the CXCR4 affinity. *EJNMMI Res.* **2016**, *6*, 36. [CrossRef]
25. Ahn, B.C. Personalized medicine based on theranostic radioiodine molecular imaging for differentiated thyroid cancer. *BioMed Res. Int.* **2016**, *2016*, 1680464. [CrossRef]
26. Miller, C.; Rousseau, J.; Ramogida, C.F.; Celler, A.; Rahmim, A.; Uribe, C.F. Implications of physics, chemistry and biology for dosimetry calculations using theranostic pairs. *Theranostics* **2022**, *12*, 232. [CrossRef]
27. Ehlerding, E.B.; Ferreira, C.A.; Aluicio-Sarduy, E.; Jiang, D.; Lee, H.J.; Theuer, C.P.; Engle, J.W.; Cai, W. $^{86/90}$Y-based theranostics targeting angiogenesis in a murine breast cancer model. *Mol. Pharm.* **2018**, *15*, 2606–2613. [CrossRef]
28. Ferreira, C.A.; Ehlerding, E.B.; Rosenkrans, Z.T.; Jiang, D.; Sun, T.; Aluicio-Sarduy, E.; Engle, J.W.; Ni, D.; Cai, W. $^{86/90}$Y-Labeled monoclonal antibody targeting tissue factor for pancreatic cancer theranostics. *Mol. Pharm.* **2020**, *17*, 1697–1705. [CrossRef]
29. Ming-qi, L.I.; Qi-min, D.; Zuo-yong, C.; Mao-liang, L.I. Production and application of medical radionuclide: Status and urgent problems to be resolved in China. *J. Isot.* **2013**, *26*, 186. (In Chinese)
30. Mushtaq, A. Reactors are indispensable for radioisotope production. *Ann. Nucl. Med.* **2010**, *24*, 759–760. [CrossRef]
31. Xoubi, N.; Primm, R.T., III. *Modeling of the High Flux Isotope Reactor Cycle 400*; ORNL/TM-2004/251; Oak Ridge National Laboratory: Oak Ridge, Tennessee, USA, 2005.
32. Ruth, T.J. The medical isotope crisis: How we got here and where we are going. *J. Nucl. Med. Technol.* **2014**, *42*, 245–248. [CrossRef]
33. Koleška, M.; Lahodová, Z.; Šoltés, J.; Viererbl, L.; Ernest, J.; Vinš, M.; Stehno, J. Capabilities of the LVR-15 research reactor for production of medical and industrial radioisotopes. *J. Radioanal. Nucl. Chem.* **2015**, *305*, 51–59. [CrossRef]
34. OECD-NEA. *The Supply of Medical Radioisotopes: 2019 Medical Isotope Supply and Capacity Projection for the 2019–2024 Period*; OECD-NEA: Paris, France, 2019.
35. Gao, F.; Lin, L.; Liu, Y.; Ma, X. Production situation and technology prospect of medical isotopes. *J. Isot.* **2016**, *29*, 116–120. (In Chinese)
36. Kurenkov, N.V.; Shubin, Y.N. Radionuclides for nuclear medicine. Медицинская Радиология И Радиационная Безопасность **1996**, *41*, 54–63.
37. IAEA. *Nuclear Research Reactors in the World*; IAEA: New York, NY, USA, 1997; 120p, ISBN 92-0-100298-X.
38. Hoedl, S.A.; Updegraff, W.D. The production of medical isotopes without nuclear reactors or uranium enrichment. *Sci. Glob. Secur.* **2015**, *23*, 121–153. [CrossRef]
39. Van der Keur, H. Medical radioisotopes production without a nuclear reactor. 2010. Available online: http://www.laka.org/info/publicaties/2010-medical_isotopes.pdf (accessed on 20 June 2022).
40. Ziwei, L.; Yuncheng, H.; Xiaoyu, W.; Jiachen, Z.; Yongfeng, W.; Qunying, H. Production Status and Technical Prospects of Medical Radioisotope 99 Mo/99m Tc. *Nucl. Phys. Rev.* **2019**, *36*, 170–183. (In Chinese)
41. Starovoitova, V.N.; Tchelidze, L.; Wells, D.P. Production of medical radioisotopes with linear accelerators. *Appl. Radiat. Isot.* **2014**, *85*, 39–44. [CrossRef] [PubMed]
42. Kaur, C.D.; Mishra, K.K.; Sahu, A.; Panik, R.; Kashyap, P.; Mishra, S.P.; Kumar, A. Theranostics: New era in nuclear medicine and radiopharmaceuticals. In *Medical Isotopes*; Naqvi, S.A.R., Imrani, M.B., Eds.; IntechOpen: London, UK, 2020.
43. Zhang, T.; Fan, M.; Wei, S.; Chen, S.; Yang, F. The present situation and the prospect of medical cyclotrons in China. *Sci. China Phys. Mech. Astron.* **2011**, *54*, 260–265. [CrossRef]
44. Sunderland, J.; Erdahl, C.; Bender, B.; Sensoy, L.; Watkins, G. Considerations, measurements and logistics associated with low-energy cyclotron decommissioning. In Proceedings of the AIP Conference Proceedings, Playa del Carmen, México, 26–29 August 2012; American Institute of Physics: New York, NY, USA, 2012; Volume 1509, pp. 16–20.
45. International Atomic Energy Agency. *Alternative Radionuclide Production with a Cyclotron*; IAEA Radioisotopes and Radiopharmaceuticals Reports No. 4; IAEA: Vienna, Austria, 2021.
46. Chernyaev, A.P.; Varzar, S.M. Particle accelerators in modern world. *Phys. At. Nucl.* **2014**, *77*, 1203–1215. [CrossRef]
47. Goethals, P.E.; Zimmermann, R.G. *Cyclotrons used in Nuclear Medicine World Market Report & Directory*; MEDraysintell: Louvain-la-Neuve, Belgium, 2015.
48. Available online: https://www.machinedesign.com/learning-resources/whats-the-difference-between/article/21832184/what-are-the-differences-between-linear-accelerators-cyclotrons-and-synchrotrons (accessed on 1 June 2022).

49. Available online: https://www.iaea.org/newscenter/news/cyclotrons-what-are-they-and-where-can-you-find-them (accessed on 10 July 2022).
50. Available online: https://www.iaea.org/sites/default/files/gc/gc65-inf2.pdf (accessed on 10 July 2022).
51. Available online: https://www.iaea.org/sites/default/files/gc/gc64-inf2.pdf (accessed on 10 July 2022).
52. Available online: https://www.iaea.org/sites/default/files/gc/gc61inf-4_en.pdf (accessed on 10 July 2022).
53. Available online: https://www.iaea.org/sites/default/files/ntr2015.pdf (accessed on 10 July 2022).
54. Available online: https://www.iaea.org/sites/default/files/ntr2014.pdf (accessed on 10 July 2022).
55. Available online: https://www-legacy.iaea.org/OurWork/ST/NE/Pess/assets/13-25751_rep_ntr_2013_web.pdf (accessed on 10 July 2022).
56. Available online: https://www-legacy.iaea.org/OurWork/ST/NE/Pess/assets/ntr2012_web.pdf (accessed on 10 July 2022).
57. Available online: https://www-legacy.iaea.org/OurWork/ST/NE/Pess/assets/ntr2011.pdf (accessed on 10 July 2022).
58. Chao, A.W.; Chou, W. (Eds.) *Reviews of Accelerator Science and Technology-Volume 3: Accelerators as Photon Sources*; World Scientific: Chiyoda City, Tokyo, Japan, 2011.
59. Leo, K.W.K.; Hashim, S. Accelerator Selection for Industry and Medical Applications. (This is a preprint article, it offers immediate access but has not been peer reviewed).
60. Synowiecki, M.A.; Perk, L.R.; Nijsen, J.F.W. Production of novel diagnostic radionuclides in small medical cyclotrons. *EJNMMI Radiopharm. Chem.* **2018**, *3*, 3. [CrossRef]
61. Zuoyuan, D.; Bin, W. *Securities Research Report-In-Depth Discussion Series-Nuclear Medicine*; Pacific Securities: Guangdong, China, 2019.
62. Yuan, Z. FDA approved radiopharmaceuticals. In *Foreign Medical Sciences*; Section of Radiation Medicine and Nuclear Medicine: Tianjin, China, 2000; Volume 24, pp. 161–163, ISSN 1001-098X.
63. Racette, B.A.; Antenor, J.A.; McGee-Minnich, L.; Moerlein, S.M.; Videen, T.O.; Kotagal, V.; Perlmutter, J.S. [$^{18}$F] FDOPA PET and clinical features in parkinsonism due to manganism. *Mov. Disord.* **2005**, *20*, 492–496. [CrossRef]
64. Zhao, P.; Zhang, B.; Gao, S.; Li, X. Clinical features, MRI, and $^{18}$F-FDG-PET in differential diagnosis of Parkinson disease from multiple system atrophy. *Brain Behav.* **2020**, *10*, e01827. [CrossRef]
65. Rahmim, A.; Zaidi, H. PET versus SPECT: Strengths, limitations and challenges. *Nucl. Med. Commun.* **2008**, *29*, 193–207. [CrossRef]
66. Ruth, T.J.; Wolf, A.P. Absolute cross sections for the production of $^{18}$F via the $^{18}$O (p, n) $^{18}$F reaction. *Radiochim. Acta* **1979**, *26*, 21–24. [CrossRef]
67. Hess, E.; Blessing, G.; Coenen, H.H.; Qaim, S.M. Improved target system for production of high purity [18F] fluorine via the $^{18}$O (p, n) $^{18}$F reaction. *Appl. Radiat. Isot.* **2000**, *52*, 1431–1440. [CrossRef]
68. Kambali, I.; Parwanto; Suryanto, H.; Huda, N.; Listiawadi, F.D.; Astarina, H.; Ismuha, R.R.; Kardinah. Dependence of $^{18}$F Production Yield and Radioactive Impurities on Proton Irradiation Dose. *Phys. Res. Int.* **2017**, *2017*, 2124383. [CrossRef]
69. P Perini, E.A.; Skopchenko, M.; Hong, T.T.; Harianto, R.; Maître, A.; Rodríguez, M.R.R.; de Oliveira Santos, N.; Guo, Y.; Qin, X.; Zeituni, C.A.; et al. Pre-feasibility study for establishing radioisotope and radiopharmaceutical production facilities in developing countries. *Curr. Radiopharm.* **2019**, *12*, 187–200. [CrossRef]
70. Barnhart, T.E.; Nickles, R.J.; Roberts, A.D. Revisiting Low Energy Deuteron Production of [18F] Fluoride and Fluorine for PET. In Proceedings of the AIP Conference Proceedings, Denton, Texas, USA, 12-16 November 2002; American Institute of Physics: New York, NY, USA, 2003; Volume 680, pp. 1086–1089.
71. Haug, A.R.; Cindea-Drimus, R.; Auernhammer, C.J.; Reincke, M.; Wängler, B.; Uebleis, C.; Schmidt, G.P.; Göke, B.; Bartenstein, P.; Hacker, M. The role of $^{68}$Ga-DOTATATE PET/CT in suspected neuroendocrine tumors. *J. Nucl. Med.* **2012**, *53*, 1686–1692. [CrossRef]
72. Kręcisz, P.; Czarnecka, K.; Królicki, L.; Mikiciuk-Olasik, E.b.; Szymański, P. Radiolabeled peptides and antibodies in medicine. *Bioconjugate Chem.* **2020**, *32*, 25–42. [CrossRef]
73. Vaughn, B.A. *Chelation Approaches for the Theranostic Radioisotopes of Copper, Scandium and Lutetium*; State University of New York at Stony Brook: York, NE, USA, 2021.
74. Krebs, S.; O'Donoghue, J.A.; Biegel, E.; Beattie, B.J.; Reidy, D.; Lyashchenko, S.K.; Lewis, J.S.; Bodei, L.; Weber, W.A.; Pandit-Taskar, N. Comparison of $^{68}$Ga-DOTA-JR11 PET/CT with dosimetric $^{177}$Lu-satoreotide tetraxetan ($^{177}$Lu-DOTA-JR11) SPECT/CT in patients with metastatic neuroendocrine tumors undergoing peptide receptor radionuclide therapy. *Eur. J. Nucl. Med. Mol. Imaging* **2020**, *47*, 3047–3057. [CrossRef]
75. Maffey-Steffan, J.; Scarpa, L.; Svirydenka, A.; Nilica, B.; Mair, C.; Buxbaum, S.; Bektic, J.; von Guggenberg, E.; Uprimny, C.; Horninger, W.; et al. The $^{68}$Ga/$^{177}$Lu-theragnostic concept in PSMA-targeting of metastatic castration–resistant prostate cancer: Impact of post-therapeutic whole-body scintigraphy in the follow-up. *Eur. J. Nucl. Med. Mol. Imaging* **2020**, *47*, 695–712. [CrossRef]
76. Scarpa, L.; Buxbaum, S.; Kendler, D.; Fink, K.; Bektic, J.; Gruber, L.; Decristoforo, C.; Uprimny, C.; Lukas, P.; Horninger, W.; et al. The $^{68}$Ga/$^{177}$Lu theragnostic concept in PSMA targeting of castration-resistant prostate cancer: Correlation of SUVmax values and absorbed dose estimates. *Eur. J. Nucl. Med. Mol. Imaging* **2017**, *44*, 788–800. [CrossRef]
77. Sartor, O.; Herrmann, K. Prostate Cancer Treatment: $^{177}$Lu-PSMA-617 Considerations, Concepts, and Limitations. *J. Nucl. Med.* **2022**, *63*, 823–829. [CrossRef]
78. Velikyan, I. $^{68}$Ga-based radiopharmaceuticals: Production and application relationship. *Molecules* **2015**, *20*, 12913–12943. [CrossRef]

79. Lin, M.; Waligorski, G.J.; Lepera, C.G. Production of curie quantities of $^{68}$Ga with a medical cyclotron via the $^{68}$Zn (p, n) $^{68}$Ga reaction. *Appl. Radiat. Isot.* **2018**, *133*, 1–3. [CrossRef]
80. Razbash, A.A.; Sevastianov, Y.u.G.; Krasnov, N.N.; Leonov, A.I.; Pavlekin, V.E. Germanium-68 row of products. In Proceedings of the 5th International Conference on Isotopes, 5ICI, Brussels, Belgium, 25–29 April 2005; Medimond: Bologna, Italy; p. 147.
81. Rösch, F. Past, present and future of $^{68}$Ge/$^{68}$Ga generators. *Appl. Radiat. Isot.* **2013**, *76*, 24–30. [CrossRef]
82. Engle, J.; Lopez-Rodriguez, V.; Gaspar-Carcamo, R.; Valdovinos, H.; Valle-Gonzalez, M.; Trejo-Ballado, F.; Severin, G.W.; Barnhart, T.; Nickles, R.; Avila-Rodriguez, M.A. Very high specific activity $^{66/68}$Ga from zinc targets for PET. *Appl. Radiat. Isot.* **2012**, *70*, 1792–1796. [CrossRef]
83. Sadeghi, M.; Kakavand, T.; Rajabifar, S.; Mokhtari, L.; Rahimi-Nezhad, A. Cyclotron production of 68Ga via proton-induced reaction on 68Zn target. *Nukleonika* **2009**, *54*, 25–28.
84. Nelson, B.J.; Wilson, J.; Richter, S.; Duke, M.J.M.; Wuest, M.; Wuest, F. Taking cyclotron $^{68}$Ga production to the next level: Expeditious solid target production of $^{68}$Ga for preparation of radiotracers. *Nucl. Med. Biol.* **2020**, *80*, 24–31. [CrossRef]
85. Mardon, A.; Saleem, H.; Parish, G.; Syed, M.; Inayat, E.; Henry, J.; Heinen, L.; Amiscaray, D.E.; Dong, F.; Mak, E.; et al. *What in the World is Medical Isotope Production?* Golden Meteorite Press: Edmonton, Alberta, Canada, 2021.
86. Pandey, M.K.; Byrne, J.F.; Jiang, H.; Packard, A.B.; DeGrado, T.R. Cyclotron production of $^{68}$Ga via the $^{68}$Zn (p, n) $^{68}$Ga reaction in aqueous solution. *Am. J. Nucl. Med. Mol. Imaging* **2014**, *4*, 303.
87. Keinänen, O.; Fung, K.; Brennan, J.M.; Zia, N.; Harris, M.; van Dam, E.; Biggin, C.; Hedt, A.; Stoner, J.; Donnelly, P.S.; et al. Harnessing $^{64}$Cu/$^{67}$Cu for a theranostic approach to pretargeted radioimmunotherapy. *Proc. Natl. Acad. Sci. USA* **2020**, *117*, 28316–28327. [CrossRef]
88. Pfeifer, A.; Knigge, U.; Mortensen, J.; Oturai, P.; Berthelsen, A.K.; Loft, A.; Binderup, T.; Rasmussen, P.; Elema, D.; Klausen, T.L.; et al. Clinical PET of neuroendocrine tumors using $^{64}$Cu-DOTATATE: First-in-humans study. *J. Nucl. Med.* **2012**, *53*, 1207–1215. [CrossRef]
89. Avila-Rodriguez, M.A.; Nye, J.A.; Nickles, R.J. Simultaneous production of high specific activity $^{64}$Cu and $^{61}$Co with 11.4 MeV protons on enriched $^{64}$Ni nuclei. *Appl. Radiat. Isot.* **2007**, *65*, 1115–1120. [CrossRef]
90. Szelecsényi, F.; Kovács, Z.; Nagatsu, K.; Zhang, M.R.; Suzuki, K. Excitation function of (p, α) nuclear reaction on enriched $^{67}$Zn: Possibility of production of $^{64}$Cu at low energy cyclotron. *Radiochim. Acta* **2014**, *102*, 465–472. [CrossRef]
91. Obata, A.; Kasamatsu, S.; McCarthy, D.W.; Welch, M.J.; Saji, H.; Yonekura, Y.; Fujibayashi, Y. Production of therapeutic quantities of $^{64}$Cu using a 12 MeV cyclotron. *Nucl. Med. Biol.* **2003**, *30*, 535–539. [CrossRef]
92. McCarthy, D.W.; Shefer, R.E.; Klinkowstein, R.E.; Bass, L.A.; Margeneau, W.H.; Cutler, C.S.; Anderson, C.J.; Welch, M.J. Efficient production of high specific activity $^{64}$Cu using a biomedical cyclotron. *Nucl. Med. Biol.* **1997**, *24*, 35–43. [CrossRef]
93. Hilgers, K.; Stoll, T.; Skakun, Y.; Coenen, H.H.; Qaim, S.M. Cross-section measurements of the nuclear reactions natZn (d, x) $^{64}$Cu, $^{66}$Zn (d, α) $^{64}$Cu and $^{68}$Zn (p, αn) $^{64}$Cu for production of $^{64}$Cu and technical developments for small-scale production of $^{67}$Cu via the $^{70}$Zn (p, α) $^{67}$Cu process. *Appl. Radiat. Isot.* **2003**, *59*, 343–351. [CrossRef]
94. Elomaa, V.V.; Jurttila, J.; Rajander, J.; Solin, O. Automation of $^{64}$Cu production at Turku PET Centre. *Appl. Radiat. Isot.* **2014**, *89*, 74–78. [CrossRef]
95. Zeisler, S.K.; Pavan, R.A.; Orzechowski, J.; Langlois, R.; Rodrigue, S.; Van Lier, J.E. Production of $^{64}$Cu on the Sherbrooke TR-PET cyclotron. *J. Radioanal. Nucl. Chem.* **2003**, *257*, 175–177. [CrossRef]
96. Thieme, S.; Walther, M.; Pietzsch, H.J.; Henniger, J.; Preusche, S.; Mäding, P.; Steinbach, J. Module-assisted preparation of $^{64}$Cu with high specific activity. *Appl. Radiat. Isot.* **2012**, *70*, 602–608. [CrossRef]
97. Van der Meulen, N.P.; Hasler, R.; Blanc, A.; Farkas, R.; Benešová, M.; Talip, Z.; Müller, C.; Schibli, R. Implementation of a new separation method to produce qualitatively improved $^{64}$Cu. *J. Label. Compd. Radiopharm.* **2019**, *62*, 460–470. [CrossRef]
98. Xie, Q.; Zhu, H.; Wang, F.; Meng, X.; Ren, Q.; Xia, C.; Yang, Z. Establishing reliable Cu-64 production process: From target plating to molecular specific tumor micro-PET imaging. *Molecules* **2017**, *22*, 641. [CrossRef]
99. Ohya, T.; Nagatsu, K.; Suzuki, H.; Fukada, M.; Minegishi, K.; Hanyu, M.; Fukumura, T.; Zhang, M.-R. Efficient preparation of high-quality $^{64}$Cu for routine use. *Nucl. Med. Biol.* **2016**, *43*, 685–691. [CrossRef]
100. Verel, I.; Visser GW, M.; Boellaard, R.; Stigter-van Walsum, M.; Snow, G.B.; Van Dongen, G.A. $^{89}$Zr immuno-PET: Comprehensive procedures for the production of $^{89}$Zr-labeled monoclonal antibodies. *J. Nucl. Med.* **2003**, *44*, 1271–1281.
101. Vento, J.; Mulgaonkar, A.; Woolford, L.; Nham, K.; Christie, A.; Bagrodia, A.; de Leon, A.D.; Hannan, R.; Bowman, I.; McKay, R.M.; et al. PD-L1 detection using $^{89}$Zr-atezolizumab immuno-PET in renal cell carcinoma tumorgrafts from a patient with favorable nivolumab response. *J. Immunother. Cancer* **2019**, *7*, 144. [CrossRef]
102. Taghilo, M.; Kakavand, T.; Rajabifar, S.; Sarabadani, P. Cyclotron production of $^{89}$Zr: A potent radionuclide for positron emission tomography. *Int. J. Phys. Sci.* **2012**, *7*, 1321–1325. [CrossRef]
103. Ciarmatori, A.; Cicoria, G.; Pancaldi, D.; Infantino, A.; Boschi, S.; Fanti, S.; Marengo, M. Some experimental studies on $^{89}$Zr production. *Radiochim. Acta* **2011**, *99*, 631–634. [CrossRef]
104. Sadeghi, M.; Enferadi, M.; Bakhtiari, M. Accelerator production of the positron emitter zirconium-89. *Ann. Nucl. Energy* **2012**, *41*, 97–103. [CrossRef]
105. Tang, Y.; Li, S.; Yang, Y.; Chen, W.; Wei, H.; Wang, G.; Yang, J.; Liao, J.; Luo, S.; Liu, N. A simple and convenient method for production of $^{89}$Zr with high purity. *Appl. Radiat. Isot.* **2016**, *118*, 326–330. [CrossRef] [PubMed]

106. Deri, M.A.; Zeglis, B.M.; Francesconi, L.C.; Lewis, J.S. PET imaging with $^{89}$Zr: From radiochemistry to the clinic. *Nucl. Med. Biol.* **2013**, *40*, 3–14. [CrossRef]
107. Kandil, S.A.; Spahn, I.; Scholten, B.; Saleh, Z.A.; Saad, S.M.M.; Coenen, H.H.; Qaim, S.M. Excitation functions of (α, xn) reactions on natRb and natSr from threshold up to 26 MeV: Possibility of production of $^{87}$Y, $^{88}$Y and $^{89}$Zr. *Appl. Radiat. Isot.* **2007**, *65*, 561–568. [CrossRef]
108. Liqiang, L.; Feng, W.; Teli, L.; Hua, Z.; Zhi, Y. Production of Iodine-124 and Its Application in PET Molecular Imaging. *J. Isot.* **2018**, *31*, 188. (In Chinese)
109. Freudenberg, L.S.; Jentzen, W.; Stahl, A.; Bockisch, A.; Rosenbaum-Krumme, S.J. Clinical applications of $^{124}$I-PET/CT in patients with differentiated thyroid cancer. *Eur. J. Nucl. Med. Mol. Imaging* **2011**, *38*, 48–56. [CrossRef]
110. Aboian, M.S.; Huang, S.-y.; Hernandez-Pampaloni, M.; Hawkins, R.A.; VanBrocklin, H.F.; Huh, Y.; Vo, K.T.; Gustafson, W.C.; Matthay, K.K.; Seo, Y. $^{124}$I-MIBG PET/CT to monitor metastatic disease in children with relapsed neuroblastoma. *J. Nucl. Med.* **2021**, *62*, 43–47. [CrossRef]
111. Lewis, J.S. *Production, Use and Applications of* $^{124}$*I*. PowerPoint Slides; Memorial–Sloan Kettering Cancer Center: New York, NY, USA, 2020.
112. Braghirolli AM, S.; Waissmann, W.; da Silva, J.B.; dos Santos, G.R. Production of iodine-124 and its applications in nuclear medicine. *Appl. Radiat. Isot.* **2014**, *90*, 138–148. [CrossRef]
113. Bzowski, P.; Borys, D.; Gorczewski, K.; Chmura, A.; Daszewska, K.; Gorczewska, I.; Kastelik-Hryniewiecka, A.; Szydło, M.; d'Amico, A.; Sokół, M. Efficiency of $^{124}$I radioisotope production from natural and enriched tellurium dioxide using $^{124}$Te (p, xn) $^{124}$I reaction. *EJNMMI Phys.* **2022**, *9*, 41. [CrossRef]
114. Bastian, T.; Coenen, H.H.; Qaim, S.M. Excitation functions of $^{124}$Te (d, xn) $^{124,125}$I reactions from threshold up to 14 MeV: Comparative evaluation of nuclear routes for the production of $^{124}$I. *Appl. Radiat. Isot.* **2001**, *55*, 303–308. [CrossRef]
115. Payolla, F.B.; Massabni, A.C.; Orvig, C. Radiopharmaceuticals for diagnosis in nuclear medicine: A short review. *Eclética Química* **2019**, *44*, 11–19.
116. Schaffer, P.; Bénard, F.; Bernstein, A.; Buckley, K.; Celler, A.; Cockburn, N.; Corsaut, J.; Dodd, M.; Economou, C.; Eriksson, T.; et al. Direct production of $^{99m}$Tc via $^{100}$Mo (p, 2n) on small medical cyclotrons. *Phys. Procedia* **2015**, *66*, 383–395. [CrossRef]
117. Takacs, S.; Hermanne, A.; Ditroi, F.; Tárkányi, F.; Aikawa, M. Reexamination of cross sections of the $^{100}$Mo (p, 2n) $^{99m}$Tc reaction. *Nucl. Instrum. Methods Phys. Res. Sect. B Beam Interact. Mater. At.* **2015**, *347*, 26–38. [CrossRef]
118. Scholten, B.; Lambrecht, R.M.; Cogneau, M.; Ruiz, H.V.; Qaim, S.M. Excitation functions for the cyclotron production of $^{99m}$Tc and $^{99}$Mo. *Appl. Radiat. Isot.* **1999**, *51*, 69–80. [CrossRef]
119. Rodrigue, S.; van Lier, J.E.; van Lier, M.A.S.E. Cyclotron production of $^{99m}$Tc: An approach to the medical isotope crisis. *J. Nucl. Med.* **2010**, *51*, 13N.
120. Hoehr, C.; Bénard, F.; Buckley, K.; Crawford, J.; Gottberg, A.; Hanemaayer, V.; Kunz, P.; Ladouceur, K.; Radchenko, V.; Ramogida, C.; et al. Medical isotope production at TRIUMF–from imaging to treatment. *Phys. Procedia* **2017**, *90*, 200–208. [CrossRef]
121. Pillai, M.R.A.; Dash, A.; Knapp, F.F.R. Sustained availability of $^{99m}$Tc: Possible paths forward. *J. Nucl. Med.* **2013**, *54*, 313–323. [CrossRef]
122. Lebeda, O.; van Lier, E.J.; Štursa, J.; Rális, J.; Zyuzin, A. Assessment of radionuclidic impurities in cyclotron produced $^{99m}$Tc. *Nucl. Med. Biol.* **2012**, *39*, 1286–1291. [CrossRef]
123. Pupillo, G.; Esposito, J.; Gambaccini, M.; Haddad, F.; Michel, N. Experimental cross section evaluation for innovative $^{99}$Mo production via the (α, n) reaction on $^{96}$Zr target. *J. Radioanal. Nucl. Chem.* **2014**, *302*, 911–917. [CrossRef]
124. Hagiwara, M.; Yashima, H.; Sanami, T.; Yonai, S. Measurement of the excitation function of $^{96}$Zr (α, n) $^{99}$Mo for an alternative production source of medical radioisotopes. *J. Radioanal. Nucl. Chem.* **2018**, *318*, 569–573. [CrossRef]
125. Jacobson, A.F.; Deng, H.; Lombard, J.; Lessig, H.J.; Black, R.R. $^{123}$I-meta-iodobenzylguanidine scintigraphy for the detection of neuroblastoma and pheochromocytoma: Results of a meta-analysis. *J. Clin. Endocrinol. Metab.* **2010**, *95*, 2596–2606. [CrossRef]
126. Eslami, M.; Kakavand, T.; Mirzaii, M. Simulation of Proton beam using the MCNPX code; A prediction for the production of $^{123}$I via $^{124}$Xe (p, x) $^{123}$I reaction. In Proceedings of the DAE-BRNS symposium on nuclear physics, Mumbai, India, 2–6 December 2013; Volume 58, p. 860.
127. EXFOR. Experimental Nuclear Reaction Data. 2011. Available online: http://www-nds.iaea.org/exfor (accessed on 15 June 2022).
128. Kakavand, T.; Sadeghi, M.; Kamali Moghaddam, K.; Shokri Bonab, S.; Fateh, B. Computer simulation techniques to design Xenon-124 solid target for iodine-123 production. *Iran. J. Radiat. Res.* **2008**, *5*, 207–212.
129. Tárkányi, F.; Qaim, S.M.; Stöcklin, G.; Sajjad, M.; Lambrecht, R.M.; Schweickert, H. Excitation functions of (p, 2n) and (p, pn) reactions and differential and integral yields of $^{123}$I in proton induced nuclear reactions on highly enriched $^{124}$Xe. *Int. J. Radiat. Appl. Instrumentation. Part A Appl. Radiat. Isot.* **1991**, *42*, 221–228. [CrossRef]
130. Hupf, H.B.; Beaver, J.E.; Armbruster, J.M.; Pendola, J.P. Production of ultra-pure I-123 from the $^{123}$Te (p, n) $^{123}$I reaction. *AIP Conf Proc* **2001**, *576*, 845–848.
131. Mertens, J. New Development in Radio-Iodinated Radiopharmaceuticals for SPECT and Radionuclide Therapy. In Proceedings of the IAEA-CN-130 International Symposium on Trends in Radiopharmaceuticals ISTR-2005, Vienna, Austria, 14–18 November 2005; IAEA: New York, NY, USA; pp. 101–103.

132. Scholten, B.; Qaim, S.M.; Stöcklin, G. Excitation functions of proton induced nuclear reactions on natural tellurium and enriched $^{123}$Te: Production of $^{123}$I via the $^{123}$Te (p, n) $^{123}$I-process at a low-energy cyclotron. *Int. J. Radiat. Appl. Instrumentation. Part A Appl. Radiat. Isot.* **1989**, *40*, 127–132. [CrossRef]
133. Kratochwil, C.; Haberkorn, U.; Giesel, F.L. 225Ac-PSMA-617 for therapy of prostate cancer. In *Seminars in Nuclear Medicine*; WB Saunders: Philadelphia, PA, USA, 2020; Volume 50, pp. 133–140.
134. Kratochwil, C.; Bruchertseifer, F.; Giesel, F.L.; Weis, M.; Verburg, F.A.; Mottaghy, F.; Kopka, K.; Apostolidis, C.; Haberkorn, U.; Morgenstern, A. $^{225}$Ac-PSMA-617 for PSMA-targeted α-radiation therapy of metastatic castration-resistant prostate cancer. *J. Nucl. Med.* **2016**, *57*, 1941–1944. [CrossRef]
135. Zacherl, M.J.; Gildehaus, F.J.; Mittlmeier, L.; Böning, G.; Gosewisch, A.; Wenter, V.; Unterrainer, M.; Schmidt-Hegemann, N.; Belka, C.; Kretschmer, A. First clinical results for PSMA-targeted α-therapy using $^{225}$Ac-PSMA-I&T in advanced-mCRPC patients. *J. Nucl. Med.* **2021**, *62*, 669–674.
136. Królicki, L.; Kunikowska, J.; Bruchertseifer, F.; Koziara, H.; Królicki, B.; Jakuciński, M.; Pawlak, D.; Rola, R.; Morgenstern, A.; Rosiak, E.; et al. 225Ac-and 213Bi-substance P analogues for glioma therapy. In *Seminars in Nuclear Medicine*; WB Saunders: Philadelphia, PA, USA, 2020; Volume 50, pp. 141–151.
137. Nagatsu, K.; Suzuki, H.; Fukada, M.; Ito, T.; Ichinose, J.; Honda, Y.; Minegishi, K.; Higashi, T.; Zhang, M.-R. Cyclotron production of $^{225}$Ac from an electroplated $^{226}$Ra target. *Eur. J. Nucl. Med. Mol. Imaging* **2021**, *49*, 279–289. [CrossRef]
138. Lee, K.C. 225Ac production at KIRAMS. In Proceedings of the IAEA Workshop on the Supply of 225Ac, Vienna, Austria, 9–10 October 2018, (unpublished).
139. Bruchertseifer, F.; Kellerbauer, A.; Malmbeck, R.; Morgenstern, A. Targeted alpha therapy with bismuth-213 and actinium-225: Meeting future demand. *J. Label. Compd. Radiopharm.* **2019**, *62*, 794–802. [CrossRef]
140. Ermolaev, S.; Zhuikov, B.; Kokhanyuk, V.; Matushko, V.; Kalmykov, S.N.; Aliev, R.A.; Tananaev, I.G.; Myasoedov, B.F. Production of actinium, thorium and radium isotopes from natural thorium irradiated with protons up to 141 MeV. *Radiochim. Acta* **2012**, *100*, 223–229. [CrossRef]
141. Chen, D.; Liu, W.; Huang, Q.; Cao, S.; Tian, W.; Yin, X.; Tan, C.; Wang, J.; Chu, J.; Jia, Z.; et al. Accelerator Production of the Medical Isotope $^{211}$At and Monoclonal Antibody Labeling. *Acta Chim. Sin.* **2021**, *79*, 1376–1384. [CrossRef]
142. Watabe, T.; Kaneda-Nakashima, K.; Shirakami, Y.; Liu, Y.; Ooe, K.; Teramoto, T.; Toyoshima, A.; Shimosegawa, E.; Nakano, T.; Kanai, Y.; et al. Targeted alpha therapy using astatine ($^{211}$At)-labeled phenylalanine: A preclinical study in glioma bearing mice. *Oncotarget* **2020**, *11*, 1388. [CrossRef]
143. Zalutsky, M.R.; Reardon, D.A.; Akabani, G.; Coleman, R.E.; Friedman, A.H.; Friedman, H.S.; McLendon, R.E.; Wong, T.Z.; Bigner, D.D. Clinical experience with α-particle–emitting $^{211}$At: Treatment of recurrent brain tumor patients with $^{211}$At-labeled chimeric antitenascin monoclonal antibody 81C6. *J. Nucl. Med.* **2008**, *49*, 30–38. [CrossRef] [PubMed]
144. Lindegren, S.; Albertsson, P.; Bäck, T.; Jensen, H.; Palm, S.; Aneheim, E. Realizing clinical trials with astatine-211: The chemistry infrastructure. *Cancer Biother. Radiopharm.* **2020**, *35*, 425–436. [CrossRef]
145. Cederkrantz, E.; Andersson, H.; Bernhardt, P.; Bäck, T.; Hultborn, R.; Jacobsson, L.; Jensen, H.; Lindegren, S.; Ljungberg, M.; Magnander, T.; et al. Absorbed doses and risk estimates of $^{211}$At-MX35 F (ab′) 2 in intraperitoneal therapy of ovarian cancer patients. *Int. J. Radiat. Oncol. Biol. Phys.* **2015**, *93*, 569–576. [CrossRef]
146. Washiyama, K.; Oda, T.; Sasaki, S.; Aoki, M.; Gomez, F.L.G.; Taniguchi, M.; Nishijima, K.-i.; Takahashi, K. At-211 production using the CYPRIS MP-30. *J. Med. Imaging Radiat. Sci.* **2019**, *50*, S42. [CrossRef]
147. Alfarano, A.; Abbas, K.; Holzwarth, U.; Bonardi, M.; Groppi, F.; Alfassi, Z.; Menapace, E.; Gibson, P. Thick target yield measurement of $^{211}$At through the nuclear reaction $^{209}$Bi (α, 2n). In *Journal of Physics: Conference Series*; IOP Publishing: Bristol, UK, 2006; Volume 41, p. 009.
148. Feng, Y.; Zalutsky, M.R. Production, purification and availability of 211At: Near term steps towards global access. *Nucl. Med. Biol.* **2021**, *100*, 12–23. [CrossRef]
149. Guérard, F.; Gestin, J.F.; Brechbiel, M.W. Production of [$^{211}$At]-astatinated radiopharmaceuticals and applications in targeted α-particle therapy. *Cancer Biother. Radiopharm.* **2013**, *28*, 1–20. [CrossRef]
150. Cullinane, C.; Jeffery, C.M.; Roselt, P.D.; van Dam, E.M.; Jackson, S.; Kuan, K.; Jackson, P.; Binns, D.; van Zuylekom, J.; Harris, M.; et al. Peptide receptor radionuclide therapy with $^{67}$Cu-CuSarTATE is highly efficacious against a somatostatin-positive neuroendocrine tumor model. *J. Nucl. Med.* **2020**, *61*, 1800–1805. [CrossRef]
151. DeNardo, S.J.; DeNardo, G.L.; Kukis, D.L.; Shen, S.; Kroger, L.A.; DeNardo, D.A.; Goldstein, D.S.; Mirick, G.R.; Salako, Q.; Mausner, L.F.; et al. $^{67}$Cu-21T-BAT-Lym-1 pharmacokinetics, radiation dosimetry, toxicity and tumor regression in patients with lymphoma. *J. Nucl. Med.* **1999**, *40*, 302–310.
152. Pupillo, G.; Sounalet, T.; Michel, N.; Mou, L.; Esposito, J.; Haddad, F. New production cross sections for the theranostic radionuclide $^{67}$Cu. *Nucl. Instrum. Methods Phys. Res. Sect. B Beam Interact. Mater. At.* **2018**, *415*, 41–47. [CrossRef]
153. Katabuchi, T.; Watanabe, S.; Ishioka, N.S.; Iida, Y.; Hanaoka, H.; Endo, K.; Matsuhashi, S. Production of $^{67}$Cu via the $^{68}$Zn (p, 2p) $^{67}$Cu reaction and recovery of $^{68}$Zn target. *J. Radioanal. Nucl. Chem.* **2008**, *277*, 467–470. [CrossRef]
154. Mou, L.; Martini, P.; Pupillo, G.; Cieszykowska, I.; Cutler, C.S.; Mikołajczak, R. $^{67}$Cu production capabilities: A mini review. *Molecules* **2022**, *27*, 1501. [CrossRef] [PubMed]
155. Hovhannisyan, G.H.; Stepanyan, A.V.; Saryan, E.R.; Amirakyan, L.A. Methods of Production the Isotope $^{67}$Cu. *J. Contemp. Phys. (Armen. Acad. Sci.)* **2020**, *55*, 183–190. [CrossRef]

156. Kin, T.; Nagai, Y.; Iwamoto, N.; Minato, F.; Iwamoto, O.; Hatsukawa, Y.; Segawa, M.; Harada, H.; Konno, C.; Ochiai, K.; et al. New production routes for medical isotopes $^{64}$Cu and $^{67}$Cu using accelerator neutrons. *J. Phys. Soc. Jpn.* **2013**, *82*, 034201. [CrossRef]
157. Szelecsényi, F.; Steyn, G.F.; Dolley, S.G.; Kovács, Z.; Vermeulen, C.; Van der Walt, T.N. Investigation of the $^{68}$Zn (p, 2p) $^{67}$Cu nuclear reaction: New measurements up to 40 MeV and compilation up to 100 MeV. *Nucl. Instrum. Methods Phys. Res. Sect. B Beam Interact. Mater. At.* **2009**, *267*, 1877–1881. [CrossRef]
158. Pandey, M.K.; DeGrado, T.R. Cyclotron production of PET radiometals in liquid targets: Aspects and prospects. *Curr. Radiopharm.* **2021**, *14*, 325–339. [CrossRef]
159. Deng, X.; Rong, J.; Wang, L.; Vasdev, N.; Zhang, L.; Josephson, L.; Liang, S.H. Chemistry for positron emission tomography: Recent advances in $^{11}$C-, $^{18}$F-, $^{13}$N-, and $^{15}$O-labeling reactions. *Angew. Chem. Int. Ed.* **2019**, *58*, 2580–2605. [CrossRef]
160. McQuade, P.; Rowland, D.J.; Lewis, J.S.; Welch, M.J. Positron-emitting isotopes produced on biomedical cyclotrons. *Curr. Med. Chem.* **2005**, *12*, 807–818. [CrossRef]
161. Schmitz, J. The production of [$^{124}$I] iodine and [$^{86}$Y] yttrium. *Eur. J. Nucl. Med. Mol. Imaging* **2011**, *38*, 4–9. [CrossRef]
162. van der Meulen, N.P.; Bunka, M.; Domnanich, K.A.; Müller, C.; Haller, S.; Vermeulen, C.; Türler, A.; Schibli, R. Cyclotron production of $^{44}$Sc: From bench to bedside. *Nucl. Med. Biol.* **2015**, *42*, 745–751. [CrossRef]
163. van der Meulen, N.P.; Hasler, R.; Talip, Z.; Grundler, P.V.; Favaretto, C.; Umbricht, C.A.; Müller, C.; Dellepiane, G.; Carzaniga, T.S.; Braccini, S. Developments toward the implementation of $^{44}$Sc production at a medical cyclotron. *Molecules* **2020**, *25*, 4706. [CrossRef]
164. Lagunas-Solar, M.C.; Jungerman, J.A.; Paulson, D.W. Cyclotron production of Thallium-201 via the 205 Tl (p, 5n) 201 Pb→ 201 Tl reaction. In Proceedings of the International Symposium on Radiopharmaceuticals, Seattle, WA, USA, 18–23 March 1979; Society of Nuclear Medicine: New York, NY, USA; pp. 779–789.
165. Misiak, R.; Walczak, R.; Wąs, B.; Bartyzel, M.; Mietelski, J.W.; Bilewicz, A. $^{47}$Sc production development by cyclotron irradiation of $^{48}$Ca. *J. Radioanal. Nucl. Chem.* **2017**, *313*, 429–434. [CrossRef]
166. Abel, E.P.; Domnanich, K.; Clause, H.K.; Kalman, C.; Walker, W.; Shusterman, J.A.; Greene, J.; Gott, M.; Severin, G.W. Production, collection, and purification of $^{47}$Ca for the generation of $^{47}$Sc through isotope harvesting at the national superconducting cyclotron laboratory. *ACS Omega* **2020**, *5*, 27864–27872. [CrossRef]
167. Lawrence, J.H. Nuclear physics and therapy: Preliminary report on a new method for the treatment of leukemia and polycythemia. *Radiology* **1940**, *35*, 51–60. [CrossRef]
168. Hupf, H.B.; Beaver, J.E. Cyclotron production of carrier-free gallium-67. *Int. J. Appl. Radiat. Isot.* **1970**, *21*, 75–76. [CrossRef]
169. do Carmo, S.J.C.; Scott, P.J.H.; Alves, F. Production of radiometals in liquid targets. *EJNMMI Radiopharm. Chem.* **2020**, *5*, 2. [CrossRef]
170. Skliarova, H.; Cisternino, S.; Cicoria, G.; Marengo, M.; Cazzola, E.; Gorgoni, G.; Palmieri, V. Medical Cyclotron Solid Target Preparation by Ultrathick Film Magnetron Sputtering Deposition. *Instruments* **2019**, *3*, 21. [CrossRef]
171. McNeil, B.L.; Robertson, A.K.; Fu, W.; Yang, H.; Hoehr, C.; Ramogida, C.F.; Schaffer, P. Production, purification, and radiolabeling of the $^{203}$Pb/$^{212}$Pb theranostic pair. *EJNMMI Radiopharm. Chem.* **2021**, *6*, 6. [CrossRef]
172. Gracheva, N.; Carzaniga, T.S.; Schibli, R.; Braccini, S.; van der Meulen, N.P. $^{165}$Er: A new candidate for Auger electron therapy and its possible cyclotron production from natural holmium targets. *Appl. Radiat. Isot.* **2020**, *159*, 109079. [CrossRef]
173. Nelson, B.J.B.; Wilson, J.; Andersson, J.D.; Wuest, F. High yield cyclotron production of a novel $^{133/135}$La theranostic pair for nuclear medicine. *Sci. Rep.* **2020**, *10*, 22203. [CrossRef] [PubMed]
174. Dey, M.K.; Gupta, A.D.; Chakrabarti, A. Design of ultra-light superconducting proton cyclotron for production of isotopes for medical applications. *Proc. Cyclotr.* **2013**, *2013*, 447–450.
175. Waites, L.H.; Alonso, J.R.; Conrad, J. IsoDAR: A cyclotron-based neutrino source with applications to medical isotope production. *AIP Conf. Proc* **2019**, *2160*, 040001.
176. Waites, L.H.; Alonso, J.; Conrad, J.M.; Koser, D.; Winklehner, D. Tools for the Development and Applications of the IsoDAR Cyclotron. *Energy (MeV/Nucl.)* **2021**, *60*, 30.
177. Pramudita, A. Linacs for medical isotope production. In *Proceeding on the scientific meeting and presentation on accelerator technology and its applications: Physics, nuclear reactor, Yogyakarta, Indonesia, 13 December 2011*; National Nuclear Energy Agency: Tangerang, Jawa Barat, Indonesia, 2012; Volume 47, pp. 11–16.
178. Griswold, J.R.; Medvedev, D.G.; Engle, J.W.; Copping, R.; Fitzsimmons, J.; Radchenko, V.; Cooley, J.; Fassbender, M.; Denton, D.; Murphy, K.; et al. Large scale accelerator production of $^{225}$Ac: Effective cross sections for 78–192 MeV protons incident on $^{232}$Th targets. *Appl. Radiat. Isot.* **2016**, *118*, 366–374. [CrossRef]
179. Zhuikov, B.L.; Ermolaev, S.V. Radioisotope research and development at the Linear Accelerator of the Institute for Nuclear Research of RAS. *Phys.-Uspekhi* **2021**, *64*, 1311. [CrossRef]
180. Antipov, K.; Ayzatsky, M.; Akchurin, Y.I.; Boriskin, V.; Beloglasov, V.; Biller, E.; Demidov, N.; Dikiy, N.; Dovbnya, A.; Dovbush, L.; et al. Electron linacs in NSC KIPT: R&D and application. Вопросы Атомной Науки И Техники **2001**, *37*, 40–47.
181. Danagulyan, A.S.; Hovhannisyan, G.H.; Bakhshiyan, T.M.; Avagyan, R.H.; Avetisyan, A.E.; Kerobyan, I.A.; Dallakyan, R.K. Formation of medical radioisotopes $^{111}$In, $^{117m}$Sn, $^{124}$Sb, and $^{177}$Lu in photonuclear reactions. *Phys. At. Nucl.* **2015**, *78*, 447–452. [CrossRef]
182. Courtney, W.; Sowder, K.; McGyver, M.; Stevenson, N.; Brown, D. The challenges of commercial isotope production on a linear accelerator. *Nucl. Instrum. Methods Phys. Res. Sect. B Beam Interact. Mater. At.* **2007**, *261*, 739–741. [CrossRef]

183. Matthews, M.; Saey, P.; Bowyer, T.; Vandergrift, G.; Cutler, N.R.C.; Ponsard, B.; Mikolajczak, R.; Tsipenyuk, Y.; Solin, L.; Fisher, D.; et al. *Workshop on Signatures of Medical and Industrial Isotope Production: A Review*; Pacific Northwest National Laboratory: Richland, Washington, USA, 2010.
184. Dikiy, N.P.; Dovbnya, A.N.; Medvedyeva, E.P.; Tur, Y.D. Experience of Technetium-99m Generation for Nuclear Medicine on Electron Linac; VANT: Kharkov, Ukraine, 1997; pp. 165–167.
185. Uvarov, V.L.; Dikiy, N.P.; Dovbnya, A.N.; Medvedyeva, Y.P.; Pugachov, G.D.; Tur, Y.D. Electron Accelerator's Based Production of Technetium-99m for Nuclear Medicine. *Bull. Amer. Phys. Soc.* **1997**, *42*, 1338.
186. De Jong, M. Producing medical isotopes using X-rays. *Sin Proc. IPAC* **2012**, *12*, 3177.
187. Chemerisov, S.; Bailey, J.; Heltemes, T.; Jonah, C.; Makarashvili, V.; Tkac, P.; Rotsch, D.; Virgo, M.; Vandegrift, G. *Results of the Six-and-a-Half Day Electron-Accelerator Irradiation of Enriched Mo-100 Targets for the Production of Mo-99*; Argonne National Lab.(ANL): Argonne, IL, USA, 2016.
188. Takeda, T.; Fujiwara, M.; Kurosawa, M.; Takahashi, N.; Tamura, M.; Kawabata, T.; Fujikawa, Y.; Suzuki, K.N.; Abe, N.; Kubota, T.; et al. $^{99m}$Tc production via the (γ, n) reaction on natural Mo. *J. Radioanal. Nucl. Chem.* **2018**, *318*, 811–821. [CrossRef]
189. Szpunar, B.; Rangacharyulu, C.; Date, S.; Ejiri, H. Estimate of production of medical isotopes by photo-neutron reaction at the Canadian light source. *Nucl. Instrum. Methods Phys. Res. Sect. A Accel. Spectrometers Detect. Assoc. Equip.* **2013**, *729*, 41–50. [CrossRef]
190. Babcock, C.; Goodacre, T.D.; Amani, P.; Au, M.; Bricault, P.; Brownell, M.; Cade, B.; Chen, K.; Egoriti, L.; Johnson, J.; et al. Offline target and ion source studies for TRIUMF-ARIEL. *Nucl. Instrum. Methods Phys. Res. Sect. B Beam Interact. Mater. At.* **2020**, *463*, 464–467. [CrossRef]
191. Danon, Y.; Block, R.C.; Testa, R.; Testa, R.; Moore, H. Medical isotope production using a 60 MeV linear electron accelerator. *Trans.-Am. Nucl. Soc.* **2008**, *98*, 894.
192. Yagi, M.; Kondo, K. Preparation of carrier-free $^{67}$Cu by the $^{68}$Zn (γ, p) reaction. *Int. J. Appl. Radiat. Isot.* **1978**, *29*, 757–759. [CrossRef]
193. Hovhannisyan, G.H.; Bakhshiyan, T.M.; Dallakyan, R.K. Photonuclear production of the medical isotope $^{67}$Cu. *Nucl. Instrum. Methods Phys. Res. Sect. B Beam Interact. Mater. At.* **2021**, *498*, 48–51. [CrossRef]
194. Gopalakrishna, A.; Suryanarayana, S.; Naik, H.; Nayak, B.; Patil, B.; Devraju, S.; Upreti, R.; Kinhikar, R.; Deshpande, D.; Maletha, P.; et al. Production of $^{99}$Mo and $^{64}$Cu in a mixed field of photons and neutrons in a clinical electron linear accelerator. *J. Radioanal. Nucl. Chem.* **2018**, *317*, 1409–1417. [CrossRef]
195. Maslov, O.D.; Sabel'nikov, A.V.; Dmitriev, S.N. Preparation of $^{225}$Ac by $^{226}$Ra (γ, n) photonuclear reaction on an electron accelerator, MT-25 microtron. *Radiochemistry* **2006**, *48*, 195–197. [CrossRef]
196. Robertson, A.K.H.; Ramogida, C.F.; Schaffer, P.; Radchenko, V. Development of $^{225}$Ac radiopharmaceuticals: TRIUMF perspectives and experiences. *Curr. Radiopharm.* **2018**, *11*, 156–172. [CrossRef]
197. Inagaki, M.; Sekimoto, S.; Tanaka, W.; Tadokoro, T.; Ueno, Y.; Kani, Y.; Ohtsuki, T. Production of $^{47}$Sc, $^{67}$Cu, $^{68}$Ga, $^{105}$Rh, $^{177}$Lu, and $^{188}$Re using electron linear accelerator. *J. Radioanal. Nucl. Chem.* **2019**, *322*, 1703–1709. [CrossRef]
198. Radel, R.; Sengbusch, E.; Piefer, G. Recent Progress on the PNL Accelerator-Based Intense Fusion Neutron Source. *Trans. Am. Nucl. Soc.* **2016**, *114*, 11–12.
199. Rotsch, D.A.; Brown, M.A.; Nolen, J.A.; Brossard, T.; Henning, W.F.; Chemerisov, S.D.; Gromov, R.G.; Greene, J. Electron linear accelerator production and purification of scandium-47 from titanium dioxide targets. *Appl. Radiat. Isot.* **2018**, *131*, 77–82. [CrossRef] [PubMed]
200. Melville, G.; Allen, B.J. Cyclotron and linac production of Ac-225. *Appl. Radiat. Isot.* **2009**, *67*, 549–555. [CrossRef]
201. Deshpande, A.; Dixit, T.; Bhat, S.; Jadhav, P.; Kottawar, A.; Krishnan, R.; Thakur, K.; Vidwans, M.; Waingankar, A. Design of High Energy Linac for Generation of Isotopes for Medical Applications. In Proceedings of the IPAC 2021-12th International Particle Accelerator Conference, Campinas, SP, Brazi, 24-28 May 2021; JACoW Publishing: Geneva, Switzerland, 2021; pp. 2472–2474.
202. Leung, K.N. New compact neutron generator system for multiple applications. *Nucl. Technol.* **2020**, *206*, 1607–1614. [CrossRef]
203. Kononov, V.N.; Bokhovko, M.V.; Kononov, O.E.; Soloviev, N.A.; Chu, W.T.; Nigg, D. Accelerator-based fast neutron sources for neutron therapy. *Nucl. Instrum. Methods Phys. Res. Sect. A Accel. Spectrometers Detect. Assoc. Equip.* **2006**, *564*, 525–531. [CrossRef]
204. Cloth, P.; Conrads, H. Neutronics of a dense-plasma focus—An investigation of a fusion plasma. *Nucl. Sci. Eng.* **1977**, *62*, 591–600. [CrossRef]
205. Csikai, J.; Dóczi, R. Applications of neutron generators. *Handb. Nucl. Chem.* **2011**, *3*, 363.
206. Reijonen, J. Compact neutron generators for medical, home land security, and planetary exploration. In Proceedings of the 2005 Particle Accelerator Conference, Knoxville, TN, USA, 16–20 May 2005; pp. 49–53.
207. Dovbnya, A.N.; Kuplennikov, E.L.; Tsymba, V.A.; Krasil'nikov, V.V. Possibility of $^{99m}$Tc production at neutron generator. Вопросы Атомной Науки И Техники **2009**, *5*, 64–66.
208. Pagdon, K.; Gentile, C.; Cohen, A.; Ascione, G.; Baker, G. Production of Tc-99m from naturally occurring molybdenum absent uranium. In Proceedings of the 2011 IEEE/NPSS 24th Symposium on Fusion Engineering, Chicago, IL, USA, 26–30 June 2011; pp. 1–4.
209. Mausolf, E.J.; Johnstone, E.V.; Mayordomo, N.; Williams, D.L.; Guan, E.Y.Z.; Gary, C.K. Fusion-Based Neutron Generator Production of Tc-99m and Tc-101: A Prospective Avenue to Technetium Theranostics. *Pharmaceuticals* **2021**, *14*, 875. [CrossRef]

210. National Academies of Sciences, Engineering, and Medicine. *Molybdenum-99 for Medical Imaging*; National Academies Press: Washington, DC, USA, 2016.
211. Youker, A.J.; Chemerisov, S.D.; Tkac, P.; Kalensky, M.; Heltemes, T.A.; Rotsch, D.A.; Vandegrift, G.F.; Krebs, J.F.; Makarashvili, V.; Stepinski, D.C. Fission-produced Mo-99 without a nuclear reactor. *J. Nucl. Med.* **2017**, *58*, 514–517. [CrossRef]
212. Leung, K.N.; Leung, J.K.; Melville, G. Feasibility study on medical isotope production using a compact neutron generator. *Appl. Radiat. Isot.* **2018**, *137*, 23–27. [CrossRef] [PubMed]
213. Badwar, S.; Ghosh, R.; Lawriniang, B.M.; Vansola, V.; Sheela, Y.; Naik, H.; Naik, Y.; Suryanarayana, S.V.; Jyrwa, B.; Ganesan, S. Measurement of formation cross-section of $^{99}$Mo from the $^{98}$Mo (n, γ) and $^{100}$Mo (n, 2n) reactions. *Appl. Radiat. Isot.* **2017**, *129*, 117–123. [CrossRef] [PubMed]
214. Auditore, L.; Amato, E.; Baldari, S. Theoretical estimation of 64Cu production with neutrons emitted during 18F production with a 30 MeV medical cyclotron. *Appl. Radiat. Isot.* **2017**, *122*, 229–234. [CrossRef] [PubMed]
215. Pandit-Taskar, N.; Batraki, M.; Divgi, C.R. Radiopharmaceutical therapy for palliation of bone pain from osseous metastases. *J. Nucl. Med.* **2004**, *45*, 1358–1365.
216. Kim, S.K.; Choi, H.D. New technique for Producing Therapeutic Radioisotope $^{89}$Sr. In Proceedings of the Korean Nuclear Society Conference, jeju, Korea, 26–26 May 2005; Korean Nuclear Society: Seoul, Korea; pp. 751–752.
217. Molla, N.I.; Basunia, S.; Miah, M.R.; Hossain, S.M.; Rahman, M.M.; Spellerberg, S.; Qaim, S.M. Radiochemical Study of $^{45}$Sc (n, p) $^{45}$Ca and $^{89}$Y (n, p) $^{89}$Sr Reactions in the Neutron Energy Range of 13.9 to 14.7 MeV. *Radiochim. Acta* **1998**, *80*, 189–192. [CrossRef]
218. Capogni, M.; Capone, M.; Pietropaolo, A.; Fazio, A.; Dellepiane, G.; Falconi, R.; Colangeli, A.; Palomba, S.; Valentini, G.; Fantuzi, M.; et al. $^{64}$Cu production by 14 MeV neutron beam. *J. Neutron Res.* **2020**, *22*, 257–264. [CrossRef]
219. Voyles, A.; Basunia, M.; Batchelder, J.; Bauer, J.; Becker, T.; Bernstein, L.; Matthews, E.; Renne, P.; Rutte, D.; Unzueta, M.; et al. Measurement of the $^{64}$Zn, $^{47}$Ti (n, p) cross sections using a DD neutron generator for medical isotope studies. *Nucl. Instrum. Methods Phys. Res. Sect. B Beam Interact. Mater. At.* **2017**, *410*, 230–239. [CrossRef]
220. Mellard, S.C.; Biegalski, S.R. MCNP based simulations for the optimization of radioxenon via DD and DT neutron generators from $^{132}$Xe. *J. Radioanal. Nucl. Chem.* **2018**, *318*, 313–322. [CrossRef]
221. Weicheng, Z. Short-lived medical isotopes produced by 14MeV neutron generator. *At. Energy Sci. Technol* **1981**, *03*, 366–368. (In Chinese)
222. Yuntao, L.; Shizhong, A.; Jixin, L. Current Situation and Development Trend of Nuclear Technology Application. *Sci. Technol. Rev.* **2022**, *40*, 88–97. (In Chinese)
223. Huang, Z.; Wang, J.; Ma, Z.; Lu, X.; Wei, Z.; Zhang, S.; Liu, Y.; Zhang, Z.; Zhang, Y.; Yao, Z. Design of a compact D–D neutron generator. *Nucl. Instrum. Methods Phys. Res. Sect. A Accel. Spectrometers Detect. Assoc. Equip.* **2018**, *904*, 107–112. [CrossRef]
224. Jixin, L.; Yuqing, C.; Guang, L.; Xuesong, D.; Yijia, S.; Laicheng, Q.; Yuping, L.; Hua, J.; Guiqun, L. Production process of 64Cu by C-30 cyclotron. In *Annual Report for China Institute of Atomic Energy*; China Institute of Atomic Energy: Beijing, China, 2013.

Review

# Radiopharmaceutical Treatments for Cancer Therapy, Radionuclides Characteristics, Applications, and Challenges

Suliman Salih [1,2], Ajnas Alkatheeri [1], Wijdan Alomaim [1] and Aisyah Elliyanti [3,*]

[1] Radiology and Medical Imaging Department, Fatima College of Health Sciences, Abu Dhabi 3798, United Arab Emirates
[2] National Cancer Institute, University of Gezira, Wad Madani 2667, Sudan
[3] Nuclear Medicine Division of Radiology Department, Faculty of Medicine, Universitas Andalas, Padang 25163, Indonesia
* Correspondence: aelliyanti@med.unand.ac.id

**Abstract:** Advances in the field of molecular biology have had an impact on biomedical applications, which provide greater hope for both imaging and therapeutics. Work has been intensified on the development of radionuclides and their application in radiopharmaceuticals (RPs) which will certainly influence and expand therapeutic approaches in the future treatment of patients. Alpha or beta particles and Auger electrons are used for therapy purposes, and each has advantages and disadvantages. The radionuclides labeled drug delivery system will deliver the particles to the specific targeting cell. Different radioligands can be chosen to uniquely target molecular receptors or intracellular components, making them suitable for personal patient-tailored therapy in modern cancer therapy management. Advances in nanotechnology have enabled nanoparticle drug delivery systems that can allow for specific multivalent attachment of targeted molecules of antibodies, peptides, or ligands to the surface of nanoparticles for therapy and imaging purposes. This review presents fundamental radionuclide properties with particular reference to tumor biology and receptor characteristic of radiopharmaceutical targeted therapy development.

**Keywords:** alpha particles; auger electron; beta particles; nanotargeted therapy; radioligand therapy

## 1. Introduction

In the early 1900s, Henri Becquerel and Marie Curie discovered radioactivity. Therapeutic applications immediately followed this discovery [1,2]. For many years radionuclide therapy was limited to the use of Iodide-131 ($^{131}$I) for thyroid cancer and hyperthyroidism and phosphate-32 ($^{32}$P) for polycythemia vera [2–6]. Radionuclides labeled molecules such as a drug, a protein, or a peptide that operate as a delivery vehicle that accumulates and binds to specific targets such as tumors or other undesirable cell proliferation [3,7,8]. The development of radionuclide use has been growing exponentially with the introduction of more new radiopharmaceuticals (RPs) for therapy and imaging.

In recent times, RPs use in nuclear medicine has become popular in theranostics. These are used in therapeutic interventions after imaging verifies the presence of a biological target [6,9,10]. Unlike radiotherapy, RPs are administrated intravenously to be delivered to a target tumor or associated structure. RPs have advantages in treating systemic malignancy in areas such as the bone or brain, which are impossible to treat using external radiotherapy [2]. The targeted tumor cell absorbs a dose of radiation from an RP which exponentially decreases over time (Figure 1a). On the other hand, in external radiotherapy, radiation beams are directed at tumor tissue and cannot avoid healthy cells (Figure 1b).

Radiopharmaceutical therapy (RPT) is a novel modality that can be effective with minimal toxicity [6,7]. The advantages of RPT are, firstly, it can be targeted at tumors, including metastasis sites. The RPs can be used in radiotracer imaging to determine the uptake of the RP in the target tissues before administering a therapeutic dose. Secondly, a wide variety of radionuclides are now available emitting different types of radiation at different energies. For instance, high linear energy transfer (LET) radionuclides are used effectively to kill resistant hypoxic cells. Thirdly, this therapy allows for a relatively lower whole-body absorbed dose [7,10–13].

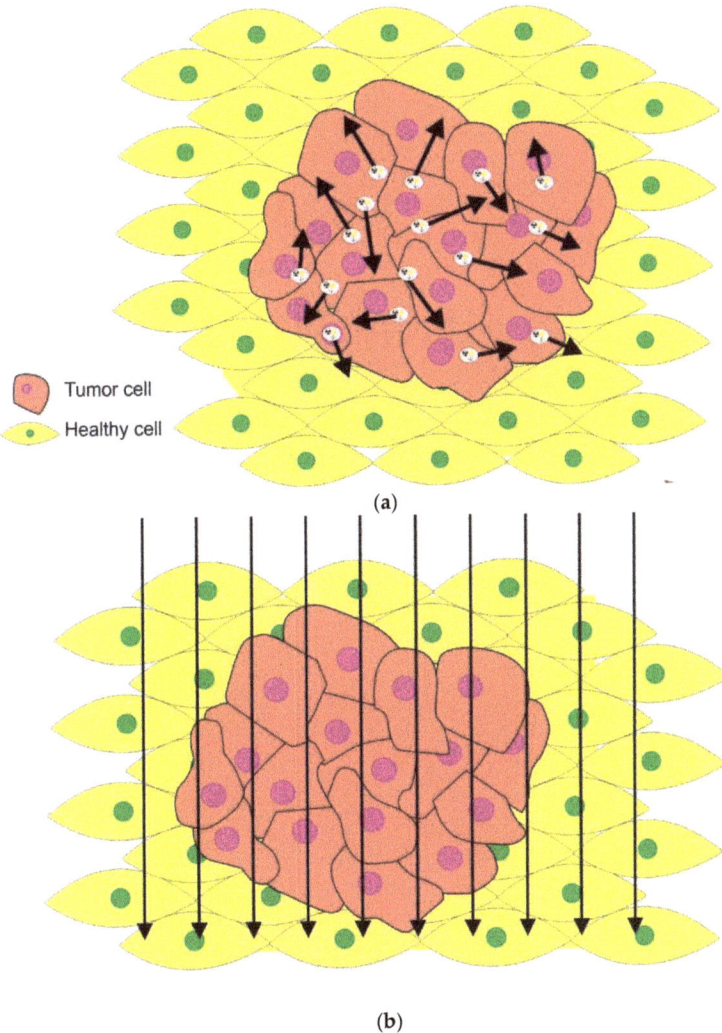

**Figure 1.** The cell's radiation distribution by RPT (**a**) and external radiotherapy (**b**). Radiopharmaceuticals are administrated intravenously to be delivered to a target tumor. The targeted tumor cell absorbs a dose of radiation which exponentially decreases over time. The tumor mass's periphery cells will receive absorbed and crossfire doses from other target cells (**a**). Radiation beams are directed at tumor tissue in external radiotherapy and can also affect healthy cells (**b**).

RPT can be used as adjuvant therapy with or after other treatment options such as chemotherapy and surgery [2]. It is being used to control symptoms and shrink and stabilize tumors in systemic metastatic cancer, where conventional therapy or chemotherapy is impossible. RPT can be a good choice, especially for patients who no longer respond to other treatments [2,3,7,10]. This review describes some fundamental radionuclide properties with particular reference to tumor biology and the receptor characteristics of radiopharmaceutical targeted therapy development.

## 2. Radionuclide Emission Properties

The physical characteristics of a radionuclide should be considered when selecting it for therapy purposes. These include physical half-life, radiation energy, type of emissions, daughter product(s), production method, and radionuclide purity [2,9]. Ideally, the physical half-life of the radionuclide should be between 6 h and seven days [14]. The RPs with a long half-life will expose the target tumor and surrounding environment to radiation for longer. However, RPs with a very short physical half-life have limitations due to the delivery time. There must be sufficient retention time for the emission to be delivered to the tumor target [15].

Furthermore, in vivo stability, toxicity, and the biological half-life within the patient's body must be considered [7,16], along with the type and size of the tumor, method of administration, and uptake mechanism [1,2,6,15]. The tumor uptake mechanism is specific to the target cell. It depends on processes such as antigen–antibody reactions, physical particle trapping, receptor binding sites, removal of damaged cells from circulation, and transportation of a chemical species across a cell membrane and metabolic cycle [2,17]. The condition will influence the ratio of the concentration of radionuclides in the tumor to that in normal tissues. This ratio should be optimized [2]. The other factors that must be considered are radionuclide particle size, toxicity, specific gravity for optimal flow and distribution, and clearance rate [2,6,18–22].

Radionuclides used in RPT are primarily beta ($\beta$)-particle (0.2 keV/$\mu$m) or alpha ($\alpha$)-particle (50–230 keV/$\mu$m) emitters [2,9,11,15,23], and Auger electrons (AE) (4–26 keV/$\mu$m) [2,9,11,15,23,24]. Various radionuclides and their characteristics are summarized in Table 1. Each of these radiation types results in ionization along the travel length, and they are fully deposited in the cell [16]. The radiation destroys the cell directly and indirectly [6,25]. The distance traveled by particles and the energy deposited in cells must be considered to ensure optimal targeted cell destruction and minimize ionization interaction with healthy cells [2,6,7,15].

Table 1. Characteristics of radionuclides used in radiotherapy.

| Radionuclides | Emitting | Physical Half-Life | Mean Eα/β- (MeV) Max | Mean Eα/β- (MeV) Min | Mean Eα/β- (MeV) Mean | Primary Eα/β- (MeV) (%) | Mean Range in Soft Tissue (mm) | Indication | References |
|---|---|---|---|---|---|---|---|---|---|
| 131I | β | 8.02 d | 0.606 MeV | 0.069 MeV | 0.356 MeV | 0.3645 MeV (81%) | 0.4 mm | Hyperthyroid, thyroid cancer, Radioimmunotherapy (RIT) for NHL and neuroblastoma, pheochromocytoma, carcinoid, medullary thyroid cancer | [2,3,6,8,22,24] |
| 32P | β | 14.26 d | 1.71 MeV | 0.695 MeV | 1.015 MeV | - | 2.6 mm | Polycythemia vera, keloid, cystic craniopharyngioma | [2,3,23] |
| 89Sr | β | 50.53 d | 1.491 MeV | 0.583 MeV | 0.908 MeV | 0.91 MeV (0.01%) | 2.4 mm | Bone pain palliation | [2,3,6,8,23] |
| 90Y | β | 64.10 d | 2.284 MeV | 0.935 MeV | 1.349 MeV | (0.01%) | 3.6 mm | Liver metastasis, hepatocellular carcinoma, RIT for NHL, neuroendocrine tumor | [2,3,6,8,22,23] |
| 153Sm | β | 46.50 h | 0.8082 MeV | - | - | 0.1032 MeV (29.8%) | 0.7 mm | Bone pain palliation, synovitis | [2,3,6,8] |
| 169Er | β | 9.4 d | 0.35 MeV | - | - | 0.084 MeV (0.16%) | 0.3 mm | Synovitis | [2,3] |
| 177Lu | β | 6.73 d | 0.497 MeV | 0.047 MeV | 0.208 MeV | 0.208 MeV (11%) | 0.28 mm | Synovitis and RIT for various cancer | [2,6,8,22–24] |
| 186Re | β | 3.72 d | 1.077 MeV | 0.308 MeV | 0.769 MeV | 0.137 MeV (9.4%) | 1.2 mm | Bone pain palliation, arthritis. | [2,6,8,23] |
| 188Re | β | 17 h | 2.12 MeV | 0.528 MeV | 1.592 MeV | 0.155 MeV (15%) | 2.1 mm | Bone pain palliation, RIT for various cancer, rheumatoid arthritis | [2,3,8,22,23] |
| 223Ra | α | 11.44 d | 5.9792 MeV | - | 6.59 MeV | 0.154 MeV (5.59%) | 0.054 mm | Bone pain palliation | [2,5,13] |
| 211At | α | 7.2 h | - | - | 6.79 MeV | (5.87%) | 0.057 mm | RIT leukemia, brain tumor, RLT prostate cancer | [2,3,23,25] |
| 213Bi | α | 46 mins | - | - | 8.32 MeV | (26%) | 0.078 mm | RIT leukemia, brain tumor | [3,22,23,25] |
| 225Ac | α | 10 d | - | - | 0.218 MeV | (11.4%) | 0.05–0.08 mm | Radioligand (RLT) prostate cancer | [2,8,24] |

## 2.1. Beta Particles

Beta particles have been used in cancer therapy over the last 40 years [6]. They are the product of the β decay process, wherein an unstable nucleus is converted to a proton, and a β particle, a high-energy electron [7,26]. β particles are the most frequently used radiation in RPT agents and are widely available [7]. β particles are negatively charged. They have a relatively long path from 0.0 to 12 mm, and some emit a gamma (γ) ray such as $^{32}P$, $^{89}Sr$, $^{90}Y$, and $^{169}Er$ [3]. They emit γ ray <10%, which is acceptable for imaging to confirm the tumor uptake and biodistribution and dosimetric calculations [2,3]. They have a low linear energy transfer (LET) of approximately 0.2 keV/μm, so more β particles are required to deliver a similar absorbed dose compared to alpha particles.

The most familiar and frequently used β particle is iodine-131 ($^{131}I$). Hertz and Roberts used radioiodine I-130 ($^{130}I$) for hyperthyroid therapy in 1941, which rose at the birth of nuclear medicine [27–29]. In August 1946, $^{130}I$ was replaced by $^{131}I$ because it was much cheaper [27,29]. $^{131}I$ is a β and γ emitter with a half-life of 8.05 days. The β particle has a peak energy of 0.606 MeV, with a maximum range of ~3 mm in the tissue, and it is used for therapy. The peak energy of the γ ray is 0.364 MeV and is used for imaging [27]. Since then, $^{131}I$ has been used countless times for therapy for hyperthyroid and thyroid cancer [3,6,27–30]. In 1981, $^{131}I$-iobenguane (meta-iodobenzylguanidine, MIBG) was introduced as a diagnostic agent, and in 1984, it was used for treating malignant phaeochromocytoma [31]. Monoclonal antibodies are used to label with $^{131}I$, and, in 2003, FDA approved $^{131}I$-tositumomab (Bexxar) for the treatment of refractory non-Hodgkin's lymphoma (NHL) [2,6,7]. Several studies have reported the monoclonal antibodies labeled on other beta particle emitters, including Yttrium-90 ($^{90}Y$) and Lutetium-177 ($^{177}Lu$), for more effective therapy purposes [2,7,31–33].

The high-energy β from Yttrium-90 ($^{90}Y$) or Rhenium-188 ($^{188}Re$) is preferable for treating higher volume solid and poorly perfused tumors and is less suited for targeting micro-metastases to avoid crossfire doses to neighbor cells [9,11,34]. $^{90}Y$, widely available like $^{131}I$, is a popular radionuclide for liver cancer and metastases [35,36]. Neuroendocrine tumors (NETs) have been treated with radionuclide therapy (PPRT) targeting peptide receptors with radiopharmaceuticals labeled with $^{90}Y$. Antibodies also labeled with $^{90}Y$, have been introduced for ovarian and hematological cancers [7,26,37–39]. Low-energy SS, like those seen with lutetium-177 ($^{177}Lu$), is more efficient for small tumors [1,9]; hence, $^{177}Lu$ is becoming a popular SS-particle source for treating small tumors [7,9]. $^{177}Lu$ has a half-life of 6.73 days and is compatible with antibodies and peptides [40,41]. Furthermore, it also emits gamma-rays and can be detected externally as a theranostic agent [1,7,40]. Samarium-153 ($^{153}Sm$) is used to treat palliative bone metastases and other primary cancers [3,42,43]. Ethylenediamine-tetra-methylene-phosphonic acid (EDTMP) chelator binds with $^{153}Sm$ through six ligands (four phosphate groups and two amines). It has been widely used since FDA approval in various osteoblastic metastatic lesions, especially in prostate and breast cancer [44]. However, not all possible β particle sources have been widely adopted because of the complexity of the radiochemistry or the absence of commercial availability. The decision to use one β-particle source over another must consider the absorbed dose ratio between tumor to non-tumor tissue [7].

## 2.2. Alpha Particles

The application of targeted α particle therapy (TAT) gained approval in 2013 [19]. Alpha particles are high energy and have shorter path lengths, resulting in higher efficacy in some applications [2,8,15,25,26]. TAT is an attractive therapeutic option for multiple micro-metastases. It is easy to administer and can be used to treat multiple lesions simultaneously. It is also possible to combine it with other therapeutic approaches, primarily for cancer treatment [45,46].

An alpha particle is a $^4He$ nucleus without its surrounding electrons (sometimes denoted as ($He^{2+}$)) [26,45]. Alpha radiation is emitted from the nucleus of a radioactive atom undergoing decay with an energy is 4–9 MeV, and the particles travel only

1–3 cell diameters (40–100 μm) in tissue [7,15,32,45,46]. The particles have high LET (60–230 keV/μm) throughout their range, peaking to three times the initial value at the end of the path range (the Bragg peak) [16,26,32]. Most alpha particles also emit gamma-ray. However, treatment planning or post-therapeutic imaging using alpha particles is not performed yet in clinical settings due to technical limitations [45].

Furthermore, intracellular accumulation of the α particles effectively creates double-strand breaks (DSBs) in DNA, and numerous clusters of DSBs in target cells, making cellular repair systems ineffectual [7,32,47]. The cytotoxicity of α-particles is much higher than that of β-particles due to the particle deposit energy per unit path length, which is 1500 times more than beta particles [45,48]. In addition, the short travel distance of α particles reduces the damage to surrounding healthy tissue [15,49]. The particle radiation has been demonstrated to be independent of cell oxygen concentration [15,32,45,50]. The physical and biological characteristics of alpha, beta particles, and Auger electrons are summarized in Table 2, and DNA damage by that radiations are illustrated in Figure 2a,b.

Improvements in understanding molecular tumor biology, labeling techniques, technology development, and other related disciplines have paved the way for significant new clinical applications of α radiation as a novel therapeutic agent [7,15,51]. Alpha particle-labeled biological molecules such as monoclonal antibodies (mAb) allow close radiation targeting and selectively deliver high radiation to the target, with limited toxicity to normal tissues [15]. The mAbs are labeled radionuclides that bind to the extracellular domain of PSMA, demonstrating promising results in imaging and therapy of prostate cancers [9]. The monoclonal antibodies are labeled with bismuth-213 ($^{213}$Bi) and astatine-211 ($^{211}$At) and are used to treat leukemia and brain tumors [11,52]. The monoclonal antibody MX35 labeled $^{213}$Bi successfully treated ovarian cancer in animal models with no signs of toxicity [53]. $^{213}$Bi has a short half-life and is produced using a generator and labeling to produce TAT compounds is therefore completed on-site [26,54]. Because of its short half-life, $^{213}$Bi needs to be delivered directly into tumor tissue, and it can be given at a high dose over a short period, which is more effective than low dose rates given over a more extended period [26,32,55]. $^{213}$Bi has been used to label DOTA peptides in preclinical and clinical trials with >99% purity [15,26]. In preclinical and clinical studies, $^{213}$Bi and $^{225}$Ac have been used to label somatostatin receptors [15,26,32].

Radium-223 dichloride (Xofigo), a α particle emitter used for bone pain palliation in prostate and breast cancer patients, was approved by FDA in 2013 [7,11,26,32,45]. The emission energy of $^{223}$Ra can generate irreparable DNA double-strand breaks in the adjacent osteoblasts and osteoclasts, which has a detrimental effect on the adjacent cells and inhibits abnormal bone formation [7]. $^{223}$Ra is being studied as a radioactive label for other cytotoxic agents such as poly (ADP-ribose) polymerase inhibitors (olaparib), docetaxel (DORA trial), and new androgen axis inhibitors as enzalutamide and abiraterone citrate. The recently high number of $^{223}$Ra and in combination with other therapeutics, showed promising results [7].

Another alpha particle attracting increasing interest is $^{225}$Ac, the parent of $^{213}$Bi, which is relatively long-lived, with a half-life of 9.9 days [54]. $^{255}$Ac is produced via the neutron transmutation of $^{225}$Ra or decay of $^{233}$U [26,54,55]. $^{225}$Ac can be used to treat neuroendocrine tumors. It has been used to label PSMA with a radiochemical purity of >98% for prostate cancer therapy [26,54,56]. It also labeled antibodies to test for myeloid malignancy [9] and shows a potential for therapy, and post-therapy imaging, even though the images are suboptimal [26,32,55]. Results of clinical trials using TAT results indicate that this treatment strategy presents a promising alternative to targeted cancer therapy [52]. Lately, $^{225}$Ac-labeled PSMA-ligands have gained popularity as an alternative to $^{177}$Lu-PSMA [26,54,56]. However, $^{225}$Ac may damage the healthy cells due to daughter radionuclides such as $^{221}$Fr, $^{217}$At, and $^{213}$Bi [47]. Danger from radiation from daughter radionuclides needs to be carefully evaluated.

**Table 2.** Physical and biological characteristics of α, β particles, and Auger electron.

| | Alpha Particle | Beta Particle | Auger Electron |
|---|---|---|---|
| Type of particles | $^4$He nucleus | Energetic electron | Low energy electron; electron capture (ec) and/or internal conversion (ic) |
| Particle energy | 4–9 MeV | 50–2300 keV | 25–80 keV |
| Particle path length | 40–100 μm | 0.05–12 mm | Nanomicrometers |
| Linear energy transfer | ~80 keV/μm | ~0.2 keV/μm | 4–26 keV/μm |
| Hypoxic tumors | Effective | Less effective | Effective |
| Toxicity | Effective in creating double-strand breaks in DNA | High dose rates (tumor survival rates close to linear exponential). Low dose rates (single-strand breaks), repairable with shouldering the dose-response curve | Potential creation of double-strand breaks DNA, and cell membrane |
| Bystander effect/crossfire | Yes/low | Yes | Yes |
| Tumor size | Micro/small | Higher volume solid tumor | Micro |

Ref: [7,8,13,24–26,32,47,50,55,57].

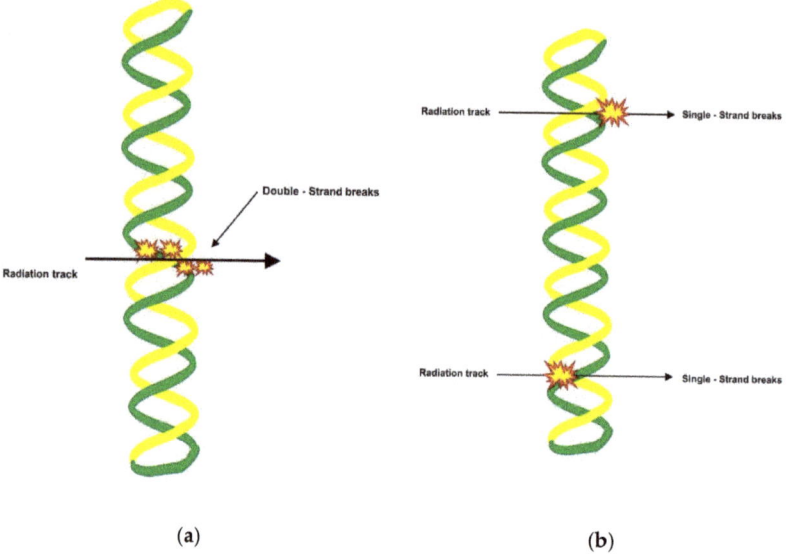

(a)         (b)

**Figure 2.** High and intermediate LET radiation (alpha particle and Auger electron, respectively), cause double-strand breaks in DNA (**a**). Single-strand breaks in DNA due to radiation by low LET (beta particle) (**b**).

### 2.3. Auger Electrons

Auger electrons (AE) have an even shorter range than alpha particles delivering radiation of only 1–1000 nm through the tissue causing potent tumor cell death if they can be conjugated with suitable ligands that effectively target micro-metastasis, particularly of DNA and cell membranes [2,11,24,32,34,52]. AEs are generated from suborbital transitions, and their range depends on their energy. They have an intermediate LET (4–26 keV/μM) [4,32,58]. Bromine-77 ($^{77}$Br), indium-111 ($^{111}$In), iodine-123 ($^{123}$I), and iodine-125 ($^{125}$I) are the most commonly used radionuclides sources [24,59,60]. Human studies using locoregional administration have shown promising results in therapy [7].

Despite the short range of AE, local cross-dose effects occur in cells adjacent to the radionuclide decay mediated by the several micrometer ranges of higher energy AEs and internal conversion (IC) that causes the death of distant non-irradiated cells through the bystander effect [16,24,25,27,32]. Lethally damaged tumor cells may release mediators that cause the death of distant non-irradiated cells [24,25]. Radiation also releases heat shock protein-70 and high mobility group box-1, which activate the dendritic cells (DCs). The activated DCs activate cytotoxic T cells that result in tumor regression at distant sites [11,61]. It has been observed that the effects of ionizing radiation can work synergistically with targeted immune treatment observed at the site(s) distant from targeted tissues/organs. This phenomenon is suggestive of the role of the abscopal effect [11,25,61,62]. Attention has been focused on delivering AE to the nucleus/DNA as the primary cellular radiation target to maximize toxic effects. However, cell membrane targeting has also been proven to be an effective strategy for killing cancer cells [24,63]. Cell membrane damage further induces γH2AX foci in the nucleus of the cells exposed to $^{125}$I-anti-CEA mAbs and in non-exposed cells through a bystander effect. $^{125}$I-labeled anti-CEA 35A7 was also found to be effective in vivo for treating small peritoneal tumors in mice [24,63]. Toxicity may also be induced indirectly by free radical-mediated pathways [24,25,57].

So, AE nuclear targeting is essential but not always required for RPT [24]. The abscopal and indirect killing effects suggest that targeting cell surface antigens overexpressed on tumor cells that are recognized by monoclonal antibodies (Mabs) or other ligands may be effective [25,64]. AE therapy has not been widely adopted yet. The fact that auger electron agents are incorporated into the DNA, and the cell membrane results in unfavorable pharmacokinetics, might be the reason for the lack of efficacy. Technological developments could overcome obstacles and increase interest in AE for therapy development [7].

## 3. Therapy Application

Radiopharmaceutical or radioligand therapy includes systemic radiation therapy, molecular radiotherapy, targeted radiation therapy, or peptide receptor radionuclide therapy (PRRT) and there are examples of where RPT is applied in optimizing and balancing the therapeutic index (TI). Various radioligands are being developed and investigated to target molecular receptors or intracellular components in personal therapy [58].

### 3.1. Antibodies

The monoclonal antibodies (mAbs) are labeled with radionuclides. The smaller fragments and new fusion proteins are directed against tumor antigens to deliver radionuclides to the targeted tumor [2,9]. The FDA has approved these agents for the clinical management of liquid malignancies (ibritumomab tiuxetan (Zevalin) labeled with $^{90}$Y, and tositumomab (Bexxar) labeled with $^{131}$I) is used for lymphoma therapy [2,6,7,55,64,65], and some the RPT in optimizing and balancing the therapeutic index (TI) [65]. The therapeutic benefit is achieved when the cells absorb continuous radiation emitted by radionuclides tagged to mAbs while minimizing toxicities in non-target tissues.

The effect of the RPT depends on the radiation's energy and the antibody's affinity, antigen target concentration on the cells, tissue vascularity, and antibody/antigen rate constants [64]. Novel antibody engineering techniques have enabled the development of antibodies that bind to antigens expressed in target cancer cells. An antibody that binds to a particular antigen will allow for a higher RP uptake within tumor tissue. However, antibodies are larger molecules, limiting the tumor penetration and distribution of the radiolabeled antibody within the tumor. Furthermore, antibodies have a prolonged circulation time and slow biological clearance, leading to larger radiation-absorbed doses to healthy organs and blood. Pre-infusion of a certain mass of non-radiolabeled antibody (cold antibody) may be used before the infusion of radiolabeled (hot antibody) to saturate antigenic sites in normal cells to avoid unnecessary radiation to healthy cells [64], reducing the binding of the hot antibody and decreasing the radiation doses to healthy organs. However, a pre-infusion time before administration of the hot antibody must be determined and optimized for every therapy [64].

Patient selection for RPT should be based on the predetermined expression of specific tumor antigens or diagnostic results [65]. Several antigens or receptors are expressed on the surface of the membrane of tumor cells, such as human epidermal growth factor receptor 2+ (HER2+), epidermal growth factor receptor (EGFR), CD20, prostate-specific membrane antigen (PSMA), vascular endothelial growth factor (VEGF), mucin 1 (MUC1) and tumor necrosis factor (TNF). Any of these can be labeled with various radionuclides [9]. Beta particle emitters have often been labeled with antibodies because they emit β and γ rays and have a longer half-life of 8 days. Lately, alpha particles have rapidly gained interest and have been used to label antibodies to deliver radiation to tumors, such as $^{227}$Th-anti CD22 and $^{225}$Ac-PSMA-617 [55]. However, α particles cannot be imaged unless they emit γ rays as $^{223}$Ra and $^{227}$Th do [64]. Unfortunately, these radionuclides only emit γ rays in low concentrations, which is not optimal for assessment. This imaging limitation may lead to noncompliance, and other radionuclides imaging may be required to establish lesion targeting and dosimetry [45,64].

Radionuclide-labeled mAbs demonstrate more efficacy in inducing cancer remissions than unlabeled molecules and are also more effective than chemotherapy [9]. They have been shown to benefit lung, pancreatic, stomach, ovarian, breast, colorectal, leukemia, and high-grade brain glioma cancers [2]. Fortunately, the application of the RPT in giant solid tumors is less successful than in small volume tumors such as malignant lymphoma due to poor perfusion, increased intratumoral hydrostatic pressure, and various radionuclides uptakes by the cells [8,64,65].

### 3.2. Prostate-Specific Membrane Antigen (PSMA)

$^{131}$I-labeled prostate-specific membrane antigen (PSMA) ligands showed promise for prostate cancer therapy and were further developed to "the $^{177}$Lu-PSMA" introduced in 2015 [60]. PSMA is a transmembrane protein that is over-expressed in prostate cancer (PC) cells, and its expression increases progressively in higher-grade cancers such as metastatic castration-resistant prostate cancer (mCRPC) PC [56,66–68]. Its benefits remain high even after multiple lines of therapy [56,66]. Radionuclide PSMA is a promising therapeutic approach for mCRPC patients for whom chemotherapy has been ineffective [55,56,66,69]. Early reports show that $^{177}$Lu-PSMA is safe and effectively reduces the tumor burden. It has low toxicity [69] and has become popular, with more than a thousand therapy cycles performed [66,69]. Severe hematological side effects are rare. Organs at risk after treatment with $^{177}$Lu-PSMA, including the salivary glands and the kidneys. However, the radiation dose to bone marrow, spleen, and liver is below critical limits [68].

Currently, the two most frequently used PSMA ligands are PSMA-617 and PSMA-I&T (imaging and therapy), labeled with $^{177}$Lu [68]. PSMA- targeting ligands using $^{225}$Ac maybe have an advantage compared to PSMA-targeting ligands using β particles. Clinical studies using $^{225}$Ac-labeled PSMA-ligands (PSMA-617 or PSMA-I&T) have demonstrated remarkable therapeutic results recently. Data on treatment with $^{225}$Ac-PSMA-617 indicate an excellent effect on tumor control in both early and late-stage mCRPC [70]. A novel α particle treatment with a $^{227}$Th-PSMA has shown potency in in vitro studies and efficacy in xenograft models of prostate cancer [8,67]. However, α particles have a more significant radiobiological effect on the organs at risk [56]. Concerns have been raised about treatment-associated, mostly permanent xerostomia, frequently leading to treatment discontinuation in many patients [56,68]. Combining α particles with β particle emitters is called "tandem therapy" and may reduce these significant adverse effects compared to using α particles alone [56,71,72].

### 3.3. Peptide Receptor Radionuclide Therapy (PRRT)

Receptor-based radionuclide therapies (PRRT) targeting the somatostatin receptor (SSTR), have since early 1990 been an important treatment modality for neuroendocrine tumors [7,26]. The efficacy of PPRT therapy might be due to the somatostatin receptor ligand that binds the specific receptor (SSTR1–5) [30,73]. Peptide receptors expressed in various tumor cells, including NETs, are significantly higher than in normal tissues or cells. NETs overexpress the SSTR2 potential for SSTR2 targeted therapies such as synthetic somatostatin analogs (SSAs) and radio-peptides or PRRT [30,73], and SSTR2 is primarily targeted by PRRT [73]. Octreotide and lanreotide are two SSAs developed and employed for clinical practice, which bind primarily to SSTR2 and SSTR5 [73]. Peptides have been labeled with several radionuclides, such as beta particles emitter $^{177}$Lu and $^{90}$Y. $^{177}$Lu–SSTR ligand is more effective in small-sized tumors, whereas, for larger tumor volumes, $^{90}$Y might be a better choice [30,73]. The first agent used was $^{90}$Y-labeled DOTATOC and DOTATATE. However, significant permanent kidney damage has been reported [34,74]. $^{177}$Lu-labeled DOTATATE or DOTATOC was the next PRRT radiopharmaceutical, causing less nephrotoxicity compared to $^{90}$Y [26] and a more negligible crossfire effect, particularly on small and metastatic tumors [74]. $^{177}$Lu-DOTATATE (Lutathera®) has also become the most widely used PRRT radiopharmaceutical at present [34].

Overall, α-emitters PRRT has shown good results. However, crossfire effects on small-size tumors have a significant impact. Additionally, hypoxia tumor tissue could be resistant to β-emitters treatment. α particles with high LET over a short range can minimize damage to surrounding healthy tissue. $^{213}$Bi and $^{225}$Ac have been clinically tested for brain tumors, neuroendocrine tumors, and prostate cancer therapy [26]. $^{213}$Bi and $^{225}$Ac-DOTA chelated peptides have been developed for peptide receptor radiotherapies, such as DOTA-Substance P targeting the neurokinin-1 receptor and somatostatin-analogs (e.g., DO-TATOC, DOTATATE) [74]. However, the results from these agents need to be confirmed in further studies.

### 3.4. Radioiodine Concentration via Sodium Iodide Symporter

$^{131}$I has been used for adjuvant therapy to manage well-differentiated thyroid cancer (DTC) for more than 60 years. It is used to destroy remaining thyroid cells post-thyroidectomy, including in metastases, and is relatively inexpensive and widely available [14,75]. It increases the 10-year survival rate to 80% and decreases mortality by 12% [75]. One-third of advanced DTC metastases show low uptake of iodine. Losing the ability to accumulate iodine can occur during the progression of the disease due to dysfunction and loss of sodium iodide symporter (NIS) expression [75,76], indicating a status of dedifferentiation known as a radioiodine refractory disease [75–77].

A sodium iodide symporter (NIS) transports iodine through the cell membrane. Iodine is transported into follicular thyroid cells against the electrochemical gradient [27,75,76]. In a normal condition, the gradient between a thyroid cell and the extracellular environment is 100:1 [27,75]. The expression of NIS provides the molecular basis of radioiodine for diagnostic and therapeutic use in patients with thyroid disease [76–78]. It resides in the thyroid in the basolateral membrane of epithelial cells and transports two cations of sodium (Na$^+$) and one anion of iodide (I-) into the cells. This process is facilitated by an enzyme Na$^+$/K$^+$ ATPase [27,29,75,76].

Genetic alteration causes the mitogen-activated protein kinase (MAPK) and phosphoinositide 3-kinase (PI3K) pathways associated with the silencing of solute carrier family five-member 5 (SLC5A5), which encodes NIS. The condition causes the cancer cell failure to take radioiodine [77]. A clinical trial of kinase inhibitors targeting the MAPK or PI3K pathways has shown promising effects in redifferentiation therapy. It brings hope to future therapy using either kinase inhibitors with different targets or kinase inhibitors and $^{131}$I in managing radioiodine refractory disease in DTC [77].

Furthermore, NIS transgene has been successfully transferred selectively into extra-thyroidal tumor cells or cells in the tumor environment using various gene delivery systems [78]. An advanced endogenous PDAC mouse model study indicated genetically engineered mesenchymal stem cells (MSC) as NIS gene delivery vehicles demonstrate high stromal targeting of NIS by selective recruitment of NIS-MSCs after systemic application resulting in an impressive $^{131}$I therapeutic effect [78].

*3.5. Nanotargeted Radionuclides*

In the last three decades, there has been a rapid increase in the use of new nanomaterials and radionuclides to enhance cancer diagnosis and therapies [2,79]. Many organic and inorganic materials can be used as nanoparticles [58,80,81]. Nanoparticle (NP) delivery systems have enhanced imaging and therapeutic efficacy by targeting the delivery of radio-labeled drugs to the tumor site and reducing their toxic side effects [79,81,82]. The significant advantages of nanoparticles are that they can be prepared in sizes <100 nm. This increases the localization of the drugs and radionuclides and the permeability and retention (EPR) effect of passive targeting tumor cells and facilitates uptake by active targeting tumor cells [81,82]. The surface of nanomaterials is usually coated with polymers or ligands to improve biocompatibility and the selection of specific targets [80]. A nanomaterial's final size and structure depend on the salt concentrations, surfactant additives, reactant concentrations, reaction temperatures, and solvent conditions used during synthesis [79,80]. Two mechanisms of nanoparticle delivery system for diagnostics and therapy to tumor sites are (i) specific passive targeting cells and (ii) specific active targeting cells [81,82].

Nanotargeted radionuclides have three main components, the nanoparticle core, the targeting biomolecule (which must be able to recognize a specific biological target), and the radionuclide [80]. Nanoparticles drug delivery systems can be made from polymers (polymeric nanoparticles, micelles, or dendrimers), lipids (liposomes), viruses (viral nanoparticles), organometallic compounds (nanotubes), inorganic nanoparticles (fullerenes, carbon nanotubes, quantum dots, or magnetic nanoparticles) [47,80–82]. The physical and chemical properties of nanoparticles play a critical role in determining particle–cell interactions, cellular trafficking mechanisms, biodistribution, pharmacokinetics, and optical properties [80]. Each nanoparticle type shows certain advantages and disadvantages that are inherent features of a particular material, such as solubility, thermal conductivity, ability to bind biomolecules or linkers, chemical stability, and capacity to incorporate and release compounds, as well as biocompatibility, toxicity, immunogenicity, and controlled drug release rate [47,80].

The targeting biomolecule must have a high affinity for the targeted epitopes. For ligands to bind effectively, each radionuclide can be conjugated directly on the nanoparticle surface, with or without a spacer, or can be attached to the nanoparticle during chemical synthesis. The spacer groups between the nanoparticle surface, the radionuclide, or the biomolecule can be a simple hydrocarbon chain, a peptide sequence, or a PEG linker [80–82]. In some cases, a bifunctional chelating group (BFC), such as 1, 4, 7, 10-tetraazadodecane-DOTA, must be conjugated to the nanoparticle, and then a radioactive metal needs to be attached. This requires modification of nanoparticles before radiolabeling [80].

There are many passive and active targeted nanoparticle therapies being developed. Most development is still at the in vitro or animal study stage. The most significant development is of $^{131}$I labeled nanoparticles for targeted therapy of different tumor types, according to the targeting strategy of the prepared NPs in which $^{131}$I is incorporated. The targeting strategy of these NPs depends on either passive or active targeting. $^{131}$I labeled NPs (silver or polymeric) shows $^{131}$I accumulation in different tumor types [79]. $^{131}$I labeled NPs targeting integrin have been studied. This protein is essential in regulating angiogenesis processes and tumor progression. Radionuclide labeled arginine–glycine–aspartate (RGD) can specifically target tumor integrin receptors [79]. Other β-particles such as $^{188}$Re, Holmium-166 ($^{166}$Ho), $^{90}$Y, and gold-198 ($^{198}$Au)-NPs have also been investigated for tumor-targeted radiotherapy [79,82]. $^{88}$Re-liposome has been shown to have a therapeutic effect in various animal models and translational clinical research [83]. $^{166}$Ho nanoparticles have also been prepared and studied in radionuclide tumor therapy for skin cancer and ovarian cancer metastases [80,82]. Liposome labeled with $^{90}$Y has been investigated for colon and melanoma tumors in animal models [79]. Gum arabic, functionalized peptide and protein, coated $^{198}$Au NP have been shown to be potential prostate cancer therapeutic agents in an animal model [79].

Nanoparticles labeled with α-particle emitters have been synthesized to enhance therapeutic efficiency with minimum danger to healthy tissues. $^{211}$At has been studied as a prospective NP's alpha particle emitter, but the main disadvantage of $^{211}$At for NTR is low in vivo stability [79]. The sodium form of A-type nano-zeolites targeting peptides has been labeled with $^{223}$Ra, showing a cytotoxic effect on glioma cells [79]. A preliminary study reported that $^{225}$Ac-Au@TADOTAGA administrated intratumoral delayed tumor growth in glioma xenografts, and it is the first reported study using $^{225}$Ac-labeled gold nanoparticles [84]. However, $^{225}$Ac use remains challenging for insufficient retention of its daughter's products due to the α recoil effect observed upon release of an α particle [84]. So, the use of α particles still has challenges related to incorporation into useful targeting vectors, such as in vivo stability, the weakness of α emitter–biomolecule bond, organ toxicity of inappropriate leakage of radionuclides from the bioconjugate, and uncoupling (or trans-chelated) than distribution to off-target areas [47].

## 4. Challenges in Radiopharmaceutical Therapy

In the last three decades, there has been a growing interest in radioiodine, a beta and gamma emitter, as new RPs are introduced for therapy and imaging (theranostics) for specific target tumor cells. However, one must be aware of issues related to the crossfire effect and toxicity of ß particles. The high LET and short range of α particles enable effective and rapid cancer therapies but are hindered by short half-lives and rarely emit gamma radiation for imaging [9,15,55]. Combining α and β particle emitters may reduce some of these obstacles [56,64]. Some issues with RPs are related to tumor-targeting uptake, biocompatibility, side effects, nonspecific uptake and distribution, and the radiation exposure effect in healthy tissues.

The development of intelligent drug delivery agents such as peptides, small molecules, mAb, and mAb fragments, and especially nanoparticle cores offer the promise of better diagnostic and therapeutic options [55,81]. However, the heterogeneity of RP uptake by tumor cells is challenging when using radiolabeled antibodies. The larger size of whole antibodies may limit penetration into the tumor tissue and crossfire effects, which occur when radiation interacts with cells away from the site of actual binding of the antibody agent [64].

Radiopharmaceuticals provide effective cancer treatment, particularly when other standard therapeutic approaches have failed. However, even after more than four decades of clinical investigation, RPs have still not become a standard part of cancer management therapy, which is peculiar, especially in light that other "targeted therapy" have clinical trial failure rates of 97% and is more popular than RPs that [7,8,85]. Furthermore, changing the fear of the public perception of radioactivity and the perceived complexity of the treatment are challenges in developing and applying RPs for therapy and imaging.

## 5. Conclusions and Future Direction

Radiopharmaceutical therapy can be a safe and effective targeted approach to treating many types of cancer. RPT has shown high efficacy with minimal toxicity compared to other systemic cancer treatment options. Different radioligands can be chosen to uniquely target molecular receptors or intracellular components, making them suitable for personal patient-tailored therapy in modern cancer therapy management. Further research is still needed regarding specific targets, radioligand stability in vivo, toxic effects, crossfire, dosimetry, and bond stability with daughter nuclides, particularly for alpha emitters.

However, new particle drug delivery systems continue to enhance targeted therapy efficacy and safety, including the use of nanoparticles. The number of successful studies exploring new drug delivery agents' different delivery systems of radionuclide particles will probably increase the effectiveness and range of applicants. With a growing positive track record, public understanding and perception of the safety and success of RPT may improve.

If this occurs, then RPT will be adopted as an increasingly mainstream cancer therapy approach and the investment needed to resolve issues of radionuclides supply. In the coming decades, RPT may provide an increasing variety of rapid, personalized, practical, effective, and affordable treatments that offer new hope to cancer patients.

**Author Contributions:** Conceptualization, A.E.; writing—draft preparation, S.S., A.A. and W.A.; writing—review, A.E. and S.S.; editing A.E.; funding acquisition, S.S. All authors have read and agreed to the published version of the manuscript.

**Funding:** Article publication fund from Fatima College of Health Sciences, the United Arab Emirates.

**Institutional Review Board Statement:** Not applicable.

**Informed Consent Statement:** Not applicable.

**Data Availability Statement:** Not applicable.

**Acknowledgments:** Thanks to Fatima College of Health Sciences, the United Arab Emirates, for fund of the article publication processing and Fay Farley for English editing assistance.

**Conflicts of Interest:** The authors declare no conflict of interest.

## References

1. Goldsmith, S.J. Targeted Radionuclide Therapy: A Historical and Personal Review. *Semin. Nucl. Med.* **2020**, *50*, 87–97. [CrossRef] [PubMed]
2. Yeong, C.H.; Cheng, M.H.; Ng, K.H. Therapeutic Radionuclides in Nuclear Medicine: Current and Future Prospects. *J. Zhejiang Univ. Sci. B* **2014**, *15*, 845–863. [CrossRef] [PubMed]
3. Ercan, M.T.; Caglar, M. Therapeutic Radiopharmaceuticals. In *Current Pharmaceutical Design*; Bentham Science Publishers: Sharjah, United Arab Emirates, 2000; Volume 6, pp. 1085–1121.
4. Gabriel, M. Radionuclide therapy beyond radioiodine. *Wien. Med. Wochenschr.* **2012**, *162*, 430–439. [CrossRef] [PubMed]
5. Hillegonds, D.J.; Franklin, S.; Shelton, D.K.; Vijayakumar, S.; Vijayakumar, V. The management of painful bone metastases with an emphasis on radionuclide therapy. *J. Natl. Med. Assoc.* **2007**, *99*, 785–794. [PubMed]
6. Asadian, S.; Mirzaei, H.; Kalantari, B.A.; Davarpanah, M.R.; Mohamadi, M.; Shpichka, A.; Nasehi, L.; Es, H.A.; Timashev, P.; Najimi, M.; et al. β-radiating radionuclides in cancer treatment, novel insight into promising approach. *Pharmacol. Res.* **2020**, *160*, 105070. [CrossRef] [PubMed]
7. Sgouros, G.; Bodei, L.; McDevitt, M.R.; Nedrow, J.R. Radiopharmaceutical therapy in cancer: Clinical advances and challenges. *Nat. Rev. Drug Discov.* **2020**, *19*, 589–608. [CrossRef] [PubMed]
8. Elliyanti, A. Radiopharmaceuticals in Modern Cancer Therapy. In *Radiopharmaceutical Current Research for Better Diagnosis, Therapy, Environmental and Pharmaceutical Applications*, 1st ed.; Badria, F.A., Ed.; Intechopen: London, UK, 2021. [CrossRef]
9. Kramer-Marek, G.; Capala, J. The role of nuclear medicine in modern therapy of cancer. *Tumor Biol.* **2012**, *33*, 629–640. [CrossRef] [PubMed]
10. Herrmann, K.; Schwaiger, M.; Lewis, J.S.; Solomon, S.B.; McNeil, B.J.; Baumann, M.; Gambhir, S.S.; Hricak, H.; Weissleder, R. Radiotheranostics: A roadmap for future development. *Lancet Oncol.* **2020**, *21*, e146–e156. [CrossRef]
11. Kumar, C.; Shetake, N.; Desai, S.; Kumar, A.; Samuel, G.; Pandey, B.N. Relevance of radiobiological concepts in radionuclide therapy of cancer. *Int. J. Radiat. Biol.* **2016**, *92*, 173–186. [CrossRef]
12. Wulbrand, C.; Seidl, C.; Gaertner, F.C.; Bruchertseifer, F.; Morgenstern, A.; Essler, M.; Senekowitsch-Schmidtke, R. Alpha-particle emitting 213Bi-anti-EGFR immunoconjugates eradicate tumor cells independent of oxygenation. *PLoS ONE* **2013**, *8*, e64730. [CrossRef]
13. Calais, P.J.; Turner, J.H. Outpatient 131I-rituximab radioimmunotherapy for non-Hodgkin lymphoma: A study in safety. *Clin. Nucl. Med.* **2012**, *37*, 732–737. [CrossRef] [PubMed]
14. Qaim, S.M. Therapeutic radionuclides and nuclear data. *Radiochim. Acta* **2001**, *89*, 297–304. [CrossRef]
15. Ferrier, M.G.; Radchenko, V. An appendix of radionuclides used in targeted alpha therapy. *J. Med. Imaging Radiat. Sci.* **2019**, *50*, S58–S65. [CrossRef] [PubMed]
16. Baskar, R.; Dai, J.; Wenlong, N.; Yeo, R.; Yeoh, K.W. Biological response of cancer cells to radiation treatment. *Front. Mol. Biosci.* **2014**, *1*, 24. [CrossRef]
17. Komal, S.; Nadeem, S.; Faheem, Z.; Raza, A.; Sarwer, K.; Umer, H.; Roohi, S.; Naqvi, S.A.R. Localization Mechanisms of Radiopharmaceuticals. In *Medical Isotopes*, 1st ed.; Naqvi, S.A.R., Imran, M.B., Eds.; Intechopen: London, UK, 2021. [CrossRef]
18. Elliyanti, A. An introduction to nuclear medicine in oncological molecular imaging. In Proceedings of the AIP Conference Proceedings, Padang, Indonesia, 10 December 2019. [CrossRef]

19. Arslan, N.; Emi, M.; Alagöz, E.; Üstünsöz, B.; Oysul, K.; Arpacı, F.; Uğurel, Ş.; Beyzadeoğlu, M.; Ozgüven, M.A. Selective intraarterial radionuclide therapy with Yttrium-90 (Y-90) microspheres for hepatic neuroendocrine metastases: Initial experience at a single center. *Vojnosanit. Pregl.* **2011**, *68*, 341–348. [CrossRef]
20. Kucuk, O.N.; Soydal, C.; Lacin, S.; Ozkan, E.; Bilgic, S. Selective intraarterial radionuclide therapy with yttrium-90 (Y-90) microspheres for unresectable primary and metastatic liver tumors. *World J. Surg. Oncol.* **2011**, *9*, 86. [CrossRef]
21. Houle, S.; Yip, T.K.; Shepherd, F.A.; Rotstein, L.E.; Sniderman, K.W.; Theis, E.; Cawthorn, R.H.; Richmond-Cox, K. Hepatocellular carcinoma: Pilot trial of treatment with Y-90 microspheres. *Radiology* **1989**, *172*, 857–860. [CrossRef]
22. Thamboo, T.; Tan, K.B.; Wang, S.C.; Salto-Tellez, M. Extrahepatic embolisation of Y-90 microspheres from selective internal radiation therapy (SIRT) of the liver. *Pathology* **2003**, *35*, 351–353.
23. Widel, M.; Przybyszewski, W.M.; Cieslar-Pobuda, A.; Saenko, Y.V.; Rzeszowska-Wolny, J. Bystander normal human fibroblasts reduce damage response in radiation targeted cancer cells through intercellular ROS level modulation. *Mutat. Res.* **2012**, *731*, 117–124. [CrossRef]
24. Ku, A.; Facca, V.J.; Cai, Z.; Reilly, R.M. Auger electrons for cancer therapy—A review. *EJNMMI Radiopharm. Chem.* **2019**, *4*, 27. [CrossRef]
25. Elliyanti, A. Molecular Radiobiology and Radionuclides Therapy Concepts. In *The Evolution of Radionanotargeting towards Clinical Precission Oncology: A Festschrift in Honor of Kalevi Kairemo*; Jekunen, A., Ed.; Bentham Science: Sharjah, United Arab Emirates, 2022; pp. 395–408. [CrossRef]
26. Navalkissoor, S.; Grossman, A. Targeted alpha particle therapy for neuroendocrine tumours: The next generation of peptide receptor radionuclide therapy. *Neuroendocrinology* **2019**, *108*, 256–264. [CrossRef] [PubMed]
27. Elliyanti, A. Radioiodine for Graves' Disease Therapy. In *Graves' Diseas*, 1st ed.; Gensure, R., Ed.; Intechopen: London, UK, 2021. [CrossRef]
28. Slonimsky, E.; Tulchinsky, M. Radiotheragnostics Paradigm for Radioactive Iodine (Iodide) Management of Differentiated Thyroid Cancer. *Curr. Pharm. Des.* **2020**, *26*, 3812. [CrossRef] [PubMed]
29. Luster, M.; Pfestroff, A.; Hänscheid, H.; Verburg, F.A. Radioiodine Therapy. *Semin. Nucl. Med.* **2017**, *47*, 126–134. [CrossRef]
30. Kendi, A.T.; Moncayo, V.M.; Nye, J.A.; Galt, J.R.; Halkar, R.; Schuster, D.M. Radionuclide therapies in molecular imaging and precision medicine. *PET Clin.* **2017**, *12*, 93–103. [CrossRef] [PubMed]
31. Koziorowski, J.; Ballinger, J. Theragnostic radionuclides: A clinical perspective. *Q. J. Nucl. Med. Mol. Imaging* **2021**, *65*, 306–314. [CrossRef] [PubMed]
32. Kassis, I.A.; Adelstein, S.J. Radiobiologic principles of radionuclide therapy. *J. Nucl. Med.* **2005**, *46*, 4S–12S. [PubMed]
33. Gholami, Y.H.; Maschmeyer, R.; Kuncic, Z. Radio-enhancement effects by radiolabeled nanoparticles. *Sci. Rep.* **2019**, *9*, 14346. [CrossRef]
34. Pouget, J.P.; Navarro-Teulon, I.; Bardiès, M.; Chouin, N.; Cartron, G.; Pèlegrin, A.; Azria, D. Clinical Radioimmunotherapy—The role of radiobiology. *Nat. Rev. Clin. Oncol.* **2011**, *8*, 720–734. [CrossRef]
35. Jia, Z.; Wang, W. Yttrium-90 radioembolization for unresectable metastatic neuroendocrine liver tumor: A systematic review. *Eur. J. Radiol.* **2018**, *100*, 23–29. [CrossRef]
36. Filippi, L.; Schillaci, O.; Cianni, R.; Bagni, O. Yttrium-90 resin microspheres and their use in the treatment of intrahepatic cholangiocarcinoma. *Future Oncol.* **2018**, *14*, 809–818. [CrossRef]
37. Oei, A.L.; Verheijen, R.H.; Seiden, M.V.; Benigno, B.B.; Lopes, A.D.B.; Soper, J.T.; Epenetos, A.A.; Massuger, L.F. Decreased intraperitoneal disease recurrence in epithelial ovarian cancer patients receiving intraperitoneal consolidation treatment with yttrium-90-labeled murine HMFG1 without improvement in overall survival. *Int. J. Cancer* **2007**, *120*, 2710–2714. [CrossRef] [PubMed]
38. Waldmann, T.; White, J.; Carrasquillo, J.; Reynolds, J.; Paik, C.; Gansow, O.; Brechbiel, M.; Jaffe, E.; Fleisher, T.; Goldman, C. Radioimmunotherapy of interleukin-2R alpha-expressing adult T-cell leukemia with yttrium-90-labeled anti-Tac. *Blood* **1995**, *86*, 4063–4075. [CrossRef] [PubMed]
39. Nisa, L.; Savelli, G.; Giubbini, R. Yttrium-90 DOTATOC therapy in GEP-NET and other SST2 expressing tumors: A selected review. *Ann. Nucl. Med.* **2011**, *25*, 75–85. [CrossRef] [PubMed]
40. Kang, J.; Li, C.; Rosenkrans, Z.T.; Huo, N.; Chen, Z.; Ehlerding, E.B.; Huo, Y.; Ferreira, C.A.; Barnhart, T.E.; Engle, J.W.; et al. CD38-Targeted Theranostics of Lymphoma with $^{89}Zr/^{177}Lu$-Labeled Daratumumab. *Adv. Sci.* **2021**, *8*, 2001879. [CrossRef]
41. Da Silva, T.N.; van Velthuysen, M.L.F.; van Eijck, C.H.J.; Teunissen, J.J.; Hofland, J. Successful neoadjuvant peptide receptor radionuclide therapy for an inoperable pancreatic neuroendocrine tumour. *Endocrinol. Diabetes Metab. Case Rep.* **2018**, *11*, 18-0015. [CrossRef]
42. Sartor, O. Overview of samarium Sm 153 lexidronam in the treatment of painful metastatic bone disease. *Rev. Urol.* **2004**, *6*, S3–S12.
43. Sgouros, G. Alpha-particles for targeted therapy. *Adv. Drug Deliv. Rev.* **2008**, *60*, 1402–1406. [CrossRef]
44. Manafi-Farid, R.; Masoumi, F.; Divband, G.; Saidi, B.; Ataeinia, B.; Hertel, F.; Schweighofer-Zwink, G.; Morgenroth, A.; Beheshti, M. Targeted Palliative Radionuclide Therapy for Metastatic Bone Pain. *J. Clin. Med.* **2020**, *9*, 2622. [CrossRef]
45. Guerra Liberal, F.D.C.; O'Sullivan, J.M.; McMahon, S.J.; Prise, K.M. Targeted Alpha Therapy: Current Clinical Applications. *Cancer Biother. Radiopharm.* **2020**, *35*, 404–417. [CrossRef]

46. Jurcic, J.G.; Levy, M.; Park, J.; Ravandi, F.; Perl, A.; Pagel, J.; Smith, B.D.; Orozco, J.; Estey, E.; Kantarjian, H.; et al. Trial in progress: A phase I/II study of lintuzumab-Ac225 in older patients with untreated acute myeloid leukemia. *Clin. Lymphoma Myeloma Leuk.* **2017**, *17*, S277. [CrossRef]
47. Kleynhans, J.; Sathekge, M.; Ebenhan, T. Obstacles and Recommendations for Clinical Translation of Nanoparticle System-Based Targeted Alpha-Particle Therapy. *Materials* **2021**, *14*, 4784. [CrossRef] [PubMed]
48. Goyal, J.; Antonarakis, E.S. Bone-targeting radiopharmaceuticals for the treatment of prostate cancer with bone metastases. *Cancer Lett.* **2012**, *323*, 135. [CrossRef] [PubMed]
49. Filippi, L.; Chiaravalloti, A.; Schillaci, O.; Cianni, R.; Bagni, O. Theranostic approaches in nuclear medicine: Current status and future prospects. *Expert Rev. Med. Devices* **2020**, *17*, 331–343. [CrossRef] [PubMed]
50. Bertolet, A.; Ramos-Méndez, J.; Paganetti, H.; Schuemann, J. The relation between microdosimetry and induction of direct damage to DNA by alpha particles. *Phys. Med. Biol.* **2021**, *66*, 155016. [CrossRef] [PubMed]
51. Liberini, V.; Huellner, M.W.; Grimaldi, S.; Finessi, M.; Thuillier, P. The Challenge of Evaluating Response to Peptide Receptor Radionuclide Therapy in Gastroenteropancreatic Neuroendocrine Tumors: The Present and the Future. *Diagnostics* **2020**, *10*, 1083. [CrossRef]
52. McDevitt, M.R.; Sgouros, G.; Sofou, S. Targeted and nontargeted α-particle therapies. *Annu. Rev. Biomed. Eng.* **2018**, *20*, 73–93. [CrossRef]
53. Gustafsson-Lutz, A.; Bäck, T.; Aneheim, E.; Hultborn, R.; Palm, S.; Jacobsson, L.; Morgenstern, A.; Bruchertseifer, F.; Albertsson, P.; Lindegren, S. Therapeutic efficacy of α-radioimmunotherapy with different activity levels of the $^{213}$Bi-labeled monoclonal antibody MX35 in an ovarian cancer model. *EJNMMI Res.* **2017**, *7*, 38. [CrossRef]
54. Ahenkorah, S.; Cassells, I.; Deroose, C.; Cardinaels, T.; Burgoyne, A.; Bormans, G.; Ooms, M.; Cleeren, F. Bismuth-213 for Targeted Radionuclide Therapy: From Atom to Bedside. *Pharmaceutics* **2021**, *3*, 599. [CrossRef]
55. Silindir-Gunay, M.; Karpuz, M.; Ozer, A.Y. Targeted alpha therapy and Nanocarrier approach. *Cancer Biother. Radiopharm.* **2020**, *35*, 446–458. [CrossRef]
56. Rosar, F.; Krause, J.; Bartholomä, M.; Maus, S.; Stemler, T.; Hierlmeier, I.; Linxweiler, J.; Ezziddin, S.; Khreish, F. Efficacy and safety of [$^{225}$Ac] Ac-PSMA-617 augmented [$^{177}$Lu] Lu-PSMA-617 Radioligand therapy in patients with highly advanced mCRPC with poor prognosis. *Pharmaceutics* **2021**, *13*, 722. [CrossRef]
57. Reissig, F.; Wunderlich, G.; Runge, R.; Freudenberg, R.; Lühr, A.; Kotzerke, J. The effect of hypoxia on the induction of strand breaks in plasmid DNA by alpha-, beta- and Auger electron-emitters $^{223}$Ra, $^{188}$Re, $^{99m}$Tc and DNA-binding $^{99m}$Tc-labeled pyrene. *Nucl. Med. Biol.* **2020**, *80–81*, 65–70. [CrossRef] [PubMed]
58. Stéen, E.J.L.; Edem, P.E.; Nørregaard, K.; Jørgensen, J.T.; Shalgunov, V.; Kjaer, A.; Herth, M.M. Pretargeting in nuclear imaging and radionuclide therapy: Improving efficacy of theranostics and nanomedicines. *Biomaterials* **2018**, *179*, 209–245. [CrossRef] [PubMed]
59. Kennel, S.J.; Mirzadeh, S.; Eckelman, W.C.; Waldmann, T.A.; Garmestani, K.; Yordanov, A.T.; Stabin, M.G.; Brechbiel, M.W. Vascular-targeted radioimmunotherapy with the alpha- particle emitter 211At. *Radiat. Res.* **2002**, *157*, 633–641. [CrossRef]
60. Persson, L. The Auger electron effect in radiation dosimetry. *Health Phys.* **1994**, *67*, 471–476. [CrossRef] [PubMed]
61. Widel, M. Radionuclides in radiation-induced bystander effect; may it share in radionuclide therapy? *Neoplasma* **2017**, *64*, 641–654. [CrossRef] [PubMed]
62. Kirsch, D.G.; Diehn, M.; Kesarwala, A.; Maity, A.; Morgan, M.A.; Schwarz, J.K.; Bristow, R.; DeMaria, S.; Eke, I.; Griffin, R.J.; et al. The Future of Radiobiology. *J. Natl. Cancer Inst.* **2018**, *110*, 329–340. [CrossRef]
63. Paillas, S.; Ladjohounlou, R.; Lozza, C.; Pichard, A.; Boudousq, V.; Jarlier, M.; Sevestre, S.; Le Blay, M.; Deshayes, E.; Sosabowski, J.; et al. Localized irradiation of cell membrane by auger electrons is cytotoxic through oxidative stress-mediated nontargeted effects. *Antioxid. Redox Signal.* **2016**, *25*, 467–484. [CrossRef]
64. Pandit-Taskar, N. Targeted Radioimmunotherapy and Theranostics with Alpha Emitters. *J. Med. Imaging Radiat. Sci* **2019**, *50*, S41–S44. [CrossRef]
65. White, J.M.; Escorcia, F.E.; Viola, N.T. Perspectives on metals-based radioimmunotherapy (RIT): Moving forward. *Theranostics* **2021**, *11*, 6293–6314. [CrossRef]
66. Fendler, W.P.; Rahbar, K.; Herrmann, K.; Kratochwil, C.; Eiber, M. 177Lu-PSMA Radioligand Therapy for Prostate Cancer. *J. Nucl. Med.* **2017**, *58*, 1196–1200. [CrossRef]
67. Nevedomskaya, E.; Baumgart, S.J.; Haendler, B. Recent advances in prostate Cancer treatment and drug discovery. *Int. J. Mol. Sci.* **2018**, *19*, 1359. [CrossRef] [PubMed]
68. Lunger, L.; Tauber, R.; Feuerecker, B.; Gschwend, J.E.; Eiber, M. Narrative review: Prostate-specific membrane antigen-radioligand therapy in metastatic castration-resistant prostate cancer. *Transl. Androl. Urol.* **2021**, *10*, 3963–3971. [CrossRef] [PubMed]
69. Kairemo, K.; Joensuu, T. Lu-177-PSMA treatment for metastatic prostate cancer: Case examples of major responses. *Clin. Transl Imaging* **2018**, *6*, 223–237. [CrossRef]
70. Sathekge, M.; Bruchertseifer, F.; Knoesen, O.; Reyneke, F.; Lawal, I.; Lengana, T.; Davis, C.; Mahapane, J.; Corbett, C.; Vorster, M.; et al. $^{225}$Ac-PSMA-617 in chemotherapy-naive patients with advanced prostate cancer: A pilot study. *Eur. J. Nucl. Med. Mol. Imaging* **2019**, *46*, 129–138. [CrossRef] [PubMed]

71. Khreish, F.; Ebert, N.; Ries, M.; Maus, S.; Rosar, F.; Bohnenberger, H.; Stemler, T.; Saar, M.; Bartholomä, M.; Ezziddin, S. $^{225}$Ac-PSMA-617/$^{177}$Lu-PSMA-617 tandem therapy of metastatic castration-resistant prostate cancer: Pilot experience. *Eur. J. Nucl. Med. Mol. Imaging* **2020**, *47*, 721–728. [CrossRef]
72. Kratochwil, C.; Haberkorn, U.; Giesel, F.L. $^{225}$Ac-PSMA-617 for Therapy of Prostate Cancer. *Semin. Nucl. Med.* **2020**, *50*, 133–140. [CrossRef]
73. Basu, S.; Parghane, R.V.; Chakrabarty, S. Peptide Receptor Radionuclide Therapy of Neuroendocrine Tumors. *Semin. Nucl. Med.* **2020**, *50*, 447–464. [CrossRef]
74. Kunikowska, J.; Królicki, L. Targeted α-Emitter Therapy of Neuroendocrine Tumors. *Semin. Nucl. Med.* **2020**, *50*, 171–176. [CrossRef]
75. Elliyanti, A.; Rustam, R.; Tofrizal, T.; Yenita, Y.; Susanto, Y.D.B. Evaluating the Natrium iodide Symporter expressions in thyroid Tumors. *Open Access Maced. J. Med. Sci.* **2021**, *9*, 18–23. [CrossRef]
76. Elliyanti, A.; Rusnita, D.; Afriani, N.; Susanto, Y.D.B.; Susilo, V.Y.; Setiyowati, S.; Harahap, W.A. Analysis natrium iodide symporter expression in breast cancer subtypes for radioiodine therapy response. *Nucl. Med. Mol. Imaging* **2020**, *54*, 35–42. [CrossRef]
77. Liu, J.; Liu, Y.; Lin, Y.; Liang, J. Radioactive Iodine-Refractory Differentiated Thyroid Cancer and Redifferentiation Therapy. *Endocrinol. Metab.* **2019**, *34*, 215–225. [CrossRef]
78. Schug, C.; Gupta, A.; Urnauer, S.; Steiger, K.; Cheung, P.F.Y.; Neander, C.; Savvatakis, K.; Schmohl, K.A.; Trajkovic-Arsic, M.; Schwenk, N. A Novel Approach for Image-Guided $^{131}$I Therapy of Pancreatic Ductal Adenocarcinoma Using Mesenchymal Stem Cell-Mediated NIS Gene Delivery. *Mol. Cancer Res.* **2019**, *17*, 310–320. [CrossRef]
79. Bayoumi, N.A.; El-Kolaly, M.T. Utilization of nanotechnology in targeted radionuclide cancer therapy: Monotherapy, combined therapy and radiosensitization. *Radiochim. Acta* **2021**, *109*, 459–475. [CrossRef]
80. Mirshojaei, S.F.; Ahmadi, A.; Morales-Avila, E.; Ortiz-Reynoso, M.; Reyes-Perez, H. Radiolabelled nanoparticles: Novel classification of radiopharmaceuticals for molecular imaging of cancer. *J. Drug Target.* **2016**, *24*, 91–101. [CrossRef]
81. Ting, G.; Chang, C.H.; Wang, H.E.; Lee, T.W. Nanotargeted radionuclides for cancer nuclear imaging and internal radiotherapy. *J. Biomed. Biotechnol.* **2010**, *2010*, 953537. [CrossRef]
82. Farzin, L.; Sheibani, S.; Moassesi, M.E.; Shamsipur, M. An overview of nanoscale radionuclides and radiolabeled nanomaterials commonly used for nuclear molecular imaging and therapeutic functions. *J. Biomed. Mater. Res. A* **2019**, *107*, 251–285. [CrossRef]
83. Chang, C.H.; Chang, M.C.; Chang, Y.J.; Chen, L.C.; Lee, T.W.; Ting, G. Translating Research for the Radiotheranostics of Nanotargeted $^{188}$Re-Liposome. *Int. J. Mol. Sci.* **2021**, *22*, 3868. [CrossRef]
84. Salvanou, E.-A.; Stellas, D.; Tsoukalas, C.; Mavroidi, B.; Paravatou-Petsotas, M.; Kalogeropoulos, N.; Xanthopoulos, S.; Denat, F.; Laurent, G.; Bazzi, R.; et al. A Proof-of-Concept Study on the Therapeutic Potential of Au Nanoparticles Radiolabeled with the Alpha-Emitter Actinium-225. *Pharmaceutics* **2020**, *12*, 188. [CrossRef]
85. Wong, C.H.; Siah, K.W.; Lo, A.W. Estimation of clinical trial success rates and related parameters. *Biostatistics* **2018**, *20*, 273–286. [CrossRef]

Article

# Production and Quality Control of [$^{177}$Lu]Lu-PSMA-I&T: Development of an Investigational Medicinal Product Dossier for Clinical Trials

Valentina Di Iorio [1,*], Stefano Boschi [2], Cristina Cuni [1], Manuela Monti [1], Stefano Severi [1], Giovanni Paganelli [1] and Carla Masini [1]

[1] IRCCS Istituto Romagnolo per lo Studio dei Tumori "Dino Amadori" IRST, 47014 Meldola, Italy; cristina.cuni@irst.emr.it (C.C.); manuela.monti@irst.emr.it (M.M.); stefano.severi@irst.emr.it (S.S.); giovanni.paganelli@irst.emr.it (G.P.); carla.masini@irst.emr.it (C.M.)

[2] Department of Pharmacy and Biotechnologies, University of Bologna, 47921 Rimini, Italy; stefano.boschi@unibo.it

* Correspondence: valentina.diiorio@irst.emr.it; Tel.: +39-0543739930

**Abstract:** Since prostate cancer is the most commonly diagnosed malignancy in men, the theranostic approach has become very attractive since the discovery of urea-based PSMA inhibitors. Different molecules have been synthesized starting from the Glu-urea-Lys scaffold as the pharmacophore and then optimizing the linker and the chelate to improve functional characteristics. This article aimed to highlight the quality aspects, which could have an impact on clinical practice, describing the development of an Investigational Medicinal Product Dossier (IMPD) for clinical trials with [$^{177}$Lu]Lu-PSMA-I&T in prostate cancer and other solid tumors expressing PSMA. The results highlighted some important quality issues of the final preparation: radiolabeling of PSMA-I&T with lutetium-177 needs a considerably longer time compared with the radiolabeling of the well-known [$^{177}$Lu]Lu-PSMA-617. When the final product was formulated in saline, the stability of [$^{177}$Lu]Lu-PSMA-I&T was reduced by radiolysis, showing a decrease in radiochemical purity (<95% in 24 h). Different formulations of the final product with increasing concentrations of ascorbic acid have been tested to counteract radiolysis and extend stability. A solution of 20 mg/mL of ascorbic acid in saline prevents radiolysis and ensures stability over 30 h.

**Keywords:** prostate cancer; [$^{177}$Lu]Lu-PSMA-I&T; IMPD; quality assurance

## 1. Introduction

Prostate cancer (PCa) is the most commonly diagnosed malignancy in men worldwide and remains one of the leading causes of cancer-related deaths. Prostate-specific membrane antigen (PSMA) is a type II membrane glycoprotein with an extensive extracellular domain (44–750 amino acids) and plays a significant role in prostate carcinogenesis and progression [1–4].

PSMA expression correlates with the malignancy of the disease, being further increased in metastatic and hormone-refractory patients [5]. As a consequence, PSMA has attracted attention as a target for molecular imaging and targeted radioligand therapy, especially in metastatic castration-resistant prostate cancer (mCRPC).

Since the discovery of urea-based PSMA inhibitors in 2001 [6], a variety of PSMA-targeted radioligands for imaging prostate cancer was developed. Briefly, most of the relevant molecules are structured by three main components: a pharmacophore, usually X-urea-Glu (XuE)-scaffold, in our case X = K; a linker to enhance affinity by interaction with receptor hydrophobic pocket; and a chelator to bind the radionuclide. Other types of PSMA inhibitors are the ones labeled with fluorine-18 ([$^{18}$F]F-PSMA-1007, ([$^{18}$F]F-DCMPyL, and others), the chelator is in those cases replaced by organic structures with a leaving group to support an SN2 reaction. The "gold standard" ligand for PSMA imaging is Glu-NH-CO-NH-Lys(Ahx)-$^{68}$Ga-HBED-CC ([$^{68}$Ga]Ga-PSMA-HBED-CC) characterized by a high-affinity chelator for Gallium-68 (logK = 38.5) that seems to interact advantageously with the lipophilic part of the PSMA binding pocket [7]. Unfortunately, the HBED-CC chelator is unsuitable for radiolabeling with therapeutic radiometals such as yttrium-90 and lutetium-177, and therefore new theranostic compounds were designed to bind gallium-68, as well as yttrium-90 and lutetium-177.

Substitution of HBED-CC with DOTA and systematical modification of the side chain with the introduction of a naphtylic linker have led to PSMA-617 [8] with excellent pharmacokinetic properties, high binding affinity and internalization, prolonged tumor uptake, and high tumor-to-background ratio, which are extremely important for both imaging quality and therapy.

In parallel, another theranostic PSMA-targeted radioligand, PSMA-I&T, was explored [9]. The DOTAGA-FFK(SubKuE)-scaffold represents a flexible and adjustable backbone for the development of KuE-based PSMA inhibitors. Additionally, the DOTA chelator was substituted with DOTAGA (1,4,7,10-tetraazacyclododecane-1-(glutaric acid)-4,7,10-triacetic acid) to facilitate the yttrium-90 and lutetium-177-labeling procedure, improve pharmacokinetics and, potentially, affinity to the receptor. DOTAGA derivatives showed higher hydrophilicity (logP = $-3.9 \pm 0.1$ for DOTAGA 177Lu derivative compared with $-2.7 \pm 0.02$ for that DOTA derivative) and improved affinity to PSMA, compared with DOTA-coupled counterpart, resulting in about a twofold-increased specific internalization of the $^{68}$Ga- and $^{177}$Lu-labelled DOTAGA analogue [10]. The substitution of one of the D-phenylalanine residues in the peptidic linker by 3-iodo-D-tyrosine improved the interaction of the tracer molecule with a remote binding site. These modifications, together with increasing the lipophilic interaction of the tracer with the PSMA enzymes, led to the second-generation theranostic tracers DOTAGA-(I-y), fk(Sub-KuE), and the PSMA I&T [9].

In order to use a radiopharmaceutical in human applications, it should be manufactured under Good Manufacturing Practice (GMP) or National Regulations. In Italy, the reference quality assurance system is "Norme di Buona Preparazione dei Radiofarmaci per Medicina Nucleare (NBP-MN)" [11], which is a GMP-like quality assurance system dealing with no-profit clinical trials.

To submit the clinical study to the Italian Medicines Agency (AIFA), an Investigational Medicinal Product Dossier (IMPD) needs to be produced for [$^{177}$Lu]Lu-PSMA-I&T according to European Medicines Agency EMA guideline [12]. This guideline aims to address the documentation on the chemical and pharmaceutical quality of investigational medicinal products (IMPs), including radiopharmaceuticals, to ensure their quality, safety, and efficacy.

The aim of this paper is to describe the process validation as well as the analytical methods, along with establishing acceptance criteria for [$^{177}$Lu]Lu-PSMA-I&T according to the purpose of obtaining an IMPD.

At the moment the ligand PSMA-I&T is commercially available from ABX and there are no patents that prevent us from working with PSMA-I&T.

## 2. Results

IMPD for [$^{177}$Lu]Lu-PSMA-I&T was prepared according to the EMA guideline [12]. IMPD includes the most up-to-date information relevant to the clinical trial available at the time of submission of the clinical trial application. It essentially consists of two parts, the first dedicated to the drug substance and the second dedicated to the investigational medicinal product under test.

### 2.1. Drug Substance

Two drug substances have been identified: the ligand PSMA-I&T and the precursor [$^{177}$Lu]LuCl$_3$.

2.1.1. PSMA-I&T

Nomenclature:

(R)-DOTAGA-D-Tyr(3-I)-D-Phe-D-Lys[Sub-Lys-CO-Glu]-OH (supplied as acetate salt)
Synonyms (R)-DOTAGA-(I-y)fk(Sub-KuE) Structure
Molecular formula: $C_{63}H_{92}IN_{11}O_{23}$
Molecular weight: 1498.37 g/mol

The chemical structure of PSMA-I&T is showed in Figure 1.

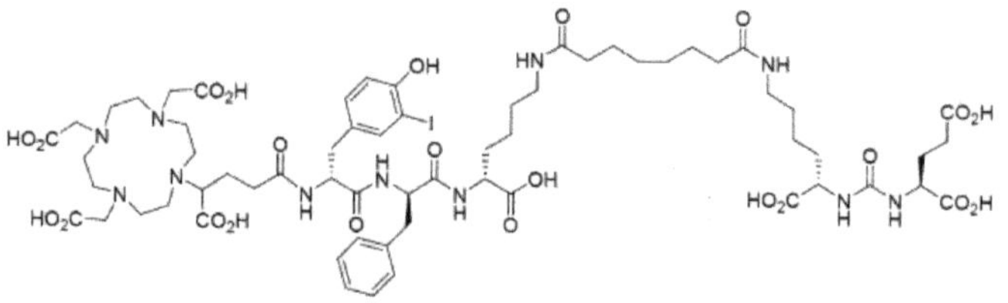

Figure 1. The chemical structure of PSMA- I&T.

2.1.2. Lutetium-177

PSMA-I&T was radiolabeled with lutetium-177 chloride. Lutetium-177 chloride was produced under a marketing authorization (MA) by irradiation of highly enriched (>99%) ytterbium-176 by a neutron source.

Lutetium-177 was produced according to the following nuclear reaction:

$$[^{176}Yb(n,\gamma)^{177}Yb \to (\beta^-) \to {}^{177}Lu]$$

The neutron thermal flux was between 1013 and 1016 $cm^{-2}\,s^{-1}$. The nuclear reaction is no-carried added (n.c.a.). The n.c.a. reaction results in a very high specific activity (≥3000 GBq/mg), in comparison with the lower specific activity (500 GBq/mg) when the Lutetium-177 is obtained by neutron irradiation of Lutetium-176. Moreover, the "direct production" leads to formation of a radionuclidic impurity (Lutetium-177m), not present in the Lutetium-177 produced by ytterbium-176. Metallic impurities concentration was very low, the sum of both metallic as well as other radionuclidic impurities cannot interfere with labeling nor with radiochemical and radionuclidic purity.

The specifications for the release of lutetium-177 chloride are indicated in Table 1.

Table 1. Specifications of lutetium-177 chloride.

| Test | Method | Specification | Unit |
|---|---|---|---|
| [$^{177}$Lu]LuCl$_3$ in HCl 0.04M pH 1-2 Activity per Vial value decay corrected to ART | n.a. | 90–110 of the activity stated in the label | % |
| Volume delivered | n.a. | 0.4–0.8 mL According to the radioactivity ordered | mL |
| Appearance | Visual test | Clear and colorless solution | n.a. |
| Identity Lu-177 | Gamma Spectrometry | 113 KeV gamma line 208 KeV gamma line | n.a. |
| Identity Chloride | Eu. Phar. | White precipitate visible | n.a. |
| Specific activity value decay corrected to ART | ICP-MS | ≥3000 | GBq/mg |
| Radionuclidic purity Radiochemical purity | Gamma Spectrometry TLC | Yb-175 ≤ 0.01 Sum of impurities ≤ 0.01 ≥99.0 as $^{177}$LuCl$_3$ | % % % |

Table 1. Cont.

| Test | Method | Specification | Unit |
|---|---|---|---|
| Chemical purity | ICP-MS | Fe ≤ 0.25<br>Cu ≤ 0.5<br>Zn ≤ 0.5<br>Pb ≤ 0.5<br>Yb-176 ≤ 0.1<br>Sum of impurities ≤ 0.5 | µg/GBq |
| Radiolabeling yield | TLC | ≥99.0 | % |
| Sterility | Eu. Phar. | Sterile | n.a. |
| Bacterial endotoxins | Eu. Phar. | ≤20 | EU/mL |

ART = Activity reference time, ICP-MS = Inductively coupled plasma–mass spectrometry, TLC = Thin layer chromatography, Eu.Phar. = European Pharmacopoeia. Radiolabeling yield (TLC): based on radiolabeling Lu-177 of DOTA-derivated molar ratio 1:4 (CoA).

### 2.2. Investigational Medicinal Product (IMP) under Test

Description and Composition of the IMP

The IMP consists of a description of the [$^{177}$Lu]Lu-PSMA-I&T solution, among other things stating the range of radioactivity (17,980–29,790 MBq), at the end of synthesis (EOS), which, in this case, is also considered Activity Reference Time (ART); the final volume is 17–25 mL.

The radioactive concentration is between 1057 and 1192 MBq/mL. IMP is formulated as a multidose drug with the components described in Table 2.

Table 2. Batch formula of [$^{177}$Lu]Lu-PSMA-I&T.

| Components | Function | Amount/Activity |
|---|---|---|
| [$^{177}$Lu]LuCl$_3$ | Active Pharmaceutical Ingredient (API) | 18,350–31,040 MBq<br>Activity Reference Time (ART) |
| PSMA-I&T | Precursor | 500–800 µg (334–534 nmol) |
| Water for injection | For reconstitution of PSMA-I&T | 0.5–1 mL |
| Gentisic/ascorbic buffer composition: | | |
| Gentisic acid | Radical scavenger | 16.8 mg (109 µmol) |
| Sodium acetate | Buffer solution | 32.4 mg (395 µmol) |
| Sodium hydroxide | pH balance buffer | 9.6 mg (240 µmol) |
| Ascorbic acid | Radical scavenger | 31.2 mg (177 µmol) |
| Ascorbic acid solution in NaCl 0.9% 20 mg/mL | Diluent and radical scavenger | 17–25 ml |

Typical radiometric and UV chromatograms of [$^{177}$Lu]Lu-PSMA-I&T syntheses are shown in Figure 2, which shows the difference between the retention time of [$^{177}$Lu]Lu-PSMA-I&T and the precursor PSMA-I&T (6.72 and 7.01, respectively). The largest peak in the UV chromatogram is gentisic acid. The radiometric detector evidences the high radiochemical purity of the product, since only the peak of [$^{177}$Lu]Lu-PSMA-I&T is present.

### 2.3. Quality Controls

2.3.1. Acceptance Criteria

Acceptance criteria, specifications, and release timing were chosen in compliance with the general texts and monographs of the current European Pharmacopoeia and are summarized in Table 3. The product should meet the acceptance criteria for all the established quality parameters. The administered patient dose ranged from 5500 to 7400 MBq according to the patient conditions.

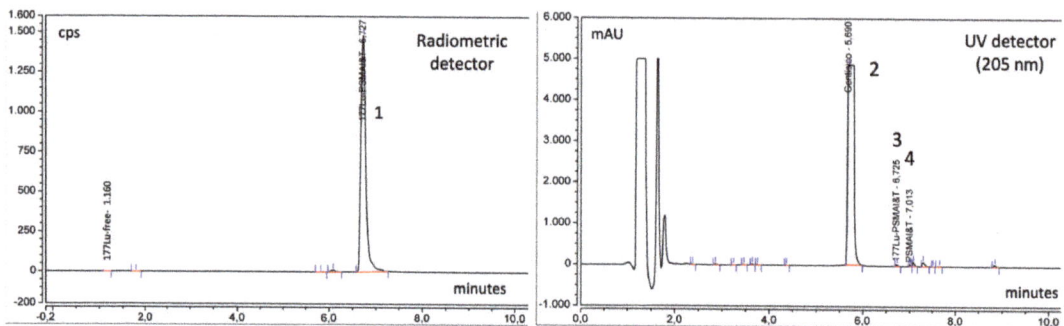

**Figure 2.** Relevant chromatograms of the final products of [$^{177}$Lu]Lu-PSMA-I&T with radiometric and UV detector: Peak 1 = [$^{177}$Lu]Lu-PSMA-I&T, Peak 2 = Gentisic acid, Peak 3 = [$^{177}$Lu]Lu-PSMA-I&T, Peak 4 = PSMA-I&T.

**Table 3.** Recommended test for the quality controls.

| Parameter | Method | Acceptance Criteria |
|---|---|---|
| [$^{177}$Lu]Lu-PSMA-I&T activity | Dose calibrator | 17,980–29,790 MBq |
| Radioactive concentration | Dose calibrator | 1057–1192 MBq/mL |
| Volume |  | 17–25 mL |
| Appearance | Visual test | Clear and colorless solution |
| Identification | HPLC | Rt [$^{177}$Lu]Lu-PSMA-I&T ± 0.2 min vs. Rt $^{nat}$Lu-PSMA-I&T reference standard |
| Radionuclidic identity | Gamma Spectrometry | 113 KeV gamma line 208 KeV gamma line |
| Yb-175 content | Gamma Spectrometry | Yb-175 ≤ 0.01% |
| Radiochemical purity | HPLC | [$^{177}$Lu]Lu ≤ 3% [$^{177}$Lu]Lu-PSMA-I&T ≥ 97% |
| Radiochemical purity | TLC | [$^{177}$Lu]Lu colloids ≤ 3% [$^{177}$Lu]Lu-PSMA-I&T ≥ 97% |
| pH | pH strips | 4.5–5.5 |
| Filter integrity | Bubble Point Test | ≥50 psi |
| Sterility | Sterility Test (Eur. Ph.) | Sterile |
| Bacterial endotoxins | Eur. Ph. | ≤175 EU/V |

PSMA-I&T ≤ 0.1 mg/V$_{max}$ where Vmax is the maximum injectable volume of the preparation. Volume is determined by the sum of the volume of the reagents and the volume of Ascorbic acid solution in NaCl 0.9% 20 mg/mL added to dilute the final product.

All the tests, except sterility, were carried out before the release.

2.3.2. Validation of the Analytical Procedures

Validation is the act of proving that any procedure, process, equipment, material, activity, or system actually leads to the expected results, with the aim to contribute to and guarantee the quality of a radiopharmaceutical.

The objective of validation of an analytical procedure is to demonstrate that it is suitable for its intended purpose.

The validation of the analytical procedures, the acceptance limits, and the parameters (specificity, linearity, range, accuracy, precision, quantification, and detection limit) for performing validation of analytical methods was carried out according to the ICH guideline Q2(R1) [13]. For the HPLC determination of chemical purity, $^{nat}$Lu-PSMA-I&T and PSMA-I&T were used. The high concentration and absorbance of gentisic acid cause it to be by far the highest peak in the UV (205 nm) spectrum of the PSMA-I&T product; we, therefore, chose to perform the validation of the PSMA-I&T analytical method in presence of gentisic acid to simulate closely the condition of the final product formulation. Parameters and acceptance criteria for the validation of the radio-HPLC method and the results obtained are shown in Table 4. Radionuclidic purity, [$^{177}$Lu]Lu-PSMA-I&T is verified on the basis of the certificate of analysis attached by the supplier of the [$^{177}$Lu]LuCl$_3$.

**Table 4.** Parameters and acceptance criteria for the validation of the radio-HPLC method and the obtained results.

| Parameters | Acceptance Criteria | Results |
|---|---|---|
| **Chemical Purity UV Detector** | | |
| Specificity | Rs $^{nat}$LuPSMA I&T and PSMA-I&T Rs $\geq$ 1.5 | Comply |
| Precision | CV% PSMA-I&T $\leq$ 5% CV% $^{nat}$LuPSMA-I&T $\leq$ 5% | <4% <3% |
| Linearity | R$^2$ PSMA-I&T $\geq$ 0.99 R$^2$ $^{nat}$LuPSMA-I&T $\geq$ 0.99 | $\geq$0.999 >0.999 |
| LOQ (µg/mL) | Experimental | PSMA-I&T = 6.8 $^{nat}$LuPSMA-I&T = 13.2 |
| LOD (µg/mL) | Experimental | PSMA-I&T = 2.2 $^{nat}$LuPSMA-I&T = 4.3 |
| Range | 80–120% | Comply |
| Accuracy | Average bias < 5% | Comply |
| **Radiochemical Purity Radiodetector** | | |
| Parameters | Acceptance Criteria | Results |
| Specificity | Difference t$_R$ ± 5% RT compared with RT $^{nat}$LuPSMA-I&T | ±4% |
| Precision | CV% $\leq$ 5% | $\leq$3.2% |
| Linearity | R$^2$ $\geq$ 0.99 | $\geq$0.999 |
| LOQ | n.a. | n.a. |
| LOD | n.a. | n.a. |
| Range | n.a. | n.a. |

Rs = Resolution, CV= coefficient of variation, R$^2$ = correlation coefficient, LOQ = quantitation limit, LOD = detection limit, n.a. = not applicable.

Chromatograms of standards $^{nat}$Lu-PSMA-I&T and PSMA-I&T are shown in Figure 3.

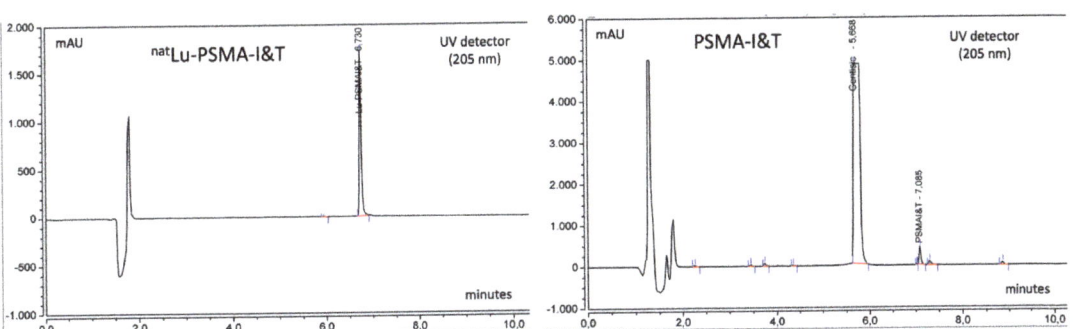

**Figure 3.** Chromatograms of standard solutions of $^{nat}$Lu-PSMA-I&T (20 µL of a 0.1 mg/mL solution) and PSMA-I&T (20 µL of a 0.04 mg/mL solution). Rt of $^{nat}$Lu-PSMA-I&T is slightly lower than that Rt of [$^{177}$Lu]Lu-PSMA-I&T with radiometric detection because it is positioned after the UV detector.

### 2.3.3. Bioburden

The pre-filtrated product was sent for a bioburden test, using 1 mL for each test sample. (Eurofins Laboratory Biolab Srl, Vimodrone, Milan Italy.)

The results were:

Total aerobic microbial count (TAMC) < 1 cfu/mL,
Total yeast and mold count (TYMC) < 1 cfu/mL,
where <1 cfu/mL means absence of colonies.

### 2.3.4. Batch Analysis and Process Validation

Process Validation should be intended as a means to establish that all the process parameters that bring to the preparation of the intended radiopharmaceutical and their quality characteristics are consistently and reproducibly met.

Process validation was carried out by producing three different batches of [$^{177}$Lu]Lu-PSMA-I&T on three different days, in the same conditions set for typical routine preparations and in the activity range reported in the acceptance criteria. Each batch was prepared accordingly the validation protocol should be fully characterized from the analytical point of view, with the aim to verify that the product meets the acceptance criteria as for all the established quality parameters.

Parameters were measured by the tests described in Table 3.

The results for three representative batches are shown in Table 5.

**Table 5.** Results of [$^{177}$Lu]Lu-PSMA-I&T representative batches.

| Parameter | Method | Acceptance Criteria | Batch 08/04/2021 | Batch 15/04/2021 | Batch 07/05/2021 |
|---|---|---|---|---|---|
| [$^{177}$Lu]Lu-PSMA-I&T Activity | Dose Calibrator | 17,980–29,790 MBq | 17,980 MBq | 21,720 MBq | 29,790 MBq |
| Radioactive concentration | Dose Calibrator | 1057–1192 MBq/mL | 1058 MBq/mL | 1086 MBq/mL | 1192 MBq/mL |
| Volume | - | 17–25 ml | 17 mL | 20 mL | 25 mL |
| Appearance | Visual test | Clear and Colorless Solution | Complies | Complies | Complies |
| Identification | HPLC | Rt [$^{177}$Lu]Lu-PSMA-I&T ± 0.2 min vs. Rt $^{nat}$Lu-PSMA-I&T reference standard | +0.012 | +0.03 | +0.01 |
| Radionuclidic identity | Gamma Spectrometry | 113 KeV gamma line 208 KeV gamma line | Comply | Comply | Comply |
| Yb-175 content | Gamma Spectrometry | Yb-175 ≤ 0.01% | Comply | Comply | Comply |
| Radiochemical purity | TLC | [$^{177}$Lu]Lu colloids ≤ 3% [$^{177}$Lu]Lu-PSMA-I&T ≥ 97% | 0 100% | 0 100% | 0 100% |
| Radiochemical purity | HPLC | [$^{177}$Lu]Lu ≤ 3% [$^{177}$Lu]Lu-PSMA-I&T ≥ 97% | 0.02% 99.3% | 0.02% 99.4% | 0.01% 99.4% |
| Chemical purity | HPLC | PSMA-I&T ≤ 0.1 mg/Vmax Sum of impurities ≤ 0.5 mg/Vmax | Complies | Complies | Complies |
| pH | pH Strips | 4.5–5.5 | 5 | 5 | 5 |
| Filter integrity | Bubble Point Test | ≥50 psi | ≥50 psi | ≥50 psi | ≥50 psi |
| Sterility | Sterility Test (Eur. Ph.) | Sterile | Sterile | Sterile | Sterile |
| Bacterial endotoxins | Eur. Ph. | ≤175 EU/V | ≤10 EU/V | ≤10 EU/V | ≤10 EU/V |

Rt = retention time.

All the batches used for process validation complied with the acceptance criteria.

### 2.3.5. Stability

Stability was assessed at 0, 24, and 30 h after the end of the synthesis. The three batches used for process validation were kept at room temperature and then parameters that could change over time such as appearance, radiochemical purity, and pH were reanalyzed after 24 and 30 h. The synthesis does not affect the radionuclidic purity so the radionuclidic purity is the same as observed on the sheet of [$^{177}$Lu]LuCl$_3$.

The analysis of the radiochemical purity was carried out by HPLC and TLC because the TLC method is used to detect the presence of colloidal lutetium-177 [14] that is not detectable with HPLC analysis.

For organizational reasons, it is not possible to administer the drug to the patient, later than 24 h after preparation. Chromatograms are shown in Figure 4; the stability data are shown in Table 6.

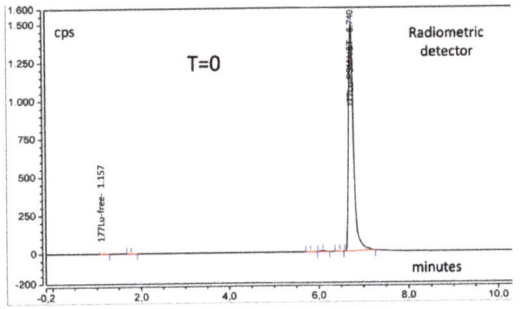
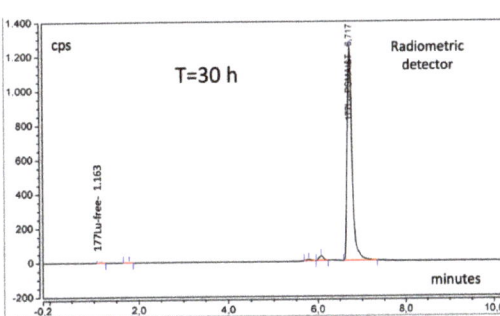
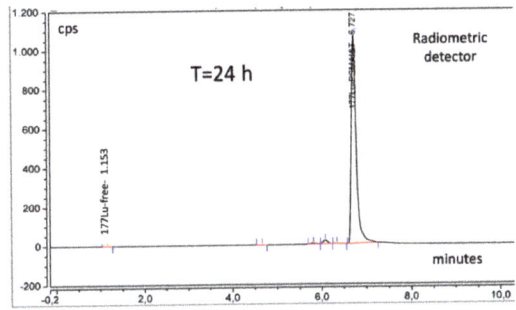

| [177Lu]Lu-PSMA-I&T Relative area | |
|---|---|
| T=0 | 99.9 % |
| T=24 | 97.73% |
| T=30 | 97.31% |

**Figure 4.** [$^{177}$Lu]Lu-PSMA-I&T (radiometric detector) during stability studies, at T = 0, T = 24 h, T = 30. In the table inside the figure are reported relative areas of [$^{177}$Lu]Lu-PSMA-I&T as a percentage of the total areas.

**Table 6.** Stability data of [$^{177}$Lu]Lu-PSMA-I&T at T0 and after 24 and 30 h at room temperature.

| | | | T0 Stability Test | | | |
|---|---|---|---|---|---|---|
| Parameter | Method | | Acceptance Criteria | Batch 08/04/2021 | Batch 15/04/2021 | Batch 07/05/2021 |
| Appearance | Visual test | | Clear and Colorless Solution | Complies | Complies | Complies |
| Radiochemical purity | TLC | | [$^{177}$Lu]Lu colloids ≤ 3%<br>[$^{177}$Lu]Lu-PSMA-I&T ≥ 97% | 0<br>100% | 0<br>100% | 0<br>100% |
| Radiochemical purity | HPLC | | [$^{177}$Lu]Lu ≤ 3%<br>[$^{177}$Lu]Lu-PSMA-I&T ≥ 97% | 0.02%<br>99.3% | 0.02%<br>99.4% | 0.01%<br>99.4% |
| pH | pH Strips | | 4.5–5.5 | 5 | 5 | 5 |
| | | | 24 h stability test | | | |
| Parameter | Method | | Acceptance Criteria | Batch 08/04/2021 | Batch 15/04/2021 | Batch 07/05/2021 |
| Appearance | Visual test | | Clear and Colorless Solution | Complies | Complies | Complies |
| Radiochemical purity | TLC | | [$^{177}$Lu]Lu colloids ≤ 3%<br>[$^{177}$Lu]Lu-PSMA-I&T ≥ 97% | 0.1%<br>99.9% | 0.1%<br>99.9% | 0.1%<br>99.9% |

Table 6. Cont.

| | | 24 h stability test | | | |
|---|---|---|---|---|---|
| Parameter | Method | Acceptance Criteria | Batch 08/04/2021 | Batch 15/04/2021 | Batch 07/05/2021 |
| Radiochemical purity | HPLC | [$^{177}$Lu]Lu ≤ 3%<br>[$^{177}$Lu]Lu-PSMA-I&T ≥ 97% | 0.01%<br>98% | 0.02%<br>98% | 0.04%<br>97.7% |
| pH | pH Strips | 4.5–5.5 | 5 | 5 | 5 |
| | | 30 h stability test | | | |
| Parameter | Method | Acceptance Criteria | Batch 08/04/2021 | Batch 15/04/2021 | Batch 07/05/2021 |
| Appearance | Visual Test | Clear and Colorless Solution | Complies | Complies | Complies |
| Radiochemical purity | TLC | [$^{177}$Lu]Lu colloids ≤ 3%<br>[$^{177}$Lu]Lu-PSMA-I&T ≥ 97% | 0.1%<br>99.9% | 0.2%<br>99.8% | 0.1%<br>99.9% |
| Radiochemical purity | HPLC | [$^{177}$Lu]Lu ≤ 3%<br>[$^{177}$Lu]Lu-PSMA-I&T ≥ 97% | 0.04%<br>97.5% | 0.04%<br>97.4% | 0.07%<br>97.3% |
| pH | pH Strips | 4.5–5.5 | 5 | 5 | 5 |

## 3. Discussion

[$^{68}$Ga]Ga-PSMA-HBED-CC [7] represents a breakthrough in the imaging and staging of PCa. To meet the clinical need for a therapeutic agent for treatment of PCa, some promising urea-based candidates have been investigated, [$^{177}$Lu]Lu-PSMA-617 and [$^{177}$Lu]Lu-PSMA-I&T [8,9]. The two molecules have the same pharmacophore but different linkers and chelators.

The effect of replacing the DOTA chelator with DOTAGA led to an increase in radiolabeling reaction time: PSMA-I&T was incubated with [$^{177}$Lu]LuCl$_3$ at 95 °C for 30 min. to obtain [$^{177}$Lu]Lu-PSMA-I&T instead of 8 min at 100 °C for [$^{177}$Lu]Lu-PSMA-617. The longer reaction time for the incorporation of Lutetium-177 in the DOTAGA is in agreement with Weineisen et al. [9] and could probably be ascribed to a slower kinetic of incorporation due to different conformational changes in the chelators along with the overall molecular structure of the linker.

The two molecules showed slight differences in solubility characteristics, which, however, did not require substantial changes in the buffer solutions.

The longer incubation time needed to prepare [$^{177}$Lu]Lu-PSMA-I&T did not affect the impurities profile. In terms of quality parameters, the experimental results from the three batches of [$^{177}$Lu]Lu-PSMA-I&T fulfilled the specifications. Furthermore, the radiolabeling conditions always led to a very high radiochemical yield.

To prevent radiolysis of the radiopharmaceutical, the radiolabeling was carried out in presence of ascorbic and gentisic acid.

The tests performed for quality controls are commonly used in radiochemical and radiopharmaceutical methods.

HPLC allows the use of two detectors in series for example UV or a mass detector coupled with a radiometric detector.

Acceptance criteria are based on the Eu. Pharm. general monograph "Radiopharmaceutical Preparation" [15] and on National Regulations for preparation of radiopharmaceuticals [11].

Gamma-ray spectroscopy is used for the measurement of the radionuclidic purity of radiopharmaceuticals. The Pharmacopoeia generally states that radiopharmaceuticals should have a radionuclidic purity of at least 99.9% throughout their shelf-life. High purity germanium detectors are required to detect impurities of less than 0.1%. For identification, the same approach was used to compare γ energies lines characteristics of the radionuclide.

The most significant quality issue in the preparation of [$^{177}$Lu]Lu-PSMA-I&T is the poor stability of the finished product when diluted with saline to a final volume of 17–25 mL, a general practice with other radiopharmaceuticals such as [$^{177}$Lu]Lu-PSMA-617. Comparative results on the stability of both radiopharmaceuticals are reported in Table 7.

**Table 7.** Comparative stability data of [$^{177}$Lu]Lu-PSMA-I&T and [$^{177}$Lu]Lu-PSMA-617, diluted with saline, without ascorbic acid, at T = 0 and after 24 and 30 h at room temperature.

| Parameter | Method | Acceptance Criteria | [$^{177}$Lu]Lu-PSMA-I&T | [$^{177}$Lu]Lu-PSMA-617 |
|---|---|---|---|---|
| | | **T0 Stability Test** | | |
| Radiochemical purity | TLC | [$^{177}$Lu]Lu colloids ≤ 3% <br> [$^{177}$Lu]Lu-PSMA-I&T ≥ 97% | 0.3% <br> 99.7% | 0 <br> 100% |
| Radiochemical purity | HPLC | [$^{177}$Lu]Lu ≤ 3% <br> [$^{177}$Lu]Lu-PSMA-I&T ≥ 97% | 0.2% <br> 99.1% | 0.02% <br> 99.8% |
| | | **24 h stability test** | | |
| Radiochemical purity | TLC | [$^{177}$Lu]Lu colloids ≤ 3% <br> [$^{177}$Lu]Lu-PSMA-I&T ≥ 97% | 3.3% <br> 96.7% | 0.1% <br> 99.9% |
| Radiochemical purity | HPLC | [$^{177}$Lu]Lu ≤ 3% <br> [$^{177}$Lu]Lu-PSMA-I&T ≥ 97% | 0.2% <br> 94.6% | 0.04% <br> 97.4% |
| | | **30 h stability test** | | |
| Radiochemical purity | TLC | [$^{177}$Lu]Lu colloids ≤ 3% <br> [$^{177}$Lu]Lu-PSMA-I&T ≥ 97% | 3.5% <br> 96.5% | 0.2% <br> 99.8% |
| Radiochemical purity | HPLC | [$^{177}$Lu]Lu ≤ 3% <br> [$^{177}$Lu]Lu-PSMA-I&T ≥ 97% | 0.2% <br> 93.2% | 0.06% <br> 97.2% |

[$^{177}$Lu]Lu-PSMA-I&T was found to be instable in saline solution (<95% radiochemical purity after 24 h). These data emphasize that what is reported in the Supplemental data by Weineisen et al. [9] either does not consider the long-term stability of the preparation or establishes less stringent acceptance criteria compared with this paper (≥97% radiochemical purity). Another aspect to consider is the large difference between the final radioactivities reported in the literature [9] and the radioactivities used in this study. Long-term radiolysis due to high radioactivity concentration should be taken into account. [$^{177}$Lu]Lu-PSMA-617, in our experience, is stable in saline; this behavior can probably be ascribed to [$^{177}$Lu]Lu-PSMA-I&T and [$^{177}$Lu]Lu-PSMA-617 having different linkers that are affected differently by radiolysis.

To overcome this problem, [$^{177}$Lu]Lu-PSMA-I&T was formulated in an ascorbic 20mg/mL saline solution to a final volume of 17–25 mL, resulting in a shelf life of 30 h, which is the optimal time interval for the management of patients.

## 4. Materials and Methods

### 4.1. [$^{177}$Lu]Lu-PSMA-I&T Manufacturing Process and Process Controls

For the radiolabeling of [$^{177}$Lu]Lu-PSMA-I&T, a manual synthesis was used. The radiosynthesis was carried out in a shielded isolator offering a class A environment with class B pre-chambers (Manuela Beta, COMECER S.p.A, Castelbolognese, Italy) located in a class C cleanroom.

4.1.1. Reagents

[$^{177}$Lu]LuCl$_3$.EndolucinBeta, radiopharmaceutical precursor with a MA was obtained from ITM Medical Isotope GmbH, Germany—and was supplied by Gamma Servizi S.r.l, Borgarello, Italy

PSMA-I&T GMP precursor vials 1 mg purchased from ABX Advanced Biochemical Compounds Biomedizinische Forschungsreagenzien GmbH Radeberg, Germany.

Water for Injectable Preparations (100 mL bottles) with a MA was purchased from Monico S.p.A. Venezia/Mestre, Italy.

Gentisic acid, (97.5–102.5% purity), was supplied by Merck KGaA, Darmstadt, Germany.
Ascorbic acid, (99.0–100.5% purity), was supplied by VWR International, Leuven, Belgium.
NaOH, ($\geq$99% purity), was supplied by Merck KgaA, Darmstadt, Germany.
Anhydrous sodium acetate, ($\geq$99% purity), was supplied by Merck KgaA, Darmstadt, Germany.

Gentisic acid/ascorbic acid buffer pH = 5 was prepared by IRST Radiopharmacy prior to the radiolabeling. The buffer was prepared with 3.1 g of sodium acetate, 1.6 g of gentisic acid, and 3.0 g of ascorbic acid, which were dissolved in 46 mL of water for injectable preparations measured with a 60 mL sterile syringe. This solution was pH adjusted by the addition of 11.5 mL of a 2N NaOH solution. The final pH was $5.2 \pm 0.1$ and was verified by pH meter. This solution was sterile filtered through a single-use, sterile, pyrogen-free 0.22 μm ventilated filter (Vented Millex-GV SLGV255F Merck Millipore Ltd.). The buffer was aliquoted into ten 1 mL fractions, which were stored at $-20\ °C$. Shelf life of the laboratory-prepared labeling buffer solution was 1 month.

Ascorbic acid-Vitamin C SALF 1000 mg/5 mL solution for injection vials with an MA was purchased from S.A.L.F Laboratorio Farmacologico, Cenate Sotto Bergamo, Italy.

Sodium Chloride 0.9% 100 mL, with a MA was purchased from Fresenius Kabi S.r.l., Isola della Scala, Italy.

Sterile glass vials under partial vacuum—manufactured by Eckert&Ziegler GmbH, Berlin, Germany—were purchased by Radius S.r.l. Budrio, Italy.

4.1.2. Manufacturing of [$^{177}$Lu]Lu-PSMA-I&T

The precursor was incubated with lutetium-177 at 95 °C for 30 min in the presence of ascorbic and gentisic acid. A typical synthesis time for the complexation reaction yielding [$^{177}$Lu]Lu-PSMA I&T in 60 min with a radiochemical yield of $96.8 \pm 0.9\%$ ($n = 5$). The flow chart of the radiolabeling of PSMA-I&T is shown in Figure 5.

Step A—Verification of dose calibrator response using a certified source of cesium-37 (NuklearMedizin, Dresden, Germany). The deviation between the measured and expected value should never be greater than 5%.

Step B—The radioactivity of the received [$^{177}$Lu]LuCl$_3$ vial using dose calibrator.

Step C—The reaction mixture was prepared with 0.5–0.8 mL of PSMA-I&T (in water) and 0.5 mL of gentisic–ascorbic acid solution. The amount of PSMA-I&T should be adequate to obtain the specific activity (SA) of 36–39 GBq/mg. The reaction mixture is set up in a 2.5 mL syringe in the shielded isolator immediately before radiolabeling.

Step D—PSMA-I&T solution was added to the [$^{177}$Lu]LuCl$_3$ vial (0.4–0.8 mL). The vial was placed in the heater, at 95 °C with continuous temperature monitoring for 30 min, to help the complexation reaction to take place.

Step E—The vial containing a volume of about 1.5 mL of [$^{177}$Lu]Lu-PSMA-I&T was diluted by adding 15–23 mL of an ascorbic acid solution (20 mg/mL) to the reaction vial. The reactor vial is connected to the product vial via a sterile needle, tubing, and a 0.22 μm ventilated sterile filter. The needle in the reaction vial needs to be at the bottom of the reaction vial. The pressure created in the reaction vial pushed the solution ca. 16–25 mL into the product vial.

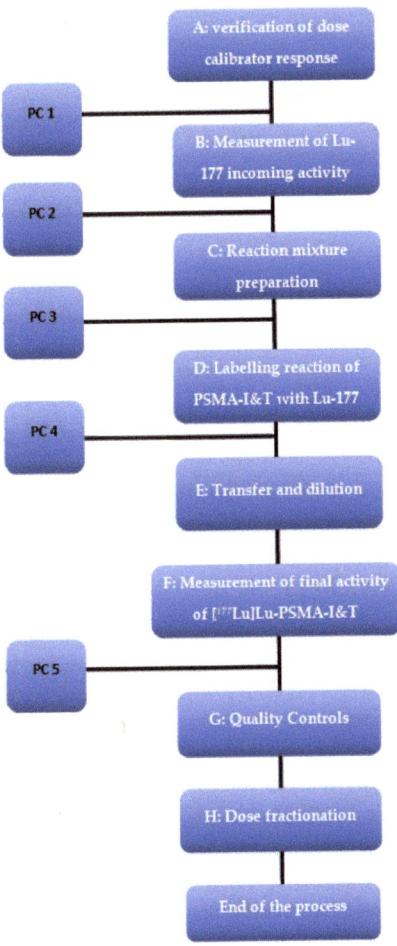

**Figure 5.** Flow chart of the radiolabeling of PSMA-I&T. PC = In-Process Control.

4.1.3. In Process Controls (PC)

PC 1: Accuracy testing of emitting β-sources is performed prior to the introduction of a new measured geometry (e.g., new vial size of lutetium-177).

PC 2: A daily radioactivity check is performed before each production run and the deviation between the read and calculated activity, according to the calibration certificate, must be $<\pm 10\%$.

PC 3: Operator aseptic work techniques are verified by media fill test.

PC 4: Double check of the temperature.

PC 5: Control of the final product activity; the radiochemical yield is also calculated.

4.2. Quality Control

4.2.1. Standard Procedures

pH was determined by pH strips (Merck pH indicator strip, Acilit, increment 0.5 pH unit).

The Endotoxin test was performed by the Limulus amebocyte lysate test (LAL test) on an Endosafe Nexgen-PTS™ (Charles River Laboratories Italia, Calco, Italy).

Since this is a preparation that cannot be subjected to terminal sterilization, the product solution, therefore, has to be subjected to sterile filtration through a sterile filter (pores size less than 0.22 µm).

Filter integrity must be checked by bubble point test before the release. The bubble point test was performed on an Integritest 4 system (Merck Millipore, Merck KgaA, Darmstadt, Germany).

A 1 mL aliquot of the product mixture was sent to the Microbiological Laboratory of the Regional Healthcare, Pievesestina, Cesena, Italy for the sterility tests. The sterility test was performed according to current European Pharmacopoeia Monograph 2.6.1 "Sterility".

#### 4.2.2. HPLC Analysis

$^{nat}$Lu-PSMA-I&T was purchased from ABX GmbH—Advanced Biochemical Compounds (Radeberg, Germany) as a reference standard.

HPLC analysis was performed on an Ultimate 3000 system equipped by a UV variable wavelength detector RS300 (Thermo Fischer Scientific, Germany) and a radiometric detector (GABI, Raytest, Germany). The system was run by Chromeleon software version 7.2 SR5 (Dionex Sunnyvale, CA, USA).

The column was an Acclaim 120 C18, 3 µm, 120Å, 3 × 150 mm (Thermo Scientific, Waltham, MA, USA).

A multi-step gradient was applied using two solvent A (0.1% TFA in water) and solvent B (0.1% TFA in acetonitrile): 92% A to 40% A in 10 min, then from 40% A to 20% A for a further 3 min, and back to 92% A in 2 min, then stable for 5 min. The flow rate was set at 0.6 mL/min. UV wavelength at 205 nm. Column Oven: 25 °C. Injection volume 20 µL. Retention times of $^{nat}$-LuPSMA-I&T and PSMA-I&T were 6.730 and 7.013, respectively.

Chemical purity was calculated by comparing the areas of the peaks of the product with a standard solution injected before the analysis of the final product, as a general practice suggested by Pharmacopoeia.

#### 4.2.3. Thin-Layer Chromatography (TLC)

TLC was performed using a TLC Silica Gel 60 (Merck KGaA, Darmstadt, Germany), Approximately 1–2 µL of IMP injection solution was spotted on the plate. The solvent for the development of the TLC plates was ammonium acetate 0.1N and methanol 50:50 $v/v$. The developed plate was analyzed by autoradiography on MS (MultiSensitive) storage phosphor screens and by a Cyclone Plus Storage Phosphor System (PerkinElmer).

Rf of lutetium-177 free and colloids was 0.2; Rf of [$^{177}$Lu]Lu-PSMA-I&T was 0.8.

### 5. Conclusions

This study demonstrates that [$^{177}$Lu]Lu-PSMA-I&T can be prepared as a radiopharmaceutical suitable for human use. Clearly defined acceptance criteria, validations plans, and methods for quality control were outlined. We compared two structural related radiopharmaceuticals, which differed in chelator and linker, but not in pharmacophore. The difference resulted in a different metal chelation kinetic, a different solubility, and, above all, a very different stability of the two radiopharmaceuticals. Ascorbic acid functions as a scavenger and can significantly prolong the shelf life of [$^{177}$Lu]Lu-PSMA-I&T by at least 30 h, and can prolong the time interval where patients can be treated.

### 6. Patents

PSMA-I&T is available by ABX and there are no patents.

**Author Contributions:** Planning and development of the study, validation, and quality documentation (IMPD), V.D.I. and S.B.; synthesis and quality controls, V.D.I. and C.C.; writing—original draft preparation, V.D.I. and S.B.; writing—review and editing of IMPD, M.M.; principal investigators of clinical study, G.P.; supervision, S.S., G.P. and C.M. All authors have read and agreed to the published version of the manuscript.

**Funding:** This research received no external funding.

**Institutional Review Board Statement:** We have not yet submitted the clinical protocol to the regulatory authority.

**Informed Consent Statement:** Not applicable.

**Data Availability Statement:** The data presented in this study are available on request from the corresponding author.

**Acknowledgments:** This work was partly supported thanks to the contribution of Ricerca Corrente by the Italian Ministry of Health within the research line "Innovative therapies, phase I–III clinical trials and therapeutic strategy trials based on preclinical models, onco-immunological mechanisms, and nanovectors".

**Conflicts of Interest:** The authors declare no conflict of interest.

**Sample Availability:** Samples of the compounds are not available from the authors.

# References

1. Siegel, R.; Ma, J.; Zou, Z.; Jemal, A. Cancer statistics, 2014. *Cancer J. Clin.* **2014**, *64*, 9–29. [CrossRef] [PubMed]
2. Ferlay, J.; Steliarova-Foucher, E.; Lortet-Tieulent, J.; Rosso, S.; Coebergh, J.W.W.; Comber, H.; Forman, D.; Bray, F. Cancer incidence and mortality patterns in Europe: Estimates for 40 countries in 2012. *Eur. J. Cancer* **2013**, *49*, 1374–1403. [CrossRef] [PubMed]
3. Schroöder, F.H.; Hugosson, J.; Roobol, M.J.; Tammela, T.L.J.; Zappa, M.; Nelen, V.; Kwiatkowski, M.; Lujan, M.; Määttänen, L.; Lilja, H.; et al. Screening and prostate cancer mortality: Results of the European Randomised Study of Screening for Prostate Cancer (ERSPC) at 13 years of follow-up. *Lancet* **2014**, *384*, 2027–2035. [CrossRef]
4. Silver, D.A.; Pellicer, I.; Fair, W.R.; Heston, W.D.; Cordon-Cardo, C. Prostate-specific membrane antigen expression in normal and malignant human tissues. *Clin. Cancer Res.* **1997**, *3*, 81–85. [PubMed]
5. Sweat, S.D.; Pacelli, A.; Murphy, G.P.; Bostwick, D.G. Prostate-specific membrane antigen expression is greatest in prostate adenocarcinoma and lymph node metastases. *Urology* **1998**, *52*, 637–640. [CrossRef]
6. Kozikowski, A.P.; Nan, F.; Conti, P.; Zhang, J.; Ramadan, E.; Bzdega, T.; Wroblewska, B.; Neale, J.H.; Pshenichkin, S.; Wroblewski, J.T. Design of remarkably simple, yet potent urea-based inhibitors of glutamate carboxypeptidase II (NAA-LADase). *J. Med. Chem.* **2001**, *44*, 298–301. [CrossRef] [PubMed]
7. Eder, M.; Schäfer, M.; Bauder-Wüst, U.; Hull, W.E.; Wängler, C.; Mier, W.; Haberkorn, U.; Eisenhut, M. 68Ga-Complex Lipophilicity and the Targeting Property of a Urea-Based PSMA Inhibitor for PET Imaging. *Bioconj. Chem.* **2012**, *44*, 1014–1024. [CrossRef] [PubMed]
8. Benešova, M.; Schäfer, M.; Baüder-Wust, U.; Afshar-Oromieh, A.; Kratochwil, C.; Mier, W.; Haberkorn, U.; Kopka, K.; Eder, M. Preclinical Evaluation of a Tailor-Made DOTA-Conjugated PSMA Inhibitor with Optimized Linker Moiety for Imaging and Endoradiotherapy of Prostate Cancer. *J. Nucl. Med.* **2015**, *56*, 914–920. [CrossRef] [PubMed]
9. Weineisen, M.; Schottelius, M.; Simecek, J.; Baum, R.P.; Yildiz, A.; Beykan, S.; Kulkarni, H.R.; Lassmann, M.; Klette, I.; Eiber, M.; et al. 68Ga- and 177Lu-labeled PSMA I&T: Optimization of a PSMA-targeted theranostic concept and first proof-of-concept human studies. *J. Nucl. Med.* **2015**, *56*, 1169–1176. [CrossRef] [PubMed]
10. Weineisen, M.; Simecek, J.; Schottelius, M.; Schwaiger, M.; Wester, H.J. Synthesis and preclinical evaluation of DOTAGA-conjugated PSMA ligands for functional imaging and endoradiotherapy of prostate cancer. *EJNMMI Res.* **2014**, *4*, 63. Available online: http://www.ejnmmires.com/content/4/1/63 (accessed on 13 April 2022). [CrossRef] [PubMed]
11. Norme di Buona Preparazione dei Radiofarmaci per Medicina Nucleare. *Ital. Pharm.* **2005**, *146* (Suppl. S11), 168. Available online: https://www.sifoweb.it/images/pdf/attivita/attivita-scientifica/aree_scientifiche/radiofarmacia/normativa/NBP_Radiofarmaci.pdf (accessed on 2 May 2021).
12. Guideline on the Requirements for the Chemical and Pharmaceutical Quality Documentation Concerning Investigational Medicinal Products in Clinical Trials—EMA/CHMP/QWP/545525/2017. Available online: https://www.ema.europa.eu/en/documents/scientific-guideline/guideline-requirements-chemical-pharmaceutical-quality-documentation-concerning-investigational_en.pdf (accessed on 13 April 2022).
13. ICH Topic Q 2 (R1) Validation of Analytical Procedures: Text and Methodology, CPMP/ICH/381/95. Available online: https://www.ema.europa.eu/en/ich-q2-r1-validation-analytical-procedures-text-methodology (accessed on 13 April 2022).

14. Sosabowski, J.K.; Mather, S.J. Conjugation of DOTA-like chelating agents to peptides and radiolabeling with trivalent metallic isotopes. *Nat. Protoc.* **2006**, *1*, 972–976. [CrossRef] [PubMed]
15. *The European Pharmacopoeia: Radiopharmaceuticals Preparation*; General Monograph 07/2016:0125; European Directorate for the Quality of the Medicines (EDQM): Strasbourg, France, 2020.

*Article*

# Issues with the European Pharmacopoeia Quality Control Method for $^{99m}$Tc-Labelled Macroaggregated Albumin

Svend Borup Jensen [1,2,*], Lotte Studsgaard Meyer [1], Nikolaj Schandorph Nielsen [1] and Søren Steen Nielsen [1]

[1] Department of Nuclear Medicine, Aalborg University Hospital, 9100 Aalborg, Denmark; lsn@rn.dk (L.S.M.); nsn@rn.dk (N.S.N.); ssn@rn.dk (S.S.N.)
[2] Department of Chemistry and Biochemistry, Aalborg University, 9220 Aalborg, Denmark
* Correspondence: svbj@rn.dk

**Abstract:** Technetium-99m macroaggregated albumin ([$^{99m}$Tc]Tc-MAA) is an injectable radiopharmaceutical used in nuclear medicine for lung perfusion scintigraphy. After changing to a new batch of macroaggregated albumin (MAA), we saw unwanted uptake in the liver and spleen. The batch was therefore tested by both the supplier and us and we found it to comply with the requirements of the European Pharmacopoeia (Ph. Eur.). However, a simple comparison between the problematic batch and a batch supplied by another manufacturer showed that there was a significant difference. The quality testing showed a higher number of small particles in the problem encumbered MAA batch with unwanted in vivo uptake. In this article we present a simple method of testing for particle size of [$^{99m}$Tc]Tc-MAA, which gives a good indication of how the radioactive drug performs in vivo. We argue that the quality control method described in the Ph. Eur. should be changed. The changes will improve concordance between the laboratory analyzes and what is seen in vivo in human lung perfusion scintigraphy. Furthermore, we hope that the MAA suppliers without delay will replace their release procedure to be in accordance with the method described in this article.

**Keywords:** technetium-99m labelled macroaggregated albumin [$^{99m}$Tc]Tc-MAA; European Pharmacopoeia; lung perfusion scintigraphy; quality control; EANM recommendation

## 1. Introduction

Technetium-99m labelled macroaggregated albumin ([$^{99m}$Tc]Tc-MAA) is, at our hospital, used to perform lung perfusion scintigraphy on suspicion of pulmonary embolism and prior to operation for lung cancer to be able to calculate postoperative lung function. Our customary supplier of MAA labelling kits notified us that they could not supply us with labeling kits for preparation of [$^{99m}$Tc]Tc-MAA, due to production issues.

We applied and were granted a compassionate user permission for MAA from Medi-Radiopharma, Hungary, from the Danish Medical Agency. The first batch was received and performed as expected. However, the second batch of the MAA labelling kits, received in July 2020, resulted in unwanted radioactivity in the liver and spleen in many patients.

Additional information on this second MAA batch: the temperature data logger that accompanied the parcel revealed that the batch had been too warm during the shipment to Denmark. The maximum temperature should not have exceeded 8 °C but had been 10 °C for 12 h. We consulted Medi-Radiopharma who replied that based on their stability data for MAA, they could conclude that the temperature exceedance did not affect the product quality. Our standard TLC quality control of the [$^{99m}$Tc]Tc-MAA batch revealed nothing abnormal, and it complied with specification. The [$^{99m}$Tc]Tc-MAA was therefore released for clinical use. Shortly after, our doctors detected unusual uptake in the liver and spleen and started questioning the quality of the product.

The TLC analysis used for the release of the batch was repeated and again the batch complied with our release specification. The common TLC release analysis only differentiates between [$^{99m}$Tc]Tc-MAA and free [$^{99m}$Tc]pertechnetate. However, looking at the in vivo uptake pattern, it seemed likely that it could be an unusually higher number of small particles, which had been labeled and then caused the abnormal activity pattern in the perfusion scintigraphy. Generally, we rely on the company analysis for the particle size analysis. The release document stated that 90.34% of the MAA particles were between 10–100 μm and that no particles were above 150 μm, measured by optical microscope as described in the monograph on Tecchnetium ($^{99m}$Tc) macrosalb injection of the European Pharmacopoeia (Ph. Eur.).

The monograph says that 90% of the particles typically should have diameters between 10–100 μm [1]. However, Ph. Eur. does not describe an exact method by which one can measure this. Ph. Eur suggests that one can use an optical microscope to verify whether the particles are not too big, but not that one can use an optical microscope to determine if the particles are too small, as Medi-Radiopharma did. Ph. Eur. suggests an indirect test method for particles size, it is a filter method (pore size of 3 μm). This method should reveal if there are too many small radiolabelled particles. The requirements of the United States Pharmacopeia on $^{99m}$Tc-labeled macroaggregated albumin (MAA) are very similar to those of the European Pharmacopoeia it states that 90% of the particle size should have a diameter between 10 and 90 μm and none of the observed particles have a diameter greater than 150 μm [2]. This range was chosen to ensure that particles smaller than the 7–8 μm diameter of capillaries would be absent, eliminating the possibility of localization in the brain, kidneys, and other internal organs. The limit value at the high end, was chosen to minimize the risk of occlusion of larger vessels, especially in the patient with pulmonary artery hypertension, in which case vasoconstriction is present [3].

MAA is prepared from human serum albumin under carefully controlled conditions of pH, time, temperature, agitation, and reagent concentration to insure correct particle size formation [3].

Reviewing the literature on macroaggregated albumin particle sizes revealed a couple of studies which examined the particle size distributions in commercially available MAA brands [4,5]. And an older study which compared the pharmaceutical quality of a laboratory made MAA with commercially available products [6]. All three studies applied microscope methods for particle size determination. The conclusion was that the mean particle size was similar for all the five macroaggregated albumin preparations tested, but the actual particle size distribution varied considerably among the products tested [4,5]. In the study from Hung and coworkers [5], they also examined and found particle sizes intervariance among the best-performing kit and they concluded that the particle sizes may affect the accuracy and the reproducibility of their patient studies.

We decided to examine the [$^{99m}$Tc]Tc-MAA by filtration, not just with one size filter as described in Ph. Eur., but with four different pore size filters. We found the problem encumbered [$^{99m}$Tc]Tc-MAA, called [$^{99m}$Tc]Tc-Makro-Albumon, to comply with the Ph. Eur. However, comparing the particle sizes between this batch and a different batch supplied by a different vendor of MAA, it was noticeable that the distribution of particle size was significantly different.

On the basis of the in vivo human data and of the results of our particle size experiments, we decided to discard the problem encumbered batch. We have informed the Danish Medicines Agency that we believe that the quality control described in the Ph. Eur. should be changed, so that it will identify batches which will result in unwanted uptake in the liver and spleen. The Danish Medicines Agency has agreed to look at our data (private mail communication between us and the Danish Medicines Agency), but since it may take some time to change the Ph. Eur., we hope that the MAA producers will comply with the proposed changes for the benefit of the patients.

## 2. Results

### 2.1. Labeling

Radioactive labeling with [$^{99m}$Tc]pertechnetate is a one pot chelation reaction with less than 2% unreacted [$^{99m}$Tc]pertechnetate detected after chelation. For comparison two MAA precursors were labeled, both the problem encumbered batch from Medi-Radiopharma and a batch from CIS Bio International France, called Pulmocis The radiolabeled [$^{99m}$Tc]Tc-MAA preparations are designated as respectively [$^{99m}$Tc]Tc-Makro-Albumon and [$^{99m}$Tc]Tc-Pulmocis.

### 2.2. Particle Size Determined by a Microscope

Determination of particle size by a microscope was performed by Medi-Radiopharma Hungary and revealed that an average of 92.8% of the particles have a size between 10–100 μm (10 vials examined), see Table 1.

**Table 1.** Particle size determination by Medi-Radiopharma, Hungary on the problem cumbered [$^{99m}$Tc]Tc-Makro-Albumon batch by a microscopic analysis of the contents of ten labelled kits.

| Test | 10–100 μm | 100–150 μm | >150 μm |
|---|---|---|---|
| 1 | 92.4% | 0 | 0 |
| 2 | 92.5% | 2 | 0 |
| 3 | 91.3% | 1 | 0 |
| 4 | 93.8% | 2 | 0 |
| 5 | 91.2% | 2 | 0 |
| 6 | 91.8% | 1 | 0 |
| 7 | 93.5% | 5 | 0 |
| 8 | 92.3% | 7 | 0 |
| 9 | 93.9% | 8 | 0 |
| 10 | 94.9% | 4 | 0 |
| Mean | 92.8% | 3.2 | 0 |
| Specification | >90% | max:10 pcs * | 0 pcs * |

* A minimum of 5000 particles (pcs) examined.

### 2.3. Indirect Particle Size Determination

An indirect determination of particle size by a filtration method applying four different pore size filters (pore size of 3, 5, 8 and 10 μm) was performed. The test results from three examinations of the problem encumbered [99mTc]Tc-Makro-Albumon preparation is given in Table 2. The test results from three examinations of the [99mTc]Tc-MAA preparations from Pulmocis labelling kits is given in Table 3.

**Table 2.** Table showing filter test results from three examinations of the problem encumbered [$^{99m}$Tc]Tc-Makro-Albumon preparation.

| Filter Pore Size | Filter (MBq) | Filtered Solution (MBq) | Filter/(Filter + Filtered Solution) (%) |
|---|---|---|---|
| 3 μm | 84.3 | 9.26 | 90.1 |
| 5 μm | 76.3 | 10.51 | 87.9 |
| 8 μm | 67.1 | 18.26 | 78.6 |
| 10 μm | 65.0 | 21.27 | 75.3 |
| **Filter Pore Size** | **Filter (MBq)** | **Filtered Solution (MBq)** | **Filter/(Filter + Filtered Solution) (%)** |
| 3 μm | 83.8 | 8.39 | 90.9 |
| 5 μm | 69.9 | 9.11 | 88.5 |
| 8 μm | 56.4 | 13.62 | 80.5 |
| 10 μm | 56.1 | 16.07 | 77.7 |
| **Filter Pore Size** | **Filter (MBq)** | **Filtered Solution (MBq)** | **Filter/(Filter + Filtered Solution) (%)** |
| 3 μm | 71.2 | 12.29 | 85.2 * |
| 5 μm | 71.3 | 16.86 | 80.8 * |
| 8 μm | 66.3 | 20.90 | 76.0 * |
| 10 μm | 65.3 | 22.90 | 74.0 * |

* The expiration date was exceeded by about one month.

**Table 3.** Table showing filter test results from three [$^{99m}$Tc]Tc-MAA preparations from Pulmocis labelling kits.

| Filter Pore Size | Filter (MBq) | Filtered Solution (MBq) | Filter/(Filter + Filtered Solution) (%) |
|---|---|---|---|
| 3 μm | 113.0 | 1.46 | 98.7 |
| 5 μm | 112.2 | 1.76 | 98.5 |
| 8 μm | 111.7 | 2.07 | 98.2 |
| 10 μm | 111.4 | 2.11 | 98.1 |
| **Filter Pore Size** | **Filter (MBq)** | **Filtered Solution (MBq)** | **Filter/(Filter + Filtered Solution) (%)** |
| 3 μm | 114.2 | 2.66 | 97.7 |
| 5 μm | 109.5 | 2.97 | 97.4 |
| 8 μm | 93.6 | 3.43 | 96.5 |
| 10 μm | 91.6 | 3.05 | 96.8 |
| **Filter Pore Size** | **Filter (MBq)** | **Filtered Solution (MBq)** | **Filter/(Filter + Filtered Solution) (%)** |
| 3 μm | 91.9 | 1.13 | 98.8 |
| 5 μm | 92.0 | 1.31 | 98.6 |
| 8 μm | 86.0 | 1.46 | 98.3 |
| 10 μm | 88.6 | 1.49 | 98.3 |

## 3. Discussion

This study started because our doctors noticed abnormal uptake in the lung perfusion scintigraphy when using a new batch of [$^{99m}$Tc]Tc-MAA, Figure 1.

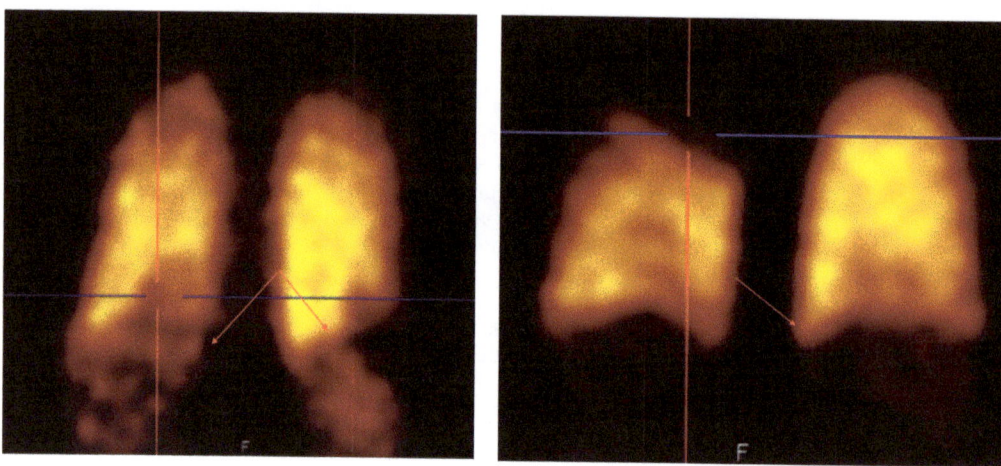

**Figure 1.** Two examples of abnormal hepatic and spleen uptake seen in lung perfusion scintigraphy when using the problematic [$^{99m}$Tc]Tc-MAA batch, called [$^{99m}$Tc]Tc-Makro-Albumon.

To evaluate this observation, an experienced nuclear medicine specialist randomly selected 55 [$^{99m}$Tc]Tc-MAA perfusion scintigrams made after injection of [$^{99m}$Tc]Tc-Makro-Albumon perfusion reviewed them. None of the patients had known history of cardiac shunt or liver cirrhosis which can be physiological causes of right-to-left shunting causing trapping of particles in organs like the liver and spleen. In 78% of the scintigraphies, he identified liver or spleen uptake which is not normally seen using [$^{99m}$Tc]Tc-MAA prepared from labelling kits supplied by a different vendor without the higher number of small particles. In fact, three of the 55 reviewed [$^{99m}$Tc]Tc-Makro-Albumon scanned patients had an earlier scintigram with a [$^{99m}$Tc]Tc-MAA prepared using a labeling kit supplied from a different vendor, and none of these earlier scintigrams showed unwanted activity in the liver or spleen.

The patients with abnormal uptake (78%) were divided into 3 groups: visible, moderate and clearly increased uptakes. In 33% of all the scintigrams, the uptake was classified as a moderate uptake so in one out of three patients the unwanted uptake slightly affected the study interpretation. Furthermore, a clearly increased uptake in the liver or spleen was found in four out of the 55 of the patients (7%) and this uptake affected the interpretation of the study.

We could not explain the unwanted uptake using our commonly applied quality control method, which is a radio-TLC. The radio-TLC revealed that less than 2% was free [$^{99m}$Tc]pertechnetate, which is in line with what we usually see. These results indicaed that more than 98% of the radioactivity could be in the form of [$^{99m}$Tc]Tc-MAA.

Our attention, therefore, turned to the particle size of the problem encountered MAA batch. Too many small particles could probably explain the unwanted lung and spleen uptake [7]. Was the [$^{99m}$Tc]Tc-Makro-Albumon from Medi-Radiopharma, Hungary within specification? Our suspicion was increased by the fact that the kit, during shipment, exceeded the threshold temperature value and also by the fact that the certificate of analysis stated that only 90.3% of the particles was between 10–100 μm. According to the certificate of analysis 90% or more of the particles must be between 10–100 μm, so the batch was only just within specification.

Due to the unwanted uptake in the liver and spleen in human scans, we started a dialog with the supplier, who asked us to return the problematic batch (no. MA-200405-1) for reanalysis. We returned 12 kits. They tested ten kits of the troublesome batch. On the original certificate of analysis for MA-200405-1, they stated that 90.3% of the particles were within 10–100 μm. For results of the reanalysis, see Table 2. The figures were between 91.2% and 94.9% (mean 92.8%), so all the ten vials reexamined performed better than the original vial which were used for batch release. The Ph. Eur. also states that on a total of 5000 particles examined no more than ten particles must have a dimension bigger than 100 μm and that no particles must have a dimension above 150 μm. Medi-Radiopharma found that batch no. MA-200405-1 complies with Ph. Eur.'s specifications on all accounts.

We decided to perform our own particle size test. The Ph. Eur. states that 90% of the radioactivity must be retained on a polycarbonate membrane filter, pore size 3 μm [1]. We hypothesized that we could gain more information on the particle size of the denatured albumin particles in the MAA labelling kits if we used several filters with different pore sizes. We, therefore, decided to apply four different pore size filters (3, 5, 8, and 10 μm) in the experiment. All the filters had a diameter of 25 mm.

For comparison, we tested two different [$^{99m}$Tc]Tc-MAA batches, both the problematic batch from Medi-Radiopharma here called [$^{99m}$Tc]Tc-Makro-Albumon and a batch from CIS Bio International France here called [$^{99m}$Tc]Tc-Pulmocis.

If we look at [$^{99m}$Tc]Tc-Makro-Albumon (named MAA in Figure 2) first, 90.1% of the radioactivity is retained on polycarbonate membrane filter with a pore size 3 μm in the first experiment and 90.9% in the second experiment. That is in both cases within the acceptance limit according to the Ph. Eur.

In the last experiment, only 85.2% of the radioactivity was retained on a 3 μm pore size filter. That was below the required 90%. But the Macro-Albumon kits had exceeded their expiration date by about a month when we did experiment 3, so the kit would not have been used in humans anyway.

In other words, the [$^{99m}$Tc]Tc-Makro-Albumon batch still complied with the requirements set out in the Ph. Eur [1]. We therefore cannot claim compensation because the [$^{99m}$Tc]Tc-Makro-Albumon batch meets the requirements described by the Ph. Eur., even though we do not think the batch should be used in humans.

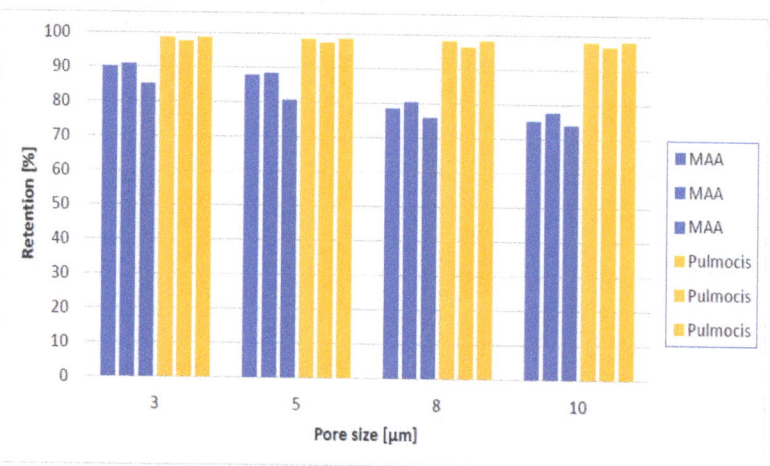

**Figure 2.** Filter test results from applying the four pore size filters on three [$^{99m}$Tc]Tc-Makro-Albumon (MAA) and three [$^{99m}$Tc]Tc-Pulmocis (Pulmocis) products.

The recommendation described in the European Association of Nuclear Medicine, (EANM) guidelines for ventilation/perfusion scintigraphy from 2009 [8] or 2019 [9] both say that the size of the aggregates of a [$^{99m}$Tc]Tc-MAA preparation should be within 15–100 μm not within 10–100 μm as the Ph. Eur. suggests. We find it peculiar that the Ph. Eur. and EANM do not recommend the same interval for particle size.

An examination of our human data, clearly showed that particle size is important in connection with the MAA particles to lodge in the pulmonary capillaries and in the precapillary arterioles. The question is, would the [$^{99m}$Tc]Tc-Makro-Albumon batch with the unwanted liver and spleen uptake have failed the quality control if it was required that that 90% of the particles had to be in-between 15–100 μm instead of 10–100 μm? We cannot say for sure, but it is likely since the batch was close to failing. The certificate of analysis says that only 90.3% of the particles were between 10–100 μm. If we look at the filtration analysis of the [$^{99m}$Tc]Tc-Makro-Albumon for the two experiments where the batch still meets the expiration date (experiment 1 and 2, Table 2, it was found that approximately 90.5% of the radioactivity was retained applying a 3 μm pore size filter, 88.2% for a 5 μm pore size filter, 79.5% for a 8 μm pore size filter and 76.5% for a 10 μm pore size filter. Thus, 90.5% − 88.2% = 2.3% of the particles were retained on a 3 μm pore size but not on a 5 μm pores, 11% were retained by 3 μm but not by 8 μm pore, and 14% were retained by 3 μm but not by 10 μm pores. Regardless of which pore size filters were used more than 90% of the [$^{99m}$Tc]Tc-Pulmocis particles were retained.

Looking at the four filters and the six experiments in Figure 2, the two products have a measurable difference in particle size composition. In this article we present a simple analytical method which can distinguish between a [$^{99m}$Tc]Tc-MAA batch that results in unwanted lung and spleen uptake and a batch that does not. The method is so simple that the test can be performed at any nuclear medicine department and does not require expensive equipment. To do the test, one just needs a dose calibrator and a polycarbonate membrane filter. Instead of using a polycarbonate membrane filter with a pore size of 3 μm, one could do the test with a filter having a slightly bigger pore size. In our case one could use either a 5, 8 or 10 μm with the same results. The batch with the unwanted liver and spleen uptake would have failed, and one could avoid using a [$^{99m}$Tc]Tc-MAA batch with unwanted uptake in the lung and spleen in humans.

## 4. Materials and Methods

### 4.1. Materials

Macro-Albumon labelling kits were purchased from Medi-Radiopharma, Érd, Hungary through Wiik Pharma, Hinnerup, Denmark. Pulmocis labelling kits were purchased from CIS Bio International, Saclay, France via Dupharma, Kastrup, Denmark. The technetium-99m generator was an Ultra-TechneKow FM Generator from Curium, Petten, Netherland bought through GE, Brøndby, Denmark. Acetone (GC quality, purity ≥ 99.5%) was purchased from Sigma-Aldrich Søborg, Denmark. Silica gel 60 Aluminum TLC plates and the four different types of Merck Isopre Polycarbonate membrane filters, all 25 mm in diameter, were purchased from Merck Life Science A/S, Søborg, Denmark with pore size diameters of respectively 3.0 µm, 5.0 µm, 8.0 µm and 10.0 µm (Merck IsopreTM Membrane Filter, 3.0 µm, Item number: TSTP02500/Lot R9MA87321), (Merck IsopreTM Membrane Filter, 5.0µm Item number.: TMTPO2500/Lot.ROEB65842), (Merck IsopreTM Membrane Filter, 8.0µm Item number. TETPO2500/LotROMB21842), (Merck IsopreTM Membrane Filter, 10.0µm Item number. TOTPO2500/LotROHB80056). Radio-TLC was performed using a Scan-RAM radio-TLC scanner from Lablogic, Sheffield, UK. Microscopic analysis of the particle size was done by Medi-Pharma, Érd, Hungary. For measurement of radioactivity a gamma counter (Wizard 2/2480, Perkin Elmer, Skovlunde, Denmark) and a Capintec (CRC®-55tR), dose calibrator (Hoy Scientific, Hadsund, Denmark) were used.

### 4.2. The Preparation of [$^{99m}$Tc] Macroaggregated Albumin ([$^{99m}$Tc]Tc-MAA)

The production of [$^{99m}$Tc]Tc-Makro-Albumon and [$^{99m}$Tc]Tc-Pulmocis followed the manufacturer's package inserts.

The $^{99}$Mo/$^{99m}$Tc generator had been eluted within 24 h before it was used for preparation of [$^{99m}$Tc]Tc-Makro-Albumon or [$^{99m}$Tc]Tc-Pulmocis. The generator eluate had been checked for molybdenum-99 breakthrough prior to labeling. The $^{99}$Mo/$^{99m}$Tc generator was eluted, the radioactivity was measured in a dose calibrator and an appropriate part of the eluate was then used for labeling shortly after (see below).

#### 4.2.1. [$^{99m}$Tc]Tc-Makro-Albumon

[$^{99m}$Tc]Tc-Makro-Albumon was prepared with 3700 MBq Tc-99m in 8 mL saline solution (9 g/L NaCl) without a breather needle. After adding the [$^{99m}$Tc]pertechnetate solution to the macroaggregated albumin labeling vial, an equivalent volume of nitrogen was withdrawn to avoid excess pressure in the vial. The kit was turned upside down after the radioactivity had been added and placed on a tilting device for two minutes. Then the kit was left standing at room temperature for 20 min before it was ready for use.

#### 4.2.2. [$^{99m}$Tc]Tc-Pulmocis

[$^{99m}$Tc]Tc-Pulmocis was prepared with 3700 MBq Tc-99m in a 10 mL saline solution (9g/L NaCl) without a breather needle. After adding the [$^{99m}$Tc]pertechnetate solution to the macroaggregated albumin labeling vial, an equivalent volume of nitrogen was withdrawn to avoid excess pressure in the vial. The kit was then gently turned upside down for two minutes, and it was then left at room temperature for 15 min, after which it was stored refrigerated until usage.

### 4.3. Quality Control of the [$^{99m}$Tc] Macroaggregated Albumin ([$^{99m}$Tc]Tc-MAA)

#### 4.3.1. Microscopic Analysis

Determination of particle size by microscopic analysis was performed by Medi-Radiopharma, following Ph. Eur. [1], after dilution of the preparations to a level where the number of particles was just low enough for individual particles to be distinguished. Using a syringe fitted with a needle having a caliber not less than 0.35 mm, a suitable volume was placed in a counting chamber. The preparation was allowed to settle for one minute, a cover-slide was applied carefully without squeezing the sample. The examined area covered at least 5000 particles.

The Ph. Eur. [1] explains how to determine large molecules using a microscope. However, Medi-Radiopharma uses this test to identify both particles that are too big and those which are too small.

4.3.2. Non-Filterable Radioactivity

A polycarbonate membrane filter with a 13–25 mm diameter and a pore size of 3 µm was used as prescribed by the Ph. Eur. However we decided to use four different pore size filters (3, 5, 8, and 10 µm) in the experiment. All 4 polycarbonate membrane filters had a diameter of 25 mm.

For comparison and to test the method, we tested both the [$^{99m}$Tc]Tc-Makro-Albumon from Medi-Radiopharma and the similar radiopharmaceutical [$^{99m}$Tc]Tc-Pulmocis from Dupharma. Both products where gently turned upside down a couple of times before taking $4 \times 0.2$ mL out for the four analyses. The 0.2 mL product solution was added to each of the above-described filters followed by 20 mL saline (9 g/L). The solution was passed gently through the filter. To calculate the retained fraction, both the filtered solvent and the filter were measured in a dose calibrator. The results are presented in Tables 2 and 3.

4.3.3. Radio-TLC Methods for the Quantitative Determination of [$^{99m}$Tc]Tc-MAA

The product mixture was turned gently upside down a couple of times before transferring 5 µL of the product mixture onto a 10 cm long Silica gel-60 TLC plate one cm from the bottom. The TLC plate was placed in a beaker, the bottom of the beaker was covered with acetone (0.5 cm). When the solvent front reached about 9cm, the TLC-plate was removed from the beaker and examined by a radio TLC scanner.

The [$^{99m}$Tc]Tc-MAA does not move, whereas free [$^{99m}$Tc]pertechnetate does.

There were found below 2% free [$^{99m}$Tc]pertechnetate in both the [$^{99m}$Tc]Tc-Makro-Albumon and the [$^{99m}$Tc]Tc-Pulmocis production.

4.4. In Vivo Evaluation in Patients

An experienced nuclear medicine specialist reviewed 55 [$^{99m}$Tc]Tc-MAA perfusion scintigraphy. These studies represented consecutive examinations from 15 October 2020 until 19 November 2020. We also randomly retrieved lung-perfusion scintigraphy of the month prior to the unwanted uptake in the liver and spleen.

## 5. Conclusions

We have shown that the limit values described in Ph. Eur. are too wide and are unable to identify a [$^{99m}$Tc]Tc-macroaggregated albumin batch with unwanted lung and spleen uptake in more than 75% of patients examined. The requirement that at least 90% of the particles must be larger in size than 10 was met for the batch with the unwanted liver and spleen uptake, as shown by the microscopic analysis of the Hungarian manufacturer. The aim must be to identify a batch like that before it gets into patients. There is therefore a need for tighten the requirement. A cautious recommendation for the guidelines could be that 95% of the particles must be larger in size than 10 µm. Moreover, we have shown that a simple and inexpensive filter analysis using polycarbonate membrane filter (pore size 5, 8 or 10 µm) in our case can distinguish between the batches performing as it should in vivo from the batch with the unwanted lung and spleen uptake.

This article will be sent to the Danish Medicines Agency with the aim of changing the European Pharmacopoeia Quality Control Method for 99mTc-labeled of macroaggregated albumin. We suggest changing the wording of the Ph. Eur. to that 95% of the particles should be between 10–100 µm, and that 95% of the particles should be retained on a polycarbonate membrane filter with a pore size of 3 µm (or alternatively that 90% must be retained on a filter with pore size 5, 8 or 10 µm).

**Author Contributions:** Idea Origin, S.S.N. and S.B.J.; Investigation, N.S.N., L.S.M., S.S.N. and S.B.J.; Laboratory Experiment L.S.M. and N.S.N.; Reading Scintigrams S.S.N.; Writing and original draft preparation S.B.J. All authors have read and agreed to the published version of the manuscript.

**Funding:** This research received no external funding.

**Institutional Review Board Statement:** Not applicable.

**Informed Consent Statement:** Not applicable.

**Data Availability Statement:** Not applicable.

**Acknowledgments:** Pia Afzelius and Lars Jødal for constructive criticism of the article.

**Conflicts of Interest:** The authors declare no conflict of interest.

# References

1. European Pharmacopoeia 8.0; 1100-1101 (01/2009:0296, Corrected 7.4.): Technetium (99mTc) Macrosalb Injection. Available online: https://www.edqm.eu/en/european-pharmacopoeia-ph-eur-10th-edition- (accessed on 19 April 2022).
2. United States Pharmacopeia. Technetium Tc 99m Albumin Macroaggregated Injection. Available online: http://www.pharmacopeia.cn/v29240/usp29nf24s0_m80590.html (accessed on 19 April 2022).
3. NucMedTutorials. (Online Continuing Education for Nuclear Medicine): Tutorial on "Pulmonary Perfusion Imaging with Tc-99m MAA". Available online: https://nucmedtutorials.files.wordpress.com/2018/03/pulmonary-perfusion-imaging-with-tc-99m-maa.pdf (accessed on 19 April 2022).
4. Callahan, R.J.; Swanson, D.P.; Petry, N.A.; Beightol, R.W.; Vaillancourt, J.; Dragotakes, S.C. Multi-institutional in vitro evaluation of commercial/sup 99m/Tc macroaggregated albumin kits. *J. Nucl. Med. Technol.* **1986**, *14*, 206–209.
5. Hung, J.C.; Redfern, M.G.; Mahoney, D.W.; Thorson, L.M.; Wiseman, G.A. Evaluation of macroaggregated albumin particle sizes for use in pulmonary shunt patient studies. *J. Am. Pharm. Assoc.* **2000**, *40*, 46–51. [CrossRef]
6. Moerlin, S.; Colombetti, L.G.; Patel, G. Comparison of pharmaceutical quality of laboratory made macroaggregated albumin with commercially available products. *Radiobiol. Radiother.* **1977**, *18*, 45–51.
7. Küçüker, K.A.; Güney, I.B.; Aikimbaev, K.; Paydaş, S. Unexpected Hepatic Uptake of Tc-99m-MAA in Lung Perfusion Scintigraphy in a Patient with End-stage Renal Disease. *Mol. Imaging Radionucl. Ther.* **2019**, *28*, 27–29. [CrossRef] [PubMed]
8. Bajc, M.; Neilly, J.B.; Miniati, M.; Schuemichen, C.; Meignan, M.; Jonson, B. EANM guidelines for ventilation/perfusion scintigraphy. *Eur. J. Nucl. Med. Mol. Imaging* **2009**, *36*, 1356–1370. [CrossRef] [PubMed]
9. Bajc, M.; Schümichen, C.; Grüning, T.; Lindqvist, A.; Roux, P.L.; Alatri, A.; Bauer, R.W.; Dilic, M.; Neilly, B.; Verberne, H.J.; et al. EANM guideline for ventilation/perfusion single-photon emission computed tomography (SPECT) for diagnosis of pulmonary embolism and beyond. *Eur. J. Nucl. Med. Mol. Imaging* **2019**, *46*, 2429–2451. [CrossRef] [PubMed]

Article

# Analysis of Pros and Cons in Using [$^{68}$Ga]Ga-PSMA-11 and [$^{18}$F]PSMA-1007: Production, Costs, and PET/CT Applications in Patients with Prostate Cancer

Costantina Maisto [1], Michela Aurilio [2], Anna Morisco [1], Roberta de Marino [1], Monica Josefa Buonanno Recchimuzzo [1], Luciano Carideo [1], Laura D'Ambrosio [1], Francesca Di Gennaro [1], Aureliana Esposito [1], Paolo Gaballo [1], Valentina Pirozzi Palmese [1], Valentina Porfidia [1], Marco Raddi [1], Alfredo Rossi [1], Elisabetta Squame [1] and Secondo Lastoria [1,*]

[1] Nuclear Medicine Division, Istituto Nazionale Tumori-IRCCS-Fondazione G. Pascale, 80127 Napoli, Italy; c.maisto@istitutotumori.na.it (C.M.); a.morisco@istitutotumori.na.it (A.M.); roberta.demarino@istitutotumori.na.it (R.d.M.); monica.buonanno@istitutotumori.na.it (M.J.B.R.); luciano.carideo@istitutotumori.na.it (L.C.); l.dambrosio@istitutotumori.na.it (L.D.); f.digennaro@istitutotumori.na.it (F.D.G.); aureliana.esposito@istitutotumori.na.it (A.E.); p.gaballo@istitutotumori.na.it (P.G.); valentina.pirozzipalmese@istitutotumori.na.it (V.P.P.); valentina.porfidia@istitutotumori.na.it (V.P.); marco.raddi@istitutotumori.na.it (M.R.); alfredo.rossi@istitutotumori.na.it (A.R.); e.squame@istitutotumori.na.it (E.S.)
[2] Hospital Pharmacy Department, ASL 1 "Ospedale del Mare" Hospital, 80127 Napoli, Italy; michela.aurilio@aslnapoli1centro.it
* Correspondence: s.lastoria@istitutotumori.na.it

**Abstract:** The aim of this work is to compare [$^{68}$Ga]Ga-PSMA-11 and [$^{18}$F]PSMA-1007 PET/CT as imaging agents in patients with prostate cancer (PCa). Comparisons were made by evaluating times and costs of the radiolabeling process, imaging features including pharmacokinetics, and impact on patient management. The analysis of advantages and drawbacks of both radioligands might help to make a better choice based on firm data. For [$^{68}$Ga]Ga-PSMA-11, the radiochemical yield (RCY) using a low starting activity (**L**, average activity of 596.55 ± 37.97 MBq) was of 80.98 ± 0.05%, while using a high one (**H**, average activity of 1436.27 ± 68.68 MBq), the RCY was 71.48 ± 0.04%. Thus, increased starting activities of [$^{68}$Ga]-chloride negatively influenced the RCY. A similar scenario occurred for [$^{18}$F]PSMA-1007. The rate of detection of PCa lesions by Positron Emission Tomography/Computed Tomography (PET/CT) was similar for both radioligands, while their distribution in normal organs significantly differed. Furthermore, similar patterns of biodistribution were found among [$^{18}$F]PSMA-1007, [$^{68}$Ga]Ga-PSMA-11, and [$^{177}$Lu]Lu-PSMA-617, the most used agent for RLT. Moreover, the analysis of economical aspects for each single batch of production corrected for the number of allowed PET/CT examinations suggested major advantages of [$^{18}$F]PSMA-1007 compared with [$^{68}$Ga]Ga-PSMA-11. Data from this study should support the proper choice in the selection of the PSMA PET radioligand to use on the basis of the cases to study.

**Keywords:** PCa; [$^{68}$Ga]Ga-PSMA-11; [$^{18}$F]PSMA-1007; PSMA; [$^{68}$Ge]/[$^{68}$Ga] generator; cyclotron RCY; RCP

## 1. Introduction

Prostate adenocarcinoma (PCa) is the second most commonly diagnosed cancer among males and the fifth cause of death worldwide related to disease progression [1]. Advanced prostate cancer represents the ultimate challenge in terms of reducing its specific mortality. Accurate diagnosis and efficient staging of recurrences in metastatic castration-resistant prostate cancer (mCRPC) are greatly required.

In the last years, it has been demonstrated that prostate-specific membrane antigen (PSMA) plays a pivotal role in the detection, at diagnosis, of primary PCa as well as of recurrent and metastatic sites of disease, being an optimal target highly overexpressed in poorly differentiated, metastatic, and hormone-refractory PCa [2,3]. In fact, PSMA PET/CT has been recently included in international guidelines for imaging in biochemical recurrence (BCR) [4]. Furthermore, there is a high demand for monitoring patients with relapsed or metastasized prostate cancer because PSMA-specific PET tracers are not negatively affected by androgen deprivation therapy (ADT) once metastases became castration resistant. Conversely, for castration-resistant patients, the impact of ADT on PSMA expression in hormone-sensitive prostate cancer (HSPC) remains unclear and is under investigation. Several radiolabeled PSMA-inhibitors have been developed and, among these, two classes of probes gained particular relevance for clinical use as PET/CT radiopharmaceuticals in PCa: those labeled with gallium-68 [$^{68}$Ga] including PSMA-11, PSMA I&T, PSMA-617, etc., and those labeled with Fluoride-18 [$^{18}$F] including PSMA-1007, DCFPyL, and JK-PSMA-7, etc. All of these synthetic peptides, independently from the radionuclide used for PET imaging, are characterized by high affinities for PSMA, although different, and ensure excellent performances in terms of the detection rate and diagnostic accuracy. Different patterns of biodistribution and variable levels of uptake in normal organs as well as pharmacokinetic properties were found. For such characteristics, these agents are in Pharmacopeia or under evaluation for authorization by regulatory agencies worldwide and they rapidly gained great relevance in the management of PCa patients. Thus, the knowledge of pros and cons of $^{68}$Ga-labeled and $^{18}$F-labeled-PSMA is mandatory for the proper choice. In this setting, we have analyzed our data of the past years at NCI of Napoli, by reviewing all of the parameters involved in the preparation and use of [$^{68}$Ga]Ga-PSMA-11 and [$^{18}$F]-PSMA-1007. In detail, we considered the radiochemical yields, the imaging performances, and the impact of PET findings in the management of patients with PCa, including the selection of eligible patients for radioligand therapy (RLT) [5]. Finally, we evaluated the cost of production, the amount of the final product along with the number of allowed PET/CT exams. To evaluate the costs of production of [$^{68}$Ga]Ga-PSMA-11 and [$^{18}$F]PSMA-1007 in a nuclear medicine center with its own cyclotron/radiopharmacy facility, we adopted the same procedure previously published for [$^{18}$F]PSMA-1007 [6]. The knowledge obtained from all of these data will support the choice of the best option to adopt in a nuclear medicine center.

## 2. Results

*2.1. Production/Labeling*

Two starting activities of [$^{68}$Ga]Gallium chloride, low (L) (596.55 ± 37.97 MBq) and high (H) (1436.27 ± 68.68 MBq), were used for the labeling of PSMA-11 (20 productions, 10 for each activity range). This choice was driven by the eluted radioactivity in two different periods of [$^{68}$Ge]/[$^{68}$Ga] generator shelf life. In detail, H activities correspond to the first seven weeks of generator life from the calibration date, and L activities correspond to the last seven weeks of generator life.

For all produced batches, the quality controls, performed on the final products, were in compliance with the acceptance criteria described by Ph. Eur. XII.

The values of radiochemical purity (RCP), evaluated by radio-HPLC and radio-iTLC, were ≥99%. Residual ethanol was ≤10% $v/v$.

Radionuclide purity, assessed by $\gamma$-spectrometry, ranged from 490 to 531 KeV, while the half-life measurement was between 62 and 74 min, both in the normal range.

The endotoxin value, determined by the Limulus Amebocyte Lysate test (LAL), was ≤2.5 EU/mL.

The RCP at the end of synthesis (EOS) and the related RCY are summarized in Table 1. The average value of RCY for L activity, not corrected for decay, was 80.98 ± 0.05%, decreasing to 71.48 ± 0.04% for H activity. The RCP was unaffected by the initial [$^{68}$Ga]Gallium chloride activity, resulting in 99.91% (99.75–100%) for L and 99.96% (99.81–100%) for H activities.

Table 1. Comparison of [$^{68}$Ga]Gallium chloride starting activities (low and high) and related RCY and RCP.

| [$^{68}$Ga] Low Activity (L) | | | [$^{68}$Ga] High Activity (H) | | |
|---|---|---|---|---|---|
| $^{68}$Ga Starting Activities (MBq) | RCY (%) | RCP (%) | $^{68}$Ga Starting Activities (MBq) | RCY (%) | RCP (%) |
| 640.84 | 77.70 | 100 | 1323.12 | 67 | 100 |
| 637.88 | 80.70 | 100 | 1536.61 | 71.40 | 99.99 |
| 572.39 | 83.10 | 100 | 1519.96 | 74.50 | 99.95 |
| 648.61 | 79.40 | 100 | 1499.98 | 76.10 | 99.91 |
| 572.39 | 79.20 | 100 | 1434.49 | 67.20 | 99.95 |
| 598.29 | 78.60 | 100 | 1448.18 | 72.10 | 99.98 |
| 582.38 | 78.50 | 100 | 1387.13 | 64.80 | 99.99 |
| 526.51 | 94.60 | 99.75 | 1365.67 | 74 | 100 |
| 580.90 | 77.60 | 100 | 1441.89 | 72.90 | 99.81 |
| 605.32 | 80.40 | 99 | 1405.63 | 74.80 | 100 |
| 596.55 * | 80.98 * | 99.91 * | 1436.27 * | 71.48 * | 99.96 * |
| 37.97 ** | 0.05 ** | 0.01 ** | 68.68 ** | 0.04 ** | 0.01 ** |

RCY (%) not corrected for decay and RCP (%) reported for every batch production of [$^{68}$Ga]Ga-PSMA-11; for low starting activities (range L) RCY score is 80.98%, for high starting activities (range H) RCY decreased to 71.48%. * Average ** Standard deviation.

The costs for each production of [$^{68}$Ga]Ga-PSMA-11, besides the starting activity, was approximately 1.830€; ranging from 66€/37 MBq to 140€/37 MBq for high and low activities of [$^{68}$Ga]Gallium chloride, respectively. These differences in terms of costs per 37 MBq were obtained by considering the number of allowed PET/CT exams: two PET/CT exams for L activities and up to five for H activities.

The labeling procedure, quality controls, and cost of production for preparing [$^{18}$F]PSMA-1007 in our institution have been analyzed and published [6]. Considering a starting activity of about 90 GBq of [$^{18}$F]Fluoride, a final activity of 45–50 GBq of [$^{18}$F]PSMA-1007 was obtained, allowing 25–30 PET/CT exams per day. The costs for the production of [$^{18}$F]PSMA-1007 was approximately 5.450€ (4.31€/37 MBq), as shown in Table 2.

Table 2. Costs of production and PET/CT exams for [$^{18}$F]PSMA-1007 and [$^{68}$Ga]Ga-PSMA-11.

| Radioligand | Costs of Production | | N° PET/CT Performed | | Costs for Single PET/CT Exam | |
|---|---|---|---|---|---|---|
| [$^{18}$F]PSMA-1007 | 5.454€ | | 25 | | 30€ | |
| [$^{68}$Ga]Ga-PSMA-11 | L activity 1.831€ | H activity 1.831€ | L activity 2 | H activity 5 | L activity 583€ | H activity 275€ |

Costs of each, single, production of [$^{18}$F]PSMA-1007 and [$^{68}$Ga]Ga-PSMA-11 correlated to the allowed number of PET/CT. The cost of each single PET/CT exam considered a standard dose of 259 MBq for [$^{18}$F]PSMA-1007 and of 154 MBq for [$^{68}$Ga]Ga-PSMA-11. For [$^{18}$F]PSMA-1007 the presented range of starting activity is the most cost/effective, as previously shown [6].

### 2.2. PET/CT Imaging

[$^{18}$F]PSMA-1007 and [$^{68}$Ga]Ga-PSMA-11 have different patterns of uptake in normal organs, as shown in Figure 1A,B. The "physiological", hepatic, uptake is higher for [$^{18}$F]PSMA-1007 than for [$^{68}$Ga]Ga-PSMA-11. Increased hepatic extraction of [$^{18}$F]PSMA-1007 causes increased excretion throughout the intestine, and often prolonged retention within the gall bladder is observed. This finding is related to the higher lipophilicity of fluorinated radioligand. Such biodistribution might interfere with the detection of the involved abdominal lymph nodes.

Increased uptake and elevated excretion via the urinary system are prevalent for [$^{68}$Ga]Ga-PSMA-11. The prolonged and persistent presence of radioactive urine in the bladder might musk recurrences in the prostate and/or pelvic lymph nodes.

Nevertheless, the detection of PCa deposits, in our experience, was not significantly affected by the behavior of the two radioligands in normal organs/tissues. In the wide majority of cases, both radioligands depicted the same lesions with different levels of uptake as shown in Figure 1. This patient had an initial PET/CT exam with [$^{18}$F]PSMA-1007, performed before RLT with [$^{177}$Lu]Lu-PSMA-617, and the radioligand was concentrated within skeletal and lymph nodal metastases. After the first two cycles of RLT, the patient was reevaluated by PET/CT with [$^{68}$Ga]Ga-PSMA-11, that documented the same lesions although with different uptake.

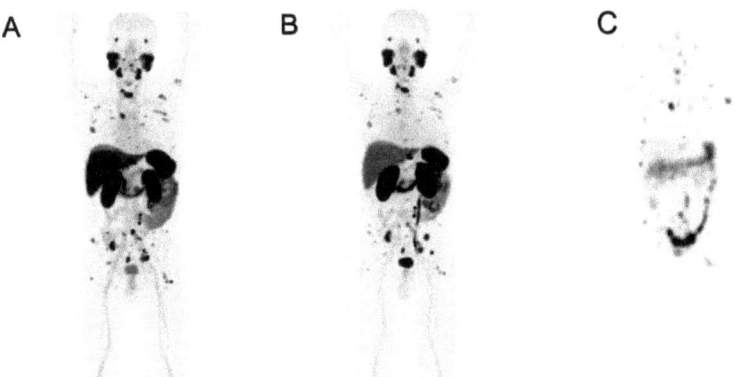

**Figure 1.** Positive PET/CT scans performed with (**A**) [$^{18}$F]PSMA-1007 (70–90 min p.i.) and with (**B**) [$^{68}$Ga]Ga-PSMA-11 (45–60 min p.i.) for the same patient. Planar images (**C**) acquired with SPECT/CT γ-camera 96 h after administration of [$^{177}$Lu]Lu-PSMA-617.

PET/CT images of both radioligands were also compared with planar, whole-body SPECT/CT images, acquired for dosimetric purposes 96 h after the first cycle (7.4 GBq of [$^{177}$Lu]Lu-PSMA-617). All of the known lesions were seen by the three different PSMA-radioligands. Such a finding indicates that the selection of candidates for RLT may be reliable, independently from the used PET-PSMA radioligand.

## 3. Discussion

In current clinical practice, PSMA-PET/CT is gaining a key role in the management of patients with PCa. It has been shown to be effective in the diagnosis, follow-up, namely, for patients with biochemical recurrence, and in the selection of patients to submit to RLT [7].

According to our experience with more than 2000 patients using either $^{68}$Ga-labeled or $^{18}$F-labeled PSMA, we have tried to summarize the advantages and the limitations of the two PET radioligands and their application in the clinical routine at our site.

The wide majority of published papers used [$^{68}$Ga]Ga-PSMA-11 and emphasized, as pro, the elevated diagnostic sensitivity (greater than 90%), and as a con the relatively low amounts of radioligand available at the end of each synthesis [8]. In fact, the amount of $^{68}$Ga-labeled compounds in general, and of PSMA in this specific case, is always defined by the activity eluted by the [$^{68}$Ge]/[$^{68}$Ga] generator, that ranges between 370 and 1110 MBq.

The RCY was affected by the activities of [$^{68}$Ga]Gallium chloride, decreasing by about 10% for higher activities (1300–1500 MBq), usually obtained in the first seven weeks of generator life from the calibration date.

In the near future, the production of [$^{68}$Ga]Gallium chloride by cyclotron and a liquid target will likely improve the limited amounts of injectable radioligands by starting with activities 20–30 fold greater than those obtained with a generator [9,10]. RCY should be carefully evaluated in this new scenario. Previous reports have shown that larger amounts of [$^{68}$Ga]Ga-PSMA-11 were obtained using cyclotron and solid targets. Activity as high as 72 GBq at EOS was obtained, with no evidence of negative effects on the quality of the final product (i.e., colloid or unreacted [$^{68}$Ga]Gallium chloride). In terms of patient doses, this activity will allow 12–15 studies to be performed in a center with two PET cameras [11,12].

Conversely, the limit of low starting activity does not apply to the production of fluorinated PSMA inhibitors. The starting activity range of [$^{18}$F]Fluoride plays a key role in the final yield for [$^{18}$F]PSMA-1007; thus, the optimization of the variables influencing the yields of the synthesis is mandatory as we have previously demonstrated [6]. In fact, using different starting activities of [$^{18}$F]Fluoride: **low** (55.91 ± 6.69 GBq), **medium** (89.06 ± 4.02 GBq), and **high** (162.38 ± 6.46 GBq), we found that the medium one was the most compliant for clinical needs, ensuring the best match of RCY, costs, and number of performed PET/CT exams [6].

For both radioligands, the measured RCP was ≥99% in all batches and it was unaffected by the different starting activities.

Moreover, we considered the production costs of the two radioligands (including cyclotron beam/generator elution, ligand, disposables for dispensation, reagents, personnel and maintenance, and radioactive waste dismission including the wasting of disposables used for the synthesis processes). According to the actual volume of PSMA PET/CT exams performed in our institution, which ranges between 25–30 exams on a weekly basis, the costs of production for [$^{68}$Ga]Ga-PSMA-11 will be much higher than those for [$^{18}$F]PSMA-1007. They will range from 9.155€ to 18.310€ according to the [$^{68}$Ge]/[$^{68}$Ga] generator shelf life and the number of required syntheses (5 or 6) to prepare 25–30 doses. All of the required syntheses (5–6) could not be performed in one day because the [$^{68}$Ge]/[$^{68}$Ga] generator needs time to regenerate after elution [13]. As a reminder, according to our analysis, the cost of a [$^{18}$F]PSMA-1007 production to perform the same number of PET/CT exams is approximatively 5.450€.

From a clinical point of view, the different chemical properties of the two diagnostic radioligands somehow influence their biodistribution; the timing of PET/CT acquisition does not significantly affect the diagnostic accuracy and their use in the various clinical scenarios of PCa patients, including the definition of eligibility for RLT.

Our experience indicated that the highest uptake within lesions was observed by [$^{18}$F]-PSMA-1007 while the overall rate of detection was not significantly affected by the PSMA probe. The highest uptake of [$^{18}$F]-PSMA-1007 may be explained by the highest injected activity (up to 3-fold) of [$^{68}$Ga]Ga-PSMA-11 and the relatively high number of emitted positrons vs. [$^{68}$Ga]Ga-PSMA-11, as shown in Figure 1.

At least two significant differences were detectable on PET images and were strictly related to the pharmacokinetic properties of these two radioligands as shown in Figure 2. A greater physiological hepatic retention occurs when using [$^{18}$F]PSMA-1007, and it is related to the highest lipophilicity [14,15]. In kidneys and the urinary tract, [$^{68}$Ga]Ga-PSMA-11 accumulates highly because of its greater hydrophilicity. Among the differences between the two radiopharmaceuticals, a greater [$^{18}$F]PSMA-1007 uptake in the skeleton is frequently observed, which according to the extensive and prolonged quality controls we have performed in our experience, is not related to free [$^{18}$F]Fluorine. The nature of the isotope may be a possible explanation: the lower positron energy and the higher rate of photon emission (photon flux density) of [$^{18}$F]Fluorine compared to [$^{68}$Ga]Gallium contribute to the detection of more positive benign lesions in the skeleton, related to increased osteoblastic activity (i.e., osteoarthritis, degenerative changes, fractures, etc.) [16,17].

The analysis of images might be in favor of fluorinated PSMA because its positron energy emission (0.65 MeV) that is lower than that of [$^{68}$Ga]Gallium (1.90 MeV) enables a better spatial resolution on PET/CT images (Figure 1) and a lower radiation burden [13,14]. Furthermore, the positron yield of [$^{68}$Ga]Gallium is lower than that of [$^{18}$F]Fluoride (89.14% vs. 96.86%), increasing image noise by possible prompt gamma contamination in PET data and negatively impacting the detection sensitivity [14].

The anatomical sites of "normal" uptake are the same for both diagnostic PSMA ligands, and the different contrast on PET/CT images is due to the density of $\beta^+$ generated by the decay, which is major for [$^{18}$F]Fluoride [14,18].

A higher uptake of [$^{18}$F]PSMA-1007 than [$^{68}$Ga]Ga-PSMA-11 within recurrent metastases has been reported as well as a biodistribution similar to PSMA-617, currently used for the RLT of mCRPC [19]. Despite the differences in the structure of the molecules, in the radiolabeling strategy (a prosthetic rather than a chelator group), PSMA-1007 and PSMA-617 show reduced kidney uptake compared with PSMA-11. Thus, [$^{18}$F]PSMA-1007 and [$^{177}$Lu]Lu-PSMA-617 seem to be a perfect theragnostic tandem [20] as confirmed in our experience.

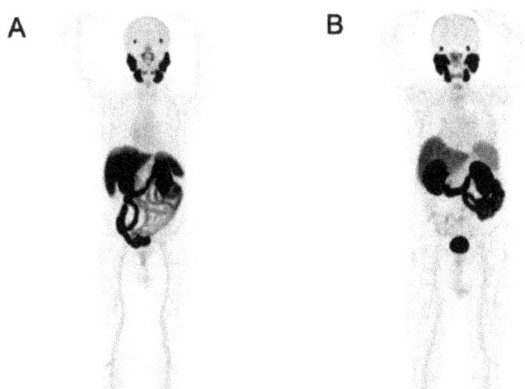

**Figure 2.** Negative PET/CT scans: patient (**A**) was examined with [$^{18}$F]PSMA-1007, and patient (**B**) was examined with [$^{68}$Ga]Ga-PSMA-11.

## 4. Materials and Methods

Radiosyntheses and quality controls of [$^{18}$F]PSMA-1007 were performed as previously described [5].

### 4.1. Radiosynthesis of [$^{68}$Ga]Ga-PSMA-11

[$^{68}$Ga]Gallium was obtained by a $^{68}$Ge/$^{68}$Ga generator (0.74–1.85 GBq) (Galliapharm, Eckert-Ziegler, Berlin, Germany). Radiosyntheses were performed using an All In One mini automated synthesizer (Trasis, Ans, Belgium) and cassettes without pre-purification after generator elution. The reagents kit (ABX, Radeberg, Germany) included PSMA-11 (GMP grade), acetate buffer (0.7 M), HCl (0.1 M), ethanol, and a saline bag.

Radiochemical yield was evaluated by radio-high performance liquid chromatography (HPLC) analyses that were carried out using an LC 20AD Pump with a SPD-20AV UV/VIS detector (Shimadzu, Kyoto, Japan) equipped with a GABI radiometric detector (Raytest, Elysia, Straubenhardt, Germany). A 5 μm C18 300 Å, 250 × 4.6 mm column (Jupiter®, Phenomenex, Bologna, Italy) was used with a flow rate of 1 mL/min and the following gradient (acetonitrile 0.1% Trifluoroacetic acid (TFA) as solvent A, water 0.1% TFA as solvent B): 100% A for 5 min, A 25% and B 75% for 3 min, the same gradient for 4 min, and then 100% B for 3 min. The UV/VIS detector was set at 220 nm and 254 nm.

Thin-layer chromatography (TLC) was carried out using 1 M ammonium acetate/methanol (1:1) as the mobile phase and iTLC-SG glass microfiber chromatography paper impregnated with silica gel (Agilent Technologies, Santa Clara, CA, USA) as the stationary phase. TLCs were analyzed by a storage phosphor system (Cyclon Plus, Perkin Elmer, Oxford, UK).

Residual ethanol was quantified by a gas-chromatography system (GC 2010 Plus, Shimadzu, Japan) equipped with a flame-ionization detector (FID) and a capillary column (Elite-1301, 6% cyanopropylphenyl 94% dimethyl polysiloxane, L 30 m, ID 0.53; Perkin Elmer, UK). The temperature for the split was 240 °C and 280 °C for the FID.

Radionuclidic purity was determined using a multi-channel analyzer (Mucha Star, Raytest, Elysia, Germany) and by half-life measurement with a dose calibrator (Atomlab 500, Biodex, New York, NY, USA).

A radiotracer was also tested for bacterial endotoxin through the kinetic chromogenic Limulus Amebocyte Lysate (LAL) method (Endosafe Nexgen-PTS, Charles River, Wilmington, MA, USA).

All quality controls of the final products followed the acceptance criteria described in Ph. Eur. XII.

*4.2. Imaging Protocols*

PET/CT images were performed on a time-of-flight PET/CT digital scanner (Discovery MI, GE Healthcare, Waukesha, WI, USA) with a 25 cm axial field of view. A low-dose CT scan (120 kV; automated current modulation; 0.98 pitch; 3.75 mm slice thickness; 0.5-s rotation time) from the vertex of the skull to the mid-thighs was used for anatomic localization and attenuation correction. PET scanning followed at 2 min/bed position with a 27-slice overlap. Images were reconstructed using ordered-subset expectation maximization (OSEM, four iterations, eight subsets, cutoff 6 mm, 256 × 256 matrix) applying all appropriate corrections for dead time, randoms, scatter, coincidence, and detector normalization.

Post therapy images were acquired on a SPECT/CT Discovery NM/CT 670 system (GE Healthcare) equipped with MEGP (medium energy general purpose) collimators and 3/8″ NaI(Tl) crystal thickness. A 20% energy window centered on the 208 keV photopeak and a 10% scatter correction window centered on 178 keV were applied. Whole-body images were acquired in continuous mode (15 cm/min), with a 1.0 zoom, 2.21 mm pixel size, and an automatic body contour.

## 5. Conclusions

[$^{18}$F]PSMA-1007 is in many ways an ideal radiotracer for PCa imaging; it can be produced with purity and in high yield and it can be transported to sites that do not have cyclotron and radiopharmacy facilities.

[$^{68}$Ga]Ga-PSMA-11 can also be produced in high purity but in much lower yield than [$^{18}$F]PSMA-1007 unless it is produced by a cyclotron; moreover, [$^{68}$Ga]Ga-PSMA-11 has to be produced on site due the short half-life and low yield.

The costs of [$^{68}$Ga]Ga-PSMA-11 production are the same for both starting activity ranges. However, comparing the two diagnostic radiopharmaceuticals, [$^{18}$F]PSMA-1007 is the most compliant one, according to the aspects previously evaluated.

From a clinical point of view, our experience suggests that [$^{18}$F]PSMA-1007 is characterized by a better specificity in detecting recurrences and a similar behavior to the therapeutical analogue [$^{177}$Lu]Lu-PSMA-617. This makes the radioligand a better probe for patients eligible for RLT.

To optimize the production of [$^{68}$Ga]Gallium chloride, it is mandatory to overcome the limited scalability of the generator, by introducing a liquid target for the production of [$^{68}$Ga]Gallium by a cyclotron. This will enable a significant reduction of costs due to the synthesis process of [$^{68}$Ga]Ga-PSMA-11 improving the number of PET/CT exams performed at once in a similar range/fashion to [$^{18}$F]PSMA-1007.

Our experience also shows that among the different PSMA radioligands used to image or to treat PCa, there are no significant differences in their binding within tumor lesions. As shown in Figure 1, no differences in the biodistribution in lesions were identified among patients, which has been studied before starting RLT with [$^{177}$Lu]Lu-PSMA-617 and imaged, after two cycles of RLT, with [$^{68}$Ga]Ga-PSMA-11. The relevant aspect is the similar distribution and levels of uptake within malignant deposits among diagnostic ligands and therapeutic companions while some differences were observed in normal organs, mainly related to greater lipophilicity of [$^{18}$F]PSMA-1007 vs. other radioligands.

In conclusion, the choice between [$^{68}$Ga]Ga-PSMA-11 and [$^{18}$F]PSMA-1007 is not influenced by significant differences in the rate of detection, while it might reflect economical and/or the number of PET/CT studies to perform on a daily or weekly basis. In our analysis, [$^{18}$F]-PSMA-1007 seems to add several advantages in routine clinical applications, related to economic convenience, greater availability, and consequent higher number of performed

PET/CT exams, as well as a pattern of distribution similar to [$^{177}$Lu]Lu-PSMA-617 within PCa lesions, which supports its clinical use for proper selection of patients eligible for RLT.

**Author Contributions:** Conceptualization, C.M., A.M., R.d.M. and S.L.; methodology, M.A., E.S., M.J.B.R., V.P. and A.E.; clinical data analysis, F.D.G., L.C., M.R., A.R. and S.L.; investigation, L.D. and V.P.P.; data curation, P.G.; writing—original draft preparation, C.M., A.M. and R.d.M.; writing—review and editing, S.L.; supervision, S.L.; funding acquisition, S.L. All authors have read and agreed to the published version of the manuscript.

**Funding:** This research received no external funding.

**Institutional Review Board Statement:** Not applicable.

**Informed Consent Statement:** Not applicable.

**Data Availability Statement:** Raw data is available on https://zenodo.org/record/6563187#.YoY_uS98qX1, accessed on 19 May 2022.

**Conflicts of Interest:** The authors declare no conflict of interest.

**Sample Availability:** Not applicable.

# References

1. Eeles, R.; Ni Raghallaigh, H. Men with a susceptibility to prostate cancer and the role of genetic based screening. *Transl. Androl. Urol.* **2018**, *7*, 61–69. [CrossRef] [PubMed]
2. Lengana, T.; Lawal, I.O.; Rensburg, C.V.; Mokoala, K.M.G.; Moshokoa, E.; Ridgard, T.; Vorster, M.; Sathekge, M.M. A comparison of the diagnostic performance of $^{18}$F-PSMA-1007 and $^{68}$GA-PSMA-11 in the same patients presenting with early biochemical recurrence. *Hell. J. Nucl. Med.* **2021**, *24*, 178–185. [CrossRef] [PubMed]
3. Umbricht, C.A.; Benešová, M.; Schmid, R.M.; Türler, A.; Schibli, R.; van der Meulen, N.P.; Müller, C. $^{44}$Sc-PSMA-617 for radiotheragnostics in tandem with $^{177}$Lu-PSMA-617-preclinical investigations in comparison with $^{68}$Ga-PSMA-11 and $^{68}$Ga-PSMA-617. *EJNMMI Res.* **2017**, *7*, 9. [CrossRef] [PubMed]
4. Parker, C.; Castro, E.; Fizazi, K.; Heidenreich, A.; Ost, P.; Procopio, G.; Tombal, B.; Gillessen, S. Prostate cancer: ESMO Clinical Practice Guidelines for diagnosis, treatment and follow-up. *Ann. Oncol.* **2020**, *31*, 1119–1134. [CrossRef] [PubMed]
5. Fendler, W.P.; Rahbar, K.; Herrmann, K.; Kratochwil, C.; Eiber, M. $^{177}$Lu-PSMA Radioligand Therapy for Prostate Cancer. *J. Nucl. Med.* **2017**, *58*, 1196–1200. [CrossRef] [PubMed]
6. Maisto, C.; Morisco, A.; de Marino, R.; Squame, E.; Porfidia, V.; D'Ambrosio, L.; Di Martino, D.; Gaballo, P.; Aurilio, M.; Buonanno, M.; et al. On site production of [$^{18}$F]PSMA-1007 using different [$^{18}$F]fluoride activities: Practical, technical and economical impact. *EJNMMI Radiopharm. Chem.* **2021**, *6*, 36. [CrossRef] [PubMed]
7. Kaewput, C.; Vinjamuri, S. Update of PSMA Theranostics in Prostate Cancer: Current Applications and Future Trends. *J. Clin. Med.* **2022**, *11*, 2738. [CrossRef] [PubMed]
8. Amor-Coarasa, A.; Schoendorf, M.; Meckel, M.; Vallabhajosula, S.; Babich, J.W. Comprehensive Quality Control of the ITG 68Ge/68Ga Generator and Synthesis of 68Ga-DOTATOC and 68Ga-PSMA-HBED-CC for Clinical Imaging. *J. Nucl. Med.* **2016**, *57*, 1402–1405. [CrossRef] [PubMed]
9. Jussing, E.; Milton, S.; Samén, E.; Moein, M.M.; Bylund, L.; Axelsson, R.; Siikanen, J.; Tran, T.A. Clinically Applicable Cyclotron-Produced Gallium-68 Gives High-Yield Radiolabeling of DOTA-Based Tracers. *Biomolecules* **2021**, *11*, 1118. [CrossRef] [PubMed]
10. Siikanen, J.; Jussing, E.; Milton, S.; Steiger, C.; Ulin, J.; Jonsson, C.; Samén, E.; Tran, T.A. Cyclotron-produced $^{68}$Ga from enriched $^{68}$Zn foils. *Appl. Radiat. Isot.* **2021**, *176*, 109825. [CrossRef] [PubMed]
11. Tieu, W.; Hollis, C.A.; Kuan, K.W.; Takhar, P.; Stuckings, M.; Spooner, N.; Malinconico, M. Rapid and automated production of [$^{68}$Ga]gallium chloride and [$^{68}$Ga]Ga-DOTA-TATE on a medical cyclotron. *Nucl. Med. Biol.* **2019**, *74–75*, 12–18. [CrossRef] [PubMed]
12. Thisgaard, H.; Kumlin, J.; Langkjær, N.; Chua, J.; Hook, B.; Jensen, M.; Kassaian, A.; Zeisler, S.; Borjian, S.; Cross, M.; et al. Multi-curie production of gallium-68 on a biomedical cyclotron and automated radiolabelling of PSMA-11 and DOTATATE. *EJNMMI Radiopharm. Chem.* **2021**, *6*, 1. [CrossRef] [PubMed]
13. Amor-Coarasa, A.; Kelly, J.M.; Gruca, M.; Nikolopoulou, A.; Vallabhajosula, S.; Babich, J.W. Continuation of comprehensive quality control of the itG $^{68}$Ge/$^{68}$Ga generator and production of $^{68}$Ga-DOTATOC and $^{68}$Ga-PSMA-HBED-CC for clinical research studies. *Nucl. Med. Biol.* **2017**, *53*, 37–39. [CrossRef] [PubMed]
14. Piron, S.; Verhoeven, J.; Vanhove, C.; De Vos, F. Recent advancements in $^{18}$F-labeled PSMA targeting PET radiopharmaceuticals. *Nucl. Med. Biol.* **2022**, *106–107*, 29–61. [CrossRef] [PubMed]
15. Foley, R.W.; Redman, S.L.; Graham, R.N.; Loughborough, W.W.; Little, D. Fluorine-18 labelled prostate-specific membrane antigen (PSMA)-1007 positron-emission tomography-computed tomography: Normal patterns, pearls, and pitfalls. *Clin. Radiol.* **2020**, *75*, 903–913. [CrossRef] [PubMed]

16. Kroenke, M.; Mirzoyan, L.; Horn, T.; Peeken, J.C.; Wurzer, A.; Wester, H.J.; Makowski, M.; Weber, W.A.; Eiber, M.; Rauscher, I. Matched-Pair Comparison of $^{68}$Ga-PSMA-11 and $^{18}$F-rhPSMA-7 PET/CT in Patients with Primary and Biochemical Recurrence of Prostate Cancer: Frequency of Non–Tumor-Related Uptake and Tumor Positivity. *J. Nucl. Med.* **2021**, *62*, 1082–1088. [CrossRef] [PubMed]
17. Sanchez-Crespo, A. Comparison of Gallium-68 and Fluorine-18 imaging characteristics in positron emission tomography. *Appl. Radiat. Isot.* **2013**, *76*, 55–62. [CrossRef] [PubMed]
18. Hoffmann, M.A.; von Eyben, F.E.; Fischer, N.; Rosar, F.; Müller-Hübenthal, J.; Buchholz, H.G.; Wieler, H.J.; Schreckenberger, M. Comparison of [$^{18}$F]PSMA-1007 with [$^{68}$Ga]Ga-PSMA-11 PET/CT in Restaging of Prostate Cancer Patients with PSA Relapse. *Cancers* **2022**, *14*, 1479. [CrossRef] [PubMed]
19. Conti, M.; Eriksson, L. Physics of pure and non-pure positron emitters for PET: A review and a discussion. *EJNMMI Phys.* **2016**, *3*, 8. [CrossRef] [PubMed]
20. Farolfi, A.; Calderoni, L.; Mattana, F.; Mei, R.; Telo, S.; Fanti, S.; Castellucci, P. Current and Emerging Clinical Applications of PSMA PET Diagnostic Imaging for Prostate Cancer. *J. Nucl. Med.* **2021**, *62*, 596–604. [CrossRef] [PubMed]

Article

# Synthesis and Evaluation of [11]C-Labeled Triazolones as Probes for Imaging Fatty Acid Synthase Expression by Positron Emission Tomography

James M. Kelly [1,2,*], Thomas M. Jeitner [1], Nicole N. Waterhouse [2], Wenchao Qu [2,†], Ethan J. Linstad [3,4], Banafshe Samani [3], Clarence Williams, Jr. [1], Anastasia Nikolopoulou [1,2,‡], Alejandro Amor-Coarasa [1,§], Stephen G. DiMagno [3,∥] and John W. Babich [1,2,5,∥]

[1] Molecular Imaging Innovations Institute (MI3), Department of Radiology, Weill Cornell Medicine, New York, NY 10065, USA; tmj4001@med.cornell.edu (T.M.J.); clw2012@med.cornell.edu (C.W.J.); ann2010@med.cornell.edu (A.N.); alejandro.amor@einsteinmed.edu (A.A.-C.); jwbabich@gmail.com (J.W.B.)
[2] Citigroup Biomedical Imaging Center, Weill Cornell Medicine, New York, NY 10021, USA; niw2014@med.cornell.edu (N.N.W.); weq2002@med.cornell.edu (W.Q.)
[3] Departments of Medicinal Chemistry & Pharmacognosy and Chemistry, University of Illinois-Chicago, Chicago, IL 60612, USA; ethanl@uic.edu (E.J.L.); bsaman4@uic.edu (B.S.); sdimagno@uic.edu (S.G.D.)
[4] Department of Chemistry, University of Nebraska-Lincoln, Lincoln, NE 68588, USA
[5] Sandra and Edward Meyer Cancer Center, Weill Cornell Medicine, New York, NY 10065, USA
* Correspondence: jak2046@med.cornell.edu
† Present address: Department of Psychiatry, Stony Brook University, Stony Brook, NY 11794, USA.
‡ Present address: The Janssen Pharmaceutical Companies of Johnson & Johnson, Spring House, PA 19477, USA.
§ Present address: Department of Radiology, Albert Einstein College of Medicine, Bronx, NY 10461, USA.
∥ Present address: Ratio Therapeutics, LLC, Boston, MA 02210, USA.

**Abstract:** Cancer cells require lipids to fulfill energetic, proliferative, and signaling requirements. Even though these cells can take up exogenous fatty acids, the majority exhibit a dependency on de novo fatty acid synthesis. Fatty acid synthase (FASN) is the rate-limiting enzyme in this process. Expression and activity of FASN is elevated in multiple cancers, where it correlates with disease progression and poor prognosis. These observations have sparked interest in developing methods of detecting FASN expression in vivo. One promising approach is the imaging of radiolabeled molecular probes targeting FASN by positron emission tomography (PET). However, although [[11]C]acetate uptake by prostate cancer cells correlates with FASN expression, no FASN-specific PET probes currently exist. Our aim was to synthesize and evaluate a series of small molecule triazolones based on GSK2194069, an FASN inhibitor with IC$_{50}$ = 7.7 ± 4.1 nM, for PET imaging of FASN expression. These triazolones were labeled with carbon-11 in good yield and excellent radiochemical purity, and binding to FASN-positive LNCaP cells was significantly higher than FASN-negative PC3 cells. Despite these promising characteristics, however, these molecules exhibited poor in vivo pharmacokinetics and were predominantly retained in lymph nodes and the hepatobiliary system. Future studies will seek to identify structural modifications that improve tumor targeting while maintaining the excretion profile of these first-generation [11]C-methyltriazolones.

**Keywords:** fatty acid synthase; cancer metabolism; positron emission tomography; carbon-11

## 1. Introduction

Increased lipogenesis is a phenotypic hallmark of many cancer cells [1,2]. Fatty acids support a number of essential processes in cancer, including proliferation, energy, oncogenic signaling pathways, and resistance to therapy [1,3]. The fatty acid pool in cancer cells is fed by both de novo fatty acid synthesis and the uptake of exogenous fatty acids [4]. However, the majority of cancer cells overexpress lipogenic enzymes, including fatty acid synthase (E.C. 2.3.1.85; FASN) and exhibit a dependency on de novo fatty acid synthesis [5]. FASN synthesizes palmitate from acetyl-CoA and malonyl-CoA, using NADPH as a reducing equivalent [2], and is the rate-limiting enzyme of de novo fatty acid synthesis [6]. The products of de novo fatty acid synthesis in cancer cells are predominantly esterified to phospholipid aggregates that participate in signal transduction, intracellular trafficking, and cell migration [7]. FASN is also proposed to support post-translational modification and provide palmitate as a source of fuel [8]. FASN expression is regulated by the PI3K/Akt axis [9], which is frequently dysregulated in cancer [10]. Consequently, expression and activity of FASN is elevated in multiple cancers [11–15], where it correlates with disease progression and poor prognosis [13,16,17].

In contrast to cancerous cells, FASN expression in normal tissue is typically low [5]. This facilitates a therapeutic strategy based on inhibition of FASN activity. FASN exists as a homodimer comprised of identical subunits of seven domains, many of which have been targeted by pharmacological inhibitors [2,18]. Early studies using broader spectrum inhibitors targeting specific domains of FASN demonstrated decreased viability of cancer cells and restrained tumor growth in xenograft models as a consequence of FASN inhibition [19–26] but these compounds were limited by poor solubility, poor bioavailability, and off-target toxicity [2,18,27]. More recently, multiple classes of high affinity small molecule inhibitors have been developed and advanced into preclinical [28–32] and clinical evaluation [33,34].

In spite of the prevalence of small molecule FASN inhibitor platforms, no FASN-specific agent exists for in vivo imaging of FASN expression by positron emission tomography (PET) [34]. Our aims were to develop a family of $^{11}$C-labeled FASN inhibitors based on high affinity triazolone-containing compounds related to GSK2194069 [35,36] and to evaluate these compounds in prostate cell lines and xenograft models characterized by high and low FASN expression. We report the synthesis of three [$^{11}$C]methyltriazolones that are rapidly taken up by lymph node carcinoma of the prostate (LNCaP) cells in an FASN-specific manner, but are limited in their imaging utility by poor in vivo pharmacokinetics. These compounds may pave the way for radiolabeled FASN inhibitors with better pharmacokinetics in the future.

## 2. Results

### 2.1. Lead Compound Identification

Our strategy for developing $^{11}$C-labeled FASN inhibitors was to N-alkylate triazolone FASN inhibitors with [$^{11}$C]CH$_3$I or [$^{11}$C]CH$_3$OTf. This approach was based on the observation that N-methylation of the triazolone moiety of GSK2194069 (1), a commercially-available, high affinity FASN ligand, does not substantially decrease FASN affinity (10 nM vs. 20 nM) [35]. We selected two fluorine-containing triazolones (2 and 3) from a library of previously published library of GSK2194069 analogues [35] for this purpose since they could be labeled with fluorine-18 as an alternative to carbon-11 should our initial studies prove successful. Fluorine-18 has superior physical properties (e.g., higher positron branching, lower positron $E_{max}$) for PET imaging than carbon-11, but the chemistry of fluoride incorporation on electron-rich aromatic rings is challenging. Our target compounds are shown in Figure 1.

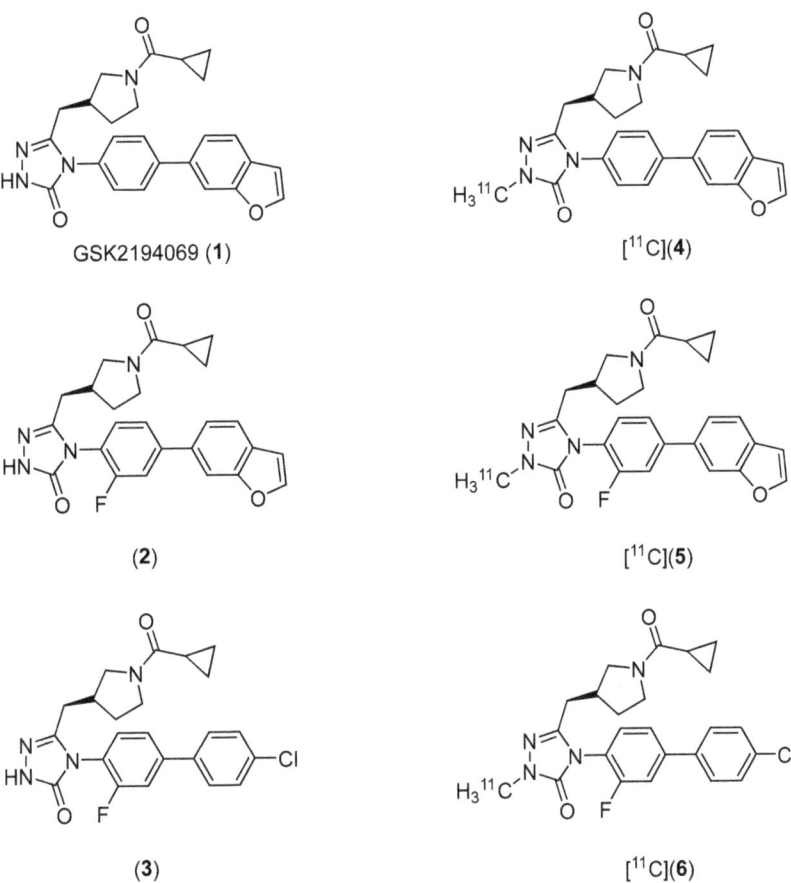

**Figure 1.** Target compounds. Fluorine-containing compounds **2** and **3**, and their derivatives [$^{11}$C]**5** and [$^{11}$C]**6**, were selected to allow the possibility of $^{18}$F-fluorination as an alternative labeling strategy.

## 2.2. Radiosynthesis

The compounds were synthesized by base-catalyzed alkylation of the triazolone precursor. Each compound was isolated in greater than 97% radiochemical purity (Figures S1–S3).

(S)-[*methyl*-$^{11}$C]4-(4-(Benzofuran-6-yl)phenyl)-5-((1-(cyclopropanecarbonyl)pyrrolidin-3-yl)methyl)-2-methyl-2,4-dihydro-3H-1,2,4-triazol-3-one ([$^{11}$C]**4**) was synthesized from GSK2194069 (Figure 2) in 44 ± 2 min from end of bombardment (EOB). Production and delivery of [$^{11}$C]CH$_3$I was accomplished in 14 ± 1 min, and the synthesis of [$^{11}$C]**4** was therefore completed in approximately 30 min from delivery of [$^{11}$C]CH$_3$I. The radiochemical yield of [$^{11}$C]**4**, decay-corrected (dc) to the delivery of [$^{11}$C]CH$_3$I, was 46.8 ± 9.0% (*n* = 12) when NaOH was used as base and DMSO as the solvent. The molar activity, $A_m$, was 420 ± 320 GBq/µmol. When K$_2$CO$_3$/DMF were used, [$^{11}$C]**4** was isolated in 52.5 ± 1.5% decay-corrected radiochemical yield (dc RCY) (*n* = 2) and $A_m$ = 240 GBq/µmol.

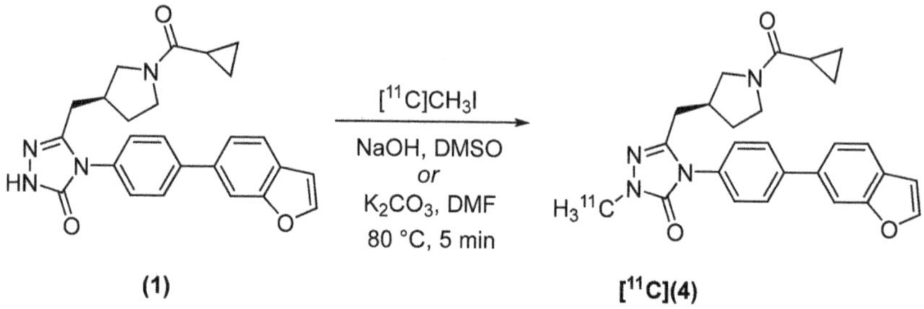

**Figure 2.** Radiosynthesis of [$^{11}$C](4).

The synthesis of (S)-[methyl-$^{11}$C]4-(4-(benzofuran-6-yl)-2-fluorophenyl)-5-((1-(cyclopropanecarbonyl)pyrrolidin-3-yl)methyl)-2-methyl-2,4-dihydro-3H-1,2,4-triazol-3-one ([$^{11}$C]5) was accomplished by two strategies. Our first method employed [$^{11}$C]CH$_3$I as the alkylating agent (Figure 3) and resulted in a dc RCY, corrected to delivery of [$^{11}$C]CH$_3$I, of 43.2 ± 17.8% ($n$ = 3) and A$_m$ = 322 ± 96 GBq/µmol. Synthesis time was 48 ± 2 min from EOB. Alkylation of (2) was also accomplished using [$^{11}$C]CH$_3$OTf in 50 ± 1 min from EOB, of which 16 ± 1 min were required to produce and deliver [$^{11}$C]CH$_3$OTf. Under these reaction conditions, the dc RCY, corrected to delivery of [$^{11}$C]CH$_3$OTf, was 65.3 ± 18.5% ($n$ = 3) and A$_m$ = 1280 GBq/µmol.

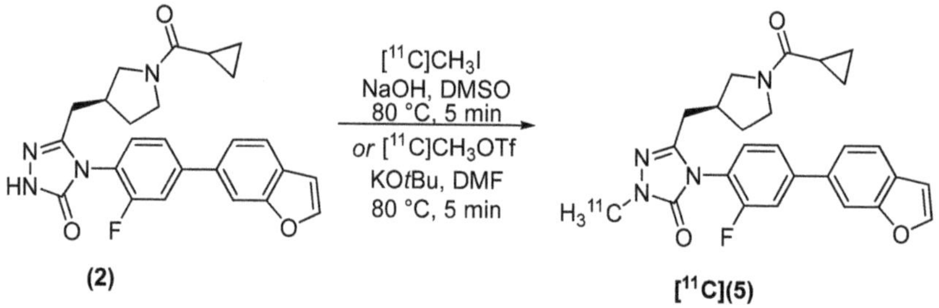

**Figure 3.** Radiosynthesis of [$^{11}$C](5).

(S)-[methyl-$^{11}$C]-4-(4'-Chloro-3-fluoro-[1,1'-biphenyl]-4-yl)-5-((1-(cyclopropanecarbonyl)pyrrolidin-3-yl)methyl)-2-methyl-2,4-dihydro-3H-1,2,4-triazol-3-one ([$^{11}$C]6) was synthesized using [$^{11}$C]CH$_3$I (Figure 4). The final product was isolated in 61.8 ± 15.5% dc RCY ($n$ = 3) and A$_m$ = 49 ± 27 GBq/µmol ($n$ = 3). Synthesis time was 33 ± 6 min from EOB.

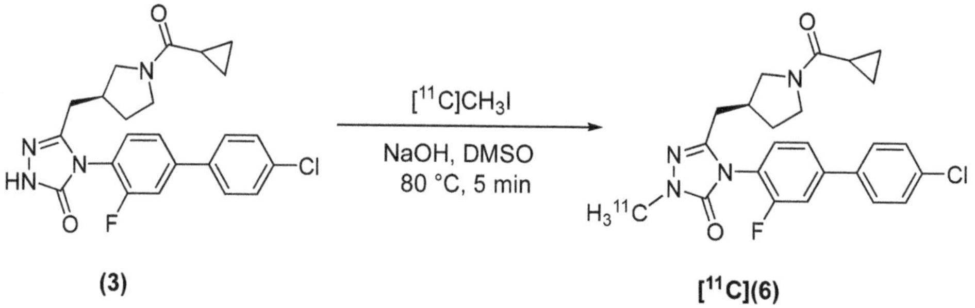

**Figure 4.** Radiosynthesis of [$^{11}$C](6).

## 2.3. In Vitro Cell Binding

The activity of the probes was determined against LNCaP cells, which highly express FASN (Figure S4), and PC3 prostate cancer cells, which express FASN at much lower levels and served as our negative controls. Each of the compounds exhibited higher binding to LNCaP cells than to PC3 cells after the harvested counts were normalized to protein content. Binding was characterized by a rapid initial rate of uptake, followed by a period of equilibrium (Figure 5). Peak cell binding was reached after 10–15 min and was two-fold higher in LNCaP cells. Efflux of radioactivity was negligible up to 50 min after addition of activity. Peak binding of [$^{11}$C]5 and [$^{11}$C]6 to LNCaP cells was comparable (4.23 ± 0.14 vs. 5.06 ± 0.18% added activity), while peak binding of [$^{11}$C]4 was slightly lower (2.37 ± 0.21% added activity). Binding of [$^{11}$C]5 to LNCaP cells was significantly higher than PC3 cells ($p = 0.03$).

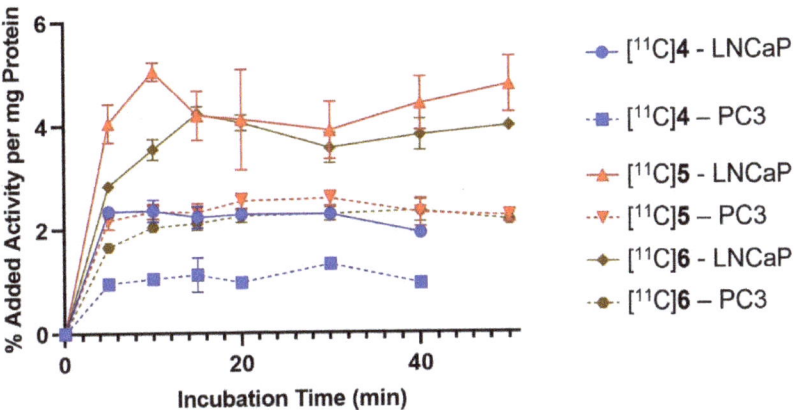

**Figure 5.** Time course binding of [$^{11}$C]4 (blue), [$^{11}$C]5 (red), and [$^{11}$C]6 (brown) to LNCaP cells or PC3 cells incubated at 37 °C for up to 50 min. The counts were corrected for decay and normalized to protein content.

## 2.4. In Vitro Growth Inhibition Assays

The ability of the compounds to inhibit the growth of LNCaP and PC3 cells was determined for ligand concentrations ranging from 1 nM to 100 µM (Figure 6). The EC$_{50}$ in LNCaP cells of GSK2194069 is 16 nM. By comparison, the EC$_{50}$ of GSK2194069 in A549 cells, another cell line with high FASN expression, is 15 ± 0.5 nM [36]. Analogues **5** and **4** substantially reduced the viability of LNCaP cells, with EC$_{50}$ = 14 nM and EC$_{50}$ = 22 nM, respectively. By contrast, compound **6** did not inhibit LNCaP cells as successfully, with EC$_{50}$ = 2.9 µM.

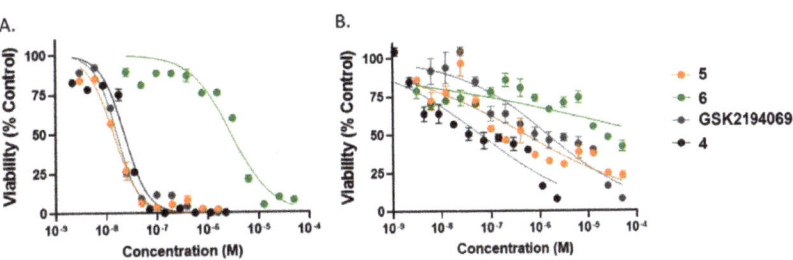

**Figure 6.** Growth inhibition of (**A**) LNCaP cells or (**B**) PC3 cells following treatment with GSK2194069 (gray), or the non-radioactive standards **4** (black), **5** (orange), and **6** (green) for 4 d. Studies were performed in triplicate.

Each of the compounds preferentially kills LNCaP cells as compared to PC3 cells. The effectiveness in PC3 cells is reduced by a factor of 90 for GSK2194069 (EC$_{50}$ = 1.5 µM) and a factor of nearly 40 for **5** (EC$_{50}$ = 551 nM) (Figure S4).

*2.5. MicroPET/CT Imaging*

In spite of promising in vitro activities in LNCaP cells, none of the $^{11}$C-labeled FASN inhibitors accumulates in LNCaP xenograft tumors. [$^{11}$C]**4** distributes to the hepatobiliary system (Figure 7A). In addition, accumulation in thymus is evident at early time points but clears by 60 min p.i. [$^{11}$C]**5** predominantly accumulates in the liver and intestines (Figure 7B), with uptake in lungs at early time points evident. In addition, uptake is evident in the brachial, cervical, and lumbar lymph nodes. The tumor is indistinguishable from background. To account for the possibility that the uptake in liver and nodes is due to self-aggregation of the lipophilic probe, imaging was performed using a pre-mixed solution of [$^{11}$C]**5** and Captisol® (Ligand, San Diego, CA, USA), a modified cyclodextrin solubilizing agent [37]. There was no apparent change in the tissue distribution (data not shown). [$^{11}$C]**6** distributes similarly to both [$^{11}$C]**4** and [$^{11}$C]**5** (Figure 7C). FASN expression was confirmed in the xenografted tumors by Western blot (Figure S4).

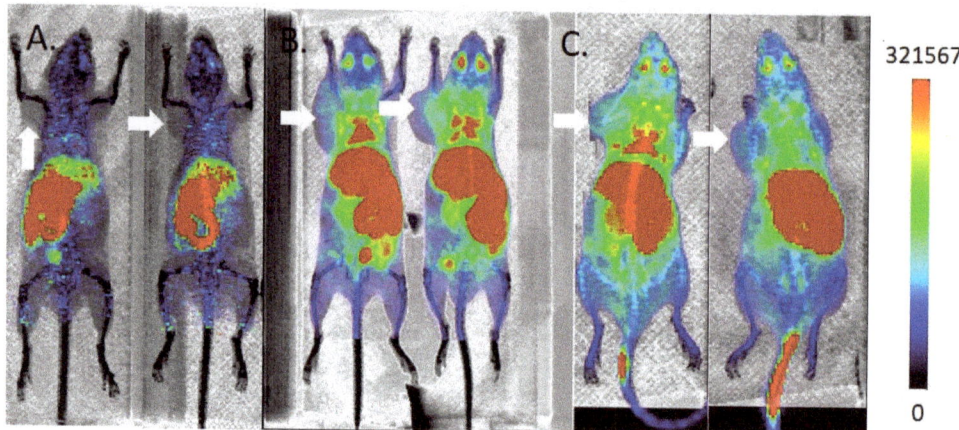

**Figure 7.** Maximum intensity projection microPET/CT images of $^{11}$C-labeled triazolones in male inbred athymic nu/nu mice bearing LNCaP xenografts (*n* > 4 per compound). Mice were imaged in groups of 4, with 2 representative examples shown per compound. Tumors are indicated with white arrows. The mice were administered (**A**) 8–10 MBq [$^{11}$C]**4** in 10% EtOH/saline; (**B**) 9–11 MBq [$^{11}$C]**5** in 10% EtOH/saline; (**C**) 9–11 MBq [$^{11}$C]**6** in 10% EtOH/saline.

## 3. Discussion

In contrast to many cancers, prostate cancer exhibits highly variable uptake of [$^{18}$F]FDG [38]. This variability likely reflects the biological and clinical heterogeneity of disease [39]. [$^{18}$F]FDG uptake has prognostic value, particularly in metastatic prostate cancer [40–42], but uptake in androgen-sensitive disease is generally low [43]. [$^{11}$C]Choline and [$^{18}$F]fluorocholine exploit upregulation of choline metabolism in prostate cancer, which leads to increased phosphatidylcholine levels and turnover [44,45]. Similarly to [$^{18}$F]FDG PET, [$^{11}$C]choline and [$^{18}$F]fluorocholine PET suffer from diminished sensitivity in primary cancer, but may be of greater value in detecting recurrent disease [46]. Sensitive detection of prostate cancer by metabolic imaging therefore requires other metabolic pathways to be considered.

Lipid metabolism is upregulated even at early stages of prostate cancer [47], rendering this pathway an attractive target for diagnostic and prognostic imaging. FASN catalyzes the de novo synthesis of long-chain fatty acids and is an important regulator of lipid metabolism [48]. In spite of the emerging importance of FASN as a marker of tumor aggressiveness [11], a possible pharmacological target [31,32], and a prognostic indicator [11,49], there are no probes for direct in vivo PET imaging of FASN expression. [$^{11}$C]Acetate is proposed to be a probe for FASN on the basis that its uptake correlates with FASN expression in multiple prostate cancer cell lines [50,51]. More recent studies in hepatocellular carcinoma cell lines do not find similar correlation [52]. This discrepancy can be explained by the fact that [$^{11}$C]acetate uptake is only an indirect measure of lipid synthesis. [$^{11}$C]Acetate is thought to be taken up by monocarboxylate transporters [53]. In the cell, [$^{11}$C]acetate may be oxidized in mitochondria, resulting in the liberation of [$^{11}$C]CO$_2$ [54], or act as a substrate for Acetyl-CoA synthetase [55]. As [$^{11}$C]CO$_2$ rapidly effluxes from cells and tissue, retention of radioactivity in tumors is likely to be due to formation of acetyl-CoA [54]. However, acetyl-CoA is not only a building block for fatty acid synthesis, but for sterols, ketone bodies, and the TCA cycle as well [56]. In this light, increased uptake of [$^{11}$C]acetate in tumors may indicate upregulation of alternative mechanisms to glycolysis for preserving the pool of acetyl-CoA [57].

[$^{18}$F]Fluoroacetate is a potential analogue of [$^{11}$C]acetate on the basis of structural similarity [58], but in vivo defluorination is significant in preclinical models [58,59], and [$^{18}$F]fluoroacetate does not appear to be a functional analogue of [$^{11}$C]acetate [60]. 2-[$^{18}$F]Fluoropropionic acid is another short chain fatty acid analogue that shows good uptake and retention in prostate cancer [60] and liver cancer xenografts [61,62]. However, to date, the significance of 2-[$^{18}$F]fluoropropionic acid tumor uptake remains poorly understood. Uptake of 2-[$^{18}$F]fluoropropionic acid appears to correlate with FASN expression [62], but the precise biochemical pathways and cellular entities targeted by this tracer have yet to be elucidated.

In contrast to prior approaches to imaging FASN, we chose small molecule inhibitors of FASN as parent structures for radiotracer development. We reasoned that these radiotracers would be specific for FASN expression. Our compound library was based on one of the most potent and selective inhibitors of FASN reported to date. GSK2194069 targets the β-ketoacyl reductase domain of human FASN and inhibits the enzyme with IC$_{50}$ < 10 nM [36]. Methylation of the triazolone moiety of GSK2194069 (4; Figure 1) retains its potency, with IC$_{50}$ = 20 nM [35]. Similarly, introduction of a fluorine atom at the 2-position of the phenyl ring, results in compounds with a potency as high as 3 nM depending on the selection of (hetero)aromatic substituent at the 4-position of the phenyl ring. Our structural modifications, which consisted of the addition of a fluorine atom to 4 (5; Figure 1), and substitution of a 4-chlorophenyl moiety for the benzofuran moiety of 5 (6; Figure 1), had variable effects on potency. Analogue 5 retained the potency of GSK2194069 in LNCaP cells, but the potency of 6 was substantially compromised.

Our quantification of FASN expression by Western blot confirms high expression in LNCaP cells and much lower expression in PC3 cells (Figure S4), in agreement with literature reports [50]. In this light, the greater binding seen in LNCaP cells is consistent with FASN targeting. The difference in binding between the two cell lines is approximately two-fold but binding to both cell lines is relatively low when expressed as a percentage of the added dose. It is possible that the difference would be accentuated at higher levels of cell binding. When LNCaP cells were cultured for longer than 48 h, we achieved greater total cell binding (17% vs. 7% added activity of [$^{11}$C]5 at peak binding; Figure S6), but cells no longer grew as a monolayer. Consequently, we adjusted our assay conditions to increase reproducibility at the expense of total cell binding.

Notwithstanding our promising findings in vitro, the tumor targeting of [$^{11}$C]4, [$^{11}$C]5, and [$^{11}$C]6 in vivo was poor. Given that we did not observe a loss in FASN expression in the tumors relative to the parent cell line (Figure S4), we investigated alternative explanations. Alkylation of triazolones on the oxygen atom under basic conditions has been reported [63]. In the absence of reference standards for O-methylation, we could not be certain that our analytical method could distinguish between the N-[$^{11}$C]methyl and O-[$^{11}$C]methyl compounds. Therefore, we explored alternative radiolabeling strategies, including variation of base (NaOH, K$_2$CO$_3$, KOtBu), solvent (DMF, DMSO), reaction temperature (25 °C,

80 °C), and alkylating agent ([$^{11}$C]CH$_3$I, [$^{11}$C]CH$_3$OTf). Small differences in decay corrected radiochemical yield were evident, but the purified final products behaved identically in analytical assays and in vivo imaging studies. These findings led us to conclude that the poor tumor targeting is a consequence of poor pharmacokinetics. Our images indicated high retention of signal in the hepatobiliary pathway. We did not observe significant renal clearance, and consequently activity in the bladder was low. This contrasts with [$^{18}$F]FDG, for which accumulation of activity in the bladder can obscure the primary tumor in the prostate gland. Our next generation FASN ligands will seek to retain the excretion profile of these first-generation compounds while improving the tumor targeting.

## 4. Materials and Methods

### 4.1. Synthesis of Precursors

(S)-4-(4-(Benzofuran-6-yl)phenyl)-5-((1-(cyclopropanecarbonyl) pyrrolidin-3-yl)methyl)-2,4-dihydro-3H-1,2,4-triazol-3-one (GSK2194069) was purchased from Tocris Bioscience (Minneapolis, MN, USA) and used without further purification. The remaining precursors and non-radioactive standards were synthesized according to published procedures [35]. The characterization of final compounds is described herein, with full experimental details available in the Supporting Information Scheme S1.

(S)-4-(4-(Benzofuran-6-yl)phenyl)-5-((1-(cyclopropanecarbonyl)pyrrolidin-3-yl)methyl)-2-methyl-2,4-dihydro-3H-1,2,4-triazol-3-one (**4**).

$^1$H-NMR (CDCl$_3$, 500 MHz, 25 °C): δ 7.83 (d, $J$ = 1.5 Hz, 1H), 7.76 (d, $J$ = 8.5 Hz, 2H), 7.71 (d, $J$ = 2.5 Hz, 1H), 7.61 (d, $J$ = 8.5 Hz, 1H), 7.55 (dd, $J$ = 8.5 Hz, 2.0 Hz, 1H), 7.38 (d, $J$ = 8.0 Hz, 2H), 6.86 (d, $J$ = 2.0 Hz, 0.5 Hz, 1H), 3.85 (m, 1H), 3.70 (m, 1H), 3.53 (m, 1H), 3.18 (m, 1H), 3.55 (s, 3H), 2.64–2.42 (br m, 2.5H), 2.26–2.15 (br m, 1H), 2.10–2.02 (m, 0.5H), 1.77–1.68 (m, 0.5H), 1.61–1.56 (m, 1.5H), 1.02–0.96 (m, 2H), 0.79–0.74 (m, 2H). $^{13}$C-NMR (CDCl$_3$, 150 MHz, 25 °C): δ 172.3, 154.8, 153.5, 145.9, 142.8, 137.2, 135.1, 131.5, 128.9, 128.8, 128.1, 127.4, 127.3, 123.9, 119.9, 111.8, 106.8, 51.8, 45.3, 42.8, 32.5, 29.7, 22.8, 12.2, 7.6, 7.5. LCMS calc. for C$_{26}$H$_{26}$N$_4$O$_3$ [M + H]$^+$: 443.20. Found: 443.24. [α]$^{20}_D$ = −8.04 ± 0.59° (CH$_2$Cl$_2$).

(R)-4-(4-(Benzofuran-6-yl)-2-fluorophenyl)-5-((1-(cyclopropanecarbonyl)pyrrolidin-3-yl)methyl)-2,4-dihydro-3H-1,2,4-triazol-3-one (**2**).

$^1$H-NMR (CDCl$_3$, 400 MHz, 25 °C): δ 10.24 (br s, 0.5H), 10.12 (br s, 0.5H), 7.80 (br s, 1H), 7.69 (br s, 1H), 7.60 (dd, $J$ = 8.5 Hz, 2.8 Hz, 1H), 7.56–7.50 (m, 3H), 7.42 (dt, $J$ = 11.7 Hz, 2.0 Hz, 1H), 6.84 (br s, 1H), 3.96–3.92 (m, 0.5H), 3.78–3.69 (m, 1H), 3.66–3.58 (m, 1H), 3.42–3.35 (m, 0.5H), 3.30–3.26 (m, 0.5H), 3.10–3.06 (m, 0.5H), 2.74–2.64 (m, 0.5H), 2.64–2.42 (br m, 2.5H), 2.26–2.15 (br m, 0.5H), 2.10–2.02 (m, 0.5H), 1.77–1.68 (m, 0.5H), 1.62–1.54 (m, 1.5H), 1.02–0.93 (m, 2H), 0.78–0.69 (m, 2H). $^{13}$C-NMR (CDCl$_3$, 100 MHz, 25 °C): δ 172.4, 159.1, 156.6, 155.3, 155.2, 155.1, 146.6, 146.3, 146.2, 146.0, 145.93, 145.90, 145.8, 134.02, 133.95, 130.1, 130.0, 128.39, 128.37, 124.4, 123.9, 120.1, 118.6, 118.52, 118.48, 118.39, 116.1, 115.9, 112.13, 112.09, 106.9, 51.8, 51.2, 51.0, 45.9, 45.3, 36.0, 34.3, 31.6, 30.5, 29.5, 29.3, 12.6, 12.4, 7.8, 7.7, 7.6. $^{19}$F-NMR (CDCl$_3$, 376 MHz, 25 °C): δ -120.2 (br s, 1F). HRMS (ESI+) calculated for C$_{25}$H$_{23}$FN$_4$O$_3$: 446.1754. Found: 446.1685. [α]$^{20}_D$ = −8.35 ± 0.30° (CH$_2$Cl$_2$).

(R)-4-(4-(Benzofuran-6-yl)-2-fluorophenyl)-5-((1-(cyclopropanecarbonyl)pyrrolidin-3-yl)methyl)-2-methyl-2,4-dihydro-3H-1,2,4-triazol-3-one (**5**).

$^1$H-NMR (CDCl$_3$, 400 MHz, 25 °C): δ 7.80 (s, 1H), 7.69 (s, 1H), 7.59 (d, $J$ = 8.4 Hz, 1H), 7.54–7.48 (m, 3H), 7.39 (dd, $J$ = 8.7 Hz, 7.7 Hz, 1H), 6.84 (s, 1H), 3.94–3.90 (m, 0.5H), 3.79–3.69 (m, 1H), 3.65–3.56 (m, 1H), 3.52 (d, $J$ = 6.9 Hz, 3H), 3.42–3.35 (m, 0.5H), 3.28–3.24 (m, 0.5H), 3.08–3.04 (m, 0.5H), 2.72–2.63 (m, 0.5H), 2.63–2.45 (br m, 2.5H), 2.26–2.15 (br m, 0.5H), 2.10–2.02 (m, 0.5H), 1.75–1.66 (m, 0.5H), 1.60–1.51 (m, 1.5H), 1.05–0.92 (m, 2H), 0.78–0.69 (m, 2H). $^{13}$C-NMR (CDCl$_3$, 100 MHz, 25 °C): δ 172.3, 159.1, 156.6, 155.2, 146.24, 146.20, 145.84, 145.76, 145.73, 145.66, 144.5, 134.03, 133.98, 130.1, 130.0, 128.4, 124.3,

123.9, 120.1, 119.1, 119.02, 118.95, 118.89, 116.0, 115.9, 112.09, 112.07, 106.9, 51.8, 51.2, 51.0, 45.9, 45.3, 36.0, 34.3, 32.7, 31.6, 30.5, 29.5, 29.3, 12.6, 12.4, 7.8, 7.7, 7.6. $^{19}$F-NMR (CDCl$_3$, 376 MHz, 25 °C): δ-120.3 (br s, 1F). HRMS (ESI+) calculated for C$_{26}$H$_{25}$FN$_4$O$_3$: 460.1911. Found: 460.1846. [α]$^{20}_D$ = −2.99 ± 0.28° (CH$_2$Cl$_2$).

(R)-4-(4′-Chloro-3-fluoro-[1,1′-biphenyl]-4-yl)-5-((1-(cyclopropanecarbonyl)pyrrolidin-3-yl)methyl)-2,4-dihydro-3H-1,2,4-triazol-3-one (3).

$^1$H-NMR (CDCl$_3$, 400 MHz, 25 °C): δ 10.94 (br s, 1H), 7.50 (dd, J = 8.5 Hz, 1.6 Hz, 2H), 7.47–7.39 (m, 5H), 3.93–3.89 (m, 0.5H), 3.75–3.67 (m, 1H), 3.63–3.53 (m, 1H), 3.40–3.33 (m, 0.5H), 3.27–3.23 (m, 0.5H), 3.07–3.02 (m, 0.5H), 2.70–2.63 (m, 0.5H), 2.63–2.40 (m, 2.5H), 2.23–2.14 (m, 0.5H), 2.08–2.00 (m, 0.5H), 1.73–1.64 (m, 0.5H), 1.59–1.49 (m, 1.5H), 1.01–0.91 (m, 2H), 0.78–0.69 (m, 2H). $^{13}$C-NMR (CDCl$_3$, 100 MHz, 25 °C): δ 172.5, 159.1, 156.5, 155.15, 155.09, 146.2, 144.1, 144.0, 143.9, 137.2, 137.1, 135.0, 134.9, 130.3, 130.2, 129.41, 129.38, 128.5, 123.9, 119.31, 119.23, 119.18, 119.10, 115.7, 115.5, 51.8, 51.2, 45.9, 45.3, 35.9, 34.3, 31.5, 30.4, 29.4, 29.2, 12.6, 12.3, 7.74, 7.67, 7.57. $^{19}$F-NMR (CDCl$_3$, 376 MHz, 25 °C): δ-119.7 (br s, 1F). HRMS (ESI+) calculated for C$_{23}$H$_{22}$ClFN$_4$O$_2$: 440.1415. Found: 440.1339. [α]$^{20}_D$ = −9.37 ± 0.34° (CH$_2$Cl$_2$).

(R)-4-(4′-Chloro-3-fluoro-[1,1′-biphenyl]-4-yl)-5-((1-(cyclopropanecarbonyl)pyrrolidin-3-yl)methyl)-2-methyl-2,4-dihydro-3H-1,2,4-triazol-3-one (6).

$^1$H-NMR (CDCl$_3$, 400 MHz, 25 °C): δ 7.52 (br d, J = 8.4 Hz, 2H), 7.47–7.39 (m, 5H), 3.93–3.89 (m, 0.5H), 3.75–3.67 (m, 1H), 3.63–3.56 (m, 1H), 3.52 (s, 1.5H), 3.51 (s, 1.5H), 3.40–3.33 (m, 0.5H), 3.27–3.23 (m, 0.5H), 3.07–3.02 (m, 0.5H), 2.70–2.63 (m, 0.5H), 2.63–2.40 (m, 2.5H), 2.23–2.14 (m, 0.5H), 2.08–2.00 (m, 0.5H), 1.73–1.64 (m, 0.5H), 1.59–1.49 (m, 1.5H), 1.01–0.91 (m, 2H), 0.78–0.69 (m, 2H). $^{13}$C-NMR (CDCl$_3$, 100 MHz, 25 °C): δ 172.3, 159.1, 156.6, 153.2, 144.3, 137.3, 137.2, 135.1, 135.0, 130.33, 130.28, 129.50, 129.48, 128.5, 123.9, 115.8, 115.6, 51.8, 51.2, 45.9, 45.3, 34.3, 32.63, 32.60, 31.5, 30.4, 29.4, 29.2, 12.6, 12.3, 7.74, 7.67, 7.57. $^{19}$F-NMR (CDCl$_3$, 376 MHz, 25 °C): δ-119.7 (br s, 1F). HRMS (ESI+) calculated for C$_{24}$H$_{24}$ClFN$_4$O$_2$: 454.1572. Found: 454.1470. [α]$^{20}_D$ = −8.74 ± 0.16° (CH$_2$Cl$_2$).

## 4.2. Radiosynthesis

[$^{11}$C]CO$_2$ (11–22 GBq) was produced by a $^{14}$N(p,α)$^{11}$C reaction using a TR19 cyclotron (Advanced Cyclotron Systems, Inc). The [$^{11}$C]CO$_2$ was subsequently converted to [$^{11}$C]CH$_3$I using a TracerLab FX$_C$ Pro unit (GE Healthcare, Chicago, IL, USA). The transformation was typically accomplished within 10 min and resulted in a starting activity of 5.5–14.8 GBq [$^{11}$C]CH$_3$I.

All solvents and reagents were purchased from Sigma Aldrich and were used without further purification unless otherwise noted.

The chemical and radiochemical purity of the final product was determined by analytical HPLC using a ProStar HPLC (Varian, Palo Alto, CA, USA) with a Luna C18(2) column (5 μm, 250 × 4.6 mm$^2$, 100 Å) (Phenomenex, Torrance, CA, USA) and a 50:50 0.3 M NH$_4$COOH (pH 4.2):MeCN mobile phase at a flow rate of 1.5 mL/min. The HPLC was coupled to single wavelength UV-Vis detector and an Eckert & Ziegler FlowCount NaI(Tl) radio-HPLC scintillation detector (Bioscan, Washington DC, USA). Absorbance was measured at 254 nm. The retention time, $t_R$, of the radiolabeled product was compared to the $t_R$ of a non-radioactive standard.

4.2.1. (S)-4-(4-(benzofuran-6-yl)phenyl)-5-((1-(cyclopropanecarbonyl)pyrrolidin-3-yl)methyl)-2-[methyl-$^{11}$C]-2,4-dihydro-3H-1,2,4-triazol-3-one ([$^{11}$C]4)

The [$^{11}$C]CH$_3$I was trapped on ascarite and released as a bolus into a capped reaction vial containing a pre-mixed solution of either 1 mg GSK2194069 and 3.0 μL 2.5 M NaOH in 350 μL DMSO or 1 mg (2.3 μmol) GSK2194069 and 5 mg (36 μmol) K$_2$CO$_3$ in 350 μL DMF. The reaction was stirred for 5 min at 80 °C before it was quenched by addition of 2 mL 60:40 0.3 M NH$_4$COOH (pH 4.2):MeCN.

The quenched reaction mixture was purified using a Phenomenex Luna C18 column (5 µm, 250 × 10 mm$^2$, 100 Å) and a 60:40 0.3 M NH$_4$COOH (pH 4.2):MeCN mobile phase at a flow rate of 6 mL/min. The fraction containing the product was collected, diluted with H$_2$O, and passed through a pre-conditioned Sep-Pak® C18 Plus Light cartridge (Waters, Milford, MA, USA). The cartridge was washed with 10 mL H$_2$O and dried under a flow of N$_2$. [$^{11}$C]4 was eluted using 700 µL EtOH followed by 9.3 mL normal saline (0.9% NaCl). The purity of the final product was determined as described above.

4.2.2. (S)-4-(4-(benzofuran-6-yl)-2-fluorophenyl)-5-((1-(cyclopropanecarbonyl)pyrrolidin-3-yl)methyl)-2-[methyl-$^{11}$C]-2,4-dihydro-3H-1,2,4-triazol-3-one [$^{11}$C]5

A solution of 1 mg (2.2 µmol) (S)-4-(4-(benzofuran-6-yl)-2-fluorophenyl)-5-((1-(cyclopropanecarbonyl)pyrrolidin-3-yl)methyl)-2,4-dihydro-3H-1,2,4-triazol-3-one (2) and 5.0 µL 2 M NaOH in 350 µL DMSO was stirred for 2 min at 80 °C. Then [$^{11}$C]CH$_3$I was introduced as a bolus into the sealed reaction vessel. The reaction was stirred for 5 min at 80 °C, then quenched by addition of 650 µL H$_2$O + 0.01% TFA.

The quenched reaction product was purified using a Waters Bondapak®, C18 column (15–20 µm, 7.8 × 300 mm$^2$, 125 Å) and an isocratic mobile phase of 50:50 H$_2$O + 0.01 TFA/90% MeCN/H$_2$O + 0.01% TFA at a flow rate of 4 mL/min. The retention time of the product was 10–13 min. The fraction corresponding to the product was collected and diluted to 30 mL with H$_2$O. The diluted fraction was passed through a pre-conditioned Sep Pak® C18 Plus Light cartridge (Waters). The cartridge was washed with 10 mL H$_2$O and dried with air. [$^{11}$C]5 was eluted with 100 µL EtOH followed by 900 µL saline (0.9% NaCl). The purity of the final compound was determined as described above.

4.2.3. (S)-4-(4'-chloro-3-fluoro-[1,1'-biphenyl]-4-yl)-5-((1-(cyclopropanecarbonyl)pyrrolidin-3-yl)methyl)-2-[methyl-$^{11}$C]-2,4-dihydro-3H-1,2,4-triazol-3-one ([$^{11}$C]6)

[$^{11}$C]CH$_3$I was introduced into a sealed reaction vial containing a premixed solution of 1 mg (2.3 µmol) (S)-4-(4'-chloro-3-fluoro-[1,1'-biphenyl]-4-yl)-5-((1-(cyclopropanecarbonyl)pyrrolidin-3-yl)methyl)-2,4-dihydro-3H-1,2,4-triazol-3-one (3) and 3.0 µL 2.5 M NaOH in 350 µL DMSO. The reaction was stirred for 5 min at 80 °C, then quenched with 600 µL 0.3 M NH$_4$COOH (pH 4.2).

The quenched reaction mixture was purified using a Phenomenex Luna C18 column (5 µm, 250 × 10 mm$^2$, 100 Å) and a 50:50 0.3 M NH$_4$COOH (pH 4.2):MeCN mobile phase at a flow rate of 6 mL/min. The fraction containing the product was collected, diluted with H$_2$O, and passed through a pre-conditioned Sep-Pak® C18 Plus Light cartridge (Waters). The cartridge was washed with 10 mL H$_2$O and dried under a flow of N$_2$. [$^{11}$C]6 was eluted using 1 mL EtOH followed by 13 mL normal saline (0.9% NaCl). The purity of the final compound was determined as described above.

*4.3. Cell Lines*

The FASN-expressing human prostate cancer cell line LNCaP was obtained from the American Type Culture Collection (ATCC, Manassas, VA, USA). The low-expressing human prostate cancer cell line PC3 was also obtained from the ATCC. Cell culture supplies were purchased from Corning unless otherwise noted. Both cell lines were cultured in RPMI-1640 media supplemented with 10% heat-inactivated fetal bovine serum (FBS; GE Healthcare), 100 U/mL penicillin, 100 U/mL streptomycin, 1 mM sodium pyruvate (Gibco, ThermoFisher Scientific, Waltham, MA, USA), 0.45% glucose, MEM non-essential amino acids, and 2 mM glutamine (Gibco) at 37 °C and in 5% CO$_2$. Cells were removed from flasks for passage or for transfer to 24-well assay plates by incubating them with 0.25% trypsin/ethylenediaminetetraacetic acid (EDTA).

### 4.4. Western Blotting

The LNCaP and PC3 cells were grown to confluence in RPMI-1640 media supplemented with 10% FBS, pen/strep, glutamine, pyruvate, glucose, and non-essential amino acids. The cells were lysed with 50 mM Tris.HCl (pH 7.4), 0.5 mM EDTA, and Pierce Halt™ Protease Inhibitor Cocktail (ThermoFisher) with 30–60 s grinding using Molecular Grinding Resin™ (Sigma Aldrich, St. Louis, MO, USA). Excised tumors were immediately flash frozen and stored at −80 °C. The samples were thawed and sliced, and the tissue slices were minced and then homogenized as described for the cell lines. A ratio of 4 parts lysis buffer to 1 part cellular material was maintained for all tumor samples. Cell protein content was determined using the BCA method (ThermoFisher) with bovine serum albumin as the standard. The blot was stained with either a rabbit polyclonal anti-FASN antibody from Bethyl Laboratories (1:1000, Bethyl Laboratories, Montgomery, TX, USA) or a mouse monoclonal anti-β-actin antibody (1:100, Santa Cruz Biotechnology, Dallas, TX, USA) followed by HRP-conjugated goat anti-mouse IgG (1:1000, Bio-Rad Laboratories, Hercules, CA, USA).

### 4.5. Binding to Prostate Cancer Cell Lines

LNCaP cells were seeded at a density of $2.5 \times 10^5$ cells/mL in 24-well plates and grown for 24–48 h in RPMI-1640 media supplemented with 10% fetal bovine serum (FBS). The cells were a confluent monolayer at the time of the assay. Immediately prior to the assay, the growth medium was removed from each well under suction and replaced with 480 µL fresh RPMI-1640 media supplemented with 0.05% bovine serum albumin (BSA). A stock solution was prepared by adding 111–222 MBq of the $^{11}$C-labeled tracer in 1 mL 10% EtOH/saline to 25 mL of the RPMI-1640 media. From this stock solution were added 20 µL aliquots, containing approximately 111 kBq at time of addition, to each well. The experiment was performed in triplicate. The plates were incubated at 37 °C for 5, 10, 15, 20, 30, 40, or 50 min. Then the media was removed under suction, the wells were washed three times with PBS, and the cells were detached with 0.8 mL RIPA buffer. Non-specific binding was determined by immediately removing the media after addition to the wells. The cells were transferred to tubes and counted on a Wizard $^2$ automated γ-counter (Perkin Elmer, Waltham, MA, USA). A 10% added activity standard was counted for quantification. Counts were corrected for decay, the background and non-specific binding were subtracted. The resulting values were expressed as a percentage of added activity ± standard error of the mean (SEM) and normalized to the protein content of each corresponding well.

PC3 cells were seeded at a density of $5 \times 10^4$ cells/mL in 24-well plates and grown for 24–48 h in RPMI-media supplemented with 10% FBS. The cells were a confluent monolayer at the time of the assay. The binding was determined as described above. The resulting values were expressed as a percentage of added activity ± SEM and normalized to the protein content of each corresponding well.

The protein content of each sample was determined by analysis of a 5 µL aliquot of the harvested cell mixture after activity decayed to background.

### 4.6. Growth Inhibition Assay

The growth inhibition of LNCaP and PC3 following treatment with FASN inhibitors was carried out according to recently published methods [64]. Briefly, LNCaP and PC3 cells were seeded at 1 mL/well on 24-well plates at a density of $1 \times 10^6$ cells/mL or $1.25 \times 10^4$ cells/mL, respectively. The cells were seeded in RPMI-1640 media supplemented with 1% heat-inactivated FBS, 100 U/mL penicillin, 100 U/mL streptomycin, 1 mM sodium pyruvate, 0.45% glucose, MEM non-essential amino acids, and 2 mM glutamine. One day after seeding, when the cells had grown to approximately 50% confluence, the seeding media was removed and replaced with 2 mL of the culture media described above containing either 0.1% $v/v$ DMSO (vehicle) or solutions of GSK2194069, **4**, **5**, **6** in DMSO. The concentration of DMSO in the culture media was 0.1% $v/v$ for all inhibitor concentrations. The final concentration of inhibitor in the well ranged from $10^{-4}$ M to $10^{-9}$ M. The experiments were performed in triplicate. The treated cells were cultured for four days at 37 °C and in

5% $CO_2$. Cell viability was then assessed by replacing the treatment media with 0.25 mM of a solution of Presto Blue (ThermoFisher), prepared as recommended by the supplier, and incubation for a further 40 min at 37 °C and in 5% $CO_2$. After incubation, 0.2 mL was removed from the wells for quantification at $660_{ex}$ and $690_{em}$ nm. The resulting data was compared to the values obtained using the vehicle and expressed as the percentage of control viability. The results were analyzed and plotted as mean ± SEM using GraphPad Prism 8.4.3.

### 4.7. Animal Models

Animals were housed under standard conditions in approved facilities with 12 h light–dark cycles. Food and water were provided ad libitum throughout the course of the studies. Eight-week-old male inbred athymic nu/nu mice were purchased from The Jackson Laboratory. LNCaP cells were suspended at $4 \times 10^7$ cells/mL in a 1:1 mixture of phosphate-buffered saline:Matrigel (BD Biosciences, San Jose, CA, USA). Each mouse was injected in the left flank with 0.25 mL of the cell suspension. Imaging studies were performed once tumors reached a volume of 150–500 mm³, approximately 4 weeks post inoculation.

### 4.8. MicroPET/CT Imaging

*Method A*: Tumor-bearing mice were injected intravenously with 100 µL of the final product solution, containing 8–10 MBq [$^{11}$C]4, 9–11 MBq [$^{11}$C]5, or 9–11 MBq [$^{11}$C]6. The mice were then placed on the imaging bed and a 60 min dynamic acquisition was performed by microPET/CT (Siemens Inveon™, Siemens Medical Solutions USA, Malvern, PA, USA). The acquisition typically began 15 min after injection of the radiotracer. A CT scan was performed immediately upon conclusion of the microPET scan for attenuation correction and anatomical co-registration. The images were processed using open-source image processing software (AMIDE).

*Method B*: A 500 µL aliquot containing approximately 111 MBq (38 ng) [$^{11}$C]5 in 10% EtOH/saline was added to a vial containing 50 mg Captisol® in 0.5 mL saline, prepared 4 h previously. The mixture was shaken vigorously for 10 min. A control solution was prepared by diluting a 500 µL aliquot containing approximately 111 MBq (38 ng) [$^{11}$C]5 in 10% EtOH/saline with 0.5 mL saline.

Two male athymic nu/nu mice bearing LNCaP xenograft tumors (150–500 mm³) were injected intravenously with 100 µL of the solution, containing 9–11 MBq [$^{11}$C]5 and 5 mg Captisol®. In parallel, two mice were injected intravenously with 100 µL of the control solution, containing 9–11 MBq. The mice were then placed on the imaging bed and a 60 min dynamic acquisition was performed by microPET/CT (Siemens Inveon™). The acquisition typically began 15 min after injection of [$^{11}$C]5. A CT scan was performed immediately upon conclusion of the microPET scan for attenuation correction and anatomical co-registration. The images were processed using AMIDE.

### 4.9. Statistics

Statistical comparisons were drawn using the Tukey's multiple comparison test (GraphPad Prism) with a significance level of $p < 0.05$.

## 5. Conclusions

Small molecule triazolones structurally related to GSK2194069, a potent FASN inhibitor, are readily $^{11}$C-methylated in good yield and high purity and molar activity under mild conditions. Of these compounds, [$^{11}$C]5 shows considerable promise in vitro as it reduces the viability of LNCaP cells with an $EC_{50}$ comparable to GSK2194069 and is taken up to a significantly higher extent in FASN-positive LNCaP cells than in FASN-negative PC3 cells in vitro. However, the probes demonstrate remarkably poor tumor uptake in vivo, with the majority of the activity retained in lymph nodes and the hepatobiliary system.

Future studies should target structural modifications that improve the pharmacokinetics of these triazolone FASN inhibitors.

**Supplementary Materials:** The following supporting information can be downloaded at: online, Full experimental details, including Scheme S1: Synthesis of 4 from GSK2194069, and Scheme S2: Synthesis of non-radioactive standards 5 and 6 and their precursors 2 and 3; Figure S1: Radio (top) and UV (bottom) chromatograms of purified [$^{11}$C]4; Figure S2: Radio (top) and UV (bottom) chromatograms of purified [$^{11}$C]5; Figure S3: Radio (top) and UV (bottom) chromatograms of purified [$^{11}$C]6; Figure S4. Analysis of FASN expression in cell lines and xenograft tumors by Western blot; Figure S5. Comparison of viability of LNCaP and PC3 cells treated with A. GSK2194069 or B. 2 (right); Figure S6. Comparison of [$^{11}$C]5 binding to LNCaP cells cultured for >48 h (black) and <48 h (blue); and $^1$H-NMR, $^{13}$C-NMR, and $^{19}$F-NMR spectra of all final compounds and intermediates.

**Author Contributions:** Conceptualization, J.M.K., A.A.-C., S.G.D. and J.W.B.; methodology, J.M.K., T.M.J. and J.W.B.; data analysis, J.M.K. and T.M.J.; investigation, J.M.K., T.M.J., N.N.W., W.Q., E.J.L., B.S., C.W.J., A.N. and A.A.-C.; writing—original draft preparation, J.M.K.; writing—review and editing, T.M.J., S.G.D. and J.W.B.; funding acquisition, J.W.B. All authors have read and agreed to the published version of the manuscript.

**Funding:** This work was funded by the National Cancer Institute of the National Institutes of Health under award number 1R21CA213157-01. The funding agency did not influence the design of the study nor the interpretation of the results.

**Institutional Review Board Statement:** All animal studies were approved by the Institutional Animal Care and Use Committee of Weill Cornell Medicine and were undertaken in accordance with the guidelines set forth by the U.S. Public Health Service Policy on Humane Care and Use of Laboratory Animals.

**Informed Consent Statement:** Not applicable.

**Data Availability Statement:** Raw data is available from the authors upon request.

**Conflicts of Interest:** The authors have no conflict of interest.

**Sample Availability:** Samples of compounds **2–6** manuscript are available from the authors upon request.

# References

1. Zaidi, N.; Lupien, L.; Kuemmerle, N.B.; Kinlaw, W.B.; Swinnen, J.V.; Smans, K. Lipogenesis and lipolysis: The pathways exploited by the cancer cells to acquire fatty acids. *Prog. Lipid Res.* **2013**, *52*, 585–589. [CrossRef]
2. Menendez, J.A.; Lupu, R. Fatty acid synthase and the lipogenic phenotype in cancer pathogenesis. *Nat. Rev. Cancer* **2007**, *7*, 763–777. [CrossRef] [PubMed]
3. Röhrig, F.; Schulze, A. The multifaceted roles of fatty acid synthesis in cancer. *Nat. Rev. Cancer* **2016**, *16*, 732–749. [CrossRef]
4. Yellen, P.; Foster, D.A. Inhibition of fatty acid synthase induces pro-survival Akt and ERK signaling in K-Ras-driven cancer cells. *Cancer Lett.* **2014**, *353*, 258–263. [CrossRef] [PubMed]
5. Mashima, T.; Seimiya, H.; Tsuruo, T. De novo fatty-acid synthesis and related pathways as molecular targets for cancer therapy. *Br. J. Cancer* **2009**, *100*, 1369–1372. [CrossRef] [PubMed]
6. Ameer, F.; Scandiuzzi, L.; Hasnain, S.; Kalbacher, H.; Zaidi, N. De novo lipogenesis in health and disease. *Metabolism* **2014**, *63*, 895–902. [CrossRef] [PubMed]
7. Swinnen, J.V.; Veldhoven, P.P.V.; Timmermans, L.; Schrijver, E.D.; Brusselmans, K.; Vanderhoydonc, F.; Sande, T.V.d.; Heemers, H.; Heyns, W.; Verhoeven, G. Fatty Acid Synthase Drives the Synthesis of Phospholipids Partitioning Into Detergent-Resistant Membrane Microdomains. *Biochem. Biophys. Res. Commun.* **2003**, *302*, 898–903. [CrossRef]
8. Benedettini, E.; Nguyen, P.; Loda, M. The pathogenesis of prostate cancer: From molecular to metabolic alterations. *Diagn. Histopathol.* **2008**, *14*, 195–201. [CrossRef]
9. Van de Sande, T.; De Schrijver, E.; Heyns, W.; Verhoeven, G.; Swinnen, J.V. Role of the Phosphatidylinositol 3′-Kinase/PTEN/Akt Kinase Pathway in the Overexpression of Fatty Acid Synthase in LNCaP Prostate Cancer Cells. *Cancer Res.* **2002**, *62*, 642–646.
10. Yuan, T.L.; Cantley, L.C. PI3K pathway alterations in cancer: Variations on a theme. *Oncogene* **2008**, *27*, 5497–5510. [CrossRef]
11. Rossi, S.; Graner, E.; Febbo, P.; Weinstein, L.; Bhattacharya, N.; Onody, T.; Bubley, G.; Balk, S.; Loda, M. Fatty Acid Synthase Expression Defines Distinct Molecular Signatures in Prostate Cancer. *Mol. Cancer Res.* **2003**, *1*, 707–715. [PubMed]
12. Wang, Y.; Kuhajda, F.P.; Li, J.N.; Pizer, E.S.; Han, W.F.; Sokoll, L.J.; Chan, D.W. Fatty acid synthase (FAS) expression in human breast cancer culture supernatants and in breast cancer patients. *Cancer Lett.* **2001**, *167*, 99–104. [CrossRef]

13. Alo, P.L.; Amini, M.; Piro, F.; Pizzuti, L.; Sebastiani, V.; Botti, C.; Murari, R.; Zotti, G.; Tondo, U.D. Significance of Fatty Acid Synthase in Pancreatic Carcinoma. *Anticancer. Res.* **2007**, *27*, 2523–2527. [PubMed]
14. Rashid, A.; Pizer, E.S.; Moga, M.; Milgraum, L.Z.; Zahurak, M.; Pasternack, G.R.; Kuhajda, F.P.; Hamilton, S.R. Elevated Expression of Fatty Acid Synthase and Fatty Acid Synthetic Activity in Colorectal Neoplasia. *Am. J. Pathol.* **1997**, *150*, 201. [PubMed]
15. Ito, T.; Sato, K.; Maekawa, H.; Sakurada, M.; Orita, H.; Shimada, K.; Daida, H.; Wada, R.; Abe, M.; Hino, O.; et al. Elevated levels of serum fatty acid synthase in patients with gastric carcinoma. *Oncol. Lett.* **2014**, *7*, 616–620. [CrossRef]
16. Myers, R.B.; Oelschlager, D.K.; Weiss, H.L.; Frost, A.R.; Grizzle, W.E. Fatty Acid Synthase: An Early Molecular Marker of Progression of Prostatic Adenocarcinoma to Androgen Independence. *J. Urol.* **2001**, *165*, 1027–1032. [CrossRef]
17. Jones, S.F.; Infante, J.R. Molecular Pathways: Fatty Acid Synthase. *Clin. Cancer Res.* **2015**, *21*, 5434–5438. [CrossRef]
18. Flavin, R.; Peluso, S.; Nguyen, P.L.; Loda, M. Fatty acid synthase as a potential therapeutic target in cancer. *Future Oncol.* **2010**, *6*, 551–562. [CrossRef]
19. Pizer, E.S.; Wood, F.D.; Heine, H.S.; Romantsev, F.E.; Pasternack, G.R.; Kuhajda, F.P. Inhibition of Fatty Acid Synthesis Delays Disease Progression in a Xenograft Model of Ovarian Cancer. *Cancer Res.* **1996**, *56*, 1189–1193.
20. Pizer, E.S.; Jackisch, C.; Wood, F.D.; Pasternack, G.R.; Davidson, N.E.; Kuhajda, F.P. Inhibition of Fatty Acid Synthase Induces Programmed Cell Death in Human Breast Cancer Cells. *Cancer Res.* **1996**, *56*, 2745–2747.
21. Carvalho, M.A.; Zecchin, K.G.; Seguin, F.; Bastos, D.C.; Agostini, M.; Rangel, A.L.C.A.; Veiga, S.S.; Raposo, H.F.; Oliveira, H.C.F.; Loda, M.; et al. Fatty acid synthase inhibition with Orlistat promotes apoptosis and reduces cell growth and lymph node metastasis in a mouse melanoma model. *Int. J. Cancer* **2008**, *123*, 2557–2565. [CrossRef] [PubMed]
22. Kridel, S.J.; Axelrod, F.; Rozenkrantz, N.; Smith, J.W. Orlistat Is a Novel Inhibitor of Fatty Acid Synthase with Antitumor Activity. *Cancer Res.* **2004**, *64*, 2070–2075. [CrossRef] [PubMed]
23. Pizer, E.S.; Thupari, J.; Han, W.F.; Pinn, M.L.; Chrest, F.J.; Frehywot, G.L.; Townsend, C.A.; Kuhajda, F.P. Malonyl-Coenzyme-A Is a Potential Mediator of Cytotoxicity Induced by Fatty-Acid Synthase Inhibition in Human Breast Cancer Cells and Xenografts. *Cancer Res.* **2000**, *60*, 213–218. [PubMed]
24. Sadowski, M.C.; Pouwer, R.H.; Gunter, J.H.; Lubik, A.A.; Quinn, R.J.; Nelson, C.C. The fatty acid synthase inhibitor triclsoan: Repurposing an anti-microbial agent for targeting prostate cancer. *Oncotarget* **2014**, *5*, 9362–9381. [CrossRef] [PubMed]
25. Brusselmans, K.; Schrijver, E.D.; Heyns, W.; Verhoeven, G.; Swinnen, J.V. Epigallocatechin-3-gallate Is a Potent Natural Inhibitor of Fatty Acid Synthase in Intact Cells and Selectively Induces Apoptosis in Prostate Cancer Cells. *Int. J. Cancer* **2003**, *106*, 856–862. [CrossRef] [PubMed]
26. Relat, J.; Blancafort, A.; Oliveras, G.; Cufi, S.; Haro, D.; Marrero, P.F.; Puig, T. Different fatty acid metabolism effects of (-)-Epigallocatechin-3-Gallate and C75 in Adenocarcinoma lung cancer. *BMC Cancer* **2012**, *12*, 280. [CrossRef]
27. Loftus, T.M.; Jaworsky, D.E.; Frehywot, G.L.; Townsend, C.A.; Ronnet, G.V.; Lane, M.D.; Kuhajda, F.P. Reduced Food Intake and Body Weight in Mice Treated with Fatty Acid Synthase Inhibitors. *Science* **2000**, *288*, 2379–2381. [CrossRef]
28. Zaytseva, Y.Y.; Rychahou, P.G.; Le, A.-T.; Scott, T.L.; Flight, R.M.; Kim, J.T.; Harris, J.; Liu, J.; Wang, C.; Morris, A.J.; et al. Preclinical evaluation of novel fatty acid synthase inhibitors in primary colorectal cancer cells and a patient-derived xenograft model of colorectal cancer. *Oncotarget* **2018**, *9*, 24787–24800. [CrossRef]
29. Puig, T.; Aguilar, H.; Cuf, S.; Oliveras, G.; Turrado, C.; Ortega-Gutierrez, S.; Benhamú, B.; López-Rodríguez, M.L.; Urruticoechea, A.; Colomer, R. A novel inhibitor of fatty acid synthase show activity against HER2+ breast cancer xenografts and is active in anti-HER2 drug-resistant cell lines. *Breast Cancer Res.* **2011**, *13*, R131. [CrossRef]
30. Shaw, G.; Lewis, D.; Boren, J.; Montoya, A.-R.; Bielik, R.; Soloviev, D.; Brindle, K.; Neal, D. Therapeutic Fatty Acid Synthase Inhibition in Prostate Cancer and the Use of 11C-Acetate to Monitor Therapeutic Effects. *J. Urol.* **2013**, *189*, e208–e209. [CrossRef]
31. Zadra, G.; Ribeiro, C.F.; Chetta, P.; Ho, Y.; Cacciatore, S.; Gao, X.; Syamala, S.; Bango, C.; Photopoulos, C.; Huang, Y.; et al. Inhibition of de novo lipogenesis targets receptor signaling in castration-resistant prostate cancer. *Proc. Natl. Acad. Sci. USA* **2019**, *116*, 631–640. [CrossRef] [PubMed]
32. Ventura, R.; Mordec, K.; Waszczuk, J.; Wang, Z.; Lai, J.; Fridlib, M.; Buckley, D.; Kemble, G.; Heuer, T.S. Inhibition of de nov Palmitate Synthesis by Fatty Acid Synthase Induces Apoptosis in Tumor Cells by Remodeling Cell Membranes, Inhibiting Signaling Pathways, and Reprogramming Gene Expression. *EBioMedicine* **2015**, *2*, 808–824. [CrossRef] [PubMed]
33. Brenner, A.; Infante, J.; Patel, M.; Arkenau, H.-T.; Voskoboynik, M.; Borazanci, E.; Falchook, G.; Molife, L.R.; Pant, S.; Dean, E.; et al. First-in-human study of the first-in-class fatty acid synthase (FASN) inhibitor, TVB-2640 as monotherapy or in combination—Final results of dose escalation. *Mol. Cancer Ther.* **2015**, *14*, A54.
34. Falchook, G.; Patel, M.; Infante, J.; Arkenau, H.-T.; Dean, E.; Brenner, A.; Borazanci, E.; Lopez, J.; Moore, K.; Schmid, P.; et al. First in human study of the first-in-class fatty acid synthase (FASN) inhibitor TVB-2640. *Cancer Res.* **2017**, *77*, CT153.
35. Adams, N.D.; Aquino, C.J.; Chaudhari, A.M.; Ghergurovich, J.M.; Kiesow, T.J.; Parrish, C.A.; Reif, A.J.; Wiggall, K. Triazolones as Fatty Acid Synthase Inhibitors. U.S. Patent No 8,802,864 B2, 12 August 2012.
36. Hardwicke, M.A.; Rendina, A.R.; Williams, S.P.; Moore, M.L.; Wang, L.; Krueger, J.A.; Plant, R.N.; Totoritis, R.D.; Zhang, G.; Briand, J.; et al. A human fatty acid synthase inhibitor binds β-ketoacyl reductase in the keto-substrate site. *Nat. Chem. Biol.* **2014**, *10*, 774–782. [CrossRef]
37. Lockwood, S.F.; O'Malley, S.; Mosher, G.L. Improved aqueous solubility of crystalline astaxanthin (3,3′-dihydroxy-β,β-carotene-4,4′-dione) by Captisol (sulfobutyl ether β-cyclodextrin). *J. Pharm. Sci.* **2003**, *92*, 922–926. [CrossRef] [PubMed]

48. Oyama, N.; Akino, H.; Suzuki, Y.; Kanamaru, H.; Sadato, N.; Yonekura, Y.; Okada, K. The Increased Accumulation of [18F]Fluorodeoxyglucose in Untreated Prostate Cancer. *Jpn. J. Clin. Oncol.* **1999**, *29*, 623–629. [CrossRef]
49. Jadvar, H. FDG PET in Prostate Cancer. *PET Clin.* **2009**, *4*, 155–161. [CrossRef]
50. Spratt, D.E.; Gavane, S.; Tarlinton, L.; Fareedy, S.B.; Doran, M.G.; Zelefsky, M.J.; Osborne, J.R. Utility of FDG-PET in Clinical Neuroendocrine Prostate Cancer. *Prostate* **2014**, *74*, 1153–1159. [CrossRef]
51. Fricke, E.; Machtens, S.; Hofmann, M.; Hoff, J.v.d.; Bergh, S.; Brunkhorst, T.; Meyer, G.J.; Karstens, J.H.; Knapp, W.H.; Boerner, A.R. Positron Emission Tomography With $^{11}$C-acetate and $^{18}$F-FDG in Prostate Caner Patients. *Eur. J. Nucl. Med. Mol. Imaging* **2003**, *30*, 607–611. [CrossRef]
52. Meirelles, G.S.P.; Schöder, H.; Ravizzini, G.C.; Gönen, M.; Fox, J.J.; Humm, J.; Morris, M.J.; Scher, H.I.; Larson, S.M. Prognostic Value of Baseline [$^{18}$F] Fluorodeoxyglucose Positron Emission Tomography and $^{99m}$Tc-MDP Bone Scan in Progressing Metastatic Prostate Cancer. *Clin. Cancer Res.* **2010**, *16*, 6093–6099. [CrossRef] [PubMed]
53. Jadvar, H. Imaging Evaluation of Prostate Cancer with 18F-fluorodeoxyglucose PET/CT: Utility and Limitations. *Eur. J. Nucl. Med. Mol. Imaging* **2013**, *40*, 5–10. [CrossRef] [PubMed]
54. Ackerstaff, E.; Glunde, K.; Bhujwalla, Z.M. Choline Phospholipid Metabolism: A Target in Cancer Cells? *J. Cell Biochem.* **2003**, *90*, 525–533. [CrossRef] [PubMed]
55. Glunde, K.; Bhujwalla, Z.M.; Ronen, S.M. Choline metabolism in malignant transformation. *Nat. Rev. Cancer* **2011**, *11*, 835–848. [CrossRef] [PubMed]
56. Schwarzenböck, S.; Souvatzoglou, M.; Krause, B.J. Choline PET and PET/CT in Primary Diagnosis and Staging of Prostate Cancer. *Theranostics* **2012**, *2*, 318–330. [CrossRef] [PubMed]
57. Tousignant, K.D.; Rockstroh, A.; Fard, A.T.; Lehman, M.L.; Wang, C.; McPherson, S.J.; Philp, L.K.; Bartonicek, N.; Dinger, M.E.; Nelson, C.C.; et al. Lipid Uptake Is an Androgen-Enhanced Lipid Supply Pathway Associated with Prostate Cancer Disease Progression and Bone Metastasis. *Mol. Cancer Res.* **2019**, *17*, 1166–1179. [CrossRef]
58. Fhu, C.W.; Ali, A. Fatty Acid Synthase: An Emerging Target in Cancer. *Molecules* **2020**, *25*, 3935. [CrossRef]
59. Takahiro, T.; Shinichi, K.; Toshimitsu, S. Expression of Fatty Acid Synthase as a Prognostic Indicator in Soft Tissue Sarcomas. *Clin. Cancer Res.* **2003**, *9*, 2204–2212.
60. Yoshii, Y.; Furukawa, T.; Oyama, N.; Hasegawa, Y.; Kiyono, Y.; Nishii, R.; Waki, A.; Tsuji, A.B.; Sogawa, C.; Wakizaka, H.; et al. Fatty Acid Synthase is a Key Target in Multiple Essential Tumor Functions of Prostate Cancer: Uptake of Radiolabeled Acetate as a Predictor of the Targeted Therapy Outcomes. *PLoS ONE* **2013**, *8*, e64570.
61. Vāvere, A.L.; Kridel, S.J.; Wheeler, F.B.; Lewis, J.S. 1-11C-acetate as a PET Radiopharmaceutical for Imaging Fatty Acid Synthase Expression in Prostate Cancer. *J. Nucl. Med.* **2008**, *49*, 327–334. [CrossRef]
62. Li, L.; Che, L.; Wang, C.; Blecha, J.E.; Li, X.; VanBrocklin, H.F.; Calvisi, D.F.; Puchowicz, M.; Chen, X.; Seo, Y. [11C]acetate PET Imaging is not Always Associated with Increased Lipogenesis in Hepatocellular Carcinoma in Mice. *Mol. Imaging Biol.* **2016**, *18*, 360–367. [CrossRef] [PubMed]
63. Jeon, J.Y.; Lee, M.; Whang, S.H.; Kim, J.-W.; Cho, A.; Yun, M. Regulation of Acetate Utilization by Monocarboxylate Transporter 1 (MCT1) in Hepatocellular Carcinoma (HCC). *Oncol. Res.* **2018**, *26*, 71–81. [CrossRef] [PubMed]
64. Czernin, J.; Benz, M.R.; Allen-Auerbach, M.S. PET Imaging of Prostate Cancer Using $^{11}$C-Acetate. *PET Clin.* **2009**, *4*, 163–172. [CrossRef] [PubMed]
65. Yu, M.; Bang, S.-H.; Kim, J.W.; Park, J.Y.; Kim, K.S.; Lee, J.D. The Importance of Acetyl Coenzyme A Synthetase for 11C-acetate Uptake and Cell Survival in Hepatocellular Carcinoma. *J. Nucl. Med.* **2009**, *50*, 1222–1228.
66. Schug, Z.T.; Voorde, J.V.; Gottlieb, E. The metabolic fate of acetate in cancer. *Nat. Rev. Cancer* **2016**, *16*, 708–717. [CrossRef] [PubMed]
67. Gao, X.; Lin, S.-H.; Ren, F.; Li, J.-T.; Chen, J.-J.; Yao, C.-B.; Yang, H.-B.; Jiang, S.-X.; Yan, G.-Q.; Wang, D.; et al. Acetate functions as an epigenetic metabolite to promote lipid synthesis under hypoxia. *Nat. Commun.* **2016**, *7*, 11960. [CrossRef]
68. Ponde, D.E.; Dence, C.S.; Oyama, N.; Kim, J.; Tai, Y.-C.; Laforest, R.; Siegel, B.A.; Welch, M.J. $^{18}$F-Fluoroacetate: A Potential Acetate Analog for Prostate Tumor Imaging—In Vivo Evaluation of $^{18}$F-Fluoroacetate Versus $^{11}$C-Acetate. *J. Nucl. Med.* **2007**, *48*, 420–428. [PubMed]
69. Pillarsetty, N.; Punzalan, B.; Larson, S.M. 2-$^{18}$F-Fluoropropionic Acid as a PET Imaging Agent for Prostate Cancer. *J. Nucl. Med.* **2009**, *50*, 1709–1714. [CrossRef]
70. Lindhe, Ö.; Sun, A.; Ulin, J.; Rahman, O.; Långström, B.; Sörensen, J. [18F]Fluoroacetate is not a functional analogue of [11C]acetate in normal physiology. *Eur. J. Nucl. Med. Mol. Imaging* **2009**, *36*, 1453–1459. [CrossRef]
71. Zhang, Z.; Liu, S.; Ma, H.; Nie, D.; Wen, F.; Zhao, J.; Sun, A.; Yuan, G.; Su, S.; Xiang, X.; et al. Validation of R-2-[$^{18}$F]Fluoropropionic Acid as a Potential Tracer for PET Imaging of Liver Cancer. *Mol. Imaging Biol.* **2019**, *21*, 1127–1137. [CrossRef]
72. Zhao, J.; Zhang, Z.; Nie, D.; Ma, H.; Yuan, G.; Su, S.; Liu, S.; Liu, S.; Tang, G. PET Imaging of Hepatocellular Carcinomas: 18F-Fluoropropionic Acid as a Complementary Radiotracer for 18F-Fluorodeoxyglucose. *Mol. Imaging* **2019**, *18*, 1–8. [CrossRef] [PubMed]

63. Fonović, U.P.; Knez, D.; Hrast, M.; Zidar, N.; Proj, M.; Gobec, S.; Kos, J. Structure-activity relationships of triazole-benzodioxine inhibitors of cathepsin X. *Eur. J. Med. Chem.* **2020**, *193*, 112218. [CrossRef] [PubMed]
64. Singha, P.K.; Mäklin, K.; Vihavainen, T.; Laitinen, T.; Nevalainen, T.J.; Patil, M.R.; Tonduru, A.K.; Poso, A.; Laitinen, J.T.; Savinainen, J.R. Evaluation of FASN inhibitors by a versatile tookit reveals differences in pharmacology between human and rodent FASN preparations and in antiproliferative efficacy in vitro vs. in situ in human cancer cells. *Eur. J. Pharm. Sci.* **2020**, *149*, 105321. [CrossRef] [PubMed]

MDPI
St. Alban-Anlage 66
4052 Basel
Switzerland
www.mdpi.com

*Molecules* Editorial Office
E-mail: molecules@mdpi.com
www.mdpi.com/journal/molecules

Disclaimer/Publisher's Note: The statements, opinions and data contained in all publications are solely those of the individual author(s) and contributor(s) and not of MDPI and/or the editor(s). MDPI and/or the editor(s) disclaim responsibility for any injury to people or property resulting from any ideas, methods, instructions or products referred to in the content.

www.ingramcontent.com/pod-product-compliance
Lightning Source LLC
LaVergne TN
LVHW070242100526
838202LV00015B/2166